TABLE 1 THE STANDARD NORMAL DISTRIBUTION FUNCTION. $P(Z \leqq z), Z \sim N(0, 1)$.

z	$P(Z \leqq z)$	z	$P(Z \leqq z)$	z	$P(Z \leqq z)$	z	$P(Z \leqq z)$
0.00	.5000	0.44	.6700	0.88	.8106	1.32	.9066
0.01	.5040	0.45	.6736	0.89	.8133	1.33	.9082
0.02	.5080	0.46	.6772	0.90	.8159	1.34	.9099
0.03	.5120	0.47	.6808	0.91	.8186	1.35	.9115
0.04	.5160	0.48	.6844	0.92	.8212	1.36	.9131
0.05	.5199	0.49	.6879	0.93	.8238	1.37	.9147
0.06	.5239	0.50	.6915	0.94	.8264	1.38	.9162
0.07	.5279	0.51	.6950	0.95	.8289	1.39	.9177
0.08	.5319	0.52	.6985	0.96	.8315	1.40	.9192
0.09	.5359	0.53	.7019	0.97	.8340	1.41	.9207
0.10	.5398	0.54	.7054	0.98	.8365	1.42	.9222
0.11	.5438	0.55	.7088	0.99	.8389	1.43	.9236
0.12	.5478	0.56	.7123	1.00	.8413	1.44	.9251
0.13	.5517	0.57	.7157	1.01	.8437	1.45	.9265
0.14	.5557	0.58	.7190	1.02	.8461	1.46	.9279
0.15	.5596	0.59	.7224	1.03	.8485	1.47	.9292
0.16	.5636	0.60	.7257	1.04	.8508	1.48	.9306
0.17	.5675	0.61	.7291	1.05	.8531	1.49	.9319
0.18	.5714	0.62	.7324	1.06	.8554	1.50	.9332
0.19	.5753	0.63	.7357	1.07	.8577	1.51	.9345
0.20	.5793	0.64	.7389	1.08	.8599	1.52	.9357
0.21	.5832	0.65	.7422	1.09	.8621	1.53	.9370
0.22	.5871	0.66	.7454	1.10	.8643	1.54	.9382
0.23	.5910	0.67	.7486	1.11	.8665	1.55	.9394
0.24	.5948	0.68	.7517	1.12	.8686	1.56	.9406
0.25	.5987	0.69	.7549	1.13	.8708	1.57	.9418
0.26	.6026	0.70	.7580	1.14	.8729	1.58	.9429
0.27	.6064	0.71	.7611	1.15	.8749	1.59	.9441
0.28	.6103	0.72	.7642	1.16	.8770	1.60	.9452
0.29	.6141	0.73	.7673	1.17	.8790	1.61	.9463
0.30	.6179	0.74	.7704	1.18	.8810	1.62	.9474
0.31	.6217	0.75	.7734	1.19	.8830	1.63	.9485
0.32	.6255	0.76	.7764	1.20	.8849	1.64	.9495
0.33	.6293	0.77	.7794	1.21	.8869	1.65	.9505
0.34	.6331	0.78	.7823	1.22	.8888	1.66	.9515
0.35	.6368	0.79	.7852	1.23	.8907	1.67	.9525
0.36	.6406	0.80	.7881	1.24	.8925	1.68	.9535
0.37	.6443	0.81	.7910	1.25	.8944	1.69	.9545
0.38	.6480	0.82	.7939	1.26	.8962	1.70	.9554
0.39	.6517	0.83	.7967	1.27	.8980	1.71	.9564
0.40	.6554	0.84	.7995	1.28	.8997	1.72	.9573
0.41	.6591	0.85	.8023	1.29	.9015	1.73	.9582
0.42	.6628	0.86	.8051	1.30	.9032	1.74	.9591
0.43	.6664	0.87	.8079	1.31	.9049	1.75	.9599

TABLE 1 (CONTINUED)

z	$P(Z \leqq z)$	z	$P(Z \leqq z)$	z	$P(Z \leqq z)$	z	$P(Z \leqq z)$
1.76	.9608	2.12	.9830	2.48	.9934	2.84	.9977
1.77	.9616	2.13	.9834	2.49	.9936	2.85	.9978
1.78	.9625	2.14	.9838	2.50	.9938	2.86	.9979
1.79	.9633	2.15	.9842	2.51	.9940	2.87	.9980
1.80	.9641	2.16	.9846	2.52	.9941	2.88	.9980
1.81	.9649	2.17	.9850	2.53	.9943	2.89	.9981
1.82	.9656	2.18	.9854	2.54	.9945	2.90	.9981
1.83	.9664	2.19	.9857	2.55	.9946	2.91	.9982
1.84	.9671	2.20	.9861	2.56	.9948	2.92	.9983
1.85	.9678	2.21	.9864	2.57	.9949	2.93	.9983
1.86	.9686	2.22	.9868	2.58	.9951	2.94	.9984
1.87	.9693	2.23	.9871	2.59	.9952	2.95	.9984
1.88	.9699	2.24	.9875	2.60	.9953	2.96	.9985
1.89	.9706	2.25	.9878	2.61	.9955	2.97	.9985
1.90	.9713	2.26	.9881	2.62	.9956	2.98	.9986
1.91	.9719	2.27	.9884	2.63	.9957	2.99	.9986
1.92	.9726	2.28	.9887	2.64	.9959	3.00	.9987
1.93	.9732	2.29	.9890	2.65	.9960	3.01	.9987
1.94	.9738	2.30	.9893	2.66	.9961	3.02	.9987
1.95	.9744	2.31	.9896	2.67	.9962	3.03	.9988
1.96	.9750	2.32	.9898	2.68	.9963	3.04	.9988
1.97	.9756	2.33	.9901	2.69	.9964	3.05	.9989
1.98	.9761	2.34	.9904	2.70	.9965	3.06	.9989
1.99	.9767	2.35	.9906	2.71	.9966	3.07	.9989
2.00	.9773	2.36	.9909	2.72	.9967	3.08	.9990
2.01	.9778	2.37	.9911	2.73	.9968	3.09	.9990
2.02	.9783	2.38	.9913	2.74	.9969	3.10	.9990
2.03	.9788	2.39	.9916	2.75	.9970	3.20	.9993
2.04	.9793	2.40	.9918	2.76	.9971	3.30	.9995
2.05	.9798	2.41	.9920	2.77	.9972	3.40	.9997
2.06	.9803	2.42	.9922	2.78	.9973	3.50	.9998
2.07	.9808	2.43	.9925	2.79	.9974	3.60	.9998
2.08	.9812	2.44	.9927	2.80	.9974	3.70	.9999
2.09	.9817	2.45	.9929	2.81	.9975	3.80	.9999
2.10	.9821	2.46	.9931	2.82	.9976	3.90	1.0000
2.11	.9826	2.47	.9932	2.83	.9977	4.00	1.0000

MATHEMATICAL STATISTICS

Steven F. Arnold
The Pennsylvania State University

Prentice-Hall International, Inc.

A Division of Simon & Schuster
Englewood Cliffs, New Jersey 07632

Printed in the United States of America
10 9 8 7 6 5 4 3 2 1

ISBN 0-13-563099-1

Prentice-Hall International (UK) Limited, *London*
Prentice-Hall of Australia Pty. Limited, *Sydney*
Prentice-Hall Canada Inc., *Toronto*
Prentice-Hall Hispanoamericana, S.A., *Mexico*
Prentice-Hall of India Private Limited, *New Delhi*
Prentice-Hall of Japan, Inc., *Tokyo*
Simon & Schuster Asia Pte. Ltd., *Singapore*
Editora Prentice-Hall do Brasil, Ltda., *Rio de Janeiro*
Prentice-Hall, Englewood Cliffs, New Jersey

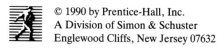

To my parents, Jean and David

CONTENTS

Sections marked with an * are optional and may be skipped without loss of continuity. Sections marked with an # contain matrices.

PREFACE

Mathematical Statistics is written as a textbook for a course in probability and statistics taken by mathematically inclined undergraduates and for a similar course taken by first-year graduate students in statistics.

In the book, I have tried to present the theory behind applied statistics. This has led to considerable emphasis on one-sample and two-sample models, regression, analysis of variance and multinomial models. There are also many realistic examples and exercises. Although these exercises are often fairly routine, a student doesn't really understand a t-test until he or she has applied it to a realistic problem.

The first six chapters of the book contain the basic probability and distribution theory needed for statistics. The central limit theorem is stated early in Chapter 4 and used there to derive normal approximations for many distributions, which are then used to solve realistic probability problems. In this way, it is hoped that the students develop an early appreciation for the theorem's usefulness.

Chapters 7 to 10 contain the basic theory of classical (non-Bayesian) parametric statistics, i.e., point and interval estimation, testing and sufficient statistics. To make the introduction to testing less intimidating, that topic is introduced through test statistics in Chapter 8. The power function is not defined until Chapter 9; UMP and likelihood ratio tests are also found there. Most of the examples in these chapters are exponential families so that the material may be covered fairly quickly.

Chapters 11 to 14 apply the theory of the previous chapters to some classical parametric models (one-sample, two-sample and paired models, regression models, analysis of variance models and multinomial models). In Chapter 13, a general approach to constructing tests for ANOVA models is given. With this approach it is possible to write down the ANOVA table for most balanced, crossed or nested models nearly immediately. It is not necessary to know the formulas before starting, as it is in the method of partitioning the sum of squares.

Chapter 15 contains non-parametric procedures. Estimators, confidence intervals and tests based on ranks are discussed for one-sample and two-sample models. Efficacies and Pitman efficiencies for sign, Wilcoxon, normal scores and t procedures are derived for the normal case. Derivations for other distributions are given in the exercises.

Chapter 16 presents Bayesian procedures. Presenting all the Bayesian procedures in a separate chapter makes clear the coherence of the Bayesian approach to statistics. The Bayesian procedures are defined in terms of the posterior distribution (as opposed to a Bayes risk perspective). This approach is compatible with the likelihood principle, which is one of the main strengths of Bayesian analysis.

In order to explain the theory behind practical statistics, it is necessary to do more asymptotics than in other books at this level. For example, the asymptotic theory is necessary to understand the asymptotic tests and confidence intervals used for discrete random variables. In addition, for two-sample models with continuous data, it is often more appealing to use the asymptotic pivotal quantity which has been correctly studentized when the variances are unequal than to use the two-sample t pivotal quantity. (The optimality of the t-test for normal models with equal variances is balanced by its incorrect studentization when the variances and sample sizes are unequal.) As a final example, the primary justifications for using maximum likelihood estimators or non-parametric procedures are asymptotic ones.

The asymptotic results are also used to indicate the sensitivity of various procedures to the assumptions under which they are derived (e.g., normality, independence, and equal variances). For example, it is shown that the confidence intervals and tests for means and differences of means are asymptotically insensitive to the normal assumption, but that an interval for the variance which is a 95% confidence interval when the observations are normally distributed is only a 68% asymptotic confidence interval for a particular non-normal model. This example shows that the usual procedures for confidence intervals and tests about the variance should only be used when we are relatively certain that the normal assumption is exactly satisfied, an unlikely occurrence in practice. Similarly, it is shown that an interval for the mean which is a 95% confidence interval when the observations are independently normally distributed is an asymptotic 0% confidence interval when there is any positive correlation between the observations.

The theory of regression, analysis of variance and multinomial models is based on the multivariate normal distribution, which is introduced in Chapter 5. To study this distribution, it is convenient to use matrices. The necessary matrix algebra can be taught to students at the level of this book in about half an hour, so that I always cover the sections using matrices. The book is written so that these sections (marked with a #) may be omitted. For this reason, there are separate sections on the bivariate and multivariate normal distributions and on simple and multiple regression. Similarly the basic results on analysis of variance and multinomial models are stated and applied without using matrices. These results are derived, using matrices, in separate sections.

To add flexibility in its use, this book contains over 30 sections marked with an *. These starred sections are optional and may be skipped without loss of continuity. Some starred sections cover such applied topics as Simpson's paradox, sampling from finite populations, correlation and causation and multicollinearity. Other starred sections contain theoretical topics such as the asymptotic properties of maximum likelihood estimators, the likelihood principle and inference for random variables which are uniformly distributed on the interval $\mu - .5$ to $\mu + .5$. (Some bizarre things happen for this model.) Further starred sections cover theoretical topics which are also of practical importance such as Scheffe' simultaneous confidence intervals and a Bayesian analysis of the two-parameter normal model. By choosing which starred sections to cover, an instructor can fine tune the course to the students. The material in the unstarred sections is at about the same level as other mathematical statistics books (e.g., DeGroot (1986) or Hogg and Craig (1978)). In some of the starred sections, the material may be at a somewhat higher level.

When I teach to graduate students (in statistics), I try to cover the first 7 chapters in the first semester and finish the book in the second semester. I cover almost all the sections (starred and unstarred) in the book. For undergraduates, I cover the unstarred sections of the first 6 chapters in the first semester. In the second semester, I cover the unstarred sections of Chapters 7–10, most of Chapter 11 and as much of Chapters 12 through 16 as possible.

This book has over 1,000 exercises. These exercises are divided into three types. Type A problems are realistic word problems, type B problems are routine calculations and derivations and type C problems are more difficult calculations and derivations. When I am teaching an undergraduate class, I assign most of the type A problems and many of the type B problems. When I am teaching graduate students, I assign fewer type A problems, and many type B and C problems. It is not necessary to use the computer to solve any of the exercises. However, a calculator would be quite useful, especially one which computes permutations and combinations, or at least factorials.

The prerequisite for this course is a sound background in calculus, including the calculus of several variables. Sections A.2 of the appendix contain a summary of the necessary calculus with examples. The sections of the text marked with an # use matrix algebra, which is summarized in Section A.3.1 of the appendix. (It is not necessary to read Section A.3.2 in order to understand the material in the text.) The sections marked with an # may be skipped if the instructor chooses not to use matrix algebra.

Professors Peter J. Brockwell, James E. Gehrmann, Harry J. Khamis, Daniel G. Martinez, Paul I. Nelson, Larry J. Ringer and William F. Stout read earlier versions of this text and made many helpful suggestions both in terms of emphasis and clarity of the text. Their contributions to this text are gratefully acknowledged.

Srinivas Emani, Laura Simon and Clint Coakley detected many typographical errors in earlier versions of this book. In particular, Laura and Clint carefully studied two versions of the book. Due to their efforts, we hope that the book is relatively free of typographical errors.

This book would not have been possible without the organizational skills of Dorothy Fletcher. Many times when I thought that I would give up on ever getting things together, I gave it to her and she rescued it for me.

I would also like to thank WordCrafters Inc., and Brian Baker, whose copy editing has improved the presentation in this manuscript considerably, and David Ostrow and the people at Prentice-Hall, who encouraged me to start this book and to continue working on it when I was about to quit.

Chapter 1

PROBABILITY

In this chapter we discuss general probability models. These models are used to describe events that occur "by chance" or "randomly." Most of us have some idea of what probability is. Surprisingly enough, however, it is difficult to define probability in a way that applies to every situation where we use the term and in a way which is agreeable to most people.

One common definition of probability is the *frequentist* definition, in which the probability of a particular outcome of an experiment is the proportion of times the outcome occurs if we perform the experiment a very large number of times. For example, if the experiment is to flip a coin, then the probability of a head is the proportion of flips on which we get heads in a large number of flips. If we say that the probability of heads is $\frac{1}{2}$, we are saying that heads should come up about half the time in a large number of flips. The weak law of large numbers (see Section 6.3) is often used to support this approach to probability.

A second common definition of probability is the *subjective* one. In this definition, the probability that a person assigns to a particular outcome in an experiment measures the person's belief as to how likely that outcome is to occur. For example, if a weatherman says that the probability of rain the next day is .4, he is telling the audience his subjective belief in the likelihood of rain the next day, given the weather conditions today. Different weather forecasters would give different probabilities from the same present conditions, which indicates the subjective nature of these probabilities.

Neither of these approaches is completely satisfactory as an interpretation of probability. One problem with the frequentist approach is that it is somewhat imprecise. Suppose we have a coin and want to check whether its probability of heads is $\frac{1}{2}$. So we toss it 100,000 times and get 49,500 heads, so that the proportion of heads is .495. It is not clear whether this number is "about half" or not. If we do the experiment 10^{10} times and get a proportion of heads of .495, is this "about half" or not? One way out of this difficulty is to define the probability of an outcome as the limit of the proportion of times it occurs as the number of trials approaches infinity.

A more severe difficulty with the frequentist approach is that it assumes that the given experiment may be replicated a very large (possibly infinite) number of times, which is often not the case. In the weather forecasting experiment, there will never be another day exactly like this one, so it is hard to picture how to replicate the experiment of predicting tomorrow's weather from today's weather. (Some people even have trouble imagining the replication of the coin-tossing experiment, since the conditions under which the coin is tossed will vary from toss to toss.)

The primary difficulty with the subjective approach to probability is that it is subjective. Each person's probability for a particular event could be different. There is no "true" probability for any event. While this approach may seem reasonable for experiments like predicting the weather, it does not seem quite so reasonable when computing probabilities for card problems or dice problems or coin-tossing problems, for which most people feel there are "true" probabilities. Due to the subjective nature of the definition, it would also be difficult for people with different subjective probabilities to communicate with each other.

The approach we take to probability in this book is to leave the interpretation of probability unspecified. This approach is similar to that taken by physicists when they do not define mass, time, or distance and rely on intuition for the meaning of these terms, but define other concepts in terms of these.

One important aspect of probability is that in defining probabilities, we are actually modeling the world. For example, we say that a coin is "fair" if the probability of heads is $\frac{1}{2}$. However, in a particular experiment we may have a highly biased coin, that is, one whose probability of heads is not near $\frac{1}{2}$. If we do probability calculations based on a fair coin when the coin is in fact biased, we get answers that are not helpful for the particular coin which we have. It should be remembered that the usefulness of any probability calculation is heavily dependent on how good the model is. If the coin is fair or nearly fair, calculations based on a fair coin are useful, but if the coin is biased, such calculations are not useful.

In Section 1.1, we discuss certain probability models which we call *equally probable* models. Such models occur when we have a finite set of possible outcomes and we think of each outcome as equally likely. These models were some of the first probability models studied. Most people have a fairly good intuitive idea of what probability means in such models. In Section 1.2, we give an axiomatic definition of a probability model. The axioms tell what a probability model must satisfy to be internally consistent. However, a probability model can satisfy the axioms and still be the wrong probability model. For example, a probability model based on tossing a fair coin would be internally consistent, but might be completely incorrect if the actual coin is biased.

In Section 1.3, we define the conditional probability that one outcome has occurred when we know that another one has. For example, we might want to know the probability that a particular football team wins on a particular day given that it rains on that day. Also in Section 1.3, we define what we mean by outcomes being independent of each other. For instance, if we flip two coins, we might reasonably expect the outcomes of the two tosses to be independent. In Section 1.4, we discuss the secretary problem, a classical problem in the study of probability.

1.1 EQUALLY LIKELY MODELS

1.1.1 Equally Likely Probability Models

We begin the discussion of probability with some models which arise in games of chance. Suppose we have a game of chance with a finite set S of possible outcomes. We call S the *sample space*. An *event* is a subset of S. If all the points of S are equally probable, then the probability of an event A is

$$P(A) = \frac{\#(A)}{\#(S)},$$

where $\#(A)$ is the number of points in A and $\#(S)$ is the number of points in S. We call this the *equally probable model* for S.

Example 1–1

A die is a cube on which one face is labeled 1, one face labeled 2, one labeled 3, one labeled 4, one labeled 5, and one labeled 6. If we roll the die and a 4 is on top, we say we have rolled a 4, etc. Now suppose that we roll two dice. Then the 36 possible outcomes in the sample space are the following:

$$\begin{matrix} (1,1) & (1,2) & (1,3) & (1,4) & (1,5) & (1,6) \\ (2,1) & (2,2) & (2,3) & (2,4) & (2,5) & (2,6) \\ (3,1) & (3,2) & (3,3) & (3,4) & (3,5) & (3,6) \\ (4,1) & (4,2) & (4,3) & (4,4) & (4,5) & (4,6) \\ (5,1) & (5,2) & (5,3) & (5,4) & (5,5) & (5,6) \\ (6,1) & (6,2) & (6,3) & (6,4) & (6,5) & (6,6) \end{matrix}.$$

We assume that these 36 points are equally probable. Let A be the event that we roll doubles (that is that the dice are the same). Then A has six points in it: ((1, 1), (2, 2), (3, 3), (4, 4), (5, 5) and (6, 6)). Therefore, the probability of A is

$$P(A) = \frac{\#(A)}{\#(S)} = \frac{6}{36} = \frac{1}{6}.$$

Now, let X be the sum of the two rolls, and let B_k be the event that $X = k$, $k = 2, \ldots, 12$. Then there are three points in B_4 (namely (1, 3), (2, 2) and (3, 1)), so that

$$P(B_4) = \frac{3}{36} = \frac{1}{12}.$$

Similarly, there are five points such that X is 8, and therefore,

$$P(B_8) = P(X \text{ is } 8) = \frac{5}{36}.$$

In fact, $P(B_k) = P(X \text{ is } k)$ is given in the following table:

k	2	3	4	5	6	7	8	9	10	11	12
$P(B_k)$	$\frac{1}{36}$	$\frac{2}{36}$	$\frac{3}{36}$	$\frac{4}{36}$	$\frac{5}{36}$	$\frac{6}{36}$	$\frac{5}{36}$	$\frac{4}{36}$	$\frac{3}{36}$	$\frac{2}{36}$	$\frac{1}{36}$

Note that the outcomes for X are not equally probable. Note also that these probabilities sum to one (as they must, since this sum is just the total number of points in S divided by the total number of points in S). Let E be the event that X is even. Since there are 18 points in the table whose sum is even, it follows that

$$P(E) = \frac{18}{36} = \frac{1}{2}.$$

We could also compute $P(E)$ by observing that X is even when X is 2, 4, 6, 8, 10 or 12. Therefore,

$$P(E) = \frac{\#(E)}{36} = \frac{(\#(B_2) + \#(B_4) + \#(B_6) + \#(B_8) + \#(B_{10}) + \#(B_{12}))}{36}$$

$$= P(B_2) + P(B_4) + P(B_6) + P(B_8) + P(B_{10}) + P(B_{12}) = \frac{1}{2}.$$

Now let F be the event that X is odd. There are 18 points in the set F, so that $P(F)$ is also $\frac{1}{2}$. But also, F occurs if and only if E does not occur, so that we could also compute $P(F)$ in the following way:

$$P(F) = \frac{\#(F)}{\#(S)} = \frac{(\#(S) - \#(E))}{\#(S)} = 1 - P(E).$$

In a similar way, we could show that the probability that the sum is divisible by three is one-third and that the probability that it is divisible by four is one-fourth, but that the probability that the sum is divisible by five is

$$P(X \text{ is divisible by } 5) = P(X \text{ is } 5) + P(X \text{ is } 10) = \frac{7}{36}.$$

Let C_j be the event that the first die is j, $j = 1, \ldots, 6$. Then, for each j, there are six points in C_j, namely, $(j, 1)$, $(j, 2)$, $(j, 3)$, $(j, 4)$, $(j, 5)$, and $(j, 6)$, so that

$$P(C_j) = \frac{6}{36} = \frac{1}{6}.$$

Therefore, if this model is correct, then the first die must be a "fair" die, i.e., one in which all six faces are equally probable. Similarly, the second die must be fair. Hence, if either of the dice is not a fair die, then the model is not correct, and any probability calculations made under it are not correct. By the same reasoning, if the dice have magnets in them so that they always come up doubles, then the probability of doubles would be one instead of $\frac{1}{6}$, so that the model would also be incorrect in this situation. However, we shall show in Section 1.3.2 that as long as the dice are fair and are "independent" of each other, then the model given is the correct model.

In the remaining examples, we use some counting formulas which may be familiar to most students. (For those students for whom they are unfamiliar, a discussion is given in Appendix 1.) For any positive integer n, we write

$$n! = n(n-1)(n-2)\ldots(3)(2)(1), \qquad 0! = 1.$$

Also, we use the symbol

$${}_nP_k = \frac{n!}{(n-k)!}$$

for the number of permutations of n things k at a time, the notation

$$\begin{bmatrix} n \\ k \end{bmatrix} = \frac{n!}{(n-k)!k!} = \frac{{}_nP_k}{k!}$$

for the number of combinations of n things k at a time, and the notation

$$\begin{bmatrix} & n & \\ i & j & k \end{bmatrix} = \frac{n!}{i!j!k!}$$

for the number of ways of dividing a set with n objects into three subsets, one with i objects, one with j objects and one with k objects. Finally, we define

$$\begin{bmatrix} & & n & \\ i & j & k & m \end{bmatrix}, \text{ etc.,}$$

similarly. Note that

$$\begin{bmatrix} & n & \\ k & & j \end{bmatrix} = \begin{bmatrix} n \\ k \end{bmatrix} = \begin{bmatrix} n \\ j \end{bmatrix}.$$

In the following examples, we often use the following basic rule.

Counting rule. If we have a k-stage experiment which has n_1 possible outcomes at the first stage, has n_2 possible outcomes at the second stage for every outcome at the first stage, has n_3 possible outcomes at the third stage for every possible pair of outcomes at the second stage, etc., then the total number of possible outcomes is

$$N = n_1 n_2 \ldots n_k = \prod_i n_i .$$

A standard deck of cards consists of 52 cards which are divided into four suits, called spades, hearts, diamonds, and clubs. There are 13 cards in each suit, which have 13 different denominations, the ace (A), the king (K), the queen (Q), the jack (J), the 10, the 9, down to the 2. (Note that there is no 1, since the ace is really a 1.) Therefore, all 52 cards are different. In typical card games, we draw a fixed number of cards from this deck or deal a certain number of cards to each of several people. The usual assumption made for computing probabilities for these games is that all possible hands are equally likely. In the next section we show that for the events we consider here, we get the same answer no matter whether we consider ordered or unordered hands. Here, we use unordered hands since the calculations are easier. Therefore, the hand in which we have the A of spades, the Q of hearts, and the 7 of diamonds is the same as the hand in which we have the Q of hearts, the 7 of diamonds, and the A of spades. (That is, we ignore the order in which we drew the cards or in which they were dealt to us.)

Example 1–2

Poker is a game which is often played by dealing five cards to each of several players. Each such set of five cards is called a "poker hand." Let us find probabilities for some poker events. The number of possible hands in the sample space S for this model is

$$\#(S) = \begin{bmatrix} 52 \\ 5 \end{bmatrix} = 2{,}598{,}960,$$

since there are 52 cards and we are choosing 5 of them. Let B be the event that all the cards are hearts. Then

$$P(B) = \frac{\#(B)}{\#(S)} = \frac{\begin{bmatrix} 13 \\ 5 \end{bmatrix}}{\begin{bmatrix} 52 \\ 5 \end{bmatrix}} = \frac{1{,}287}{2{,}598{,}960} = .0005$$

since the numerator is just the number of ways of choosing 5 cards from the 13 hearts. Now let C be the event that the cards all come from the same suit (a flush). Then $\#(C)$ is the number of hands where all the cards come from the same suit, which is four times $\#(B)$. Therefore,

$$P(C) = 4P(B) = .002.$$

Now let D be the event that we get three aces and two kings. The number of hands with three aces and two kings is the product of the number of ways of choosing three aces times the number of ways of choosing two kings (by the above counting rule). Therefore,

$$P(D) = \frac{\begin{bmatrix} 4 \\ 3 \end{bmatrix}\begin{bmatrix} 4 \\ 2 \end{bmatrix}}{\begin{bmatrix} 52 \\ 5 \end{bmatrix}} = \frac{24}{2{,}598{,}960} = .000009.$$

By similar reasoning, the probability of three aces and two queens is equal to $P(D)$, as is the probability of three fours and two tens, etc. There are 13 possibilities for three of a kind, and for each of these choices, there are 12 choices for two of a kind. Therefore, the probability of three of a kind together with a pair (a hand called a full house) is

$$(13)(12)P(D) = \frac{3{,}744}{2{,}598{,}960} = .0014.$$

Now consider the probability of the event E of getting three A's, a K, and a Q. The number of hands with three A's, a K, and a Q is the product of the number of ways of choosing three A's, the number of ways of choosing a K, and the number of ways of choosing a Q. Therefore,

$$P(E) = \frac{\begin{bmatrix} 4 \\ 3 \end{bmatrix}\begin{bmatrix} 4 \\ 1 \end{bmatrix}\begin{bmatrix} 4 \\ 1 \end{bmatrix}}{\begin{bmatrix} 52 \\ 5 \end{bmatrix}} = \frac{64}{2{,}598{,}960} = .0000246.$$

Now, the probability of three K's, a J, and a 10 is the same as $P(E)$, as is the probability of three 7's, a 5, and a 4. In general, there are $\left[\begin{smallmatrix} 13 \\ 1 \end{smallmatrix}\right]$ possibilities for three of a kind and $\left[\begin{smallmatrix} 12 \\ 2 \end{smallmatrix}\right]$ possibilities for the single cards. Therefore, the probability of a three of a kind is

$$\begin{bmatrix} 13 \\ 1 \end{bmatrix}\begin{bmatrix} 12 \\ 2 \end{bmatrix} P(E) = .0211285.$$

Finally, let F be the event that the cards form a straight, i.e., that they can be put into a sequence, for example 3, 4, 5, 6, 7. (It is assumed that A is higher than K is higher than Q is higher than J is higher than 10, so that Q, J, 10, 9, 8 is a straight, but 5, 4, 3, 2, A is not.) To compute $P(F)$, we need to find the number of possible straights. First consider the straight 2, 3, 4, 5, 6. There are four choices for the 2, four for the 3, four for the 4, four for the 5, and four for the 6. Therefore, by the previous counting rule, there are 4^5 possible straights starting with 2. There are also 4^5 possible straights starting with 3, etc. There are nine possible lowest values in a straight $(2, 3, \ldots, 10)$. Hence, there are $(9)4^5$ possible straights, and therefore,

$$P(F) = \frac{(9)4^5}{\begin{bmatrix} 52 \\ 5 \end{bmatrix}} = \frac{9{,}016}{2{,}598{,}960} = .0035.$$

Example 1–3

Now suppose that we deal two hands of five cards each from the same deck. We can consider this as a partition of the deck into one hand with 5 cards, another hand with 5 cards, and the remaining 42 cards. Therefore, the number of possible pairs of hands in the sample space S^* is

$$\#(S^*) = \begin{bmatrix} & 52 & \\ 5 & 5 & 42 \end{bmatrix} = 3.99 \times 10^{12}.$$

We now find the probability of the event A that the first player has three aces and two kings and that the second player has two kings and three queens. To count the number of ways this can occur, we first partition the four aces so that three go into the first hand, none goes into the second, and one is still in the deck. We then partition the four kings so that two go into the first hand, two go into the second, and none remains in the deck. Then we partition the four queens so that none goes to the first hand, three go to the second, and one goes to the deck. Therefore,

$$P(A) = \frac{\begin{bmatrix} & 4 & \\ 3 & 0 & 1 \end{bmatrix}\begin{bmatrix} & 4 & \\ 2 & 2 & 0 \end{bmatrix}\begin{bmatrix} & 4 & \\ 0 & 3 & 1 \end{bmatrix}}{\begin{bmatrix} & 52 & \\ 5 & 5 & 42 \end{bmatrix}} = 2.41 \times 10^{-11}.$$

This is the same as the probability that the first player gets three 7's and two 4's and the second player gets three 5's and two 4's. There are 13 choices for the first player's three of a kind, 12 remaining choices for the second player's three of a kind, and 11 choices for the common pair. Therefore, the probability of the event B that both players get full houses with a common pair is

$$P(B) = (13)(12)(11)P(A) = 4.1 \times 10^{-8}.$$

Now the probability of the event C that the first player gets three A's and two K's and the second player gets three Q's and two J's is

$$P(C) = \frac{\begin{bmatrix} & 4 & \\ 3 & 0 & 1 \end{bmatrix}\begin{bmatrix} & 4 & \\ 2 & 0 & 2 \end{bmatrix}\begin{bmatrix} & 4 & \\ 0 & 3 & 1 \end{bmatrix}\begin{bmatrix} & 4 & \\ 0 & 2 & 2 \end{bmatrix}}{\begin{bmatrix} & 52 & \\ 5 & 5 & 42 \end{bmatrix}} = 1.44 \times 10^{-10}.$$

Also, the probability of the event D that both players get a full house with different pairs is

$$P(D) = (13)(12)(11)(10)P(B) = 2.48 \times 10^{-6}.$$

Finally, since B and D have no points in comon, the probability of the event E that both players get full houses is

$$P(E) = \frac{\#(E)}{\#(S^*)} = \frac{(\#(B) + \#(D))}{\#(S^*)} = P(B) + P(D) = 2.52 \times 10^{-6}.$$

The equally likely model for cards is fairly accurate as long as the deck is fairly well shuffled before each deal and there is no cheating.

Example 1–4

Suppose we have k people in a room and want to find the probability that they all have different birthdays. We assume for simplicity that there are only 365 days a year (no leap years). Then an outcome of this experiment consists of a sequence of k birthdays. Suppose that all possible such sequences are equally likely. We have 365 choices for the first birthday, 365 choices for the second, 365 for the third, etc. So there are

$$\#(S) = 365^k$$

possible sequences of birthdays. Let E be the event that all the birthdays are different. We need to find the number of sequences in E. If all the birthdays are different, then we have 365 choices for the first birthday, 364 choices for the second birthday, 363 for the third, etc. Therefore, there are

$$\#(E) = 365(364)\ldots(366-k) = {}_{365}P_k$$

sequences in E. Consequently,

$$P(E) = \frac{\#(E)}{\#(S)} = \frac{{}_{365}P_k}{365^k}.$$

Thus, if $k = 25$, then $P(E) = .43$.

Now let D be the event that there is at least one pair with the same birthday. Then

$$P(D) = \frac{\#(D)}{\#(S)} = \frac{(\#(S) - \#(E))}{\#(S)} = 1 - P(E).$$

Therefore, if $k = 25$, then $P(D) = 1 - .43 = .57$. Hence, if this model is correct, and if we have 25 people, then the probability that at least two people have the same brithday is .57.

The result of this example is typically somewhat surprising. The student may be interested in doing the experiment several times with real people in a real room.

Exercises—A

1. **(a)** In Example 1–1, verify the results given for $P(B_k) = P(X = k)$.
   (b) In the same example, verify that the probability that X is divisible by 4 is $\frac{1}{4}$.
2. In Example 1-2, find:
   (a) The probability of four of a kind;
   (b) The probability of two pairs (but not a full house);
   (c) The probability of one pair (but not a three of a kind or two pairs).
   (d) The probability of no face cards, i.e., that all the cards in the hand are at least 2 but at most 9.
3. Six fair dice are rolled. What is the probability that each number appears once?
4. A tetrahedron is a four-sided die whose faces are marked with 1, 2, 3, and 4, respectively. (When we roll a tetrahedron there is no top face, so we read the bottom face). Suppose we roll two tetrahedrons.
   (a) List the 16 points in the sample space.

(b) Find the probability that the sum is even and the probability that the sum is divisible by three.
(c) Find the probability of doubles.
(d) Find the probability that the sum equals k, for $k = 2, 3, \ldots, 8$.

5. Suppose we draw three cards from an ordinary deck.
(a) Find the probability that all cards are of the same suit.
(b) Find the probability of three of a kind.
(c) Find the probability of a pair.

6. Suppose we play with a deck of cards which has only A's, K's, Q's, J's, and 10's in the deck, so that it has 20 cards altogether. We draw four cards from this deck.
(a) Find the probability that all the cards are of the same suit.
(b) Find the probability of three of a kind.
(c) Find the probability of two pairs.

7. Suppose we flip a coin three times. Assume that all eight possible sequences of heads (H) and tails (T) are equally probable.
(a) List the eight points in the sample space.
(b) Find the probability of all heads.
(c) Find the probability of exactly two heads.
(d) Find the probability of an odd number of heads.

8. Four couples come to a party. They are seated around a circular table. Assuming that all possible orderings are equally likely, what is the probability that each man is sitting next to his wife?

9. Find the number k of people such that the probability of finding two people with the same birthday is at least .75.

10. Bridge is a card game played by four players (say A, B, C, and D) with a standard deck of cards. All 52 cards in the deck are dealt so that each player receives 13 cards.
(a) How many possible sets of four hands are there?
(b) Find the probability that player A and player B both receive five hearts. (Partition the deck into player A, player B, and the remainder.)
(c) Find the probability that two players receive five hearts.
(d) Find the probability that players A and B receive all the A's.

11. (a) Find the probability that in a bridge game player A receives four spades and three in each of the other suits, player B receives four hearts and three in the other suits, player C receives four diamonds and three in the other suits, and player D receives four clubs and three in the other suits.
(b) Find the probability that each player receives four cards in one suit and three in each of the others.
(c) Find the probability that each player receives 13 cards in the same suit.

12. (a) Suppose that in a bridge game players A and C have 11 spades between them, including the A and Q, but that B and D have the K. Show that the probability that B and D each have one spade is .52.
(b) Suppose that players A and C have 10 spades, including the A and Q, but not the K. Find the probability that B has the K and no other spades.
(c) Show that the probability that the hand with the K has no other spades is .26.

For readers familiar with bridge and its terminology, this result implies that, in the absence of other information, playing for the drop is a better strategy than finessing when A and C have 11 spades between them, but finessing is better when they have only 10.

1.1.2 Sampling with and without Replacement

Suppose an experiment consists of drawing n objects from a set of N objects. We call the set of N objects we are drawing from the *population* and N the *population size,* and we call the n objects drawn the *sample* and n the *sample size.*

We say we are *sampling without replacement* if we draw the first object from the population, then draw the second object from the $N-1$ objects remaining after the first draw, draw the third one from the $N-2$ remaining objects, etc. We take as the sample space the set S of ${}_N P_n$ possible sequences of draws and assume an equally probable model for S.

Example 1–5

Suppose we have a set of 20 fuses of which 18 are good. We draw two fuses without replacement. We find the probability that the first one is good, the probability that the second one is good, and the probability that they are both good. To find these probabilities, we assume that the fuses are numbered and that the last two are the bad ones. We have 20 choices for the first fuse and 19 for the second, so

$$\#(S) = {}_{20}P_2 = (20)(19) = 380.$$

Let A be the event that the first fuse is good. We need to find $\#(A)$. Since A has the first fuse good and makes no statement about the second fuse, there are 18 ways to choose the first good fuse. Now suppose we drew fuse #8 on the first draw. Then there would be 19 ways to choose the second fuse (any fuse but #8). In general, for *any* good fuse we choose on the first draw, there are 19 ways to draw the second fuse. Therefore,

$$\#(A) = (18)(19) = 342, \qquad P(A) = \frac{\#(A)}{\#(S)} = \frac{342}{380} = .9.$$

(Note that this answer is 18/20, which is the obvious answer.) Now, let B be the event that the second fuse is good. We need to find $\#(B)$. We note that B can occur in two ways: if the first fuse is good and the second fuse is good, or if the first fuse is bad and the second fuse is good. There are 18 ways to draw a good fuse on the first draw. For each of those ways, there are 17 ways to draw a good fuse on the second draw. (For example, if I draw the fourth fuse on the first draw, then I can draw any of the other 17 good fuses on the second draw.) Similarly, there are two ways to draw a bad fuse on the first draw and for each of those ways there are 18 ways to draw a good fuse on the second draw (since all 18 good fuses are still in the box). Therefore,

$$\#(B) = (18)(17) + (2)(18) = 342,$$

which is the same as $\#(A)$, so that

$$P(B) = P(A) = .9.$$

Now, let C be the event that both fuses are good. There are 18 ways to choose the first good fuse and for each of those draws, there are 17 ways to draw the second fuse, so that

$$\#(C) = (18)(17) = 306, \qquad P(C) = \frac{306}{380} = .805.$$

We now return to sampling n objects from a population of N objects. We say that we are *sampling with replacement* if we draw the first object, record it, and put it back, then draw again from the original N objects, record again, and put the object back, draw again from the original N objects, etc. For sampling with replacement, we have the sample space T consisting of all possible N^n such draws. (Note that the sample space is different for sampling with replacement and sampling without replacement, since we can draw the same object several times when sampling with replacement.) We assume an equally probable model for T.

Example 1–6

Suppose again that we have 20 fuses, 18 of which are good, and that we will choose two fuses from these 20 with replacement. In this case, there are 20 choices for the first fuse and 20 choices for the second fuse (since we put the first draw back before drawing the second), so that

$$\#(T) = 20^2 = 400.$$

Let A be the event that the first fuse is good. We have 18 choices for the first fuse and 20 choices for the second fuse. Therefore, for this model,

$$\#(A) = (18)(20) = 360, \qquad P(A) = \frac{360}{400} = .9,$$

which is the same answer we got in Example 1–5 when sampling without replacement. By a similar argument, we can show that the probability that the second fuse is good is the same as for sampling without replacement. Now let C be the event that both fuses are good. For this model, there are 18 choices for the first good fuse and 18 choices for the second good fuse, so that

$$\#(C) = (18)^2 = 324, \qquad P(C) = \frac{324}{400} = .81,$$

which is different from the model of sampling without replacement. Therefore, the probabilities can be different depending on whether we sample with or without replacement.

Example 1–7

Suppose now that we have 2,000 fuses of which 1,800 are good. Suppose we draw two fuses. What is the probability of the event C that they are both good? If we sample without replacement, we have

$$\#(S) = (2{,}000)(1{,}999) = 3{,}998{,}000,$$
$$\#(C) = (1{,}800)(1{,}799) = 3{,}238{,}200,$$
$$P(C) = \frac{3{,}238{,}200}{3{,}998{,}000} = .809955.$$

If we sample with replacement, we get

$$\#(T) = (2{,}000)^2 = 4{,}000{,}000,$$
$$\#(C) = (1{,}800)^2 = 3{,}240{,}000,$$
$$P(C) = \frac{3{,}240{,}000}{4{,}000{,}000} = .81.$$

Examples 1–5 to 1–7 point up two important facts about sampling with and without replacement:

1. For sampling with replacement, we get the same answer for the probability of drawing two good fuses whether we start with 20 fuses of which 18 are good or 2,000 fuses of which 1,800 ar good. (In the exercises, you are asked to show that the answer is the same for sampling with replacement no matter how many fuses there are, as long as 90% of them are good.)
2. If n is much smaller than N, then the answer is essentially the same whether we sample with or without replacement. This fact is fairly obvious intuitively: If we are sampling a small number of objects from a huge population, it does not matter whether we put the object back or not, because we are unlikely to draw it again anyhow.

The first fact about sampling with replacement makes this method much easier to use than sampling without replacement, as we shall see in later chapters. The second fact implies that we may assume that we are sampling with replacement even when we are actually sampling without replacement, as long as N, the population size, is much larger than n, the sample size.

Example 1–8

Suppose we have a box with 25 light bulbs, 10 of which are X watts, 6 Y watts, and 9 Z watts. We draw four bulbs and want to know the probability that we get two X-watt bulbs, one Y-watt bulb, and one Z-watt bulb. Let C be this event. First, assume we sample without replacement. Then

$$\#(S) = (25)(24)(23)(22) = 303{,}600.$$

Now, C can occur in 12 ways: $XXYZ$, $XXZY$, $XYXZ$, $XZXY$, $XYZX$, $XZYX$, $YXZX$, $ZXYX$, $YXXZ$, $ZXXY$, $YZXX$, and $ZYXX$, where $XYXZ$ means we draw an X, then a Y, then another X, and then a Z. To get $XXYZ$, we have 10 choices for the first bulb, 9 for the second (since we have already got an X), 6 for the third (to get the Y-watt bulb), and 9 for the fourth. Therefore,

$$\#(XXYZ) = (10)(9)(6)(9) = 4{,}860.$$

Similarly,

$$\#(YXZX) = (6)(10)(9)(9) = 4{,}860.$$

In fact, all 12 of the events listed have the same number of points. (Why?) Therefore,

$$\#(C) = (12)(4{,}860) = 58{,}320, \qquad P(C) = \frac{58{,}320}{303{,}600} = .1921.$$

Now suppose that we sample with replacement. Then

$$\#(T) = 25^4 = 390{,}625.$$

C can still occur in the same 12 ways, so for sampling with replacement,

$$\#(XXYZ) = (10)(10)(6)(9) = 5{,}400,$$

and all the other 12 events have the same number of points. (Why?) Therefore,

$$\#(C) = (12)(5{,}400) = 64{,}800, \qquad P(C) = \frac{64{,}800}{390{,}625} = .166.$$

We next discuss an alternative method for finding probabilities for sampling n objects without replacement from a population of N objects. We say that an event A

is *unordered* if it is unchanged by a permutation of the draws. Thus, in Example 1–5, the event C that we draw two good fuses is an unordered event, but the event that the first fuse is good is an ordered event, since when the draws are permuted, this becomes the event that the second fuse is good. Similarly, in Example 1–8, the event that we draw two X-watt bulbs, one Y-watt bulb, and one Z-watt bulb is an unordered event because it does not imply anything about the ordering of the bulbs. However, the event that we draw an X-watt bulb before a Y-watt bulb is an ordered event.

Let A be an unordered event, and let $\#^*(A)$ be the number of unordered sequences in A. Since A is an unordered event, every sequence in A has n draws in it. Therefore, for every unordered sequence in A, there are $n!$ ordered sequences in A. Hence,

$$\#(A) = n!\#^*(A), \qquad \#(S) = n!\#^*(S), \qquad P(A) = \frac{\#(A)}{\#(S)} = \frac{\#^*(A)}{\#^*(S)}.$$

Consequently, for unordered events, we have a choice of two equally probable models, one which takes the set of ordered sequences as the sample space and one which takes the set of unordered sequences as the sample space. The argument just presented shows that we get the same answers for probabilities of unordered events with either model. So we typically use the space of unordered events, when possible, because the calculations are somewhat easier. (The card model we considered in Section 1.1.1 is an example of sampling without replacement, and in that section we used the sample space of unordered sequences.)

Example 1–9

We return to Example 1–8, in which we are drawing four bulbs from a box which contains 10 X-watt bulbs, 6 Y-watt bulbs, and 9 Z-watt bulbs. Since we are drawing 4 objects from 25 objects,

$$\#^*(S) = \begin{bmatrix} 25 \\ 4 \end{bmatrix} = 12{,}650.$$

Let C be the event that we draw two X-watt, one Y-watt, and one Z-watt bulb. In this case, we are drawing 2 out of the 10 X-watt bulbs, 1 out of the 6 Y-watt bulbs, and 1 out of the 9 Z-watt bulbs. Therefore,

$$\#^*(C) = \begin{bmatrix} 10 \\ 2 \end{bmatrix}\begin{bmatrix} 6 \\ 1 \end{bmatrix}\begin{bmatrix} 9 \\ 1 \end{bmatrix} = 2{,}430, \qquad P(C) = \frac{2{,}430}{12{,}650} = .1921,$$

which is the same answer we got using ordered sequences. Now let D be the event that we get two X's and two Y's in four draws. Then

$$\#^*(D) = \begin{bmatrix} 10 \\ 2 \end{bmatrix}\begin{bmatrix} 6 \\ 2 \end{bmatrix} = 675, \qquad P(D) = \frac{675}{12{,}650} = .0534.$$

Although the sample space of unordered sequences is typically easier to use when we are sampling without replacement, it is only applicable for finding probabilities of unordered events for sampling without replacement. If the event is not unordered, or if the sampling is with replacement, then it is necessary to use the sample space of ordered sequences.

Exercises—A

1. A committee with two members is to be selected (without replacement) from a collection of 30 people, of whom 10 are males and 20 are females.
   (a) Find the probability that both members are male.
   (b) Find the probability that both members are female.
   (c) Find the probability that one member is male and one is female.

   Do this calculation two ways, using ordered and unordered sequences.
2. A warehouse contains 100 tires, of which 5 are defective. Four tires are chosen at random for a new car. Find the probability that all four are good
   (a) if we sample without replacement;
   (b) if we sample with replacement.
3. The order on a ballot of six candidates for two positions is determined at random. What is the probability that the first two candidates listed are the present office holders?
4. In a shuffled poker deck,
   (a) What is the probability that the first four cards are aces?
   (b) What is the probability that the first card is not an ace, but the next four cards are aces?
   (c) What is the probability that the four aces are next to each other somewhere in the deck?

   (You must use ordered sequences to solve this problem.)
5. Suppose that we are sampling two fuses with replacement from a set of N fuses, of which $.1N$ are defective. Show that the probability of two good fuses is .81.

1.2 SETS AND PROBABILITY

1.2.1 Sets

As we have seen in Section 1.1.1, probabilities are defined on events which are subsets of a set S called the *sample space.* Let us therefore recall some basic facts about subsets. Throughout this section, we assume that all sets are subsets of the (possibly infinite) sample space S. A *subset* A of S is a collection of points in S. If x is a point in A, we say that $x \in A$. If x is not a point in A, we write $x \notin A$. The *empty set* $\emptyset$ is the set with no points in it. ($\emptyset$ is also called the *null set.*)

We now give some notation for particular sets. We use $R = R^1$ for the set of all real numbers and R^p for the set of all $\underset{\sim}{x} = (x_1, \ldots, x_p)$, where the x_i are real numbers. We write

$$B = \{x: f(x) \in A\}$$

to mean that B is the set of all x such that $f(x) \in A$, and

$$C = \{f(x): x \in A\}$$

to be that C is the set of all y such that $y = f(x)$ for some $x \in A$. Let $a < b$ and x be in R. We define

$$(a, b) = \{x: a < x < b\}, \qquad (a, b] = \{x: a < x \leqq b\}, \qquad [a, b) = \{x: a \leqq x < b\},$$

$$[a, b] = \{x: a \leqq x \leqq b\}, \qquad (a, \infty) = \{x: x > a\}, \qquad [a, \infty) = \{x: x \geqq a\},$$

$$(-\infty, a) = \{x: x < a\}, \qquad (-\infty, a] = \{x: x \leqq a), \qquad (-\infty, \infty) = R.$$

The intervals (a, b), (a, ∞), and $(-\infty, a)$ are called *open intervals*; $[a, b]$, $[a, \infty)$, and $(-\infty, a]$ are called *closed intervals.*

Let A and B be subsets of S. We say that A *is contained in* B or B *contains* A and write $A \subset B$ or $B \supset A$ if all the points which are in A are also in B. Clearly, for any subset A, $\emptyset \subset A \subset S$. Also, if $A \subset B \subset C$, then $A \subset C$. If $A \subset B$ and $B \subset A$, we say that A *equals* B and write $A = B$.

The *union* of A and B (written $A \cup B$) is the set of all points which are in either A or B or both. From this definition, it follows that

$$A \cup B = B \cup A, \qquad A \cup A = A, \qquad A \cup \emptyset = A, \qquad A \cup S = S.$$

Note that if $A \subset B$, then $A \cup B = B$. Also, the *union* of A, B, and C (written $A \cup B \cup C$) is the set of all points that are in either A, B, or C (or any two of them, or all three of them). Then

$$A \cup B \cup C = (A \cup B) \cup C = A \cup (B \cup C).$$

The *intersection* of A and B (written $A \cap B$) is the set of all points which are in both A and B. The basic properties of intersections are the following:

$$A \cap B = B \cap A, \qquad B \cap B = B, \qquad A \cap \emptyset = \emptyset, \qquad A \cap S = A.$$

Also, if $A \subset B$, then $A \cap B = A$. Finally, the *intersection* of A, B, and C (written $A \cap B \cap C$) is the set consisting of all those points which are in A and in B and in C (i.e., which are in all three sets). It follows that

$$A \cap B \cap C = A \cap (B \cap C) = (A \cap B) \cap C.$$

Two somewhat deeper properties of unions and intersections are

$$\begin{aligned} A \cap (B \cup C) &= (A \cap B) \cup (A \cap C) \\ A \cup (B \cap C) &= (A \cup B) \cap (A \cup C). \end{aligned} \quad \text{and} \qquad (1\text{–}1)$$

One method which is often used to illustrate results like these is through Venn diagrams, which are diagrams in which the sets A, B, and C are represented as ellipses or circles. To show that two sets are equal, we draw them on the Venn diagram and observe that they are the same. For example, in the Venn diagram in Figure 1–1, the shaded set is $A \cap (B \cup C)$ and also $(A \cap B) \cup (A \cap C)$, which illustrates that these two sets are equal. (In using Venn diagrams, it is very important that they be drawn showing all possible intersections.)

The *complement* of A (written A^c) is the set of all points in S which are not in A. Some basic properties of complements are the following:

$$(A^c)^c = A, \quad \emptyset^c = S, \quad S^c = \emptyset, \quad A \cup A^c = S, \quad A \cap A^c = \emptyset, \quad A \subset B \Rightarrow A^c \supset B^c.$$

Two somewhat deeper results are

$$(A \cup B)^c = A^c \cap B^c \qquad \text{and} \qquad (A \cap B)^c = A^c \cup B^c. \qquad (1\text{–}2)$$

In the Venn diagram in Figure 1–2, the shaded set is both $(A \cup B)^c$ and $A^c \cap B^c$, so that these two sets are equal.

We say that A and B are *disjoint* if there are no points which are in both A and B, i.e., if $A \cap B = \emptyset$. We say that A, B, and C are *disjoint* if, simultaneously,

$$A \cap B = \emptyset, \qquad A \cap C = \emptyset, \qquad \text{and} \qquad B \cap C = \emptyset,$$

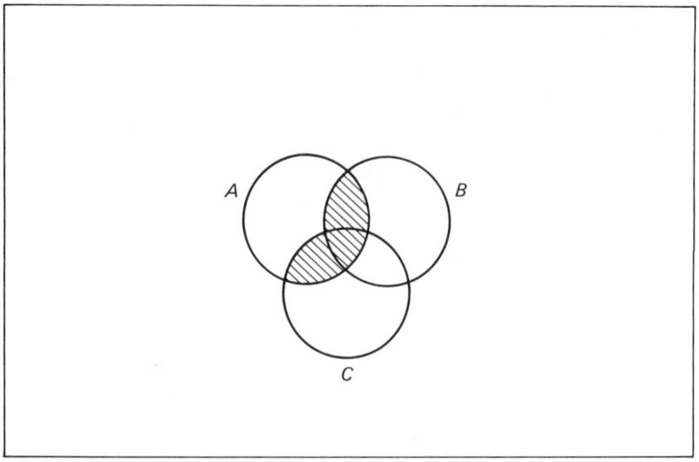

Figure 1–1 $A \cap (B \cup C) = (A \cap B) \cup (A \cap C)$.

i.e., if A and B are disjoint, A and C are disjoint, and B and C are disjoint. Note that it is not enough that $A \cap B \cap C = \emptyset$. (See the exercises.)

Example 1–10

Let $S = R$ be the set of all real numbers, let $A = (0, \infty)$ be the set of all positive numbers, let $B = [1, \infty) \cup (-\infty, -1]$ be the set of all numbers x such that $|x| \geqq 1$, and let $C = (1, \infty)$ be the set of numbers x such that $x > 1$. Since $x > 1$, x is positive, so that $C \subset A$. Similarly, $C \subset B$. $A \cup B$ is the set of all numbers such that $x > 0$ or $|x| \geqq 1$, that is, the set of all x such that $x \leqq -1$ or $x > 0$. $A \cap B$ is the set of numbers such that $x > 0$ and $|x| \geqq 1$, that is, the set of all numbers such that $x \geqq 1$. Because this set is not the null set, A and B are not disjoint. Also, since $C \subset A$, $A \cup C = A$ and $A \cap C = C$. A^c is the set of all numbers that are not positive, i.e., the set of all x such that $x \leqq 0$. C^c is the set of all numbers such that $x \leqq 1$. Note that $C^c \supset A^c$ (since $C \subset A$).

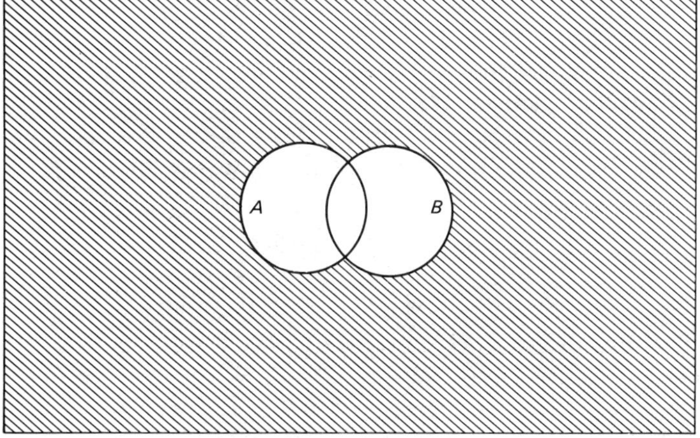

Figure 1–2 $(A \cup B)^c = A^c \cap B^c$.

We now extend some of the preceding definitions. Let B_k, $k = 1, \ldots, n$, or $k = 1, \ldots, \infty$ be a collection of sets. The *union* of the B_k is defined as the set consisting of all the points which are in at least one of the B_k, while the *intersection* of the B_k is the set of points which are in all the B_k. We write the union of the B_k as

$$\cup_{k=1}^{n} B_k \qquad \text{or} \qquad \cup_{k=1}^{\infty} B_k$$

(depending on whether the collection of sets is finite or infinite) and the intersection of the B_k as

$$\cap_{k=1}^{n} B_k \qquad \text{or} \qquad \cap_{k=1}^{\infty} B_k.$$

When there is no ambiguity, we use the simpler notation

$$\cup B_k, \qquad \cap B_k.$$

We say that the B_k are *disjoint* if B_i and B_j are disjoint for all $i \neq j$. Example 1–11 gives a collection of sets in which $\cap B_k = \emptyset$ but the B_k are not disjoint.

Example 1–11

Let C_k and B_k be the infinite sequences of sets

$$C_k = (k, k+1], \qquad B_k = (k, \infty), \qquad k = 1, 2, \ldots .$$

Then the C_k are disjoint, since $C_j \cap C_k = \emptyset$ when $j \neq k$. However, the B_j are not disjoint, since $B_j \subset B_k$, and hence $B_j \cap B_k = B_j$ if $j > k$. In addition,

$$\cup C_k = \cup B_k = (1, \infty)$$

(since all these points are in at least one of the C_k's and at least one of the B_k's). Also,

$$\cap C_k = \emptyset \qquad \text{and} \qquad \cap B_k = \emptyset$$

(since there is no point which is in all the C_k's or in all the B_k's.

We next give some results about sets which cannot be illustrated using Venn diagrams, since the number of sets may be infinite. These results are generalizations of Equations (1–1) and (1–2). We illustrate a common method for proving the equality of two sets Q and R: showing that $Q \subset R$ and $R \subset Q$.

Theorem 1–1. Let C be a subset of S, and let B_k be a (possibly infinite) collection of subsets of S. Then

a. $(\cap B_k)^c = \cup B_k^c$ and $(\cup B_k)^c = \cap B_k^c$ (DeMorgan's laws).

b. $C \cap (\cup B_k) = \cup (C \cap B_k)$ and $C \cup (\cap B_k) = \cap (C \cup B_k)$.

Proof

a. Let $x \in (\cap B_k)^c$. Then $x \notin \cap B_k$, which implies that $x \notin B_k$ for some k. Hence, $x \in B_k^c$ for some k, and $x \in \cup B_k^c$. Thus, any point in $(\cap B_k)^c$ is in $\cup B_k^c$, and it follows that

$$(\cap B_k)^c \subset \cup B_k^c.$$

Now let $y \in \cup B_k^c$. Then $y \in B_k^c$ for some k. Therefore, $y \notin B_k$ for some k, and hence $y \notin \cap B_k$ and $y \in (\cap B_k)^c$. Consequently,

$$\cup B_k^c \subset (\cap B_k)^c,$$

and it follows that $(\cap B_k)^c = \cup B_k^c$. The proof of the other equality is similar and left as an exercise.

b. Let $x \in C \cup (\cup B_k)$. Then $x \in C$ and $x \in \cup B_k$. Hence, $x \in B_k$ for some k. Therefore, $x \in C \cap B_k$ for some k, and $x \in \cup(C \cap B_k)$. Conversely, suppose that $y \in \cup(C \cap B_k)$. Then $y \in C \cap B_k$ for some k. Therefore, $y \in C$ and $y \in B_k$ for some k, so that $y \in \cup B_k$. Hence, $y \in C \cap (\cup B_k)$, and it follows that $C \cap (\cup B_k) = \cup(C \cap B_k)$. The proof of the other equality is similar and is again left as an exercise. □

Exercises—B

1. Use Venn diagrams to verify the second equations in each of Equations (1–1) and (1–2).
2. Let $A = (2, 4)$, $B = (1, 3)$, and $C = (3, 5)$. What are
   (a) $A \cap B$, $B \cap C$, and $A \cap B \cap C$?
   (b) $A \cup B$, $A \cup C$, and $A \cup B \cup C$?
   (c) $A \cup (B \cap C)$ and $A \cap (B \cup C)$?
   (d) $(A \cup B)^c$ and $A^c \cap B^c$?
3. Let $A_n = (-n, n)$, $n = 1, 2, \ldots$.
   (a) Are the A_n disjoint?
   (b) What are $\cup A_n$, $\cap A_n$, $(\cup A_n)^c$, and $(\cap A_n)^c$?
   (c) What are A_n^c and $\cap A_n^c$?
4. Let $D_n = (-n^{-1}, n^{-1})$, $n = 1, 2, \ldots$.
   (a) Are the D_n disjoint?
   (b) What are $\cap D_n$, $\cup D_n$, $(\cup D_n)^c$, and $(\cap D_n)^c$?

Exercises—C

1. (a) Prove the second DeMorgan law.
   (b) Prove the second equation in part b of Theorem 1–1.

1.2.2 General Probability Models

In discussing general probability models, we wish to generalize the equally probable model given in Sections 1.1.1 and 1.1.2 in two different directions. The first direction is to allow the possibility that not all of the points of the sample space S are equally probable. For example, when two dice are rolled, the first die may be loaded so that 1's are more likely to occur than other numbers. In that case, it would be unrealistic to assume that the point $(2, 3)$ is as probable as $(1, 3)$. The second direction we want to generalize the definition is to allow the possibility of infinitely many outcomes. For instance, suppose we choose someone at random. Let X be the person's weight, and suppose we want to find the probability that $X < 150$. It is often convenient to assume that the weight is measured on a continuum. That is, we assume that X may take on any value in some interval. Therefore, the sample space for this experiment is the set of numbers in some interval, which is an infinite set.

To define a probability space for an experiment, we first need a set S, called the *sample space*, of possible outcomes for that experiment. An *event* A is a collection of possible outcomes, or equivalently, a subset of S. We want to define the probability of A, written $P(A)$, for a suitably chosen collection of events A. As mentioned in Section 1.1.1, the probabilities are merely a model. Therefore, there are many possible probability models for a given experiment, depending on the assumptions we make. However, to be an acceptable probability model, the probabilities must satisfy certain basic assumptions. It is clear that the probabilities must at least satisfy

$$0 \leqq P(A) \leqq 1, \qquad P(\emptyset) = 0, \qquad P(S) = 1.$$

(In the dice example, the second assumption says that the probability that there is no outcome to the experiment is 0, while the third assumption says that the probability that there is some outcome is 1.)

There is one other assumption that a probability must satisfy in order to be consistent. Let A and B be two disjoint events. Then, in order for a probability to be a consistent probability measure, we must have

$$P(A \cup B) = P(A) + P(B). \tag{1–3}$$

(In the dice example, this implies that the probability that the sum of the two dice is 5 or 6 must be the probability that the sum is 5 plus the probability that the sum is 6.) Note also that for the equally likely probability model, if A and B are disjoint, then

$$P(A \cup B) = \frac{\#(A \cup B)}{\#(S)} = \frac{(\#(A) + \#(B))}{\#(S)} = P(A) + P(B).$$

We can show, by induction, that if Equation (1–3) is true for all disjoint sets A and B, then

$$P(\cup_{i=1}^{n} A_i) = \sum_{i=1}^{n} P(A_i), \tag{1–4}$$

for any finite number of disjoint sets $A_1, \ldots, A_n$. Equation (1–4) is called *finite additivity*. To make the theory work out more nicely, we make the stronger assumption that for any sequence of disjoint sets $A_1, A_2, \ldots,$

$$P(\cup_{i=1}^{\infty} A_i) = \sum_{i=1}^{\infty} P(A_i). \tag{1–5}$$

Equation (1–5) is called *countable additivity*.

We are now ready to define a probability measure.

Definition. Let S be a sample space of outcomes, and let $P(A)$ be defined on a suitably chosen collection of subsets A of S. Then $P(A)$ is a *probability measure* if it satisfies the following axioms.

Axiom 1. $0 \leqq P(A) \leqq 1$.

Axiom 2. $P(\emptyset) = 0$, $P(S) = 1$.

Axiom 3. If $A_1, A_2, \ldots$ is a sequence of disjoint subsets of S, then

$$P(\cup_{i=1}^{\infty} A_i) = \sum_{i=1}^{\infty} P(A_i).$$

(Actually, we do not need to assume that $P(A) \leqq 1$ or that $P(\emptyset) = 0$; these assertions can be derived from the other assumptions. (See Exercises C1 and C2.))

We now derive some other basic properties of probabilities which follow from the foregoing axioms. We first show that countable additivity implies finite additivity.

Theorem 1–2. Let $A_1, \ldots, A_n$ be a finite collection of disjoint subsets of S. Then

$$P(\cup_{i=1}^{n} A_i) = \sum_{i=1}^{n} P(A_i).$$

Proof. We extend the A_i into a sequence of sets by letting $A_{n+1} = \emptyset$, $A_{n+2} = \emptyset, \ldots, A_{n+k} = \emptyset, \ldots$. By hypothesis, the A_i, $i = 1, \ldots, n$, form a finite disjoint sequence of sets. In addition, $\emptyset$ is disjoint from any set, including itself. Therefore, the A_i are disjoint. Moreover,

$$\cup_{i=1}^{\infty} A_i = \cup_{i=1}^{n} A_i.$$

But then, by countable additivity,

$$P(\cup_{i=1}^{n} A_i) = P(\cup_{i=1}^{\infty} A_i) = \sum_{i=1}^{n} P(A_i) + 0 = \sum_{i=1}^{n} P(A_i)$$

(the second equality because $P(\emptyset) = 0$). □

Corollary. If A, B, and C are disjoint, then

$$P(A \cup B) = P(A) + P(B), \qquad P(A \cup B \cup C) = P(A) + P(B) + P(C).$$

We next establish some other elementary results.

Theorem 1–3

a. $P(A^c) = 1 - P(A)$.
b. If $A \subset B$, then $P(A) \leqq P(B)$.
c. Let A and B be two subsets of S (not necessarily disjoint). Then

$$P(A \cup B) = P(A) + P(A) - P(A \cap B).$$

Proof

a. We note that A and A^c are disjoint and $A \cup A^c = S$. Therefore,

$$1 = P(S) = P(A \cup A^c) = P(A) + P(A^c).$$

b. See Exercise B3.
c. Let $C_1 = A \cap B$, $C_2 = A \cap B^c$, and $C_3 = A^c \cap B$. Then C_1, C_2, and C_3 are disjoint. Also, from the Venn diagram shown in Figure 1–3, we see that

$$A = C_1 \cup C_2, \qquad B = C_1 \cup C_3, \qquad A \cup B = C_1 \cup C_2 \cup C_3.$$

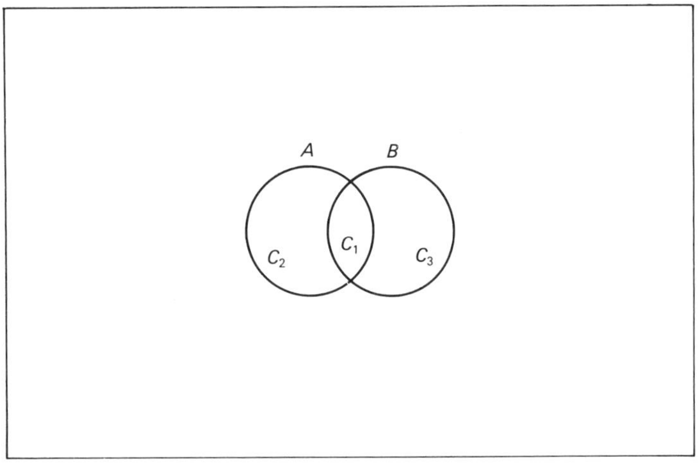

Figure 1–3 $A = C_1 \cup C_2$, $B = C_1 \cup C_3$, $A \cup B = C_1 \cup C_2 \cup C_3$.

Therefore,

$$P(A) = P(C_1) + P(C_2), \qquad P(B) = P(C_1) + P(C_3),$$
$$P(A \cap B) = P(C_1), \qquad P(A \cup B) = P(C_1) + P(C_2) + P(C_3),$$

and the result follows immediately. □

Note that parts a and b of this theorem seem like conditions that a probability should satisfy, and if we could not derive them from the given axioms, we would include them in the axioms. (Part c says that if we compute $P(A \cup B)$ by adding $P(A)$ and $P(B)$, we have double counted those points which are in both A and B, and therefore we must subtract $P(A \cap B)$.)

Example 1–12

Suppose a student has two tests on a particular day. Let the probability that he passes the first test be .6, the probability that he passes the second test be .8, and the probability that he passes both tests be .5. We find the probability that he passes at least one test. Let A be the event that he passes the first test and B be the event that he passes the second test. Then $A \cap B$ is the event that he passes both tests, and $A \cup B$ is the event that he passes at least one test. Therefore, the probability that he passes at least one test is

$$P(A \cup B) = P(A) + P(B) - P(A \cap B) = .6 + .8 - .5 = .9.$$

The event $(A \cup B)^c$ is the event that the student fails both exams. Therefore, the probability that he fails both exams is

$$P((A \cup B)^c) = 1 - P(A \cup B) = 1 - .9 = .1.$$

We now extend part c of the last theorem to unions of more than two sets.

Theorem 1–4

a. Let A, B, and C be (possibly nondisjoint) sets. Then

$$P(A \cup B \cup C) = P(A) + P(B) + P(C) - (P(A \cap B) + P(A \cap C) + P(B \cap C)) + P(A \cap B \cap C).$$

b. Let $A_1, \ldots, A_n$, $n \geqq 2$, be a finite collection of (possibly nondisjoint) subsets of S. Then

$$P(\cup A_i) = \sum_i P(A_i) - \sum\sum_{j<i} P(A_i \cap A_j) + \sum\sum\sum_{k<j<i} P(A_i \cap A_j \cap A_k) - \cdots + (-1)^{n+1}P(\cap_i A_i).$$

Proof

a. This result is a special case of part b.

b. We prove this result by induction. By Theorem 1–3, part c, it is true for $n = 2$. Suppose the result is true for $n = k - 1$. Let $A_1, \ldots, A_k$ be a collection of sets. Then

$$P(\cup_{i=1}^{k} A_i) = P((\cup_{i=1}^{k-1} A_i) \cup A_k) = P(\cup_{i=1}^{k-1} A_i) + P(A_k) - P((\cup_{i=1}^{k-1} A_i) \cap A_k).$$

By the induction hypothesis,

$$P(\cup_{i=1}^{k-1} A_i) = \sum_{i=1}^{k-1} P(A_i) - \sum_{i=1}^{k-1}\sum_{j=1}^{i-1} P(A_i \cap A_j) + \cdots + (-1)^k P(\cap_{i=1}^{k-1} A_i),$$

and

$$P((\cup_{i=1}^{k-1} A_i) \cap A_k) = P(\cup_{i=1}^{k-1}(A_i \cap A_k)) = \sum_{i=1}^{k-1} P(A_i \cap A_k) - \sum_{i=1}^{k}\sum_{j=1}^{i-1} P(A_i \cap A_j \cap A_k) + \cdots + (-1)^k P(\cap A_i),$$

(using the fact that $(A_i \cap A_k) \cap (A_j \cap A_k) = A_i \cap A_j \cap A_k$). Putting these equations together, we see that

$$P(\cup_{i=1}^{k} A_i) = \sum_{i=1}^{k} P(A_i) - \sum_{i=1}^{k}\sum_{j=1}^{i-1} P(A_i \cap A_j) + \cdots + (-1)^{k+1} P(\cap_{i=1}^{k} A_i).$$

Therefore, the result is true for $n = k$ if it is true for $n = k - 1$, and since it is also true when $n = 2$, it is true for all $n \geqq 2$. □

Theorem 1–4 is often called the inclusion-exclusion theorem. One intuitive way to understand part a is to note that when we add $P(A)$, $P(B)$, and $P(C)$ we have double counted all the points in $A \cap B$, $A \cap C$, and $B \cap C$, so we must subtract those probabilities. However, at this point we have added the points in $A \cap B \cap C$ three times and subtracted them three times, so that they have not been counted at all. We must therefore add that probability back.

Example 1–13

Suppose we have n individuals who each have a hat. Suppose they throw their hats in a room, and the hats are returned to them randomly. We find the probability that at least one person gets his or her own hat. Let A_i be the event that a person gets his or her own hat. Then we want $P(\cup A_i)$. Now, the hats are gotten at random, so that $P(A_i) = 1/n$. Therefore,

$$\sum_i P(A_i) = \frac{n}{n} = 1.$$

Furthermore, the ith person has n choices of a hat, and after that person has chosen, the jth person has $n-1$ choices. Therefore, the probability that the ith and jth individual both get the correct hat is $1/n(n-1)$; i.e., $P(A_i \cap A_j) = 1/n(n-1)$. Therefore,

$$\sum\sum_{j<i} P(A_i \cap A_j) = \frac{\begin{bmatrix} n \\ 2 \end{bmatrix}}{n\,(n-1)} = \frac{1}{2!}.$$

Similarly,

$$\sum\sum\sum_{k<j<i} P(A_i \cap A_j \cap A_k) = \frac{\begin{bmatrix} n \\ 3 \end{bmatrix}}{n\,(n-1)(n-2)} = \frac{1}{3!}.$$

This procedure can be continued, and we see that the probability p_n that at least one of the n people gets his or her own hat is

$$p_n = 1 - \frac{1}{2!} + \frac{1}{3!} - \frac{1}{4!} + \cdots + \frac{(-1)^{n+1}}{n!} = \sum_{i=1}^{n} \frac{(-1)^{i+1}}{i!}.$$

We recall from basic calculus that

$$e^{-1} = \sum_{i=0}^{\infty} \frac{(-1)^i}{i!} = 1 - \sum_{i=1}^{\infty} \frac{(-1)^{i+1}}{i!}.$$

Therefore,

$$\lim p_n = \sum_{i=1}^{\infty} \frac{(-1)^{i+1}}{i!} = 1 - e^{-1} = .6321.$$

This somewhat surprising result says that if n is large, so that we have a very large collection of people, then the probability that at least one person gets his or her own hat is about .6321. Intuitively, we might expect the probability to go to zero, since each person would have a very large number of hats to choose from, and in fact, the probability that any particular person gets his or her own hat is $1/n$, which does go to 0. However, the number of people is also increasing, so that although the probability that any *particular* person gets his or her own hat goes to zero, the probability that at least one person gets his or her own hat goes to .6321.

We say that a sequence of sets $A_1, A_2, \ldots$ is *increasing* if $A_i \subset A_{i+1}$ for all i, and that the sequence is *decreasing* if $A_i \supset A_{i+1}$ for all i.

Theorem 1–5

a. If $A_1, A_2, \ldots$ is an increasing sequence of sets, then

$$P(\cup_{n=1}^{\infty} A_i) = \lim_{n\to\infty} P(A_n).$$

b. If $B_1, B_2, \ldots$ is a decreasing sequence of sets, then

$$P(\cap_{n=1}^{\infty} B_n) = \lim_{n\to\infty} P(B_n).$$

Proof

a. Let $C_1 = A_1$, $C_i = A_i \cap (A_{i-1})^c$, $i > 1$. Then the C_i are disjoint,

$$A_n = \cup_{i=1}^{n} C_i \qquad \text{and} \qquad \cup_{n=1}^{\infty} A_n = \cup_{i=1}^{\infty} C_i.$$

Therefore,

$$P(\cup_{n=1}^{\infty} A_n) = P(\cup_{i=1}^{\infty} C_i) = \sum_{i=1}^{\infty} P(C_i) = \lim_{n\to\infty} \sum_{i=1}^{n} P(C_i)$$
$$= \lim_{n\to\infty} P(\cup_{i=1}^{n} C_i) = \lim_{n\to\infty} P(A_n).$$

b. Let $A_n = B_n^c$. Then the $A_1, A_2, \ldots$ is an increasing sequence of sets. Therefore, by part a and DeMorgan's law,

$$P(\cap_{n=1}^{\infty} B_n) = 1 - P((\cap_{n=1}^{\infty} B_n)^c) = 1 - P(\cup_{n=1}^{\infty} A_n)$$
$$= 1 - \lim_{n\to\infty} P(A_n) = \lim_{n\to\infty} P(B_n). \quad \square$$

For example, if the sample space is the space of real numbers, then

$$P((-\infty,\infty)) = \lim_{n\to\infty} P((-n,n)) \qquad \text{and} \qquad P([0]) = \lim_{n\to\infty} P((-n^{-1}, n^{-1}))$$

(where (a,b) is the open set of points between a and b and [0] is the closed set consisting of the point 0). □

Theorem 1–5 is called the continuity theorem for probabilities. The following corollary illustrates why.

Corollary 1–5. Let $C_1, C_2, \ldots$ be any sequence of sets. Then

$$P(\cup_{i=1}^{\infty} C_i) = \lim_{n\to\infty} P(\cup_{i=1}^{n} C_i) \qquad \text{and} \qquad P(\cap_{i=1}^{\infty} C_i) = \lim_{n\to\infty} P((\cap_{i=1}^{n} C_i).$$

Proof. $A_n = \cup_{i=1}^{n} C_i$ is an increasing sequence of sets and $B_n = \cap_{i=1}^{n} C_i$ is a decreasing sequence of sets, so that these results follow directly from Theorem 1–5. □

Note that the property stated in Corollary 1–5 is one that we would expect a probability measure to have, and if it did not, we might add it to our assumptions. (Note also that if we replace the assumption of countable additivity with the weaker assumption of finite additivity, the property may no longer be satisfied.)

Let $A_1, \ldots, A_n$ be a finite collection of (possibly not disjoint) sets. In the exercises, two inequalities are derived which are often useful, viz.,

$$P(\cup_i A_i) \leqq \sum_i P(A_i), \qquad P(\cap_i A_i^c) \geqq 1 - \sum_i P(A_i).$$

(The sums may be either finite or infinite.) The first inequality is called *Boole's inequality,* and the second one is called *Bonferroni's inequality.*

It should be emphasized that the three axioms given in the definition of a probability measure guarantee that probabilities are consistent. They do not guarantee, however, that probabilities are "correct." Since there are many possible probability models for a particular experiment, there are many possible models satisfying the three axioms. As mentioned earlier, the conclusions that we draw from a particular model are only as good as the model is. If we have an incorrect model, then the conclusions are likely to be incorrect. For the dice example given in Section 1.1.1, if we use the equally probable model when the dice are not balanced, we are likely to get incorrect results.

Moreover, we have been somewhat cavalier about the sets on which we define probabilities. Using the axiom of choice, we can show that it is not possible to define a consistent set of probabilities on all possible subsets for even some very simple probability models (with infinite sample spaces). However, it is possible to define probabilities for all the subsets we shall be studying. A good discussion of measure-theoretic results and their relationship to probability theory appears in Billingsley (1979).

Exercises—A

1. Suppose a die is not balanced, so that the probability of a 1 is $\frac{1}{21}$, the probability of a 2 is $\frac{2}{21}$, and in general the probability of an i is $i/21$, $i = 1,2,\ldots,6$. We roll that die one time.
   (a) Find the probability that an even number comes up.
   (b) Find the probability that the number which comes up is at most 3.
2. In a particular city, 40% of the people subscribe to magazine A, 30% to magazine B, and 50% to magazine C. However, 10% subscribe to both A and B, 25% subscribe to both A and C, and 15% subscribe to both B and C. Finally, 5% subscribe to all three magazines. A person is chosen at random.
   (a) What is the probability that the chosen person subscribes to at least one magazine?
   (b) What is the probability that the person subscribes to at least two magazines?
3. Suppose that a student takes three tests on a given day and that the probability that she passes any particular test is .6, the probability that she passes any particular pair of tests is .4, and the probability that she passes all three tests is .3. What is the probability that she passes at least one test?

Exercises—B

1. Suppose that $P(A) = .6$, $P(B) = .5$, $P(C) = .4$, $P(A \cap B) = .3$, $P(A \cap C) = .2$, $P(B \cap C) = .2$, and $P(A \cap B \cap C) = .1$.
   (a) Find $P(A \cup B \cup C)$ and $P((A \cup B \cup C)^c)$.
   (b) Find $P((A \cup B) \cap C)$ and $P(A \cup (B \cap C))$.
   (c) Find $P((A^c \cup B^c) \cap C^c)$ and $P((A^c \cap B^c) \cup C^c)$.
   (*Hint*: Draw a Venn diagram and fill in the probabilities of the various sets.)
2. Suppose, in addition to the probabilities in Exercise B1, that there is a set D such that $P(D \cap B) = P(D \cap C) = 0$, $P(A \cap D) = .1$, and $P(D) = .2$.
   (a) What are $P(B \cap C \cap D)$ and $P(A \cap C \cap D)$?
   (b) What are $P(A \cup B \cup D)$ and $P(A \cup B \cup C \cup D)$?
   (c) What is $P((A \cap B) \cup (C \cap D))$?

3. Prove part b of Theorem 1–3.
4. Suppose that a sample space consists of a finite number of points $a_1, \ldots, a_k$. Show that for any set B, $P(B) = \sum_{a_k \in B} P(a_k)$.

Exercises—C

1. Show that the assumption that $P(A) \leqq 1$ may be dropped from the assumptions, i.e., that this assumption can be derived from the other assumptions. (*Hint*: The derivation of part b of Theorem 1–3 does not use this assumption anywhere.)
2. Show that the assumption $P(\emptyset) = 0$ may be dropped from the assumptions. (*Hint*: $\emptyset = \emptyset \cup \emptyset \cup \ldots$.)
3. (a) (Boole) Use induction to show that $P(\cup_{k=1}^{n} A_k) \leqq \sum_{k=1}^{n} P(A_k)$.
   (b) (Bonferroni) Use DeMorgan's laws to show that $P(\cap_{k=1}^{n} A_k^c) \geqq 1 - \sum_{k=1}^{n} P(A_k)$.
   (c) Use Theorem 1–5 to show that we may take $n = \infty$ in parts a and b.

1.3 CONDITIONAL PROBABILITY AND INDEPENDENCE

1.3.1 Conditional Probability

Let A and B be subsets of a sample space S, and suppose that $P(B) > 0$. We define the *conditional probability of A given B*, written $P(A|B)$, by

$$P(A|B) = \frac{P(A \cap B)}{P(B)}.$$

The interpretation of $P(A|B)$ is the probability that A occurs when we know that B has occurred.

To see why this definition is sensible, we return to the equally probable model of Section 1.1.1, in which we have a finite sample space S and the probability of any subset A of S is just $\#(A)/\#(S)$. Let A and B be subsets of S. If we know that B has occurred, we can look at B as our sample space. Since all the points in B are equally probable, to find the probability that A occurs, we count the number of points in B which are also in A and divide by the number of points in B. Therefore, for the equally probable model, the conditional probability of A given B should be

$$P(A|B) = \frac{\#(A \cap B)}{\#(B)} = \frac{\#(A \cap B)/\#(S)}{\#(B)/\#(S)} = \frac{P(A \cap B)}{P(B)},$$

which matches the definition given.

To make the discussion in the previous paragraph more precise, consider again the model for rolling two dice in which each of the 36 possible outcomes is equally probable. Let A be the event that the first die is odd and B be the event that the sum of the dice is 4. Then B consists of the points (1,3), (2,2), and (3,1). Two of these points have the first die odd, so that the conditional probability of A given B should be $\frac{2}{3}$. That is, we should have

$$P(A|B) = \frac{2}{3} = \frac{2/36}{3/36} = \frac{P(A \cap B)}{P(B)}.$$

Example 1–14

Let us consider again a student who is taking two tests on a given day. Let A be the event that the student passes the first test and B be the event that he passes the second. As in Section 1.2.2, suppose that

$$P(A) = .6, \qquad P(B) = .8, \qquad P(A \cap B) = .5.$$

Then the probability that the student passes the second test given that he passes the first is

$$P(B|A) = \frac{P(A \cap B)}{P(A)} = \frac{5}{6},$$

while the probability that the student passes the first test given that he passes the second is

$$P(A|B) = \frac{P(A \cap B)}{P(B)} = \frac{5}{8}.$$

Thus, in general,

$$P(A|B) \neq P(B|A).$$

We now present a result which is often useful for finding probabilities of intersections.

Theorem 1–6

a. If $P(A) > 0$, then $P(A \cap B) = P(A)\,P(B|A)$.
b. If $P(A \cap B) > 0$, then $P(A \cap B \cap C) = P(A)\,P(B|A)\,P(C|A \cap B)$.
c. If $P(A_1 \cap A_2 \cap \cdots \cap A_{k-1}) > 0$, then $P(A_1 \cap A_2 \cap A_3 \cap \cdots \cap A_k) = P(A_1)\,P(A_2|A_1)\,P(A_3|(A_1 \cap A_2)) \ldots P(A_k|A_1 \cap A_2 \cap \cdots \cap A_{k-1})$.

Proof

a. This result follows directly from the definition of conditional probability.
b. By the definition of conditional probability,

$$P(A \cap B \cap C) = P(C|A \cap B)\,P(A \cap B) = P(C|A \cap B)\,P(B|A)\,P(A)$$

c. By the definition of conditional probability,

$$\begin{aligned} &P(A_1)\,P(A_2|A_1) \ldots P(A_k|A_1 \cap \cdots \cap A_{k-1}) \\ &= (P(A_1))\frac{P(A_1 \cap A_2)}{P(A_1)}\frac{P(A_1 \cap A_2 \cap A_3)}{P(A_1 \cap A_2)} \cdots \frac{P(A_1 \cap A_2 \cap \cdots \cap A_k)}{P(A_1 \cap A_2 \cap \cdots \cap A_{k-1})} \\ &= P(A_1 \cap A_2 \cap A_3 \cap \cdots \cap A_k). \end{aligned}$$

(Note that the condition that $P(A_1 \cap A_2 \cap \cdots \cap A_{k-1}) > 0$ guarantees that $P(A_1 \cap \cdots \cap A_j) > 0$ for $j \leqq k-1$ and hence that $P(A_{j+1}|A_1 \cap \cdots \cap A_j)$ is defined.) □

Example 1–15

We return to the model in which we find k people and check their birthdays. We want to find the probability of the event Q that no two people have the same birthday. As before, we assume that all 365 days are equally probable. Let A_1 be the event that we draw the first person, let A_2 be the event that we draw the second person and his or her birthday does not match the first person's, . . . , and let A_k be the event that we draw the kth person and his or her birthday does not match any of the previous birthdays. Then

$$Q = \bigcap_{i=1}^{k} A_i.$$

We note that $P(A_1) = 1$, $P(A_2|A_1) = \frac{364}{365}$ (since there are 364 days left to choose), and so on, i.e.,

$$P(A_3|(A_1 \cap A_2) = \frac{363}{365}, \ldots, P\left(A_k|\bigcap_{i=1}^{k-1} A_i\right) = \frac{(366-k)}{365}$$

since the conditional probability that the jth birthday is different given that the first $j-1$ are different is $(365-(j-1))/365$). Therefore,

$$P(Q) = P(A_1)P(A_2|A_1)P(A_3|A_1 \cap A_2) \ldots P(A_k|\bigcap_{i=1}^{k-1} A_i)$$

$$= \frac{365}{365}\frac{364}{365}\frac{363}{365}\cdots\frac{366-k}{365} = {}_{365}P_k/(365)^k,$$

which is the same answer we got using counting arguments.

Example 1–16

The probability that a particular football team will win the next game on a rainy day is .8, while the probability that it will win on a sunny day is .6. Suppose that the probability of a rainy day is .7 (so that the probability of a sunny day is .3). What is the probability that the football team will win its next game? Let A be the event that the team wins and B be the event that it is rainy, so that B^c is the event that it is sunny. We know that

$$P(B) = 1 - P(B^c) = .7, \qquad P(A|B) = .8, \qquad P(A|B^c) = .6.$$

We seek $P(A)$. Note that $A \cap B$ and $A \cap B^c$ are disjoint sets, and

$$(A \cap B) \cup (A \cap B^c) = A.$$

Therefore,

$$P(A) = P(A \cap B) + P(A \cap B^c) = P(B)P(A|B) + P(B^c)P(A|B^c)$$
$$= (.7)(.8) + (.3)(.6) = .74.$$

Example 1–17

It has been suggested that we implement a procedure for testing for child abuse. Suppose that when a child has been abused a doctor can correctly identify the abuse with probability .99, and that when a child has not been abused the doctor can correctly identify the nonabuse with probability .9. Suppose also that the probability that a child has been abused is .05. If the doctor says that a child has been abused, what is the probability that the child has actually been abused? Let A be the event that the child is abused, so that A^c is the event that the child is not abused. Let B be the event that the doctor decides that the child has been abused. We are given that

$$P(A) = .05, \qquad P(B|A) = .99, \qquad P(B|A^c) = .1.$$

We wish to find $P(A|B) = P(A \cap B)/P(B)$. Now,

$$P(A \cap B) = P(A)\,P(B|A) = (.05)(.99) = .0495.$$

Furthermore,

$$P(B) = P(B \cap A) + P(B \cap A^c) = P(A)\,P(B|A) + P(A^c)\,P(B|A^c)$$
$$= (.05)(.99) + (.95)(.1) = .1445.$$

Therefore,

$$P(A|B) = \frac{P(A \cap B)}{P(B)} = \frac{.0495}{.1445} = .3426.$$

We see then that even though the procedure is fairly accurate (being correct 99% of the time when the child is abused and 90% of the time when the child is not), the probability that a child who is labeled abused really is abused is only about one-third. That is, about two-thirds of the people who would be accused of child abuse with this procedure would be incorrectly accused. The reason for this is apparent in the calculations: Ninety-nine percent of the abused children is a much smaller number than 10% of the nonabused children (since only 5% of the children are abused). Phrased another way, about two-thirds of those identified as positive with this procedure would be false positives. The phenomenon of large numbers of false positives happens often when we have a test of this type (for example, lie detector tests, urinalysis tests for drugs, etc.). For this reason, such a test must be used with great caution: It is not enough to measure the performance of the test on "guilty" individuals; we must also measure performance on "innocent" individuals, as well as look at the proportion of "guilty" individuals.

Example 1–18

Suppose we have three machines for making a particular part. The first machine produces 4% defectives, the second machine produces 6% defectives, and the third machine produces 1% defectives. Suppose also that the first machine supplies 30%, the second machine 20%, and the third machine 50%, of the parts. If a part is selected at random, what is the probability that it is defective? Given that the part is defective, what is the probability that it came from the first machine? To answer these questions, let A_i be the event that the part came from the ith machine, and let D be the event that the part is defective. Then

$$P(A_1) = .3,\; P(A_2) = .2,\; P(A_3) = .5,\; P(D|A_1) = .04,\; P(D|A_2) = .06,\; P(D|A_3) = .01.$$

Now, $A_1 \cap D$, $A_2 \cap D$, and $A_3 \cap D$ are disjoint and their union is D. Therefore, the probability that a part is defective is

$$P(D) = P(D \cap A_1) + P(D \cap A_2) + P(D \cap A_3)$$
$$= (.3)(.04) + (.2)(.06) + (.5)(.01) = .029.$$

Given that the part is defective, the probability that it came from the first machine is

$$P(A_1|D) = \frac{P(D \cap A_1)}{P(D)} = \frac{(.3)(.04)}{.029} = \frac{12}{29}.$$

We next give two results which are often used for solving problems like those in the last three examples, although it usually seems just as easy to do these problems from the definition and Theorem 1–6, as we have just done. A *partition* of the sample space S is a collection of disjoint sets $C_1, \ldots, C_n$ such that $S = \cup C_i$.

Theorem 1–7. Let $C_1, \ldots, C_n$ be a partition of the sample space S, and let A be an event. Then

a. (Law of total probability) $P(A) = \sum_i P(C_i)\, P(A|C_i)$.

b. (Bayes rule) $P(C_j|A) = P(C_j)\, P(A|C_j) \big/ \sum_i P(C_i)\, P(A|C_i)$.

Proof

a. Let $B_i = A \cap C_i$. Then the B_i are disjoint sets such that

$$\cup B_i = \cup(A \cap C_i) = A \cap (\cup C_i) = A \cap S = A.$$

Therefore, using Theorem 1–5a,

$$P(A) = \Sigma P(A \cap C_i) = \Sigma P(C_i)\, P(A|C_i).$$

b. By the definition of conditional probability and part a of this theorem,

$$P(C_j|A) = \frac{P(C_j \cap A)}{P(A)} = \frac{P(C_j)\, P(A|C_j)}{\Sigma P(C_i)\, P(A|C_i)}. \quad \square$$

Exercises—A

1. Suppose that my dog has an 80% chance of detecting a burglar and, when he detects the burglar, has a 40% chance of scaring him off. If a burglar comes to the house, what is the probability that the dog will scare him off?
2. A Democratic candidate D will face the winner of a three-way primary for the Republican nomination between A, B, and C. The probability that A wins the primary is .4, that B wins it is .1, and that C wins is .5. The conditional probability that D can beat A is .4, that D can beat B is .8, and that D can beat C is .5. What is the probability that D will win the election?
3. A hitter comes to the plate two times in a game. Suppose the probability of a hit on the first at bat is .3, the conditional probability of a hit on the second at bat given a hit on the first at bat is .4, and the conditional probability of a hit on the second at bat given no hit on the first at bat is .2.
   (a) What is the probability of a hit on both at bats?
   (b) What is the probability of at least one hit?
   (c) What is the probability of a hit on the second at bat?
4. Suppose that a lie detector test has the following properties. If the suspect is telling the truth, the lie detector will correctly say so with probability .85; if the suspect is lying, it will correctly identify this with probability .99. If 95% of the people are telling the truth, find the probability that a person is actually lying when the test says that he or she is.
5. Fifty percent of the freshmen at a school favor a particular candidate, while 40% of the sophomores do, 30% of the juniors do, and 20% of the seniors do. Suppose that 20% of the students are freshmen, 30% of them are sophomores, 25% are juniors, and 25% are seniors.
   (a) What proportion of the students favor the candidate?
   (b) Given that a student favors the candidate, what is the probability that the student is a freshman?

6. Two roommates are taking the same course. Roommate A goes to class 70% of the time, and B goes 60% of the time. They both go 50% of the time.
   (a) What is the probability that at least one goes?
   (b) What are the conditional probabilities that B goes given that A goes and that A goes given that B goes?
   (c) What is the conditional probability that A is in class given that at least one of the roommates is?
7. In Example 1–1,
   (a) What is the conditional probability that the sum of the dice is even given that the first die is odd?
   (b) What is the conditional probability that the sum is at least 8 given that the first die is at least 5?
   (c) What is the conditional probability that the first die is at least 5 given that the sum is at least 8.
   (d) What is the conditional probability that the first die is at least 5 given that the sum is even?
8. Suppose that a telegraph operator is sending a series of dots and dashes. Suppose also that 60% of the message is dots and that the probability that any particular dash or dot is correctly interpreted at the other end is .9. If a particular symbol is interpreted as a dot, what is the probability that it was sent as a dot?
9. Suppose I draw five cards from an ordinary deck of cards.
   (a) Find the probabilities of at least three spades and at least four spades.
   (b) Find the conditional probability of at least four spades given at least three spades.
   (c) Find the conditional probability of four A's given at least two A's.
10. Consider Exercise A2 of Section 1.2.2.
    (a) Find the conditional probability that a person subscribes to magazine A given that he or she subscribes to magazine B.
    (b) Find the conditional probability that a person who subscribes to any magazine subscribes to magazine C.
    (c) Find the conditional probability that a person subscribes to magazines A and B given that he or she subscribes to magazine C.
    (d) Find the conditional probability that a person subscribes to magazine A given that he or she subscribes to at least one magazine, and also find the conditional probability that the person subscribes to magazine A given that he or she subscribes to at least two magazines.
    (e) Find the conditional probability that a person subscribes to at least two magazines given that he or she subscribes to at least one.
11. (Bertrand's box paradox) A man has three coins. One of the coins is a fair coin, one has two heads, and one has two tails. A blindfolded woman chooses a coin at random from these three and flips the coin. She gets a head. What is the probability that the other side is also a head?
12. (The prisoner paradox) Three prisoners, A, B, and C, are in jail. Two of the prisoners have been randomly selected to be executed, so that the probability that any one prisoner is executed is $\frac{2}{3}$. Prisoner A has a great idea. She knows that either B or C will be executed. Therefore, she asks the jailer to tell her which one of B and C will be executed, and the jailer says C. Prisoner A sleeps much better at night after hearing this, because she thinks that she and prisoner B now have equal chances of being executed, so that her probability of execution has decreased to $\frac{1}{2}$. What is wrong with this reasoning?

13. On a certain game show, there are four doors. Behind one door is a very valuable prize, but behind the other three doors there are no prizes. The contestant guesses a door. If he guesses the door with the prize, he keeps the prize, but if he guesses another door, he gets nothing. He therefore has probability $\frac{1}{4}$ of winning the prize. After he has chosen a door, the master of ceremonies (MC) tells the contestant which one of the remaining doors does not hide the prize. The MC then offers the contestant the opportunity to drop the door game and play another game in which he has probability .3 of winning the prize. The contestant reasons that since he knows one door which does not hide the prize and he has guessed one of the three remaining doors, his probability of winning the door game is now $\frac{1}{3}$, so he stays with the door game. What is wrong with his reasoning?

Exercises—B

1. Suppose that $P(A) = .6$, $P(B) = .5$, $P(C) = .4$, $P(A \cap B) = .3$, $P(A \cap C) = .2$, $P(B \cap C) = .2$, and $P(A \cap B \cap C) = .1$.
 (a) Find $P(A|(B \cap C))$, $P((A \cap B)|C)$, and $P((A \cap B)|(B \cap C))$.
 (b) Find $P((A \cup B)|C)$, $P(C|(A \cup B))$, and $P((A \cup B)|(B \cup C))$.

2. Using the probabilities given in Exercise B1 above,
 (a) Find $P((A^c \cup B)|C^c)$ and $P((A \cup B)^c|C^c)$
 (b) Find $P((A^c \cap B)|(A \cup B))$ and $P((A^c \cap B)|(B^c \cup A))$

3. Suppose that $P(A|B) < P(A)$. Show that $P(B|A) < P(B)$.

4. Show that $P(A^c|B) = 1 - P(A|B)$.

5. Show that $P(A \cup B|C) = P(A|C) + P(B|C) - P(A \cap B|C)$.

6. Let S be a sample space and $C \subset S$ be such that $P(C) > 0$. Let $Q(A) = P(A|C)$. Show that Q is a probability measure. That is, show that
 (a) $0 \leqq Q(A) \leqq 1$.
 (b) $Q(S) = 1$, $Q(\emptyset) = 0$.
 (c) If $A_1, A_2, \ldots$ is a disjoint sequence of sets, then $Q(\cup A_k) = \Sigma Q(A_k)$.

1.3.2 Independent Events

Let A and B be subsets of a sample space S. We say that A and B are *independent* if

$$P(A \cap B) = P(A)P(B).$$

We think of independent events as ones which are unrelated; that is, knowing whether A occurs tells us nothing about whether B occurs. The reason for this interpretation is the following simple result.

Theorem 1–8. If $P(B) > 0$, then A and B are independent if and only if

$$P(A|B) = P(A).$$

Proof. This follows directly from the definitions. □

(Note that the *definition* of independent events can be applied even when $P(B) = 0$, but Theorem 1.8 obtains only when $P(B) > 0$).

Example 1–19

We return again to the student who is taking two exams. As before let A and B be the events that the student passes the first and second exams, respectively. We assume that

$P(A) = .8$, $P(B) = .6$, and $P(A \cap B) = .5$. Since $P(A \cap B) \neq P(A)P(B)$, the results on the two exams are not independent. In fact, since $P(A \cap B) > P(A)P(B)$, the student is somewhat more likely to pass both exams than we would expect if the events were independent. Thus, passing the first exam may help the student's performance on the second by increasing his confidence.

Example 1–20

Consider again the roll of two dice. Let A_i be the event that we roll an i on the first die and B_j be the event that we roll a j on the second die. We assume that the dice are fair, so that

$$P(A_i) = P(B_j) = \frac{1}{6}, \qquad i = 1, \ldots, 6;\ j = 1, \ldots, 6.$$

We also assume that the dice are independent, so that

$$P(A_i \cap B_j) = P(A_i)P(B_j) = \left(\frac{1}{6}\right)\left(\frac{1}{6}\right) = \frac{1}{36}.$$

Therefore, if we assume that the dice are fair and independent, we get the model in which all 36 outcomes on the dice are equally likely. This is the model we used in Example 1–1.

We say that events A, B, and C are *independent* if they satisfy the following four conditions:

$$P(A \cap B) = \mathrm{P(A)P(B)}, \qquad P(A \cap C) = \mathrm{P(A)P(C)},$$
$$P(B \cap C) = \mathrm{P(B)P(C)}, \qquad P(A \cap B \cap C) = P(A)\mathrm{P(B)P(C)}.$$

If A, B, and C satisfy just the first three conditions, we say they are *pairwise independent.* In Exercise A2, an example is given of three events which are pairwise independent but not independent.

Example 1–21

Three tennis players, A, B, and C, are playing a round robin match. That is, A plays B once, A plays C once, and B plays C once. We say that a player wins the round robin if the player wins both of his or her matches. Suppose that the probability that A beats B is .55, the probability that A beats C is .6, and the probability that B beats C is .9. We assume that the matches are independent. Then the probability that A wins the tournament, the probability that B wins the tournament, and the probability that C wins the tournament are given respectively by

$$P(A) = (.55)(.6) = .33, \qquad P(B) = (.45)(.9) = .405, \qquad \text{and} \qquad P(C) = (.4)(.1) = .04.$$

We note that although A is more likely to beat B than B is to beat A, B is more likely to win the round robin. The probability that nobody wins the round robin is

$$\begin{aligned} P(N) &= 1 - P(A \cup B \cup C) = 1 - (P(A) + P(B) + P(C)) \\ &= 1 - (.33 + .405 + .04) = .225 \end{aligned}$$

(using the fact that A, B, and C are disjoint).

Now, let $A_1, \ldots, A_n$ be events. We say that the A_i are *independent* if, for every set of distinct subscripts $i_1, i_2, \ldots, i_k$,

$$P(A_{i_1} \cap A_{i_2} \cap \cdots \cap A_{i_k}) = P(A_{i_1})P(A_{i_2}) \ldots P(A_{i_k}).$$

(That is, the A_i are independent if the probability of any intersection of the A_i is the product of the probabilities of those A_i.)

If we have n sets and n is reasonably large, it is quite tedious to check all the 2^n possible intersections for independence. Fortunately, we shall typically assume independence, in which case it is often easy to compute probabilities because the probability of any intersection is the product of the probabilities of the sets.

We now give a result which is intuitively obvious, but is often useful in problems.

Theorem 1–9. If A and B are independent, so are A and B^c, A^c and B, and A^c and B^c.

Proof. Consider the last two sets. We note that $A^c \cap B^c = (A \cup B)^c$. Now, since A and B are independent,

$$P(A^c \cap B^c) = P((A \cup B)^c) = 1 - P(A \cup B) = 1 - (P(A) + P(B) - P(A \cap B))$$
$$= 1 - P(A) - P(B) + P(A)\,P(B) = (1 - P(A))(1 - P(B)) = P(A^c)\,P(B^c).$$

The other results are proved similarly and are left as exercises. □

Example 1–22

Consider the system of components shown in Figure 1–4. The system works if either the upper path or the lower path works. The upper path works if A and D work and either B or C works. The lower path works if E works, if either F or G works, and if either H or I works. We assume that the components each fail with probability .1 and that their failures are independent. We now find the probability that the system works. We write $P(A)$, $P(B)$, etc., for the probability that A works, the probability that B works, etc. The probability that either B or C works is one minus the probability that they both fail, i.e.,

$$P(B \cup C) = 1 - P((B \cup C)^c) = 1 - P(B^c \cap C^c) = 1 - P(B^c)\,P(C^c) = 1 - (.1)(.1) = .99$$

(using Theorem 1–9). Now, the probability that the upper path works is

$$P(A \cap (B \cup C) \cap D) = P(A)\,P(B \cup C)\,P(D) = (.9)(.99)(.9) = .802.$$

Similarly, the probability that the lower branch works is

$$P(E \cap (F \cup G) \cap (H \cup I)) = P(E)\,P(F \cup G)\,P(H \cup I) = (.9)(.99)(.99) = .882.$$

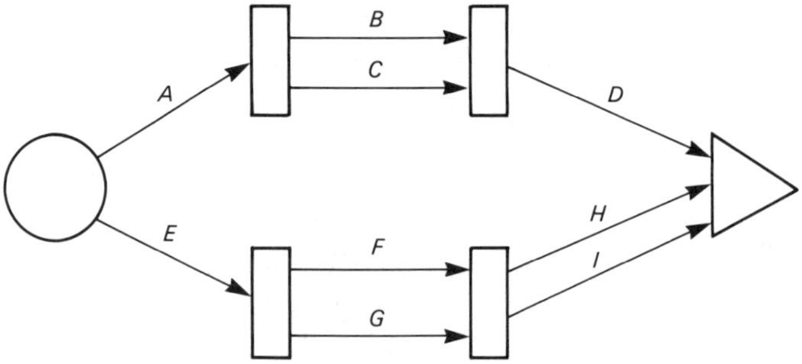

Figure 1–4 System of components for Example 1–22.

The probability $P(W)$ that the whole system works is the probability that either branch works, which is one minus the probability that they both fail, which, using the independence assumption, is

$$P(W) = 1 - (1 - .802)(1 - .882) = 1 - (.198)(.118) = .977.$$

Example 1–23

Consider the problem of drawing two cards from an ordinary deck. Let A be the event that an ace occurs on the first draw and B be the event that an ace occurs on the second draw. We first look at sampling without replacement, i.e., we draw the first card and then draw the second card without putting the first card back. Then

$$P(A) = \frac{1}{13}, \qquad P(B) = \frac{1}{13}, \qquad P(A \cap B) = \frac{1}{(13)(17)}$$

(see Exercise A1). In this case, we see that A and B are not independent.

Now suppose we sample with replacement, that is, we put the first card back before drawing the second card. In this case,

$$P(A) = \frac{1}{13}, \qquad P(B) = \frac{1}{13}, \qquad P(A \cap B) = \frac{1}{(13)^2},$$

so that these events are independent when we sample with replacement.

The result of the preceding example can be generalized: Whenever we sample with replacement the draws are independent, but when we sample without replacement they are not. If we replace the object we have drawn on the first draw before drawing another one, then knowing what we drew the first time does not help us at all in knowing what we will draw the second time. By contrast, if we sample without replacement, knowing what we drew the first time does help us in knowing what we will draw the second time. If we get an ace the first time, we are less likely to get an ace the second time, because there are only three aces left in the deck to draw. It is the independence of the draws in sampling with replacement which makes it easier to use as a model than sampling without replacement. For this reason, we often assume we are sampling with replacement when we are actually sampling without replacement. As mentioned in Section 1.1.2, we can make this assumption as long as N, the population size, is much larger than n, the sample size.

Example 1–24

Suppose that a person rolls a fair die until she gets a six. Let B_k be the event that the first six occurs on the kth roll. Then B_1 is the event in which she gets a six on the first roll, so that

$$P(B_1) = \frac{1}{6}.$$

Now, B_2 is the event in which she doesn't get a six on the first roll, but does get a six on the second roll. Since the rolls are independent,

$$P(B_2) = P(\text{no 6 on first})\, P(\text{6 on second}) = \left(\frac{5}{6}\right)\left(\frac{1}{6}\right).$$

Similarly, B_k is the event that she doesn't get a six on the first $k-1$ rolls but gets a six on the kth. Using independence again, we obtain

$$P(B_k) = \left(\frac{5}{6}\right)^{k-1}\left(\frac{1}{6}\right).$$

Now let C be the event that she eventually gets a six, that is, she doesn't go on rolling the die forever. Then $C = \cup B_k$, and the B_k are disjoint. Therefore,

$$P(C) = P(\cup B_k) = \sum_{k=1}^{\infty}\left(\frac{5}{6}\right)^{k-1}\left(\frac{1}{6}\right) = \left(\frac{1}{6}\right)\sum_{j=0}^{\infty}\left(\frac{5}{6}\right)^{j} = 1,$$

using the formula for the sum of a geometric series. Therefore, she will eventually get a six with probability one.

Example 1–25

A box contains two coins, one fair coin and one coin with heads on both sides. One of the coins is selected at random and flipped twice, independently. Let A be the event that the first toss is a heads and B be the event that the second toss is a heads. Most people initially believe that A and B are independent, since the flips are independent. However, A and B are not independent, as we now show.

Let C be the event that the fair coin is chosen. Then

$$P(A) = P(C)P(A|C) + P(C^c)P(A|C^c) = \left(\frac{1}{2}\right)\left(\frac{1}{2}\right) + \left(\frac{1}{2}\right)(1) = \frac{3}{4} = P(B).$$

Now,

$$P(A \cap B) = P(C)P(A \cap B|C) + P(C^c)P(A \cap B|C^c).$$

If the fair coin is drawn, then the probability of two heads is $\frac{1}{4}$, and if the two-headed coin is drawn, then the probability of two heads is 1. Therefore,

$$P(A \cap B) = \left(\frac{1}{2}\right)\left(\frac{1}{4}\right) + \left(\frac{1}{2}\right)(1) = \frac{5}{8} \neq \frac{9}{16} = P(A)P(B).$$

It is fairly easy to see why these events should not be independent. A heads on the first toss is some evidence that we have drawn the two-headed coin. Therefore, if we get a heads on the first toss, we would expect the probability of a heads on the second toss to be a little higher than if we get a tails on the first toss. In fact,

$$P(B|A) = \frac{P(A \cap B)}{P(A)} = \left(\frac{5}{8}\right)\Big/\left(\frac{3}{4}\right) = \frac{5}{6} > \frac{3}{4} = P(B).$$

We now give an interesting application of independence to genetics.

Example 1–26

(Hardy-Weinberg law) Suppose a gene is either dominant (A) or recessive (a). Each individual in the population has two genes, so the possible gene combinations are AA, Aa, and aa. Suppose that in the first generation, the proportions of each type are $p = P(\text{AA})$, $q = P(\text{Aa})$, and $r = P(\text{aa})$, for both males and females. (Note that $p + q + r = 1$.) Suppose that all matings are random, so the contribution from the mother is independent of the contribution from the father. Then the probability that the father transmits an A to the child is $p + q/2$ (since he is sure to transmit an A if he is AA, and he transmits it with probability $\frac{1}{2}$ if he is Aa). Similarly, the probability that the mother transmits an A is $p + q/2$. Since the mother's and father's contributions are independent, the probability that the offspring is AA is

$$p^* = \left(p + \frac{q}{2}\right)^2.$$

Similarly, the probabilities that the offspring is Aa and that it is aa are given, respectively, by

$$q^* = 2\left(p + \frac{q}{2}\right)\left(r + \frac{q}{2}\right) \qquad \text{and} \qquad r^* = \left(r + \frac{q}{2}\right)^2.$$

Now let p^{**}, q^{**}, and r^{**} be the proportions of AA, Aa, and aa in the third generation. Then, replacing p, q, and r by p^*, q^*, and r^* in the preceding formulas, we see that

$$p^{**} = \left(p^* + \frac{q^*}{2}\right)^2 = \left[\left(p + \frac{q}{2}\right)^2 + \left(p + \frac{q}{2}\right)\left(r + \frac{q}{2}\right)\right]^2$$

$$= \left[\left(p + \frac{q}{2}\right)(p + q + r)\right]^2 = \left(p + \frac{q}{2}\right)^2 = p^*,$$

Similarly,

$$q^{**} = q^* \qquad \text{and} \qquad r^{**} = r^*.$$

Therefore, the proportions in all future generations are also p^*, q^*, and r^*. Surprisingly enough, the proportions have converged in only one generation. This result is called the Hardy-Weinberg law. (The assumption of random mating, and hence independence of genetic contributions, may often be suspect in practice. For example, if the gene has an effect on height, then the mating would probably not be random, since it may be that taller people tend to marry each other, as may shorter people.)

Exercises—A

1. Consider the problem of drawing two cards from a deck of cards. Verify that if we draw without replacement the probability of 2 A's is $1/(13)(17)$ and that if we draw with replacement the probability of 2 A's is $1/(13)^2$.
2. A graduate school has four applicants for a fellowship. Suppose that one applicant is a woman, one is black, one is handicapped, and one is a black handicapped woman. Suppose that the recipient of the fellowship is chosen at random. Let A be the event that the recipient is a woman, B the event that the recipient is black, and C be the event that the recipient is handicapped. Show that A, B, and C are pairwise independent, but are not independent.
3. Two physicians A and B are on call during nonworking hours. Physician A is within earshot of his beeper 85% of the time, while physician B is within earshot of his beeper 70% of the time, and they are independent of each other.
   (a) Find the probability that at least one of the physicians is within earshot of his beeper.
   (b) The medical service is thinking of adding a third physician, C, who is within earshot of her beeper 80% of the time. If they do this, what is the probability that at least one of the three physicians is within earshot of his or her beeper?
4. Three hunters, A, B, and C, fire at a deer. The probability that A hits the deer is .4, the probability that B does is .3, and the probability that C does is .2. Suppose that the shots

are independent. Find the conditional probability that A killed the deer given that exactly one shot hit the deer.

5. A mountain climber has been lost on a mountain on either slope A or slope B. The head of the rescue mission believes that the mountain climber is on A with probability .6. Suppose that she has 12 rescue parties which are all equally competent and will locate the person with probability .3 if they are on the correct slope. Suppose further that they work independently of each other.

(a) If the mission leader sends eight parties to slope A and four to slope B, what is the probability that they rescue the lost mountain climber? (*Hint*: First find the conditional probability that they rescue him given that he is on slope A and the conditional probability that they rescue him given that he is on slope B.)

(b) Suppose the leader sends k parties to slope A and $12-k$ to slope B. What is the probability that they rescue the climber?

(c) How many rescue parties should she send to slope A in order to maximize the probability that the skier is rescued?

6. Suppose we roll two independent dice such that the probability that face i appears is $i/21$. (For example, the probability of a 6 is $\frac{6}{21}$.)

(a) What is the probability that we get doubles?

(b) What is the probability that the sum is 7?

7. Suppose two dice are in a box. One is a fair die, so that the probability of any face is $\frac{1}{6}$. The other die is not fair, and the probability of an i is $i/21$. Suppose a die is chosen at random from the box and rolled twice. Let A be the event that the first roll gives a 6 and B the event that the second roll gives a 6.

(a) Find $P(A)$, $P(B)$, and $P(A \cap B)$.

(b) Are A and B independent?

(c) What is the conditional probability of a 6 on the second roll given a 6 on the first?

8. A system works if any of its three branches works. The first branch works if A works, B or C works, and D works. The second branch works if E or F works or G works, and if H works. The third branch works if I works, either J, K, or L works, and either M or N works. If each component works with probability .95, what is the probability that the system works?

9. Suppose that someone flips a fair coin until a head occurs. Let A_k be the event that the first head occurs on the kth toss.

(a) Find $P(A_k)$.

(b) Show that $P(\cup A_k) = 1$. (That is, a head eventually occurs with probability one.)

10. Verify the formulas for q^*, r^*, q^{**}, and r^{**} in Example 1–26.

11. To win a contest in a fast food chain, a contestant must collect at least one of each of the letters A, B, C, D and E. Suppose that a contestant has collected 10 coupons, each containing one of the above letters, with each letter equally likely on each coupon. What is the probability that the contestant has won the contest? (*Hint*: What is the probability that he is missing a particular letter? the probability that he is missing a particular pair of letters? Use Theorem 1.4.b.)

12. Twenty people get on an elevator in the basement of a building. If each person is equally likely to choose any of the 10 floors, independently, what is the probability that the elevator stops at all 10 floors? (Use Theorem 1.4.b.)

13. Show that if A and B are independent, then A and B^c are independent.

14. **(a)** Suppose that $P(A) = 0$. Show that A is independent of any other event B.

(b) Suppose that $P(C) = 1$. Show that C is independent of any other event B.

1.4 A FAMOUS PROBABILITY PROBLEM

*1.4.1 The Secretary Problem

Suppose an executive needs to hire a new secretary. She finds n possible candidates for the job. Naturally, she wants to hire the best candidate, so she interviews them one at a time. After each interview, she knows only whether that candidate is the best of the ones she has already interviewed. She has no idea how the candidate compares to the remaining candidates. However, she must decide at that time whether or not to hire the candidate. She wants a strategy which maximizes her chance of hiring the best candidate. We assume that the candidates come in random order, so that the probability of any sequence of candidates is equally likely. (This problem has many other forms and names. For example, it is often phrased in terms of a prince choosing his wife in order to maximize his dowry. When he sees a candidate, he has no knowledge of the dowries to be offered by future princesses. In this form, the problem is called the dowry problem.)

We consider strategies of the following form. The executive interviews the first r candidates, but does not hire any of them. We might call these the training sample. She then hires the next candidate who is better than any candidate in the training sample. (Using an indifference argument, it can be shown that the optimal strategy must have this form. See Mosteller (1965), pp. 46–48; readers should be cautioned that our r is his $r - 1$.)

Using this strategy, the executive loses the best candidate if that candidate is in the training sample. Even if the best candidate is not in the training sample, that candidate is lost if another candidate who is better than any candidate in the training sample comes before the best candidate does. For example, if the training sample has 20 candidates, the 25th candidate is the second best one, and the best candidate is the 30th one, the executive would choose the second best candidate and therefore would lose out on the best one.

From the preceding discussion, we want to choose the training sample large enough to get good information about the secretaries (so that we do not hire an inferior secretary by stopping too soon), but not so large that the best secretary is likely to be in the training sample.

Accordingly, consider the strategy in which we take a training sample of size r from a population of n secretaries. Let $p_{r,n}$ be the probability that we get the best secretary with this strategy. In order to find $p_{r,n}$, let B be the event that the best candidate is hired with this strategy, and let A_i be the event that the best candidate is the ith candidate interviewed. Then

$$p_{r,n} = P(B) = \sum_i P(A_i)\, P(B \mid A_i).$$

Since the candidates can come in any order,

$$P(A_i) = \frac{1}{n},$$

and since the executive is not going to hire any of the first r candidates,

$$P(B|A_i) = 0, \qquad i = 1, \ldots, r.$$

We note first that if the best candidate is in the $(r+1)$th position, we are sure to hire him or her, so that

$$P(B|A_{r+1}) = 1.$$

We now find $P(B|A_{r+2})$, that is, the probability that we hire the best candidate given that he or she is in the $(r+2)$th position. We shall hire this candidate as long as we don't hire the one in the $(r+1)$th position, that is, as long as the best candidate in the first $r+1$ positions is not in the $(r+1)$th position. Therefore,

$$P(B|A_{r+2}) = \frac{r}{r+1}.$$

Now, consider the case that the best candidate is in the $(r+k)$th position. The probability that we hire that candidate is the same as the probability that the best candidate among the first $r+k$ candidates is in the training sample (i.e., that the $(r+k)$th candidate is the best candidate after the training sample). Therefore,

$$P(B|A_{r+k}) = \frac{r}{r+k-1}, \qquad k = 1, \ldots n-r.$$

Hence,

$$p_{r,n} = P(B) = \frac{1}{n}\left(1 + \frac{r}{r+1} + \frac{r}{r+2} + \cdots + \frac{r}{n-1}\right).$$

We now find the optimal choice for r. Let

$$a_{r,n} = \frac{np_{r,n}}{n} = \left(\frac{1}{r} + \frac{1}{r+1} + \frac{1}{r+2} + \cdots + \frac{1}{n-1}\right).$$

Then

$$p_{r,n} - p_{r-1,n} = \frac{a_{r,n} - 1}{n}$$

(see Exercise B2). Therefore, $p_{r,n}$ increases as long as $a_{r,n} > 1$ and decreases when $a_{r,n} < 1$. Hence, $p_{r,n}$ is maximized by the largest r such that $a_{r,n} > 1$. Call this value r_n^*. By the preceding equations, the optimal probability is

$$q_n = p_{r_n^*,n} = \frac{r^* a_{r_n^*}}{n}.$$

For example, if $n = 5$, $a_{4,5} = \frac{1}{4}$, $a_{3,5} = (\frac{1}{4} + \frac{1}{3}) = \frac{7}{12}$, and $a_{2,5} = (\frac{1}{4} + \frac{1}{3} + \frac{1}{2}) = \frac{13}{12}$. Therefore, the largest r such that $a_{r,5} > 1$ is 2, and hence, $r^* = 2$. The optimal probability in this case is

$$q_5 = \frac{2(13/12)}{5} = .43.$$

Other values of r^* and p_n are given in the following table:

TABLE 1.1

n	3	4	5	6	7	8	9	10	20	30	50	100	1,000
r^*	1	1	2	2	2	3	3	3	7	11	18	37	368
q_n	.5	.46	.43	.43	.41	.41	.41	.40	.38	.38	.37	.37	.37

The table is rather surprising. Even if there are 1,000 candidates for the position, we have a strategy for which we can find the best candidate 37% of the time: We interview the first 368 as a training sample and then take the next candidate who is better than all those in the training sample.

We now investigate the behavior of $a_{r,n}$, r_n^*, and p_n as $n \to \infty$. Note that

$$\log\left[\frac{n-1}{r-1}\right] \geqq a_{r,n} \geqq \log\left[\frac{n}{r}\right]$$

(see Exercise B3). This implies that for large n,

$$a_{r,n} \doteq \log\left[\frac{n}{r}\right].$$

Since r_n^* is the largest r such that $a_{r,n} > 1$, for large n,

$$1 \doteq a_{r_n^*,n} \doteq \log\left[\frac{n}{r^*}\right] = -\log\left[\frac{r^*}{n}\right],$$

and we see that

$$r^* \doteq n\exp(-1) = \frac{n}{e} \doteq .368n,$$

and

$$p_n = \frac{r^* a_{r^*,n}}{n} \doteq e^{-1} = .368.$$

Therefore, for large n, the optimal strategy is to use the first $.368n$ as a training sample and take the next candidate who is better than anyone in the training sample. With this strategy, we have a probability of .368 of finding the optimal secretary. Most people are rather surprised that for such large samples there is a strategy that permits them to find the best candidate almost 37% of the time.

Exercises—B

1. Verify the entries for $n = 7$ given in Table 1.1.
2. Verify that $p_{r,n} - p_{r-1,n} = (a_{r,n} - 1)/n$.
3. Suppose that $x \in (r, r+1)$.
   (a) Show that $x^{-1} \leqq r^{-1} \leqq (x-1)^{-1}$.
   (b) Show that $\log[(r+1)/r] \leqq r^{-1} \leqq \log[r/(r-1)]$. (Note that $\int_r^{r+1} x^{-1}\,dx = \log[(r+1)/r]$ and $\int_r^{r+1}(x-1)^{-1}\,dx = \log[r/(r-1)]$.)
   (c) Show that $\log(n/r) \leqq a_r \leqq \log[(n-1)/(r-1)]$.

Chapter 2

RANDOM VARIABLES AND RANDOM VECTORS

The study of probability theory is much simplified by using random variables. A *random variable* X is a function from a probability space to the set of real numbers. The *range* of X is the set of possible values for X. In the last chapter we defined a probability space for the outcomes of the experiment of throwing two independent dice. The sum of the two dice was denoted by Z, a random variable with range $\{2, 3, \ldots, 12\}$. As another example, suppose we choose a male student at random and let X be his weight. Then the probability space is the set of all male students with the equally likely probability model, and X is a random variable (since it is a function of the student chosen). It is somewhat difficult to define exactly what the range of X is. We often take the range to be all positive numbers, with extremely small probabilities for very large weights.

Most of the random variables we study are either *discrete* or *continuous*. A discrete random variable X is used to model an experiment when we are counting things, such as the number of aces in a poker hand, or the number of people in a locale who get cancer, or the number of people on welfare in a particular city. Typically, the range of a discrete random variable contains only nonnegative integers.

A continuous random variable Y is used when we are measuring something, such as weight or blood pressure or intelligence or the time till an accident. Typically, continuous random variables take on values in an interval, say $(0, \infty)$. One interesting property of a continuous random variable Y is that $P(Y = c) = 0$ for all c. For example, the probability that a randomly chosen male is exactly 72″ tall (as opposed to 72.000001″, say) is assumed to be zero. Continuous random variables are often much easier to analyze than discrete ones and thus are often used for modeling data which are discrete but whose range contains many points, such as SAT scores.

Many times, we want to make two or more observations on a particular probability space. For example, we might want to know a randomly chosen freshman's high school GPA and SAT score. A random vector $\mathbf{X}$ is a vector whose components are random variables. For instance, if we choose a student at random and let X_1 be the number of courses the student is taking and X_2 be the student's GPA, then $\mathbf{X} = (X_1, X_2)$ is a bivariate random vector.

In Section 2.2, the theory is extended to bivariate random vectors. In Section 2.3, the theory is further extended to multivariate random vectors.

2.1 RANDOM VARIABLES

2.1.1 Discrete Density Functions

We say that a random variable X is a *discrete* random variable if X takes on a finite set $a_0 < a_1 < \cdots < a_k$ or a sequence $a_0 < a_1 < a_2 < \cdots$ of possible values. (In most of the examples we study, $a_0 = 0, a_1 = 1, \ldots$.) We define the *density function* of a discrete random variable X by

$$f(t) = P(X = t).$$

Note that $f(t)$ is defined for all real t, but that $f(t)$ is 0 when $t \notin \{a_0, a_1, \ldots\}$. We define the *range* S of X as those points t for which $f(t) > 0$. That is, the range of X is the set of possible values for X. If the range of X contains only integers, we say that X is an *integer-valued* random variable.

Now let X be a discrete random variable with range S. Then, in order to be an acceptable density function for X, $f(t)$ must satisfy the following conditions:

$$f(t) > 0 \text{ for } t \in S, \qquad f(t) = 0 \text{ for } t \notin S, \qquad \sum_S f(t) = 1. \tag{2–1}$$

(See Exercise C1.)

Example 2–1

Let Z be the sum of the two faces of two independently tossed fair dice. In the last chapter, we showed that $P(X = k) = (6 - |7 - k|)/36, k = 2, \ldots, 12$. Therefore, the range of this random variable is $S = \{2, \ldots, 12\}$ and

$$f(t) = \begin{cases} (6 - |7 - t|)/36, & t = 2, \ldots, 12 \\ 0 & \text{elsewhere.} \end{cases}$$

Note that $f(t) > 0$ for $t \in S$, $f(t) = 0$ for $t \notin S$, and $\sum_{t \in S} f(t) = 1$.

Example 2–2

Suppose we have a fair coin which we shall flip independently until we get a head. Let X be the number of flips till we get that head. That is, if we get a head on the first flip then $X = 1$, if we get the first head on the second flip then $X = 2$, etc. Therefore, the range of this random variable is the set $\{1, 2, \ldots\}$ of positive integers, i.e., X is integer valued. Since $X = 1$ means we get a head on the first flip, $f(1) = \frac{1}{2}$. Also, since $X = 2$ means that we get a tail on the first flip and a head on the second flip, $f(2) = \left(\frac{1}{2}\right)\left(\frac{1}{2}\right) = \frac{1}{4}$, using the independence of the flips. In general, $f(k) = \left(\frac{1}{2}\right)^k, k = 1, \ldots$. Therefore, for this model, we have

$$f(t) = \begin{cases} \left(\frac{1}{2}\right)^t, & t = 1, 2, \ldots \\ 0, & \text{elsewhere.} \end{cases}$$

Using the countable additivity of probabilities, we see that the probability that X is even is

$$P(X \text{ is even}) = P(X = 2 \text{ or } X = 4 \text{ or} \ldots) = P(\cup\{X = 2j\}) = \sum_j P(X = 2j) = \sum_j f(2j)$$

$$= \sum_j \left(\frac{1}{2}\right)^{2j} = \sum_j \left(\frac{1}{4}\right)^j = \frac{\frac{1}{4}}{1 - \frac{1}{4}} = \frac{1}{3}.$$

Also,

$$P(X \text{ is odd}) = 1 - P(X \text{ is even}) = \frac{2}{3}.$$

Similarly,

$$P(3 \leqq X \leqq 5) = f(3) + f(4) + f(5) = \frac{1}{8} + \frac{1}{16} + \frac{1}{32} = \frac{7}{32},$$

and

$$P(X \geqq 4) = 1 - P(X < 4) = 1 - [f(1) + f(2) + f(3)] = 1 - \left[\frac{1}{2} + \frac{1}{4} + \frac{1}{8}\right] = \frac{1}{8}.$$

We could also calculate this probability by noticing that

$$P(X \geqq 4) = P(\text{all tails on first 3 tosses}) = \left(\frac{1}{2}\right)^3.$$

We finish the example by calculating some conditional probabilities:

$$P(X = 2 | X < 4) = \frac{P(\{X = 2\} \cap \{X < 4\})}{P(X < 4)} = \frac{P(X = 2)}{P(X < 4)} = \frac{\frac{1}{4}}{1 - \frac{1}{8}} = \frac{2}{7}.$$

$$P(X > 2 | X < 4) = \frac{P(\{X > 2\} \cap \{X < 4\})}{P(X < 4)} = \frac{P(X = 3)}{P(X < 4)} = \frac{\frac{1}{8}}{\frac{7}{8}} = \frac{1}{7}.$$

From the preceding example, we see that if we know the density function of a discrete random variable, we can compute the probabilities of any events having to do with that variable (at least if we can sum the series involved). For this reason, we often define the density function of a random variable without reference to the underlying probability space. Typically, the density function of X is all we need to know.

In later sections, we shall often be considering several random variables at the same time. In such situations, it is often convenient to replace t in the definition of f with x. We therefore define

$$f(x) = P(X = x).$$

If we have a random variable Y, we use y for the associated variable. (Note that we can call the variable in the definition of f by essentially any name we choose. By whatever such name, $f(3) = P(X = 3)$.)

In future examples we often shall not write down the "elsewhere" part of the density function, writing just the range S and the formula for $f(x)$ where $x \in S$. Thus, for the random variable of Example 2–2, we shall write

$$f(x) = \left(\frac{1}{2}\right)^x, \qquad x = 1, 2, 3, \ldots.$$

It may be assumed that densities are 0 elsewhere. In this form of the density function, it is quite important to include the range $x = 1, 2, \ldots$. The density function is not defined without it.

We next present a basic theorem about discrete density functions which is fairly obvious from Examples 2–1 and 2–2.

Theorem 2–1. Let X be a discrete random variable with range S. Let A be a set in S. Then

$$P(X \in A) = \sum_{x \in A} f(x).$$

Proof. We note that $f(x) > 0$ for at most countably many points in A, say, $b_1, b_2, \ldots$. Therefore, by the countable additivity of probabilities and the fact that $\{x = b_i\}$ are disjoint sets,

$$P(X \in A) = P(\cup \{X = b_i\}) = \sum_i P(X = b_i) = \sum_i f(b_i) = \sum_{x \in A} f(x). \quad \square$$

Exercises—A

1. (a) Suppose we draw an unordered hand of five cards from an ordinary deck, with the probability space being the one in which all possible unordered hands are equally likely. Let X be the number of aces. Find the density function of X. How many points are there in the probability space of unordered hands?
   (b) Suppose we draw five cards in sequence without replacement from an ordinary deck. Let Y be the number of kings. Find the density function of Y. How many points are there in the probability space of possible sequences of cards?

   Note that the preceding two densities are the same, even though the probability spaces are quite different. Therefore, for most purposes, it does not matter which probability space is used; all we need to know is the density function of X (or Y).
2. Suppose we draw three cards from an ordinary deck with replacement. Let Z be the number of queens drawn. Find the density function of Z.
3. Suppose a box contains four red balls and six blue balls. Suppose further that we draw three balls from the box. Let X be the number of red balls drawn.
   (a) Find the density function of X, and verify that it satisfies the conditions given in Equation (2–1).
   (b) Find $P(X = 2)$ and $P(X > 2)$.
   (c) Find $P(X = 2 | X \geq 2)$ and $P(X \geq 2 | X = 2)$.
4. Suppose a fair coin is tossed three times. Let X be the number of heads tossed.
   (a) Find the density function of X. (*Hint*: list the eight points of the sample space.)
   (b) Find $P(X = 2)$ and $P(X < 3)$.
   (c) Find $P(X = 2 | X \leqq 2)$.
5. Suppose a fair die is rolled until a six appears. Let X be the number of rolls necessary (including the one with the six).
   (a) Find the density function of X.
   (b) Show that this density function satisfies Equation (2–1).
   (c) Find $P(X \leqq 5)$ and $P(4 \leqq X \leqq 6)$.
   (d) Find $P(X \text{ is odd})$.

Exercises—B

1. Let X be a discrete random variable with density function $f(x) = cx, x = 1, 2, 3, 4$.
   (a) What is c?
   (b) Find $P(X \text{ is odd})$ and $P(X \text{ is even})$.
   (c) Find $P(X = 2 \text{ or } x = 3)$.
   (d) Find $P(X = 2|X \text{ is even})$ and $P(X = 3|X \text{ is even})$.
2. Let X be a random variable with discrete density function

$$f(x) = c\begin{bmatrix}3\\x\end{bmatrix}, \qquad x = 0, 1, 2, 3.$$

   (a) What is c?
   (b) Find $P(X = 2)$ and $P(X > 1)$.
3. Let X be a random variable having density function $cx, x = .1, .2, .3, .4, .5$.
   (a) What is c?
   (b) Is X integer-valued?
   (c) Find $P(X > .2)$.
4. Show that the density function in Example 2–2 in fact satisfies the conditions for being a density function.

Exercises—C

1. (a) Let S be a finite or countable set. Let $f(x)$ be a function that $f(x) > 0$ for $x \in S$, $f(x) = 0$ for $x \notin S$, and $\sum_{x \in S} f(x) = 1$. Let $P(A) = \sum_{x \in A} f(x)$. Show that $P(A)$ is a probability measure.
   (b) Suppose that $P(A) = \sum_{x \in A} f(x)$ is a probability measure. Let the set of points S where f is positive be a countable set. Show that $f(x) \geqq 0$ and $\sum_{x \in S} f(x) = 1$.

2.1.2. Continuous Density Functions

A random variable X is a *continuous* random variable if there exists a function $f(x) \geqq 0$ such that

$$P(X \in A) = \int_A f(x)dx. \tag{2–2}$$

We call $f(x)$ the *density function* of X and the set of x such that $f(x) > 0$ the *range* of X. We think of the range as the set of possible values for X. In order to be a continuous density function for a random variable X with range S, $f(x)$ must satisfy

$$f(x) > 0 \text{ for } x \in S, \qquad f(x) = 0 \text{ for } x \notin S, \qquad \int_S f(x)dx = 1. \tag{2–3}$$

(See Section 2.1.3., Exercise C3.) If X is a continuous random variable, then

$$P(X = a) = \int_a^a f(x)dx = 0$$

for any a.

It seems intuitive that if $P(X = a) = 0$ for all a, then all events have to have probability zero. However, this is not true. The interval $\{c < x < d\}$ contains uncountably many points, and probability measures are only countably additive. Thus, $P(c \leqq X \leqq d)$ is often positive, even though the probability of each point in the interval is zero.

Example 2–3

Let X be a continuous random variable with density

$$f(x) = \begin{cases} 0, & x < 0 \\ 2e^{-2x}, & x \geqq 0. \end{cases}$$

Then the range of X is the set $[0, \infty) = \{x: 0 \leqq x < \infty\}$, and

$$P(X = 3) = \int_3^3 2e^{-2x}\,dx = -e^{-2x}\Big|_3^3 = 0,$$

$$P(X < 1) = \int_0^1 2e^{-2x}\,dx = -e^{-2x}\Big|_0^1 = 1 - e^{-2},$$

$$P(X > 2) = \int_2^\infty 2e^{-2x}\,dx = -e^{-2x}\Big|_2^\infty = e^{-4},$$

$$P(-1 < X < 1) = P(X < 1) = 1 - e^{-2}.$$

To find some conditional probabilities, we use the probabilities just calculated, together with the facts that $P(X > 1) = 1 - P(X < 1)$ and $P(1 < X < 2) = P(X > 1) - P(X > 2)$ (since $P(X = a) = 0$). We obtain

$$P(X > 2 | X > 1) = \frac{P(\{X > 2 \cap \{X > 1\})}{P(X > 1)} = \frac{P(X > 2)}{P(X > 1)} = \frac{e^{-4}}{e^{-2}} = e^{-2},$$

$$P(X < 2 | X > 1) = \frac{P(1 < X < 2)}{P(X > 1)} = \frac{(e^{-2} - e^{-4})}{e^{-2}} = 1 - e^{-2}.$$

We conclude the example with a general result about probabilities of intervals:

$$P(a \leqq X \leqq a + b) = \int_a^{a+b} 2e^{-2x}\,dx = -e^{-2x}\Big|_a^{a+b}$$

$$= e^{-2a} - e^{-2(a+b)} \leqq 1 - e^{-2b} = P(0 \leqq X \leqq b).$$

From the last calculation in this example, we can see that $h_b(a) = P(a \leqq X \leqq a + b)$ is maximized for $a = 0$, which is also the maximum for the density function. If the density function of a continuous random variable is high over a point, then intervals containing that point have a higher probability than intervals of equal length which do not contain the point. (We would like to say *points* where the density function is high have higher probability than those where the density is low, but of course, all points have probability zero for a continuous random variable.)

We also note from the preceding example that $f(0) = 2$, so that continuous density functions can be greater than one. In fact, there are cases in which the density function is unbounded. (Note that discrete density functions must be at most one. Why?)

In subsequent examples, we shall often omit the set where the density is zero. It may then be assumed that the density is zero everywhere other than the set listed.

For example, we would write the density of the function in Example 2–3 as

$$f(x) = 2e^{-2x}, \qquad x \geqq 0.$$

In using this form, it is obviously important to include the range ($x \geqq 0$): The density function is not defined without it.

We shall often define the density function of a random variable without saying whether the random variable is discrete or continuous. However, if we look at the range of the random variable, it is clear which type it is. If the range has finitely or countably many points, then X is a discrete random variable; if the range is an interval, then X is a continuous random variable.

The discrete density function is the probability of a particular event and is thus uniquely defined. However, the continuous density function is a function $f(x)$ which satisfies Equation (2–2). Unfortunately, we can change $f(x)$ on any finite or countable set and not change the integral in Equation (2–2). Thus, in Example 2–3, $f^*(x) = e^{-x}, x > 0$, satisfies Equation (2–2) just as well as $f(x)$ ($f^*(x) = f(x)$, except when $x = 0$). Therefore, $f(x)$ is not uniquely defined. Similarly, since the range is the set on which $f(x) > 0$, the range of a continuous random variable is also not uniquely defined. (In Example 2–3, the range of X using $f(x)$ is the closed set $[0, \infty) = \{x: x \geqq 0\}$, while the range for $f^*(x)$ is the open set $(0, \infty) = \{x: x > 0\}$.) In this book, however, we shall essentially ignore the nonuniqueness of f and S for two reasons. The first is that whichever choice we make for f does not have much effect on our inferences, since any two choices are equal with probability one. The second reason is that there is often only one essentially continuous choice for f and that is the choice we make. The only ambiguity in the choice is at the end points, which can be either included in S or excluded from S. Typically, the density is not continuous there. (In Example 2–3, the limit of $f(x)$ from the left is zero and the limit from the right is two.)

Example 2–4 (Uniform distribution)

A continuous random variable X has a *uniform distribution* on the interval $[a, b] = \{a \leqq x \leqq b\}$ if X has density function

$$f(x) = c, \qquad a \leqq x \leqq b.$$

Since this density is constant over the interval $[a, b]$, we use this model for situations in which we feel that all the observations in the interval are "equally probable." (Since X is a continuous random variable, $P(X = d) = 0$ for any d, so that the notion of equal probability is not too precise here.) Note that the conditions for $f(x)$ to be a density function imply that

$$1 = \int_a^b c\,dx = cx \Big|_a^b = c(b - a),$$

so that

$$c = (b - a)^{-1}.$$

Therefore, we could write the uniform density in the form

$$f(x) = (b - a)^{-1}, \qquad a \leqq x \leqq b.$$

Now, let $a \leqq h \leqq g \leqq b$. Then

$$P(h \leqq X \leqq g) = \int_h^g (b-a)^{-1}\,dx = \frac{g-h}{b-a}.$$

In other words, if X is uniformly distributed, then the probability of any interval contained in its range is the ratio of the length of that interval to the length of the range. (This model, therefore, seems like the natural analogue to the equally likely model discussed in Section 1.1.1.) One application of the uniform distribution is the random number generator in a computer, which is designed to give numbers which are uniformly distributed on the interval $[0, 1]$.

Exercises—A

1. Suppose that my quitting time on a given day is uniformly distributed between 4:00 and 5:00.
   (a) Find the probability that I leave before 4:15.
   (b) Find the probability that I leave after 4:30.
   (c) Find the conditional probability that I leave after 4:30 given that I leave after 4:15.
   (d) Find the conditional probability that I leave before 4:15 given that I leave before 4:30.
2. Let X be the lifetime in weeks of a particular light bulb. Suppose that X has density $f(x) = 10^{-1}\exp[-x/10], x > 0$.
   (a) Find the probability that the light bulb lasts at least 10 weeks.
   (b) Find the number a such that $P(X \leqq a) = .5$.

Exercises—B

1. Show that $f(x)$ in Example 2–3 is a density function.
2. Let X be a continuous random variable with density function $f(x) = 3x^2/2, -1 \leqq x \leqq 1$.
   (a) Find $P(X=0)$, $P(X>0)$, $P\left(X>\frac{1}{2}\right)$, $P\left(|X|>\frac{1}{2}\right)$, $P\left(\frac{1}{2}<X<\frac{3}{4}\right)$, $P\left(\frac{1}{2}<X<2\right)$.
   (b) Find $P\left(X<\frac{3}{4}\middle|X>\frac{1}{2}\right)$.
3. Let X have density function $f(x) = e^{-|x|}/2, -\infty < x < \infty$.
   (a) Find $P(X=2)$, $P(X>1)$, $P(X>-1)$, $P(|X|>1)$, $P(-1<X<2)$, $P(-2<X<2)$. (*Hint*: $\int_{-1}^{2} f(x)dx = \int_{-1}^{0} f(x)dx + \int_0^2 f(x)dx$.)
   (b) Find $P(X<2|X>1)$ and $P(X>4|X^2>1)$.
4. Let X be a random variable with density function $f(x) = cx^{-1/2}, 0<x<1$.
   (a) What is c?
   (b) Find $P\left(X=\frac{1}{2}\right)$, $P\left(X \geqq \frac{1}{4}\right)$, $P\left(\frac{1}{9} \leqq X \leqq \frac{1}{4}\right)$.
   (c) Find $P\left(X>\frac{1}{9}\middle|X \leqq \frac{1}{4}\right)$.
5. Let X be a random variable with density function $f(x)$ which satisfies $f(-x) = f(x)$ for all x. Show that $P(a<X<b) = P(-b<X<-a)$ for all $0<a<b$.
6. (a) Show that $f(x) = c/x, x \geqq 0$ is not a density function for any c.
   (b) Show that $f(x) = c(x-1), 0 \leqq x \leqq 2$ is not a density function for any c.
   (c) For what c is $f(x) = c(x-1), 0 \leqq x \leqq 1$ a density function?

Exercises—C

1. Let X be a random variable which is uniformly distributed on the interval $[0, 1]$. Let Y be the first digit of X.
   (a) Show that $P(Y=0) = .1$. (*Hint*: $P(Y=0) = P(0 \leqq X < .1)$. Why?)

(b) Show that Y has the discrete density function $g(y) = .1, y = 0, 1, \ldots, 9$.
(c) Let Z be the second digit of X. What is the density function of Z?

2.1.3 Distribution Functions

Let X be a random variable. Then the *distribution function* of X is

$$F(x) = P(X \leqq x).$$

The distribution function is defined for any random variable, whereas the density function is defined only for discrete and continuous random variables.

We next show how to find probabilities from the distribution function. If $F(x)$ is not continuous at $x = a$, we define the height of the jump of $F(x)$ at a to be the limit of $F(x)$ as x goes to a from above minus the limit of $F(x)$ as X goes to a from below. (These limits exist because $F(x)$ is nondecreasing, as shown in Theorem 2–5.)

Theorem 2–2

a. If $F(x)$ is continuous at a, then $P(X = a) = 0$. If $F(x)$ is not continuous at a, then $P(X = a)$ is the height of the jump of $F(x)$ at a.
b. If $a < b$, then

$$\begin{aligned} P(a < X \leqq b) &= F(b) - F(a), \\ P(a \leqq X \leqq b) &= F(b) - F(a) + P(X = a), \\ P(a < X < b) &= F(b) - F(a) - P(X = b), \\ P(a \leqq X < b) &= F(b) - F(a) + P(X = a) - P(X = b). \end{aligned}$$

If F is continuous at a and b, then these probabilities are all equal.
c. For any a and b,

$$\begin{aligned} P(X \leqq b) &= F(b), \qquad & P(X < b) &= F(b) - P(X = b), \\ P(X > a) &= 1 - F(a), \qquad & P(X \geqq a) &= 1 - F(a) + P(X = a). \end{aligned}$$

Proof

a. Let $b_1 > b_2 > b_3, \ldots$ be a sequence of points which converges to zero from above. Let

$$A_i = (a - b_i, a + b_i].$$

Then by part b whose proof does not depend on that of part a,

$$P(A_i) = P(a - b_i < X \leqq a + b_i) = F(a + b_i) - F(a - b_i).$$

Now, the A's form a decreasing sequence of sets whose intersection is the point a. Therefore, by Theorem 1–5,

$$\begin{aligned} P(X = a) &= P(\cap_{i=1}^{\infty} A_i) \\ &= \lim_{n\to\infty} P(A_n) = \lim_{n\to\infty} (F(a + b_n) - F(a - b_n)) \end{aligned}$$

which is the height of the jump at a. If $F(x)$ is continuous at a, then this limit is zero.

b. The event $X \leqq b$ is the disjoint union of the events $X \leqq a$ and the events $a < X \leqq b$. Therefore,

$$F(b) = P(X \leqq b) = P(X \leqq a) + P(a < X \leqq b) = F(a) + P(a < X \leqq b),$$

and hence $P(a < X \leqq b) = F(b) - F(a)$. Similarly, the event $a \leqq X \leqq b$ is the disjoint union of the events $a < X \leqq b$ and the event $X = a$. Therefore,

$$P(a \leqq X \leqq b) = P(X = a) + P(a < X \leqq b) = P(X = a) + F(b) - F(a).$$

The other equalities follow in a similar manner.

c. The event $X > a$ is the complement of the event $X \leqq a$, and therefore,

$$P(X > a) = 1 - P(X \leqq a) = 1 - F(a).$$

The other equations follow similarly. □

Example 2–5

Let the random variable X have the distribution function

$$F(x) = \begin{cases} 0 & \text{if } x < -1 \\ (x+1)/4 & \text{if } -1 \leqq x < 0 \\ (x+3)/4 & \text{if } 0 \leqq x < 1 \\ 1 & \text{if } x \geqq 1. \end{cases}$$

This function is graphed in Figure 2–1. From the graph, we see that $F(x)$ is continuous except at $x = 0$. Therefore, $P(X = a) = 0$ for all $x \neq 0$. However, the limits of $F(x)$ as x goes to zero from above and below are

$$F(0^+) = \lim_{x \to 0} \frac{x+3}{4} = \frac{3}{4}$$

and

$$F(0^-) = \lim_{x \to 0} \frac{x+1}{4} = \frac{1}{4}$$

and hence,

$$P(X = 0) = F(0^+) - F(0^-) = \frac{3}{4} - \frac{1}{4} = \frac{1}{2}.$$

(It is also apparent from the graph that the height of the jump at zero is $\frac{1}{2}$.) Now,

$$P(\tfrac{1}{4} \leqq X \leqq \tfrac{3}{4}) = F(\tfrac{3}{4}) - F(\tfrac{1}{4}) + P(X = \tfrac{1}{4}) = \frac{15}{16} - \frac{13}{16} + 0 = \frac{1}{8}.$$

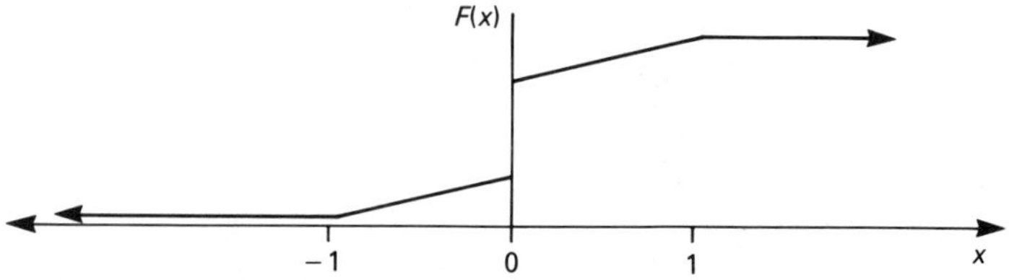

Figure 2–1 Graph of $F(x)$ in Example 2–5.

Finally,

$$P(X \geqq \tfrac{1}{2}) = 1 - F(\tfrac{1}{2}) + P(X = \tfrac{1}{2}) = 1 - \frac{7}{8} + 0 = \frac{1}{8},$$

$$P(X \geqq 0) = 1 - F(0) + P(X = 0) = 1 - \frac{3}{4} + \frac{1}{2} = \frac{3}{4}.$$

From this example, we see that if we know the distribution function of a random variable X, we can find the probability that X equals any point or lies in any interval.

We now establish the relationship between the density function and the distribution function for discrete random variables. Let $a_0 < a_1 < \cdots$ be a finite set or an infinite sequence of points in R. We say that $F(x)$ is a *step function* with jumps at the a_i if $F(x)$ is flat except at the a_i (i.e., if $F(c) - F(b) = 0$ for any points b and c which are contained in any interval $a_i < b < c < a_{i+1}$). One particular step function is graphed in Figure 2–2.

Theorem 2–3. Let X be a discrete random variable with distribution function $F(x)$, density function $f(x)$, and range $a_0 < a_1 < a_2, \ldots$.

a. $F(x)$ is a step function with jumps at the a_i.
b. $F(x) = \sum_{a_i \leqq x} f(a_i)$, and $f(a_i) = F(a_i) - F(a_{i-1})$.
c. Let X be an integer-valued random variable, so that $a_i = i$. Then

$$f(k) = F(k) - F(k-1)$$

$$P(j < X \leqq k) = F(k) - F(j), \qquad P(j \leqq X < k) = F(k-1) - F(j-1),$$

$$P(j < X < k) = F(k-1) - F(j), \qquad P(j \leqq X \leqq k) = F(k) - F(j-1).$$

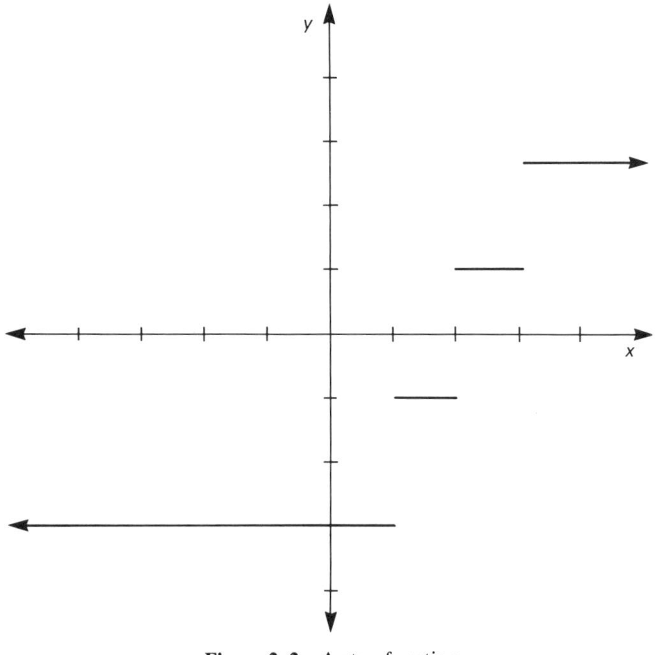

Figure 2–2 A step function.

Proof

a. This result follows from the definitions.

b. The first equality follows directly from Theorem 2–1. Since X only takes on the values $\{a_0, a_1, \ldots\}$, and since these values are ordered,

$$F(a_i) = P(X \leqq a_i) = P(X \leqq a_{i-1}) + P(X = a_i) = F(a_{i-1}) + f(a_i).$$

c. The first equality follows directly from part b. In addition,

$$P(j < X < k) = P(j < k \leqq k - 1) = F(k-1) - F(j).$$

The other equations follow similarly. □

If X is an an integer-valued random variable, then the distribution function $F(x)$ is a step function which is flat between integers and has a jump of height $f(k)$ at the integer k.

The next two examples illustrate how to find the distribution function from the density function, and the density function from the distribution function, for discrete random variables.

Example 2–6

If X has density function $f(x) = (\frac{1}{2})^x$, $x = 1, 2, \ldots$, then

$$F(k) = P(X \leqq k) = \sum_{x=1}^{k} (\tfrac{1}{2})^x = 1 - (\tfrac{1}{2})^k, \qquad k = 1, 2, \ldots.$$

Therefore, the distribution function of X is

$$F(x) = \begin{cases} 1 - (\frac{1}{2})^k, & k \leqq x < k+1, \\ 0, & x < 1. \end{cases} \qquad k = 1, 2, \ldots.$$

Example 2–7

Suppose X is a random variable with distribution function

$$F(x) = \begin{cases} 0 & \text{if } x < 0 \\ .4 & \text{if } 0 \leqq x < 1 \\ 1 & \text{if } x \geqq 1. \end{cases}$$

Then X is a discrete random variable (since F is a step function) which takes on the values 0 and 1 (since the distribution function jumps at those points). The density function of X is

$$f(x) = \begin{cases} .4 & \text{if } x = 0 \\ .6 & \text{if } x = 1 \\ 0 & \text{elsewhere.} \end{cases}$$

(Note that .4 and .6 are the heights of the jumps at 0 and 1, respectively.)

If we know both the density function and the distribution function for a discrete random variable X, it is usually easier to use the distribution function to find probabilities. For most of the events we want to consider, finding the probabilities from the distribution function involves taking one difference, while finding probabilities from the density function can involve many additions. For example

$$P(3 < X < 675) = F(674) - F(3) = \sum_{x=4}^{674} f(x).$$

For this reason, the distribution functions rather than the density functions are tabled in the back of the book. However, for most other purposes, we shall see that the density function is the more important function.

We now establish the relationship between the density function and the distribution function for continuous random variables.

Theorem 2–4. Let X be a continuous random variable with density function $f(x)$ and distribution function $F(x)$. Then

a. $F(x) = \int_{-\infty}^{x} f(t)dt$, and hence $F(x)$ is continuous.
b. $f(x) = F'(x)$ at all points where $f(x)$ is continuous.
c. $P(a < X \leqq b) = P(a < X < b) = P(a \leqq X < b) = P(a \leqq X \leqq b) = F(b) - F(a)$ for any $b > a$.

Proof

a. By definition,

$$F(x) = P(X \leqq x) = \int_{t \leqq x} f(t)dt = \int_{-\infty}^{x} f(t)dt.$$

b. This result follows directly from part a and the fundamental theorem of calculus.
c. This result follows directly from Theorem 2–2, since $F(x)$ is a continuous function by part a. □

In later sections we shall tacitly assume that $F'(x) = f(x)$ for all x, although for some examples it can happen that the right derivative is not equal to the left derivative. In those examples, we can define the density function using either derivative. The distribution of a continuous random variable is unaffected by changing the density on any finite (or countable) set of points.

We next give two examples which indicate how to find the distribution function from the density function, and the density function from the distribution function, for continuous random variables.

Example 2–8

Let X be a continuous random variable with density $f(x) = 1, 0 \leqq x \leqq 1$. Let $a < 0$, $0 \leqq b \leqq 1, c > 1$. Then

$$F(a) = \int_{-\infty}^{a} 0\,dx = 0, \qquad F(b) = \int_{-\infty}^{0} 0\,dx + \int_{0}^{b} 1\,dx = b,$$

$$F(c) = \int_{-\infty}^{0} 0\,dx + \int_{0}^{1} 1\,dx + \int_{1}^{c} 0\,dx = 1.$$

Therefore, the distribution function for this example is

$$F(x) = \begin{cases} 1 & \text{if } x > 1 \\ x & \text{if } 0 \leqq x \leqq 1 \\ 0 & \text{if } x < 0. \end{cases}$$

Example 2–9

Let X be a random variable with distribution function

$$F(x) = \begin{cases} 0, & x < 0 \\ x^2, & 0 \leqq x < 1 \\ 1, & x \geqq 1. \end{cases}$$

Then $f(x) = F'(x) = 2x, 0 \leqq x \leqq 1$, and $f(x) = 0$ elsewhere. (Note that $F'(x)$ does not exist at the point 1. However, we can define $f(x)$ arbitrarily at that point.)

For a continuous random variable, a table of the density function would be of very little help in computing probabilities (since we must integrate the density to find probabilities). Therefore, we have tabled several continuous distribution functions in the back of the book. We have not tabled density functions. However, we shall see that for most other purposes, the density function is the more important function. (The density function tells us which points are "more probable" for the random variable.)

As mentioned in previous sections, we shall often define the distribution of a random variable without reference to the probability space on which it is defined. Therefore, it is quite useful to know the conditions which a function $F(x)$ must satisfy in order to be a distribution function. The following theorem gives those conditions; we define $F(a^+)$ to be the limit of $F(x)$ as x goes to a from above.

Theorem 2–5. A function $F(x)$ is a distribution function if and only if it satisfies the following conditions:

a. $F(x)$ is a nondecreasing function of x.

b. $F(x^+) = F(x)$.

c. $\lim_{x\to\infty} F(x) = 1$, and $\lim_{x\to-\infty} F(x) = 0$.

Proof. We prove only that a distribution function satisfies these conditions. For a proof of the converse, see Billingsley (1979), pp. 159–160.

a. Let $F(x)$ be a distribution function for a random variable X. If $a < b$, then

$$F(b) - F(a) = P(a < X \leqq b) \geqq 0.$$

Therefore, F is nondecreasing.

b. Let $b_1 > b_2 > b_3, \ldots$ be a sequence of constants which converge to zero from the positive side, and let $B_i = (-\infty, x + b_i]$. Then $\cap B_i = (-\infty, x]$ and the B_i are decreasing. Therefore, by Theorem 1–4,

$$F(x) = P\left(\bigcap_{i=1}^{\infty} B_i\right) = \lim_{n\to\infty} P(B_n) = \lim_{n\to\infty} F(x + b_n) = F(x^+).$$

c. Let $x_1 < x_2, \ldots$ be a sequence of constants which converge to ∞, and let $C_i = (-\infty, x_i)$. Then $\cup C_i = (-\infty, \infty)$ and the C_i are increasing. Now, since $P(-\infty < X < \infty) = 1$,

$$1 = P\left(\bigcup_{i=1}^{\infty} C_i\right) = \lim_{n\to\infty} P(C_n) = \lim_{n\to\infty} F(x_n)$$

(by Theorem 1–4 again). This limit obtains for all sequences x_n which go to infinity, and therefore, $\lim_{x\to\infty} F(x) = 1$. The other limit follows similarly. □

Theorem 2–5 says that any function $F(x)$ which satisfies its three conditions is an acceptable distribution function for X. However, a given $F(x)$ may not be the correct function for a particular application. If we are using the wrong distribution function to model some random phenomenon, even if that function is acceptable, our probability calculations are likely to be incorrect.

Exercises—B

1. For the distribution function in Example 2–5, find $P(X = \frac{3}{4})$, $P(0 < X < \frac{3}{4})$, and $P(0 \leqq X \leqq \frac{3}{4})$.
2. Let X be a random variable with distribution function $F(x)$ such that $F(x) = 0$ for $x < 0$, $F(x) = (2x + 1)/4$ for $0 \leqq x < 1$, and $F(x) = 1$ for $x \geqq 1$.
   (a) Graph $F(x)$. Is X a continuous or discrete random variable?
   (b) Find $P(X = 0)$, $P(X = \frac{1}{2})$, and $P(X = 1)$.
   (c) Find $P(0 < X < 1)$, $P(0 \leqq X \leqq 1)$, and $P(X \geqq 1)$.
3. Let X be a discrete random variable with density function $f(x) = x/10$ for $x = 1, 2, 3, 4$.
   (a) Find and graph the distribution function $F(x)$.
   (b) Find $P(1 < X \leqq 3)$ and $P(1 \leqq X < 3)$ using the density function and using the distribution function.
4. Let X be a continuous random variable with density function $f(x) = 3x^2/2$ for $-1 \leqq x \leqq 1$.
   (a) Find and graph the distribution function $F(x)$.
   (b) Use the distribution function to find $P(-\frac{1}{2} < X < \frac{1}{3})$ and $P(-\frac{1}{3} \leqq X \leqq \frac{1}{2})$.
5. Let X be a discrete random variable with distribution function $F(x)$ such that $F(x) = 0$ if $x < 0$, $F(x) = [x]/4$ if $0 \leqq x < 5$, and $F(x) = 1$ for $x \geqq 5$ (where $[x]$ is the largest integer less than or equal to x).
   (a) Graph $F(x)$ and find the range of X.
   (b) What is the density function of X?
6. Let X be a continuous random variable with distribution function $F(x)$ such that $F(x) = 0$ for $x < 0$, $F(x) = x^3/8$ for $0 \leqq x \leqq 2$, and $F(x) = 1$ for $x > 2$.
   (a) Graph $F(x)$. What is the range of X?
   (b) Find $P(\frac{1}{2} < X < 1)$ and $P(\frac{1}{2} \leqq X \leqq 1)$.
   (c) Find the density function of X, and use it to find $P(\frac{1}{2} < X < 1)$.
7. (a) For any distribution function F, show that $P(a < X < b) = F(b) - F(a) - P(X = b)$.
   (b) Show that $P(X \geqq a) = 1 - F(a) + P(X = a)$.

Exercises—C

1. Show that $F(x)$ in Example 2–5 satisfies the conditions for being a distribution function.
2. Show that $\lim_{x \to -\infty} F(x) = 0$. (*Hint*: Let $d_1 > d_2, \ldots$ converge to $-\infty$, and let $D_i = (-\infty, d_i]$. What is $\bigcap D_i$?)
3. (a) Let $f(x)$ be the density function of a continuous random variable with range S (so that $f(x)$ satisfies Equation (2–3)). Let $F(x) = \int_{-\infty}^{x} f(x)dx$. Show that $F(x)$ satisfies the conditions of Theorem 2–5.
   (b) Let $F(x)$ be the distribution function of a continuous random variable (so that F satisfies the conditions of Theorem 2–5). Let $f(x) = F'(x)$, and suppose that f is continuous. Show that $f(x)$ satisfies Equation (2–3).

   (This result establishes that Equation (2–3) is a necessary and sufficient condition for $f(x)$ to be a continuous density function.)

2.1.4. Functions of Random Variables

Let X be a random variable with density function $f(x)$. Let $U = h(X)$. Then U is also a random variable, which we call the *induced random variable*. In this section, we discuss methods for deriving the density function $f_U(u)$ of U from the density function $f_X(x)$ of X.

We first assume that X is a discrete random variable. If so, then $f(x)$ is a probability, and we typically derive the density function directly.

Example 2–10

Let X be a discrete random variable with range $S = \{-1, 0, 1, 2\}$ and density function $f_X(x)$, where $f_X(-1) = .2, f_X(0) = .3, f_X(1) = .4$, and $f_X(2) = .1$. Let $U = X^2$. Then U has range $\{0, 1, 4\}$ and density $f_U(u)$, where

$$f_U(0) = P(U = 0) = P(X = 0) = .3, \qquad f_U(4) = P(U = 4) = P(X = 2) = .1,$$
$$f_U(1) = P(U = 1) = P(X = 1 \text{ or } X = -1) = P(X = 1) + P(X = -1) = .2 + .4 = .6.$$

In the continuous case, the density function is not a probability, so that it is more difficult to derive the induced density function. Typically, we first find the distribution function of the induced random variable and then differentiate it to find the density function. We only need to find the distribution function $F_U(u)$ for u in the range of U.

Example 2–11

Let X be a continuous random variable with density function $f_X(x) = 3x^2/2, -1 \leqq x \leqq 1$. Let $U = X^2$. Then U has range $0 \leqq u \leqq 1$, and the distribution function for U is given by

$$F_U(u) = P(U \leqq u) = P(X^2 \leqq u) = P(-u^{1/2} \leqq X \leqq u^{1/2})$$
$$= \int_{-u^{1/2}}^{u^{1/2}} \frac{3x^2}{2} dx = \frac{x^3}{2}\bigg|_{-u^{1/2}}^{u^{1/2}} = u^{3/2}, \qquad 0 \leqq u \leqq 1.$$

Therefore, the density function for U is

$$f_U(u) = F'_U(u) = \frac{3u^{1/2}}{2}, \qquad 0 \leqq u \leqq 1.$$

(Note that it is very important to give the range of U; the definition is not complete without it.)

Example 2–12

Let X have density $f_X(x) = \frac{1}{3}, -1 \leqq x \leqq 2$. Let $U = X^2$ again. Then U has range $0 \leqq u \leqq 4$. To find the distribution function of U, we must break this range into two pieces, $0 \leqq u \leqq 1$ and $1 < u \leqq 4$. (Note that on the first set both $-u^{1/2}$ and $u^{1/2}$ contribute to the density, while on the second set only $u^{1/2}$ contributes.) If $0 \leqq u \leqq 1$, then

$$F_U(u) = P(X^2 \leqq u) = \int_{-u^{1/2}}^{u^{1/2}} \frac{1}{3} dx = \frac{2u^{1/2}}{3}.$$

If $1 < u \leqq 4$, then

$$F_U(u) = P(X^2 \leqq u) = \int_{-u^{1/2}}^{-1} 0\, dx + \int_{-1}^{u^{1/2}} \frac{1}{3} dx = \frac{(u^{1/2} + 1)}{3}.$$

(F is continuous at $u = 1$.) Therefore, the density function $f_U(u)$ is given by

$$f_U(u) = F'_U(u) = \begin{cases} u^{-1/2}/3 & \text{if } 0 \leqq u \leqq 1 \\ u^{-1/2}/6 & \text{if } 1 < u \leqq 4. \end{cases}$$

(Note that we could also define $f_U(1)$ to be $\frac{1}{6}$ instead of $\frac{1}{3}$. It simply doesn't matter what choice we make for $f(x)$ at a point like $x = 1$, where f is discontinuous.)

We now derive a general method which is often helpful for deriving the density function without first finding the distribution function. Let $y = h(x)$ be a function. We say that h is *invertible* if there exists an *inverse function* $x = g(y)$ such that

$$y = h(x) \Leftrightarrow x = g(y).$$

For example $y = e^x$ is invertible because $x = \log(y)$. Also, $y = x^2$ is an invertible function if x is nonnegative; because $x = y^{1/2}$, but is not an invertible function if x is allowed to be any real number, because we would then have $x = \pm y^{1/2}$, which is not a function. Note that the only continuous functions which are invertible are either increasing functions or decreasing functions. (Nondecreasing and nonincreasing functions are not good enough, because flat spots make a function noninvertible.)

In many calculus courses, various criteria are given for establishing whether a function $h(x)$ is invertible (one to one and onto, $h'(x) > 0$, etc.). In this book, we always need to find the inverse anyway, so such criteria are not relevant. For our purposes, a function is invertible if we can invert it.

Let X be a random variable with range S_X. Then the range of $U = h(X)$ is the set S_U such that

$$X \in S_X \Leftrightarrow U \in S_U.$$

We are now ready for the main theorem of this section.

Theorem 2–6

a. Let X be a discrete random variable with density function $f_X(x), x \in S_X$. Let $U = h(X)$, where h is an invertible function with inverse function $x = g(u)$. Then U has the density function

$$f_U(u) = f_X(g(u)), \qquad u \in S_U.$$

b. Let X be a continuous random variable with density function $f_X(x), x \in S_X$. Let $U = h(X)$, where h is an invertible function with inverse $x = g(u)$, where $g'(u)$ exists for all $u \in S_X$. Then U has density function

$$f_U(u) = f_X(g(u))|g'(u)|, \qquad u \in S_U.$$

Proof

a. In the discrete case,

$$f_U(u) = P(U = u) = P(h(X) = u) = P(X = g(u)) = f_X(g(u)).$$

b. Since $g'(u)$ exists, $g(u)$ is continuous. Since g is invertible, g and h are either both increasing or both decreasing. If they are decreasing, then

$$F_U(u) = P(U \leqq u) = P(h(X) \leqq u) = P(X \geqq g(u)) = 1 - F(g(u)).$$

Therefore, by the chain rule and the fact that g' is negative (since g is decreasing),

$$f_U(u) = F_U'(u) = -f(g(u))g'(u) = f(g(u))|g'(u)|$$

The proof for increasing g is left as an exercise. □

Part a of the theorem seems quite intuitive: Since X is $g(U)$, we should be able to substitute $g(u)$ for x in the density function for x. The result for the continuous case is not quite so simple: We must multiply by the absolute value of the derivative of g. Note that in order to apply this theorem, h only need be an invertible function from S_X to S_U. For example, if X is a nonnegative random variable, then $Y = X^2$ is an invertible function, but it is not if X takes on both positive and negative values.

Example 2–13

Let X be a continuous random variable with density function $f_X(x) = e^{-x}, x > 0$. Let $U = X^{1/2}$, so that $X = U^2 = g(U)$. Then the range of U is the same as the range of X, and $g'(u) = 2u$. Therefore,

$$f_U(u) = f_X(u^2)|2u| = 2ue^{-u^2}, \qquad u > 0.$$

Let $V = e^{-X}$. Then $X = -\log(V) = g^*(V)$. It follows that the range of V is $[0, 1]$, and $g^{*\prime}(v) = -1/v$. Therefore,

$$f_V(v) = f_X(-\log(v))\left|\frac{-1}{v}\right| = \frac{v}{v} = 1.$$

Therefore, V is uniformly distributed on $[0, 1]$.

It sometimes happens that X is a continuous random variable and U is a discrete random variable. In that case, we typically derive the density function of U directly, since it is again a probability.

Example 2–14

Let X be distributed uniformly on the interval $[0, 1]$. Let U be the first digit in X. Then U has range $S_U = \{0, 1, 2, \ldots, 9\}$. Also,

$$f_U(3) = P(U = 3) = P(.3 \leqq X < .4) = .1.$$

Similarly,

$$f_U(u) = .1, \qquad u = 1, 2, 3, \ldots, 9.$$

Now, let V be the second digit of X. Then V also has range $\{0, 1, \ldots, 9\}$, and

$$P(V = 2) = P(.02 \leqq X < .03) + P(.12 \leqq X < .13) + \cdots + P(.92 \leqq X < .93)$$
$$= 10(.01) = .1.$$

In a similar way, we could show that all the digits have this same distribution.

Now,

$$P(V = 2 \text{ and } U = 3) = P(.23 \leqq X < .24) = .01 = P(V = 2)P(U = 3).$$

Therefore, the events $V = 2$ and $U = 3$ are independent, and in general, the events $U = j$ and $V = k$ are independent for all j and k. In a similar way, we could show that all the digits of X are independent.

This discussion suggests a method for simulating a random variable X which is uniform on the interval $[0, 1]$. First, choose a digit U randomly from $0, \ldots, 9$. Make this digit the first digit of X. Then simulate a second digit V randomly from $0, \ldots, 9$, independently of the first digit, and make this the second digit of X. Then simulate a third digit W randomly from $0, \ldots, 9$, independently of both U and V, and make this the third digit of X. Continue in this fashion forever, and X is uniformly distributed. (Actually, many software packages stop with six digits.)

We next prove a basic result about the uniform distribution.

Theorem 2–7. Let $F(x)$ be a distribution function such that $u = F(x)$ is invertible. Let $x = G(u)$ be the inverse of $F(x)$.

a. Let U be uniformly distributed on the interval $[0, 1]$, and let $X = G(U)$. Then X has distribution function $F(x)$.
b. Let X be a random variable with distribution function $F(x)$, and let $U = F(X)$. Then U is uniformly distributed on $[0, 1]$.

Proof. We note first that both $F(x)$ and $G(u)$ are increasing functions.

a. Let $X = G(U)$, and let $F_X(x)$ be the distribution function of X. We wish to show that $F_X(x) = F(x)$. To do so, we merely need note that
$$F_X(x) = P(X \leqq x) = P(G(U) \leqq x) = P(U \leqq F(x)) = F(x).$$
b. Let $U = F(X)$, where X has distribution function $F(x)$. Then
$$F_U(u) = P(U \leqq u) = P(F(X) \leqq u) = P(X \leqq G(u))$$
$$= F(G(u)) = u, \qquad 0 \leqq u \leqq 1. \quad \square$$

Part a of the theorem can be extended to distribution functions which are not invertible by an appropriate choice of $G(x)$. (See Exercise C1.) Also, part a is often used for simulating random variables from a nonuniform distribution. We first simulate a uniform $[0, 1]$ random variable U and let $X = G(U)$. Then X has distribution function F.

Example 2–15

Suppose we want to simulate a random variable X having density function $f(x)$ and distribution function $F(x)$, given below
$$f(x) = 10 \exp(-10x), \qquad F(x) = 1 - \exp(-10x), \qquad x \geqq 0.$$
Then
$$u = 1 - \exp(-10x) \Leftrightarrow x = \frac{-\log(1-u)}{10}.$$
Therefore, if U is uniformly distributed on the interval $[0, 1]$, then
$$X = G(U) = \frac{-\log(1-U)}{10}$$
has distribution function $F(x)$ by Theorem 2–7. (Note that $1 - U \leqq 1$, so that $X \geqq 0$.) Therefore, to simulate X, we first simulate U from a uniform $[0, 1]$ distribution and let $X = G(U)$.

Unfortunately, although the above method of simulation is quite general, it is typically not too useful because it is often quite difficult for the computer to compute $G(U)$.

We shall not discuss simulations any further in this text. An interesting reference for learning about simulations is Ross (1985), pp. 433–493.

Exercises—B

1. Let X be a discrete random variable with range $S = \{0, 1, 2, 3, 4\}$ and density function $f(0) = .1, f(1) = .3, f(2) = .6$.

(a) Let $U = X^2$. Find the density function of U.
(b) Let $V = (X - 1)^2$. Find the density function of V.
(c) Let $W = e^X$. Find the density function of W.

2. Let X be a continuous random variable with density function $f(x) = 4x^3, 0 \leqq x \leqq 1$.
   (a) Let $U = X^2$. Find the density function of U.
   (b) Let $V = e^X$. Find the density function of V.
3. Let X be a continuous random variable with density function $f(x) = 5x^4/64, -2 \leqq x \leqq 2$.
   (a) Let $U = X^3$. Find the density function of U.
   (b) Let $V = e^X$. Find the density function of V.
   (c) Let $W = X^4$. Find the density function of W.
4. Let X be a continuous random variable which is uniformly distributed on the interval $[-2, 3]$.
   (a) Let $U = e^X$. Find the density function of U.
   (b) Let $V = X^3$. Find the density function of V.
   (c) Let $W = X^4$. Find the density function of W.
5. Let X be a continuous random variable with range $-\infty < x < \infty$ and density function $f_X(x)$. Let $U = X^2$. Show that U has density function $f_U(u) = (\frac{1}{2})u^{-1/2}(f_X(u^{1/2}) + f_X(-u^{1/2})), u > 0$.
6. Prove Theorem 2–6, part b, in the case where g is an increasing function.

Exercises—C

1. Let $F(x)$ be a distribution function (not necessarily invertible). Let $G(u)$ be the smallest x such that $F(x) = u$.
   (a) Show graphically that $G(u) \leqq x$ if and only if $u \leqq F(x)$.
   (b) Let U be uniformly distributed on $[0, 1]$, and let $X = G(U)$. Show that X has distribution function $F(x)$.

2.2 BIVARIATE RANDOM VECTORS

2.2.1 Bivariate Density Functions

Suppose we have two random variables X and Y defined on a sample space. (For example, suppose we choose a student at random and let X be her weight and let Y be her height.) We often call (X, Y) a (bivariate) *random vector*.

We say that (X, Y) is a *discrete* random vector if X and Y are both discrete random variables and that (X, Y) is a *nonnegative integer-valued* random vector if X and Y are nonnegative integer-valued random variables. If (X, Y) is a discrete random vector, we define the *joint density function* $f(x, y)$ of X and Y by

$$f(x, y) = P(X = x \text{ and } Y = y)$$

and the *range* of (X, Y) to be those points where $f(x, y) > 0$. That is, the range of (X, Y) is the set of possible values for (X, Y). Note that if (x, y) is in the range of (X, Y), then x is in the range of X and y is in the range of Y, and the ranges of X and Y are finite or countable sets. Therefore, the range of (X, Y) is also a finite or countable set.

The function $f(x, y)$ is a possible density function for a discrete random vector (X, Y) with range S if and only if

$$f(x,y) > 0 \text{ if } (x,y) \in S, \qquad f(x,y) = 0 \text{ if } (x,y) \notin S, \qquad \text{and} \qquad \sum_{(x,y)\in S}\sum f(x,y) = 1.$$

By a proof similar to that in Section 2.1.1, we can show that for any set A,

$$P((X,Y)\in A) = \sum_{(x,y)\in A}\sum f(x,y).$$

Example 2–16

Let (X, Y) be a random vector with range

$$S = \{(0,1), (0,2), (1,0), (1,1), (2,0)\}$$

and density function

$$f(x,y) = c(x + 2y), \qquad (x,y) \in S.$$

By the conditions for a density function,

$$1 = c \sum_{(x,y)\in S}\sum (x+2y) = c(2+4+1+3+2) = 12c,$$

so that $c = \frac{1}{12}$. It then follows that

$$P(X > Y) = f(1,0) + f(2,0) = \frac{3}{12}, \qquad P(X \geqq Y) = f(1,0) + f(2,0) + f(1,1) = \frac{6}{12},$$

$$P(X = 1) = f(1,0) + f(1,1) = \frac{4}{12}, \qquad P(X + Y \text{ is odd}) = f(0,1) + f(1,0) = \frac{3}{12}.$$

Example 2–17

Suppose we are drawing hands of five cards from a standard deck. Let X be the number of aces and Y be the number of kings. Then the range S of (X, Y) is given by $x = 0, 1, \ldots, 4, y = 0, 1, \ldots, 4, x + y \leqq 5$ (since there are only four aces and four kings in the deck, and we can draw at most a total of five aces and kings in five draws). Now,

$$f(2,1) = P(X = 2, Y = 1) = P(2 \text{ aces, } 1 \text{ king, } 2 \text{ others}) = \frac{\begin{bmatrix}4\\2\end{bmatrix}\begin{bmatrix}4\\1\end{bmatrix}\begin{bmatrix}44\\2\end{bmatrix}}{\begin{bmatrix}52\\5\end{bmatrix}},$$

and by a similar argument,

$$f(x,y) = \frac{\begin{bmatrix}4\\x\end{bmatrix}\begin{bmatrix}4\\y\end{bmatrix}\begin{bmatrix}44\\5-x-y\end{bmatrix}}{\begin{bmatrix}52\\5\end{bmatrix}}, \qquad (x,y) \in S.$$

We say that (X, Y) is a *continuous random vector* if there exists an $f(x, y) \geqq 0$ such that

$$P((X,Y)\in A) = \iint_A f(x,y)\,dx\,dy.$$

(The right side is the double integral computed over the set $(x,y) \in A$.) We call $f(x,y)$ the *joint density function* of (X, Y), and the set S where $f(x,y) > 0$ we call the *range* of (X, Y). As in the discrete case, we think of S as the set of possible values of (X, Y). In order to be a joint density function for a continuous random vector (X, Y) with range S, $f(x,y)$ must satisfy

$$f(x,y) > 0 \text{ for } (x,y) \in S, \qquad f(x,y) = 0 \text{ for } (x,y) \notin S, \qquad \iint_S f(x,y)\,dxdy = 1.$$

In the next section, we shall show that if (X, Y) is a continuous random vector, then X and Y are continuous random variables. However, in Exercise C2 we give an example in which X and Y are continuous random variables, but (X, Y) is not a continuous random vector.

If (X, Y) is a continuous random vector, then

$$P(X = Y) = \int_{S^*} \int_y^y f(x,y)dxdy = \int_{S^*} 0dy = 0$$

(where S^* is the range of Y). In a similar way, we can show that $P(X = Y^2) = 0$, $P(X = 4Y) = 0$, $P(Y = e^X) = 0$, and $P(X^2 + Y^2 = 1) = 0$. That is if (X, Y) is a continuous random vector, then the probability of any one-dimensional curve is zero.

Before looking at some examples, we note that

$$P(X \in A) = \iint_A f(x,y)dxdy = \iint_{A \cap S} f(x,y)dxdy,$$

since the density is zero when $(x,y) \notin S$.

Example 2–18

Let (X, Y) be a continuous random vector with joint density function

$$f(x,y) = cxy, \qquad 0 < x < y, 0 < y < 1.$$

(We follow the earlier convention of giving the density function only on the range of (X, Y) and assuming that the density is zero elsewhere.) Then

$$1 = \iint_S f(x,y)dxdy = c\int_0^1 \int_0^y xy\,dxdy = c\int_0^1 \left(\frac{x^2y}{2}\bigg|_0^y\right)dy$$

$$= c\int_0^1 \frac{y^3}{2}dy = \frac{cy^4}{8}\bigg|_0^1 = \frac{c}{8},$$

and therefore, $c = 8$. We find $P(X + Y < 1)$. Let $A = \{(x,y\}: x + y < 1\}$. From Figure 2–3, we see that $A \cap S$ is given by the union of the sets $C = \{(x,y): 0 < x < y, 0 < y < \frac{1}{2}\}$ and $D = \{(x,y): 0 < x < 1 - y, \frac{1}{2} \leq y < 1\}$. Since C and D are disjoint,

$$P((X,Y) \in A) = P((X, Y \in C) + P((X,Y) \in D) = \int_0^{.5} \int_0^y 8xy\,dxdy + \int_{.5}^1 \int_0^{1-y} 8xy\,dxdy$$

$$= \int_0^{.5} 4y^3\,dy + \int_{.5}^1 (4y - 8y^2 + 4y^3)\,dy = y^4\bigg|_0^{.5} + 2y^2 - \frac{8}{3}y^3 + y^4\bigg|_{.5}^{.1} = \frac{1}{6}.$$

Actually, $A \cap S$ also equals $E = \{(x,y): x < y < 1 - x, 0 < x < .5\}$. Therefore, integrating in the reverse order, we obtain

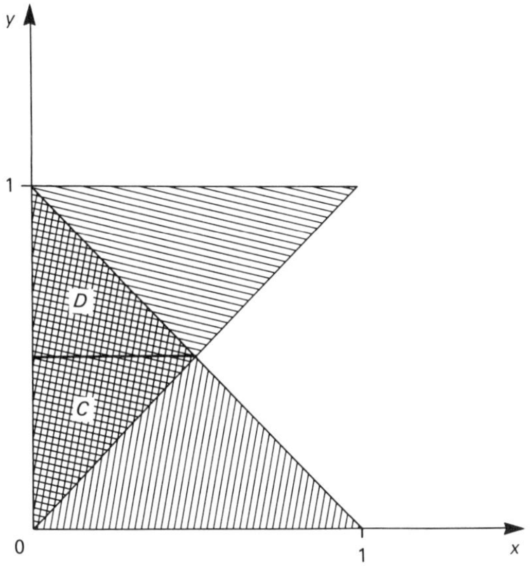

Figure 2–3 Graph of $A \cap S$ in Example 2-18 ■ $S = \{(x, y); 0 < x < y < 1\}$; ▨ $A = \{(x, y): x + y < 1\}$.

$$P((X, Y) \in A) = P((X, Y) \in E) = \int_0^{.5} \int_x^{1-x} 8xy\, dydx = \int_0^{.5} 4x((1-x)^2 - x^2)dx$$

$$= \int_0^{.5} (4x - 8x^2)dx = 2x^2 - \frac{8}{3}x^3 \Big|_0^{.5} = \frac{1}{6}.$$

Example 2–18 suggests that it is often useful to interchange the order of integration in computing double integrals. Note, however, that when we interchange the order of integration, we do not merely interchange the limits. We have to look back at the graph to find the new limits. In particular, there should not be x's or y's in the limits for the outside integrals.

Example 2–19

Let (X, Y) be a continuous random vector with range S. We say that (X, Y) is uniformly distributed over S if $f(x, y) = c$ for $(x, y) \in S$. Invoking the condition for being a density function, we see that $c = (\text{area of } S)^{-1}$, and

$$P(A) = \iint_{A \cap S} c\, dxdy = \frac{\text{area}(A \cap S)}{\text{area}(S)}.$$

Suppose, for example, that (X, Y) is uniformly distributed on the circle with radius 2, so that the area of S is 4π. The circle of radius 1 has area π, so that

$$P(X^2 + Y^2 \leqq 1) = \frac{\pi}{4\pi} = \frac{1}{4}.$$

Similarly, the square $-1 \leqq x \leqq 1, -1 \leqq y \leqq 1$ has area 4, so that

$$P(-1 \leqq X \leqq 1, -1 \leqq Y \leqq 1) = \frac{4}{4\pi} = \frac{1}{\pi}.$$

Now let A be the set in which $x > y$. If we intersect this set with the range of (X, Y), we get a half-circle with area 2π, and therefore,

$$P(X > Y) = \frac{2\pi}{4\pi} = \frac{1}{2}.$$

Let (X, Y) be a bivariate random vector (not necessarily discrete or continuous). Then the *joint distribution function* of (X, Y) is defined as

$$F(x, y) = P(X \leqq x \text{ and } Y \leqq y).$$

If (X, Y) is an integer-valued discrete random vector with joint density function $f(x, y)$ and joint distribution function $F(x, y)$, then

$$f(i, j) = F(i, j) - F(i, j-1) - F(i-1, j) + F(i-1, j-1)$$

and

$$F(x, y) = \sum_{i \leqq x} \sum_{j \leqq y} f(i, j).$$

If (X, Y) is a continuous random variable, then

$$f(x, y) = \frac{\partial^2}{\partial x\, \partial y} F(x, y) \qquad \text{and} \qquad F(x, y) = \int_{-\infty}^{x} \int_{-\infty}^{y} f(s, t) dt ds.$$

Typically, bivariate distribution functions are rather complicated. Fortunately, we shall not need to compute them often (although we shall use the definition several times).

Exercises—A

1. Suppose a friend and I are to meet at a bar sometime between 5:00 and 6:00. Let X be the time I arrive, and let Y be the time my friend arrives. Suppose (X, Y) is uniformly distributed on the square $5{:}00 \leqq X \leqq 6{:}00, 5{:}00 \leqq Y \leqq 6{:}00$.
   (a) Find the probability that I have to wait at least 10 minutes for my friend.
   (b) Find the probability that at least one of us waits at least 10 minutes.
2. Suppose I draw a hand of five cards from an ordinary deck with all hands equally likely. Let X be the number of spades and Y be the number of hearts drawn. Find the joint density function of (X, Y).
3. Suppose I draw a hand of five cards from a normal deck with all hands equally likely. Let X be the number of aces and Y be the number of hearts I draw.
   (a) Find $P(X = 1, Y = 2)$. (*Hint*: This can occur in either of two ways: the ace of hearts and one other heart, or two hearts which are not aces.)
   (b) Find the joint density of (X, Y).

Exercises—B

1. Let (X, Y) be a discrete bivariate random vector with range $(1, 1)$, $(1, 2)$, $(1, 3)$, $(2, 1)$, $(2, 2)$, $(2, 3)$, $(3, 1)$, $(3, 2)$, $(3, 3)$ and density function $f(x, y) = cxy$, $(x, y) \in S$.
   (a) What is c?
   (b) Find $P(X + Y \leqq 4)$ and $P(X + Y < 4)$.
   (c) Find $P(X = 1)$, $P(X = 2)$, and $P(X = 3)$.

2. Let (X, Y) be a continuous random vector with joint density function $f(x,y) = cx^2y, 0<x<1, 0<y<1$.
   (a) What is c?
   (b) Find $P(X \leqq .5, Y \leqq .5)$ and $P(X \leqq .5)P(Y \leqq .5)$. Are the events $\{X \leqq .5\}$ and $\{Y \leqq .5\}$ independent?
   (c) Find $P(X + Y \leqq \frac{1}{3})$.
   (d) Find $P(X + Y \leqq \frac{2}{3})$. (*Hint*: Find the probability of the complement.)
3. Let (X, Y) be a continuous random vector with joint density function $f(x,y) = c(x^2 + y^2)$, $-1 \leqq x \leqq 1, -1 \leqq y \leqq 1$.
   (a) What is c?
   (b) Find $P(X \geqq Y)$ and $P(X \geqq 2Y)$.
   (c) Find $P(X^2 + Y^2 \leq 1)$. (*Hint*: Use polar coordinates.)
4. Let (X, Y) be a continuous random vector with joint density function $f(x,y) = ce^{-x}e^{-2y}$, $0 \leqq x \leqq 2y < \infty$.
   (a) Find c.
   (b) Find $P(X \leqq Y)$.
   (c) Find $P(X \leqq 1, Y \leqq 1)$ and $P(X \leqq 1)P(Y \leqq 1)$. Are these events independent?
5. Let (X, Y) be a discrete bivariate random vector with density function $f(x,y)$. Show that $P(X \in A) = \sum\sum_{(x,y) \in A} f(x,y)$.
6. Let (X, Y) be a random vector with distribution function $F(x,y)$.
   (a) Show that $P(a < X \leqq b, c < Y \leqq d) = F(b,d) - F(b,c) - F(a,d) + F(a,c)$.
   (b) Let X and Y be integer valued. Show that $f(i,j) = F(i,j) - F(i,j-1) - F(i-1,j) + F(i-1,j-1)$.

Exercises—C

1. (a) Show that the joint distribution function for (X, Y) in Example 2–18 is $F(x,y) = x^2(2y^2 - x^2), 0<x<y<1$. (*Hint*: First find $F(\frac{1}{3}, \frac{1}{2})$.)
   (b) Show that $F(x,y) = F(y,y), 0<y<x<1$. (First find $F(\frac{1}{2}, \frac{1}{3})$.)
2. Let X be a continuous random variable with density function $f(x) = e^{-x}, x>0$. Let $Y = X$.
   (a) Show that Y is a continuous random variable. What is its density function?
   (b) Show that (X, Y) is not a continuous random vector.
3. Let (X, Y) be a continuous random vector with joint density function $f(x,y)$ and joint distribution function $F(x,y)$.
   (a) Show that $F(x,y) = \int_{-\infty}^{x} \int_{-\infty}^{y} f(s,t)\, dt ds$.
   (b) Show that $f(x,y) = (\partial^2/\partial x\, \partial y)F(x,y)$ at points where $f(x,y)$ is continuous.

2.2.2 Marginal and Conditional Density Functions

If (X, Y) is a bivariate random vector, then X and Y are random variables. The *marginal density of* X, denoted $f_1(x)$, is defined to be the density function of X (ignoring Y), and the *conditional density of X given $Y = y$*, denoted $f_1(x|y)$, is defined to be the density of X when we know that $Y = y$. For example, suppose X is a randomly chosen female's weight and Y is her height. Then the marginal density of X is the density function for the weight of a randomly chosen female, and the conditional density of X given $Y = 60$ in. is the density function for the weight of a female randomly chosen from the class of 60 in.-tall females. Similarly, the *marginal density of Y*, denoted $f_2(y)$, is the density of Y (ignoring X), and the *conditional density of Y given X* $= x$, denoted $f_2(y|x)$, is the density of Y when we know that $X = x$. (Note that when f_1 is a function of only one variable, as in $f_1(x)$, then it is a marginal density function, but when it is a function of two variables, as in $f_1(x|y)$, then it is a conditional density function.)

The *range* S_1 of X is the set of all x's such that $f(x, y) > 0$ for some y. That is, S_1 is the set of possible values of X. Similarly, the *range* S_2 of Y is the set of all y's such that $f(x, y) > 0$ for some x.

Theorem 2–8. Let (X, Y) be a random vector with joint density function $f(x, y)$. Let S_1 and S_2 be the ranges of X and Y.

a. If (X, Y) is a discrete random vector, then X and Y are discrete random variables, and the marginal density functions are given by

$$f_1(x) = \sum_{y \in S_2} f(x, y), \qquad x \in S_1; \qquad f_2(y) = \sum_{x \in S_1} f(x, y), \qquad y \in S_2.$$

b. If (X, Y) is a continuous random vector, then X and Y are continuous random variables, and the marginal density functions are given by

$$f_1(x) = \int_{S_2} f(x, y)\, dy, \qquad x \in S_1; \qquad f_2(y) = \int_{S_1} f(x, y)\, dx, \qquad y \in S_2.$$

c. If (X, Y) is a discrete or continuous random vector, then the conditional density functions are given by

$$f_1(x|y) = \frac{f(x, y)}{f_2(y)}, \qquad (x, y) \in S; \qquad f_2(y|x) = \frac{f(x, y)}{f_1(x)}, \qquad (x, y) \in S.$$

Proof

a. See Exercise B5.

b. We first find the distribution function of X:

$$F_1(x) = P(X \leqq x) = P(X \leqq x, Y < \infty) = \int_{-\infty}^{x} \int_{-\infty}^{\infty} f(s, t)dtds.$$

Therefore, by the fundamental theorem of calculus,

$$f_1(x) = F_1'(x) = \int_{-\infty}^{\infty} f(x, t)dt = \int_{-\infty}^{\infty} f(x, y)dy = \int_{S_2} f(x, y)dy.$$

(Note that it does not matter whether we call the variable of integration t or y and that $f(x, y) = 0$ if $y \notin S_2$.) The proof for $f_2(y)$ is similar.

c. We prove this result only in the discrete case. For a discussion of conditional densities in the continuous case, see Billingsley (1979), pp. 354–407. If (X, Y) is a discrete random vector, then

$$f_1(x|y) = P(X = x|Y = y) = \frac{P(X = x \text{ and } Y = y)}{P(Y = y)} = \frac{f(x, y)}{f_2(y)}.$$

The proof for $f_2(y|x)$ is similar. $\square$

For discrete random vectors, part a of Theorem 2–8 says that to find the marginal distribution $f(x)$ of X, we just sum out $f(x, y)$ over all values y such that $f(x, y) > 0$. In the continuous case, we integrate out over all such values. The conditional distribution of X given $Y = y$ in the continuous case is actually rather difficult to define carefully (since the set $Y = y$ has probability zero). However, part c of the theorem is true for the continuous random vectors we shall study in this book.

Observe that the marginal density functions are density functions, and for each fixed y, $f(x|y)$ is a density function in x. For example if (X, Y) is a continuous random vector (so that X is a continuous random variable), then

$$f_1(x) > 0 \text{ for } x \in S_1, \qquad f_1(x) = 0 \text{ for } x \notin S_1, \qquad \int_{S_1} f_1(x)dx = 1,$$

$$f_1(x|y) > 0 \text{ for } (x, y) \in S, \qquad \int_{S_1} f(x|y)dx = 1$$

(see Exercise B6).

Example 2–20

Let (X, Y) be a discrete random vector with range $S = \{(0, 1), (0, 2), (1, 0), (1, 1), (2, 0)\}$ and density function $f(x, y) = (x + 2y)/12$. We note that X can take on the values 0, 1, 2, so that $S_1 = \{0, 1, 2\}$. Similarly, $S_2 = \{0, 1, 2\}$. Now, suppose that $x = 0$. Then $f(x, y) = 0$ unless $y \in \{1, 2\}$. Therefore,

$$f_1(0) = f(0, 1) + f(0, 2) = \frac{(2 + 4)}{12} = \frac{6}{12}.$$

Similarly, if $x = 1$, then y can only be 0 or 1, and if $x = 2$, then y must be 0. Therefore,

$$f_1(1) = f(1, 0) + f(1, 1) = \frac{4}{12}, \qquad f_1(2) = f(2, 0) = \frac{2}{12}.$$

(Note that $f_1(0) + f_1(1) + f_1(2) = 1$.) Consequently, for this model, X is 0 with probability $\frac{6}{12}$, 1 with probability $\frac{4}{12}$, and 2 with probability $\frac{2}{12}$.

In a similar manner, we have

$$f_2(0) = \frac{3}{12}, \qquad f_2(1) = \frac{5}{12}, \qquad f_2(2) = \frac{4}{12}.$$

Now,

$$f_1(0|1) = \frac{f(0, 1)}{f_2(1)} = \frac{2/12}{5/12} = \frac{2}{5}, \qquad f_1(1|1) = \frac{f(1, 1)}{f_2(1)} = \frac{3}{5}.$$

Therefore, if $Y = 1$, then X is 0 with probability $\frac{2}{5}$ and 1 with probability $\frac{3}{5}$. (Note again that $f_1(0|1) + f_1(1|1) = 1$.) Also, we have

$$f_1(0|2) = \frac{f(0, 2)}{f_2(2)} = 1,$$

so that if $Y = 2$, then $X = 0$ (which is fairly obvious from the joint range). Note that $f_1(0|1) + f_1(0|2) \neq 1$. Finally,

$$f_2(0|1) = \frac{f(1, 0)}{f_1(1)} = \frac{1}{4} \neq f_1(0|1).$$

When $Y = 1$, X must be either 0 or 1, so that the range of X when $Y = 1$ is not the same as the range of X alone, which is 0, 1, and 2.

Example 2–21

Suppose we draw five cards and let X be the number of aces and Y the number of kings. In the last section, we saw that

$$f(x,y) = \frac{\begin{bmatrix}4\\x\end{bmatrix}\begin{bmatrix}4\\y\end{bmatrix}\begin{bmatrix}44\\5-x-y\end{bmatrix}}{\begin{bmatrix}52\\5\end{bmatrix}}, \qquad (x,y) \in S,$$

where $S = \{(x,y)\text{: } x = 0,1,\ldots,4, y = 0,1,\ldots,4, x+y \leqq 5\}$. Now, since we can get anywhere from zero to four aces, $S_1 = \{0,1,\ldots,4\}$. Suppose, then, that $Y = 3$, so that we know that the hand has three kings. Then we can get at most two aces. Similarly, if $Y = 2$, then X can only be in $\{0,1,2,3\}$.

The marginal density function of X is

$$f_1(x) = \sum_{y \in S_2} f(x,y).$$

Unfortunately, this sum is nontrivial to compute. However, we can find the marginal density of X easily from the definition. For example,

$$f_1(2) = P(X=2) = P(2 \text{ aces and 3 other cards}) = \frac{\begin{bmatrix}4\\2\end{bmatrix}\begin{bmatrix}48\\3\end{bmatrix}}{\begin{bmatrix}52\\5\end{bmatrix}},$$

and similarly,

$$f_1(x) = P(X=x) = \frac{\begin{bmatrix}4\\x\end{bmatrix}\begin{bmatrix}48\\5-x\end{bmatrix}}{\begin{bmatrix}52\\5\end{bmatrix}}, \qquad x = 0,1,\ldots,4.$$

This example points out that it is occasionally easier to find the marginal density for a random variable from the definition than using the theorem. Now, the conditional density of X given that $Y = 2$ is

$$f_1(x|2) = \frac{f(x,2)}{f_2(2)} = \frac{\begin{bmatrix}4\\x\end{bmatrix}\begin{bmatrix}4\\2\end{bmatrix}\begin{bmatrix}44\\3-x\end{bmatrix}}{\begin{bmatrix}52\\5\end{bmatrix}} \Bigg/ \frac{\begin{bmatrix}4\\2\end{bmatrix}\begin{bmatrix}44\\3\end{bmatrix}}{\begin{bmatrix}52\\5\end{bmatrix}}$$

$$= \frac{\begin{bmatrix}4\\x\end{bmatrix}\begin{bmatrix}44\\3-x\end{bmatrix}}{\begin{bmatrix}48\\3\end{bmatrix}}, \qquad x = 0,1,2,3.$$

Note that this answer also has a very natural interpretation. If we know we have exactly two kings, we can picture drawing the other three cards from a deck with only 48 cards, four of which are aces, and this leads to the conditional probability given. Similarly,

$$f_1(x|y) = \frac{\begin{bmatrix}4\\x\end{bmatrix}\begin{bmatrix}44\\5-y-x\end{bmatrix}}{\begin{bmatrix}48\\5-y\end{bmatrix}}, \qquad x = 0, 1, \ldots, \min(4, 5-y); y = 0, 1, \ldots, 4.$$

(If we have y kings, then the number of aces must be at most $5 - y$, because there are only $5 - y$ cards left in the hand. In addition, there cannot be more than four aces in the hand.)

Example 2–22

Let (X, Y) be a continuous random vector with $f(x, y) = 8xy, 0 < x < y < 1$. Then the range S_1 of X is $0 < x < 1$. In addition, unless $y > x$, $f(x, y) = 0$. Therefore, the marginal density of X is

$$f_1(x) = \int_x^1 8xy\, dy = 4xy^2 \Big|_{y=x}^{y=1} = 4x(1 - x^2), \qquad 0 < x < 1.$$

Similarly, the marginal density of Y is

$$f_2(y) = \int_0^y 8xy\, dx = 4x^2 y \Big|_{x=0}^{x=y} = 4y^3, \qquad 0 < y < 1.$$

The conditional density of X given $Y = y$ is

$$f_1(x|y) = \frac{f(x, y)}{f_2(y)} = \frac{8xy}{4y^3} = \frac{2x}{y^2}, \qquad 0 < x < y, 0 < y < 1.$$

Note that for any y in $(0, 1)$,

$$\int_0^y \frac{2x}{y^2}\, dx = \frac{x^2}{y^2}\Big|_{x=0}^{x=y} = \frac{y^2}{y^2} = 1,$$

so that $f_2(x|y)$ is a density function for X for any fixed y between zero and one. Similarly, the conditional density of Y given $X = x$ is

$$f_2(y|x) = \frac{f(x, y)}{f_1(x)} = \frac{8xy}{4x(1 - x^2)} = \frac{2y}{(1 - x^2)}, \qquad x < y < 1, 0 < x < 1,$$

which is a density function for Y for any x in the interval $(0, 1)$.

Example 2–23

Let (X, Y) be uniformly distributed in the circle of radius 2, so that $f(x, y) = 1/4\pi$, $x^2 + y^2 \le 4$. Then X takes on values between -2 and 2, and hence, $S_1 = (-2, 2)$. If $X = x$, then Y takes on values between $-(4 - x^2)^{1/2}$ and $(4 - x^2)^{1/2}$. Hence, the marginal density of X is

$$f_1(x) = \int_{-(4-x^2)^{1/2}}^{(4-x^2)^{1/2}} \left(\frac{1}{4\pi}\right) dy = \frac{(4 - x^2)^{1/2}}{2\pi}, \qquad -2 < x < 2.$$

Thus, X is not uniformly distributed. This is not surprising. If x is near 0, the interval of possible y's is larger than if x is near -2 or 2.

Similarly,

$$f_2(y) = \frac{(4 - y^2)^{1/2}}{2\pi}, \qquad -2 < y < 2.$$

Therefore, the conditional density of X given that $Y = y$ is

$$f_1(x|y) = \frac{1/4\pi}{(4 - y^2)^{1/2}/2\pi} = (2(4 - y^2)^{1/2})^{-1}, \qquad -(4 - y^2)^{1/2} < x < (4 - y^2)^{1/2}.$$

We thus see that conditionally on $Y = y$, X is uniformly distributed on the interval $(-(4-y^2)^{1/2}, (4 - y^2)^{1/2})$.

Examples 2–20 through 2–23 indicate that often the most difficult part of finding marginal and conditional density functions is finding the regions of integration or summation.

We often turn around the process in these examples and use the marginal and conditional density to determine the joint density. By part c of Theorem 2–8,

$$f(x, y) = f_1(x)f_2(y|x) = f_2(y)f_1(x|y).$$

(Note that there is no way we can use $f_1(x)$ and $f_1(x|y)$ to find the joint density of X and Y.)

Example 2–24

Suppose $f_1(x) = 3x^2$ for $0 < x < 1$ and $f_2(y|x) = 2y/x^2$ for $0 < y < x$. Then

$$f(x, y) = 6y, \qquad 0 < y < x < 1.$$

Therefore,

$$f_2(y) = \int_y^1 6y\,dx = 6y(1-y), \quad 0 < y < 1; \qquad f_1(x|y) = \frac{1}{1-y}, \quad 0 < y < x < 1.$$

Thus, given that $Y = y$, X is uniformly distributed on the interval $(y, 1)$.

Exercises—A

1. For the model in Exercise A1 of the last section, find the marginal density function of Y and the conditional density function of X given $Y = y$.
2. For Exercise A2 of the last section, find the marginal density function of Y and the conditional density function of X given $Y = y$.
3. For the model in Exercise A3 of the last section, find the marginal density function of X and the conditional density function of Y given X.

Exercises—B

1. For the joint density function of Exercise B1 of the last section, find both marginal density functions. Find the conditional density function of Y given $X = x$.
2. Let (X, Y) have the joint density function $6x^2y, 0 < x < 1, 0 < y < 1$. Show that the marginal density of X and the conditional density of X given Y are the same.
3. Let (X, Y) have the joint density function $f(x, y) = 3(x^2 + y^2)/2, 0 < x < 1, 0 < y < 1$. Find the marginal density function of X and the conditional density function of Y given X.
4. Let (X, Y) have the joint density function $f(x, y) = 4e^{-x-2y}, 0 < x < 2y < \infty$.
   (a) Find the marginal density of X and the conditional density of Y given $X = x$.
   (b) Verify that the marginal density of X is a density function over the range of X, and that the conditional density of Y given $X = x$ is a density function. (Be careful of the range.)
5. Prove part a of Theorem 2–8. (*Hint*: There is no need to use the distribution function in the discrete case.)
6. Let (X, Y) be a continuous random vector.
   (a) Show that $\int_{S_1} f_1(x)\,dx = 1$.

(b) Show that $\int_{S_1} f_1(x|y)\,dx = 1$ for all $y \in S_2$.

7. Let X be uniformly distributed on the interval $(0, 1)$, and let the conditional distribution of Y given $X = x$ be uniform on the interval $(-x, x)$.
   (a) Find the joint density of (X, Y).
   (b) Find the marginal density of Y and the conditional density of X given $Y = y$.

2.2.3 Independent Random Variables

Let (X, Y) be a bivariate random vector. We say that X and Y are *independent* if

$$P(X \in A \text{ and } Y \in B) = P(X \in A)P(Y \in B)$$

for all sets A and B. That is, X and Y are independent random variables if the events $X \in A$ and $Y \in B$ are independent events. Intuitively, X and Y are independent if information about one variable doesn't tell anything about the second variable. For example, it might be reasonable to suppose that a randomly chosen person's height and SAT score are independent.

We now give several conditions which are equivalent to independence.

Theorem 2–9. Let (X, Y) be a continuous or discrete random vector with joint density function $f(x, y)$ and range S. Let X and Y have marginal density functions $f_1(x)$ and $f_2(y)$ and ranges S_1 and S_2.

a. X and Y are independent if and only if $f(x, y) = f_1(x)f_2(y)$.
b. X and Y are independent if and only if $S = \{(x, y)\colon x \in S_1, y \in S_2\}$ and there exist functions $g(x)$ and $h(y)$ such that $f(x, y) = g(x)h(y)$ for all $x \in S_1$, $y \in S_2$.

Proof. We prove the theorem for the continuous case. The discrete case is in Exercise B4.

a. First suppose that $f(x, y) = f_1(x)f_2(y)$. Then

$$\begin{aligned} P(X \in A, Y \in B) &= \int_A \int_B f(x, y)\,dydx = \int_A \int_B f_1(x)f_2(y)\,dydx \\ &= \int_A f_1(x)\left(\int_B f_2(y)\,dy\right)dx \\ &= \int_A f_1(x)P(Y \in B)\,dx = P(Y \in B)\int_A f_1(x)dx \\ &= P(Y \in B)P(X \in A) \end{aligned}$$

(since $P(Y \in B)$ does not involve x). Therefore X and Y are independent.

Now suppose that X and Y are independent. Then the joint distribution function is given by

$$F(x, y) = P(X \leqq x, Y \leqq y) = P(X \leqq x)P(Y \leqq y).$$

Therefore

$$f(x, y) = \frac{\partial^2}{\partial x\,\partial y}F(x, y) = F_1'(x)F_2'(y) = f_1(x)f_2(y).$$

b. Let X and Y be independent. Then $f(x, y) = f_1(x)f_2(y)$. This in turn implies that the range $S = \{(x, y): x \in S_1, x \in S_2\}$ and that $f(x, y)$ factors into $g(x) = f_1(x), h(y) = f_2(y)$.

Now suppose that the range and density function satisfy the latter conditions, and let

$$c = \int_{S_1} h(x)\,dx, \qquad d = \int_{S_2} g(y)dy.$$

Then

$$f_1(x) = \int_{S_2} f(x, y)\,dy = h(x)\int_{S_2} g(y)\,dy = dh(x).$$

Similarly, $f_2(y) = cg(y)$. Finally,

$$1 = \int_{S_1}\int_{S_2} h(x)g(y)\,dx\,dy = \left(\int_{S_1} h(x)\,dx\right)\left(\int_{S_2} g(y)\,dy\right) = cd.$$

Therefore,

$$f(x, y) = h(x)g(y) = (dh(x))(cg(y)) = f_1(x)f_2(y), \qquad (x, y) \in S,$$

and X and Y are independent by part a. □

Part b of Theorem 2–9 is typically used to see whether two random variables are independent. It implies that we do not need to find the marginal densities of X and Y to find out if they are independent. All we have to do is verify that the joint density function and the joint range factor in the sense given in the theorem. On the other hand, part a is used to find the joint density of X and Y when we assume that they are independent.

We next give the following corollary of Theorem 2–9, which is also helpful in understanding independence.

Corollary 2–9

a. X and Y are independent if and only if the conditional density of X given $Y = y$ is the same as the marginal density of X.
b. X and Y are independent if and only if the conditional density of X given $Y = y$ does not depend on y.

This corollary implies that knowledge of Y does not tell us anything about the distribution of X. (We could obviously also show that X and Y are independent if and only if the conditional distribution of Y given $X = x$ is the same as the marginal distribution of Y.)

Example 2–25

Let us check several joint density functions to see whether X and Y are independent. By part b of Theorem 2–9, we need to verify that the range is a "rectangle" (i.e., that it has the form $x \in Q, y \in R$) and that the density function factors into a product of a function of x and a function of y.

a. Let $f(x, y) = 2e^{-x-2y}$ for $x > 0, y > 0$. Then the range of x and y is a rectangle, and $f(x, y) = (2e^{-x})(e^{-2y})$, so that X and Y are independent.

b. Let $f(x, y) = 2e^{-x-y}$ for $0 < x < y < \infty$. Then X and Y are not independent, because the range does not factor as it should. (If we know that $Y = 1$, then we know that $X < 1$.)

c. Let $f(x, y) = 3(x^2 + y^2)/8$, $-1 < x < 1$, $-1 < y < 1$. Then the range factors appropriately, but the density function is not a product of a function of x and a function of y, so that X and Y are not independent.

Example 2–26

Suppose we know that X and Y are independent random variables with marginal density functions $f(x) = x/6, x = 1, 2, 3$, and $f(y) = (y+1)/15, y = 0, 1, 2, 3, 4$. Then $f(x, y) = x(y+1)/90, x = 1, 2, 3, y = 0, 1, 2, 3, 4$.

We conclude the subsection with a simple but important theorem.

Theorem 2–10. Let X and Y be independent random variables, and let $U = g(x)$ and $V = h(y)$. Then U and V are independent random variables.

Proof. Let A and B be sets, and let A^* and B^* be defined by

$$U \in A \Leftrightarrow X \in A^*, \qquad V \in B \Leftrightarrow Y \in B^*.$$

Then

$$\begin{aligned} P(U \in A \text{ and } V \in B) &= P(X \in A^* \text{ and } Y \in B^*) = P(X \in A^*)P(Y \in B^*) \\ &= P(U \in A)P(V \in B). \quad \square \end{aligned}$$

Exercises—A

1. Let X and Y be the lifetimes of two light bulbs, and suppose that X and Y are independent, with densities $f(x) = (100)^{-1} \exp(-X/100), x > 0$, and $g(y) = (50)^{-1} \exp(-y/50), y > 0$.
   (a) Find the probability that X lasts longer than Y.
   (b) Find the probability that X lasts at least twice as long as Y.
2. A person is throwing darts at a circular target. Let (X, Y) be the point he hits, and suppose that this point is uniformly distributed on the target. Are X and Y independent? Why or why not?

Exercises—B

1. For which of the following densities are X and Y independent?
   (a) $f(x, y) = c_1 x^2 y^3, -1 < x < 1, 0 < y < 4$.
   (b) $f(x, y) = c_2 x^2 y^3, -y < x < y, 0 < y < 4$.
   (c) $f(x, y) = c_3 \log(x^2 y^3), 1 < x < 2, 2 < y < 4$.
2. Let X and Y be independent, and let $f_1(x) = e^{-x}, x > 0$, and $f_2(y) = e^{-y}, y > 0$.
   (a) Find the joint density of X and Y.
   (b) Find $P(X + Y \leqq 1)$.
   (c) Find $P(X + Y \leqq z)$ for all $z > 0$.
   (d) Let $Z = X + Y$. Find the density function of Z.
3. Let X and Y be independent with discrete marginal densities $f(x) = x/10, x = 1, 2, 3, 4$, and $g(y) = \frac{1}{5}, y = 1, 2, 3, 4, 5$. Find $P(X = Y)$ and $P(X < Y)$.

4. Prove Theorem 2–9 for the discrete case. (*Hint*: There is no need to use the distribution function in the discrete case.)
5. Prove Corollary 2–9.

Exercises—C

1. Show that if X and Y are independent continuous random variables, then (X, Y) is a continuous random vector.

2.2.4 Functions of Bivariate Random Vectors

Let (X, Y) be a random vector with density function $f(x, y)$, and let $U = h(X, Y)$ for some function h. We often desire to find the density function of U. If U is discrete, we typically just find the density function directly.

Example 2–27

Let $f(x,y) = (x + 2y)/12, (x,y) \in \{(0,1),(0,2),(1,0),(1,1),(2,0)\}$. Let $U = X + Y$. Then U has range $\{1, 2\}$. Now,

$$f_U(1) = P(U = 1) = f(0,1) + f(1,0) = \frac{3}{12}, \qquad f_U(2) = 1 - f_U(1) = \frac{9}{12}.$$

For continuous random variables, we typically find the distribution function first and then differentiate it.

Example 2–28

Let X and Y be independent, and let $f_1(x) = e^{-x}, x > 0$, and $f_2(y) = e^{-y}, y > 0$. Suppose $U = X + Y$. We wish to find the density function of U. We note that the range of U is $u > 0$. The distribution function of U is given by

$$F_U(u) = P(X + Y \leqq u) = \int_0^u \int_0^{u-x} e^{-x-y}\, dy\, dx = \int_0^u e^{-x}(-e^{-y})\Big|_{y=0}^{y=u-x} dx$$

$$= \int_0^u e^{-x}(1 - e^{-u+x})\, dx = \int_0^u (e^{-x} - e^{-u})\, dx = -e^{-x} - xe^{-u}\Big|_{x=0}^{x=u} = 1 - e^{-u} - ue^{-u}.$$

Therefore, the density function is

$$f_U(u) = F_U'(u) = e^{-u} - e^{-u} + ue^{-u} = ue^{-u}, \qquad u > 0.$$

Now let h be a two-dimensional function of X and Y, and let

$$(U, V) = h(X, Y) = (h_1(X, Y), h_2(X, Y))$$

(so that $U = h_1(X, Y), V = h_2(X, Y)$). We say that h is *invertible* if there exists a two-dimensional function $g(U, V) = (g_1(U, V), g_2(U, V))$ such that

$$(U, V) = h(X, Y) \Leftrightarrow (X, Y) = g(U, V).$$

We call g the *inverse function* of h. If g is differentiable, we define the *Jacobian* of g by

$$J_g = \frac{\partial}{\partial u} g_1(u, v) \frac{\partial}{\partial v} g_2(u, v) - \frac{\partial}{\partial v} g_1(u, v) \frac{\partial}{\partial u} g_2(u, v).$$

An easy way to remember this formula is in matrix notation:

$$J_g = \det\begin{Bmatrix} \dfrac{\partial}{\partial u} g_1(u, v) & \dfrac{\partial}{\partial v} g_1(u, v) \\ \dfrac{\partial}{\partial u} g_2(u, v) & \dfrac{\partial}{\partial v} g_2(u, v) \end{Bmatrix}$$

(where $\det \mathbf{A}$ is the determinant of the matrix $\mathbf{A}$).

Theorem 2–11. Let (X, Y) be a random vector with joint density function $f_{X,Y}(x, y)$. Let $(U, V) = h(X, Y)$ be an invertible function with inverse $(X, Y) = g(U, V)$.

a. If (X, Y) is discrete, then (U, V) has the joint density function

$$f_{U,V}(u, v) = f_{X,Y}(g(u, v)).$$

b. If (X, Y) is continuous and g has Jacobian $J_g(u, v)$, then (U, V) has the joint density function

$$f_{U,V}(u, v) = f_{X,Y}(g(u, v))|J_g(u, v)|$$

(where $|J_g(u, v)|$ is the absolute value of the Jacobian).

Proof

a. See the exercises.

b. Let $A \subset R^2$ be a set and let A^* satisfy $(U, V) \in A \Leftrightarrow (X, Y) \in A^*$. By the theorem from calculus on change of variables in a multiple integral,

$$P((U, V) \in A) = P((X, Y) \in A^*) = \iint_{A^*} f(x, y)\, dxdy$$

$$= \iint_A f(g(u, v))|J_g(u, v)|\, dudv.$$

The desired result follows by the definition of the continuous density function.

□

Example 2–29

Let X and Y be independent, and let $f_1(x) = e^{-x}, x > 0$, and $f_2(y) = e^{-y}, y > 0$. Suppose

$$U = X + Y, \qquad V = \frac{X}{X + Y}.$$

To find the range of (U, V), we first note that $U > 0$. Now, for any $U > 0$, V can take any value between 0 and 1. Hence, the range of (U, V) is $\{(u, v): u > 0, 0 < v < 1\}$.

We now find the inverse of this function. Since $X = UV$ and $Y = U - X = U(1 - V)$, it follows that

$$g(u, v) = (uv, u(1 - v)), \qquad g_1(u, v) = uv, \qquad g_2(u, v) = u(1 - v).$$

Therefore, the Jacobian of g is

$$J_g(u, v) = v(-u) - u(1 - v) = -u, \qquad |J_g(u, v)| = u.$$

Hence,

$$f_{U,V}(u, v) = e^{-uv - u(1 - v)} u = ue^{-u}, \qquad u > 0, 0 < v < 1.$$

This result implies that U and V are independent (since both the density function and the range factor appropriately). Also, from Example 2–28, we see that U has density $ue^{-u}, u > 0$, so that v is uniformly distributed in $(0, 1)$.

Now, let

$$U = X + Y, \qquad V = X.$$

U can take on any positive number, and for each U, V can take on any positive number less than U. Hence the range of (U, V) is $\{(u, v)\colon u > 0, 0 < v < u\}$. Since the range is not a rectangle, U and V cannot be independent. The inverse function is $X = V$, $Y = U - X = U - V$, so that

$$g_1(u, v) = v, \qquad g_2(u, v) = u - v.$$

The Jacobian for this transformation is

$$J_g(u, v) = 0(-1) - 1(1) = -1, \qquad |J_g(u, v)| = 1.$$

Therefore, the joint density of (U, V) is

$$f_{U,V}(u, v) = e^{-v-(u-v)} = e^{-u}, \qquad u > 0, 0 < v < u.$$

From Example 2–28, we see that the marginal density of U is $f_U(u) = ue^{-u}$, $u > 0$. Therefore, the conditional density of V given $U = u$ is

$$f_{U,V}(v|u) = \frac{f_{U,V}(u, v)}{f_U(u)} = \frac{e^{-u}}{ue^{-u}} = \frac{1}{u}, \qquad 0 < v < u,$$

and the conditional distribution of V given $U = u$ is uniform on the interval $(0, u)$.

Finally, let

$$U = \frac{Y}{X}, \qquad V = X.$$

Then the range of V is $v > 0$. If $V = v$, then U can still take on any positive value. Therefore, the range of (U, V) is $\{(u, v)\colon u > 0, v > 0\}$. The inverse function for this problem is then

$$x = g_1(u, v) = v, \qquad y = g_2(u, v) = uv,$$

and the Jacobian of this transformation is

$$J_g(u, v) = 0(u) - 1(v) = -v, \qquad |J_g(u, v)| = v.$$

Therefore,

$$f_{U,V}(u, v) = e^{-v-uv}v = ve^{-v(1+u)}, \qquad u > 0, v > 0.$$

The marginal density of $V = X$ is $f_V(v) = e^{-v}, v > 0$. Therefore, the conditional density of U given $V = v$ is

$$f_{U,V}(u|v) = \frac{f_{U,V}(u, v)}{f_V(v)} = ve^{-vu}, \qquad v > 0, u > 0.$$

Using integration by parts with $s = v$ and $dt = e^{-v(1+u)}\,dv$, we obtain the marginal density of U as

$$\int_0^\infty ve^{-v(1+u)}\,dv = \frac{-ve^{-v(1+u)}}{1+u}\Bigg|_{v=\infty}^{v=\infty} - \int_0^\infty \frac{-e^{-v(1+u)}}{1+u}\,dv$$

$$= 0 - 0 - \frac{e^{-v(1+u)}}{(1+u)^2}\Bigg|_{v=\infty}^{v=\infty} = \frac{1}{(1+u)^2}, \qquad u > 0.$$

(Note that $\lim_{v\to\infty}(v/e^{v(1+u)}) = 0$ by L'Hospital's rule.)

From this example, we see that often the most difficult part of the calculation is finding the range of (U, V) and the inverse function g.

Exercises—B

In the following problems, be sure to indicate the ranges for all the induced densities.

1. Let X and Y have the joint density function given in Exercise B1 of Section 2.2.1.
  (a) Let $U = X + Y$. Find the density function of U.
  (b) Let $V = XY$. Find the density function of V.

2. Let X and Y be independently uniformly distributed on $(0, 1)$. Let $U = X + Y$. Find the density function of U. (*Hint*: Break this into two cases, $0 \leq u < 1$ and $1 < u < 2$.)

3. Let X and Y be independently uniformly distributed on $(0, 1)$. Let $U = X/Y$. Find the density function of U.

4. Let (X, Y) be (jointly) uniformly distributed on the triangle $0 < x < y < 1$ (so that X and Y are not uniformly distributed). Let $U = X/Y$. Find the density of U.

5. Let X and Y be independent, and let $f_1(x) = e^{-x}, x > 0$, and $f_2(y) = e^{-y}, y > 0$.
  (a) Let U be the larger of X and Y. Find the density function of U. (*Hint*: $U \leqq u$ if and only if $X \leqq u$ and $Y \leqq u$.)
  (b) Let V be the smaller of X and Y. Find the density function of V.

6. Let X and Y be independent, with $f_1(x) = e^{-x}, x > 0$, and $f_2(y) = e^{-y}, y > 0$. Let $U = X + Y, V = X - Y$. Find the joint density of U and V and the marginal density of V.

7. Let X and Y be independent, with $f_1(x) = e^{-x}, x > 0$, and $f_2(y) = e^{-y}, y > 0$.
  (a) Let $U = (X + Y)^{1/2}, V = X$. Find the joint density of U and V.
  (b) Let $U = \log[XY], V = \log[X/Y]$. Find the joint density of U and V.

8. Prove Theorem 2–11, part a.

9. Let X and Y be independent continuous random variables with marginal density functions $f_1(x)$ and $f_2(y)$. Let $U = X + Y$.
  (a) Show that U has density function $f_U(u) = \int_{-\infty}^{\infty} f_1(v) f_2(u - v)\, dv$. (*Hint*: Let $V = X$. What is the joint density of (U, V)?)
  (b) Show that when X and Y are positive random variables, U has density function $f_U(u) = \int_0^u f_1(v) f_2(u - v)\, dv$. (The density of U derived in this problem is called the *convolution* of f_1 and f_2.)

2.3 MULTIVARIATE RANDOM VECTORS

2.3.1 Multivariate Density Functions

In many probability problems, we have more than two random variables defined on a probability space. Let $X_1, X_2, \ldots, X_n$ be n such random variables. We say that $\mathbf{X} = (X_1, \ldots, X_n)$ is a *multivariate random vector*. For simplicity, we shall write vectors as row vectors, except in those sections which involve matrices. We shall also write $\mathbf{x} = (x_1, \ldots, x_n)$ and use the shorthand notation

$$\sum_{\mathbf{x} \in A} h(\mathbf{x}) = \sum\sum \cdots \sum_{(x_1, \ldots, x_n) \in A} h(x_1, \ldots, x_n),$$

$$\int_A h(\mathbf{x})\, d\mathbf{x} = \int\int \cdots \int_A h(x_1, \ldots, x_n)\, dx_1\, dx_2 \ldots dx_n.$$

For example,

$$\int_{R^n} f(\mathbf{x})\,d\mathbf{x} = \int_{-\infty}^{\infty}\int_{-\infty}^{\infty}\cdots\int_{-\infty}^{\infty} f(x_1, x_2, \ldots, x_n)\,dx_1\,dx_2\ldots dx_n.$$

The results in this section are fairly straightforward extensions of the results in Sections 2.2.1 and 2.2.2 and so are stated without proof.

Now let $\mathbf{X} = (X_1, \ldots, X_n)$ be an n-dimensional random vector. We say that $\mathbf{X}$ is a *discrete random vector* if $X_1, \ldots, X_n$ are all discrete random variables and that $\mathbf{X}$ is a *nonnegative integer-valued* random vector if the X_i are nonnegative integer-valued random variables. The *joint density function* of the discrete random vector $\mathbf{X}$ is defined as

$$f(\mathbf{x}) = P(\mathbf{X} = \mathbf{x}) = P(X_1 = x_1, \ldots, X_n = x_n).$$

The *range* S of $\mathbf{X}$ is the set of points $\mathbf{x}$ such that $f(\mathbf{x}) > 0$. As before, the range consists of finitely or countably many points. Typically, all the components of $\mathbf{x}$ are integer valued for $\mathbf{x} \in S$. If $\mathbf{X}$ is a discrete random vector with range S, $f(\mathbf{x})$ is a possible density function for $\mathbf{X}$ if and only if

$$f(\mathbf{x}) > 0 \text{ for } \mathbf{x} \in S, \qquad f(\mathbf{x}) = 0 \text{ for } x \notin S, \qquad \sum_{x \in S} f(\mathbf{x}) = 1.$$

As before,

$$P(\mathbf{X} \in A) = \sum_{x \in A} f(\mathbf{x}).$$

A random vector $\mathbf{X}$ is a *continuous random vector* if there exists an $f(\mathbf{x}) \geqq 0$ such that

$$P(\mathbf{X} \in A) = \int_A f(\mathbf{x})d\mathbf{x}.$$

We call $f(\mathbf{x})$ the *joint density function* of $\mathbf{X}$. The *range* S of $\mathbf{X}$ is the set of $\mathbf{x}$'s such that $f(\mathbf{x}) > 0$. If $\mathbf{X}$ is a random vector with range S, then $f(\mathbf{x})$ is a possible density function for $\mathbf{X}$ if

$$f(\mathbf{x}) > 0 \text{ for } \mathbf{x} \in S, \qquad f(\mathbf{x}) = 0 \text{ for } \mathbf{x} \notin S, \qquad \int_S f(\mathbf{x})\,d\mathbf{x} = 1.$$

Let $\mathbf{X} = (X_1, \ldots, X_n)$ be a random vector. Then the *marginal density function* of X_i is defined as the density of X_i alone (ignoring the other variables). Now let $\mathbf{Y}$ be a subvector of $\mathbf{X}$. Then the *marginal density function* of $\mathbf{Y}$ is the (joint) density of the components of $\mathbf{Y}$ alone. For example, if $\mathbf{Y} = (X_2, X_4, X_5)$, then the marginal density function of $\mathbf{Y}$ is the joint density of (X_2, X_4, X_5). Finally, let $\mathbf{Z}$ be the subvector of $\mathbf{X}$ which contains all of the components of $\mathbf{X}$ which are not in $\mathbf{Y}$. Then the *conditional density function* of $\mathbf{Z}$ given $\mathbf{Y} = \mathbf{y}$ is defined to be the density function of $\mathbf{Z}$ when we know that $\mathbf{Y} = \mathbf{y}$.

As an example, suppose we randomly select a female and let X_1 be her age, X_2 her salary, X_3 her height, and X_4 her intelligence. Let $\mathbf{Y} = (X_1, X_4), Z = (X_2, X_3)$. Then the marginal density of $\mathbf{Y}$ is the joint density of age and intelligence across all the women in the population, and the conditional density of $\mathbf{Y}$ given that $X_2 =$

\$40,000 and $X_4 = 120$ is the density function of a female who is randomly chosen from those women whose salary is \$40,000 and intelligence is 120.

We now give the basic result for finding marginal and conditional density functions.

Theorem 2–12. Let $\mathbf{X}$ be a random vector with density function $f(\mathbf{x})$. Let $\mathbf{Y}$ be a subvector of $\mathbf{X}$, and let $\mathbf{Z}$ contain the remaining components of $\mathbf{X}$. Let S_Y and S_Z be the ranges of $\mathbf{Y}$ and $\mathbf{Z}$, respectively.

a. If $\mathbf{X}$ is a discrete random vector, then the marginal density of $\mathbf{Y}$ is

$$f_Y(\mathbf{y}) = \sum_{\mathbf{z} \in S_Z} f(\mathbf{x}), \qquad \mathbf{y} \in S_Y.$$

b. If $\mathbf{X}$ is a continuous random vector, then the marginal density of $\mathbf{Y}$ is

$$f_Y(\mathbf{y}) = \int_{S_Z} f(\mathbf{x})\, d\mathbf{z}, \qquad \mathbf{y} \in S_Y.$$

c. The conditional density of $\mathbf{Z}$ given $\mathbf{Y} = \mathbf{y}$ is

$$f_{Z|Y}(\mathbf{z}|\mathbf{y}) = \frac{f(\mathbf{x})}{f_Y(\mathbf{y})}, \qquad (\mathbf{y}, \mathbf{z}) \in S,$$

and hence $f(\mathbf{x}) = f_Y(\mathbf{y}) f_{Z|Y}(\mathbf{z}|\mathbf{y})$.

Proof. The proof is essentially the same as that for Theorem 2–8 and is omitted. □

In order to keep our notation within reasonable bounds, for the remainder of the section we shall drop the subscripts from the density functions, using $f(\mathbf{y})$ instead of $f_Y(\mathbf{y})$ and $f(\mathbf{z}|\mathbf{y})$ instead of $f_{Z|Y}(\mathbf{z}|\mathbf{y})$.

If $\mathbf{X} = (X_1, X_2, X_3, X_4)$ is a discrete random vector with joint density function $f(x_1, x_2, x_3, x_4)$, then the marginal density of (X_1, X_3) is

$$f(x_1, x_3) = \sum\sum_{(x_2, x_4) \in S_{2,4}} f(x_1, x_2, x_3, x_4), \qquad (x_1, x_3) \in S_{1,3},$$

where $S_{1,3}$ and $S_{2,4}$ are the ranges of (X_1, X_3) and (X_2, X_4), respectively. In addition, the conditional density of (x_2, x_4) given $X_1 = x_1$ and $X_3 = x_3$ is

$$f(x_2, x_4|x_1, x_3) = \frac{f(x_1, x_2, x_3, x_4)}{f(x_1, x_3)}, \qquad (x_1, x_2, x_3, x_4) \in S.$$

We can also find the conditional density of X_1 and X_3 given X_4. We first find the marginal joint density $f(x_1, x_3, x_4)$ of (X_1, X_2, X_3) and the marginal density $f(x_4)$ of X_4. Then the conditional density of X_1 and X_3 given $X_4 = x_4$ is

$$f(x_1, x_3|x_4) = \frac{f(x_1, x_3, x_4)}{f(x_4)}.$$

Similar comments apply to continuous random vectors, as is seen in Example 2–30.

Example 2–30

Let $\mathbf{X} = (X_1, X_2, X_3)$ be a continuous random vector having joint density

$$f(x_1, x_2, x_3) = 6\exp(-x_1 - x_2 - x_3), \qquad 0 < x_1 < x_2 < x_3.$$

Note that $f(x_1, x_2, x_3) = 0$ unless $0 < x_1 < x_2$ and $x_2 < x_3$. Therefore, X_2 has marginal density function

$$\begin{aligned} f(x_2) &= \int_0^{x_2} \int_{x_2}^{\infty} 6\exp(-x_1 - x_2 - x_3)\, dx_3\, dx_1 \\ &= 6\exp(-x_2) \int_0^{x_2} \exp(-x_1) \left[-\exp(-x_3) \Big|_{x_3 = x_2}^{x_3 = \infty} \right] dx_1 \\ &= 6\exp(-2x_2)[-\exp(-x_1)] \Big|_{x_1 = 0}^{x_1 = x_2} \\ &= 6\exp(-2x_2)(1 - \exp(-x_2)), \qquad x_2 > 0. \end{aligned}$$

The conditional density of $(X_1, X_3) | X_2$ is

$$\begin{aligned} f(x_1, x_3 | x_2) &= \frac{6\exp(-x_1 - x_2 - x_3)}{6\exp(-2x_2)(1 - \exp(-x_2))} \\ &= \frac{\exp(x_2 - x_1 - x_3)}{(1 - \exp(-x_2))}, \qquad 0 < x_1 < x_2 < x_3. \end{aligned}$$

The marginal (joint) density of (X_1, X_2) is

$$f(x_1, x_2) = \int_{x_2}^{\infty} 6\exp(-x_1 - x_2 - x_3)\, dx_3 = 6\exp(-x_1 - 2x_2), \qquad 0 < x_1 < x_2.$$

Therefore, the conditional density of X_3 given $(X_1, X_2) = (x_1, x_2)$ is

$$f(x_3 | x_1, x_2) = \frac{6\exp(-x_1 - x_2 - x_3)}{6\exp(-x_1 - 2x_2)} = \exp(x_2 - x_3), \qquad 0 < x_1 < x_2 < x_3.$$

(Note that this conditional density does not depend on x_1.) Hence,

$$P(X_3 < 5 | x_1 = 1, x_2 = 2) = \int_2^5 \exp(2 - x_3)\, dx_3 = -\exp(2 - x_3) \Big|_2^5 = 1 - \exp(-3).$$

Finally, the conditional density of X_1 given $X_2 = x_2$ is

$$\begin{aligned} f(x_1 | x_2) &= \frac{6\exp(-x_1 - 2x_2)}{6\exp(-2x_2)(1 - \exp(-x_2))} \\ &= \frac{\exp(-x_1)}{(1 - \exp(-x_2))}, \qquad 0 < x_1 < x_2, x_2 > 0. \end{aligned}$$

The *joint distribution function* $F(\mathbf{x})$ for an n-dimensional random vector $\mathbf{X}$ is defined as

$$F(\mathbf{x}) = F(x_1, \ldots, x_n) = P(X_1 \leqq x_1, \ldots, X_n \leqq x_n).$$

If X is a continuous random vector with joint density function $f(\mathbf{x})$, then

$$f(\mathbf{x}) = f(x_1, \ldots, x_n) = \frac{\partial^n}{\partial x_1 \ldots \partial x_n} F(x_1, \ldots, x_n).$$

Exercises—A

1. Suppose we draw five cards without replacement from an ordinary deck. Let X_1 be the number of aces, X_2 the number of kings, and X_3 the number of queens.
   (a) What is the joint density of (X_1, X_2, X_3)?
   (b) What is the marginal density of (X_1, X_2)?
   (c) What is the marginal density of X_1?
   (d) What is the conditional density of (X_2, X_3) given X_1?
   (e) What is the conditional density of X_3 given (X_1, X_2)?
   (See Example 2–21.)

Exercises—B

1. For Example 2–30:
   (a) Find the marginal density of X_1 and the conditional density of (X_2, X_3) given $X_1 = x_1$.
   (b) Find the marginal joint density of (X_1, X_3) and the conditional density of X_2 given $(X_1, X_3) = (x_1, x_3)$. Find $P(X_2 > 1 | X_1 = .4, X_3 = 2)$.
   (c) Find the conditional density of X_3 given $X_1 = x_1$.

2. Let $\mathbf{X} = (X_1, X_2, X_3)$ be a continuous random vector with joint density function $f(x_1, x_2, x_3) = c(1 - x_1 - x_2 - x_3), x_1 > 0, x_2 > 0, x_3 > 0, x_1 + x_2 + x_3 < 1$.
   (a) Find c for this density function.
   (b) Find the marginal density of X_3 and the conditional density of (X_1, X_2) given $X_3 = x_3$.
   (c) Find the marginal (joint) density of (X_1, X_3) and the conditional density of X_2 given $(X_1, X_3) = (x_1, x_3)$.
   (d) Find the conditional density of X_1 given $X_3 = X_3$.

3. Let $\mathbf{X} = (X_1, X_2, X_3)$ be a discrete random vector with joint density $f(x_1, x_2, x_3) = c$, $x_1 = 0, 1, x_2 = 0, 1, x_3 = 0, 1, x_1 + x_2 + x_3 \leqq 1$.
   (a) What are the points of the range of X? Find c.
   (b) Find the marginal density of X_1, the conditional density of (X_1, X_2) given $X_3 = 0$, and $P(X_1 = 0, X_2 = 0 | X_3 = 0)$.
   (c) Find the marginal joint density of (X_1, X_2), the conditional density of X_3 given $(X_1, X_2) = (0, 0)$, and $P(X_3 = 1 | X_1 = 0, X_2 = 0)$.
   (d) Find the conditional density of X_2 given $X_1 = 0$, and find $P(X_2 = 1 | X_1 = 0)$.

4. Let $\mathbf{X} = (X_1, X_2, X_3)$ be a discrete random vector with joint density function $f(x_1, x_2, x_3) = c(x_1 + 2x_2 + 3x_3), (x_1, x_2, x_3) = (1, 1, 1), (1, 1, 2), (1, 2, 1), (2, 1, 1)$.
   (a) What is c?
   (b) Find the conditional density of X_2 given (X_1, X_3).
   (c) Find the conditional density of X_2 given X_1.
   (d) Find the conditional density of (X_1, X_2) given x_3.

5. Let X_1 and X_2 be independent random variables uniformly distributed on the interval $(0, 2)$. After observing $X_1 = x_1$ and $X_2 = x_2$, let X_3 be uniformly distributed on the interval $(-x_1, x_2)$.
   (a) What is the joint density of (X_1, X_2), and what is the conditional density of X_3 given $(X_1, X_2) = (x_1, x_2)$?
   (b) What is the joint density of (X_1, X_2, X_3)?

2.3.2 Independence

Let $\mathbf{X} = (X_1, \ldots, X_n)$ be a random vector. Then the X_i are *independent* if

$$P(X_1 \in A_1, X_2 \in A_2, \ldots, X_n \in A_n) = P(X_1 \in A_1)P(X_2 \in A_2) \ldots P(X_n \in A_n)$$

for all appropriate sets $A_1, \ldots, A_n$. That is, the X_i are independent random variables if the events $X_i \in A_i$ are independent events for any sets A_i.

Theorem 2–13. Let $\mathbf{X} = (X_1, \ldots, X_n)$ be a random vector with discrete or continuous joint density function $f(\mathbf{x})$ and range S. Let X_i have marginal density function $f_i(x_i)$ and range S_i.

a. The X_i are independent if and only if $f(x_1, \ldots, x_n) = f_1(x_1)f_2(x_2) \ldots f_n(x_n)$.
b. The X_i are independent if and only if $S = \{(x_1, \ldots, x_n)\colon x_1 \in S_1,\ x_2 \in S_2, \ldots, x_n \in S_n\}$ and $f(x_1, \ldots, x_n) = h_1(x_1)h_2(x_2) \ldots h_n(x_n)$ for some functions h_i.

Proof. The proof is essentially the same as that for Theorem 2–9 and is left as an exercise for the interested reader. □

Part b of Theorem 2–13 is used in the same way that the analogous result for two variables is used to tell quickly from the density function whether variables are independent. Part a is used often in this text to construct joint density functions for independent random variables.

Example 2–31

Let X_i be independent with marginal density function $f_i(x_i) = \exp(-x_i), x_i > 0$. Then $\mathbf{X} = (X_1, \ldots, X_n)$ has joint density function

$$f(\mathbf{x}) = f(x_1, \ldots, x_n) = \prod f_i(x_i = \prod \exp(-x_i) = \exp\left(-\sum x_i\right), x_i > 0.$$

Example 2–32

Let X_i be independent with marginal density functions $f_i(x_i) = i\ \exp(-ix_i), x_i > 0$. Then $\mathbf{X} = (X_1, \ldots, X_n)$ has joint density function

$$f(\mathbf{x}) = f(x_1, \ldots, x_n) = \prod i\ \exp(-ix_i) = n!\ \exp\left(-\sum ix_i\right)x_i > 0.$$

Example 2–33

Let X_i be independent continuous random variables with the same marginal density function $f_X(x)$ and marginal distribution function $F_X(x)$. Let Y be the largest of the X_i. We find the density function of Y. As usual, we first find the distribution function. We have

$$F_Y(y) = P(Y \leqq y) = P(\text{all } X_i \leqq y) = \prod P(X_i \leqq y) = (P(X_1 \leqq y))^n = (F_X(y))^n$$

(where the third equality follows from the independence of the X_i and the fourth one from their having the same distribution). Therefore, Y has density function

$$f_Y(y) = F'_Y(y) = n(F_X(y))^{n-1}f_X(x).$$

Example 2–34

Suppose that a system will work as long as any of three components work. Suppose also that the lifetimes of the three components are independent and that each component has density function $f(x) = (200)^{-1} \exp(-x/200), x > 0$. We seek the probability that each component lasts for 300 hours and the probability that the system lasts for at least 300 hours. Note that

$$F(x) = 1 - \exp(-x/200).$$

Let X_i be the lifetime of the ith component. Then the probability that a component lasts 300 hours is

$$P(X_i > 300) = 1 - F(300) = \exp(-300/200) = .22.$$

The system works as long as any component works, so that the time Y till failure is the maximum of the X_i. Therefore, the probability that the system works at least 300 hours is (using Example 2–33)

$$1 - F_Y(300) = 1 - (F(300))^3 = .53.$$

The density function of the time Y till failure is

$$f_Y(y) = 3[1 - \exp(-y/200)]^2(200)^{-1} \exp(-y/200), \qquad y > 0.$$

Let $\mathbf{X} = (X_1, \ldots, X_n)$ be a random vector. We say that the X_i are *pairwise independent* if X_i and X_j are independent for $i \neq j$. If the X's are independent, then they are pairwise independent. In Exercise B5, an example is given of three random variables which are pairwise independent but not independent. In subsequent sections, we shall assume that the X_i are independent and shall basically ignore the concept of pairwise independence.

We next extend the notion of independence to random vectors. Two random vectors $\mathbf{Y}$ and $\mathbf{Z}$ are *independent* if

$$P(\mathbf{Y} \in A, \mathbf{Z} \in B) = P(\mathbf{Y} \in A)P(\mathbf{Z} \in B)$$

for all appropriate sets A and B. Now let $\mathbf{X} = (\mathbf{Y}, \mathbf{Z})$. Then by the joint density of $\mathbf{Y}$ and $\mathbf{Z}$, we mean the joint density of $\mathbf{X}$, and by the joint range of $(\mathbf{Y}, \mathbf{Z})$, we mean the range of $\mathbf{X}$.

Theorem 2–14. Let $\mathbf{Y}$ and $\mathbf{Z}$ be random vectors with joint density $f(\mathbf{y}, \mathbf{z})$, marginal densities $f_Y(\mathbf{y})$ and $f_Z(\mathbf{z})$, joint range S, and marginal ranges S_1 and S_2.

a. $\mathbf{Y}$ and $\mathbf{Z}$ are independent if and only if $f(\mathbf{y}, \mathbf{z}) = f_Y(\mathbf{y})f_Z(\mathbf{z})$.
b. $\mathbf{Y}$ and $\mathbf{Z}$ are independent if and only if $S = \{(\mathbf{y}, \mathbf{z})\colon \mathbf{y} \in S_1, \mathbf{z} \in S_2\}$ and $f(\mathbf{y}, \mathbf{z}) = h(\mathbf{y})g(\mathbf{z})$ for some functions h and g.

Proof. Same as for Theorem 2–9. □

The definition of independent random vectors can be extended in the obvious way to k random vectors $\mathbf{X}_1, \ldots, \mathbf{X}_k$.

Example 2–35

Suppose that (X_1, X_2) is independent of X_3 and that X_1 and X_2 are independent. Then

$$P(X_1 \in A_1, X_2 \in A_2, X_3 \in A_3) = P(X_1 \in A_1, X_2 \in A_2)P(X_3 \in A_3)$$
$$= P(X_1 \in A_1)P(X_2 \in A_2)P(X_3 \in A_3).$$

(The first equality follows from the independence of (X_1, X_2) from X_3, the second from the independence of X_1 and X_2.) Therefore, X_1, X_2, and X_3 are independent.

We conclude the section with an important theorem.

Theorem 2–15. Let $\mathbf{X}_1, \ldots, \mathbf{X}_k$ be independent random variables or vectors, and let $\mathbf{U}_i = g_i(\mathbf{X}_i)$ for some functions g_i. Then the $\mathbf{U}_i$ are independent random variables or vectors.

Proof. The proof is left as an exercise. □

Note that we do not need to make any assumptions about the functions g_i for Theorem 2–15 to hold.

Exercises—A

1. Suppose that a system of three components fails when any component fails. Suppose also that the component lifetimes are independent with density function $f(x) = (200)^{-1} \exp(-x/200)$.
(a) Find the probability that the system lasts at least 150 hours.
(b) Find the density function of the time till the system fails.
(See Exercise B2 below.)

2. Suppose three light bulbs have independent lifetimes X, Y, and Z with densities $f(x) = (100)^{-1} \exp(-x/100)$ for $x > 0$, $g(y) = (200)^{-1} \exp(-y/200)$ for $y > 0$, and $h(z) = (300)^{-1} \exp(-z/300)$ for $z > 0$.
(a) Find the probability that Z lasts longer than Y and Y lasts longer than X.
(b) Find the probability that Z lasts the longest.

Exercises—B

1. Let $X_1, \ldots, X_n$ be independent. Find the joint density of the X_i if their marginal densities are as follows.
(a) $f_i(x_i) = 1, 0 < x_i < 1$.
(b) $f_i(x_i) = (.25)^{x_i}(.75)^{1-x_i}$, $x_i = 0$ or $x_i = 1$.
(c) $f_i(x_i) = 2^{x_i} e^{-2}/x_i!$, $x_i = 0, 1, \ldots$.
(d) $f_i(x_i) = i^{x_i} e^{-i}/x_i!$, $x_i = 0, 1, \ldots$.
(e) $f_i(x_i) = 4x_i e^{-2x_i}$, $x_i > 0$.

2. In Example 2–33, find the density function of Z, the smallest of the X_i.

3. Prove Theorem 2–13 for continuous random vectors when $n = 3$.

4. Suppose that (X_1, X_2) is independent of (X_3, X_4), that X_1 and X_2 are independent, and that X_3 and X_4 are independent. Show that the X_i are independent.

5. Let $\mathbf{X} = (X_1, X_2, X_3)$ be a discrete random vector having joint density $f(x_1, x_2, x_3) = \frac{1}{4}$ for the four points $(1,0,0), (0,1,0), (0,0,1), (1,1,1)$.
   (a) Find the marginal densities of the X_i, and show that they are not independent.
   (b) Find the joint density of (X_1, X_2), and show that they are independent. (Similar arguments show that X_1 and X_3 are independent and that X_2 and X_3 are also.)
6. Prove Theorem 2–15. Do not use density functions. (See the proof of Theorem 2–10.)

Exercises—C

1. In Example 2–33, let Z be the minimum of the X_i.
   (a) Find the joint distribution function of (Y, Z). (*Hint*: $P(Y \leqq y, Z \leqq z) = P(\text{all } X\text{'s} \leqq y) - P(z \leqq \text{all } X\text{'s} \leqq y)$. Why?)
   (b) Show that the joint density function of (Y, Z) is

$$f(y,z) = n(n-1)(F_x(y) - F_x(z))^{n-2} f_X(y) f_X(z).$$

*# *2.3.3 Multivariate Transformations*

Let $\mathbf{X} = (X_1, \ldots, X_n)$ be a multivariate continuous random vector with joint density function $f(\mathbf{x})$. Let $\mathbf{U} = (U_1, \ldots, U_n) = h(\mathbf{X})$ be an invertible function of $\mathbf{X}$ such that $\mathbf{X} = g(\mathbf{U})$. (That is, g is the inverse function of h.) Note that $\mathbf{X}$ and $\mathbf{U}$ have the same dimension. Let $g(\mathbf{u}) = (g_1(\mathbf{u}), \ldots, g_n(\mathbf{u}))$ (so that $x_i = g_i(\mathbf{u})$). Then the *Jacobian* $J_g(\mathbf{u})$ of $g(\mathbf{u}) = (g_1(\mathbf{u}), \ldots, g_n(\mathbf{u}))$ is the determinant of the $n \times n$ matrix $\mathbf{J}(\mathbf{u})$ whose (i, j)th component is

$$J_{ij}(\mathbf{u}) = \frac{\partial}{\partial u_j} g_i(\mathbf{u}).$$

That is, $J_g(\mathbf{u})$ is the determinant of the $n \times n$ matrix whose (ij)th element is the derivative of $g_j(\mathbf{u})$ with respect to u_i.

We now give an extension of Theorem 2–11 to the case of n-dimensional random vectors.

Theorem 2–16. Let $\mathbf{X}$ be an n-dimensional random vector with joint density function $f_X(\mathbf{x})$. Let $h(\mathbf{x})$ be an invertible n-dimensional function with inverse $\mathbf{x} = g(\mathbf{u})$. Let $\mathbf{U} = h(\mathbf{X})$.

a. If $\mathbf{X}$ is a discrete random vector, then $\mathbf{U}$ is a discrete random vector with joint density function $f_U(\mathbf{u}) = f_X(g(\mathbf{u}))$.
b. If $\mathbf{X}$ is a continuous random vector, then $\mathbf{U}$ is a continuous random vector with joint density function $f_U(\mathbf{u}) = f_X(g(\mathbf{u}))|J_g(\mathbf{u})|$.

Proof. The proof is essentially the same as that for Theorem 2–11. □

(Note in the above theorem that $|J_g|$ is the absolute value of J_g.)

Corollary 2–16. Let $\mathbf{X}$ be an $n \times 1$ continuous random vector with joint density function $f_X(\mathbf{x})$. Let $\mathbf{A}$ be an invertible $n \times n$ matrix, and let $\mathbf{b}$ be an $n \times 1$ vector. Let $\mathbf{U} = \mathbf{AX} + \mathbf{b}$. Then $\mathbf{U}$ has density function $f_U(\mathbf{u}) = |\det(\mathbf{A})|^{-1} f_X(\mathbf{A}^{-1}(\mathbf{u} - \mathbf{b}))$.

Proof. Let $\mathbf{B} = \mathbf{A}^{-1}$ have components B_{ij}. Then $\mathbf{X} = \mathbf{B}(\mathbf{U} - \mathbf{b}) = g(\mathbf{U})$, so that

$$g_i(\mathbf{u}) = \sum_j B_{ij}(u_j - b_j), \qquad \frac{\partial}{\partial u_j} g_i(u) = B_{ij}.$$

Hence, $J_g = \det(\mathbf{B}) = (\det(\mathbf{A}))^{-1}$, and the desired result follows. □

Example 2–36

Let X_1, X_2, X_3 be independent with X_i having density $f(x_i) = \exp(-x_i), x_i > 0$. Let $U_1 = X_1 + X_2 + X_3$, $U_2 = X_2/U_1$, and $U_3 = X_3/U_1$. We find the joint density of U_1, U_2, U_3. Note first that U_1 can be any positive number. Then, for any fixed value of U_1, U_2 and U_3 can be any positive numbers such that $U_2 + U_3 < 1$. Therefore, the range of $\mathbf{U} = (U_1, U_2, U_3)$ is $u_1 > 0, u_2 > 0, u_3 > 0,\ u_2 + u_3 < 1$. Since $U_3 U_1 = X_3$, $U_2 U_1 = X_2$, and $U_1(1 - U_2 - U_3) = X_1$, the inverse functions are

$$x_1 = g_1(u_1, u_2, u_3) = u_1(1 - u_2 - u_3),$$

$$x_2 = g_2(u_1, u_2, u_3) = u_1 u_2, \qquad x_3 = g_3(u_1, u_2, u_3) = u_1 u_3.$$

The Jacobian of the function g is

$$J_g = \det\begin{Bmatrix} 1 - u_2 - u_3 & -u_1 & -u_1 \\ u_2 & u_1 & 0 \\ u_3 & 0 & u_1 \end{Bmatrix} = u_1^2.$$

(Hence, $|J_g| = u_1^2$.) Now,

$$f_X(\mathbf{x}) = f(x_1)f(x_2)f(x_3) = \exp(-x_1 - x_2 - x_3), \qquad x_1 > 0, x_2 > 0, x_3 > 0.$$

Therefore,

$$f_U(u_1, u_2, u_3) = u_1^2 \exp(-u_1), \qquad u_1 > 0, u_2 > 0, u_3 > 0, u_2 + u_3 < 1.$$

Note that u_1 is independent of (u_2, u_3), but that u_2 and u_3 are not independent, because $u_2 + u_3 < 1$. Therefore, the marginal density of U_1 and the joint density of (U_2, U_3) must satisfy

$$f_1(u_1) = \frac{u_1^2 e^{-u_1}}{c}, \ u_1 > 0; \qquad f_2(u_2, u_3) = c, \quad u_2 > 0, u_3 > 0, u_2 + u_3 < 1.$$

Hence, (U_1, U_2) is uniformly distributed over the triangle $u_2 > 0, u_3 > 0, u_2 + u_3 < 1$, and c must be $\frac{1}{2}$.

Exercises—B

1. Let X_1, X_2, X_3 be independent with density $f(x_i) = \exp(-x_i), x_i > 0$. Let $U_1 = X_1 + X_2 + X_3$, $U_2 = (X_1 + X_2)/(X_1 + X_2 + X_3)$, and $U_3 = X_1/(X_1 + X_2 + X_3)$. Find the joint density of $\mathbf{U} = (U_1, U_2, U_3)$.
2. Let $\mathbf{X} = (X_1, X_2, X_3)$ be uniformly distributed on the unit sphere. That is, $f(x_1, x_2, x_3) = 3/4\pi, x_1^2 + x_2^2 + x_3^2 \leq 1$. Let $X_1 = U_1 \cos(U_2) \sin(U_3)$, $X_2 = U_1 \sin(U_2) \sin(U_3)$, and $X_3 = U_1 \cos(U_3)$. (That is, (U_1, U_2, U_3) gives the spherical coordinates for $\mathbf{X}$.) Find the joint density $\mathbf{U} = (U_1, U_2, U_3)$.

Chapter 3

EXPECTATION

Consider a simple game in which a player will be paid \$1 with probability $\frac{1}{2}$, \$2 with probability $\frac{1}{4}$, and \$3 with probability $\frac{1}{4}$. What is a fair price to play this game? One fairly intuitive answer is that a fair price is

$$\$1\left(\frac{1}{2}\right) + \$2\left(\frac{1}{4}\right) + \$3\left(\frac{1}{4}\right) = \$1.75,$$

which represents the player's "average winning" for playing the game. In other words, if she plays the game many times, she will win about \$1.75 per game. Therefore, if she pays less than \$1.75 per game, she will make money in the long run, and if she pays more than \$1.75 per game, she will lose money in the long run. (This seemingly obvious statement is a consequence of the weak law of large numbers derived in Section 6.2.1.)

Now consider another game in which a player wins x dollars with probability $f(x)$, for x in some set S. Let X be the actual amount he will win, so that X is a random variable with range S. We define his expected winnings as

$$EX = \sum_{x \in S} xf(x).$$

Again, we think of EX as the player's average winnings if he plays the game many times, or as a fair price to play the game. We call EX the *expectation* of X.

In this chapter, we examine the expectation of discrete and continuous random variables. From the interpretations just given, we can think of the expectation of X as a measure of the middle of the distribution of X. Notice from the example in the first paragraph that the expectation need not in fact be a possible value of X.

In Section 3.1, we define the expectations of discrete and continuous random variables and give the most important property of expectation, which we call the law of expectation. We prove this law in the discrete case, derive an equivalent expression for expectation, and then use this expression to prove the law of expectation in the continuous case. Subsequently, we give the basic facts about expectations, most of which follow from the law of expectation. In Section 3.2, we consider some particular types of expectations. Section 3.3 examines moment-generating functions; we use these in Chapters 4 through 6 to simplify many derivations. At the end of the section, we look at characteristic functions, which often can be used to give more rigorous proofs than those based on moment-generating functions. Since characteristic functions use complex variables, we do not make use of them in this text. In Section 3.4 we define conditional expectation, and in Section 3.5 we discuss a measure of the center of a distribution called the median. The median has some desirable properties, but expectation is much easier to study because the law of expectation is so useful.

3.1 DEFINITION AND PROPERTIES OF EXPECTATION

3.1.1 Definition and Law of Expectation

Let X be a discrete random variable with density function $f(x)$ and range S. The *expectation* of X is defined as

$$EX = \sum_{x \in S} xf(x),$$

provided this sum is finite. We think of EX as the average value of X—e.g., the average payoff in a game of chance.

Example 3–1

Let X be a discrete random variable with density function $f(x) = \frac{1}{6}$, $x = -2, -1, 0, 1, 2, 3$. Then

$$EX = \left(\frac{1}{6}\right)((-2) + (-1) + 0 + 1 + 2 + 3) = \frac{1}{2}.$$

(Note that EX is not a possible value of X in this case.) Now, let $U = X^2$. Then U has range $0, 1, 4, 9$ and density function

$$f_U(0) = f_U(9) = \frac{1}{6}, \qquad f_U(1) = f_U(4) = \frac{2}{6}.$$

(Note that $P(U = 1) = P(X = 1) + P(X = -1) = \frac{2}{6}$.) Therefore,

$$EU = 0\left(\frac{1}{6}\right) + 1\left(\frac{2}{6}\right) + 4\left(\frac{2}{6}\right) + 9\left(\frac{1}{6}\right) = \frac{19}{6}.$$

Let X be a continuous random variable with density function $f(x)$ and range S. The *expectation* of X is defined as

$$EX = \int_S xf(x)\,dx$$

(provided the integral is finite). We again think of the expectation of X as the average value of X. One way to understand expectation is to picture a (possibly infinitely long) rod which has mass density $f(x)$. Then EX is the center of mass of this rod. That is, if we put a fulcrum under the rod at the point EX, the rod would balance, not tipping to either side. Again, EX is an indicator of the "center" of the distribution of X.

Example 3–2

Let X be uniformly distributed on the interval (a, b). Then $f(x) = 1/(b-a), a < x < b$. Therefore,

$$EX = \int_a^b \frac{x}{b-a}\,dx = \frac{x^2}{2(b-a)}\bigg|_a^b = \frac{b^2-a^2}{2(b-a)} = \frac{b+a}{2},$$

which is a quite natural choice for the middle of this distribution.

Example 3–3

Let $f(x) = 2x, 0 < x < 1$. Then

$$EX = \int_0^1 2x^2\,dx = \frac{2x^3}{3}\bigg|_0^1 = \frac{2}{3}.$$

(Note that the density function puts more weight on higher values of x than on lower values, so that it is not surprising that EX is in the upper half of the interval.) Now, let $U = X^3$. We seek EU. The range of U is $0 < u < 1$, and the distribution function of U is

$$F_U(u) = P(X^3 \leqq u) = P(X \leqq u^{1/3}) = \int_0^{u^{1/3}} 2x dx = x^2\bigg|_0^{u^{1/3}} = u^{2/3}, \qquad 0 < u < 1.$$

Therefore, the density function of U is

$$f_U(u) = F_U'(u) = \frac{2u^{-1/3}}{3}, \qquad 0 < u < 1.$$

Finally, we can find EU from the definition:

$$EU = \int_0^1 u\frac{(2u^{-1/3})}{3}\,du = \frac{2u^{5/3}}{5}\bigg|_0^1 = \frac{2}{5}.$$

Example 3–4

Let X and Y have joint density function $f(x, y) = 8xy, 0 < x < y < 1$. Let $U = XY$. We want to find EU. We note that U has range $0 < u < 1$. The distribution function of U is $F_U(u) = P(U \leqq u) = P(XY \leqq u)$. Now, the set $xy \leqq u$, and $0 < x < y < 1$ is the union of the sets

$$\{(x, y)\colon u < x \leqq u^{1/2},\ x < y \leqq u/x\} \quad \text{and} \quad \{(x, y)\colon 0 < x \leqq u,\ x < y < 1\}.$$

(Draw a picture.) Therefore,

$$\begin{aligned} F_U(u) &= \int_u^{u^{1/2}}\int_x^{u/x} 8xy\,dydx + \int_0^u\int_x^1 8xydydx = \int_u^{u^{1/2}} 4x(u^2x^{-2} - x^2)dx + \int_0^u 4x(1-x^2)dx \\ &= 4u^2(\log(u^{1/2}) - \log(u)) - u^2 + u^4 + 2u^2 - u^4 \\ &= -2u^2\log(u) + u^2, \qquad 0 < u < 1. \end{aligned}$$

Hence, U has density function

$$f_U(u) = F_U'(u) = -4u\log(u) - 2u + 2u = -4u\log(u),$$

and, using integration by parts,

$$EU = \int_0^1 -4u^2\log(u)du = \frac{-4u^3\log(u)}{3}\bigg|_0^1 + \int_0^1 \frac{4u^2}{3}du = 0 - 0 + \frac{4u^3}{9}\bigg|_0^1 = \frac{4}{9}.$$

In Examples 3–1, 3–3, and 3–4, we have found the expected value of the random variables $U = X^2$, $U = X^3$, and $U = XY$ by finding the density function of U and then using the definition of expectation. Fortunately, there is a much simpler way to find the expected value of $U = h(X)$ without finding the density function of U. Recall that if $\mathbf{X} = (X_1, \ldots, X_n)$ is a random vector, then

$$\sum_{\mathbf{x} \in S} h(\mathbf{x}) f(\mathbf{x}) \quad \text{and} \quad \int_S h(\mathbf{x}) f(\mathbf{x}) d\mathbf{x}$$

are an n-dimensional sum and an n-dimensional integral, respectively. In particular, if $\mathbf{X} = (X, Y)$, then

$$\sum_{\mathbf{x} \in S} h(\mathbf{x}) f(\mathbf{x}) = \sum\sum_{(x,y) \in S} h(x, y) f(x, y)$$

and

$$\int_S h(\mathbf{x}) f(\mathbf{x}) d\mathbf{x} = \iint_S h(x, y) f(x, y) dx dy.$$

We then have the following theorem.

Theorem 3–1. (The law of expectation)

a. Let $\mathbf{X}$ be a discrete random variable or vector with range S and density function $f(\mathbf{x})$. Let $U = h(\mathbf{X})$. Then

$$EU = \sum_{\mathbf{x} \in S} h(\mathbf{x}) f(\mathbf{x}).$$

b. Let $\mathbf{X}$ be a continuous random variable or vector with range S and density function $f(\mathbf{x})$. Let $U = h(\mathbf{X})$. Then

$$EU = \int_S h(\mathbf{x}) f(\mathbf{x}) d\mathbf{x}.$$

Proof

a. Let U have range S_U, and for $u \in S_U$, let $A_u = \{\mathbf{x}: h(\mathbf{x}) = u\}$. Then

$$f_U(u) = P(U = u) = \sum_{\mathbf{x} \in A_u} f(\mathbf{x}).$$

Now, if $\mathbf{x} \in A_u$, then $u = h(\mathbf{x})$. Furthermore, the A_u are disjoint, and $\cup A_u = S$, the range of $\mathbf{X}$. Therefore,

$$EU = \sum_{u \in S_U} u f_U(u) = \sum_{u \in S_U} \left(u \sum_{\mathbf{x} \in A_u} f(\mathbf{x}) \right)$$

$$= \sum_{u \in S_U} \sum_{\mathbf{x} \in A_u} h(\mathbf{x}) f(\mathbf{x}) = \sum_{\mathbf{x} \in \cup A_u} h(\mathbf{x}) f(\mathbf{x}) = \sum_{\mathbf{x} \in S} h(\mathbf{x}) f(\mathbf{x}).$$

b. The proof of this result is somewhat difficult and is presented in Section 3.1.2. Some special cases are given in the exercises. □

Note that although $\mathbf{X}$ can be a random vector in Theorem 3–1, U must be a random variable.

Although the proof of part a of the theorem is quite elementary, it is often difficult to follow at first. It may help to look back at Example 3–1, with $U = X^2$. In that example, $A_0 = \{0\}$, $A_1 = \{1, -1\}$, $A_4 = \{2, -2\}$, and $A_9 = \{3\}$. Therefore,

$$\begin{aligned}
EU &= 0f_X(0) + 1(f_X(1) + f_X(-1)) + 4(f_X(2) + f_X(-2)) + 9f_X(3) \\
&= 0^2 f_X(0) + (1^2 f_X(1) + (-1)^2 f_X(-1)) + (2^2 f_X(2) + (-2)^2 f_X(-2)) \\
&\quad + 3^2 f_X(3) \\
&= (-2)^2 f_X(-2) + (-1)^2 f_X(-1) + 0^2 f_X(0) + 1^2 f_X(1) + 2^2 f_X(2) \\
&\quad + 3^2 f_X(3) \\
&= \sum_S x^2 f(x).
\end{aligned}$$

Because of this theorem, when $U = h(\mathbf{X})$, we ofen use the notation $Eh(\mathbf{X})$ for EU.

Example 3–5

We return to Example 3–3, where X has density function $f(x) = 2x, 0 < x < 1$, and $U = X^3$. Then

$$EU = EX^3 = \int_0^1 x^3(2x)dx = \left.\frac{2x^5}{5}\right|_0^1 = \frac{2}{5}.$$

which is the same answer we obtained the hard way before.

Example 3–6

As in Example 3–4, let X and Y have joint density $f(x, y) = 8xy$ for $0 < x < y < 1$, and let $U = XY$. By the law of expectation,

$$EU = EXY = \int_0^1 \int_0^y xy\; 8xy\; dxdy = \int_0^1 8y^2\left(\left.\frac{x^3}{3}\right|_{x=0}^{x=y}\right)dy = \int_0^1 \frac{8y^5}{3}dy = \left.\frac{4y^6}{9}\right|_0^1 = \frac{4}{9},$$

which is the same answer we obtained before by much messier calculations. This result indicates clearly what a simplification the law of expectation gives.

Example 3–7 shows that expectations sometimes are not finite.

Example 3–7

Let X be a continuous random variable with density function

$$f(x) = (\pi(1 + x^2))^{-1}, \qquad -\infty < x < \infty.$$

Then

$$f(x) > 0, \qquad \int_{-\infty}^{\infty} \frac{1}{\pi(1 + x^2)}dx = \left.\frac{\arctan(x)}{\pi}\right|_{-\infty}^{\infty} = 1,$$

so that $f(x)$ is a density function. Now,

$$EX = \int_{-\infty}^{\infty} \frac{x}{\pi(1 + x^2)}dx = \left.\frac{\log(1 + x^2)}{\pi}\right|_{-\infty}^{\infty} = \infty - (-\infty),$$

which is indeterminate. In this case, we say that EX does not exist. (If a rod had density $f(x) = 1/(\pi(1 + x^2))$, $-\infty < x < \infty$, then there is no point at which a fulcrum can be put to make the bar balance. If X has the density $f(x) = (\pi(1 + x^2))^{-1}$, we say that X has a *Cauchy* distribution.

We next present an interesting story related to infinite expectations, called the St. Petersburg game paradox.

Example 3–8

Consider a game in which you are going to flip a coin until you get a head. If the first head comes on the first toss, you get \$2. If it comes on the second toss, you get \$4, on the third toss, you get \$8, . . . , on the kth toss, you get $\$2^k$. How much should you pay to make this game a fair game? To answer the question, let X be the number of dollars you will win. Now, $X = 2$ with probability $\frac{1}{2}$, $X = 4$ with probability $\frac{1}{4}$, . . . , $X = 2^k$ with probability $1/2^k$, so that

$$EX = 2\left(\frac{1}{2}\right) + 4\left(\frac{1}{4}\right) + 8\left(\frac{1}{8}\right) + \cdots = 1 + 1 + 1 + \cdots = \infty.$$

Therefore, the only fair amount for you to pay to play this game is ∞. In other words, no matter how much you pay to play the game, you can expect to make money! However, before you go out looking for a sucker to play this game with you, consider the following fact. Suppose that your opponent only has a maximum of 2^s dollars that he can pay you. Then your payoff is $h(X)$, where $h(x) = x$ when $x < 2^s$ and $h(x) = 2^s$ when $x \geqq 2^s$. Therefore,

$$Eh(X) = 2\left(\frac{1}{2}\right) + \cdots + 2^{s-1}\left(\frac{1}{2^{s-1}}\right) + 2^s\left(\frac{1}{2^s} + \frac{1}{2^{s+1}} + \frac{1}{2^{s+2}} + \cdots\right) = s - 1 + 2 = s + 1.$$

For example, if your opponent only has $\$2^{20} = \$1{,}048{,}576$, then your expected payoff is only \$21.00. In this case, if you pay more than \$21 to play the game, you are paying too much.

Now, let X be an arbitrary continuous random variable. We say that EX is *finite* as long as both

$$I_1 = \int_0^\infty xf(x)dx < \infty \quad \text{and} \quad I_2 = -\int_{-\infty}^0 xf(x)dx < \infty,$$

or, equivalently, as long as

$$E|X| = \int_{-\infty}^\infty |x| f(x)dx = I_1 + I_2 < \infty.$$

We say that $EX = \infty$ if $I_1 = \infty$ and $I_2 < \infty$, and that $EX = -\infty$ if $I_1 < \infty$ and $I_2 = \infty$. If both I_1 and I_2 are infinite, we say that EX is indeterminate.

If $U = h(\mathbf{X})$, $V = g(\mathbf{X})$, $|g(\mathbf{x})| \leqq |h(\mathbf{x})|$, and EU is finite, then EV is finite, because

$$E|V| = \int_S |g(\mathbf{x})| f(\mathbf{x})d\mathbf{x} \leqq \int_S |h(\mathbf{x})| f(\mathbf{x})d\mathbf{x} = EU < \infty.$$

This relationship can be used to show that if EX^k is finite for some k, then EX^j is finite for $j < k$. (See Exercise C3.)

Finally, if a random variable U satisfies $a \leqq U \leqq b$, then EU is finite. (See Exercise B7.) In fact, most practical random variables are bounded and hence have finite expectations. (For example, if U is the weight of a person, we may not know a sharp upper bound for U, but we can be fairly certain that $0 < U < 10{,}000{,}000{,}000$ tons.) However, it is often convenient to approximate these bounded variables by unbounded ones, whose expectation may not be finite. Fortunately, density functions with infinite expectations are rarely used for modeling practical random events, and therefore, they will be infrequently encountered in the remainder of this book.

The preceding discussion on continuous random variables can be extended to discrete random variables in the obvious way. Since most discrete random variables are nonnegative, typically the only possibilities are that EX is finite or $EX = \infty$.

Exercises—A

1. In a particular dice game, a player picks a number between one and six. Then three dice are rolled. If the number comes up once, the player wins \$1, if it comes up twice, he wins \$2, and if it comes up three times, he wins \$3.
   (a) Find the probability that the player wins \$1.
   (b) Find the player's expected winnings.
2. A light bulb has a lifetime X which has density function $f(x) = (200)^{-1} \exp(-x/200)$, $x > 0$. Find the expected lifetime of the bulb.
3. A slot machine has three dials, each with 20 "faces." Each face contains one of six objects which are cherries, oranges, lemons, plums, bells, and bars. If you get bars on each face, you receive \$60; if you get three bells, you receive \$20; if you get bell, bell, bar, you receive \$18; for three oranges you receive \$10; for orange, orange, bar, you receive \$8; if you get cherries on the first two dials, you receive \$2; and if you get cherries on the first dial but not on the second, you receive \$1. Otherwise, you receive no money.

   Of the 20 faces on the first dial, 7 are cherries, 3 are oranges, 3 are lemons, 4 are plums, 2 are bells, and 1 is a bar. On the second dial, 7 are cherries, 7 are oranges, none are lemons, 1 is a plum, 2 are bells, and 3 are bars. On the third dial, none are cherries, 6 are oranges, 4 are lemons, 6 are plums, 3 are bells, and 1 is a bar.

   You pull a lever, spinning the three dials. Assuming that the three dials are independent of each other and that all 20 faces are equally likely on each dial,
   (a) What is the probability that you win \$60? That you win \$20?
   (b) What is your expected winning? (Note how much less this is than \$1, which is what casinos charge you to play the slot machine.)
4. Suppose a system contains three components, A, B, and C, and the component lives are independent with common density function $f(x) = (200)^{-1} \exp(-x/200)$.
   (a) What is the expected time until component A fails?
   (b) Suppose the system works as long as at least one component works. What is the expected time until the system fails?
   (c) Suppose the system works as long as all the components work. What is the expected time till the system fails?

   (*Hint*: The law of expectation is no help in this problem. See Example 2–35 and Exercise A1, Section 2.3.2, where the density functions for these times until failure are derived.)

Exercises—B

1. Let X be a random variable with density function $f(x) = 4x^3, 0 < x < 1$.
   (a) Find EX, EX^2, EX^{-1}, and $EX^{1/2}$.
   (b) Find $E(X \log[X])$. (Use integration by parts.)
2. Let X be a random variable with density function $f(x) = .2, x = 0, 1, 2, 3, 4$.
   (a) Find EX, EX^2, $E(1/(X + 1))$, and $EX^{1/2}$.

3. Let (X, Y) be a bivariate random vector with joint density function $f(x, y) = 6xy^2$, $0 < x < 1, 0 < y < 1$.
   (a) Find $E(XY)$, $E(X + Y)$, EX, and EY.
   (b) Does $E(XY) = EXEY$? Does $E(X + Y) = EX + EY$?
   (c) Find $E(X/Y)$.
4. Let (X, Y) be a bivariate random vector with joint density function $f(x, y) = 10xy^2$, $0 < x < y < 1$.
   (a) Find $E(XY), E(X + Y), EX, EY$.
   (b) Does $E(XY) = EXEY$? Does $E(X + Y) = EX + EY$?
   (c) Find $E(XY^{1/2})$.
5. Let (X, Y) be a bivariate random vector with joint density function $f(x, y) = (x + y)/18$, $x = 0, 1, 2, y = 0, 1, 2$.
   (a) Find $E(XY)$, $E(X + Y)$, EX, and EY.
   (b) Does $E(XY) = EXEY$? Does $E(X + Y) = EX + EY$?
6. Let (X, Y) be a continuous random vector. Show that $E(X + Y) = EX + EY$.
7. Let X be a continuous random variable whose range is contained in the set $a \leqq x \leqq b$. Show that $a \leqq EX \leqq b$. (*Hint*: For any x in the range of X, $af(x) \leqq xf(x) \leqq bf(x)$.)
8. Let U be uniformly distributed on $(-\pi/2, \pi/2)$. Let $Y = \tan(U)$. Show that Y has a Cauchy distribution.
9. Let X be a continuous random variable with density function $f(x)$ and range $a \leqq x \leqq b$. Let $U = h(X)$, where h is an increasing function. (Hence, h is invertible.)
   (a) What are the density function $f_U(u)$ and range S_U of U?
   (b) Show that $\int_{S_U} uf_U(u)du = \int_a^b h(x)f(x)dx$. (Make the change of variable $u = h(x)$ in the right-hand integral.)
10. Let (X, Y) be a continuous random vector with range $a \leqq x \leqq b, c \leqq y \leqq d$.
    (a) What is the marginal density $f_1(x)$ of X?
    (b) Show that $\int_a^b xf_1(x)dx = \int_a^b \int_c^d xf(x, y)dydx$.

Exercises—C

1. Let X be a continuous random variable with range $-\infty < x < \infty$ and density function $f(x)$ and distribution function $F(x)$. Let $U = X^2$.
   (a) What is the range of U?
   (b) Show that U has distribution function $F_U(u) = F(u^{1/2}) - F(-u^{1/2})$ and density function $f_U(u) = (f(u^{1/2}) + f(-u^{1/2}))/2u^{1/2}$.
   (c) Show, without using the law of expectation, that $EU = \int_{-\infty}^{\infty} x^2 f(x)dx$.
2. Let X be a positive continuous random variable with density function $f(x)$, and let $U = [X]$ (where $[x]$ is the largest integer in X, e.g., $[2.87] = 2$).
   (a) Show that U has the discrete density function $f_U(u) = \int_u^{u+1} f(x)dx, u = 0, 1, \ldots$.
   (b) Without using the law of expectation, show that $\sum_{u=0}^{\infty} uf_U(u) = \int_0^{\infty} [x]f(x)dx$.
3. Suppose that EX^k exists for some k. Let $j < k$.
   (a) Show that $E|X^k| + 1$ exists.
   (b) Show that $|x^j| \leqq |x^k| + 1$.
   (c) Show that EX^j exists.

* 3.1.2 Derivation of the Law of Expectation

In this section, we give a rather ingenious proof of the law of expectation in the continuous case. On the way, we derive an alternative expression for expectation which can be used as a definition for random variables which are neither discrete nor continuous.

Theorem 3–2. Let X be a discrete or continuous random variable. Then

$$EX = \int_0^\infty P(X > y)dy - \int_{-\infty}^0 P(X < y)dy.$$

Proof

a. (Continuous case) Since $f(x) = 0$ if $x \notin S$,

$$EX = \int_S f(x)dx = \int_0^\infty xf(x)dx - \int_{-\infty}^0 -xf(x)dx.$$

Now,

$$\int_0^\infty xf(x)dx = \int_0^\infty \left(\int_0^x dy\right)f(x)dx = \int_0^\infty \left(\int_y^\infty f(x)dx\right)dy = \int_0^\infty P(X > y)dy$$

(by interchanging the order of integration). Similarly,

$$\int_{-\infty}^0 -xf(x)dx = \int_{-\infty}^0 P(X < y)dy, \tag{3–1}$$

which finishes the proof in the continuous case.

b. (Discrete case) Although the theorem is true for arbitrary discrete random variables, for simplicity we assume that X is a nonnegative integer-valued random variable. In that case,

$$EX = \sum_{j=0}^\infty jf(j) = \sum_{j=1}^\infty jf(j) = \sum_{j=1}^\infty \sum_{k=1}^j f(j) = \sum_{k=1}^\infty \sum_{j=k}^\infty f(j) = \sum_{k=1}^\infty P(X \geqq k). \tag{3–2}$$

Now, $P(X > y) = P(X \geqq k)$ for $y \in (k-1, k]$. Therefore,

$$\int_0^\infty P(X > y)dy - \int_{-\infty}^0 P(X < y)dy = \sum_{k=1}^\infty \int_{k-1}^k P(X \geqq k)dy + 0 = \sum_{k=1}^\infty P(X \geqq k),$$

and the theorem is proved. □

In the previous section, we defined expectation only for random variables which are discrete or continuous. Now, using Theorem 3–2, we can define EX for any random variable X by

$$EX = \int_0^\infty P(X > y)dy - \int_{-\infty}^0 P(X < y)dy.$$

Note that for this definition X does not need a density function, but the definition agrees with that in the last section when X is discrete or continuous, by Theorem 3–2.

Corollary 3–2. If X is a nonnegative integer-valued random variable, then

$$EX = \sum_{k=0}^\infty P(X > k).$$

Proof. From Equation (3–2), we see that

$$EX = \sum_{k=1}^{\infty} P(X \geqq k) = \sum_{k=1}^{\infty} P(X > k - 1) = \sum_{k=0}^{\infty} P(X > k). \quad \square$$

The formula given in this corollary is occasionally useful for finding expectations, as the following example illustrates.

Example 3–9

Let X be a discrete random variable with density function $f(x) = 2^{-x}, x = 1, 2, \ldots$. In this case, it is somewhat difficult to find EX from the definition in the previous section. However, we have shown that $P(X > k) = 2^{-k}$. Therefore,

$$EX = \sum_{k=0}^{\infty} P(X > k) = \sum_{k=0}^{\infty} 2^{-k} = 2,$$

by the usual formula for summing a geometric series.

We are now ready for the main result of this section.

Theorem 3–3. (Law of expectation) Let $\mathbf{X}$ be a continuous random variable or random vector with density function $f(\mathbf{x})$ and range S. Let $U = h(\mathbf{X})$ be a random variable. Then

$$EU = \int_S h(\mathbf{x}) f(\mathbf{x}) d\mathbf{x}.$$

Proof. For all $y > 0$, let $A_y = \{\mathbf{x}: h(\mathbf{x}) > y\}$. Let $A = A_0 = \{\mathbf{x}: h(\mathbf{x}) > 0\}$. Now $y > 0$, $\mathbf{x} \in A_y$, if and only if $\mathbf{x} \in A, y < h(\mathbf{x})$. (Seeing this may be the hardest part of the proof.) Therefore,

$$\int_0^{\infty} P(U > y) dy = \int_0^{\infty} \int_{A_y} f(\mathbf{x}) d\mathbf{x} dy = \int_A \int_0^{h(\mathbf{x})} f(\mathbf{x}) dy d\mathbf{x} = \int_A h(\mathbf{x}) f(\mathbf{x}) d\mathbf{x}.$$

Now, let $B = \{\mathbf{x}: h(\mathbf{x}) < 0\}$. By a similar argument

$$-\int_{-\infty}^{0} P(U < y) dy = \int_B h(\mathbf{x}) f(\mathbf{x}) d\mathbf{x}. \tag{3–3}$$

Finally, let $C = \{\mathbf{x}: h(\mathbf{x}) = 0\}$. Then A, B, and C are disjoint, $A \cup B \cup C = R^k$, and

$$\int_C h(\mathbf{x}) f(\mathbf{x}) d\mathbf{x} = 0.$$

(Why?) Therefore, by Theorem 3–2,

$$\begin{aligned} EU &= \int_0^{\infty} P(U > y) dy - \int_{-\infty}^{0} P(U < y) dy + 0 \\ &= \int_A h(\mathbf{x}) f(\mathbf{x}) d\mathbf{x} + \int_B h(\mathbf{x}) f(\mathbf{x}) d\mathbf{x} + \int_C h(\mathbf{x}) f(\mathbf{x}) d\mathbf{x} \\ &= \int_{R^k} h(\mathbf{x}) f(\mathbf{x}) d\mathbf{x} = \int_S h(\mathbf{x}) f(\mathbf{x}) d\mathbf{x} \end{aligned}$$

(since $f(\mathbf{x}) = 0$ if $x \notin S$). $\square$

The preceding proof is quite subtle. To illustrate it, we let X be a continuous random variable with range $S = (0, \infty)$ and let $h(x) = \log(x)$. Then $A_y = \{x: \log(x) > y\} = (e^y, \infty)$. Now,

$$P(\log(X) > y) = P(X \in A_y) = P(X > e^y).$$

Therefore,

$$\int_0^\infty P(\log(X) > y)dy = \int_0^\infty \int_{e^y}^\infty f(x)dxdy = \int_1^\infty \int_0^{\log(x)} f(x)dydx = \int_1^\infty \log(x)f(x)dx,$$

and by a similar argument,

$$\int_{-\infty}^0 P(X < y)dy = \int_{-\infty}^1 \log(x)f(x)dx.$$

Note that U in Theorem 3–3 may be a discrete random variable, a continuous random variable, or neither discrete nor continuous. The proof works in all three cases.

Example 3–10

Let X be a continuous random variable with density function $f(x) = e^{-x}, x > 0$. Let

$$U_1 = h_1(X) = e^{-X}, \qquad U_2 = h_2(X) = \begin{cases} 1 & \text{if } X \leqq 1 \\ 0 & \text{if } X > 1 \end{cases}, \qquad U_3 = h_3(X) = \begin{cases} 1 & \text{if } X \leqq 1 \\ X & \text{if } X > 1 \end{cases}.$$

Then U_1 is a continuous random variable, and U_2 is a discrete random variable. U_3 is not discrete, because it takes on an interval of values, but neither is it continuous, because $P(U_3 = 1) = P(X \leqq 1) = 1 - e^{-1}$. However, Theorem 3–3 implies that we can use the law of expectation to show that

$$EU_1 = \int_0^\infty e^{-x}e^{-x}\,dx = \frac{1}{2}, \qquad EU_2 = \int_0^1 1e^{-x}\,dx + \int_1^\infty 0e^{-x}\,dx = 1 - e^{-1},$$

$$EU_3 = \int_0^1 1e^{-x}\,dx + \int_1^\infty xe^{-x}\,dx = 1 + e^{-1}.$$

This example indicates that in order to prove the law of expectation in the continuous case, we need a new definition of expectation to cover the case of random variables, like U_3, which are neither discrete nor continuous.

Exercises—B

1. Let X have density $f(x) = e^{-x}, x > 0$. Use Theorem 3–2 to derive EX.
2. Let X be uniformly distributed on $(0, 1)$. Let $U = .5$ if $X \leqq .5$ and $U = X$ if $X > .5$.
   (a) Show that U is neither a discrete nor a continuous random variable.
   (b) Find EU.
3. Prove Equations (3–1) and (3–3).

3.1.3 Properties of Expectation

In this section, we present the basic results regarding expectations. Most of these results are based on the law of expectation. All the proofs are given in the continuous case; proofs in the discrete case are left for the reader. (For the results in this section, we just replace the integrals in the continuous case by sums in the discrete case. In the previous section and in earlier chapters, we often needed quite different proofs for the discrete and continuous cases.)

Theorem 3–4

a. Let X be a random variable such that EX is finite. Then $E(aX + b) = aEX + b$.
b. Let $X_1, \ldots, X_n$ be random variables such that EX_i is finite. Then $E(\Sigma_i X_i) = \Sigma_i EX_i$.
c. Let X_i be *independent* random variables such that EX_i is finite. Then $E(\Pi_i X_i) = \Pi_i E(X_i)$.

Proof

a. This is an exercise.
b. We give the proof when $\mathbf{X} = (X_1, \ldots, X_n)$ is a continuous random vector with range S and joint density function $f(\mathbf{x})$. The proof for the discrete case is similar. Using the linearity of integrals, we have

$$E\left(\sum_i X_i\right) = \int_S \sum_i x_i f(\mathbf{x}) d\mathbf{x} = \sum_i \int_S x_i f(\mathbf{x}) d\mathbf{x} = \sum_i E(X_i).$$

c. We assume that the X_i are continuous random vectors with range S_i and marginal density function $f_i(x_i)$, so that $\mathbf{X} = (X_1, \ldots, X_n)$ has joint density function $f(\mathbf{x}) = \Pi f_i(x_i), x_1 \in S_1, \ldots, x_n \in S_n$. Then

$$E\left(\prod_i X_i\right) = \int_{S_1} \cdots \int_{S_n} \prod_i (x_i f_i(x_i) dx_i) = \prod_i \int_{S_i} x_i f_i(x_i) dx_i = \prod_i EX_i,$$

and the theorem is proved. □

Part a of Theorem 3–4 implies that the expected value of any linear function of a random variable is the same linear function of its expectation. It should be emphasized that this result is only true for linear functions. For example, for any random variable X which is not constant, we shall see that $EX^2 > (EX)^2$. In part b, it is shown that the expectation of any sum is the sum of the expectations. However, in part c, it is shown that the expectation of a product is the product of the expectations as long as the terms are independent. (In the exercises for Section 3.1.1, several problems are given in which $EXY \neq EXEY$, and hence X and Y are dependent.)

Following are some simple corollaries of Theorem 3–4.

Corollary 3–4(a). Let $\mathbf{X}$ be a random variable or vector.

a. If $Eh(\mathbf{X})$ is finite, then $E(ah(\mathbf{X}) + b) = aE(h(\mathbf{X})) + b$.
b. If $Eh_i(\mathbf{X})$ are finite, then $E(\Sigma_i a_i h_i(\mathbf{X}) + b) = \Sigma_i a_i Eh_i(\mathbf{X}) + b$.

Proof

a. $U = h(\mathbf{X})$ is a random variable, so that the desired result follows from part a of Theorem 3–4.
b. Let $U_i = h_i(\mathbf{X})$. By parts a and b of Theorem 3–4,

$$E\left(\sum_i a_i h_i(\mathbf{X}) + b\right) = E\left(\sum_i a_i U_i + b\right) = \sum_i a_i EU_i + b = \sum a_i Eh(\mathbf{X}) + b. \quad \square$$

Corollary 3–4(b). Let $\mathbf{X}_i$ be independent random variables or vectors. If $E(h_i(\mathbf{X}_i))$ is finite, then $E(\Pi_i h_i(\mathbf{X}_i)) = \Pi_i Eh_i(\mathbf{X}_i)$.

Proof. Let $U_i = h_i(\mathbf{X}_i)$. By Theorem 2–15, the U_i are independent, so that this result follows from part c of Theorem 3–4. □

Example 3–11

Let X be a random variable with $EX = 5$ and $EX^2 = 30$. Then

$$E(3X+4) = 3EX + 4 = 19, \qquad EX(X-1) = E(X^2 - X) = EX^2 - EX = 25,$$
$$E(X-1)^2 = E(X^2 - 2X + 1) = EX^2 - 2EX + 1 = 30 - 10 + 1 = 21.$$

Example 3–12

Suppose that X and Y are random variables such that $EX = 5$, $EX^2 = 30$, $EY = 3$, $EY^2 = 20$, and $EXY = 9$. Then

$$E(X - 3Y)^2 = EX^2 - 6EXY + 9EY^2 = 30 - 54 + 180 = 156,$$
$$E(X + 2Y + 3)^2 = EX^2 + 4EY^2 + 9 + 4EXY + 6EX + 12EY$$
$$= 30 + 80 + 9 + 36 + 30 + 36 = 221.$$

Example 3–13

Let X and Y be independent, and let $EX = 5$, $EY = 3$, $EX^2 = 30$, and $EY^2 = 10$. Then $EXY = EXEY = 15$, and $E(X^2Y^2) = EX^2EY^2 = 300$.

Example 3–14

Suppose a person draws five cards from an ordinary deck without replacement. Let X be the number of hearts she draws, and let Y be the number of aces. Let $U = X + Y$. Then $EU = EX + EY$. To find the EX, let X_i be the number of hearts drawn on the ith draw. Then $X = \Sigma X_i$, and $X_i = 1$ with probability $\frac{1}{4}$ and $X_1 = 0$ with probability $\frac{3}{4}$. Therefore,

$$EX_i = 1\left(\frac{1}{4}\right) + 0\left(\frac{3}{4}\right) = \frac{1}{4}, \qquad EX = \sum EX_i = 5\left(\frac{1}{4}\right) = \frac{5}{4}.$$

Similarly, let Y_i be the number of aces drawn on the ith draw. Then

$$EY_i = 1\left(\frac{1}{13}\right) + 0\left(\frac{12}{13}\right) = \frac{1}{13}, \qquad EY = \sum EY_i = \frac{5}{13}.$$

Hence,

$$EU = EX + EY = \frac{5}{4} + \frac{5}{13} = \frac{85}{52}.$$

We now consider a notational convention we make for the rest of the text. Let X be a random variable. If $P(X = a) = 1$, we say that X is *degenerate* at a and write

$$X \equiv a.$$

If X is not degenerate at a for any a, we say that X is *nondegenerate*. For all practical purposes, when $X \equiv a$, then $X = a$. In particular,

$$X \equiv a \Rightarrow g(X) \equiv g(a) \quad \text{and} \quad Eg(X) = g(a)$$

for any function g.

Let X and Y be two random variables. We write

$$X \equiv Y \Leftrightarrow P(X = Y) = 1$$

(that is, if $X - Y \equiv 0$). Then

$$Y \equiv X \Rightarrow g(Y) \equiv g(X) \quad \text{and} \quad Eg(Y) = Eg(X).$$

Informally, when $Y \equiv X$, we think of Y as being equal to X.

We next derive several important inequalities pertaining to expectations.

Theorem 3–5. Let X be a random variable such that $X \geqq 0$.

a. (Markov's inequality) For any $t > 0$, $P(X \geqq t) \leqq EX/t$.
b. $EX \geqq 0$ with equality only if $X \equiv 0$.

Proof

a. We again work only with the continuous case. In that case,

$$EX = \int_0^\infty xf(x)dx \geqq \int_t^\infty xf(x)dx \geqq t\int_t^\infty f(x)dx = tP(X \geqq t)$$

(where the first inequality follows because $xf(x) \geqq 0$, and the second one because $x \geqq t$ on the region of integration).

b. If $X \geqq 0$, then

$$EX = \int_0^\infty xf(x)dx \geqq \int_0^\infty 0dx = 0.$$

Now, suppose $EX = 0$ and $X \geqq 0$. Let A_n be the set where $X > 1/n$, and let B be the set where $X > 0$. Then $B = \bigcup A_n$. By Markov's inequality, $P(X \in A_n) \leqq 0$, and hence $P(X \in A_n) = 0$. Therefore, by Theorem 1–5,

$$P(X \neq 0) = P(X \in B) = P\left(X \in \bigcup_{k=1}^{\infty} A_k\right) = \lim_{n\to\infty} P(A_n) = 0,$$

and hence $X \equiv 0$. □

Corollary 3–5

a. If $X \geqq a$, then $EX \geqq a$ with equality only if $X \equiv a$.
b. If $X \leqq b$, then $EX \leqq b$ with equality only if $X \equiv b$.

Proof

a. Let $Y = X - a \geqq 0$. Then $EX - a = EY \geqq 0$ with equality only if $Y \equiv 0$. Therefore, $EX \geqq a$ with equality only if $X \equiv a$. Part b follows similarly, with $Z = b - X$. □

In particular, Corollary 3–5 implies that if $a \leqq X \leqq b$ then $a \leqq EX \leqq b$. For example, if the payoff in a game is between 2 and 5, then the expected payoff for the game must also be between 2 and 5. It should be emphasized that these results are true only when a and b are finite. In the example of the St. Petersburg game in Section 3.1.1, there is a random variable which is sure to be finite whose expectation is infinite.

We now present an equality which will be useful in later sections.

Theorem 3–6. (Cauchy-Schwarz inequality) Let X and Y be random variables such that

$$0 < EX^2 < \infty, \qquad 0 < EY^2 < \infty.$$

Then EXY is finite, and

$$(EXY)^2 \leqq EX^2 EY^2$$

with equality if and only if $X \equiv aY$ for some a.

Proof. Observe that $|xy| \leqq (x^2 + y^2)/2$ and $E(X^2 + Y^2)/2$ is finite. Therefore, EXY is finite. Now, $(X - aY)^2 \geqq 0$. Therefore, by Theorem 3–5, for all a,

$$0 \leqq E(X - aY)^2 = EX^2 + a^2 EY^2 - 2aEXY.$$

We choose $a = EXY/EY^2$ so that

$$0 \leqq EX^2 + \frac{(EXY)^2}{EY^2} - \frac{2(EXY)^2}{EY^2} = EX^2 - \frac{(EXY)^2}{EY^2}.$$

Therefore, $(EXY)^2 \leqq EX^2 EY^2$. Again by Theorem 3–5, the inequality in this equation is an equality if and only if $(X - aY)^2 \equiv 0$ if and only if $X \equiv aY$. □

The last theorem in this section is somewhat of a digression and may be skipped without loss of continuity. Let I be an interval on the real line. We allow I to be half-infinite or infinite, open or closed. Examples of such intervals are $(-1, 3]$, $[2, \infty)$, $(-\infty, 3)$, and $(-\infty, \infty)$. We say that a function $h(x)$ from I to the real line is *convex* if

$$h(au + (1 - a)v) \leqq ah(u) + (1 - a)h(v), \qquad \text{for all } a \in (0, 1), u \in I, v \in I,$$

and *strictly convex* if the inequality is a strict inequality (i.e., if we replace $\leqq$ by $<$). Suppose we connect the points $(u, h(u))$ and $(v, h(v))$ with a straight line. If this line is no lower than the graph of $y = h(x)$ for all x between u and v and for all u and v in I, then h is convex. If the straight line is always higher than the graph, then h is strictly convex. In many calculus courses, it is shown that any function $h(x)$ such that $h''(x) \geqq 0$ for all $x \in I$ is convex. Similarly, if $h''(x) > 0$ for all $x \in I$, then f is strictly convex.

Theorem 3–7. (Jensen's inequality)

a. Let X be a random variable whose range is contained in an interval I. If $h(x)$ is convex on I and EX is finite, then

$$E(h(X)) \geqq h(EX).$$

b. If $h(X)$ is strictly convex on I, X is not a degenerate random variable, and EX is finite, then

$$E(h(X)) > h(EX).$$

Proof

a. Let $\mu = EX$. By Corollary 3–5, $\mu \in I$. In many advanced calculus books, it is shown that there then exists a constant a such that

$$h(x) - h(\mu) - a(x - \mu) \geqq 0$$

for all $x \in I$. (This theorem is often called the *supporting hyperplane theorem*.) Therefore,

$$0 \leqq E(h(X) - h(\mu) - a(X - \mu)) = E(h(X)) - h(\mu) - a(\mu - \mu)$$
$$= E(h(X)) - h(\mu),$$

which proves part a.

b. Now suppose that h is strictly convex. Then

$$h(x) - h(\mu) - a(x - \mu) > 0$$

for all $x \neq \mu$. By Theorem 3–5, this inequality is an equality if and only if

$$h(X) - h(\mu) - a_\mu(X - \mu) \equiv 0,$$

which happens only when $X \equiv \mu$. □

Corollary 3–7. If X is nondegenerate and EX is finite, then $EX^2 > (EX)^2$, $E|X|^3 > |EX|^3$, $E(\exp(X)) > \exp(EX)$, and $E|X| \geqq |EX|$. If, in addition, $X > 0$, then $E(\log(X)) > \log(EX)$ and $EX^{-1} > 1/EX$.

Proof. The functions x^2, $|x|^3$, and $\exp[x]$ are strictly convex functions for the whole line, the function $|x|$ is convex for the whole line, and the functions $\log[x]$ and x^{-1} are strictly convex for the set of $x > 0$. Therefore, the desired results follow from Jensen's inequality. □

Exercises—A

1. Suppose I draw a card from a deck without replacement. I will be paid \$4 if I draw a spade, \$3 if I draw a heart, \$2 if I draw a diamond, and \$1 if I draw a club. In addition, I will be paid \$$k$ if I draw a card with k spots on it, \$11 if I draw a jack, \$12 if I draw a queen, and \$13 if I draw a king. For example, if I draw the seven of hearts, I get \$3 + \$7 = \$10. How much is a fair price to play this game? (*Hint*: Let X be the payoff from the suit drawn and Y be the payoff from the number on the card. Then the total payoff is $X + Y$.)
2. Suppose I draw five cards without replacement, with the same payoff on each card as given in problem 1. What is my total expected payoff?
3. Suppose a student takes two tests on a day. The probability that she passes the first one is .5, and the probability that she passes the second one is .8. Find the expected number of tests that she passes. (Note that we do not need to know the probability that she passes both tests in order to do this problem.) (*Hint*: Let $X = 1$ if she passes the first test and $X = 0$ if she fails, and let Y be similarly defined for the second test. Then the number of tests she passes is $X + Y$.)

Exercises—B

1. Let X and Y be random variables such that $EX = 2$, $EY = 3$, $EX^2 = 5$, $EY^2 = 10$, and $EXY = 7$.

(a) Are X and Y independent?
(b) Find $E([X+Y][X-Y]), E([X+3Y+1][X+2Y+4])$.

2. Let X be a random variable such that $\mu = EX$ is finite. Show that $a = \mu$ minimizes $q(a) = E(X-a)^2$.
3. Prove Theorem 3–4, part a for continuous and discrete random variables.
4. (a) Prove part b of Theorem 3–4 in the discrete case.
   (b) Prove part c of Theorem 3–4 in the discrete case.
   (c) Prove Markov's inequality in the discrete case.
5. Show that $|xy| \leqq (x^2+y^2)/2$. (*Hint*: $(x \pm y)^2 \geqq 0$.)
6. Let $h(a) = EX^2 - 2aEXY + a^2EY^2$. Show that $h(a)$ is minimized for $a = EXY/EY^2$.

Exercises—C

1. Let X be a continuous random variable whose density satisfies $f(c+x) = f(c-x)$ for some c and for all x. Then we say that X is *symmetric* about c. Show that if EX is finite, then $EX = c$. (*Hint*: Show that $E(X-c) = 0$ by breaking the integral into two pieces, one where $x < c$ and the other where $x > c$. On the first piece, let $u = x - c$, and on the second let $u = -(x-c)$.)
2. Let X and Y be random variables. Show that $\equiv$ is an equivalence relation. That is, show that
   (a) $X \equiv X$ (reflexivity);
   (b) If $X \equiv Y$, then $Y \equiv X$ (symmetry);
   (c) If $X \equiv Y$ and $Y \equiv Z$, then $X \equiv Z$ (transitivity).
   (*Hint*: $P(X \neq Z) \leqq P(X \neq Y) + P(Y \neq Z)$. Why?)
3. (a) Show that if $X \equiv a$, then $h(X) \equiv h(a)$ for any h.
   (b) Show that if $X^2 \equiv 0$, then $X \equiv 0$.
   (c) Show that if $Y \equiv X$, then $h(Y) \equiv h(X)$ for any h.
4. Suppose that $EX^2 = 0$. Show that $E(XY) = 0$, so that the Cauchy-Schwarz inequality is true when $EX^2 = 0$. (*Hint*: If $EX^2 = 0$, then $XY \equiv 0$. Why?)
5. (a) Show that x^2, $|x|^3$, and e^{-x} are strictly convex functions for all x.
   (b) Show that $\log[x]$, $1/x$, and $1/x^2$ are strictly convex functions for $x > 0$.
   (c) Argue, by means of a graph, that $|x|$ is a convex function, but not a strictly convex one.

3.2 SPECIAL EXPECTATIONS

3.2.1 The Mean, Variance, and Standard Deviation

Let X be a random variable. We respectively define the *mean* and *variance* of X as

$$\mu = EX \quad \text{and} \quad \sigma^2 = E(X-\mu)^2.$$

We assume here that both of these expectations are finite. We call $\sigma = (\sigma^2)^{1/2} = (\text{var}(X))^{1/2}$ the *standard deviation*. The mean is just another name for the expectation of X and measures where the center of the distribution of X lies, while the standard deviation is a measure of how spread out the distribution is. If σ is small, then $(X-\mu)^2$ is small with high probability, and hence X must be near μ "most of the time."

Following are some elementary facts about σ^2.

Theorem 3–8

a. $\sigma^2 = \text{var}(X) = EX^2 - \mu^2$.

b. $\text{var}(X) \geqq 0$ with equality only if $X \equiv \mu$.

Proof

a. $\sigma^2 = E(X-\mu)^2 = EX^2 - 2\mu EX + \mu^2 = EX^2 - 2\mu^2 + \mu^2 = EX^2 - \mu^2$.

b. $(X-\mu)^2 \geqq 0$, and therefore, $\sigma^2 = E(X-\mu)^2 \geqq 0$ with equality only if $(X-\mu)^2 \equiv 0$, in turn only if $X \equiv \mu$. □

Note that Theorem 3–8 implies that $EX^2 - \mu^2 > 0$ unless $X \equiv \mu$.

Example 3–15

Let X be a discrete random variable with density function

$$f(x) = \frac{x}{10}, \qquad x = 1, 2, 3, 4.$$

Then

$$\begin{aligned} \mu = EX &= 1(.1) + 2(.2) + 3(.3) + 4(.4) = 3, \\ EX^2 &= 1(.1) + 4(.2) + 9(.3) + 16(.4) = 10, \\ \sigma^2 &= 10 - (3)^2 = 1, \\ \sigma &= 1^{1/2} = 1. \end{aligned}$$

Example 3–16

Let X be a continuous random variable with density function

$$f(x) = 2x, \qquad 0 < x < 1.$$

Then

$$\mu = EX = \int_0^1 2x^2\,dx = \frac{2}{3}, \qquad EX^2 = \int_0^1 2x^3\,dx = \frac{1}{2}, \qquad \sigma^2 = \frac{1}{2} - \left(\frac{2}{3}\right)^2 = \frac{1}{18},$$

and hence $\sigma = \left(\frac{1}{18}\right)^{1/2}$. Now, let $U = X^2$. We can find the mean and variance of U without finding its density:

$$EU = EX^2 = \frac{1}{2}, \qquad EU^2 = EX^4 = \int_0^1 2x^5\,dx = \frac{1}{3}, \qquad \sigma^2 = \frac{1}{3} - \left(\frac{1}{2}\right)^2 = \frac{1}{12},$$

and hence $\sigma = \left(\frac{1}{12}\right)^{1/2}$.

We next present some other elementary properties of means and variances.

Theorem 3–9

a. $E(aX+b) = aEX + b$, $\text{var}(aX+b) = a^2\,\text{var}(X)$, and the standard deviation of $aX+b$ is $|a|$ times the standard deviation of X.

b. Let X and Y be independent random variables. Let $V = aX + bY + c$. Then

$$EV = aEX + bEY + c, \qquad \text{var}(V) = a^2\,\text{var}(X) + b^2\,\text{var}(Y).$$

c. Let X_i be independent random variables. Let $W = \sum_i a_i X_i + c$. Then

$$EW = \sum_i a_i EX_i + c, \qquad \text{var}(W) = \sum_i a_i^2 \text{ var}(X_i).$$

Proof

a. Let $\mu_X = EX$ and $U = aX + b$. We showed in the previous section that $\mu_U = EU = a\mu_X + b$. Therefore,

$$\begin{aligned}\text{var}(U) &= E(U - \mu_U)^2 = E(aX + b - (a\mu_X + b))^2 = Ea^2(X - \mu_X)^2 \\ &= a^2 E(X - \mu_X)^2 = a^2 \text{ var}(X).\end{aligned}$$

b. This is a special case of part c.

c. In the last section, we showed that $EW = \Sigma a_i EX_i + c$ (even if the X_i are not independent). If the X_i are independent with $EX_i = \mu_i$, then for $i \neq j$,

$$E(X_i - \mu_i)(X_j - \mu_j) = E(X_i - \mu_i)E(X_j - \mu_j) = (\mu_i - \mu_i)(\mu_j - \mu_j) = 0.$$

Therefore,

$$\begin{aligned}\text{var}(W) &= E\left(\sum a_i X_i + c - \left(\sum a_i \mu_i + c\right)\right)^2 = E\left(\sum a_i(X_i - \mu_i)\right)^2 \\ &= E\left(\sum_i \sum_j a_i a_j (X_i - \mu_i)(X_j - \mu_j)\right) \\ &= \sum_i \sum_j a_i a_j E(X_i - \mu_i)(X_j - \mu_j) \\ &= \sum_i a_i^2 E(X_i - \mu_i)^2 = \sum a_i^2 \text{ var}(X_i). \quad \square\end{aligned}$$

Part c implies that the variance of a sum of independent random variables is the sum of the variances. It should be emphasized, however, that the standard deviation of a sum is *not* the sum of the standard deviations. Also, part b implies that if X and Y are independent, then

$$E(X - Y) = EX - EY \quad \text{and} \quad \text{var}(X - Y) = \text{var}(X) + \text{var}(Y).$$

(Many students incorrectly assume that the variance of a difference is the difference of the variances.)

Example 3–17

Let X, Y, and Z be independent random variables with expectations 1, 2, and 3, and standard deviations 4, 5, and 6, respectively. Then $E(-3X + 7) = -3(1) + 7 = 4$, $\text{var}(-3X + 10) = 9(4^2) = 144$, and the standard deviation of $-3X + 7$ is $(144)^{1/2} = 12$. Now, let $W = X + Y + Z$. Then $EW = 1 + 2 + 3 = 6$, $\text{var}(W) = 4^2 + 5^2 + 6^2 = 77$, and the standard deviation of W is $(77)^{1/2}(\neq 4 + 5 + 6)$. Finally, let $V = 2X + Y - 3Z + 4$. Then $EV = 2(1) + 2 - 3(3) + 4 = -1$ and $\text{var}(V) = 4(4^2) + 5^2 + 9(6^2) = 413$.

We conclude the section with an inequality which is occasionally useful.

Theorem 3–10. (Chebyshev's inequality) Let X be a random variable with mean μ and variance σ^2. Then, for any $a > 0$,

$$P(|X - \mu| \geqq a) \leqq \sigma^2/a^2.$$

Proof. See Exercise B8. $\square$

Exercises—A

1. Let Z be the sum of the faces on three independent dice. What are the mean and variance of Z?
2. Suppose that a student takes two tests on a day. He has probability .8 of passing the first test and probability .6 of passing the second. Let p be the probability that he passes both tests. Find the mean and variance of the number of tests he passes. (Note that the mean does not depend on p, but the variance does.)
3. A wall is to be built with 25 cement blocks. The length of the cement blocks has mean 20″ and standard deviation .20″.
   (a) What are the mean and standard deviation of the length of the wall?
   (b) Use Chebyshev's inequality to find a lower bound on the probability that the length of the wall is within 3″ of 500″.

Exercises—B

1. Let X be a discrete random variable with density function $f(x) = x^2/19, x = -2, -1, 1, 2, 3$. Find the mean, variance, and standard deviation of X.
2. Let X be a continuous random variable with density $f(x) = x^2/3, -1 < x < 2$. Find the mean, variance, and standard deviation of X.
3. Let X be a continuous random variable with density function $f(x) = e^{-x}, x > 0$. Find the mean, variance, and standard deviation of X.
4. Let X be a continuous random variable with density function $f(x) = 2x^{-3}, x > 1$. Find the mean of X, and show that the variance of X is infinite.
5. Let X be uniformly distributed on the interval (a, b). Show that $\text{var}(X) = (b - a)^2/12$.
6. Let X and Y be independent. Show that $\text{var}(X - Y) = \text{var}(X) + \text{var}(Y)$.
7. Let X, Y, and Z be independent random variables with means -3, 4, and 1, and variances 6, 16, and 4, respectively. Let $U = 3X - 2$, $W = X + Y + Z$, and $V = X - 2Y + Z$. Find the means, variances, and standard deviations of U, V, and W. Find the mean and variance of $U + V + W$.
8. Prove Theorem 3–10. (*Hint*: Let $U = (X - \mu)^2 \geqq 0$. Then $P(|X - \mu| > a) = P(U > a^2)$.)

3.2.2 The Covariance and the Correlation Coefficient

Let (X, Y) be a random vector with $EX = \mu$, $EY = \nu$, $\text{var}(X) = \sigma^2 > 0$, and $\text{var}(Y) = \tau^2 > 0$ (so that X and Y are not degenerate). We define the *covariance* δ between X and Y, and the *correlation coefficient* ρ between X and Y, by

$$\delta = \text{cov}(X, Y) = E(X - \mu)(Y - \nu) \quad \text{and} \quad \rho = \text{corr}(X, Y) = \text{cov}(X, Y)/\sigma\tau,$$

respectively. Note that

$$\text{cov}(X, Y) = \text{cov}(Y, X) \quad \text{and} \quad \text{cov}(X, X) = \text{var}(X).$$

We first give some simple results about covariances and correlation coefficients.

Theorem 3–11

a. $\mathrm{cov}(X, Y) = EXY - EXEY$.

b. If X and Y are independent, then $\mathrm{cov}(X, Y) = \mathrm{corr}(X, Y) = 0$.

c. If $Y \equiv aX + b$ for some $a > 0$ and b, then $\mathrm{corr}(X, Y) = 1$. If $Y \equiv aX + b$ for some $a < 0$ and b, then $\mathrm{corr}(X, Y) = -1$.

d. If $Y \not\equiv aX + b$ for any a and b, then

$$-1 < \mathrm{corr}(X, Y) < 1.$$

e. If $\mathrm{var}(X)$ and $\mathrm{var}(Y)$ are finite, then $\mathrm{cov}(X, Y)$ is finite.

Proof. The proofs of a and b are left as exercises.

c. Let $\mu = EX$. If $Y \equiv aX + b$, then $\mathrm{var}(Y) = a^2\ \mathrm{var}(X)$, and

$$\mathrm{cov}(X, Y) = E(X - \mu)(aX + b - (a\mu + b)) = aE(X - \mu)^2 = a\ \mathrm{var}(X).$$

Now, $(a^2)^{1/2} = |a|$. Therefore,

$$\mathrm{corr}(X, Y) = \frac{a\ \mathrm{var}(X)}{|a|\ \mathrm{var}(X)} = \frac{a}{|a|} = 1 \text{ if } a > 0 \quad \text{and} \quad -1 \text{ if } a < 0.$$

d. Let $U = X - \mu$, $V = Y - \nu$. By the Cauchy-Schwarz inequality, if $Y \not\equiv aX + b$ for any a and b (so that $V \not\equiv aU$ for any a), then

$$(\mathrm{cov}(X, Y))^2 = (EUV)^2 < EU^2\,EV^2 = \mathrm{var}(X)\ \mathrm{var}(Y),$$

so that

$$(\mathrm{corr}(X, Y))^2 = \frac{(\mathrm{cov}(X, Y))^2}{\mathrm{var}(X)\ \mathrm{var}(Y)} < 1,$$

and hence, $-1 < \mathrm{corr}(X, Y) < 1$.

e. The desired result follows again from the Cauchy-Schwarz inequality applied to U and V. □

Parts b and c of Theorem 3–11 imply that the correlation coefficient ρ lies between -1 and 1. If it is 1, then $Y \equiv aX + b$ for some $a > 0$, so that there is an exact linear relationship between X and Y. Similarly, if $\rho = -1$, then $Y \equiv aX + b$ for some $a < 0$, and there is an exact linear relationship between X and Y which has negative slope. Finally, if X and Y are independent (so that there is no relationship between them), then $\rho = 0$. If ρ is near 1, we say that there is a strong positive correlation between X and Y. This means that high values for X tend to go with high values for Y. If ρ is near -1, we say that there is a strong negative correlation between X and Y, which means that high values of Y go with low values of X. If ρ is near 0, we say that there is a weak correlation between X and Y.

As an example, let X, Y, and Z respectively be a person's height, weight, and speed in a race. We would expect X and Y to be positively correlated, since tall people tend to weigh more. We might expect Y and Z to be negatively correlated, since lighter people can often run faster than heavier ones. Finally, we might expect X and Z to be weakly correlated, since height and speed might well be independent.

Example 3–18

Let (X, Y) have joint density function $f(x,y) = 8xy, 0 < x < y < 1$. Then

$$EX = \int_0^1 \int_0^y 8x^2 y dx dy = \frac{8}{15}, \qquad EY = \int_0^1 \int_0^y 8xy^2 dx dy = \frac{12}{15},$$

$$EX^2 = \int_0^1 \int_0^y 8x^3 y dx dy = \frac{1}{3}, \qquad EY^2 = \int_0^1 \int_0^y 8xy^3 dx dy = \frac{2}{3},$$

$$EXY = \int_0^1 \int_0^y 8x^2 y^2 dx dy = \frac{4}{9}.$$

Therefore,

$$\text{var}(X) = \frac{1}{3} - \left(\frac{8}{15}\right)^2 = \frac{11}{225}, \qquad \text{var}(Y) = \frac{2}{3} - \left(\frac{12}{15}\right)^2 = \frac{6}{225},$$

$$\text{cov}(X, Y) = \frac{4}{9} - \left(\frac{8}{15}\right)\left(\frac{12}{15}\right) = \frac{4}{225},$$

$$\text{corr}(X, Y) = \frac{4/225}{((11/225)(6/225))^{1/2}} = \frac{4}{(66)^{1/2}}.$$

Example 3–19

Let X be uniformly distributed on $(-1, 1)$, and let $Y = X^2$. Then $\text{cov}(X, Y) = \text{corr}(X, Y) = 0$. See the exercises.

Example 3–19 indicates that the correlation coefficient does not measure the strength of the relationship between X and Y. In the example, X and Y have zero correlation coefficient, even though Y is identically X^2. The reason this happens here is that Y is a decreasing function of X for negative X and an increasing function of X for positive X. If Y is always a decreasing function of X, then the correlation coefficient is negative, while if Y is always an increasing function of X, then the correlation coefficient is positive. In this case, Y is decreasing half the time and increasing the other half.

We say that X and Y are *uncorrelated* if $\text{corr}(X, Y) = 0$ and *correlated* otherwise. That is, X and Y are uncorrelated if

$$EXY = EXEY.$$

From Theorem 3–11, we see that if X and Y are independent, then they are uncorrelated. However, Example 3–19 presents a case in which uncorrelated random variables are not independent.

Let $\mathbf{X} = (X_1, \ldots, X_n)$ be a random vector. Then the X_i are *uncorrelated* if $\text{cov}(X_i, X_j) = 0$ for all $i \neq j$. Note that if the X_i are pairwise independent, then X_i and X_j are independent for all $i \neq j$, and hence the X_i are uncorrelated.

We now present the basic theorem for finding the variance of a sum of dependent random variables.

Theorem 3–12

a. Let X and Y be (possibly dependent) random variables, and let $V = aX + bY + c$. Then

$$E(V) = aEX + bEY + c,$$

$$\text{var}(V) = a^2 \text{ var}(X) + b^2 \text{ var}(Y) + 2ab \text{ cov}(X, Y).$$

b. Let X_i be (possibly dependent) random variables. Let $W = \sum_i a_i X_i$. Then

$$EW = \sum_i a_i EX_i + c,$$

$$\text{var}(W) = \sum_i a_i^2 \text{ var}(X_i) + \sum\sum_{i \neq j} a_i a_j \text{ cov}(X_i, X_j)$$

$$= \sum_i a_i^2 \text{ var}(X_i) + 2 \sum\sum_{i<j} a_i a_j \text{ cov}(X_i, X_j).$$

Proof. See Exercise B9. □

Since $\text{var}(X_i) = \text{cov}(X_i, X_i)$, we could also write the equation for the variance in part b as

$$\text{var}(W) = \sum_i \sum_j a_i a_j \text{ cov}(X_i, X_j).$$

Note that the formulas for EV and EW are the same as those given in the previous section for independent X_i. When $\text{cov}(X_i, X_j) = 0$ for $i \neq j$, the formulas for $\text{var}(V)$ and $\text{var}(W)$ also reduce to those in the last section.

Example 3–20

Let W be the number of hearts we draw in 10 draws without replacement from an ordinary deck of cards. We wish to find the mean and variance of W. The density function for W is derived in Exercise B6. However, we shall find EW and $\text{var}(W)$ without the density function. Let $X_i = 1$ if we draw a heart on the ith draw, and let $X_i = 0$ if we don't. Then $W = \Sigma_i X_i$. Now, the X_i are one with probability $\frac{1}{4}$ and zero with probability $\frac{3}{4}$. Therefore,

$$EX_i = \frac{1}{4}, \qquad EX_i^2 = EX_i = \frac{1}{4}, \qquad \text{var}(X_i) = \frac{1}{4} - \left(\frac{1}{4}\right)^2 = \frac{3}{16}.$$

(Note that $X_i^2 = X_i$. Why?) Consequently,

$$EW = \sum_{i=1}^{10} EX_i = \frac{10}{4}.$$

Unfortunately, the X_i are not independent. (If we get an ace on the first draw, we are less likely to get one on the second, so that we would expect X_i and X_j to be negatively correlated.) $X_i X_j = 1$ if we draw a heart on both the ith and jth draws, and $X_i X_j = 0$ elsewhere. Now,

$$P(\text{heart on } i, \text{ heart on } j) = P(\text{heart on } i)P(\text{heart on } j | \text{heart on } i)$$

$$= \left(\frac{1}{4}\right)\left(\frac{12}{51}\right) = \frac{1}{17}.$$

Therefore,

$$EX_i X_j = \frac{1}{17}$$

and

$$\text{cov}(X_i, X_j) = \frac{1}{17} - \left(\frac{1}{4}\right)^2 = \frac{-1}{(16)(17)}.$$

Hence,

$$\begin{aligned}\text{var}(W) &= \sum_{i=1}^{10} \text{var}(X_i) + \sum_{i=1}^{10} \sum_{j \neq i} \text{cov}(X_i, X_j) = 10\ \text{var}(X_1) + 90\ \text{cov}(X_1, X_2) \\ &= \frac{30}{16} - \frac{90}{(16)(17)} = \frac{420}{(16)(17)} = \frac{105}{68}.\end{aligned}$$

We conclude the section with some further properties of covariances and variances.

Theorem 3–13

a. Let $U = aX + c$ and $V = bY + d$. Then $\text{cov}(U, V) = (ab)\ \text{cov}(X, Y)$, $\text{corr}(U, V) = \text{corr}(X, Y)$ if $ab > 0$, and $\text{corr}(U, V) = -\text{corr}(X, Y)$ if $ab < 0$.
b. Let $X_1, \ldots, X_n$ be random variables. Let $U = \sum_i a_i X_i + c$ and $V = \sum_j b_j X_j + d$. Then $\text{cov}(U, V) = \sum_i a_i b_i\ \text{var}(X_i) + \sum_i \sum_{j \neq i} a_i b_j\ \text{cov}(X_i, X_j)$.
c. Let $X_1, \ldots, X_n, Y_i, \ldots, Y_m$ be random variables. Let $U = \sum_i a_i X_i + c$ and $V = \sum_j b_j Y_j + d$. Then $\text{cov}(U, V) = \sum_i \sum_j a_i b_j\ \text{cov}(X_i, Y_j)$.

Proof. a and b are exercises.

c. Let $EX_i = \mu_i$ and $EY_j = \nu_j$. Then $\mu_U = EU = \sum_i a_i \mu_i + c$ and $\nu_V = EV = \sum_j b_j \nu_j + d$. Therefore,

$$\begin{aligned}\text{cov}(U, V) &= E\left(\sum_i a_i X_i + c - \left(\sum_i a_i \mu_i + c\right)\right)\left(\sum_j b_j Y_j + d - \left(\sum_j b_j \nu_j + d\right)\right) \\ &= E\left(\sum_i \sum_j a_i b_j (X_i - \mu_i)(Y_j - \nu_j)\right) = \sum_i \sum_j a_i b_j\ \text{cov}(X_i, Y_j). \quad \square\end{aligned}$$

Part a of Theorem 3–13 implies that the correlation coefficient is unchanged when we change the units of X or Y.

Exercises—A

1. Suppose n people each have a hat which they check at a hatroom. Suppose the hats are returned randomly. Let Y be the number of people who get their own hats. Find EY and $\text{var}(Y)$. (*Hint*: Let $X_i = 1$ if the ith person gets his or her own hat and $X_i = 0$ if not. Then $Y = \Sigma_i X_i$. It may be assumed that $EX_i = EX_1$, $\text{var}(X_i) = \text{var}(X_1)$, and $\text{cov}(X_i, X_j) = \text{cov}(X_1, X_2)$.)

Exercises—B

1. Let (X, Y) be a discrete random vector with joint density function $f(x, y) = (x + y)/8$, $(x, y) \in \{(0, 1), (0, 2), (1, 0), (1, 1), (2, 0)\}$. Find the correlation coefficient between X and Y.
2. Let (X, Y) have joint density function $(x + 2y)/4, 0 < x < 2, 0 < y < 1$. Find the correlation coefficient between X and Y.

3. Let (X, Y) have joint density function $f(x, y) = 1/2y, 0 < y < 1, -y < x < y$. Find the correlation coefficient between X and Y.
4. For Example 3–19, show that $\text{corr}(X, Y) = 0$.
5. Let X, Y, and Z be random variables with $\text{var}(X) = 7$, $\text{var}(Y) = 8$, $\text{var}(Z) = 9$, $\text{cov}(X, Y) = 3$, $\text{cov}(X, Z) = 4$, and $\text{cov}(Y, Z) = 5$.
   (a) Find the correlation coefficient between X and Y.
   (b) Find the covariance and the correlation coefficient between $U = 2X + 3Y$ and $V = X - 2Z$.
   (c) Find the correlation coefficient between $2X + 2Y + 2Z + 2$ and $X - Y - Z + 3$.
6. Find the density function of V in Example 3–20.
7. Show that if X and Y have the same variance, then $X + Y$ and $X - Y$ are uncorrelated.
8. Prove parts a and b of Theorem 3–11.
9. Prove Theorem 3–12.
10. Prove parts a and b of Theorem 3–13.

Exercises—C

1. In Example 3–20, let U be the number of hearts drawn on the first three draws
   (a) What is the density function of U?
   (b) Show that $P(X_4 = 1) = \frac{1}{4}$. (*Hint*: $P(X_4 = 1) = \Sigma_j P(U = j)P(X_4 = 1|U = j)$.)
   (c) Show that $P(X_i = 1) = \frac{1}{4}, i = 1, \ldots, 10$, and hence that $EX_i = \frac{1}{4}$ and $\text{var}(X_i) = \frac{3}{16}$.
2. In Example 3–20, let U be the number of hearts drawn on the first three draws, and let V be the number drawn on the fifth and sixth draws.
   (a) Show that $P(X_4 = 1, X_7 = 1) = \frac{1}{17}$. (*Hint*: $P(X_4 = 1, X_7 = 1) = \Sigma\Sigma_{jk} P(U = j) \cdot P(X_4 = 1|U = j)P(V = k|U = j, X_4 = 1)P(X_7 = 1|U = j, X_1 = 1, V = k)$.
   (b) Show that $P(X_j = 1, X_k = 1) = P(X_1 = 1, X_2 = 1)$, and hence that $\text{cov}(X_i, X_j) = \text{cov}(X_1, X_2)$, and hence that $\text{cov}(X_i, X_j) = -1/(16)(17)$.

3.3 GENERATING FUNCTIONS

3.3.1 The Moment-Generating Function

Let X be a random variable. We define the kth *moment* of X by EX^k, and the *moment-generating function* of X by

$$M(t) = E(e^{tX}), \qquad t \in A,$$

where A is the set such that $M(t)$ is finite. (Note that $M(0) = 1$, so that $0 \in A$.) We say that $M(t)$ *exists* if A contains an interval $(-a, a)$ for some a (that is, if $M(t)$ is finite on an interval of points on either side of zero).

In applied mathematics courses, the moment-generating function is called the (two-tailed) Laplace transform of the density function of X. The moment-generating function does not have any obvious meaning by itself, but we shall see that it is very useful for doing distribution theory. In fact, the next two chapters might appropriately be called "applied moment generating functions."

The most basic property of the moment-generating function is that it generates the moments.

Theorem 3–14. Let

$$M^{(k)}(t) = \frac{d^k}{dt^k} M(t).$$

If $M(t)$ exists, then EX^k is finite for all k, and

$$EX = M'(0), \qquad EX^2 = M''(0),$$

and, in general,

$$EX^k = M^{(k)}(0).$$

Proof. We assume that X is a continuous random variable with range S. Note that

$$\frac{\partial^k}{\partial t^k} e^{tx} = x^k e^{tx}.$$

It may be shown that we may pass the derivative through the integral sign in this case. Therefore,

$$M^{(k)}(t) = \frac{d^k}{dt^k}\int_S e^{tx} f(x)dx = \int_S \frac{\partial^k}{\partial t^k}(e^{tx} f(x))dx = \int_S x^k e^{tx} f(x)dx = EX^k e^{tX}.$$

Therefore,

$$M^{(k)}(0) = E(X^k e^0) = EX^k. \quad \square$$

Corollary 3–14. Let $\psi(t) = \log(M(t))$. Then the mean μ and variance σ^2 of X are respectively given by

$$\mu = \psi'(0) \quad \text{and} \quad \sigma^2 = \psi''(0).$$

Proof. The derivative $\psi'(t) = M'(t)/M(t)$, and the second derivative $\psi''(t) = (M''(t)M(t) - M'(t))^2/(M(t))^2$. Now, $M(0) = 1$, $M'(0) = \mu$, and $M''(0) = EX^2$. Therefore, $\psi'(0) = \mu/1 = \mu$, and $\psi''(0) = (EX^2(1) - \mu^2)/1^2 = EX^2 - \mu^2 = \sigma^2$. $\square$

Example 3–21

Let X have density function $f(x) = e^{-x}, x > 0$. We can find EX^k using integration by parts, but let's use the moment-generating function of X instead. We have

$$M(t) = Ee^{Xt} = \int_0^\infty e^{xt} e^{-x}\, dx = (1 - t)^{-1}, \qquad t < 1.$$

Now, $M^{(k)}(t) = k!(1 - t)^{-k-1}$. Therefore,

$$EX^k = M^{(k)}(0) = k!, \qquad \mu = EX = 1, \qquad EX^2 = 2, \qquad \sigma^2 = EX^2 - \mu^2 = 2 - 1 = 1.$$

Now, let $\psi(t) = \log[M(t)] = -\log(1 - t)$. Then $\psi'(t) = (1 - t)^{-1}$, and $\psi''(t) = (1 - t)^{-2}$. Therefore, we see again that $\mu = \psi'(0) = 1$ and $\sigma^2 = \psi''(0) = 1$.

From this example, it is plain that we can compute the mean and variance using either $M(t)$ or $\psi(t)$. In later sections we shall use $\psi(t)$, because it is typically easier to differentiate than $M(t)$. ($\psi(t)$ is often called the *cumulant-generating function.*)

Generating moments is only one of the uses we shall have for moment-generating functions. Most of the other uses depend on the following important theorem.

Theorem 3–15. (Uniqueness theorem) Let X and Y be random variables that have the same moment-generating functions. Then X and Y have the same distribution.

Proof. A general proof of this result is beyond the scope of this book. We give instead a proof for the case where X and Y are nonnegative and integer valued. (For a general proof, see Billingsley (1979), p. 300.) Let X and Y be nonnegative integer-valued random variables with $p_k = P(X = k)$, $q_k = P(Y = k)$, $M(t) = Ee^{tX}$, and $N(t) = Ee^{tY}$. Let $R(t) = M[\log(t)] = Et^X$ and $S(t) = N[\log(t)] = Et^Y$. If $M(t) = N(t)$, it follows that $S(t) = R(t)$, and

$$\sum_{k=1}^{\infty} p_k t^k = R(t) = S(t) = \sum_{k=1}^{\infty} q_k t^k.$$

In advanced calculus, it is often shown that if two power series are equal, then their coefficients are equal. Therefore, $p_k = q_k$, and X has the same distribution as Y. □

Theorem 3–15 implies that it is not possible for two different distributions to have the same moment-generating function. That is, a random variable's moment-generating function completely determines its distribution. We can now describe the distribution of a random variable X in one of four ways: by defining $P(X \in A)$ for all A, by giving the distribution function of X, by giving the density of X (if X is discrete or continuous), or by giving the moment-generating function of X.

We now give some elementary results about moment-generating functions which shall be quite useful in later sections.

Theorem 3–16

a. Let X be a random variable with moment-generating function $M_X(t)$. Let $U = aX + b$. Then U has moment-generating function

$$M_U(t) = e^{bt} M_X(at).$$

b. Let $X_1, \ldots, X_n$ be independent random variables such that X_i has moment-generating function $M_i(t)$. Let $V = \sum_i X_i$ and $W = \sum_i a_i X_i + b$. Then V and W have moment-generating functions

$$M_V(t) = \prod_i M_i(t) \quad \text{and} \quad M_W(t) = e^{bt} \prod_i M_i(a_i t).$$

c. Let $X_1, \ldots, X_n$ be independent random variables with the same distribution and with common moment-generating function $M(t)$. Then $V = \sum_i X_i$ has moment-generating function

$$M_V(t) = (M(t))^n.$$

Proof

a. $M_U(t) = Ee^{tU} = Ee^{taX + tb} = e^{tb} Ee^{(at)X} = e^{tb} M_X(at)$.

b. V is a special case of W. So we have

$$M_W(t) = Ee^{tW} = Ee^{t(\sum a_i X_i + b)} = e^{tb} E \prod e^{ta_i X_i} = e^{tb} \prod Ee^{(a_i t)X_i} = e^{tb} \prod M_i(a_i t).$$

c. This result follows directly from part b. □

It is not necessary to remember Theorem 3–16: If its derivation is understood, it can be rederived very quickly whenever it is needed.

Example 3–22

Let X have moment-generating function $M_X(t) = (1 - t)^{-1}$. Then $3X + 4$ has moment-generating function $e^{4t}(1 - 3t)^{-1}$.

Example 3–23

Let X and Y be independent, and let X have moment-generating function $M_X(t) = (1 - t)^{-2}$ and Y have moment-generating function $M_Y(t) = (1 - t)^{-5}$. Let $V = X + Y$. Then V has moment-generating function

$$M_V(t) = M_X(t)M_Y(y) = (1 - t)^{-7}.$$

Now let $W = 3X + 4Y + 2$. Then W has moment-generating function

$$M_W(t) = e^{2t} M_X(3t)M_Y(4t) = e^{2t}(1 - 3t)^{-2}(1 - 4t)^{-5}.$$

Example 3–24

Let X_j be independent, with moment-generating function $M_j(t) = (1 - t)^{-j}, j = 1, \ldots, n$. Let $V = \sum X_j$. Then V has moment-generating function

$$M_V(t) = \prod (1 - t)^{-j} = (1 - t)^{-\sum j} = (1 - t)^{-n(n+1)/2}.$$

Let $W = \sum jX_j$. Then W has moment-generating function

$$M_W(t) = \prod (1 - jt)^j.$$

From the foregoing three examples, we see that it is often easy to find the moment-generating function of a linear combination W from the individual moment-generating functions. If we recognize the moment-generating function, then we have found the distribution of W, because of the uniqueness theorem. As we shall see in the next chapter, when the moment-generating function approach to finding an induced distribution works, it is the easiest method to use.

Exercises—B

1. Let X have density $f(x) = x/10, x = 1, 2, 3, 4$. Find the moment-generating function of X.
2. Let X have density function $f(x) = (\frac{1}{2})^x, x = 1, 2, \ldots$.
   a. Show that X has moment-generating function $M(t) = e^t/(2 - e^t)$.
   b. Find the mean and variance of X.
3. Suppose that X has moment-generating function $M(t) = (1 - 2t)^{-5}$. Find the mean and variance of X.
4. Let X be uniformly distributed on the interval (a, b). Find the moment-generating function of X.

5. Let X and Y be independent with moment-generating functions $M_X(t) = e^{t^2}$ and $M_Y(t) = e^{2t^2+t}$.
   a. Find the mean and variance of X.
   b. Find the moment-generating function of $X + Y$ and $X + 2Y + 3$.
6. Let $X_1, \ldots, X_n$ be independent with moment-generating functions $M_j(t) = e^{t^2+jt}$.
   a. Find the mean and variance of X_j.
   b. Let $V = \sum X_j$ and $W = \sum X_j/i$. Find the moment-generating functions of V and W. (Do not simplify the one for W.)
7. Let X be degenerate at a. Find the moment-generating function of X.
8. Show that $P(X \geqq c) \leqq (M(t)/e^{tc}$ for any $t > 0$ and that $P(X \leqq c) \leqq M(t)/e^{tc}$ for any $t < 0$.

3.3.2 The Joint Moment-Generating Function

Let $\mathbf{X} = (X_1, \ldots, X_n)$ be a random vector. Let $\mathbf{t} = (t_1, \ldots, t_n)$. The *joint moment-generating function* of $\mathbf{X}$ is

$$M(\mathbf{t}) = M(t_1, \ldots, t_n) = E\left(\exp\left[\sum_i t_i X_i\right]\right), \qquad \mathbf{t} \in A.$$

We say that $M(\mathbf{t})$ *exists* if the set A where $M(\mathbf{t})$ is finite contains a set of the form $\{\mathbf{t}: \sum t_i^2 < a\}$ for some a—that is, if A contains a ball of some positive radius about the origin. Since many of the results in this section are extensions of results from the last section, we state them without proof.

We first state how the moment-generating function generates the moments of $\mathbf{X}$.

Theorem 3–17. Let

$$M_i(\mathbf{t}) = \frac{\partial}{\partial t_i} M(\mathbf{t}), \qquad M_{ii}(\mathbf{t}) = \frac{\partial^2}{\partial t_i^2} M(\mathbf{t}), \qquad M_{ij}(\mathbf{t}) = \frac{\partial^2}{\partial t_i\, \partial t_j} M(\mathbf{t}).$$

Let $\mathbf{X} = (X_1, \ldots, X_n)$. If the joint moment-generating function $M(\mathbf{t})$ of $\mathbf{X}$ exists, then EX_i, EX_i^2, and EX_iX_j are finite, and

$$EX_i = M_i(0), \qquad EX_i^2 = M_{ii}(0),$$

and, in general,

$$EX_iX_j = M_{ij}(0).$$

In words, Theorem 3–17 says that to find EX_i, we differentiate $M(\mathbf{t})$ once with respect to t_i and then substitute zero for all the t_j. Similarly, to find EX_iX_j, we differentiate $M(\mathbf{t})$ once with respect to t_i and once with respect to t_j and put in zero for all the t's.

In a similar way, we could find $E\Pi_i X_i^{k(i)}$ for any nonnegative integers $k(i)$.

The following corollary follows directly from Theorem 3–17.

Corollary 3–17. Let $\psi(\mathbf{t}) = \log[M(\mathbf{t})]$. Then

$$EX_i = \psi_i(0), \qquad \operatorname{var}(X_i) = \psi_{ii}(0), \qquad \operatorname{cov}(X_i, X_j) = \psi_{ij}(0)$$

(where ψ_i, ψ_{ii}, and ψ_{ij} are defined analogously to M_i, M_{ii}, and M_{ij}).

($\psi(\mathbf{t})$ is called the cumulant-generating function of $\mathbf{X}$.) We use $\psi(\mathbf{t})$ for finding means, variances, and covariances because it is typically easier to differentiate than $M(\mathbf{t})$.

Example 3–25

Let $\mathbf{X} = (X_1, X_2, X_3)$ have joint moment-generating function

$$M(t_1, t_2, t_3) = (1 - t_1 + 2t_2)^{-4}(1 - t_1 + 3t_3)^{-3}(1 - t_1)^{-2}.$$

Then

$$\psi(\mathbf{t}) = -4\log(1 - t_1 + 2t_2) - 3\log(1 - t_1 + 3t_3) - 2\log(1 - t_1).$$

Therefore,

$$\psi_2(t) = -8(1 - t_1 + 2t_2)^{-1} \quad \text{and} \quad \psi_{22}(t) = 16(1 - t_1 + 2t_2)^{-2},$$

so that

$$EX_2 = \psi_2(0) = -8 \quad \text{and} \quad \text{var}(X_2) = \psi_{22}(0) = 16.$$

Similarly, we can show that $EX_1 = 9$ and $\text{var}(X_1) = 9$. Now,

$$\psi_{12}(\mathbf{t}) = -8(1 - t_1 + 2t_2)^{-2},$$

so that

$$\text{cov}(X_1, X_2) = \psi_{12}(0) = -8 \quad \text{and} \quad \text{corr}(X_1, X_2) = \frac{-8}{((16)(9))^{1/2}} = \frac{-8}{12}.$$

Hence, X_1 and X_2 have a moderately strong negative correlation.

We next state the most important theorem in this section, which is the obvious generalization of Theorem 3–15 of the last section.

Theorem 3–18. (Uniqueness theorem) Let $\mathbf{X}$ and $\mathbf{Y}$ be n-dimensional random vectors. If $\mathbf{X}$ and $\mathbf{Y}$ have the same joint moment-generating function, then $\mathbf{X}$ and $\mathbf{Y}$ have the same distribution.

Now, let $\mathbf{Y}$ be a subvector of $\mathbf{X}$, and let $\mathbf{Z}$ contain the remaining components of $\mathbf{X}$. Then the *marginal moment-generating function* of $\mathbf{Y}$ is the moment-generating function of $\mathbf{Y}$ ignoring the variables in $\mathbf{Z}$. For simplicity in stating the following theorem, we assume that $\mathbf{X} = (\mathbf{Y}, \mathbf{Z})$.

Theorem 3–19. Let $\mathbf{X} = (\mathbf{Y}, \mathbf{Z})$, where $\mathbf{Y}$ and $\mathbf{Z}$ are random variables or vectors. Let $\mathbf{t} = (\mathbf{u}, \mathbf{v})$, where $\mathbf{u}$ and $\mathbf{v}$ have the same dimension as $\mathbf{Y}$ and $\mathbf{Z}$. Suppose that $\mathbf{X}$ has moment-generating function $M_X(\mathbf{t})$. Then

a. $\mathbf{Y}$ and $\mathbf{Z}$ have marginal moment-generating functions

$$M_Y(\mathbf{u}) = M_X(\mathbf{u}, \mathbf{0}) \quad \text{and} \quad M_Z(\mathbf{v}) = M_X(\mathbf{0}, \mathbf{v}).$$

b. $\mathbf{Y}$ and $\mathbf{Z}$ are independent if and only if

$$M_X(\mathbf{u}, \mathbf{v}) = M_Y(\mathbf{u})M_Z(\mathbf{v}).$$

Proof

a. $M_X(\mathbf{u}, \mathbf{v}) = E(\exp[\sum u_i Y_i + \sum v_i Z_i])$. Therefore,

$$M_Y(\mathbf{u}) = E\left(\exp\left[\sum u_i Y_i\right]\right) = E\left(\exp\left[\sum u_i Y_i + \sum 0 Y_j\right)\right) = M_X(\mathbf{u}, \mathbf{0}).$$

The proof for M_Y is similar.

b. Suppose that **Y** and **Z** are independent. Then

$$M_X(\mathbf{u}, \mathbf{v}) = E\left(\exp\left[\sum u_i Y_i + \sum v_j Z_j\right]\right) = E\left(\exp\left[\sum u_i Y_i\right]\right)\left(\exp\left[\sum v_j Z_j\right]\right)$$

$$= \left(E\left(\exp\left[\sum u_i Y_i\right]\right)\right)\left(E\left(\exp\left[\sum v_j Z_j\right]\right)\right) = M_Y(\mathbf{u}) M_Z(\mathbf{v}),$$

by Theorem 3–4.

For the converse, suppose that $M_X(\mathbf{u}, \mathbf{v}) = M_Y(\mathbf{u}) M_Z(\mathbf{v})$. Let **S** and **T** be independent random vectors such that **S** has the same distribution as **Y** and **T** has the same distribution as **Z.** Then **S** has the same moment-generating function as **Y,** and **T** has the same moment-generating function as **Z.** Now, let $\mathbf{R} = (\mathbf{S}, \mathbf{T})$. Then, since **S** and **T** are independent, by what we have just proved, **R** has joint moment-generating function

$$M_R(\mathbf{u}, \mathbf{v}) = M_S(\mathbf{u}) M_T(\mathbf{v}) = M_Y(\mathbf{u}) M_Z(\mathbf{v}) = M_X(\mathbf{u}, \mathbf{v}).$$

Hence, **R** has the same moment-generating function as **X,** and by the uniqueness theorem, **R** and **X** have the same distribution. But then, since **S** and **T** are independent, so must **Y** and **Z** be. □

Part a of Theorem 3–19 says that to find the marginal moment-generating function of a subset of **X,** we just put zeros in the moment-generating function for the t's associated with the variables that are not in the subset. Note that to find the marginal density function from the joint density function, we have to integrate or sum out the variables that we do not want. Obviously, it is considerably easier to substitute zeros into a function than to integrate or sum out variables, so that we shall often use the joint moment-generating functions to find marginal distributions.

Part b of the theorem says that **X** and **Y** are independent if and only if the moment-generating function factors into the product of the marginal moment-generating functions. We have already seen that **X** and **Y** are independent if and only if the probability function factors into the product of the marginal probability functions and that they are independent if and only if the joint density function factors into the product of the marginal density functions. (We could also show that they are independent if and only if the joint distribution function factors into the product of the marginal distribution functions.)

The proof of the second part of part b often seems circular at first inspection. In words, what we are doing in this proof is showing that if the joint moment-generating function is the product of the marginal moment-generating functions, then **Y** and **Z** have the same joint moment-generating function as if they were independent. Therefore, by the uniqueness theorem, they are independent.

Before looking at an example, we state another theorem.

Theorem 3–20. Let $\mathbf{X} = (X_1, \ldots, X_n)$ be a random vector with joint moment-generating function $M(\mathbf{t}) = M(t_1, \ldots, t_n)$.

a. Let $V = \sum_i X_i$ and $W = \sum a_i X_i + b$. Then V and W have (marginal) moment-generating functions

$$M_V(s) = M_X(s, s, \ldots, s) \quad \text{and} \quad M_W(s) = e^{bs} M_X(a_1 s, \ldots, a_n s).$$

b. Let $W = \sum_i a_i X_i + b$ and $U = \sum_i c_i X_i + d$. Then W and U have joint moment-generating function

$$M_{W,U}(s, r) = e^{bs + dr} M_X(a_1 s + c_1 r, \ldots, a_n s + c_n r).$$

Proof

a. This result is left as an exercise.

b. We have

$$M_{W,U}(s, r) = E\left(\exp\left[s\left(\sum_i a_i X_i + b\right) + r\left(\sum_i c_i X_i + d\right)\right]\right)$$

$$= e^{sb + rd} E\left(\exp\left[\sum (a_i s + c_i r) X_i\right]\right)$$

$$= e^{sb + rd} M_X(a_1 s + c_1 r, \ldots, a_n s + c_n r). \quad \square$$

Example 3–26

As in Example 3–25, let $\mathbf{X} = (X_1, X_2, X_3)$ have moment-generating function

$$M(t_1, t_2, t_3) = (1 - t_1 + 2t_2)^{-4}(1 - t_1 + 3t_3)^{-3}(1 - t_1)^{-2}.$$

To find the moment-generating function of (X_1, X_3), we merely substitute $t_2 = 0$ to get

$$M_{X_1, X_3}(t_1, t_3) = M(t_1, 0, t_3) = (1 - t_1)^{-6}(1 - t_1 + 3t_3)^{-3}.$$

Note that X_1 and X_3 are not independent, since M_{X_1, X_3} does not factor into a product of a function of t_1 and a function of t_3. Similarly, to get the moment-generating function of X_1 alone, we let

$$M_{X_1}(t_1) = M(t_1, 0, 0) = (1 - t_1)^{-9}.$$

The moment-generating function of (X_2, X_3) is

$$M_{X_2, X_3}(t_2, t_3) = M(0, t_2, t_3) = (1 + 2t_2)^{-4}(1 + 3t_3)^{-3},$$

which does factor so that X_2 and X_3 are independent. Now, let $V = X_1 + X_2 + X_3$ and $W = 2X_1 - X_2 + 4$. Then V and W have moment-generating functions

$$M_V(s) = M(s, s, s) = (1 + s)^{-4}(1 + 2s)^{-3}(1 - s)^{-2}$$

and

$$M_W(r) = e^{4r} M(2r, -r, 0) = e^{4r}(1 - 4r)^{-4}(1 - 2r)^{-5},$$

and have joint moment-generating function

$$M_{V,W}(s, r) = e^{4r} M(s + 2r, s - r, s) = e^{4r}(1 + s - 4r)^{-4}(1 + 2s - 2r)^{-3}(1 - s - 2r)^{-2}.$$

From this example, we see that it is often quite easy to find moment-generating functions of linear functions of the X's once we know the joint moment-generating function of $\mathbf{X}$. Again, it is not necessary to remember results like Theorem 3–20: Once we understand proofs of this type, we can rederive the results almost as quickly as we could write them down if we had memorized them.

Exercises—B

1. Let X and Y be discrete random variables such that $f(x, y) = \frac{1}{6}$, $(x, y) = (0, 0)$, $(0, 1)$, $(1, 0)$, $(2, 0)$, $(1, 1)$, $(0, 2)$.
   - **(a)** Find the joint moment-generating function of (X, Y).
   - **(b)** Find EX, $\text{var}(X)$, and $\text{cov}(X, Y)$ in two ways:
     1. By using Theorem 3–17 or Corollary 3–17;
     2. By using the definitions.
   - **(c)** Find the marginal moment-generating function of X in two ways:
     1. By using Theorem 3–19;
     2. By finding the marginal density of X.
   - **(d)** Find the moment-generating function of $U = X + Y$ in two ways:
     1. By using Theorem 3–20;
     2. By first finding the density of U.
2. Let (X, Y) be continuous random variables such that $f(x, y) = \exp(-y), 0 < x < y < \infty$. Do parts a–d of problem 1.
3. Let (X, Y, Z) have joint moment-generating function $M(r, s, t) = (1 + r)^{-1}(1 + r - s)^{-2} \times (1 + r - s + t)^{-3}$, $r > -1$, $r - s > -1$, $r - s + t > -1$.
   - **(a)** What are EY, $\text{var}(Y)$, and $\text{cov}(X, Y)$?
   - **(b)** What are the marginal moment-generating functions of X, Y, and Z? Are X, Y, and Z independent?
   - **(c)** What is the joint moment-generating function of $U = X - Y$ and $V = Z - Y$? Are U and V independent?
4. **(a)** Prove Theorem 3–17 when $n = 2$.
   (b) Prove the Corollary 3–17 when $n = 2$.
5. Prove part a of Theorem 3–20.

Exercises—C

1. Let **X**, **Y**, and **Z** have joint moment-generating function $M(\mathbf{r}, \mathbf{s}, \mathbf{t})$.
   - **(a)** Show that **X**, **Y**, and **Z** are independent if and only if
   $$M(\mathbf{r}, \mathbf{s}, \mathbf{t}) = M(\mathbf{r}, \mathbf{0}, \mathbf{0})M(\mathbf{0}, \mathbf{s}, \mathbf{0})M(\mathbf{0}, \mathbf{0}, \mathbf{t}).$$
   - **(b)** Under what conditions are **X**, **Y**, and **Z** pairwise independent?

3.3.3 The Characteristic Function

Many of the theorems in the previous sections depend on the assumption that the moment-generating function $M(t)$ of a random variable X is finite on an interval $-a < t < a$. In particular, the uniqueness theorem and the theorem on moments depend on this assumption. However, we have seen that if $M(t)$ is finite on some such interval, then EX^k is finite for all k. Therefore, if EX^k is not finite for some k, then $M(t)$ is not finite on such an interval. In particular, if EX does not exist or is infinite, then $M(t)$ does not exist on such an interval.

To handle such distributions, we often use the characteristic function. Let i be the imaginary number $(-1)^{1/2}$. Then the *characteristic function* of a random variable

X is defined by

$$c(t) = E(e^{iXt}) = E(\cos(Xt)) + iE(\sin(Xt))$$

(since $e^{ia} = \cos(a) + i\ \sin(a)$ for any real a, by DeMoivre's theorem). Now,

$$-1 \leqq \cos(Xt) \leqq 1, \qquad -1 \leqq \sin(Xt) \leqq 1.$$

Therefore, $E(\cos(Xt))$ and $E(\sin(Xt))$ are both finite, and hence, $c(t)$ is finite for all t for any random variable X. In applied mathematics, the characteristic function is often called the Fourier transform of the density function.

If X has moment-generating function $M(t)$, then X has characteristic function

$$c(t) = M(it),$$

so that we can find the characteristic function of many random variables from the moment-generating function.

Let $c^{(k)}(t)$ be the kth derivative of $c(t)$ (if it exists). One interesting property of characteristic functions is that EX^k exists if and only if $c^{(k)}(0)$ exists, and

$$EX^k = \frac{c^{(k)}(0)}{i^k}.$$

Therefore, if EX^2 exists but EX^3 does not, then $c^{(2)}(0)$ exists but $c^{(3)}(0)$ does not.

Example 3–27

Let X have the Cauchy density $f(x) = (\pi(1+x^2))^{-1}$. It can be shown that X then has characteristic function

$$c(t) = \pi^{-1}\int_{-\infty}^{\infty} \frac{\cos(tx)}{1+x^2}\,dx + i\pi^{-1}\int_{-\infty}^{\infty} \frac{\sin(tx)}{1+x^2}\,dx = e^{-|t|}.$$

(These integrals are actually quite hard to evaluate.) Note that the right derivative of $c(t)$ at $t = 0$ is -1 and the left derivative is 0, so that $c(t)$ is not differentiable at $t = 0$, which we expect, because EX does not exist for the Cauchy distribution.

The characteristic function also satisfies the uniqueness theorem. That is, if X and Y have the same characteristic function, then they have the same distribution. In later sections, we shall occasionally start proofs of fairly general theorems by letting $M(t)$ be the moment-generating function of an arbitrary random variable X. However, X may not in fact have a finite moment-generating function. To make those proofs rigorous, we should substitute the characteristic function for the moment-generating function. We use the moment-generating function instead of the characteristic function in deference to those students who have not studied complex variables.

Exercises—B

1. Let X_1 and X_2 be independent and have the Cauchy density given in Example 3–27. Show that $(X_1 + X_2)/2$ has that same distribution.

3.4 CONDITIONAL EXPECTATION

3.4.1 Conditional Expectation

Conditional expectation is one of the most useful concepts in probability. Let $\mathbf{X}$ and $\mathbf{Y}$ be random variables or vectors with ranges S and T, respectively. Let $U = h(\mathbf{X}, \mathbf{Y})$ be a (univariate) random variable. Then the *conditional expectation* of U given $\mathbf{Y} = \mathbf{y}$ is the expectation of $h(\mathbf{X}, \mathbf{y})$ computed with respect to the conditional density $f(x|y)$ of $\mathbf{X}$ given $\mathbf{Y} = \mathbf{y}$. That is, if $(\mathbf{X}, \mathbf{Y})$ is a discrete random vector, then the conditional expectation of U given $\mathbf{Y} = \mathbf{y}$ is

$$E(U|\mathbf{Y} = \mathbf{y}) = E(h(\mathbf{X}, \mathbf{Y})|\mathbf{Y} = \mathbf{y}) = \sum_{x \in S} h(\mathbf{x}, \mathbf{y}) f(\mathbf{x}|\mathbf{y}), \qquad \mathbf{y} \in T.$$

If $(\mathbf{X}, \mathbf{Y})$ is a continuous random vector, then

$$E(U|\mathbf{Y} = \mathbf{y}) = E(h(\mathbf{X}, \mathbf{Y})|\mathbf{Y} = \mathbf{y}) = \int_S h(\mathbf{x}, \mathbf{y}) f(\mathbf{x}|\mathbf{y})\, d\mathbf{x}, \qquad \mathbf{y} \in T.$$

Note that $E(h(\mathbf{X}, \mathbf{Y})|\mathbf{Y} = \mathbf{y})$ is a function of $\mathbf{y}$, but not of $\mathbf{X}$. (Note also that the law of expectation applied to the conditional density of $\mathbf{X}$ given $\mathbf{Y} = \mathbf{y}$ implies that this definition is consistent.)

We think of $E(h(\mathbf{X}, \mathbf{Y})|\mathbf{Y} = \mathbf{y})$ as the expected value of $h(\mathbf{X}, \mathbf{Y})$ when we know that $\mathbf{Y} = \mathbf{y}$. For example, suppose we randomly choose a male student. Let X be his weight and Y his height. Then $E(X|Y = 70)$ is the expected weight of a student who is 70″ tall, and $E((X/Y)|Y = 70)$ is the expected ratio of weight to height for a student who is 70″ tall.

Theorem 3–21

a. $E[g(\mathbf{Y})h(\mathbf{X}, \mathbf{Y})|\mathbf{Y} = \mathbf{y}] = g(\mathbf{y})E[h(\mathbf{X}, \mathbf{Y})|\mathbf{Y} = \mathbf{y}]$.

b. $E[g(\mathbf{Y})|\mathbf{Y} = \mathbf{y}] = g(\mathbf{y})$.

c. $E[(h(\mathbf{X}, \mathbf{Y}) + k(\mathbf{X}, \mathbf{Y}))|\mathbf{Y} = \mathbf{y}] = E[h(\mathbf{X}, \mathbf{Y})|\mathbf{Y} = \mathbf{y}] + E[k(\mathbf{X}, \mathbf{Y})|\mathbf{Y} = \mathbf{y}]$.

Proof

a. We prove the result in the continuous case; the discrete case is similar.

$$\begin{aligned} E[g(\mathbf{Y})h(\mathbf{X}, \mathbf{Y})|\mathbf{Y} = \mathbf{y}] &= \int_S g(\mathbf{y})h(\mathbf{x}, \mathbf{y})f(\mathbf{x}|\mathbf{y})d\mathbf{x} \\ &= g(\mathbf{y})\int_S h(\mathbf{x}, \mathbf{y})f(\mathbf{x}|\mathbf{y})d\mathbf{x} \\ &= g(\mathbf{y})E[h(\mathbf{X}, \mathbf{Y})|\mathbf{Y} = \mathbf{y}]. \end{aligned}$$

b. and c. See Exercise B5. □

Note that parts a and b of Theorem 3–21 are intuitively clear when we note that $\mathbf{Y} = \mathbf{y}$ is a constant in the conditional distribution of $\mathbf{X}$ given $\mathbf{Y} = \mathbf{y}$. Part c is just a restatement of the linearity of expectation.

Now, let X be a univariate random variable. Let $m(\mathbf{y}) = E(X|\mathbf{Y} = \mathbf{y})$. We often call $m(\mathbf{y})$ the *conditional mean* of X given $\mathbf{Y} = \mathbf{y}$, and we define the *conditional variance* of X given $\mathbf{Y} = \mathbf{y}$ to be the variance of the conditional distribution of X given $\mathbf{Y} = \mathbf{y}$. That is,

$$v(\mathbf{y}) = \text{var}(X|\mathbf{Y} = \mathbf{y}) = E((X - m(\mathbf{y}))^2|\mathbf{Y} = \mathbf{y}) = E(X^2|\mathbf{Y} = \mathbf{y}) - (m(\mathbf{y}))^2.$$

For example, if X is the weight and Y is the height of a randomly chosen male student, then $m(70)$ and $v(70)$ are the mean and variance of the weights of male students who are 70″ tall.

Of course, we can define

$$m^*(x) = E(Y|X = x) \quad \text{and} \quad v^*(x) = \text{var}(Y|X = x)$$

in a similar fashion. In fact, if $\mathbf{Q}$ is a random vector, then we can find the conditional mean of any of its components with respect to any other set of components. We illustrate these calculations with several examples.

Example 3–28

Let (X, Y) have joint density function $f(x, y) = 8xy, 0 < x < y < 1$. Then the marginal density function of Y is $f_2(y) = 4y^3, 0 < y < 1$, so that the conditional density function of X given $Y = y$ is $f(x|y) = 2x/y^2, 0 < x < y < 1$. Also, $f(x|y)$ is 0 if $x > y$. Hence,

$$m(y) = E(X|Y = y) = \int_0^y (x)\left(\frac{2x}{y^2}\right) dx = \frac{2y}{3},$$

$$E(X^2|Y = y) = \int_0^y (x^2)\left(\frac{2x}{y^2}\right) dx = \frac{y^2}{2},$$

$$v(y) = \text{var}(X|Y = y) = \frac{y^2}{2} - \left(\frac{2y}{3}\right)^2 = \frac{y^2}{18}.$$

Note that $m(y)$ and $v(y)$ are functions of y only. The marginal density of X is $f_1(x) = 4x(1 - x^2), 0 < x < 1$, and hence,

$$f(y|x) = \frac{2y}{(1 - x^2)}, \qquad 0 < x < y < 1.$$

Therefore,

$$m^*(x) = E(Y|X = x) = \int_x^1 (y)\left(\frac{2y}{(1 - x^2)}\right) dy = \frac{2(1 - x^3)}{3(1 - x^2)},$$

$$E(Y^2|X = x) = \int_x^1 (y^2)\left(\frac{2y}{1 - x^2}\right) dy = \frac{1 - x^4}{2(1 - x^2)} = \frac{1 + x^2}{2},$$

$$v^*(x) = \text{var}(Y|X = x) = \frac{1 + x^2}{2} - \left(\frac{2(1 - x^3)}{3(1 - x^2)}\right)^2.$$

Note that $m^*(x)$ and $v^*(x)$ are functions of x only. Finally,

$$E(XY|Y = y) = yE(X|Y = y) = \frac{2y^2}{3},$$

$$E(X^2Y^2|X = x) = x^2E(Y^2|X = x) = \frac{x^2(1 + x^2)}{2}.$$

Example 3–29

Let X and Y have joint density function $f(x, y) = y/10, (x, y) = (1, 1), (1, 2), (1, 3), (2, 1), (2, 2), (3, 1)$. Then X has marginal density function $f_1(1) = \frac{6}{10}, f_1(2) = \frac{3}{10}, f_1(3) = \frac{1}{10}$.

Therefore, the conditional density function of Y given $x = 1$ is

$$f(1|1) = \frac{1/10}{6/10} = \frac{1}{6}, \qquad f(2|1) = \frac{2}{6}, \qquad f(3|1) = \frac{3}{6}.$$

Hence,

$$m(1) = E[Y|X = 1] = 1\left(\frac{1}{6}\right) + 2\left(\frac{2}{6}\right) + 3\left(\frac{3}{6}\right) = \frac{7}{3},$$

$$E[Y^2|X = 1] = 1\left(\frac{1}{6}\right) + 4\left(\frac{2}{6}\right) + 9\left(\frac{3}{6}\right) = 6, \qquad v(1) = 6 - \left(\frac{7}{3}\right)^2 = \frac{5}{9}.$$

Similarly, the conditional density of Y given $X = 2$ is

$$f(1|2) = \frac{1}{3}, \qquad f(2|2) = \frac{2}{3}, \qquad m(2) = \frac{5}{3}, \qquad v(2) = \frac{2}{9}.$$

Finally, if $X = 3$, then Y must be 1, and hence,

$$m(3) = 1 \quad \text{and} \quad v(3) = 0.$$

Example 3–30

Let $f(x, y, z) = 48xyz, 0 < x < y < z < 1$. Then the marginal density of (X, Z) is

$$\int_x^z 48xyz\,dy = 24xz(z^2 - x^2), \qquad 0 < x < z < 1,$$

and the conditional density of Y given $(X, Z) = (x, z)$ is

$$f(y|(x, z)) = \frac{48xyz}{24xz(z^2 - x^2)} = \frac{2y}{(z^2 - x^2)}, \qquad 0 < x < y < z < 1.$$

Hence,

$$m(x, z) = E(Y|(X, Z) = (x, z)) = \int_x^z (y)\left(\frac{2y}{z^2 - x^2}\right) dy = \frac{2(z^3 - x^3)}{3(z^2 - x^2)},$$

$$E(Y^2|(X, Z) = (x, z)) = \int_x^z (y^2)\left(\frac{2y}{z^2 - x^2}\right) dy = \frac{z^4 - x^4}{2(z^2 - x^2)} = \frac{z^2 + x^2}{2},$$

$$v(x, z) = \text{var}(Y|(X, Z) = (x, z)) = \frac{z^2 + x^2}{2} - \left[\frac{2(z^3 - x^3)}{z^2 - x^2}\right]^2.$$

We could also find $E(Y|Z = z)$ for this model by first finding the joint density of Y and Z and then finding the marginal density of Y given $Z = z$. The joint density of Y and Z and the density of Z are, respectively,

$$f(y, z) = \int_0^y 48xyz\,dx = 24y^3 z, \qquad 0 < y < z < 1$$

and

$$f(z) = \int_0^z 24y^3 z\,dy = 6z^5, \qquad 0 < z < 1.$$

The conditional density of y given z is

$$f(y|z) = \frac{24y^3 z}{6z^5} = \frac{4y^3}{z^4}, \qquad 0 < y < z < 1.$$

Therefore,

$$E(Y|Z = z) = \int_0^z (y)\left(\frac{4y^3}{z^4}\right) dy = \frac{4z}{5}, \qquad 0 < z < 1.$$

We could find $\text{var}(Y|Z = z)$ similarly. We could also find $E(X|Z = z)$, $E(X|Y = y)$, etc.

Exercises—B

1. Let (X, Y) have joint density function $f(x, y) = x + y, 0 < x < 1, 0 < y < 1$.
   (a) Find $m(y) = \bar{r}(X|Y = y)$ and $v(y) = \text{var}(X|Y = y)$.
   (b) Find $E(XY^2|Y = y)$.
2. Let (X, Y) have joint density function $f(x, y) = e^{-y}, 0 < x < y$.
   (a) Find $m(y) = E(X|y = y)$ and $v(y) = \text{var}(X|Y = y)$.
   (b) Find $m^*(x) = E(Y|X = x)$ and $v^*(y) = \text{var}(Y|X = x)$.
3. Let (X, Y) have joint density function $f(x, y) = (x + y)/20, (x, y) = (1, 1), (1, 2), (1, 3), (2, 1), (2, 2), (3, 1)$.
   (a) Find $m(y) = E(X|Y = y)$. (That is, find $E(X|Y = 1)$, $E(X|Y = 2)$, and $E(X|Y = 3)$.)
   (b) Find $v(y) = \text{var}(X|Y = y)$.
   (c) Find $E(X^2 Y|Y = y)$.
4. Show that if X and Y are independent, then $E(X|Y = y) = EX$.
5. Prove parts b and c of Theorem 3–21 in the continuous case.
6. Verify that $E((X - m(y))^2|Y = y) = E(X^2|Y = y) - (m(y))^2$.
7. Let (X, Y, Z) have the density given in Example 3–30. Find:
   (a) $E(X)$ and $\text{var}(X)$.
   (b) $E(X|Y = y)$ and $\text{var}(X|Y = y)$.
   (c) $E(Y|Z = z)$ and $\text{var}(Y|Z = z)$.
   (d) $E(X^2 YZ|Z = z)$ and $E(X^2 Y|Z = z)$.
8. Let (X, Y, Z) have the joint density given in Example 3–30. Find:
   (a) $m(y, z) = E(X|(Y, Z) = (y, z))$.
   (b) $v(y, z) = \text{var}(X|(Y, Z) = (y, z))$.
9. Suppose that (X, Y) is independent of Z. Show that $E(X|(Y, Z) = (y, z)) = E(X|Y = y)$.

3.4.2 Expectation of Conditional Expectation

As in the previous section, we shall assume that we observe random variables or vectors $\mathbf{X}$ and $\mathbf{Y}$ with ranges S and T. We define the *conditional expectation* of $h(\mathbf{X}, \mathbf{Y})$ given $\mathbf{Y}$ by

$$E(h(\mathbf{X}, \mathbf{Y})|\mathbf{Y}) = g(\mathbf{Y}), \text{ where } g(\mathbf{y}) = E(h(\mathbf{X}, \mathbf{Y})|\mathbf{Y} = \mathbf{y}).$$

Note that $E(h(\mathbf{X}, \mathbf{Y})|\mathbf{Y})$ is a function of the random variable (or vector) $\mathbf{Y}$, rather than the dummy variable $\mathbf{y}$, and is therefore a random variable itself. Consider again the example in which X and Y are the weight and height of a randomly chosen male student. Let Y be the height of the randomly chosen student. Then $E(X|Y)$ is the mean of the weight of a male student chosen randomly from all the male students having height Y.

We now give one of the most useful results in probability.

Theorem 3–22. Let $q(\mathbf{Y}) = E(h(\mathbf{X}, \mathbf{Y})|\mathbf{Y})$. Then $Eq(\mathbf{Y}) = Eh(\mathbf{X}, \mathbf{Y})$.

Proof. We prove the theorem for the continuous case. The discrete case is left as an exercise. We have

$$Eq(\mathbf{Y}) = \int_T q(\mathbf{y})f_2(\mathbf{y})d\mathbf{y} = \int_T \left[\int_S h(\mathbf{x},\mathbf{y})f(\mathbf{x}|\mathbf{y})d\mathbf{x}\right] f_2(\mathbf{y})d\mathbf{y}$$

$$= \int_T\int_S h(\mathbf{x},\mathbf{y})f(\mathbf{x}|\mathbf{y})f_2(\mathbf{y})d\mathbf{x}d\mathbf{y}$$

$$= \int_T\int_S h(\mathbf{x},\mathbf{y})f(\mathbf{x},\mathbf{y})d\mathbf{x}d\mathbf{y} = Eh(\mathbf{X},\mathbf{Y}),$$

using the fact that $f(\mathbf{x}|\mathbf{y})f_2(\mathbf{y}) = f(\mathbf{x},\mathbf{y})$. □

In the example in which X is weight and Y is height, $q(Y) = E(X|Y)$ is the average weight of a male of height Y. Theorem 3–22 says that if we further average this average weight over all heights, we will get the overall average weight for all males. From this result, we see that the expected value for conditional expectation is unconditional expectation. The result is often written as

$$E(E[h(\mathbf{X},\mathbf{Y})|\mathbf{Y}]) = E(h(\mathbf{X},\mathbf{Y})).$$

A useful perspective on this theorem is gotten from noting that if X and Y are continuous random variables, then $E(h(X, Y))$ is the double integral of $h(x,y)f(x,y)$ over the joint range of (X, Y), but $E(E(h(X, Y)|Y))$ and $E(E(h(X, Y)|X))$ are the two iterated integrals which are often used to calculate this double integral.

Example 3–31

Let Y be uniformly distributed on the interval $(0, 1)$. Conditionally on $Y = y$ let X be uniformly distributed on the interval $(0, y)$. The mean of a uniform distribution from zero to a is $a/2$. Therefore, $EY = \frac{1}{2}$, and

$$E(X|Y) = \frac{Y}{2}, \qquad EX = E(EX|Y) = \frac{EY}{2} = \frac{1}{4}.$$

Similarly, if U is uniformly distributed on $(0, a)$, then $EU^2 = a^2/3$. Therefore, $EY^2 = \frac{1}{3}$, $\text{var}(Y) = (\frac{1}{3}) - (\frac{1}{2})^2 = \frac{1}{12}$, and

$$E(X^2|Y) = \frac{Y^2}{3}, \qquad EX^2 = E(EX^2|Y) = \frac{EY^2}{3} = \frac{1}{9},$$

$$\text{var}(X) = \frac{1}{9} - \left(\frac{1}{4}\right)^2 = \frac{7}{144}.$$

Now, by Theorem 3–22,

$$E(XY|Y) = YEX|Y = \frac{Y^2}{2}, \qquad EXY = E(EXY|Y) = \frac{EY^2}{2} = \frac{1}{6},$$

$$\text{cov}(X, Y) = \frac{1}{6} - \left(\frac{1}{4}\right)\left(\frac{1}{2}\right) = \frac{1}{24},$$

$$\text{corr}(X, Y) = \frac{1/24}{((1/12)(7/144))^{1/2}} = \left(\frac{3}{7}\right)^{1/2}.$$

Example 3–32

A rat is put into a maze in which it has to choose one of two directions. If it chooses left, it wanders around the maze for four minutes and comes back to where it started. If it chooses right, then with probability $\frac{1}{4}$ it will depart the maze in five minutes, and with probability $\frac{3}{4}$ it will come back to where it started after three minutes. We assume that at the beginning and whenever it returns to the start, it chooses to go right with probability $\frac{1}{5}$. Let X be the rat's time in the maze. We wish to find EX. Let $Y = 1$ if the rat decides to go left at the first stage and $Y = 0$ if it chooses to go right at the first stage. Therefore, if the rat chooses the left side, it wanders around for four minutes and then comes back to where it started. In this case, its expected time till it leaves the maze is $4 +$ its expected time to leave when it started, i.e.,

$$E(X|Y = 1) = 4 + EX.$$

Similarly, if it chooses the right side, then one-fourth of the time it leaves in five minutes and three-fourths of the time it returns to the start in three minutes, i.e.,

$$E(X|Y = 0) = \left(\frac{1}{4}\right)(5) + \left(\frac{3}{4}\right)(3 + EX).$$

Hence,

$$EX = E(EX|Y) = \left(\frac{4}{5}\right)(4 + EX) + \left(\frac{1}{5}\right)\left[\left(\frac{1}{4}\right)5 + \left(\frac{3}{4}\right)(3 + EX)\right] = \frac{78 + 19EX}{20},$$

and

$$EX = 78.$$

The next theorem provides the main content of the Rao-Blackwell theorem of Chapter 10.

Theorem 3–23. Let X be a random variable and let $\mathbf{Y}$ be a random variable or vector. Let $m(\mathbf{Y}) = E(X|\mathbf{Y})$ and $v(\mathbf{Y}) = \text{var}(X|\mathbf{Y}) = E([X - m(\mathbf{Y})]^2|\mathbf{Y})$. Then

$$EX = Em(Y) \quad \text{and} \quad \text{var}(X) = E(v(Y)) + \text{var}(m(Y)).$$

Proof. By Theorem 3–22, $\mu = EX = Em(\mathbf{Y})$. Since

$$v(Y) = EX^2|\mathbf{Y} - (m(\mathbf{Y}))^2 \quad \text{and} \quad \text{var}(m(Y)) = E(m(Y))^2 - \mu^2,$$

it follows that

$$\begin{aligned}\text{var}(X) = EX^2 - \mu^2 &= E([EX^2|\mathbf{Y}) - (m(\mathbf{Y}))^2 + (m(\mathbf{Y}))^2 - \mu^2]|\mathbf{Y}) \\ &= E[E(X^2|\mathbf{Y}) - (m(\mathbf{Y}))^2] + \text{var}(m(\mathbf{Y})) = Ev(\mathbf{Y}) + \text{var}(m(\mathbf{Y})). \quad \square\end{aligned}$$

We see, then, that the mean of X is the expected value of the conditional mean, but the variance of X is the expected value of the conditional variance plus the variance of the conditional mean.

We now apply Theorem 3–23 to some examples from the previous section.

Example 3–33

We return to Example 3–28, where (X, Y) has joint density function $f(x, y) = 8xy$, $0 < x < y < 1$. In Section 3.2.2, we showed that $EY = \frac{4}{5}$, $EY^2 = \frac{2}{3}$, and $\text{var}(Y) = \frac{2}{75}$ for this function. In Example 3–28, we showed that $E(X|Y) = 2Y/3$ and $\text{var}(X|Y) = Y^2/18$. Applying Theorem 3–23, we obtain

$$EX = E(EX|Y) = \frac{E2Y}{3} = \frac{2EY}{3} = \frac{8}{15},$$

$$\text{var}(X) = E(\text{var}(X|Y)) + \text{var}(EX|Y) = E\left(\frac{Y^2}{18}\right) + \text{var}\left(\frac{2Y}{3}\right)$$
$$= \frac{EY^2}{18} + \frac{4\,\text{var}(Y)}{9} = \frac{1}{27} + \frac{8}{675} = \frac{33}{675} = \frac{11}{225}.$$

These answers are the same as those we got in Section 3.2.2 for this model.

Example 3–34

We return to Example 3–29, in which (X, Y) has joint density function $f(x, y) = y/10$, $(x, y) = (1, 1), (1, 2), (1, 3), (2, 1), (2, 2), (3, 1)$. Then Y has density function $f_2(1) = \frac{3}{10}$, $f_2(2) = \frac{4}{10}$, $f_2(3) = \frac{3}{10}$. Therefore,

$$EY = 1\left(\frac{3}{10}\right) + 2\left(\frac{4}{10}\right) + 3\left(\frac{3}{10}\right) = 2,$$
$$EY^2 = 1\left(\frac{3}{10}\right) + 4\left(\frac{4}{10}\right) + 9\left(\frac{3}{10}\right) = \frac{23}{5}, \qquad \text{var}(Y) = \frac{23}{5} - 2^2 = \frac{3}{5}.$$

In Example 3–29, we saw that the marginal density function for X, the conditional expectation of x, and conditional variance of Y given $X = x$ are

$$f_1(1) = \frac{6}{10}, \qquad f_1(2) = \frac{3}{10}, \qquad f_1(3) = \frac{1}{10},$$
$$m(1) = \frac{7}{3}, \qquad m(2) = \frac{5}{3}, \qquad m(3) = 1,$$
$$v(1) = \frac{5}{9}, \qquad v(2) = \frac{2}{9}, \qquad v(3) = 0.$$

Therefore,

$$Em(X) = \left(\frac{7}{3}\right)\left(\frac{6}{10}\right) + \left(\frac{5}{3}\right)\left(\frac{3}{10}\right) + (1)\left(\frac{1}{10}\right) = 2 = EY.$$

In addition,

$$Em^2(X) = \left(\frac{49}{9}\right)\left(\frac{6}{10}\right) + \left(\frac{25}{9}\right)\left(\frac{3}{10}\right) + 1\left(\frac{1}{10}\right) = \frac{21}{5},$$
$$\text{var}(m(X)) = \frac{21}{5} - 4 = \frac{1}{5},$$
$$Ev(X) = \left(\frac{5}{9}\right)\left(\frac{6}{10}\right) + \left(\frac{2}{9}\right)\left(\frac{3}{10}\right) + 0\left(\frac{1}{10}\right) = \frac{2}{5},$$

and therefore,

$$\text{var}(m(X)) + Ev(X) = \frac{1}{5} + \frac{2}{5} = \frac{3}{5} = \text{var}(Y).$$

Exercises—A

1. Suppose that a rat is in a maze with four possible directions. If it goes in the first direction, it gets out in three minutes. If it chooses the second direction, it returns to the starting point in five minutes. If it chooses the third direction, with probability .2 it returns to the start in four minutes, and with probability .8 it returns to the start in seven minutes. If it chooses the fourth direction, it gets out of the maze in six minutes with probability .3

and returns to the maze in eight minutes with probability .7. Each time the rat returns he chooses each direction with probability 1/4. Find the expected time until the rat escapes the maze.

2. A prisoner is in a cell with three doors. He chooses a door at random (each with probability $\frac{1}{3}$). The first door leads to a long tunnel which leads to freedom in three days. The second tunnel is a trap which leads back to the cell in two days, and the third tunnel is also a trap which leads back to the prison cell, but in five days. Each time the prisoner gets back to the cell, he chooses a door at random from the three doors (again, each with probability $\frac{1}{3}$). (That is, he does not remember the door he chose the previous time.) Find the expected time till the prisoner escapes.
3. Seven customers are located in a row at a particular bar, each five feet apart. If each customer is equally likely to ask for service at any time, what is the expected distance that the bartender must travel between customers? (*Hint*: What is the conditional expected distance given that the previous customer is in the ith position.)
4. In Example 3–31, we assumed that the rat was sure to get out of the maze. Otherwise, X would not have been a random variable. Show that the probability that the rat stays in the maze forever is zero. (*Hint*: What is the probability that the rat is still in the maze after k attempts to escape?)

Exercises—B

1. Verify that $Em(Y) = EX$ for Exercise B1 of the last section.
2. For Exercise B2 of the last section,
   (a) Verify that $Em(Y) = EX$.
   (b) Verify that $\mathrm{var}(m(Y)) + E(v(Y)) = \mathrm{var}(X)$.
3. For Exercise B3 of the last section,
   (a) Verify that $E(m(Y)) = EX$.
   (b) Verify that $\mathrm{var}(X) = E(v(Y)) + \mathrm{var}(m(Y))$.
4. Prove Theorem 3–22 in the discrete case.
5. Let (X, Y) be random variables with mean μ_X and μ_Y, variances σ_X^2 and σ_Y^2, and correlation coefficient ρ. Suppose that $EX|Y = aY + b$.
   (a) Show that $b = \mu_X - a\mu_Y$.
   (b) Show that $a = \rho\sigma_X/\sigma_Y$. (*Hint*: $EXY = E(YE(X|Y))$. Why?)
   (c) Now suppose that $\mathrm{var}(X|Y) = c^2$. Show that $c^2 = \sigma_X^2(1 - \rho^2)$.
6. For Example 3–30,
   (a) Verify that $EY = Em(X, Z)$.
   (b) Verify that $EY = E(E(Y|Z))$.
   (c) Verify that $E(Z^2Y) = E(EZ^2Y|(X, Z) = E(E(Z^2Y|Z)$.

Exercises—C

1. Let (X, Y, Z) be three random variables. Let $m(Z) = EX|Z$ and $n(Z) = EY|Z$. Define the conditional covariance between X and Y given Z by $\mathrm{cov}((X, Y)|Z) = E([XY - m(Z)n(Z)]|Z)$. Show that
$$\mathrm{cov}(X, Y) = E(\mathrm{cov}((X, Y)|Z)) + \mathrm{cov}(m(Z), n(Z)).$$
2. Show that $\mathrm{var}(X) = \mathrm{var}(X|\mathbf{Y})$ if and only if $m(\mathbf{Y})$ is degenerate.

3.5 THE MEDIAN

* 3.5.1 The Median

Besides the expectation or mean of a random variable, another measure of the center of a distribution is the median. We say that a is a *median* of X if

$$P(X \leqq a) \geqq \frac{1}{2}, \qquad P(X < a) \leqq \frac{1}{2}.$$

If X is a continuous random variable, then a is a median if

$$P(X \leqq a) = F(a) = \frac{1}{2}.$$

In other words, for continuous random variables the median has half the probability above it and half the probability below it.

Example 3–35

Let X have density function $f(x) = e^{-x}, x > 0$. Then $F(x) = 1 - e^{-x}, x > 0$, so that X has median equal to log(2). Since the mean of X is $EX = 1$, it is plain that the mean and the median can be different.

Example 3–36

Let X have density function $f(x) = \frac{1}{3}, x = 1, 2, 3$. Then $P(X \leqq 2) = \frac{2}{3}$ and $P(X < 2) = \frac{1}{3}$, and therefore X has median 2, which is equal to EX, so that in this instance the median and the mean are the same.

Example 3–37

Let X have density function $f(x) = \frac{1}{4}, x = 1, 2, 3, 4$. Then $P(X \leqq 2) = \frac{1}{2}$ and $P(X < 2) = \frac{1}{4}$, so that 2 is a median for X. Similarly, $P(X \leqq 3) = \frac{3}{4}$ and $P(X < 3) = \frac{1}{2}$, so that 3 is also a median for X. (Actually, any a such that $2 \leqq a \leqq 3$ is a median for X.) This illustrates that a random variable may have several medians, in which case we say that the median is not unique.

A random variable X with density function $f(x)$ is *symmetric about b* if

$$f(y - b) = f(-y + b))$$

for all x. If X is symmetric about b, then b is a median of X, but may not be unique. (See Exercises B2 and B4.) In addition, if EX is finite, then $EX = b$. (See Exercise C1 of Section 3.1.3) Therefore, if X is symmetric about b, and if its expectation is finite and its median unique, then the median and mean of X are the same. (Recall, however, that if X has the Cauchy density $f(x) = (\pi(1 + x^2))^{-1}$, then X is symmetric about zero, but EX does not exist.)

There are several advantages to the median over the mean as a measure of the center of the distribution. One is that the median is always finite. A second is that it is not as sensitive to extreme values as the following example shows.

Example 3–38

Let X have density function $f(x) = 1$ if $0 < x < .99$ or if $9{,}999.99 < x < 10{,}000$. Then

$$EX = \int_0^{.99} x dx + \int_{9{,}999.99}^{10{,}000} x dx = 200.98.$$

It seems clear that the center of the distribution is somewhere in the interval $0 < x < .99$; however, the mean is over 200. This occurs because we have a positive (but small) chance of getting a huge number. The median of X is $\frac{1}{2}$, which seems more reasonable.

We have seen this sensitivity of the mean to extreme observations in the St. Petersburg game paradox, where X is sure to be finite but $EX = \infty$. In that case, the median of X is 2.

The mean also has several advantages. First, it is unique (when it exists). Second, it is much easier to develop a theory of expectations than of medians, because of the law of expectation: There is typically no way to find the median of $U = h(\mathbf{X})$ without finding the distribution function of U. In particular, the median of $U = X + Y$ is not the median of X plus the median of Y. Third, the mean is the correct parameter to use in the most important theorem of statistics, the central limit theorem, which we state in Chapter 4. For these reasons, for most of the remainder of this book, we shall use the mean $\mu = EX$ to measure the center of the distribution of X.

Exercises—A

1. Verify, for the St. Petersburg game (Example 3–8), that the median payoff is 2.

Exercises—B

1. Let X have density function $f(x) = 2x, 0 < x < 1$. Find the median of X.
2. Let X have density function $f(x) = 1, -1 < x < -\frac{1}{2}, \frac{1}{2} < x < 1$. Show that any point b such that $-\frac{1}{2} \leq b \leq \frac{1}{2}$ is a median of X. (This problem illustrates that even symmetric continuous random variables can have nonunique medians.)
3. Let X be a continuous random variable with density $f(x)$ such that $f(c + x) = f(c - x)$ for all x. Show that c is a median of X.

Chapter 4

UNIVARIATE PARAMETRIC FAMILIES

In the first three chapters of the text, we defined many of the fundamental ideas of probability and proved some basic theorems. In Chapters 4 and 5, we use those definitions and theorems to study some basic models in probability.

In Section 4.1, we define the most important distribution in probability, the standard normal distribution. We also state the central limit theorem, which is the result which makes the standard normal distribution so important. However, we delay its proof until Chapter 6.

A Bernoulli trial is an experiment with only two possible outcomes (heads-tails, pass-fail, cancer–no cancer, etc.). In Section 4.2, we study the distributions associated with Bernoulli trials: the Bernoulli and binomial distributions, the geometric and negative binomial distributions, and the hypergeometric distribution.

The Poisson process is a probability model that is often used for modeling rare events such as emissions from an atom, telephone calls to a switchboard, or terrorist attacks. In Section 4.3, we study distributions associated with the Poisson process: the Poisson, exponential, and gamma distributions.

We use the general normal distribution as a model for many situations when we are measuring something (e.g., weight, speed, intelligence, size). This distribution is the one that is most often used as a model for statistically analyzing data (because of the central limit theorem). In Section 4.4, we examine some properties of the general normal distribution and other distributions related to it, viz., the χ^2, t, and F distributions.

Before considering these various distributions, we shall need a definition. Let $X_1, X_2, \ldots$ be a sequence of random variables such that X_n has distribution function $F_n(x)$, and let X be a random variable with distribution function $F(x)$. We say that X_n *converges in distribution* to X as $n \to \infty$ and write

$$X_n \xrightarrow{d} X \quad \text{as } n \to \infty$$

if

$$F_n(x) \to F(x) \quad \text{as } n \to \infty$$

at all points x where $F(x)$ is continuous. We call the distribution of X the *asymptotic distribution* of X_n.

In all the models in this chapter, either X is a continuous random variable or both X_n and X are integer-valued random variables. We show in Chapter 6 that under these conditions,

$$P(X_n \leqq x) \to P(X \leqq x) \quad \text{and} \quad P(X_n < x) \to P(X < x) \tag{4–1}$$

for all x. In particular, in these cases, $F_n(x) \to F(x)$ for all x. If Equation (4–1) is satisfied, then

$$P(X_n \in I) \to P(X \in I) \quad \text{as } n \to \infty$$

for any interval I. For example, it follows from Equation (4–1) that

$$P(X_n = a) = P(X_n \leqq a) - P(X_n < a) \to P(X \leqq a) - P(X < a) = P(X = a),$$

$$P(a \leqq X_n < b) = P(X_n < b) - P(X_n < a) \to P(X < b) - P(X < a) = P(a \leqq X < b),$$

and

$$P(X_n > a) = 1 - P(X_n \leqq a) \to 1 - P(X \leqq a) = P(X > a).$$

We often write that, for large n,

$$P(X_n = a) \doteq P(X = a) \quad \text{and} \quad P(X_n \in I) \doteq P(X \in I)$$

(where $\doteq$ means "is approximately equal to").

We now give an elementary result about certain discrete random variables.

Theorem 4–1. Let $X_1, X_2, \ldots$ be a sequence of nonnegative integer-valued random variables such that X_n has density function $f_n(x)$, and let X be a nonnegative integer-valued random variable with density function $f(x)$. Then

$$X_n \xrightarrow{d} X \Leftrightarrow \lim_{n \to \infty} f_n(k) = f(k)$$

for all nonnegative integers k.

Proof. Let X_n have distribution function $F_n(x)$, and let X have distribution function $F(x)$. Suppose that $f_n(k) \to f(k)$. Then

$$F_n(x) = \sum_{i=0}^{[x]} f_n(k) \to \sum_{i=0}^{[x]} f(k) = F(x),$$

where $[x]$ is the largest integer less than or equal to x. (For fixed x, $[x]$ is a finite number, so that the sum has a finite number of terms.) Conversely, suppose that $X_n \xrightarrow{d} X$. Then $F(x)$ is continuous at $k + .5$ for any integer k. Also,

$$f_n(k) = P(X_n = k) = P(k - .5 < X_n \leqq k + .5) = F_n(k + .5) - F_n(k - .5)$$

(since X_n takes on only integer values). Therefore,

$$f_n(k) = F_n(k + .5) - F_n(k - .5) \to F(k + .5) - F(k - .5) = f(k). \quad \square$$

We could derive a result similar to Theorem 4–1 for continuous random variables (see Billingsley (1979), p. 184). However, we shall not need that result and so shall not prove it here.

One reason that we have defined convergence in distribution in terms of the distribution function rather than the density function is that we shall have examples of discrete random variables X_n which converge to a continuous random variable X and vice versa. It is not clear how to define such convergence in terms of the density function.

Another reason for defining the convergence in distribution in terms of the distribution function is that one important use for such convergence is in approximating probabilities. As mentioned in Chapter 2, the continuous density function is of little help in determining probabilities. Even the discrete density function is less useful than the distribution function for this purpose.

4.1 THE STANDARD NORMAL DISTRIBUTION

4.1.1 The Standard Normal Distribution and the Central Limit Theorem

In this section, we define the most important distribution in probability and statistics. Let Z be a continuous random variable with density function

$$f(z) = (2\pi)^{-1/2} \exp\left(\frac{-z^2}{2}\right), \qquad -\infty < z < \infty.$$

Then Z has a *standard normal distribution*. (See Figure 4–1.) Note that the density function is symmetric about zero. The curve is often called a *bell-shaped* curve.

For the remainder of this book, we reserve the symbol Z for a standard normal random variable and the expression $N(z)$ for the distribution function of Z. Unfortunately, there is no formula for $N(z)$ in terms of elementary functions. However, Table 1 at the back of the book gives selected values of $N(z)$ for $z > 0$. In the exercises, it is shown that

$$N(-z) = 1 - N(z).$$

Example 4–1

Let Z have a standard normal distribution. Using Table 1, we see that

$$P(1 < Z < 2) = N(2) - N(1) = .9773 - .8413 = .1360,$$
$$P(.5 \leqq Z \leqq 1.5) = N(1.5) - N(.5) = .9332 - .6915 = .2417,$$
$$P(-2 < Z \leqq 1) = N(1) - N(-2) = N(1) - (1 - N(2)) = .8413 + .9773 - 1 = .8186,$$
$$P(Z > -2) = 1 - N(-2) = N(2) = .9773,$$
$$P(Z > 2) = 1 - N(2) = 1 - .9773 = .0227.$$

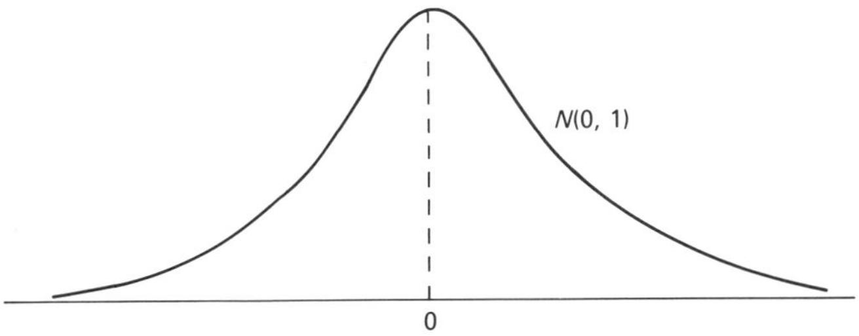

Figure 4–1 Standard normal density.

In later sections, we will often need $N(z)$ for some z which is not in the table. Typically, we shall take the nearest entry in the table. If the required z is halfway between two entries, we shall usually average the entries. (The entries in the table are close enough that it rarely makes much difference what we do.)

We next show that the density $f(z)$ given in the first paragraph of the subsection is a density function. The proof is quite ingenious; however, it does not give us much insight. In fact, there does not appear to be a natural proof of this theorem.

Theorem 4–2. The function $f(z) = (2\pi)^{-1/2}\exp(-z^2/2)$ is a density function.

Proof. Clearly, $f(z) > 0$. We therefore need only show that

$$\int_{-\infty}^{\infty} (2\pi)^{-1/2} \exp\left(\frac{-z^2}{2}\right) dz = 1.$$

Let

$$I = \int_{-\infty}^{\infty} \exp\left(\frac{-z^2}{2}\right) dz = \int_{-\infty}^{\infty} \exp\left(\frac{-x^2}{2}\right) dx = \int_{-\infty}^{\infty} \exp\left(\frac{-y^2}{2}\right) dy.$$

(Note that the integral is unchanged by renaming the variable of integration.) Then $I > 0$, so we need only show that $I^2 = 2\pi$. Now,

$$I^2 = \left(\int_{-\infty}^{\infty} \exp\left(\frac{-x^2}{2}\right) dx\right)\left(\int_{-\infty}^{\infty} \exp\left(\frac{-y^2}{2}\right) dy\right) = \int_{-\infty}^{\infty}\int_{-\infty}^{\infty} \exp\left(-\frac{(x^2+y^2)}{2}\right) dx\,dy$$

$$= \int_0^{\infty}\int_0^{2\pi} \exp\left(\frac{-r^2}{2}\right) r\,d\theta\,dr = \left(\int_0^{\infty} r \exp\left(\frac{-r^2}{2}\right) dr\right)\left(\int_0^{2\pi} d\theta\right) = 2\pi \int_0^{\infty} \exp(-u)\,du = 2\pi,$$

where the third equality follows from changing to polar coordinates, and the fifth one follows from substituting $u = r^2/2$ and $du = r\,dr$. (Note that what makes the proof work is getting the r from the change to polar coordinates.) □

We now find the moment-generating function, the mean, and the variance of the standard normal distribution.

Theorem 4–3. Let Z have a standard normal distribution. Then:

a. Z has moment-generating function

$$M(t) = \exp\left(\frac{t^2}{2}\right), \qquad -\infty < t < \infty.$$

b. $EZ = 0$ and $\mathrm{var}(Z) = 1$.

Proof

a. Let $x = z - t$. Then $dx = dz$ and $zt - z^2/2 = (t^2 - x^2)/2$, and hence,

$$M(t) = Ee^{zt} = \int_{-\infty}^{\infty} (2\pi)^{-1/2} \exp\left(zt - \frac{z^2}{2}\right) dz$$

$$= \int_{-\infty}^{\infty} (2\pi)^{-1/2} \exp\left(\frac{t^2 - x^2}{2}\right) dx = \exp\left(\frac{t^2}{2}\right) \int_{-\infty}^{\infty} (2\pi)^{-1/2} \exp\left(\frac{-x^2}{2}\right) dx$$

$$= \exp\left(\frac{t^2}{2}\right),$$

since the normal density function $(2\pi)^{-1/2} \exp(-x^2/2)$ integrates to 1.

b. Let $\psi(t) = \log(M(t)) = t^2/2$. Then $\psi'(t) = t$ and $\psi''(t) = 1$. Hence,

$$\mu = \psi'(0) = 0 \quad \text{and} \quad \sigma^2 = \psi''(0) = 1. \quad \square$$

In Section 4.4, we define a normal distribution with an arbitrary mean μ and variance σ^2. We therefore write

$$Z \sim N(0, 1)$$

to mean that Z has a normal distribution with mean 0 and variance 1, i.e., that Z has a standard normal distribution. (We shall use the symbol $\sim$ often in the remainder of this book to mean "is distributed as.")

Now, let $X_1, \ldots, X_n$ be a sequence of random variables (not necessarily normally distributed). We say that the X_i are *independently identically distributed* (i.i.d.) if the X_i are independent and have the same distribution. We write

$$T_n = \sum_{i=1}^{n} X_i, \qquad \overline{X}_n = \frac{T_n}{n}$$

to denote the total and average, respectively, of the n X_i's.

The next theorem is the most important theorem in the chapter and perhaps in the book.

Theorem 4–4. (Central limit theorem) Let $X_1, \ldots, X_n$ be i.i.d. random variables with mean μ and variance σ^2. Let

$$Z_n = \frac{n^{1/2}(\overline{X}_n - \mu)}{\sigma} = \frac{T_n - n\mu}{n^{1/2}\sigma}.$$

Then

$$Z_n \xrightarrow{d} Z \sim N(0, 1) \quad \text{as } n \to \infty.$$

Proof. See Section 6.2. $\square$

In light of Theorem 4–4, we often write, for large n,

$$Z_n \mathrel{\dot\sim} N(0, 1),$$

where the symbol $\dot\sim$ means "is approximately distributed as." Note that there are nearly no assumptions to the theorem. As long as the X_i are i.i.d. from a distribution with a finite variance, Z_n is approximately normal. The X_i could be discrete or continuous, bounded or unbounded; the theorem is still true. In fact, the theorem is true even more generally than given here: The X_i do not have to be identically distributed. The Lindeberg-Feller theorem gives fairly general conditions for sums of independent random variables to converge to normality. Other theorems also allow the possibility that the X_i are not even independent.

The size n necessary for the foregoing approximation to be useful depends, unfortunately, on the distribution of the X_i. However, for most distributions $n = 30$ is enough, and for many even $n = 25$ or $n = 20$ is enough.

In the exercises, it is shown that

$$\mu_{T_n} = ET_n = n\mu \quad \text{and} \quad \sigma^2_{T_n} = \text{var}(T_n) = n\sigma^2,$$

so that

$$Z_n = \frac{T_n - \mu_{T_n}}{\sigma_{T_n}}.$$

Similarly,

$$Z_n = \frac{\overline{X}_n - \mu_{\overline{X}_n}}{\sigma_{\overline{X}_n}}.$$

Example 4–2

An elevator sign says that it can hold 25 people or 4,000 lb. Suppose that the weight of a randomly chosen person is a random variable with mean 148 lb and standard deviation 20 lb. We wish to know the probability that the total weight of 25 randomly chosen individuals is greater than 4,000 lb. Let T be the total weight of the 25 people. Then

$$\mu_T = ET = 25(148) = 3{,}700, \qquad \sigma^2_T = 25(20)^2, \qquad \sigma_T = 100.$$

Therefore, by the central limit theorem,

$$\frac{T - 3{,}700}{100} \mathrel{\dot\sim} N(0, 1).$$

Hence,

$$P(T > 4{,}000) = P\left(\frac{T - 3{,}700}{100} > 3\right) \doteq P(Z > 3) = 1 - N(3) = .0013.$$

In later sections, we derive normal approximations for several distributions. The following theorem is often helpful in those derivations.

Theorem 4–5. Let $X_1, X_2, \ldots$ be a sequence of random variables such that X_n has mean μ_n, variance σ^2_n, and moment-generating function $M_n(t)$. If $\sigma^2_1 \neq 0$ and

$$M_n(t) = (M_1(t))^n,$$

then

$$\frac{X_n - \mu_n}{\sigma_n} \xrightarrow{d} Z \sim N(0, 1).$$

Proof. Note first that if $M_n(t) = (M_1(t))^n$, then

$$\mu_n = n\mu_1 \quad \text{and} \quad \sigma^2_n = n\sigma^2_1.$$

Now, let $V_1, V_2, \ldots$ be a sequence of independently identically distributed random variables with moment-generating function $M_1(t)$ and hence mean μ_1 and variance σ_1^2. Let

$$T_n = \sum_{i=1}^{n} V_i.$$

Then by the central limit theorem,

$$\frac{T_n - n\mu_1}{n^{1/2}\sigma_1} = \frac{T_n - \mu_n}{\sigma_n} \xrightarrow{d} Z \sim N(0,1).$$

Now, by part b of Theorem 3–16, T_n has moment-generating function $M_n(t) = (M(t))^n$. Therefore, T_n has the same distribution as X_n, and it follows that

$$\frac{X_n - \mu_n}{\sigma_n} \xrightarrow{d} Z \sim N(0,1).$$

(If X_n and T_n have the same distribution and $h_n(T_n)$ converges to Z, then $h_n(X_n)$ must also converge to Z. Why?) □

Let X_n be a sequence of random variables with $\mu_n = EX_n$ and $\sigma_n^2 = \text{var}(X_n)$. All of the normal approximations in this chapter have the form

$$Z_n = \frac{X_n - \mu_n}{\sigma_n} \xrightarrow{d} Z \sim N(0,1),$$

so that we have a simple way to remember the normal approximations to the various distributions. However, in order to derive these normal approximations, it is not enough just to find μ_n and σ_n^2; it is also necessary to use the central limit theorem or Theorem 4–5. The following example shows that finding μ_n and σ_n^2 is not sufficient to guarantee the normal approximation.

Example 4–3

Let X_n be uniformly distributed on $(0, n)$, so that $\mu_n = EX_n = n/2$ and $\sigma_n^2 = \text{var}(X_n) = n^2/12$. In this case,

$$Z_n = \frac{X_n - \mu_n}{\sigma_n} \xrightarrow{d} U$$

where U is uniformly distributed on $(-3^{1/2}, 3^{1/2})$ (see Exercise B9). Hence, Z_n does not converge to $Z \sim N(0,1)$.

Exercises—A

For the problems in this chapter, "find the probability" means "find the exact probability if possible; otherwise, find the correct approximate probability."

1. A set of 75 numbers is rounded off to the nearest integer and then averaged. Assuming that the roundoff errors are independently uniformly distributed on the interval $(-.5, .5)$, find the probability that the average of the rounded-off numbers is within .1 of the average of the original numbers.
2. A wall is to be built with 100 bricks. The lengths of the bricks are independently identically distributed with mean 1 ft. and standard deviation 1 in. What is the

probability that the length of the wall is more than 5 in. different from 100 ft.? (*Note:* 1 ft. = 12 in.)

3. Paper pads are packaged in packages of 400 by weighing them. Suppose that the weight of a pad has mean 1 oz and standard deviation .05 oz. Find the probability that a package of 400 pads weighs more than 400.5 oz (and would therefore have to have one pad removed).

Exercises—B

1. Let Z have a standard normal distribution.
   (a) Find $P(Z < -2)$, $P(Z \leqq -2)$, $P(Z = 0)$, $P(-1.5 \leqq Z < 2.5)$, $P(-2.5 \leqq Z < 1.5)$, and $P(Z > -3)$.
   (b) Find a and b such that $P(Z < a) = .95$ and $P(|Z| \leqq b) = .95$.
2. If $Z \sim N(0,1)$, show that $-Z \sim N(0,1)$.
3. Let $n(z)$ and $N(z)$ respectively be the density function and distribution function of a standard normal random variable Z.
   (a) Show that $n(-z) = n(z)$.
   (b) Show that $N(-z) = 1 - N(z)$.
   (c) Show that $P(|Z| > a) = 2(1 - N(a))$ and $P(|Z| < a) = 2N(a) - 1$.
4. Let Z be a standard normal random variable.
   (a) Use the definition of expectation to show that $EZ = 0$.
   (b) Use the definition of variance to show that $\text{var}(Z) = 1$. (Use integration by parts, with $u = z$ and $dv = z\ \exp(-z^2)dz$.)
5. Let $n(z)$ be the density function of a standard normal random variable.
   (a) Show analytically that n is maximized at $z = 0$.
   (b) Show that $n(z)$ has points of inflection at -1 and 1.
6. What is the median of a standard normal random variable?
7. Let $X_1, \ldots, X_n$ be i.i.d. with $EX_i = \mu$ and $\text{var}(X_i) = \sigma^2$. Let $T = \Sigma_i X_i$ and $\overline{X} = T/n$. Show that $ET = n\mu$, $E\overline{X} = \mu$, $\text{var}(T) = n\sigma^2$, and $\text{var}(\overline{X}) = \sigma^2/n$.
8. Show that, under the conditions of Theorem 4–5, $EX_n = n\mu_1$ and $\text{var}(X_n) = n\sigma^2$. (*Hint*: Let $\psi_n(t) = \log(M_n(t)) = n\ \log(M(t))$.)
9. Let X_n be uniformly distributed on $(0, n)$.
   (a) Show that $\mu_n = EX_n = n/2$ and $\sigma_n^2 = \text{var}(X_n) = n^2/12$.
   (b) Show that $V_n = (U_n - \mu_n)/\sigma_n$ is uniformly distributed on $(-3^{1/2}, 3^{1/2})$ and hence that $(X_n - \mu_n)/\sigma_n$ does not converge to Z.

Exercises—C

1. Let Y and Z be independent standard normal random variables. Let $X = Y/Z$. Show that X has the Cauchy density $f(x) = (\pi(1 + x^2))^{-1}$, $-\infty < x < \infty$.

4.2 DISTRIBUTIONS ASSOCIATED WITH BERNOULLI TRIALS

4.2.1 Bernoulli and Binomial Distributions

In this section, we introduce a family of distributions called binomial distributions. These distributions depend on two parameters, n and p: We get a different distribution for each possible choice of n and p.

Let X be a discrete random variable which has density function

$$f(x) = \begin{bmatrix} n \\ x \end{bmatrix} p^x(1-p)^{n-x}, \qquad x = 0, 1, \ldots, n, \tag{4–2}$$

where n is a positive integer and p is a constant such that $0 \leqq p \leqq 1$. Then we say that X has a *binomial distribution* with parameters n and p and write

$$X \sim B(n, p).$$

Table 5 at the back of the book lists distribution functions of binomial random variables for various choices of n and p.

We first derive some basic properties of the binomial distribution. (Note that although we have called the function $f(x)$ in Equation (4–2) a density function, we have not shown that it is.)

Theorem 4–6

a. The binomial density function given in Equation (4–2) is a density function for all positive integers n and p in the interval $0 \leqq p \leqq 1$.
b. If $X \sim B(n, p)$, then

$$\mu = EX = np, \qquad \sigma^2 = \mathrm{var}(X) = np(1-p),$$
$$M(t) = Ee^{Xt} = (1 - p + pe^t)^n, \qquad -\infty < t < \infty.$$

Proof

a. Note that $f(x) \geqq 0$. By the binomial theorem,

$$\sum_{x=0}^{n} \begin{bmatrix} n \\ x \end{bmatrix} p^x(1-p)^{n-x} = (p + 1 - p)^n = 1,$$

so that $f(x)$ is a density function.
b. By the binomial theorem again,

$$\begin{aligned} M(t) = Ee^{tX} &= \sum_{x=0}^{n} e^{tx} \begin{bmatrix} n \\ x \end{bmatrix} p^x(1-p)^{n-x} \\ &= \sum_{x=0}^{n} \begin{bmatrix} n \\ x \end{bmatrix} (pe^t)^x(1-p)^{n-x} = (pe^t + 1 - p)^n. \end{aligned}$$

Therefore,

$$\psi(t) = \log[M(t)] = n\, \log(pe^t + 1 - p), \quad \psi'(t) = \frac{npe^t}{pe^t + 1 - p}, \quad \mu = \psi'(0) = np,$$

$$\psi''(t) = \frac{npe^t(pe^t + 1 - p) - npe^t pe^t}{(pe^t + 1 - p)^2}, \quad \sigma^2 = \psi''(0) = np - np^2 = np(1-p). \quad \square$$

If $X \sim B(1, p)$, we say that X has a *Bernoulli distribution* with mean p. In this case,

$$f(x) = p^x(1-p)^{1-x} = \begin{cases} p & \text{if } x = 1 \\ 1 - p & \text{if } x = 0 \end{cases}, \quad EX = p, \quad \mathrm{var}(X) = p(1-p).$$

A *Bernoulli trial* is an experiment which has only two possible (incompatible) outcomes, which we shall label "success" and "failure." For example, if we are shooting baskets, we could call it a success when we make the basket and a failure when we miss it. If we have a collection of voters in an election, we could call it a success if a voter votes for candidate A and a failure if the voter does not. If we have a population of patients, we could call it a success when a patient dies and a failure when the patient lives. (This is, of course, contrary to what we might expect in such a case. But although it is traditional to use the words "success" and "failure" in describing the outcomes of a Bernoulli trial, we need not mean that the successful outcome is better than the outcome that is a failure.) In general, let $X = 1$ if the outcome of a Bernoulli trial is a success and $X = 0$ if it is a failure. Then X is a Bernoulli random variable with mean p equal to the probability of a success.

The primary application of the binomial distribution is to Bernoulli trials, as given in the next theorem.

Theorem 4–7. Suppose n independent Bernoulli trials are performed, each with probability p of success. Let X be the number of successes. Then

$$X \sim B(n,p).$$

Proof. To illustrate the proof, assume that $n = 5$. We find $P(X = 2)$. One way to get exactly two successes is to get two successes followed by three failures. Since the trials are independent, this event has probability

$$P(SSFFF) = P(S)P(S)P(F)P(F)P(F) = p^2(1-p)^3.$$

Another way to get two successes is to get $FSFSF$. This event has probability

$$P(FSFSF) = P(F)P(S)P(F)P(S)P(F) = p^2(1-p)^3$$

also. In fact, any ordering which has two successes and three failures has probability $p^2(1-p)^3$. Now, there are $\left[\begin{smallmatrix}5\\2\end{smallmatrix}\right]$ possible orderings of two S's and three F's. (We can see this by choosing the two places out of five in which to put the two successes.) Finally, the orderings are disjoint (since we cannot get both $SSFFF$ and $FSFSF$). Therefore, the probability that we get two successes and three failures is

$$f(5) = P(SSFFF) + P(FSFSF) + \cdots = \begin{bmatrix}5\\2\end{bmatrix} p^2(1-p)^3.$$

For the general problem in which X is the number of successes in n independent trials with probability p of success, note first that X could be 0 or $1, \ldots,$ or n, so that the range of X is $0, 1, \ldots, n$. Now, $X = k$ means that we get k successes and $n - k$ failures, and the probability of a particular order of successes and failures which has k successes and $n - k$ failures is

$$p^k(1-p)^{n-k}.$$

Since there are $\left[\begin{smallmatrix}n\\k\end{smallmatrix}\right]$ possible orderings of k successes and $n - k$ failures, the probability of k successes and $n - k$ failures is

$$f(k) = \begin{bmatrix}n\\k\end{bmatrix} p^k(1-p)^{n-k}. \quad \square$$

We see, therefore, that if $X \sim B(n,p)$ in a setting of Bernoulli trials, then p is the probability of a success on each trial and n is the number of trials.

Corollary 4–7. Suppose we take a sample of size n with replacement from a population of size N of which pN are successes and $(1-p)N$ are failures. Then

$$X \sim B(n,p).$$

Proof. The draws are independent when we sample with replacement, and on each draw the probability of a success is p. □

Example 4–4

Suppose we draw 10 cards with replacement from an ordinary deck. What is the probability that we draw exactly four hearts? What is the probability of at least four hearts? To answer these questions, let X be the number of hearts drawn. Then $X \sim B(10, .25)$. Therefore,

$$P(X=4) = F(4) - F(3) = .922 - .776 = .146,$$

$$P(X \geqq 4) = 1 - F(3) = 1 - .776 = .224,$$

using Table 5 in the back of the book.

Although one important use of the binomial distribution is in sampling with replacement, it has many other uses, as the next example indicates.

Example 4–5

A student is taking a multiple choice exam with 25 questions, each of which has five answers. If the student guesses randomly and independently at all the questions, what is the probability that he or she will get at most eight questions correct? What is the probability that the student will get between four and seven questions correct? To answer these questions, let X be the number of questions the student gets right. There are 25 questions, and the probability of a correct answer on each question is .2. Therefore, $X \sim B(25, .2)$, and

$$P(X \leqq 8) = F(8) = .953, \qquad P(4 \leqq X \leqq 7) = F(7) - F(3) = .891 - .234 = .657.$$

The tables of the binomial distribution are quite lengthy, since we need a different table for each choice of n and p. Therefore, the following theorem is quite useful.

Theorem 4–8. Let $X_n \sim B(n,p)$. Then

$$\frac{X_n - np}{(np(1-p))^{1/2}} \xrightarrow{d} Z \sim N(0,1) \quad \text{as } n \to \infty.$$

Proof. $\mu_n = EX_n = np$, $\sigma_n^2 = \text{var}(X_n) = np(1-p)$, and

$$M_n(t) = (pe^t + 1 - p)^n = (M_1(t))^n.$$

The desired result then follows directly from Theorem 4–5. □

A rule of thumb says that the normal approximation to the binomial distribution may be used as long as

$$np \geqq 5 \quad \text{and} \quad n(1-p) \geqq 5.$$

If $X \sim B(n,p)$, then for large n,

$$\frac{X - np}{(np(1-p))^{1/2}} \stackrel{.}{\sim} N(0,1).$$

One nice fact about this approximation is that the standard normal distribution is a single distribution and needs only one table. We are therefore able to approximate many different binomial distributions with the single standard normal distribution.

Before considering an example with the normal approximation to the binomial, let us discuss the continuity correction which we make to compensate for the fact that the binomial distribution is discrete and the normal distribution is continuous. The binomial distribution must take on integer values, so that it puts no mass on the intervals between integers. However, the normal distribution does put mass there. When we use the continuity correction, we replace the event

$$X = k$$

for the binomial random variable with the event

$$k - .5 < X < k + .5.$$

For example, we replace $P(15 < X < 20)$ with $P(15.5 < X < 19.5)$ and $P(15 \leqq X \leqq 20)$ with $P(14.5 < X < 20.5)$. Note that in the first interval we don't want to include 15 or 20, but in the second we do. (Note also that once we have made the correction, it does not matter whether we use $<$ or $\leqq$, since $P(X = 20.5) = 0$ for both the binomial distribution and the normal approximation.) In a similar manner,

$$P(X = 15) = P(14.5 < X < 15.5).$$

Example 4–6

In Example 4–5, we had $X \sim B(25, .2)$. Then $np = 5$, so this example is on the lower boundary of where the normal approximation may be used. Since

$$\mu = 25(.2) = 5, \qquad \sigma^2 = 25(.2)(.8) = 4, \qquad \sigma = 2,$$

it follows that

$$\frac{X-5}{2} \stackrel{.}{\sim} N(0,1),$$

and

$$P(X \leqq 8) = P(X \leqq 8.5) = P\left(\frac{X-5}{2} \leqq \frac{8.5-5}{2}\right) \doteq P(Z \leqq 1.75) = .9599,$$

$$P(4 \leqq X \leqq 7) = P(3.5 \leqq X \leqq 7.5) = P\left(\frac{3.5-5}{2} \leqq \frac{X-5}{2} \leqq \frac{7.5-5}{2}\right)$$

$$\doteq P(-.75 \leqq Z \leqq 1.25) = .6678.$$

In Example 4–5, we found that the exact probabilities for these events are .953 and .657. We see here, then, that the normal approximation to the binomial is fairly accurate even for n quite small.

Example 4–7

Many people cancel their reservations at hotels at the last minute. Therefore, most hotels overbook when possible. That is, they make more reservations than they have rooms. A particular hotel has 90 rooms, and, on the average, 20% of the people with reservations cancel them. Suppose that for a particular night the hotel has 100 reservations. What is the probability that more people will show up with reservations than they have rooms for? What is the probability that the people with reservations will fill at most $\frac{5}{6}$ of the hotel? What is the probability that exactly 90 guests show up? To answer these questions, let X be the number of people with reservations who show up. Then there are 100 people with reservations, and each has probability .8 of showing up. Therefore, $X \sim B(100, .8)$. Hence,

$$\mu = 100(.8) = 80, \qquad \sigma^2 = 100(.8)(.2) = 16, \qquad \sigma = 4.$$

Consequently,

$$\frac{X - 80}{4} \mathrel{\dot{\sim}} N(0, 1),$$

and

$$P(X > 90) = P(X \geqq 90.5) = P\left(\frac{X - 80}{4} \geqq \frac{90.5 - 80}{4}\right)$$
$$\doteq P(Z \geqq 2.625) = 1 - N(2.625) = .0044.$$

The probability that the people fill at most $\frac{5}{6}$ of the 90 rooms is

$$P(X \leqq 75) = P(X \leqq 75.5) = P\left(\frac{X - 80}{4} \leqq \frac{75.5 - 80}{4}\right)$$
$$\doteq P(Z \leqq -1.125) = 1 - N(1.125) = .13.$$

The probability of exactly 90 guests is

$$P(X = 90) = P(89.5 < X < 90.5) = P\left(\frac{89.5 - 80}{4} < \frac{X - 80}{4} < \frac{90.5 - 80}{4}\right)$$
$$\doteq P(2.375 < Z < 2.625) = .9956 - .9912 = .0044.$$

There is another approximation for the binomial distribution which is quite useful. We say that a discrete random variable X has a *Poisson distribution* with parameter m if X has density function

$$f(x) = \frac{m^x e^{-m}}{x!}, \qquad x = 0, 1, 2 \ldots.$$

(In Section 4.3 $f(x)$ is shown to be a density function, and other properties are derived which we do not need here.) Distribution functions of Poisson random variables for various choices of m are given in Table 6 at the back of the book.

Theorem 4–9

a. Let $X_n \sim B(n, a/n)$. Then

$$X_n \xrightarrow{d} X \sim P(a) \quad \text{as } n \to \infty.$$

b. Let $X_n \sim B(n, 1 - a/n)$. Then

$$n - X_n \xrightarrow{d} X \sim P(a) \quad \text{as } n \to \infty.$$

Proof

a. By Theorem 4–1, we need to show that the density function $f_n(x)$ of X_n converges to the density function $f(x)$ of X as $n \to \infty$ (but x is fixed). Now, X_n has density function

$$f_n(x) = \begin{bmatrix} n \\ x \end{bmatrix} \left(\frac{a}{n}\right)^x \left(1 - \frac{a}{n}\right)^{n-x} = \frac{n(n-1)\dots(n-x+1)}{n^x} \frac{a^x(1-a/n)^n}{x!}\left(1 - \frac{a}{n}\right)^{-x}.$$

Since $(1 - a/n) \to 1$, it follows that $(1 - a/n)^{-x} \to 1$ (since x is fixed). Now,

$$\left(1 - \frac{a}{n}\right)^n \to e^a.$$

(See Theorem A.9 of the Appendix.) Finally,

$$\frac{n(n-1)\dots(n-x+1)}{n^x} = 1\left(1 - \frac{1}{n}\right)\dots\left(1 - \frac{x-1}{n}\right) \to 1,$$

since x is fixed. Therefore,

$$f_n(x) \to \frac{a^x e^{-a}}{x!},$$

which is the density of $X \sim P(a)$.

b. The proof of this part of the theorem is left as an exercise. □

In part a of Theorem 4–9, we often let $p = a/n$ and say that if $X \sim B(n,p)$ and n is large but p is small, then

$$X \mathrel{\dot{\sim}} P(np).$$

Similarly, when $X \sim B(n,p)$ and both n and p are large, we say that

$$n - X \mathrel{\dot{\sim}} P(n(1-p)).$$

When we use the Poisson approximation to the binomial distribution, we do not need to make a continuity correction, since both distributions are discrete. If n is large, $np \geqq 5$, and $n(1-p) \geqq 5$, we use the normal approximation to the binomial distribution. If n is large but $np < 5$, we use the Poisson approximation in part a of Theorem 4–9, and if n is large but $n(1-p) < 5$, we use the Poisson approximation in part b of the theorem.

Example 4–8

An insurance company has 10,000 policies, each with probability .0004 of a claim. What is the probability that the company has at least nine claims? What is the probability that it has not more than three claims? Let X be the number of claims. Then $X \sim B(10{,}000, .0004)$. In this case $np = 4$, so that we cannot use the normal approximation. Instead we note that

$$X \mathrel{\dot{\sim}} P(4).$$

Therefore,

$$P(X \geqq 9) = 1 - F(8) \doteq 1 - .979 = .021, \qquad P(X \leqq 3) = F(3) \doteq .433,$$

using the distribution function for a Poisson approximation with parameter 4 given in Table 6 at the back of the book.

We now prove a theorem about the binomial distribution which we shall need later.

Theorem 4–10. Let $X_1, \ldots, X_k$ be independent, and let $X_i \sim B(n_i, p)$. Let $V = \sum_i X_i$. Then $V \sim B(N, p)$, where $N = \sum_i n_i$.

Proof. X_i has moment-generating function

$$M_i(t) = (pe^t + (1-p))^{n_i}.$$

By Theorem 3–16, V has moment-generating function

$$M_V(t) = \prod_i M_i(t) = \prod_i (pe^t + (1-p))^{n_i} = (pe^t + (1-p))^{\Sigma_i n_i} = (pe^t + (1-p))^N,$$

which is the moment-generating function of a $B(N, p)$ distribution. □

Exercises—A

1. A basketball player makes 80% of his free throws.
   - **(a)** If he shoots 10 free throws, what is the probability that he makes at least 8? exactly 8? at most 8?
   - **(b)** If he shoots 225, what is the probability that he makes at least 190? at most 190? exactly 190?
2. A certain operation is successful 75% of the time.
   - **(a)** If it is performed 10 times, what is the probability that it is successful every time? at least 8 times? exactly 8 times?
   - **(b)** If it is performed 48 times, what is the probability that it is successful at most 34 times? exactly 34 times?
3. A seed manufacturer guarantees that at least 95% of the seeds in a package will germinate. If each package contains 100 seeds and the probability that each seed germinates is .97, what is the probability that a package violates the guarantee?
4. An airline sells 300 tickets for a plane which only has 296 seats. Suppose that, on the average, 2% of the passengers cancel their seats. Find the probability that more passengers show up than there are seats.

Exercises—B

1. Let $X \sim B(25, .5)$.
   - **(a)** Find $P(10 \leqq X \leqq 13)$ and $P(X = 10)$ using Table 5 in the back of the book.
   - **(b)** Find $P(10 \leqq X \leqq 13)$ and $P(X = 10)$ using the normal approximation.
2. Let $U_1, \ldots, U_{10}$ be independently uniformly distributed on $(0, 1)$.
   - **(a)** Find the probability that at least five of the U_i are at least .6.
   - **(b)** Find the probability that at most four of the U_i are in the interval $(.25, .75)$.
3. Let X and Y be independent, $X \sim B(2, .4)$, and $Y \sim B(6, .2)$.
   - **(a)** Find $P(X = 1, Y = 1)$.
   - **(b)** Find $P(X = Y)$.
4. Let $X \sim B(5, .5)$.
   - **(a)** Find the probability that X is even.
   - **(b)** Find the conditional probability that X is even given that $X < 5$.

5. Let $X \sim B(n, .5)$.
   (a) Find the largest n such that $P(X \leqq 2) > .6$.
   (b) Find the smallest n such that $P(X \geqq 4) > .4$.

6. Prove part b of Theorem 4–9.

7. Let $X \sim B(15, .4)$. What is the median of X?

8. Let X_1 and X_2 be independent, $X \sim B(1, .3)$, and $Y \sim B(1, .2)$. Find the density of $V = X_1 + X_2$. Is it a binomial density? (Note that the moment-generating function approach is not helpful in this problem.)

Exercises—C

1. Let $X \sim B(n, p)$.
   (a) Find EX using the definition of expectation.
   (b) Find $E(X(X-1))$ using the definition, and use this to prove again that $\text{var}(X) = np(1-p)$.

2. Let X_1 and X_2 be independent, $X_1 \sim B(4, p)$, and $X_2 \sim B(3, p)$. Without using Theorem 4–10, show that $W = X_1 + X_2 \sim B(7, p)$.

4.2.2 The Geometric and Negative Binomial Distributions

In this section, we introduce a family of distributions called negative binomial distributions. Since the proofs are quite similar to those in the previous section, most of them are left as exercises.

Let X be a discrete random variable with density function

$$f(x) = \begin{bmatrix} r + x - 1 \\ x \end{bmatrix} p^r (1-p)^x, \qquad x = 0, 1, \ldots,$$

where r is a positive integer and p is a number such that $0 < p < 1$. Then X has a *negative binomial distribution* with parameters r and p, written

$$X \sim NB(r, p).$$

Note that we get a different negative binomial distribution for every such r and p.

Theorem 4–11 states some elementary facts about the negative binomial distribution.

Theorem 4–11

a. The negative binomial density function is a density function for all positive integers r and for all p in the interval $0 < p < 1$.

b. Let $X \sim NB(r, p)$. Then

$$\mu = EX = \frac{r(1-p)}{p}, \qquad \sigma^2 = \text{var}(x) = \frac{r(1-p)}{p^2},$$

$$M(t) = Ee^{Xt} = \frac{p^r}{(1-(1-p)e^t)^r}, \qquad t < -\log(1-p).$$

Proof. See Exercise B2. □

If $X \sim NB(1, p)$, we say that X has a *geometric distribution* with parameter p. In this case,

$$f(x) = p(1-p)^x, \qquad x = 0, 1, \ldots, \qquad EX = \frac{1-p}{p}, \qquad \text{var}(X) = \frac{1-p}{p^2}.$$

The main application of the geometric and negative binomial distributions is to Bernoulli trials.

Theorem 4–12. Consider a sequence of independent Bernoulli trials with constant probability p of success.

a. Let X be the number of failures before the first success. Then X has a geometric distribution with parameter p.
b. Let Y be the number of failures before the rth success. Then $Y \sim NB(r, p)$.

Proof

a. X can take on any nonnegative integer value, so the range of X is $0, 1, \ldots$. In addition, $X = x$ if and only if there are x failures followed by a success. But the probability of this event is

$$(1-p)(1-p)\ldots(1-p)p = (1-p)^x p,$$

and therefore, X has a geometric distribution.

b. The range of Y is $0, 1, \ldots$, and $Y = y$ if and only if there are y failures and $r-1$ successes in the first $y+r-1$ trials and a success on the $(y+r)$th trial. Now, the number of failures on the first $y+r-1$ trials is a binomial random variable with parameters $y+r-1$ and $1-p$. Therefore,

$$P(Y = y) = P(y \text{ failures in first } y+r-1 \text{ trials})P(\text{success on } (y+r)\text{th trial})$$
$$= \left[\begin{bmatrix} y+r-1 \\ y \end{bmatrix}(1-p)^y p^{r-1}\right]p = \begin{bmatrix} y+r-1 \\ y \end{bmatrix} p^r (1-p)^y, \qquad y = 0, 1, \ldots$$

(since the $(y+r)$th trial is independent of the first $y+r-1$ trials). Hence, $Y \sim NB(r, p)$. □

We see, then, that in a Bernoulli-trials setting, when $Y \sim NB(r, p)$, r represents the number of successes we need to get, p is the probability of a success on each trial, and Y is the number of failures before the rth success.

Now, let Y be the number of failures before the rth success, and let W be the number of trials which have occurred by the rth success. Then

$$W = Y + r \quad \text{and} \quad EW = EY + r = \frac{r(1-p)}{p+r} = \frac{r}{p}.$$

In particular, the expected number of trials which have occurred when the first success occurs is $1/p$, which is an answer we might have guessed. (If the probability of success is $\frac{1}{4}$, then we would expect four trials till we get a success.)

Example 4–9

Suppose that a basketball player makes 80% of her free throws and that the attempts are independent. Let X be the number she misses before she makes one. Then X has a geometric distribution with parameter .8, and

$$EX = \frac{.2}{.8} = .25.$$

Let $W = X + 1$ be the total number of free throws she has tried when she makes her first one. Then

$$EW = EX + 1 = .25 + 1 = 1.25 = \frac{1}{.8},$$

which is the answer we might have guessed.

Now, let Y be the number of free throws she misses before she makes five. Then $Y \sim NB(5, .8)$, and

$$EY = \frac{5}{4}$$

so that she could expect to miss 1.25 free throws before making five.

Let Z be the number she shoots before making five (including the last one). Then

$$Z = Y + 5 \quad \text{and} \quad EZ = EY + 5 = 6.25 = \frac{5}{.8}.$$

Now, let Y^* be the number she makes before missing five. Then $Y^* \sim NB(5, .2)$. (To see this, think of a miss as a success. Then the probability of a success is .2, and Y^* is the number of failures before the fifth success.)

Finally, let Z^* be the number she shoots before missing five. Then

$$EY^* = 20 \quad \text{and} \quad EZ^* = EY^* + 5 = 25 = \frac{5}{.2}.$$

Example 4–10

Teams A and B are going to play a series of baseball games until one team wins three games. Suppose that the probability that team A wins a game is .6 and the games are independent. We want to find the probability that the series ends in exactly four games. Pretend that the teams will play until team A wins three games, and let X be the number of games team A would lose before it wins its third game. Similarly, let Y be the number of games team B would lose before it wins its third game. Then

$$X \sim NB(3, .6) \quad \text{and} \quad Y \sim NB(3, .4).$$

The series ends in exactly four games if either $X = 1$ or $Y = 1$ and these events are disjoint. Therefore,

$$P(4 \text{ games}) = P(X = 1) + P(Y = 1) = \begin{bmatrix} 3 \\ 1 \end{bmatrix} .6^3 .4^1 + \begin{bmatrix} 3 \\ 1 \end{bmatrix} .4^3 \cdot .6^1 = .3744.$$

Since we have not included any tables for the distribution function of the negative binomial distribution, the following theorem is quite important.

Theorem 4–13. Let $X_n \sim NB(n, p)$. Then

$$\frac{X_n - \mu_n}{\sigma_n} \xrightarrow{d} Z \sim N(0, 1) \quad \text{as } n \to \infty,$$

where $\mu_n = EX_n = n(1-p)/p$ and $\sigma_n^2 = n(1-p)/p^2$.

Proof. See Exercise B3. □

In using Theorem 4–13, it is important to remember the continuity correction since the negative binomial distribution is discrete and the normal distribution is continuous.

The following theorem is occasionally useful.

Theorem 4–14. Let X_i be independent and $X_i \sim NB(r_i, p)$. Let $V = \sum_i X_i$. Then

$$V \sim NB(R, p), \qquad R = \sum_i r_i.$$

Proof. See Exercise B4. □

We conclude the section with an interesting property of the geometric distribution.

Theorem 4–15. Let X have a geometric distribution with parameter p. Then for any positive integers j and k,

$$P(X \geqq j + k | X \geqq j) = P(X \geqq k).$$

Proof. Using the theorem on the sum of a geometric series, we see that

$$P(X \geqq j) = \sum_{x=j}^{\infty} f(x) = \sum_{x=j}^{\infty} p(1-p)^x = p(1-p)^j \sum_{y=0}^{\infty} (1-p)^y = (1-p)^j.$$

Therefore,

$$P(X \geqq j + k | X \geqq j) = \frac{P(X \geqq j + k)}{P(X \geqq j)} = \frac{(1-p)^{j+k}}{(1-p)^j} = (1-p)^k = P(X \geqq k). \quad \square$$

The property given in Theorem 4–15 is called the *lack-of-memory property* of the geometric distribution. For example, suppose a person is playing a card game with some friends. If the games are independent and his probability of winning any game is p, then the number of games he plays till he wins his first game has a geometric distribution with parameter p. Suppose he has lost the first 10 games. Then the lack-of-memory property implies that, after losing 10 games, the time till he wins his first game has the same distribution as it did when he started playing. In other words, his loss of those first 10 games has been "forgotten." (The geometric distribution is the only discrete distribution with this lack-of-memory property; see the Exercise C1.)

Exercises—A

1. Suppose I am flipping a fair coin until I get 50 heads. Find the probability that it takes me at least 110 tosses. (*Hint*: How many tails would I flip if it takes 110 tosses?)
2. Player A has probability .4 of making a shot, while player B has probability .8 of making a shot. Player A will shoot until she makes 5 shots, while player B shoots until she makes 12 shots. Which player would expect to take fewer shots?

3. Suppose I am repeatedly playing a card game which I have probability .4 of winning. Suppose further that the games are independent.
   (a) If I have lost the last four games, what is the probability that I will win the next game?
   (b) If I have won the last five games, what is the probability that I will win the next one?
4. In Example 4–10,
   (a) Find the probability that the series ends in three games. in five games.
   (b) Find the expected number of games in the series.
5. In Example 4–10, find the probability that team A wins the series.

Exercises—B

1. Suppose that $X \sim NB(1,p)$. Find the probability that X is even.
2. Prove Theorem 4–11. (*Hint*: The negative binomial theorem says that

$$\sum_{x=0}^{\infty} \begin{bmatrix} r+x-1 \\ x \end{bmatrix} a^x = (1-a)^{-r}, \text{ as long as } |a|<1.)$$

3. Prove Theorem 4–13.
4. Prove Theorem 4–14.

Exercises—C

1. Suppose that a nonnegative integer-valued random variable X satisfies $P(X \geqq j + k | X \geqq j) = P(X \geqq k)$.
   (a) Show that $P(X = j + k | X \geqq j) = P(X = k)$.
   (b) Show that $P(X = k + 1) = P(X \geqq 1)P(X = k)$.
   (c) Show that X has a geometric distribution.

4.2.3 The Hypergeometric Distribution

In Section 4.2.1, we showed that the binomial distribution can be used to solve many probability problems involving sampling with replacement. In this section, we show how to solve those problems when we are sampling without replacement.

Suppose we have a population of N objects, pN of which are labeled successes and the remaining $(1-p)N$ failures. For example, in a population of residents of a town, the successes could be the ones who watch a particular show and the failures could be the ones who don't watch the program, or the successes could be the ones who have a particular disease and the failures the ones who don't.

Suppose we draw a sample of size n without replacement from this population. Let X be the number of successes in the sample. Then X has a *hypergeometric distribution* with parameters N, p, and n, where N is the population size, n is the sample size, and p is the proportion of successes.

We first consider the range of X. We must have $0 \leqq X \leqq n$, since we are drawing at most n objects. However, we also must have $X \leqq pN$, since there are only pN possible successes. Finally, we must have $n - X \leqq N(1-p)$, since there are only $N(1-p)$ failures to draw. Therefore, the range of X is the set S of all integers x such that

$$\max(0, n - N(1-p)) \leqq x \leqq \min(n, Np).$$

From Chapter 1, the probability of getting x successes in n draws without replacement from a population with Np successes and $N(1-p)$ failures is

$$f(x) = P(X = x) = \frac{\begin{bmatrix} Np \\ x \end{bmatrix}\begin{bmatrix} N(1-p) \\ n-x \end{bmatrix}}{\begin{bmatrix} N \\ n \end{bmatrix}}, \qquad x \in S.$$

The moment-generating function $M(t)$ of a hypergeometric distribution is finite because X has a bounded range. However, there is no formula for $M(t)$: It is quite difficult to find the mean and variance of a hypergeometric distribution from the definition. In the exercises to this section, a different method is used to prove the following theorem.

Theorem 4–16. Let X have a hypergeometric distribution with parameters n, N, and p. Then

$$EX = np \quad \text{and} \quad \text{var}(X) = \frac{np(1-p)(N-n)}{N-1}.$$

Proof. See Exercise B2. □

We now give the most important theorem about the hypergeometric distribution.

Theorem 4–17. Let X_N have a hypergeometric distribution with parameters N, p, and n. Then

$$X_N \xrightarrow{d} B(n,p) \quad \text{as } N \to \infty.$$

Proof. See Exercise C1. □

Theorem 4–17 says that if the population size N is infinite, it does not matter whether we sample with or without replacement. Even if we sample with replacement, the probability of our drawing the same individual twice is zero. The theorem also implies that if the population size N is quite large, then we may approximate the hypergeometric distribution with a binomial distribution. That is, if X has a hypergeometric distribution with parameters N, p, and n, and if N is large, then

$$X \mathrel{\dot\sim} B(n,p).$$

This result is quite important, because the hypergeometric distribution is, for most purposes, quite difficult to study. For most of the remainder of this book, we shall assume that N is enough larger than n so that we may act as though we are sampling with replacement when we are in fact sampling without replacement. We shall rarely mention the hypergeometric distribution again.

Example 4–11

Consider a city with 100,000 voting people. Suppose that 60,000 of the voters watch a particular program and that we sample five people. What is the probability that we find two watching the show? Let X be the number we find watching the show. Then X has a hypergeometric distribution with population size $N = 100{,}000$, probability of success $p = .6$, and sample size $n = 5$. Therefore,

$$P(X = 2) = \frac{\begin{bmatrix} 60{,}000 \\ 2 \end{bmatrix}\begin{bmatrix} 40{,}000 \\ 3 \end{bmatrix}}{\begin{bmatrix} 100{,}000 \\ 5 \end{bmatrix}} = .23$$

(after some messy arithmetic). However, by Theorem 4–17, $X \mathrel{\dot{\sim}} B(5, .6)$. But then, by Table 5 in the back of the book, we see that

$$P(X = 2) \doteq F(2) - F(1) = .317 - .087 = .23.$$

It can be shown, under fairly general conditions, that if $X_{N,n}$ has a hypergeometric distribution with parameters N, p, and n, then

$$\frac{X_{N,n} - \mu_{N,n}}{\sigma_{N,n}} \xrightarrow{d} Z \sim N(0, 1) \qquad \text{as } n \to \infty \quad \text{and} \quad N \to \infty,$$

where $\mu_{N,n} = EX_{N,n}$ and $\sigma^2_{N,n} = \text{var}(X_{N,n})$ are as in Theorem 4–16.

Observe that the mean of both the hypergeometric distribution and the approximating binomial distributions is np, but that the standard deviation of the hypergeometric distribution is

$$[np(1 - p)]^{1/2}\left(\frac{N - n}{N - 1}\right)^{1/2},$$

whereas that of the binomial distribution is

$$[np(1 - p)]^{1/2}.$$

We often call the ratio of these standard deviations, i.e.,

$$h_{N,n} = h = \left(\frac{N - n}{N - 1}\right)^{1/2},$$

the *finite population correction*, because it is the correction we must make in the standard deviation of the normal approximation to account for the fact that we are sampling from a finite population. If $h_{N,n}$ is near one, we may use the binomial approximation to the hypergeometric distribution with confidence in its accuracy. For instance, in Example 4–11, we have

$$h = \left(\frac{99{,}995}{99{,}999}\right)^{1/2} = .99998.$$

Example 4–12

Suppose that we want to make a survey of 40,000 people in the city of Example 4–11 and want the probability that at most 23,850 of the people sampled are watching the show. Let Y be the number of people sampled who watch the show. Then Y has a hypergeometric distribution with population size $N = 100{,}000$, proportion $p = .6$, and sample size $n = 40{,}000$. In this case, the finite population correction is

$$h = \left(\frac{60{,}000}{99{,}999}\right)^{1/2} = .775.$$

Therefore, in this setting, we should not use the binomial approximation. However, we can use the normal approximation. We have

$$\mu = 40{,}000(.6) = 24{,}000, \qquad \sigma = (40{,}000(.6)(.4))^{1/2}h = 75.9,$$

$$P(Y \leqq 23{,}850) = P\left(\frac{Y - 24{,}000}{75.9} \leqq \frac{23{,}850 - 24{,}000}{75.9}\right)$$

$$\doteq P(Z \leqq -1.98) = .024.$$

Note that if we had used the normal approximation to the binomial distribution, we would have gotten $\sigma = 98$ and approximated the probability by $P(Z \leqq -1.53) = .063$. Note also that we have not used the continuity correction, because it makes nearly no difference with numbers this large.

Example 4–12 illustrates the fact that the binomial approximation may not be appropriate even when N is large. What we need, in addition, is that n/N must be small. That is, we must be taking a relatively small sample from a large population. If the sample size is an appreciable part of the population size, then many of the procedures we shall discuss in later sections may be suspect.

Exercises—A

1. Suppose a storeroom has 1,000 tires, of which 100 are defective. I take five tires for my car. What is the probability that at least four of the tires are nondefective? (Use the binomial approximation.) What is the finite population correction for this situation?
2. A telephone directory contains 30,000 names. Ten thousand of the people favor a particular piece of legislation, and the remaining 20,000 are against it. A sample of 15,000 phone owners is chosen, and the people are asked their feelings on the legislation. If they answered truthfully, what is the probability that at most 4,900 say they like the legislation? (Use the normal approximation.) What is the finite population correction for this situation?
3. Recalculate $P(Y \leqq 23{,}850)$ in Example 4–12 for both the hypergeometric and binomial cases, using the continuity correction.

Exercises—B

1. Let X have a hypergeometric distribution with parameters $N = 20, p = 5.$, and general n. For what value of n is $\text{var}(X)$ a maximum?
2. Suppose we take a sample of size n without replacement from a population with Np successes and $N(1-p)$ failures. Let $U_i = 1$ if we get a success on the ith draw and $U_i = 0$ if we get a failure on that draw. Let $X = \Sigma U_i$ be the total number of successes drawn.
   (a) Show that $EU_i = p$ and $\text{var}(U_i) = p(1-p)$. (The U_i are Bernoulli. Why?)
   (b) Show that $\text{cov}(U_i, U_j) = -p(1-p)/(N-1)$. (See Example 3–20.)
   (c) Show that $EX = np$ and $\text{var}(X) = np(1-p)(N-n)/(N-1)$.
   (Assume that $EU_i = EU_1$, $\text{var}(U_i) = \text{var}(U_1)$, and $\text{cov}(U_i, U_j) = \text{cov}(U_1, U_2)$; see Exercise C2 below.)
3. Suppose that we sample without replacement from a population of size N with proportion of successes p.
   (a) Let U be the number of trials before we get a success. Find the density function of U.
   (b) Let V be the number of failures before the rth success. Find the density function of V.

Exercises—C

1. Let X_N have a hypergeometric density with parameters N, p, and n, and let $X \sim B(n, p)$. Argue that $f_N(x)$, the density function of X_N, converges to $f(x)$, the density function of X, as $N \to \infty$ (but p, n, and x are fixed).
2. Suppose we have a sample, without replacement, of size n from a population of size N with probability of success p. Let $U_i = 1$ if the ith trial is a success and $U_i = 0$ otherwise.
   (a) Show that $P(U_2 = 1) = p$. (The event $U_2 = 1$ is the union of the events $U_1 = 1$, $U_2 = 1$ and $U_1 = 0$, $U_2 = 1$.)
   (b) Use induction to show that $P(U_j = 1) = p$ for all p.
   (c) Show that $P(U_2 = 1, U_4 = 1) = P(U_1 = 1, U_2 = 1)$.
   (d) Show that $P(U_i = 1, U_j = 1) = P(U_1 = 1, U_2 = 1)$.

4.3 DISTRIBUTIONS ASSOCIATED WITH THE POISSON PROCESS

4.3.1 The Poisson Distribution

A discrete random variable X has a *Poisson distribution* with mean $m > 0$, denoted

$$X \sim P(m),$$

if X has density function

$$f(x) = \frac{e^{-m} m^x}{x!}, \qquad x = 0, 1, 2, \ldots.$$

Distribution functions of Poisson random variables for various choices of m are given in Table 6 at the back of the book.

We now set out the basic properties of the Poisson distribution. Note that although we have called m the mean of the Poisson distribution, we have not actually shown that it is.

Theorem 4–18

a. The Poisson density function $f(x) = e^{-m} m^x / x!$ is a density function for all $m > 0$.
b. Let $X \sim P(m)$. Then

$$\mu = EX = m, \qquad \sigma^2 = \text{var}(X) = m, \qquad M(t) = Ee^{Xt} = \exp(me^t - m).$$

Proof. We use the fact that

$$e^s = \sum_{x=0}^{\infty} \frac{s^x}{x!}.$$

a. Note that $f(x) \geqq 0$. Also,

$$\sum_{x=0}^{\infty} f(x) = \sum_{x=0}^{\infty} \frac{e^{-m} m^x}{x!} = e^{-m} \sum_{x=0}^{\infty} \frac{m^x}{x!} = e^{-m} e^m = 1.$$

b. To take the last item first,

$$M(t) = \sum_{x=0}^{\infty} (e^{tx})\left(\frac{e^{-m} m^x}{x!}\right) = e^{-m} \sum_{x=0}^{\infty} \frac{(me^t)^x}{x!} = e^{-m} \exp(me^t) = \exp(me^t - m).$$

Now, $\psi(t) = \log[M(t)] = me^t - m$. Therefore,

$$\psi'(t) = me^t, \qquad \mu = \psi'(0) = m, \qquad \psi''(t) = me^t, \qquad \sigma^2 = \psi''(0) = m. \quad \square$$

Note that for the Poisson distribution, the mean and variance are the same and are equal to the parameter m.

We next derive the normal approximation for the Poisson distribution.

Theorem 4–19. Let $X_n \sim P(na)$. Then

$$\frac{X_n - na}{(na)^{1/2}} \xrightarrow{d} Z \sim N(0, 1).$$

Proof. $\mu_n = EX_n = na$, $\operatorname{var}(X_n) = na$, and

$$M_n(t) = \exp[(na)(e^t - 1)] = (\exp[a(e^t - 1)])^n = (M_1(t))^n.$$

The desired result then follows from Theorem 4–5. $\square$

We often let $m = na$ and state Theorem 4–19 as follows: If $X \sim P(m)$ for large m, then

$$\frac{X - m}{m^{1/2}} \mathrel{\dot\sim} N(0, 1).$$

In Section 4.2.1, we discussed the use of the Poisson distribution for approximating binomial probabilities. We now discuss another application of the Poisson distribution.

Suppose we have a situation in which rare events (such as terrorist attacks) are occurring over time. We say that these occurrences are a *Poisson process* with rate λ if they satisfy the following conditions:

1. The number of occurrences in an interval I is independent of the number of occurrences in any other interval that is disjoint from I.
2. The distribution of the number of occurrences in an interval is dependent only on the length of that interval.
3. The probability of an occurrence in a short interval is approximately equal to λ times the length of the interval.
4. The probability of two or more occurrences in a very short interval is approximately zero.

(Conditions 3 and 4 are stated more precisely in the next section.) The Poisson process is used as a model for rare events such as emissions from an atom, telephone calls coming into a switchboard, accidents on a particular stretch of highway, etc.

Theorem 4–20. Consider a Poisson process with rate λ. Let X be the number of occurrences in a particular interval of length t. Then $X \sim P(\lambda t)$.

Proof. See the next section. □

Example 4–13

Suppose that accidents on a particular road occur as a Poisson process at a rate of two per month. Suppose we want to know the probability of 5 accidents in a 2-month period and the probability of 45 accidents in an 18-month period. Since both of these events imply that there are $2\frac{1}{2}$ accidents per month, many students feel that they should have the same probability. However, they do not. For let X be the number of accidents in two months. Then $X \sim P(4)$. Therefore,

$$P(5 \text{ accidents in 2 months}) = P(X = 5) = F(5) - F(4) = .785 - .629 = .156.$$

Similarly,

$$P(3 \leqq X \leqq 5) = F(5) - F(2) = .785 - .238 = .547.$$

Now, let Y be the number of accidents in 18 months. Then $Y \sim P(36)$, which is not in Table 6. We therefore use the normal approximation. Note first that

$$\mu = 36, \qquad \sigma^2 = 36, \qquad \frac{Y-36}{6} \mathrel{\dot\sim} N(0, 1).$$

Therefore,

$$\begin{aligned} P(45 \text{ accidents in 18 months}) &= P(Y = 45) = P(44.5 \leqq Y \leqq 45.5) \\ &= P\left(\frac{44.5-36}{6} \leqq \frac{Y-36}{6} \leqq \frac{45.5-36}{6}\right) \\ &\doteq P(1.42 \leqq Z \leqq 1.58) = .9429 - .9222 = .0207. \end{aligned}$$

Similarly,

$$\begin{aligned} P(30 < X < 40) = P(30.5 < X < 39.5) &= P\left(\frac{30.5-36}{6} < \frac{X-36}{6} < \frac{39.5-36}{6}\right) \\ &\doteq P(-.92 < Z < .58) = .7190 + .8212 - 1 = .5402. \end{aligned}$$

In the Poisson process, it is not necessary that "time" actually be time, as the following example indicates.

Example 4–14

Suppose that typos on a page occur as a Poisson process with rate .5 per page. Suppose also that we want to find the probability of at least 11 typos in 18 pages. Let X be the number of typos in 18 pages. Then $(.5)(18) = 9$, so that $X \sim P(9)$. Therefore,

$$P(\text{at least 11 typos in 18 pages}) = P(X \geqq 11) = 1 - F(10) = .294.$$

We compute the normal approximation to this probability. Since $\mu = 9$ and $\sigma = 3$, we have

$$P(X \geqq 11) = P(X \geqq 10.5) = P\left(\frac{X-9}{3} \geqq \frac{10.5-9}{3}\right) \doteq P(Z \geqq .5) = .3085.$$

Thus, the normal approximation is not too far off even when the mean of the Poisson distribution is as small as nine.

The Poisson distribution also arises when we are counting occurrences of some type in space, for example the number of stars in a given part of the galaxy, or the number of particles of pollutant in a volume of water. We say that these occurrences are a *spatial Poisson process* with rate λ if the following assumptions are satisfied:

1. The number of occurrences in a set S is independent of the number of occurrences in any other set that is disjoint from S.
2. The distribution of the number of occurrences in a set depends only on the volume of that set.
3. The probability of one occurrence in a small set is approximately equal to λ times the volume of that set.
4. The probability of more than one occurrence in a small set is approximately zero.

Theorem 4–21. Consider a spatial Poisson process with rate λ. Let X be the number of occurrences in a set with volume v. Then $X \sim P(\lambda v)$.

Proof. See the exercises in the next section. □

Example 4–15

Suppose the number of particles in a sample of water is a spatial Poisson process with rate 100 particles per cubic inch. Suppose we want to find the probability that there are at least two particles in a volume of .02 cubic in. Let X be the number of particles in that volume. Then X has a Poisson distribution with mean $100(.02) = 2$. Therefore,

$$P(X \geqq 2) = 1 - F(1) = 1 - .406 = .594.$$

Suppose we also want to find the probability of at least 125 particles in a cubic inch. Let Y be the number of particles per cubic inch. Then Y has a Poisson distribution with mean $100(1) = 100$. Therefore,

$$P(Y \geqq 125) = P(Y \geqq 124.5) = P\left(\frac{Y - 100}{10} \geqq \frac{124.5 - 100}{10}\right)$$

$$\doteq 1 - N(2.45) = 1 - .9929 = .0071.$$

We conclude the section with a theorem which will be useful later.

Theorem 4–22. Let X_i be independent and $X_i \sim P(m_i)$. Let $V = \sum_i X_i$. Then

$$V \sim P(M) \quad \text{and} \quad M = \textstyle\sum_i m_i.$$

Proof. From Chapter 3,

$$M_V(t) = \prod_i M_i(t) = \prod_i \exp(m_i e^t - m_i) = \exp\Big(\sum_i (m_i e^t - m_i)\Big) = \exp(Me^t - M),$$

which is the moment-generating function of a Poisson distribution with mean M. □

Exercises—A

1. Suppose that telephone calls come into a switchboard as a Poisson process at a rate of two per minute.
   (a) Find the probability of at least 15 calls in 6 minutes and the probability of at least 5 calls in 2 minutes.
   (b) Find the probability of between two and eight calls inclusive in three minutes.
   (c) Find the probability of between 50 and 60 calls inclusive in 32 minutes.
2. Suppose that terrorist attacks obey a Poisson process with a rate of one attack every two months.
   (a) Find the probability of at least eight attacks in a year and the probability of at most eight attacks in a year.
   (b) Find the probability of at least 6 attacks and at most 10 attacks in a year.
   (c) Find the probability of at least 20 but not more than 30 attacks in 50 months.
   (d) Find the probability that in three of the next four years there will be fewer than five attacks. (Think binomial!)
3. Defects occur in a roll of wallpaper at the rate of .25 per foot.
   (a) Find the probability that a 20-foot roll has no defects.
   (b) Find the probability that a 20-foot roll has between one and three defects inclusive.
   (c) Find the probability of at most 20 defects in 100 ft.
   (d) Find the probability that each of five 20-foot rolls has at most four defects.
4. Suppose the occurrence of stars in a galaxy is a spatial Poisson process with rate 1 per 1,000 cubic light years (so that $\lambda = .001$).
   (a) Find the probability of at least six stars in a given cube with side 20 light years.
   (b) Find the probability of at least 70 stars in a cube with side 40 light years.
5. Men and women arrive at a bank as independent Poisson processes, men with rate 10 per hour and women with rate 6 per hour. Find the probability that more than 70 men and women arrive in a four-hour period. (Use Theorem 4–22.)

Exercises—B

1. Suppose that $X \sim P(10)$ and the conditional distribution of Y given X is $Y|X \sim B(X, .4)$.
   (a) Find the joint density of X and Y.
   (b) Find the marginal density of Y.
2. Let $X \sim P(10)$ and $Y \sim P(8)$, with X and Y independent. Let $U = X + Y$.
   (a) How is U distributed?
   (b) Find $P(X = 2|U = 5)$.
   (c) Show that $X|U \sim B\left(U, \frac{5}{9}\right)$.

Exercises—C

1. Let $X \sim P(m)$. Without using moment-generating functions, find EX and $EX(X - 1)$, and verify the formulas given in the text for EX and $\text{var}(X)$.
2. Consider a Poisson process with rate $\lambda = 2$. Let X be the time till the first arrival.
   (a) Find $P(X > .8)$. (*Hint*: Let Y be the number of arrivals before .8. Then $P(X > .8) = P(Y = 0)$. Why?)
   (b) Show that X has density function $2e^{-2x}, x > 0$.
3. Let X and Y be independent, $X \sim P(m)$, and $Y \sim P(p)$. Let $U = X + Y$. Without using moment-generating functions or Theorem 4–22, show that $U \sim P(m + p)$.

* 4.3.2 Derivation of the Poisson Process

In this section, we derive the Poisson distribution from the assumptions of the Poisson process. In stating these assumptions, it is helpful to use "little o" notation. We say that a function $g(h) = o(h^a)$ if

$$\lim_{h\to 0} \frac{g(h)}{h^a} = 0,$$

that is, if $g(h)$ goes to zero faster than h^a does.

Suppose we have a process in which some objects are occurring randomly over time. For all a and $h > 0$, let $X(a, h)$ be the number of occurrences in the interval $(a, a + h] = \{t\colon a < t \leqq a + h\}$. (Note that $X(a, h)$ is a random variable.) Then these occurrences are a *Poisson process* with *rate* λ if

1. $X(a, h)$ and $X(a^*, h^*)$ are independent random variables whenever $(a, a + h]$ and $(a^*, a^* + h^*]$ are disjoint.
2. The distribution of $X(a, h)$ depends only on h.
3. $P(X(a, h) = 1) = \lambda h + o(h)$.
4. $P(X(a, h) > 1) = o(h)$.

These assumptions are a more precise version of the ones given in the previous section for the Poisson process.

Theorem 4–23. Suppose that the preceding assumptions are true. Then

$$X(a, t) \sim P(\lambda t).$$

Proof. Let $P_n(t) = P(X(a, t) = n)$. (See assumption 2.) We need to show that

$$P_n(t) = \frac{e^{-\lambda t}(\lambda t)^n}{n!}.$$

From assumptions 3 and 4, we obtain

$$\lim_{h\to 0} \frac{P_1(h)}{h} = \lambda, \qquad \lim_{h\to 0} \sum_{j=2}^{\infty} \frac{P_j(h)}{h} = 0,$$

$$\lim_{h\to 0} \frac{1 - P_0(h)}{h} = \lim_{h\to 0} \left(\frac{P_1(h)}{h} + \sum_{j=2}^{\infty} \frac{P_j(h)}{h} \right) = \lambda.$$

Note that $P_0(0) = 1$ and $P_n(0) = 0$ if $n > 0$. (That is, the probability of no arrivals in an interval of length zero is one, and the probability of n arrivals in that length is zero. While these assertions seem intuitively obvious, they can be derived from assumptions 1–4—see Exercise B3.) Now, $P_0(t + h)$ is the probability of no arrivals in the interval zero to $t + h$, which is the same as no arrivals in the interval zero to t and no arrivals in the interval t to $t + h$. Since these intervals are disjoint, these events are independent. Therefore, using assumption 2,

$$P_0(t+h) = P_0(t)P_0(h),$$

$$\frac{P_0(t+h) - P_0(t)}{h} = \frac{P_0(t)(P_0(h)-1)}{h},$$

$$P_0'(t) = P_0(t)\lim_{h\to 0}\frac{P_0(h)-1}{h} = -\lambda P_0(t).$$

Hence,

$$-\lambda = \frac{P_0'(t)}{P_0(t)} = \frac{d}{dt}\log[P_0(t)], \qquad \log[P_0(t)] = -\lambda t + c.$$

The condition $P_0(0) = 1$ implies that $c = 0$ and

$$P_0(t) = e^{-\lambda t}.$$

Now, the event consisting of n arrivals in the interval zero to $t + h$ is a disjoint union of the sets A_j in which there are j arrivals in the interval zero to t and $n - j$ arrivals in the interval t to $t + h$. Since these intervals are disjoint, by assumptions 1 and 2 it follows that

$$P_n(t+h) = \sum_{j=0}^{n} P_{n-j}(t)P_j(h) = P_n(t) - P_n(t)(1 - P_0(h))$$
$$+ P_{n-1}(t)P_1(h) + \sum_{j=2}^{n} P_{n-j}(t)P_j(h).$$

Now,

$$0 \leqq \sum_{j=2}^{n}\frac{P_{n-j}(t)P_j(h)}{h} \leqq \sum_{j=2}^{\infty}\frac{P_j(h)}{h} \to 0 \quad \text{as } h \to 0.$$

Hence,

$$\begin{aligned}P_n'(t) &= \lim_{h\to 0}\frac{(P_n(t+h) - P_n(t))}{h} \\ &= -P_n(t)\lim_{h\to 0}\frac{(1-P_0(h))}{h} + P_{n-1}(t)\lim_{h\to 0}\frac{P_1(h)}{h} \\ &\quad + \lim_{h\to 0}\frac{\sum_{j=2}^{n} P_{n-j}(t)P_j(h)}{h} = -\lambda P_n(t) + \lambda P_{n-1}(t) + 0.\end{aligned}$$

Therefore,

$$\lambda e^{\lambda t}P_{n-1}(t) = e^{\lambda t}(P_n'(t) + \lambda P_n(t)) = \frac{d}{dt}(e^{\lambda t}P_n(t)). \tag{4–3}$$

Hence,

$$\frac{d}{dt}(e^{\lambda t}P_1(t)) = e^{\lambda t}P_0(t) = \lambda e^{\lambda t}e^{-\lambda t} = \lambda,$$

and it follows that

$$e^{\lambda t}P_1(t) = \lambda t + c \quad \text{and} \quad P_1(t) = (\lambda t + c)e^{-\lambda t}.$$

The condition $P_1(0) = 0$ implies that

$$P_1(t) = \lambda t e^{-\lambda t}.$$

Using induction with Equation (4–3), we see that

$$P_n(t) = \frac{(\lambda t)^n e^{-\lambda t}}{n!},$$

and the theorem is proved. □

By a slightly more delicate argument than this, it can be shown that assumption 2 is a consequence of the other assumptions.

Now consider a process in which we have occurrences in three-dimensional space. For any set S, let $X(S)$ be the number of occurrences in S. We say that these occurrences are a *spatial Poisson process* with *rate* λ if

1. $X(S)$ and $X(T)$ are independent whenever S and T are disjoint.
2. The distribution of $X(S)$ depends only on v, the volume of S.
3. $P(X(S) = 1) = \lambda v + o(v)$.
4. $P(X(S) > 1) = o(v)$.

Theorem 4–24. Under conditions 1–4 just given, the number of occurrences in a set S with volume v has a Poisson distribution with mean λv.

Proof. See Exercise B2.

Exercises—B

1. Finish the induction argument at the end of the proof of Theorem 4–23.

2. Prove Theorem 4–24. (Assume that $P_0(0) = 1$ and $P_n(0) = 0$ for $n > 1$.)

3. Suppose that assumptions 1–4 just before Theorem 4–23 are satisfied.
   (a) Show that $P_0(0) = 1$. (*Hint*: $1 \geqq P_0(0) \geqq P_0(h) = 1 - (1 - P_0(h))$ for all h.)
   (b) Show that $P_n(0) = 0, n > 0$. $(0 \leqq P_n(0) \leqq \sum_{j=n}^{\infty} P_j(h).)$

4.3.3 The Exponential and Gamma Distributions

In this section, we examine some distributions which arise as waiting times for occurrences in Poisson processes.

For all $a > 0$, we define the *gamma function*

$$\Gamma(a) = \int_0^{\infty} t^{a-1}e^{-t}\,dt.$$

It can be shown that the integral converges for all $a > 0$. We now give some basic properties of $\Gamma(a)$.

Theorem 4–25

a. If $a > 1$, then $\Gamma(a) = (a-1)\Gamma(a-1)$.

b. $\Gamma(k) = (k-1)!$ for all positive integers k.

c. $\Gamma(\frac{1}{2}) = \pi^{1/2}$. $\Gamma((2k+1)/2) = (2k-1)(2k-3)\ldots(1)\pi^{1/2}/2^k$ for all positive integers k.

d. For any $c > 0$, $\int_0^\infty x^{a-1}e^{-xc}\,dx = \Gamma(a)/c^a$.

Proof

a. For all a,

$$\lim_{x\to\infty} x^a e^{-x} = 0.$$

(See Exercise B5.) Therefore, if $a > 1$,

$$\Gamma(a) = \int_0^\infty x^{a-1}e^{-x}dx = -x^{a-1}e^{-x}\Big|_0^\infty + \int_0^\infty (a-1)x^{a-2}e^{-x}dx = 0 + (a-1)\Gamma(a-1)$$

(using integration by parts with $u = x^{a-1}$ and $dv = e^{-x}\,dx$).

b. Note that

$$\Gamma(1) = \int_0^\infty e^{-x}\,dx = 1.$$

Therefore,

$$\Gamma(k) = (k-1)\Gamma(k-1) = (k-1)(k-2)\Gamma(k-2) = \cdots = (k-1)!\Gamma(1) = (k-1)!.$$

c. Since the standard normal density function integrates to one,

$$1 = 2\int_0^\infty (2\pi)^{-1/2}e^{-z^2/2}\,dz = \left(\frac{2}{\pi}\right)^{1/2}\int_0^\infty e^{-u}(2u)^{-1/2}\,du$$

$$= \pi^{-1/2}\int_0^\infty u^{-1/2}e^{-u}\,du = \pi^{-1/2}\Gamma\left(\frac{1}{2}\right),$$

and hence, $\Gamma(\frac{1}{2}) = \pi^{1/2}$. Therefore,

$$\Gamma\left(\frac{3}{2}\right) = \frac{1}{2}\Gamma\left(\frac{1}{2}\right) = \frac{\pi^{1/2}}{2},$$

$$\Gamma\left(\frac{5}{2}\right) = \frac{3}{2}\Gamma\left(\frac{3}{2}\right) = \frac{3}{4}\Gamma\left(\frac{1}{2}\right) = \frac{3}{4}\pi^{1/2}.$$

The other results follow by induction.

d. This proof is left as an exercise. □

Theorem 4–25 provides a method that is often used to define $a!$ when a is not an integer. In many books, the notation $a!$ is used for $\Gamma(a+1)$ for any $a > 0$. We shall not, however, use such a notation in this book.

A continuous random variable X has a *gamma distribution* with parameters $a > 0$ and $b > 0$, written

$$X \sim \Gamma(a, b),$$

if X has the density function

$$f(x) = \frac{x^{a-1} \exp(-x/b)}{b^a \Gamma(a)}, \qquad x > 0.$$

We next give some elementary properties of the gamma distribution.

Theorem 4–26

a. For any $a > 0$ and $b > 0$, the function $f(x)$ just given is a density function.

b. Let $X \sim \Gamma(a, b)$. Then

$$\mu = EX = ab, \qquad \sigma^2 = \text{var}(X) = ab^2, \qquad M(t) = Ee^{Xt} = (1 - bt)^{-a}, \qquad t < b^{-1}.$$

c. For any $c > -a$,

$$EX^c = \frac{\Gamma(a + c)b^c}{\Gamma(a)}.$$

d. Let $U = dX, d > 0$. Then $U \sim \Gamma(a, db)$.

Proof. a and b. See Exercise B2.

c. We cannot use the moment-generating function to prove this part of the theorem, since c may not be an integer and may be negative. So instead, we use the law of expectation and obtain

$$EX^c = (b^a \Gamma(a))^{-1} \int_0^\infty x^{a+c-1} e^{-x/b} \, dx = \frac{b^{a+c} \Gamma(a + c)}{b^a \Gamma(a)} = \frac{b^c \Gamma(a + c)}{\Gamma(a)}.$$

d. U has moment-generating function

$$M_U(t) = M_X(dt) = (1 - b(dt))^{-a} = (1 - (bd)t)^{-a},$$

which is the moment-generating function of a $\Gamma(a, db)$ distribution. The result follows from the uniqueness theorem for moment-generating functions. □

Corollary 4–26(a). If $X \sim \Gamma(a, b)$, then $EX^{1/2} = b^{1/2} \Gamma(a + \frac{1}{2}) / \Gamma(a)$. If $a > 1$, then

$$EX^{-1} = \frac{b^{-1} \Gamma(a - 1)}{\Gamma(a)} = \frac{1}{(a - 1)b}.$$

The gamma distribution is one of the few distributions for which we can calculate EX^{-1} and $EX^{1/2}$ in an easy fashion.

Corollary 4–26(b). If $X \sim \Gamma(a, b)$, then $X/b \sim \Gamma(a, 1)$.

Because of Corollary 4–26(b), it is only necessary to provide a table of the gamma distribution function when $b = 1$. However, we still need a different table, of course, for each a. Table 7 at the back of the book gives the gamma distribution function for $a = 1, 2, \ldots, 10$; $x = 1, 2, \ldots, 15$. (Note that both x and a can take noninteger values, so the table is quite limited.)

If $W \sim \Gamma(1, \mu)$, we say that W has an *exponential* distribution with mean μ. In this case,

$$f(w) = \mu^{-1} \exp\left(\frac{-w}{\mu}\right), \quad w > 0, \qquad EW = \mu, \qquad \text{var}(W) = \mu^2, \qquad M(t) = (1 - \mu t)^{-1}.$$

We note that for a Poisson distribution the mean is equal to the variance, but for an exponential distribution the mean is equal to the standard deviation.

The main application of the exponential and gamma distributions is to Poisson processes.

Theorem 4–27. Consider a Poisson process with rate λ.

a. Let W be the time till the first arrival. Then W has an exponential distribution with mean λ^{-1} (i.e., $W \sim \Gamma(1, \lambda^{-1})$).
b. Let X be the time till the kth arrival. Then $X \sim \Gamma(k, \lambda^{-1})$.

Proof

a. Fix $w > 0$, and let V be the number of arrivals by the time w. Then

$$F_W(w) = P(W \leqq w) = P(V \geqq 1) = 1 - P(V = 0).$$

That is, the event that the first arrival occurs by the time w is the same as the event that there is at least one arrival by the time w. Now, $V \sim P(\lambda u)$, by the results of the previous section. Therefore,

$$F_W(w) = 1 - P(V = 0) = 1 - \exp(-\lambda u), \quad \text{and} \quad f_U(u) = \lambda \exp(-\lambda u), \qquad u > 0,$$

which is the density function of an exponential distribution with mean λ^{-1}.

b. Fix an $x > 0$, and let Y be the number of arrivals in the interval 0 to x. Then

$$F_X(x) = P(X \leqq x) = P(Y \geqq k) = 1 - P(Y < k) = 1 - F_Y(k - 1).$$

That is, the probability that the kth arrival occurs by the time x is the same as the probability that at least k arrivals have occurred by the time x. Now, $Y \sim P(\lambda x)$. Therefore,

$$F_X(x) = 1 - P(Y < k) = 1 - \exp(-\lambda x) \sum_{j=0}^{k-1} \frac{(\lambda x)^j}{j!}.$$

Hence,

$$f_X(x) = F_X'(x) = \exp(-\lambda x)\left(\left(\lambda \sum_{j=0}^{k-1} \frac{(\lambda x)^j}{j!} - \left(\sum_{j=1}^{k-1} \frac{j(\lambda x)^{j-1}\lambda}{j!}\right)\right)\right)$$

$$= \exp(-\lambda x)\left(\sum_{i=0}^{k-1} \frac{\lambda(\lambda x)^i}{i!} - \sum_{i=0}^{k-2} \frac{\lambda(\lambda x)^i}{i!}\right) = \exp(-\lambda x)\frac{\lambda^k x^{k-1}}{\Gamma(k)},$$

which is the density function of a Γ distribution with parameters k and λ^{-1}. □

Observe that the expected time till the first arrival in a Poisson process with rate λ is $1/\lambda$. That is, if a switchboard is receiving two calls a minute, then the expected time till the first call is half a minute, as we might have guessed. This may not be true if the telephone calls are not a Poisson process.

Before looking at an application of the gamma distribution, we give its normal approximation.

Theorem 4–28. Let $X_n \sim \Gamma(nc, b)$. Then

$$\frac{X_n - ncb}{b(nc)^{1/2}} \xrightarrow{d} A \sim N(0,1) \quad \text{as } n \to \infty.$$

Proof. See Exercise B3. □

We often let $a = nc$ and say that if $X \sim \Gamma(a, b)$, then for large a,

$$\frac{X - ab}{ba^{1/2}} \mathrel{\dot{\sim}} N(0,1).$$

It is not necessary to use the continuity correction when approximating a gamma distribution with a normal distribution, since both distributions are continuous.

Example 4–16

Suppose customers arrive at a store as a Poisson process with rate 10 per hour. Then X, the time till the second customer arrives, has a $\Gamma(2, \frac{1}{10})$ distribution. Therefore, the expected time until the second customer arrives is $\frac{2}{10}$ of an hour, or 12 minutes. By Corollary 4–26(b), $10X \sim \Gamma(2, 1)$. Therefore, the probability that we have to wait at least a half hour for the second customer is

$$P\left(X \geqq \frac{1}{2}\right) = 1 - P(10X \leqq 5) = 1 - .960 = .04$$

(using Table 7). Now, let Y be the time till the 81st customer arrives. Then $Y \sim \Gamma(81, \frac{1}{10})$. Therefore, the mean, variance, and standard deviation of the time till the 81st customer arrives are, respectively,

$$\mu = EY = \frac{81}{10} = 8.1, \qquad \sigma^2 = \text{var}(Y) = .81, \qquad \sigma = .9.$$

Suppose now that we want to know the probability that the 81st customer arrives within 6.4 hours. Since 81 is large, we can use the normal approximation for Y. Therefore,

$$P(Y \leqq 6.4) = P\left(\frac{Y - 8.1}{.9} \leqq \frac{6.4 - 8.1}{.9}\right) \doteq P(Z \leqq -1.89) = .03.$$

We could also find this probability by letting W be the number of customers in the first 6.4 hours. Then $W \sim P(64)$, and

$$P(W \geqq 81) = P(W \geqq 80.5) = P\left(\frac{W - 64}{8} \geqq \frac{80.5 - 64}{8}\right) \doteq P(Z \geqq 2.06) = .02.$$

(The difference in the two answers is due to the fact that both probabilities have been approximated.)

Suppose that we have a sequence of independent Bernoulli trials with constant probability p of success. We call such a process a *Bernoulli process*. The number of successes in the first n trials of a Bernoulli process has a binomial distribution, and the number of arrivals by time t in a Poisson process has a Poisson distribution. Therefore, the Poisson distribution plays a similar role in the Poisson process as the binomial distribution does in the Bernoulli process. In a Bernoulli process the number of trials before the first success has a geometric distribution, while in a Poisson process the time till the first success has an exponential distribution, so the exponential distribution plays a similar role in the Poisson process as the geometric distribution does in the Bernoulli process. And again, the gamma distribution plays the role in a Poisson process analogous to the role of the negative binomial distribution in a Bernoulli process.

We now present a theorem about the gamma distribution similar to others derived in previous sections.

Theorem 4–29. Let X_i be independent and $X_i \sim \Gamma(a_i, b)$. Let $W = \sum_i X_i$. Then

$$W \sim \Gamma(A, b), \qquad A = \sum_i a_i.$$

Proof. See Exercise B4. □

Although there are theorems similar to Theorem 4–29 for many of the distributions in this chapter, none of them is trivial. In particular, if X_i are independent and $X_i \sim \Gamma(a, b_i)$, then the distribution of $W = \sum_i X_i$ does not have a simple form. Fortunately, for the models we consider, the version given is the one we need.

The exponential distribution is often used as a model for waiting times even when there is no Poisson process. For example, the exponential distribution is often used to model the time till failure of a light bulb or a machine. We next consider some properties of the exponential distribution which are often used in these cases. (These properties will not be used in later sections of the book.)

Theorem 4–30

a. Let X have an exponential distribution. Then

$$P(X \geqq s + t | X \geqq s) = P(X \geqq t).$$

b. If X and Y are independent exponential random variables with means μ and ν, respectively, then

$$P(X > Y) = \frac{\mu}{\mu + \nu}.$$

c. Let $X_1, \ldots, X_n$ be independent random variables, with X_i exponentially distributed with mean μ_i. Let Q be the minimum of the X_i. Then Q is exponentially distributed with mean $\theta = (\Sigma \mu_i^{-1})^{-1}$.

Proof. See Exercises B11–B13. □

Part a of Theorem 4–30 is called the lack-of-memory property of the exponential distribution. If the lifetime of a light bulb is exponentially distributed, then the probability that a 10-week-old bulb lasts at least another 15 weeks is the same as the probability that a new bulb lasts at least another 15 weeks. Similarly, suppose that telephone calls are arriving at a switchboard as a Poisson process.

Then if a person has been waiting an hour for a call, the distribution of the time till the first call is the same as if the person had just come to the switchboard. In other words, the hour he or she has already waited has been "forgotten." (The exponential distribution is the only continuous distribution which has this lack-of-memory property. See Exercise C3.)

Exercises—A

1. The time necessary to fix a car is exponentially distributed with mean four hours. What is the probability that it takes more than five hours to fix the car? What is the probability that it takes between three and five hours?
2. Suppose people move into a town as a Poisson process at a rate of five per week.
   (a) Find the mean and variance of the time till the 10th person moves into the town.
   (b) Find the probability that the time till the 10th person moves into town is at least 2 weeks.
   (c) Find the probability that the time till the 64th person moves to town is at most 15 weeks and the probability that it is between 10 and 12 weeks.
3. Suppose two people enter adjacent lines in a grocery store. Suppose the service time for person A is exponentially distributed with mean four and the service time for person B is exponentially distributed with mean six. Find the probability that person A is finished with service first. (See part b of Theorem 4–30.)
4. Consider a system with four components which will fail if any component fails. Suppose that the times to failure of components are independently exponentially distributed with means 1, 2, 4, and 8 hours.
   (a) What is the expected time till failure of the system?
   (b) What is the probability that the system lasts at least 1 hour? (*Hint*: Use part c of Theorem 4–30.)

Exercises—B

1. Prove part d of Theorem 4–25.
2. Prove parts a and b of Theorem 4–26.
3. Prove Theorem 4–28.
4. Prove Theorem 4–29.
5. (a) Use L'Hospital's rule twice to show that $\lim_{x\to\infty} x^{3/2} e^{-x} = 0$.
   (b) Show that $\lim_{x\to\infty} x^a e^{-x} = 0$ for all a.
6. Let X have an exponential distribution with mean μ. Find the median of X.
7. Let $X = \Gamma(5, 2)$.
   (a) Find $P(X \leq 12)$ and $P(4 \leq X \leq 14)$.
   (b) Find $P(X \leq 12 | X > 10)$ and $P(X \leq 10 | X \leq 12)$.
8. Let $X_1, \ldots, X_n$ be independent, with the X_i exponentially distributed with mean μ. Let $\bar{X} = \Sigma_i X_i/n$ be the sample mean. Show that $\bar{X} \sim \Gamma(n, \mu/n)$.
9. Let Y have the Weibull density function $f(y) = by^{b-1} e^{-(y/a)^b}/a^b, y > 0$. Show that $X = Y^b$ has an exponential distribution.
10. Let X and Y be independent, $X \sim \Gamma(a, c)$, and $Y \sim \Gamma(b, c)$. Let $Q = X/(X + Y)$ and $T = X + Y$.
    (a) Find the joint density of Q and T.

(b) Show that Q and T are independent.
(c) Show that the marginal density of Q is $f(q) = \Gamma(a+b)q^{a-1}(1-q)^{b-1}/\Gamma(a)\Gamma(b)$, $0 < q < 1$.
(d) Find EQ and $\mathrm{var}(Q)$. (*Hint*: Use the fact that $f(q)$ is a density function for all $a > 0$ and $b > 0$.)
(e) If $a = 1$ and $b = 1$, how is Q distributed?
(The distribution of Q is called a *beta distribution* with parameters a and b.)

11. Prove part a of Theorem 4–30. (What is $P(X \geqq u)$?)
12. Prove part b of Theorem 4–30.
13. Prove part c of Theorem 4–30. (*Hint*: $P(Q \leqq q) = 1 - \Pi_i P(X_i > q)$. Why?)
14. Let X be uniformly distributed on the interval $(0, 1)$. Let $Y = -\ln(X)$. Show that Y has an exponential distribution.

Exercises—C

1. Let Y have an exponential distribution with mean two, and let $X|Y$ have a Poisson distribution with mean Y.
   (a) What is the joint density of X and Y?
   (b) What is the marginal density of X?
   (c) What are $P(X = 0)$ and $P(X \leqq 2)$?
2. Consider a spatial Poisson process with rate λ. Fix a spot, and let U be the distance from that spot to the nearest occurrence.
   (a) Show that $P(U \geqq u) = \exp(-4\lambda\pi u^3/3)$. (*Hint*: $U \geqq u$ if and only if there are no occurrences in the sphere of radius u about the fixed spot.)
   (b) What is the density function of U?
3. (a) Suppose that X is a random variable such that $P(X \geqq s + t | X \geqq s) = P(X \geqq t)$ for all $s > 0$ and $t > 0$. Let $h(s) = \log[P(X \geqq s)]$. Show that $h(s+t) = h(s) + h(t)$.
   (b) Show that X has an exponential distribution. (*Hint*: Assume that the result of part a implies that h is a linear function, i.e., that $h(s) = cs$ for some c and for all $s > 0$.)

4.4 DISTRIBUTIONS ASSOCIATED WITH THE NORMAL DISTRIBUTION

4.4.1 The General Normal Distribution

We now present the model which is the most important in this book. We say that X is *normally distributed* with mean μ and variance $\sigma^2 > 0$ and write

$$X \sim N(\mu, \sigma^2)$$

if

$$Z = \frac{X - \mu}{\sigma} \sim N(0, 1).$$

Note that in order for X to be normally distributed, Z must be exactly normally distributed, not approximately normally distributed.

The next theorem gives some simple results about the normal distribution.

Theorem 4–31. Let $X \sim N(\mu, \sigma^2)$. Then

a. $EX = \mu$, $\text{var}(X) = \sigma^2$, $M(t) = Ee^{Xt} = \exp(\mu t + \sigma^2 t^2/2)$.

b. X has density function

$$f(x) = (2\pi)^{-1/2}\sigma^{-1} \exp\left[\frac{-(x-\mu)^2}{2\sigma^2}\right], \qquad -\infty < x < \infty.$$

Proof. We use the density function, mean, variance, and moment-generating function for the standard normal random variable Z derived in Section 4.1.1. (Note that although we have called μ and σ^2 the mean and variance, respectively, of X, that is not proof that they are.)

a. Let $Z = (X - \mu)/\sigma \sim N(0, 1)$. Then

$$X = \sigma Z + \mu, \qquad EX = \sigma EZ + \mu = \mu, \qquad \text{var}(X) = \sigma^2 \text{ var}(Z) = \sigma^2,$$

$$M_X(t) = e^{\mu t} M_Z(\sigma t) = \exp\left(\mu t + \frac{\sigma^2 t^2}{2}\right).$$

b. Let $h(x) = (x - \mu)/\sigma$. Then $h'(x) = \sigma^{-1}$ and $Z = h(X)$. Therefore,

$$f_X(x) = \sigma^{-1} f_Z\left(\frac{x-\mu}{\sigma}\right) = \sigma^{-1}(2\pi)^{-1/2} \exp\left(\frac{-(x-\mu)^2}{2\sigma^2}\right).$$

(Note that this derivation guarantees that $f_X(x)$ is a density function.) □

The densities $N(3, 4)$ and $N(5, 9)$ are graphed in Figure 4–2. From the curves, we see that the density function is "bell shaped" and centered at μ, and that if $\sigma_1 > \sigma_2$, then the $N(\mu_1, \sigma_1^2)$ curve is less concentrated about μ_1 than the $N(\mu_2, \sigma_2^2)$ is about μ_2. If $X \sim N(\mu, \sigma^2)$, we think of μ as a measure of the center of the distribution and σ^2 as a measure of the spread of the distribution. μ is often called a *location parameter*, and σ is often called a *scale parameter*.

The normal distribution is used as a model for many sorts of measurement, such as of IQ, weight, height, protein concentration, and speed. There are at least three reasons for the extensive use of the normal distribution. First, many actual data seem to have a nearly bell-shaped distribution. Second, a more extensive theory has been developed for normal distributions than for other models. And third, the central limit theorem implies that many of the procedures are not too sensitive to the normal distribution.

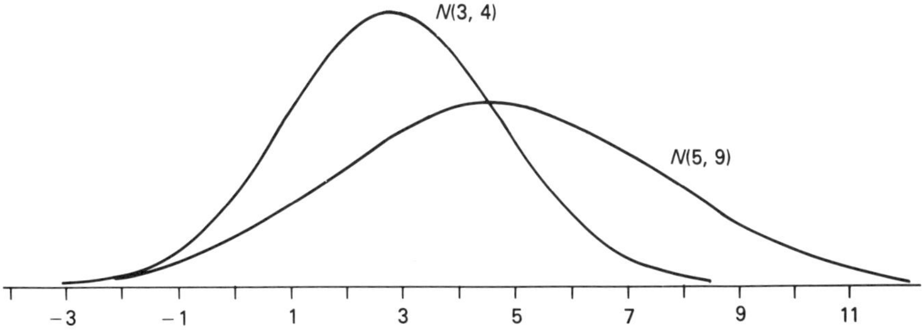

Figure 4–2 Graph of the N (3, 4) and N 5, 9) densities.

One explanation for the similarity of many actual data sets to a normal distribution is the central limit theorem, which implies that a random variable that is a sum or average of many small effects should be nearly normally distributed. (In this book, we use the central limit theorem only when the effects are i.i.d.; however, the theorem is true under much greater generality.) For example, a person's weight might be a sum of many small genetic effects as well as many small environmental factors, such as what he or she has eaten at each meal for the past few years. Similarly, a student's SAT test score is an average of his or her answers to many different questions and might reasonably be expected to be nearly normally distributed.

Example 4–17

Suppose that the height of a randomly chosen tree is normally distributed with mean 50 feet and standard deviation 10 feet. Then the probability that the height of the tree is between 40 and 70 feet is

$$P(40 < X < 70) = P\left(\frac{40-50}{10} < \frac{X-50}{10} < \frac{70-50}{10}\right) = P(-1 < Z < 2) = .8185.$$

This probability is an exact probability, since we are assuming that X is exactly normally distributed and, hence, that $(X-50)/10$ is exactly $N(0,1)$.

We conclude the section with a very useful theorem about the normal distribution.

Theorem 4–32

a. Let $X \sim N(\mu, \sigma^2)$. Let $U = aX + b$, where $a \neq 0$. Then

$$U \sim N(\mu^*, \sigma^{*2}), \qquad \mu^* = a\mu + b, \qquad \sigma^{*2} = a^2\sigma^2.$$

b. Let X_i be independent and $X_i \sim N(\mu_i, \sigma_i^2)$. Let $W = \sum_i a_i X_i + b$. Then

$$W \sim N(\nu, \tau^2), \qquad \nu = \sum_i a_i \mu_i + b, \qquad \tau^2 = \sum_i a_i^2 \sigma_i^2.$$

Proof

a. See Exercise B4.

b. X_i has moment-generating function

$$M_i(t) = \exp\left(\mu_i t + \frac{\sigma_i^2 t^2}{2}\right).$$

Therefore, W has moment-generating function

$$M(t) = e^{bt}\prod_i M_i(a_i t) = e^{bt}\prod_i \exp\left(\mu_i a_i t + \frac{\sigma_i^2 a_i^2 t^2}{2}\right)$$

$$= e^{bt}\exp\left(\sum_i \left(\mu_i a_i t + \frac{\sigma_i^2 a_i^2 t^2}{2}\right)\right) = \exp\left(\left(b + \sum_i a_i \mu_i\right)t + \frac{\left(\sum_i a_i \sigma_i^2\right)t^2}{2}\right),$$

$$= \exp\left(\nu t + \frac{\tau^2 t^2}{2}\right),$$

which is the moment-generating function of $N(\nu,\tau^2)$. The uniqueness of moment-generating functions implies that $W \sim N(\nu,\tau^2)$. □

Note that in order to prove Theorem 4–32, it is not enough to show that $EW = \nu$ and $\text{var}(W) = \tau^2$: Many different distributions have the same mean and variance. However, by the uniqueness theorem for moment-generating functions, two different distributions cannot have the same moment-generating function.

Example 4–18

Suppose that X_1, X_2, and X_3 are independent and that

$$X_1 \sim N(2,4), \qquad X_2 \sim N(3,25), \qquad X_3 \sim N(4,16).$$

Let $W = X_1 + 2X_2 - 3X_3 - 4$. Then

$$\nu = 2 + 2(3) - 3(4) - 4 = -8, \quad \text{and} \quad \tau^2 = 4 + 4(25) + 9(16) = 248.$$

Therefore,

$$W \sim N(-8, 248),$$

and it follows that

$$P(W > 0) = P\left(\frac{W+8}{15.75} > \frac{0+8}{15.75}\right) = P(Z > .51) = .305.$$

Exercises—A

1. Verbal SAT scores are assumed to be normally distributed with mean 500 and standard deviation 100.
   (a) Find the probability that a score lies between 300 and 700.
   (b) Find the conditional probability that the score is over 800 given that it is over 700.
2. Suppose that the outside diameter of a particular kind of bolt has a normal distribution with mean 1 in. and standard deviation .01 in. Suppose that the inside diameters of the nuts that go with those bolts are normally distributed with mean 1.01 in. and standard deviation .01 in. The nut and the bolt will work together as long as the diameter on the nut is bigger than the diameter of the bolt, but not more than .03 in. bigger. What is the probability that a randomly chosen nut and bolt work? (*Hint*: How is the difference distributed?)
3. Suppose that the height in inches of a randomly chosen man is normally distributed with mean 68 in. and standard deviation 4 in. If we convert the measurements to centimeters, what is the distribution of the heights?

Exercises—B

1. Let $X \sim N(3,4)$.
   (a) Find a such that $P(-a < X - 3 < a) = .90$.
   (b) Find b such that $P(X < b) = .90$.
   (c) Find $P(X > 4.5 | X > 4)$.
   (d) Find $P(X > 4.5 | 4 < X < 6)$.
2. Let X_1, X_2, and X_3 be independent, and let $X_1 \sim N(1,4)$, $X_2 \sim N(2,9)$, and $X_3 \sim N(3,16)$.
   (a) Find the probability that all the X's are greater than one.
   (b) Find the probability that exactly two of the X's are greater than one.

3. Suppose that $X \sim N(\mu, \sigma^2)$ and that $P(X \leqq 2) = .4$ and $P(X \geqq 4) = .3$. Find μ and σ^2.
4. Prove part a of Theorem 4–32.
5. Suppose that $X \sim N(\mu, \sigma)$. Let $Y = \exp(X)$.
   (a) What is the density of Y?
   (b) Find EY and EY^2. ($EY^k = M_X(k)$. Why?)
   (The distribution of Y is called the log-normal distribution.)

4.4.2 The Chi-Square, t-, and F-Distributions

In this section, we define three distributions which are quite useful in making inferences in normal models.

Let $Z_1, \ldots, Z_k$ be independent with $Z_i \sim N(0, 1)$. Let

$$U = \sum_{i=1}^{k} Z_i^2.$$

Then U has a *chi-square* distribution with k *degrees of freedom*, written

$$U \sim \chi_k^2.$$

(Note that the joint distribution of the Z_i is completely specified, so that the distribution of U depends only on k, the number of Z_i's.)

The following theorem gives the basic result about the chi-square distribution.

Theorem 4–33. Let $U \sim \chi_k^2$. Then $U \sim \Gamma(k/2, 2)$.

Proof. Let $Z \sim N(0, 1)$ and $V = Z^2$. Then V has moment-generating function

$$M_V(t) = Ee^{Vt} = Ee^{Z^2 t} = \int_{-\infty}^{\infty} (2\pi)^{-1/2} e^{z^2 t} e^{-z^2/2}\, dz$$

$$= \int_{-\infty}^{\infty} (2\pi)^{-1/2} e^{-u^2/2} (1-2t)^{-1/2}\, du = (1-2t)^{-1/2}, \qquad t < \tfrac{1}{2}.$$

(In doing the preceding, we made the change of variable $z = (1-2t)^{-1/2} u$, so that $dz = (1-2t)^{-1/2}\, du$, and used the fact that the standard normal density function integrates to one.) Now, let Z_i be independent and $Z_i \sim N(0, 1)$. Also, let

$$V_i = Z_i^2 \quad \text{and} \quad U = \sum_{i=1}^{k} V_i = \sum_{i=1}^{k} Z_i^2 \sim \chi_k^2.$$

Then the V_i are independently and identically distributed. Therefore,

$$M_U(t) = (M_V(t))^k = (1-2t)^{-k/2}, \qquad t < \tfrac{1}{2},$$

which is the moment-generating function of a $\Gamma(k/2, 2)$ random variable. But then, by the uniqueness theorem, $U \sim \Gamma(k/2, 2)$. □

Corollary 4–33. Let $U \sim \chi_k^2$. Then

a. U has density function

$$f(u) = \frac{u^{(k/2)-1} e^{-u/2}}{2^{k/2}\Gamma(k/2)},$$

and moment-generating function

$$M(t) = (1 - 2t)^{-k/2}, \qquad t < \tfrac{1}{2}.$$

b. Further

$$EU = k, \qquad \mathrm{var}(U) = 2k, \qquad EU^2 = k(k+2).$$

If $k > 1$, then $EU^{-1/2}$ is finite; if $k > 2$, then $EU^{-1} = (k-2)^{-1}$.

c. If $U_n \sim \chi_n^2$, then

$$\frac{U_n - n}{(2n)^{1/2}} \xrightarrow{d} Z \sim N(0,1) \quad \text{as } n \to \infty.$$

Proof. All three parts of the theorem follow directly from theorems proved about the gamma distribution in Section 4.3.3. □

Let Z and V be independent, and let $Z \sim N(0,1)$ and $V \sim \chi_k^2$. Also, let

$$t = \frac{Z}{(V/k)^{1/2}}.$$

Then we say that t has a *t-distribution* with k degrees of freedom and write

$$t \sim t_k.$$

(Note that the joint distribution of (Z, V) is completely determined by k, the degrees of freedom of V, and, hence, the distribution of t depends only on k.)

The preceding distribution is often called a *Student's t-distribution* after W. S. Gosset, who first described it under the pseudonym "Student." (Note that in this case we have bowed to tradition and broken our rule that random variables are denoted by capital letters.) In the exercises, it is shown that if $t \sim t_k$, then t has density function

$$f(t) = c_k\left(1 + \frac{t^2}{k}\right)^{-(k+1)/2}, \qquad -\infty < t < \infty, \qquad c_k = \frac{\Gamma((k+1)/2)}{(\pi k)^{1/2}\Gamma(k/2)}.$$

The constant c_k is merely the constant that is necessary to make $f(t)$ integrate to one. We shall not need this density function in this text.

Now, let U and V be independent, $U \sim \chi_m^2$, and $V \sim \chi_k^2$. Also, let

$$F = \frac{U/m}{V/k}.$$

Then F has an *F-distribution* with m and k degrees of freedom, written

$$F \sim F_{m,k}.$$

We call m the *numerator degrees of freedom* and k the *denominator degrees of freedom*. In the exercises, it is shown that if $F \sim F_{m,k}$, then F has density function

$$\frac{d_{m,k} f^{(m/2)-1}}{(1 + mf/k)^{(k+m)/2}}, \qquad f > 0,$$

$$d_{m,k} = \frac{\Gamma((k+m)/2)(m/k)^{m/2}}{\Gamma(k/2)\Gamma(m/2)}.$$

Note again that $d_{m,k}$ is the constant that is necessary to make the density integrate to one. Again, we shall not need this density in this text.

We now give some elementary properties of the t- and F-distributions.

Theorem 4–34

a. Let $t \sim t_k$. Then $-t \sim t_k$ and $t^2 \sim F_{1,k}$. Also, if $k > 1$, then $Et = 0$, and if $k > 2$, then $\text{var}(t) = k/(k-2)$.
b. Let $F \sim F_{m,k}$. Then $F^{-1} \sim F_{k,m}$. Also, if $k > 2$, then $EF = k/(k-2)$.

Proof

a. Let $t = Z/(U/k)^{1/2}$ as in the definition of a t-distribution. Then

$$-t = \frac{-Z}{(U/k)^{1/2}}.$$

Because $-Z$ has the same distribution as Z, we see that $-t$ has the same distribution as t and hence $-t \sim t_k$. Also,

$$t^2 = \frac{Z^2/1}{U/k} \sim F_{1,k},$$

because $V = Z^2 \sim \chi_1^2$. In addition,

$$Et = k^{1/2}(EZ)(EU^{-1/2}),$$

using the independence of Z and U. Now, $EZ = 0$, and if $k > 1$, then $EU^{-1/2}$ is finite, so that $Et = 0$ when $k > 1$. (It would be tempting to argue that the mean must be zero when $k = 1$; however, in Exercise B10 it is shown that in this case t has a Cauchy distribution, so that the mean does not exist.) If $k > 2$, then

$$\text{var}(t) = Et^2 = kEZ^2E(U^{-1}) = \frac{k(1)}{k-2} = \frac{k}{k-2}$$

(using Corollary 4–33).

b. The proof of this part is left as an exercise. □

If $t \sim t_k$, then Et^j exists if $j < k$, but does not if $j \geqq k$. Therefore, the t-distribution does not have a moment-generating function. Similarly, the F-distribution does not have a moment-generating function.

Now, let U be a random variable. Then we define the *upper* α *quantile* to be the number U^α such that

$$P(U > U^\alpha) = \alpha.$$

The χ^2, t- and F-distributions are used primarily for statistical inference. As we shall see later in the text, it is more useful for statistical analysis if a table of the distribution function is given by fixed upper quantiles than by fixed values of x. Tables 2, 3, and 4 at the back of the text give the αth upper quantiles, t_k^α, $\chi_k^{2\alpha}$, and $F_{k,n}^\alpha$, for the t_k, χ_k^2, and $F_{k,m}$ distributions for various choices of α.

We next mention some simple relations between the upper α quantiles. Since the t-distribution is symmetric about zero,

$$t_k^{1-\alpha} = -t_k^\alpha.$$

By part b of Theorem 4–34, we see that if $F \sim F_{m,k}$, then $F^* = F^{-1} \sim F_{k,m}$, and

$$P(F \geqq (F_{m,k}^{\alpha})^{-1}) = P(F^{-1} \leqq F_{m,k}^{\alpha}) = 1 - P(F^* \geqq F_{m,k}^{\alpha}) = 1 - \alpha.$$

Therefore,

$$F_{m,k}^{1-\alpha} = (F_{k,m}^{\alpha})^{-1}.$$

Finally, in Chapter 6 we shall show that if $V_n \sim t_n$, then

$$V_n \xrightarrow{d} Z \sim N(0, 1),$$

and therefore,

$$t_n^{\alpha} \rightarrow z^{\alpha},$$

the αth upper quantile for the standard normal distribution. Consequently, upper quantiles for the standard normal distribution can be read from the ∞ row of the t-table.

Example 4–19

Let $U \sim \chi_{14}^2$. From Table 3, we see that

$$\chi_{14}^{2,.9} = 7.79 \quad \text{and} \quad \chi_{14}^{2,.1} = 21.06.$$

Therefore,

$$P(U \geqq 7.79) = .9, \qquad P(U \geqq 21.1) = .1, \qquad P(7.79 \leqq U \leqq 21.06) = .9 - .1 = .8.$$

Now, suppose that $t \sim t_{18}$. Then $t_{18}^{.2} = .862$ and $t_{18}^{.05} = 1.734$. Therefore,

$$P(t \geqq .862) = .2, \qquad P(t \leqq -.862) = .2, \qquad P(-.862 < t < .862) = .6,$$

$$P(-.862 < t < 1.734) = .75, \qquad P(-1.734 < t < -.862) = .15.$$

Now, let $F \sim F_{4,6}$. Then $F_{4,6}^{.05} = 4.53$ and $F_{6,4}^{.01} = 15.2$. Therefore,

$$P(F \geqq 4.53) = .05, \qquad P(F \geqq (15.2)^{-1}) = 1 - .01 = .99,$$

$$P((15.2)^{-1} < F < 4.53) = .94.$$

Exercises—B

1. **(a)** Let $U \sim \chi_{10}^2$. Find a and b such that $P(a < U) = .9$ and $P(a \leqq U < b) = .85$.
   (b) Let $U \sim \chi_{15}^2$. Find a and b such that $P(U \leqq a) = .1$ and $P(a < U \leqq b) = .8$.
2. **(a)** Let $t \sim t_{15}$. Find a and b such that $P(-a < t \leqq a) = .9$ and $P(b \leqq t < a) = .7$.
   (b) Let $t \sim t_{12}$. Find a and b such that $P(t > a) = .8$ and $P(a < t \leqq b) = .7$.
3. **(a)** Let $F \sim F_{5,8}$. Find a and b such that $P(F < a) = .05$ and $P(a \leqq F \leqq b) = .925$.
   (b) Let $F \sim F_{7,3}$. Find a and b such that $P(a \leqq F < b) = .90$.
4. **(a)** Use the t-table (Table 2) to find the upper .05 and .01 quantiles of the standard normal distribution.
   (b) Use the normal table (Table 1) to verify that the results of part a are correct, at least to the accuracy of Table 1.
5. Show that $t_k^{1-\alpha} = -t_k^{\alpha}$.
6. Let $U \sim \chi_{32}^2$. Use the normal approximation to approximate the upper .05 and .10 quantiles of U.
7. Let $U \sim \chi_k^2$. Use the definition of the χ_k^2 distribution to find EU and $\text{var}(U)$ without using Theorem 4–33 or its corollary.

8. Prove part b of Theorem 4–34.

9. Let $F \sim F_{m,k}$. Find $\text{var}(F)$.

10. Use the t density function to show that if $t \sim t_1$, then t has the Cauchy density $f(t) = (\pi(1 + t^2))^{-1}$.

11. Show that if $t \sim t_k$ and $j < k$, then Et^j exists.

Exercises—C

1. Let Z and U be independent, $Z \sim N(0, 1)$, and $U \sim \chi_k^2$.
 (a) What is the joint density of Z and U?
 (b) Let $T = Z/(U/k)^{1/2}$ and $W = U$. Find the joint density of T and W.
 (c) Find the marginal density of T. (Use Theorem 4–25, part d.)

2. Let V and U be independent, $V \sim \chi_m^2$, and $U \sim \chi_k^2$.
 (a) What is the joint density of U and V?
 (b) Let $F = (V/m)/(U/k)$ and $W = U$. What is the joint density of F and W?
 (c) Verify the formula in the text for the marginal density of F. (Use Theorem 4–25, part d.)

4.4.3 Sampling from a Normal Distribution

Let $X_1, \ldots, X_n$ be i.i.d. with distribution function F. Then we say that $X_1, \ldots, X_n$ is a *sample* from F. Also, suppose that $n > 1$, and let

$$\overline{X} = \frac{\sum_i X_i}{n} \quad \text{and} \quad S^2 = \frac{\sum_i (X_i - \overline{X})^2}{n - 1}.$$

We call $\overline{X}$ and S^2 the *sample mean* and *sample variance*, respectively, of the X_i. We think of $\overline{X}$ as an estimator of $\mu = EX_i$ and S^2 as an estimator of $\sigma^2 = \text{var}(X_i)$. By the central limit theorem, $\overline{X}$ is approximately normally distributed for large n. In Chapter 6, we show that S^2 is also approximately normally distributed for large n.

We note the following computational formula for S^2:

$$S^2 = \frac{\sum_i X_i^2 - n\overline{X}^2}{n - 1}.$$

Example 4–20

Let $X_1 = 3$, $X_2 = 5$, $X_3 = 9$, and $X_4 = 7$. Then

$$\overline{X} = \frac{3 + 5 + 9 + 7}{4} = 6, \qquad \sum X_i^2 = (9 + 25 + 81 + 49) = 164,$$

$$S^2 = \frac{164 - (4)36}{3} = 6.67.$$

In this section, we derive the exact joint distribution of $\overline{X}$ and S^2 for a sample from a normal distribution. The rather surprising result is that for a normal distribution $\overline{X}$ and S^2 are independent, even though $\overline{X}$ is a part of the definition of S^2. This result is probably the most important result of exact distribution theory to be found in this book. In fact, we give three different proofs of it in various places in the book, one here using moment-generating functions, one in Exercise C1 using Jacobians, and one in the next chapter using matrix methods.

Theorem 4–35. Let $X_1, \ldots, X_n$ be independent, and let $X_i \sim N(\mu, \sigma^2)$, $n > 1$. Then

a. The sample mean $\overline{X}$ and the sample variance S^2 are independent.
b. $\overline{X} \sim N(\mu, \sigma^2/n)$, or equivalently, $n^{1/2}(\overline{X} - \mu)/\sigma \sim N(0, 1)$.
c. $S^2 \sim \Gamma((n-1)/2, 2\sigma^2/(n-1))$, or equivalently, $(n-1)S^2/\sigma^2 \sim \chi^2_{n-1}$.

Proof

a. $X_1, \ldots, X_n$ have joint moment-generating function

$$M_X(t_1, \ldots, t_n) = \prod_i M_i(t_i) = \exp\left(\mu \sum_i t_i + \frac{\sigma^2 \sum_i t_i^2}{2}\right).$$

Let $\mathbf{U} = (U_1, \ldots, U_n)$, $U_i = (X_i - \overline{X})$, and $V = \overline{X}$. Then with $\bar{s} = \sum_i s_i/n$, $\mathbf{U}$ and V have joint moment-generating function

$$M_{U,V}(s_1, \ldots, s_n, r) = E\ \exp\left(Vr + \sum_i U_i s_i\right) = E\ \exp\left(\overline{X}(r - n\bar{s}) + \sum_i X_i s_i\right)$$

$$= E\ \exp\left[\sum_i X_i\left(s_i - \bar{s} + \frac{r}{n}\right)\right] = M_X\left(s_1 - \bar{s} + \frac{r}{n}, \ldots, s_n - \bar{s} + \frac{r}{n}\right)$$

$$= \exp\left[\mu\left(\sum_i (s_i - \bar{s} + r/n)\right) + \sigma^2 \frac{\sum_i (s_i - \bar{s} + r/n)^2}{2}\right]$$

$$= \exp\left(\mu r + \frac{\sigma^2 r^2}{2n}\right) \exp\left[\frac{\sigma^2 \sum_i (s_i - \bar{s})^2}{2}\right]$$

$$= M_{U,V}(0, \ldots, 0, r) M_{U,V}(s_1, \ldots, s_r, 0).$$

(In these equations, we have used the identities $\sum_i (s_i - \bar{s}) = 0$ and $\sum_i (s_i - \bar{s} + r/n)^2 = \sum_i (s_i - \bar{s})^2 + r^2/n$; see the exercises.) Hence, the joint moment-generating function of $\mathbf{U}$ and V factors into the product of the moment-generating function of $\mathbf{U}$ and the moment generating function of V. Therefore, $\mathbf{U}$ and V are independent. Now,

$$\overline{X} = V \quad \text{and} \quad S^2 = \frac{\sum U_i^2}{n-1}.$$

Therefore, $\overline{X}$ and S^2 are independent.

b. $\overline{X} = \Sigma(1/n)X_i$. By results in Section 4.4.1,

$$\overline{X} \sim N(\nu, \tau^2), \qquad \nu = \sum_{i=1}^{n} \left(\frac{1}{n}\right)\mu = \mu, \qquad \tau^2 = \sum_{i=1}^{n} \left(\frac{1}{n}\right)^2 \sigma^2 = \frac{\sigma^2}{n}.$$

But then, by the definition of the normal distribution,

$$\frac{n^{1/2}(\overline{X} - \mu)}{\sigma} = \frac{\overline{X} - \mu}{(\sigma^2/n)^{1/2}} \sim N(0, 1).$$

c. Let $Z_i = (X_i - \mu)/\sigma$. Then the $Z_i \sim N(0, 1)$ and are independent. Therefore,

$$T = \sum_i Z_i^2 = \frac{\sum_i (X_i - \mu)^2}{\sigma^2} \sim \chi_n^2.$$

Furthermore, $Z = n^{1/2}(\overline{X} - \mu)/\sigma \sim N(0, 1)$. Therefore,

$$W = Z^2 = \frac{n(\overline{X} - \mu)^2}{\sigma^2} \sim \chi_1^2.$$

Now, let $Y = (n - 1)S^2/\sigma^2$. Then

$$T = W + Y$$

(see Exercise B4), and W and Y are independent. Therefore,

$$M_T(t) = M_W(t)M_Y(t),$$

$$M_Y(t) = \frac{M_T(t)}{M_W(t)} = \frac{(1 - 2t)^{-n/2}}{(1 - 2t)^{-1/2}} = (1 - 2t)^{-(n-1)/2},$$

which is the moment-generating function of a χ_{n-1}^2 distribution. Hence,

$$Y = \frac{(n - 1)S^2}{\sigma^2} \sim \chi_{n-1}^2 = \Gamma\left(\frac{n - 1}{2}, 2\right),$$

and it follows that

$$S^2 = \left(\frac{\sigma^2}{n - 1}\right)\left[\frac{(n - 1)S^2}{\sigma^2}\right] \sim \Gamma\left(\frac{n - 1}{2}, \frac{2\sigma^2}{n - 1}\right). \quad \square$$

Since $\sum_i (X_i - \overline{X}) = 0$, the numerator of S^2 is the sum of squares of $n - 1$ free terms. (The last term is the negative of the sum of the first $n - 1$ terms and, hence, is not free.) This is probably the source of the term $n - 1$ degrees of freedom.

It is interesting that if $\overline{X}$ and S^2 are the sample mean and sample variance computed from a sample from any nonnormal distribution, then $\overline{X}$ and S^2 are not independent. So this property of independence of the sample mean and sample variance characterizes the normal distribution.

Many students confuse part b of Theorem 4–35 with the central limit theorem. However, the two are quite different. The central limit theorem says that no matter what distribution the sample is from, as long as it has a finite variance, $n^{1/2}(\overline{X} - \mu)/\sigma$ is approximately a standard normal random variable for large n. Part b of the theorem says that if the X_i are a sample from a normal distribution, then $n^{1/2}(\overline{X} - \mu)/\sigma$ has exactly a standard normal distribution for any sample size n.

The following corollary gives several results about $\overline{X}$ and S^2 which follow directly from Theorem 4–35.

Corollary 4–35. If X_i are independent and $X_i \sim N(\mu, \sigma^2)$, then

a. $n^{1/2}(\overline{X} - \mu)/S \sim t_{n-1}$.

b. $ES^2 = \sigma^2$, $\text{var}(S^2) = 2\sigma^4/(n-1)$, and

$$ES = \frac{2^{1/2}\Gamma(n/2)}{(n-1)^{1/2}\Gamma((n-1)/2)}\sigma.$$

Proof

a. Let $Z = n(\overline{X} - \mu)/\sigma$ and $U = (n-1)S^2/\sigma^2$. Then, by Theorem 4–35 and the definition of the t-distribution,

$$t = \frac{Z}{(U/(n-1))^{1/2}} = \frac{n^{1/2}(\overline{X} - \mu)}{S} \sim t_{n-1}.$$

b. Let $U = S^2 \sim \Gamma((n-1)/2, 2\sigma^2/(n-1)$. Then

$$ES^2 = EU = \left(\frac{n-1}{2}\right)\left(\frac{2\sigma^2}{n-1}\right) = \sigma^2,$$

$$\text{var}(S^2) = \text{var}(U) = \left(\frac{n-1}{2}\right)\left(\frac{2\sigma^2}{n-1}\right)^2 = \frac{2\sigma^4}{n-1},$$

$$ES = EU^{1/2} = \frac{\Gamma(n/2)2^{1/2}}{(n-1)^{1/2}\Gamma((n-1)/2)}\sigma. \quad \square$$

A more intuitively appealing definition for the sample variance is

$$T^2 = \sum_i \frac{(X_i - \overline{X})^2}{n} = \frac{(n-1)S^2}{n}.$$

If we put a discrete distribution on the observations $X_1, \ldots, X_n$, each having probability n^{-1}, and then compute the variance of this distribution, we get T^2. In fact, many books use T^2 for the sample variance, and most calculators compute both T^2 and S^2. Many people prefer S^2 because $ES^2 = \sigma^2$. By contrast,

$$ET^2 = \frac{(n-1)ES^2}{n} = \frac{(n-1)\sigma^2}{n}.$$

(In later sections, we shall say that S^2 is an "unbiased" estimator of σ^2.) Most books on applied statistics use S^2 for the sample variance, so we shall do so for the remainder of this book.

Exercises—B

1. Let $X_1 = 6$, $X_2 = 4$, $X_3 = 11$, $X_4 = -3$, and $X_5 = 7$. Find $\overline{X}$ and S^2.
2. Show that $\Sigma_i (X_i - \overline{X})^2 = (\Sigma_i X_i^2) - n\overline{X}^2$.
3. Show that $\Sigma_i (s_i - \bar{s}) = 0$ and $\Sigma_i (s_i - \bar{s} + r/n)^2 = \Sigma_i (s_i - \bar{s})^2 + r^2/n$.
4. In the notation of part c of Theorem 4–35, show that $T = W + Y$.
5. Let $X_1, \ldots, X_n$ be a sample from an arbitrary distribution.
   (a) Show that $E\,\Sigma_i X_i^2 = n(\sigma^2 + \mu^2)$.
   (b) Show that $En\overline{X}^2 = \sigma^2 + n\mu^2$.
   (c) Show that $ES^2 = \sigma^2$.

6. If X_i are a sample from a normal distribution, find $\mathrm{var}(S) = \mathrm{var}(U^{1/2})$.

Exercises—C

1. Let $X_1, \ldots, X_n$ be independent and $X_i \sim N(\mu, \sigma^2)$. Let $Y_1 = \overline{X}, Y_2 = X_2 - \overline{X}, \ldots, Y_n = X_n - \overline{X}$.
   (a) Show that $\Sigma_i (X_i - \overline{X})^2 = (Y_2^2 + \cdots + Y_n)^2 + (Y_2 + \cdots + Y_n)^2 = q(Y_2 \ldots Y_n)$.
   (b) Find the joint density of the Y_i's.
   (c) Show that Y_1 is independent of $(Y_2, \ldots, Y_n)$ and, hence, that $\overline{X}$ and S^2 are independent.

Chapter 5

MULTIVARIATE PARAMETRIC FAMILIES

In this chapter, we present the basic bivariate and multivariate families used in this book. In Section 5.1 we study the trinomial and multinomial distributions, which are bivariate and multivariate extensions of the binomial distribution. In Section 5.2, we examine the bivariate extension of the normal distribution. Section 5.3 gives the basic properties of the multivariate normal distribution, including the property used in deriving the asymptotic distribution of Pearson's χ^2 in Chapter 14. We use these properties to derive the basic distribution theory for the multiple regression and analysis-of-variance models discussed in Chapters 12 and 13. It is convenient to use matrices to define and consider the multivariate extension of the normal distribution. Accordingly, a review of basic matrix algebra is given in Appendix A.3.1.

5.1 MULTINOMIAL DISTRIBUTIONS

5.1.1 The Trinomial Distribution

The binomial distribution counts the number of "successes" in n independent replications of an experiment with two possible outcomes. The trinomial distribution acts analogously when there are three possible outcomes.

Let $\mathbf{X} = (X_1, X_2)$ be a bivariate discrete random vector whose range is

$$S_n = \{(x_1, x_2): x_1 = 0, 1, \ldots, x_2 = 0, 1, \ldots, x_1 + x_2 \leqq n\}.$$

(That is, X_1 and X_2 are nonnegative integer-valued random variables such that $X_1 + X_2 \leqq n$.) Then $\mathbf{X}$ has a *trinomial distribution* with parameters n and $\mathbf{p} = (p_1, p_2)$, written

$$\mathbf{X} = (X_1, X_2) \sim T(n, (p_1, p_2)),$$

if $\mathbf{X}$ has joint density function

$$f(x_1, x_2) = \begin{bmatrix} & n & \\ x_1 & x_2 & n - x_1 - x_2 \end{bmatrix} p_1^{x_1} p_2^{x_2} (1 - p_1 - p_2)^{n - x_1 - x_2}, \qquad (x_1, x_2) \in S_n,$$

where n is a positive integer and p_1 and p_2 are nonnegative numbers such that

$$p_1 + p_2 \leqq 1.$$

Recall that if $a + b + c = n$, then

$$\begin{bmatrix} & n & \\ a & b & c \end{bmatrix} = \frac{n!}{a!b!c!}.$$

Theorem 5–1

a. The trinomial density function $f(x_1, x_2)$ just given is a joint density function for all positive integers n and p_1, p_2 such that $p_i \geqq 0$ and $p_1 + p_2 \leqq 1$.
b. Let $(X_1, X_2) \sim T(n, (p_1, p_2))$. Then

$$EX_i = np_i, \qquad \text{var}(X_i) = np_i(1 - p_i), \qquad \text{cov}(X_1, X_2) = -np_1 p_2,$$
$$M(t_1, t_2) = E\ \exp[X_1 t_1 + X_2 t_2] = (p_1 e^{t_1} + p_2 e^{t_2} + 1 - p_1 - p_2)^n.$$

Proof. The trinomial theorem says that

$$\sum_{(x_1, x_2) \in S_n}\sum \begin{bmatrix} & & n & \\ x_1 & x_2 & n - x_1 - x_2 \end{bmatrix} a_1^{x_1} a_2^{x_2} a_3^{n - x_1 - x_2} = (a_1 + a_2 + a_3)^n.$$

a. Clearly $f(\mathbf{x}) \geqq 0$. By the trinomial theorem,

$$\sum_{S_n}\sum f(\mathbf{x}) = (p_1 + p_2 + (1 - p_1 - p_2))^n = 1.$$

b. By the trinomial theorem again, we have, for the moment-generating function,

$$M(\mathbf{t}) = \sum_{S_n}\sum \begin{bmatrix} & & n & \\ x_1 & x_2 & n - x_1 - x_2 \end{bmatrix} (p_1 e^{t_1})^{x_1} (p_2 e^{t_2})^{x_2} (1 - p_1 - p_2)^{n - x_1 - x_2}$$
$$= (p_1 e^{t_1} + p_2 e^{t_2} + 1 - p_1 - p_2)^n.$$

Now, let

$$Q(\mathbf{t}) = (p_1 e^{t_1} + p_2 e^{t_2} + 1 - p_1 - p_2).$$

Then

$$\psi(\mathbf{t}) = \log[M(\mathbf{t})] = n\ \log[Q(\mathbf{t})],$$

$$\psi_i(\mathbf{t}) = \frac{np_i e^{t_i}}{Q(\mathbf{t})}, \qquad EX_i = \psi_i(\mathbf{0}) = \frac{np_i}{Q(\mathbf{0})} = np_i,$$

$$\psi_{ii}(\mathbf{t}) = \frac{np_i e^{t_i} Q(\mathbf{t}) - n(p_i e^{t_i})^2}{(Q(\mathbf{t}))^2}, \qquad \text{var}(X_i) = \psi_{ii}(\mathbf{0}) = np_i(1 - p_i),$$

$$\psi_{12}(\mathbf{t}) = \frac{-n(p_1 e^{t_1})(p_2 e^{t_2})}{(Q(\mathbf{t}))^2}, \qquad \text{cov}(X_1, X_2) = \psi_{12}(\mathbf{0}) = -np_1 p_2. \quad \square$$

The main application of the trinomial distribution is given in the following theorem.

Theorem 5–2. Suppose an experiment with three disjoint possible outcomes, A, B, and C, is repeated n independent times. Suppose outcome A has constant probability $p_1 = P(A)$, outcome B has constant probability $p_2 = P(B)$, and outcome C has constant probability $p_3 = P(C) = 1 - p_1 - p_2$. Let X_1 be the number of times in the n trials that A occurs, and let X_2 be the number of times B occurs (so that $n - X_1 - X_2$ is the number of times C occurs). Then

$$(X_1, X_2) \sim T(n, (p_1, p_2)).$$

Proof. Note first that the outcomes are disjoint and exhaust all the possibilities. Hence, in each experiment, exactly one of the events A, B, or C occurs. Therefore, $X_1 + X_2$, the number of times A or B occurs, must be less than or equal to n, and consequently, $(X_1, X_2) \in S_n$. To see that the density function has the form given, we first look at an example. Suppose that $n = 6$. Then the event $X_1 = 2, X_2 = 3$ means that we get two A's, three B's, and one C. There are many possible ways to do this. One way is to get $AABBBC$, which has probability

$$p_1 p_1 p_2 p_2 p_2 p_3 = p_1^2 p_2^3 p_3$$

(where $p_3 = 1 - p_1 - p_2$). Another is to get $ABACBB$, which again has probability

$$p_1 p_2 p_1 p_3 p_2 p_2 = p_1^2 p_2^3 p_3.$$

In fact, every possible order which leads to two A's, three B's, and one C has the same probability. Now, there are

$$\begin{bmatrix} & 6 & \\ 2 & 3 & 1 \end{bmatrix}$$

possible orderings which lead to two A's, three B's, and a C. Therefore,

$$f(2,3) = P(X_1 = 2, X_2 = 3) = \begin{bmatrix} & 6 & \\ 2 & 3 & 1 \end{bmatrix} p_1^2 p_2^3 (1 - p_1 - p_2).$$

We now return to the case of general n. The event $X_1 = x_1, X_2 = x_2$ can occur in

$$\begin{bmatrix} & & n & \\ x_1 & x_2 & n - x_1 & - x_2 \end{bmatrix}$$

different orders, and each order has probability

$$p_1^{x_1} p_2^{x_2} (1 - p_1 - p_2)^{n - x_1 - x_2}.$$

Therefore,

$$f(x_1, x_2) = \begin{bmatrix} & & n \\ x_1 & x_2 & n - x_1 - x_2 \end{bmatrix} p_1^{x_1} p_2^{x_2} (1 - p_1 - p_2)^{n - x_1 - x_2}, \qquad (x_1, x_2) \in S_n.$$

That is to say,

$$(x_1, x_2) \sim T(n, (p_1, p_2)). \quad \square$$

From Theorem 5–1, we see that X_1 and X_2 are negatively correlated, which is not surprising. If X_1 is large, then there are many occurrences of type A and, hence, few occurrences of type B (since there are only n total occurrences). Therefore, if X_1 is large, X_2 should be small.

We next find the marginal and conditional distributions for a trinomial distribution.

Theorem 5–3. Let $(X_1, X_2) \sim T(n, (p_1, p_2))$. Then

a. $$X_1 \sim B(n, p_1) \quad \text{and} \quad X_2 \sim B(n, p_2).$$

b. $$X_1|X_2 \sim B\left(n - X_2, \frac{p_1}{1 - p_2}\right) \quad \text{and} \quad X_2|X_1 \sim B\left(n - X_1, \frac{p_2}{1 - p_1}\right).$$

Proof

a. The marginal moment-generating function of X_1 is

$$M_1(t) = M(t, 0) = (p_1 e^t + p_2 e^0 + (1 - p_1 - p_2))^n = (p_1 e^t + 1 - p_1)^n,$$

which is the moment-generating function for $B(n, p_1)$, so that $X_1 \sim B(n, p_1)$. The proof for X_2 is similar.

b. Using part a, we see that

$$P(X_1 = x_1 | X_2 = x_2) = \frac{P(X_1 = x_1, X_2 = x_2)}{P(X_2 = x_2)}$$

$$\frac{\begin{bmatrix} & n & \\ x_1 & x_2 & n - x_1 - x_2 \end{bmatrix} p_1^{x_1} p_2^{x_2} (1 - p_1 - p_2)^{n - x_1 - x_2}}{\begin{bmatrix} n \\ x_2 \end{bmatrix} p_2^{x_2} (1 - p_2)^{n - x_2}}$$

$$= \begin{bmatrix} n - x_2 \\ x_1 \end{bmatrix} \left(\frac{p_1}{1 - p_2}\right)^{x_1} \left(1 - \frac{p_1}{1 - p_2}\right)^{n - x_2 - x_1},$$

which is the density function for a $B(n - x_2, p_1/(1 - p_2))$ distribution. Again, the proof for X_2 given X_1 is similar. □

Example 5–1

Suppose we take a sample of size 40 from a population of voters of whom 35% are Republicans, 45% are Democrats, and 20% are Independents. We assume that the population is large enough so that we may assume that we are sampling with replacement. Let X_1 and X_2 respectively be the number of Republicans and Democrats in the sample (so that there are $40 - X_1 - X_2$ Independents). Then

$$(X_1, X_2) \sim T(40, (.35, .45)).$$

Therefore, the probability that we get 15 Republicans, 20 Democrats, and 5 Independents in the sample is

$$\begin{bmatrix} & 40 & \\ 15 & 20 & 5 \end{bmatrix} (.35^{15})(.45^{20})(.20^5) = .01149.$$

But then, Theorem 5–3,

$$X_1 \sim B(40, .35).$$

This result is fairly obvious if we consider the sampling problem as a Bernoulli sampling problem in which Republicans are successes and Democrats and Independents are failures. (Recall our earlier statements about the meaning of "success" and "failure.") Therefore, the probability that there are at least 15 and at most 18 Republicans in the sample is

$$P(15 \leqq X_1 \leqq 18) = \sum_{x=15}^{18} \begin{bmatrix} 40 \\ x \end{bmatrix} (.35^x)(.65^{40-x}).$$

Using the normal approximation to the binomial, we obtain

$$P(15 \leqq X_1 \leqq 18) = P(14.5 \leqq X \leqq 18.5) \doteq P\left(\frac{14.5-14}{3} \leqq Z \leqq \frac{18.5-14}{3}\right)$$

$$= P(.17 \leqq Z \leqq 1.5) = .9332 - .5675 = .3657.$$

Now, the conditional distribution of X_1 given that $X_2 = 15$ is

$$X_1|(X_2 = 15) \sim B\left(40 - 15, \frac{.35}{1 - .45}\right) = B\left(25, \frac{7}{11}\right).$$

(Again, this result is fairly obvious. If there are exactly 15 Democrats in our sample of 40 voters, then there are 25 remaining voters who must be either Republicans or Independents, so that the probability that on a particular draw we draw a Republican if we know that we must draw a Republican or an Independent is $\frac{7}{11}$. In other words, when X_2 is known to be 15, then X_1 is distributed binomially with parameters 25 and $\frac{7}{11}$.) Therefore, the probability of 20 Republicans given that there are exactly 15 Democrats is

$$P(X_1 = 20|X_2 = 15) = \begin{bmatrix} 25 \\ 20 \end{bmatrix} \left(\frac{7}{11}\right)^{20} \left(\frac{4}{11}\right)^5 = .04.$$

We now derive a normal approximation for trinomial probabilities which is often useful.

Theorem 5–4. Let $\mathbf{X}_n = (X_{n1}, X_{n2}) \sim T(n, (p_1, p_2))$. Let

$$W_n = a_1 X_{n1} + a_2 X_{n2}, \qquad \nu_n = EW_n, \qquad \tau_n^2 = \text{var}(W_n).$$

Then

$$\frac{W_n - \nu_n}{\tau_n} \xrightarrow{d} Z \sim N(0, 1).$$

Proof. The moment-generating function of W_n is

$$M_n(t) = (p_1 \exp(a_1 t) + p_2 \exp(a_2 t) + 1 - p_1 - p_2)^n = (M_1(t))^n.$$

The desired result then follows from Theorem 4–5. □

The standard rule of thumb says that the normal approximation may be used as long as $np_1 \geqq 5$, $np_2 \geqq 5$, and $n(1 - p_1 - p_2) \geqq 5$.

Example 5–2

We return to Example 5–1 in which we take a sample of size 40 from a population of whom 35% are Republicans, 45% are Democrats, and 20% are Independents. As before, let X_1 be the number of Republicans and X_2 the number of Democrats. Now, suppose that we want to approximate the probability that there are more Republicans than Democrats in our sample. That is, we want the probability that $W = X_1 - X_2 > 0$.

We note that

$EW = 40(.35 - .45) = -4, \quad \text{var}(W) = 40[(.35)(.65) + (.45)(.55) + 2(.35)(.45)] = 31.6.$

Substituting into Theorem 5–4 yields

$$\frac{W+4}{(31.6)^{1/2}} \mathrel{\dot{\sim}} Z \sim N(0,1).$$

Therefore,

$P(W > 0) = P(W > .5) = P(31.6)^{-1/2}(W+4) > (31.6)^{-1/2}(.5+4)) \doteq P(Z > .8) = .2119.$

Similarly, the probability that there are as many Republicans as Democrats is

$$P(W=0) = P(-.5 \leqq W \leqq .5) \doteq P\left(\frac{3.5}{(31.6)^{1/2}} \leqq Z \leqq \frac{4.5}{(31.6)^{1/2}}\right)$$
$$= P(.62 < Z < .80) = .7881 - .7257 = .0624.$$

(Note that $X_1 - X_2$ does not have a binomial distribution, so that there is no way we could use the binomial distribution or the normal approximation to the binomial distribution to approximate this probability.)

Finally, suppose that we want to find the probability that there are at least twice as many Democrats as Republicans, i.e., that $U = X_2 - 2X_1 > 0$. Then, by Theorem 5–4 again,

$$\frac{U+10}{(71.5)^{1/2}} \mathrel{\dot{\sim}} N(0,1),$$

and hence,

$$P(U > 0) = P(U > .5) \doteq P(Z > 1.24) = .1075.$$

We now present another theorem which is sometimes useful for approximating trinomial probabilities.

Theorem 5–5. Let $(X_n, Y_n) \sim T(n, (a/n, b/n))$, and let U and V be independent with $U \sim P(a)$ and $V \sim P(b)$. Let $f_n(x,y)$ be the joint density function of (X_n, Y_n), and let $f(u,v)$ be the joint density function of (U,V). Then

$$f_n(x,y) \to f(x,y)$$

for all nonnegative integers x and y.

Proof. See Exercise C2. □

Note that $X_n \sim B(n, a/n)$ and $Y_n \sim B(n, b/n)$, so we already know that $X_n \xrightarrow{d} U$ and $Y_n \xrightarrow{d} V$. What Theorem 5–5 tells us is that U and V are independent. (X_n and Y_n are not independent for any n.) We often state this theorem in the form, If $(X, Y) \sim T(n, (p_1, p_2))$, and if n is large and p_1 and p_2 are small, then X and Y are approximately independent, and

$$X \mathrel{\dot{\sim}} P(np_1) \quad \text{and} \quad Y \mathrel{\dot{\sim}} P(np_2).$$

Example 5–3

A seed manufacturer sells seeds in packages of 500. The company guarantees that at most five of the seeds will not germinate, and that of those that do germinate, at most

10 will not live through the first year. Suppose that .4% of the seeds do not germinate and that 1% germinate but do not live for a year. Let X be the number of seeds in a packet which do not germinate and Y be the number in a packet which germinate but do not live a year. Then (X, Y) has a trinomial distribution with parameters 500, .004, and .01. Accordingly, let U and V have independent Poisson distributions with means $500(.004) = 2$ and $500(.01) = 5$. Then, using Table 6 at the back of the book, the probability that the guarantee is satisfied is

$$P(X \leqq 5, Y \leqq 10) \doteq P(U \leqq 5, V \leqq 10) = P(U \leqq 5)P(V \leqq 10) = (.983)(.986) = .969.$$

We now consider a notation for the trinomial distribution which will lead to the notation we shall use in the next section for the multinomial distribution. Let $(X_1, X_2) \sim T(n, p_1, p_2)$, and let

$$X_3 = n - X_1 - X_2 \quad \text{and} \quad p_3 = 1 - p_1 - p_2.$$

Then (X_1, X_2, X_3) has joint density function

$$f^*(x_1, x_2, x_3) = P(X_1 = x_1, X_2 = x_2, X_3 = x_3) = P(X_1 = x_1, X_2 = x_2)$$

if $x_3 = n - x_1 - x_2$. Therefore, (X_1, X_2, X_3) has joint density

$$f^*(x_1, x_2, x_3) = \begin{bmatrix} & n & \\ x_1 & x_2 & x_3 \end{bmatrix} p_1^{x_1} p_2^{x_2} p_3^{x_3}, \qquad (x_1, x_2, x_3) \in S_n^*,$$

where $S_n^* = \{(x_1, x_2, x_3) \colon x_i = 0, 1, \ldots, x_1 + x_2 + x_3 = n\}$. Using the trinomial theorem, it is straightforward to show that (X_1, X_2, X_3) has joint moment-generating function

$$M^*(t) = (p_1 e^{t_1} + p_2 e^{t_2} + p_3 e^{t_3})^n.$$

We note that the joint density function and joint moment-generating function of (X_1, X_2, X_3) are somewhat nicer than they are for (X_1, X_2). Notice also that the density functions of (X_1, X_2) and (X_1, X_2, X_3) are ways of representing the same model, in which we have n independent replications of an experiment with three possible outcomes.

When $(X_1, X_2) \sim T(n, p_1, p_2)$, the joint distribution of X_1, X_2, and $X_3 = n - X_1 - X_2$ is a special case of the multinomial distribution discussed in the next section. In this case, we often say that $\mathbf{X} = (X_1, X_2, X_3)$ has a three-dimensional multinomial distribution and write

$$(X_1, X_2, X_3) \sim M_3(n, p_1, p_2, p_3),$$

where $p_3 = 1 - p_1 - p_2$.

Exercises—A

1. Suppose a red die and a green die are rolled 20 times.

(a) Find the probability that the green die is higher eight times and they are tied three times (so that the red die is higher nine times).

(b) Find the probability that they are tied exactly three times and the probability they are tied at least three times.

(c) Find the probability that they are tied exactly three times given that they are tied at least three times.

(d) Find the conditional probability that they are tied three times given that the green die wins nine times.

2. A committee of size 15 is to be chosen from a very large high school (so that we may assume that we are sampling with replacement). Suppose the high school has 30% Hispanic students, 40% black students, and 30% white students.
   (a) Find the probability that the committee has four Hispanic students, five white students, and six black students.
   (b) Find the probability that there are equal numbers of Hispanics, whites, and blacks.
   (c) Find the probability that there are equal numbers of all three groups given that there are five blacks.
   (d) What is the covariance between the number of Hispanic students and the number of black students?
3. Answer parts a–d of Exercise 2 if the total number of students in the school is 30. (See Exercise C1.)
4. A committee of size 100 is to be chosen from a large high school in which 30% of the students are Hispanic, 40% are black, and 30% are white.
   (a) Find the probability that more than 50% of the committee is black.
   (b) Find the probability that there are more black than Hispanic students.
   (c) Find the probability that there is the same number of whites as Hispanics.
   (d) Find the probability that there are at least twice as many blacks as whites on the committee.
5. Suppose an insurance company has 1,000 life insurance policies. Suppose that .1% of the people will die of a heart attack in the next year, .2% will die of cancer, and the other 99.7% will live through the year.
   (a) Find the probability that at most three people die from a heart attack and at most five people die of cancer.
   (b) Find the probability that at most seven people die of either disease. (How is the total number of people who die distributed?)
6. Suppose we have 30 light bulbs whose lifetimes are independently exponentially distributed with mean 200 hours.
   (a) Find the probability that the number of light bulbs which burn out in the first 100 hours is the same as the number of light bulbs which burn out in the next 100 hours is the same as the number of light bulbs which last more than 200 hours.
   (b) Find the probability that the number which burn out in the first 100 hours is the same as the number which burn out after 200 hours.

Exercises—B

1. Let $X_1, \ldots, X_{25}$ be independent and $X_i \sim N(0,1)$. Let U be the number of X_i less than $-.5$, and let V be the number of X_i greater than 1.
   (a) What is the joint distribution of U and V?
   (b) Find $P(U = 10, V = 8)$.
   (c) What are EUV and $E(U - V)^2$?
2. Let $(X_1, X_2) \sim T(n, (p_1, p_2))$.
   (a) Show that $X_1 + X_2 \sim B(n, p_1 + p_2)$.
   (b) Show that $n - X_2 \sim B(n, 1 - p_2)$.
3. Verify the calculations for $\text{var}(X_i)$ and $\text{cov}(X_1, X_2)$ in Theorem 5–1.
4. Verify the formula for the moment-generating function of W in Theorem 5–4.

5. Let X_1, X_2, X_3 be independent and $X_i \sim P(m_i)$. Let $U = X_1 + X_2 + X_3$ and $q = m_1 + m_2 + m_3$.
   (a) What is the joint density function of (X_1, X_2, U)?
   (b) Show that $(X_1, X_2)|U \sim T(U, (m_1/q, m_2/q))$. (*Hint*: $U \sim P(q)$. Why?)
6. Let $\mathbf{X} = (X_1, X_2, X_3) \sim M_3(n, p_1, p_2, p_3)$. Verify the formula for the moment-generating function of $\mathbf{X}$.

Exercises—C

1. Consider a sample of size n without replacement from a population of size N of which Np_1 are type A, Np_2 are type B, and $N(1 - p_1 - p_2)$ are type C. Let X_1 be the number of type A's in the sample, and let X_2 be the number of type B's.
   (a) What is the joint density function of X_1 and X_2?
   (b) Argue that X_i has a hypergeometric distribution. What are the mean and variance of X_i?
   (c) Let $U = X_1 + X_2$. Argue that U has a hypergeometric distribution. What are the mean and variance of U?
   (d) What is $\text{cov}(X_1, X_2)$? (*Hint*: $\text{var}(U) = \text{var}(X_1) + \text{var}(X_2) + 2\,\text{cov}(X_1, X_2)$.)
   (e) Show that the correlation coefficient between X_1 and X_2 is the same whether we sample with or without replacement.
2. Prove Theorem 5–5. (See the proof of Theorem 4–9.)

5.1.2 The Multinomial Distribution

In this section, we generalize the binomial and trinomial distributions to the case in which each trial has k possible outcomes. Since the theorems presented are straightforward extensions of those in the previous section, their proofs are left to the reader.

Let $\mathbf{X} = (X_1, \ldots, X_k)$ be a k-dimensional random vector with range

$$S_n = \{(x_1, \ldots, x_k) \colon x_i = 0, 1, \ldots, x_1 + \cdots + x_k = n\}.$$

(That is, the X_i are nonnegative integer-valued random variables whose sum is n.) We say that $\mathbf{X} = (X_1, \ldots, X_k)$ has a k-dimensional *multinomial distribution* with parameters n and $\mathbf{p} = (p_1, \ldots, p_k)$ and write

$$(X_1, \ldots, X_k) \sim M_k(n, \mathbf{p})$$

if $\mathbf{X}$ has joint density function

$$f(x_1, \ldots, x_k) = \begin{bmatrix} & & n & \\ x_1 & x_2 & \cdots & x_k \end{bmatrix} p_1^{x_1} p_2^{x_2} \cdots p_k^{x_k}, \qquad \mathbf{x} \in S_n,$$

where n is a positive integer and the p_i are constants such that

$$p_1 + p_2 + \cdots + p_k = 1.$$

(Recall that if $x_1 + x_2 + \cdots + x_k = n$, then

$$\begin{bmatrix} & & n & \\ x_1 & x_2 & \cdots & x_k \end{bmatrix} = \frac{n!}{x_1!x_2! \ldots x_k!} = \frac{n!}{\prod x_i!}.\Bigg)$$

Note that

$$X_1 + X_2 + \cdots + X_k = n,$$

and hence,

$$X_k = n - X_1 - X_2 - \cdots - X_{k-1} \quad \text{and} \quad p_k = 1 - p_1 - p_2 - \cdots - p_{k-1}.$$

Note also that

$$(X_1, X_2) \sim M_2(n, (p_1, p_2)) \Leftrightarrow X_1 \sim B(n, p_1), \qquad X_2 = n - X_1$$

and

$$(X_1, X_2, X_3) \sim M_3(n, (p_1, p_2, p_3)) \Leftrightarrow (X_1, X_2) \sim T(n, (p_1, p_2)), \qquad X_3 = n - X_1 - X_2.$$

The following theorem summarizes some important facts about the multinomial distribution.

Theorem 5–6

a. The multinomial density function is a joint density for all positive integers n and $p_1, \ldots, p_k$ such that $p_i \geqq 0$ and $p_1 + \cdots + p_k = 1$.
b. If $\mathbf{X} \sim M_k(n, \mathbf{p})$, then

$$EX_i = np_i, \qquad \text{var}(X_i) = np_i(1 - p_i), \qquad \text{cov}(X_i, X_j) = -np_i p_j,$$
$$M(\mathbf{t}) = E\ \exp(X_1 t_1 + \cdots + X_k t_k) = (p_1 e^{t_1} + p_2 e^{t_2} + \cdots + p_k e^{t_k})^n.$$

c. If $\mathbf{X} \sim M_k(n, \mathbf{p})$, then

$$X_i \sim B(n, p_i) \quad \text{and} \quad (X_i, X_j) \sim T(n, p_i, p_j).$$

The next theorem indicates the primary application of the multinomial distribution.

Theorem 5–7. Suppose we have n independent replications of an experiment with k possible disjoint outcomes $A_1, \ldots, A_k$. Let

$$p_i = P(A_i),$$

and let X_i be the number of occurrences of outcome A_i. Then

$$(X_1, \ldots, X_k) \sim M_k(n, (p_1, \ldots, p_k)).$$

Example 5–4

Suppose we roll a fair die 30 times. Let X_1 be the number of times face 1 occurs, X_2 the number of times face 2 occurs, . . . , X_6 the number of times face 6 occurs. Then

$$(X_1, \ldots, X_6) \sim M_6(30, (\tfrac{1}{6}, \tfrac{1}{6}, \ldots, \tfrac{1}{6})).$$

Therefore, the probability that each face occurs five times is

$$f(5, \ldots, 5) = \begin{bmatrix} & & 30 & & & \\ 5 & 5 & 5 & 5 & 5 & 5 \end{bmatrix} \left(\frac{1}{6}\right)^{30} = .00040.$$

The probability that all faces occur five times given that face 1 occurs five times is

$$\frac{P(X_1=5,\ldots,X_6=5)}{P(X_1=5)}=\frac{\begin{bmatrix} & & 30 & & & \\ 5 & 5 & 5 & 5 & 5 & 5\end{bmatrix}(1/6)^{30}}{\begin{bmatrix}30\\5\end{bmatrix}(1/6)^5(5/6)^{25}}$$

$$=\begin{bmatrix} & & 25 & & \\ 5 & 5 & 5 & 5 & 5\end{bmatrix}\left(\frac{1}{5}\right)^{25}=.0021$$

(using the fact that $X_1 \sim B(30, \frac{1}{5})$). Actually, this answer is fairly obvious intuitively. If we know that a one occurs exactly five times in 30 rolls, then there are 25 rolls left. But then, clearly, on these 25 rolls, each of the faces 2, . . . , 6 has equal probability, $\frac{1}{5}$.

The next theorem gives a normal approximation which is often useful.

Theorem 5–8. Let $\mathbf{X}_n = (X_{1n}, \ldots, X_{kn}) \sim M_k(n, (p_1, \ldots, p_k))$. Let

$$W_n = \sum a_i X_{in}, \qquad \nu_n = EW_n, \qquad \tau_n^2 = \text{var}(W_n).$$

Then

$$\frac{W_n - \nu_n}{\tau_n} \xrightarrow{d} Z \sim N(0,1) \quad \text{as } n \to \infty.$$

A rule of thumb is that the normal approximation may be applied as long as $np_i \geqq 5$ for all i.

One standard application of the multinomial distribution is for sampling with replacement (or sampling without replacement from populations which are large enough so that the chance of being sampled twice is negligible).

Example 5–5

Suppose that a school contains 30% freshmen, 25% sophomores, 25% juniors, and 20% seniors. Suppose also that we are randomly choosing a committee of size 20. Finally, suppose that the size of the school is large enough that we may act as though we are sampling with replacement. Let X_1 be the number of freshmen, X_2 the number of sophomores, X_3 the number of juniors, and X_4 the number of seniors. Then

$$(X_1, X_2, X_3, X_4) \sim M_4(20, (.3, .25, .25, .2)).$$

Therefore,

$$EX_1 = 20(.3) = 6, \qquad EX_2 = 5, \qquad EX_3 = 5, \qquad EX_4 = 4.$$

Now, the probability of six freshmen, five sophomores, five juniors, and four seniors being on the committee is

$$P(X_1=6, X_2=5, X_3=5, X_4=4)=\begin{bmatrix} & 20 & & \\ 6 & 5 & 5 & 4\end{bmatrix}.3^6\,.25^5\,.25^5\,.2^4 = .011.$$

Let $U = X_1 + X_2$ be the number of underclassmen on the committee. Then

$$U \sim B(20, .55).$$

(Why?) Therefore, the probability that a majority of the committee is underclassmen is

$$P(U \geqq 11) = P\left(\frac{U-11}{2.22} \geqq \frac{10.5-11}{2.22}\right) \doteq P(Z \geqq -.22) = .5871,$$

using the normal approximation to the binomial distribution. Moreover,

$$(U, X_3) \sim T(20, .55, .25).$$

(Why?) Therefore,

$$P(U = 10, X_3 = 6) = \begin{bmatrix} & 20 & \\ 10 & 6 & 4 \end{bmatrix} .55^{10}\,.25^{6}\,.20^{4} = .038.$$

We now look at the probability that there are more seniors on the committee than juniors, i.e., that $W = X_4 - X_3 > 0$. Since $(X_3, X_4) \sim T(20, (.25, .20))$, the exact answer to this question is

$$\sum_{(u,v)\in T}\sum \begin{bmatrix} & 20 & \\ u & v & 20-u-v \end{bmatrix} (.25)^u(.20)^v(.55)^{20-u-v},$$

$$T = \{(u, v): 0 \leqq u < v, u + v \leqq 20\}.$$

Now,

$$EW = 20(.20 - .25) = -1,$$

and

$$\text{var}(W) = 20[(.20)(.80) + (.25)(.75) + 2(.20)(.25)] = 8.95.$$

Therefore, a normal approximation to this probability is

$$P(W > .5) \doteq P\left(Z > \frac{1.5}{(8.95)^{1/2}}\right) = P(Z > .50) = .3085.$$

We have seen that the multinomial distribution can be used for finding probabilities when we are sampling with replacement (or sampling without replacement from large populations). We next consider the appropriate distribution to use when we are sampling without replacement from small or moderate populations.

Suppose we are taking a sample of size n without replacement from a population of size N of objects which are of k types, $A_1, \ldots, A_k$. Suppose also that there are Np_i objects of type A_i (so that $\Sigma p_i = 1$). Let X_i be the number of objects of type A_i that we draw in our sample. By arguments in Chapter 1, the joint density of the X_i is

$$f^*(\mathbf{x}) = \frac{\begin{bmatrix} Np_1 \\ x_1 \end{bmatrix}\begin{bmatrix} Np_2 \\ x_2 \end{bmatrix}\cdots\begin{bmatrix} Np_k \\ x_k \end{bmatrix}}{\begin{bmatrix} N \\ n \end{bmatrix}}, \qquad \mathbf{x} \in S_n,$$

where S_n is the same set as in the multinomial distribution. (That is, the X_i are nonnegative integer-valued random variables which sum to n.) We call the joint distribution of the X_i a k-dimensional *multivariate hypergeometric* distribution with parameters N, n, and $\mathbf{p} = (p_1, \ldots, p_k)$. Now, X_i has a (univariate) hypergeometric distribution with parameters N, n, and p_i. (Why?) Therefore,

$$EX_i = np_i \quad \text{and} \quad \text{var}(X_i) = \frac{np_i(1-p_i)(N-n)}{N-1}.$$

We now find $\text{cov}(X_i, X_j)$. Let $U = X_i + X_j$. Then U has a (univariate) hypergeometric distribution with parameters N, n, and $p_i + p_j$. (Why?) Therefore,

$$\text{var}(U) = \frac{n(p_i + p_j)(1 - p_i - p_j)(N - n)}{N - 1}.$$

But $\text{var}(U) = \text{var}(X_i) + \text{var}(X_j) + 2\,\text{cov}(X_i, X_j)$. Therefore,

$$\text{cov}(X_i, X_j) = \frac{\text{var}(U) - \text{var}(X_i) - \text{var}(X_j)}{2} = \frac{-np_ip_j(N - n)}{N - 1}.$$

It is interesting that $\text{cov}(X_i, X_j)$ has the same finite population correction as $\text{var}(X_i)$ does. It is also interesting that the correlation coefficient between X_i and X_j is

$$\rho(X_i, X_j) = -\left[\frac{p_ip_j}{(1 - p_i)(1 - p_j)}\right]^{1/2}$$

which does not depend on N or n and is the same as the correlation coefficient for sampling with replacement.

Example 5–6

Suppose that we are randomly choosing a committee of 20 without replacement from a school with 100 students of whom 30% are freshmen, 25% are sophomores, 25% are juniors, and 20% are seniors. Let X_1 be the number of freshmen chosen, X_2 the number of sophomores, X_3 the number of juniors, and X_4 the number of seniors. Then (X_1, X_2, X_3, X_4) has a multivariate hypergeometric distribution with parameters $N = 100$, $n = 20$, and $\mathbf{p} = (.3, .25, .25, .2)$. Therefore,

$$P(X_1 = 6, X_2 = 5, X_3 = 5, X_4 = 4) = \frac{\begin{bmatrix}30\\6\end{bmatrix}\begin{bmatrix}25\\5\end{bmatrix}\begin{bmatrix}25\\5\end{bmatrix}\begin{bmatrix}20\\4\end{bmatrix}}{\begin{bmatrix}100\\20\end{bmatrix}} = .015.$$

Now, let $U = X_1 + X_2$ be the number of underclassmen. Then U has a hypergeometric distribution with parameters 100, 20, and .55. Therefore, the probability that the underclassmen are in the majority is

$$P(U \geqq 11) = P(U \geqq 10.5) = P\left(\frac{U - 11}{2} \geqq \frac{10.5 - 11}{2}\right) = P(Z \geqq -.25) = .5987,$$

using the normal approximation with the finite population correction. Now, (U, X_3, X_4) has a trivariate hypergeometric distribution with parameters 100, 20, and $(.55, .25, .20)$. (Why?) Therefore,

$$P(U = 10, X_3 = 6, X_4 = 4) = \frac{\begin{bmatrix}55\\10\end{bmatrix}\begin{bmatrix}25\\6\end{bmatrix}\begin{bmatrix}20\\4\end{bmatrix}}{\begin{bmatrix}100\\20\end{bmatrix}} = .047.$$

Note that these answers are somewhat different from those we got in Example 5–3 for sampling with replacement. Note for this situation, the finite population correction is

$$\left(\frac{100 - 20}{100 - 1}\right)^{1/2} = .90.$$

Exercises—A

1. Suppose that a teacher typically gives 15% A's, 25% B's, 35% C's, 15% D's, and 10% F's. Suppose further that she has a class of 50 students.
   (a) Find the probability that she gives 8 A's, 10 B's, 20 C's, and 8 D's to the class.
   (b) Find the probability that she gives 8 A's and 10 B's given that she gives 8 D's and 4 F's.
   (c) Find the probability that she gives 8 A's and 10 B's given that she gives 12 grades which are D or F.
   (d) Find the probability that she gives eight A's given that she gives four F's.
   (e) Find the variance of the difference between the number of A's and the number of F's.
2. Use the normal approximation to find approximate numerical answers to the following questions:
   (a) Find the probability that at least 40 students from the class in Exercise 1 pass (i.e., get an A, B, or C).
   (b) Find the probability that the teacher gives at most 15 A's or B's.
   (c) Find the probability that she gives more A's and B's than D's and F's.
   (d) Find the probability that she gives at most 10 A's given that she passes exactly 40 students.
3. A large forest contains 40% pine trees, 20% oak trees, 20% maple trees, 10% spruce trees, and 10% ash trees. We randomly sample 25 trees. (Assume that the forest is large enough so that we employ sampling with replacement.)
   (a) Find the probability that we get an equal number of each type of tree.
   (b) Find the probability that we get an equal number of each tree given that we get five pines and five oaks.
   (c) Find the probability that we get 10 evergreens. (Pines and spruces are evergreens.)
   (d) Find the probability that we get an equal number of each type given that we get 10 evergreen trees.
   (e) Use the normal approximation to find the probability that we get at least 10 evergreens.
4. Assume now that there are only 80 trees in the forest of Exercise 3, so that we cannot employ sampling with replacement. Answer parts a–d of Exercise 3.
5. Suppose that on a particular night, the probability that a randomly chosen person is watching CBS is .20, the probability that he or she is watching ABC is .15, the probability that the person is watching NBC is .25, the probability of another station is .15, and the probability that he or she is not watching at all is .25. Suppose that 400 such people are called.
   (a) What is the probability that more people are watching CBS than NBC?
   (b) What is the probability that more people are not watching television than are watching NBC?
   (c) What is the probability that more people are watching CBS, NBC, or ABC than are watching another station or are not watching?

Exercises—B

1. Prove parts a and b of Theorem 5–6. (The multinomial theorem says that

$$\sum \cdots \sum_{(x_1,\ldots,x_k)\in S_n} \begin{bmatrix} & n & & \\ x_1 & x_2 & \cdots & x_k \end{bmatrix} a_1^{x_1} \ldots a_k^{x_k} = (a_1 + \cdots + a_k)^n.)$$

2. Prove part c of Theorem 5–6.
3. Let $\mathbf{X} = (X_1, \ldots, X_4) \sim M_4(n, (p_1, \ldots, p_4))$.
   (a) What is $P(X_1 = 2 | X_2 = 1)$?
   (b) What is the conditional distribution of X_1 given X_2?
   (c) What is $P(X_1 = 2, X_2 = 1 | X_3 = 3)$?
   (d) What is the conditional distribution of (X_1, X_2) given X_3?
4. (a) Show that Theorem 5–7 holds in the case in which $k = 4, n = 7, x_1 = 1, x_2 = 2, x_3 = 1$, and $x_4 = 3$.
   (b) Give an argument for Theorem 5–7 in the general case.
5. Let $\mathbf{X} = (X_1, \ldots, X_6) \sim M_6(n, (p_1, \ldots, p_6))$. Let $U = X_1 + X_2$, $V = X_3 + X_4 + X_5$, and $W = X_6$.
   (a) Show that (U, V) has a trinomial distribution. What are the parameters?
   (b) Find the conditional distribution of U given V.
   (c) Find the conditional distribution of U given X_6.
6. Let $\mathbf{X} = (X_1, \ldots, X_6)$ have a six-dimensional hypergeometric distribution with parameters N, n, and $\mathbf{p} = (p_1, \ldots, p_6)$. Let $U = X_1 + X_2$, $V = X_3 + X_4 + X_5$, and $W = X_6$.
   (a) Argue that (U, V, W) has a trivariate hypergeometric distribution. What are the parameters?
   (b) What is the conditional distribution of U given V?
7. Prove Theorem 5–8.

Exercises—C

1. Let $(X_1, X_2, X_3, X_4) \sim M_4(n, (p_1, p_2, p_3, p_4))$. Let $U = X_2 + X_3$.
   (a) What is the distribution of (X_1, U)?
   (b) Show that the conditional distribution of X_1 given $X_2 = 2$ and $X_3 = 3$ is the same as the conditional distribution of X_1 given $U = 5$.

5.2 BIVARIATE NORMAL DISTRIBUTION

5.2.1 The Bivariate Normal Distribution

In this section, we extend the normal distribution to two variables. Later, we extend these results to k variables, using matrix notation. Matrix notation is somewhat more elegant than the notation used in this section, and those who are familiar with matrices may choose to skip the derivations set out here. However, the numerical examples may still be interesting.

Let $\mathbf{X} = (X_1, X_2)$. We say that $\mathbf{X}$ has a *bivariate normal distribution* with means μ_1 and μ_2, variances σ_1^2 and σ_2^2, and correlation coefficient ρ and write

$$\mathbf{X} = (X_1, X_2) \sim N_2(\mu_1, \mu_2, \sigma_1^2, \sigma_2^2, \rho)$$

if (X_1, X_2) has joint moment-generating function

$$M(t_1, t_2) = \exp\left[\mu_1 t_1 + \mu_2 t_2 + \frac{\sigma_1^2 t_1^2 + \sigma_2^2 t_2^2 + 2\rho\sigma_1\sigma_2 t_1 t_2}{2}\right],$$

where μ_1, μ_2, σ_1^2, σ_2^2, and ρ are constants satisfying

$$-\infty < \mu_i < \infty, \qquad \sigma_i^2 > 0, \qquad -1 < \rho < 1.$$

The next theorem gives the basic properties of the bivariate normal distribution. (Note that at this time we have not shown that μ_i and σ_i^2 are the means and variances, or that ρ is the correlation coefficient, of the distribution.)

Theorem 5–9

a. $M(t_1, t_2)$ is a moment-generating function for all μ_1, μ_2, $\sigma_1^2 > 0$, $\sigma_2^2 > 0$, and ρ in the interval $(-1, 1)$.
b. If $(X_1, X_2) \sim N_2(\mu_1, \mu_2, \sigma_1^2, \sigma_2^2, \rho)$, then
$$EX_i = \mu_i, \qquad \text{var}(X_i) = \sigma_i^2, \qquad \text{corr}(X_i, X_j) = \rho.$$
c. Let $(X_1, X_2) \sim N_2(\mu_1, \mu_2, \sigma_1^2, \sigma_2^2, \rho)$. Then
$$X_1 \sim N(\mu_1, \sigma_1^2) \quad \text{and} \quad X_1|X_2 \sim N\left(\mu_1 + \left(\frac{\rho\sigma_1}{\sigma_2}\right)(X_2 - \mu_2), \sigma_1^2(1-\rho^2)\right).$$
d. X_1 and X_2 are independent if and only if $\rho = 0$.

Proof

a. Let Z_1 and Z_2 be independent and $Z_i \sim N(0, 1)$. Then $\mathbf{Z} = (Z_1, Z_2)$ has joint moment-generating function
$$M_Z(s_1, s_2) = \exp\left(\frac{s_1^2}{2}\right)\exp\left(\frac{s_2^2}{2}\right) = \exp\left(\frac{s_1^2 + s_2^2}{2}\right).$$
Let
$$X_1 = \mu_1 + \sigma_1((1-\rho^2)^{1/2} Z_1 + \rho Z_2) \quad \text{and} \quad X_2 = \mu_2 + \sigma_2 Z_2. \tag{5–1}$$
Then X_1 and X_2 have joint moment-generating function
$$\begin{aligned} M_X(t_1, t_2) &= E\ \exp[t_1(\mu_1 + \sigma_1((1-\rho^2)^{1/2} Z_1 + \rho Z_2)) + t_2(\mu_2 + \sigma_2 Z_2)] \\ &= \exp(\mu_1 t_1 + \mu_2 t_2) E\ \exp[Z_1(t_1\sigma_1(1-\rho^2)^{1/2}) + Z_2(\sigma_1\rho t_1 + \sigma_2 t_2)] \\ &= \exp(\mu_1 t_1 + \mu_2 t_2) M_Z(t_1\sigma_1(1-\rho^2)^{1/2}, \sigma_1\rho t_1 + \sigma_2 t_2) \\ &= \exp(\mu_1 t_1 + \mu_2 t_2)\ \exp\left(\frac{t_1^2\sigma_1^2(1-\rho^2) + \sigma_1^2\rho^2 t_1^2 + \sigma_2^2 t_2^2 + 2\rho\sigma_1\sigma_2 t_1 t_2}{2}\right) \\ &= \exp\left[\mu_1 t_1 + \mu_2 t_2 + \frac{t_1^2\sigma_1^2 + t_2^2\sigma_2^2 + 2\rho\sigma_1\sigma_2 t_1 t_2}{2}\right], \end{aligned}$$
so that $(X_1, X_2) \sim N(\mu_1, \mu_2, \sigma_1^2, \sigma_2^2, \rho)$.
b. The means, variances, and covariance follow directly from differentiating $\psi(t_1, t_2) = \log(M(t_1, t_2))$.
c. The Z_i are independent, with $Z_i \sim N(0, 1)$. Also,
$$X_1 = \mu_1 + \sigma_1(1-\rho^2)^{1/2} Z_1 + \sigma_1\rho Z_2.$$

Therefore, by part b of Theorem 4–32,

$$X_1 \sim N(\mu_1, \sigma_1^2(1-\rho^2) + \sigma_1^2\rho^2) = N(\mu_1, \sigma_1^2).$$

We next find the conditional density of X_1 given $Z_2 = z_2$. Note that

$$Z_1 | Z_2 = z_2 \sim N(0, 1),$$

since Z_1 and Z_2 are independent. Therefore, the conditional distribution of

$$X_1 = (\mu_1 + \sigma_1 \rho Z_2) + \sigma_1(1-\rho^2)^{1/2} Z_1$$

given $Z_2 = z_2$ is

$$X_1 | Z_2 = z_2 \sim N(\mu_1 + \rho\sigma_1 z_2, \sigma_1^2(1-\rho^2))$$

(since z_2 is a constant in the conditional distribution of X_1 given $Z_2 = z_2$). Now, $X_2 = \sigma_2 Z_2 + \mu_2$. Therefore, $X_2 = x_2$ if and only if $Z_2 = (x_2 - \mu_2)/\sigma_2$, and it follows that the conditional distribution of X_1 given $X_2 = x_2$ is the same as the conditional distribution of X_1 given $Z_2 = (x_2 - \mu_2)/\sigma_2$. That is,

$$X_1 | X_2 = x_2 \sim N\left(\mu_1 + \frac{\rho\sigma_1(x_2-\mu_2)}{\sigma_2}, \sigma_1^2(1-\rho^2)\right).$$

d. The conditional distribution of X_1 given X_2 is the same as the marginal distribution of X_1 if and only if $\rho = 0$. Therefore, X_1 and X_2 are independent if and only if $\rho = 0$. □

In a similar way, we could show that

$$X_2 \sim N(\mu_2, \sigma_2^2) \quad \text{and} \quad X_2 | X_1 \sim N\left(\mu_2 + \left(\frac{\rho\sigma_2}{\sigma_1}\right)(X_1 - \mu_1), \sigma_2^2(1-\rho^2)\right).$$

Note that the conditional means are both linear functions and the conditional variances are both constants. However, even though the conditional variance of X_1 given X_2, $\sigma_1^2(1-\rho^2)$, is constant, it is not equal to the variance of X_1, i.e., σ^2, unless $\rho = 0$.

Part d of Theorem 5–9 says that if X_1 and X_2 have a bivariate normal distribution, then they are independent if and only if they are uncorrelated. Of course, independent random variables must be uncorrelated. However, earlier we saw examples of random variables which are uncorrelated but not independent. In fact, in Exercise C1, X_1 and X_2 are each marginally normally distributed and are uncorrelated, but are not independent. This problem shows that it is important that X_1 and X_2 be jointly normally distributed in order to guarantee that uncorrelated random variables are independent.

The bivariate normal distribution is the most commonly used bivariate distribution. It is used to model bivariate measurements such as height and weight, high school grade point average and college grade point average, and fertilizer and output on a field. The bivariate central limit theorem implies that any pair of measurements which is a sum of many small effects should have an approximate bivariate normal distribution. The trinomial distribution discussed in Section 5.1.1 has an approximate bivariate normal distribution when n is large. (See Chapter 6.)

Example 5–7

Suppose that the weight X_1 and height X_2 of a randomly chosen male have a bivariate normal distribution with means of 170 pounds and 68 inches, variances of 400 square pounds and 16 square inches, and correlation coefficient .8. Then $X_1 \sim N(170, (20)^2)$. Therefore,

$$P(150 < X_1 < 180) = P(-1 < Z < .5) = .5328.$$

By Theorem 5–9, the conditional distribution of weight given height is

$$X_1|X_2 \sim N(4X_2 - 102, 144).$$

Now suppose that we know the chosen male is 72 inches tall. Then

$$X_1|X_2 = 72 \sim N(186, (12)^2).$$

Hence, the probability that a randomly chosen male 72 inches tall weighs more than 198 pounds is

$$P(X_1 > 198|X_2 = 72) = P\left(\frac{X_1 - 186}{12} > \frac{198 - 186}{12}\right) = P(Z > 1) = .1587.$$

Note that the conditional standard deviation of weight for a particular height in this case is 12, while the (marginal) standard deviation is 20. This reflects the obvious fact that the distribution of weights for males of a particular height is less spread out than the distribution of weights for all males.

The next theorem gives two useful results about the bivariate normal distribution.

Theorem 5–10

a. Let $(X_1, X_2) \sim N(\mu_1, \mu_2, \sigma_1^2, \sigma_2^2, \rho)$. Let

$$W = a_1 X_1 + a_2 X_2.$$

Then

$$W \sim N(\nu, \tau^2), \qquad \nu = a_1 \mu_1 + a_2 \mu_2, \qquad \tau^2 = a_1^2 \sigma_1^2 + a_2^2 \sigma_2^2 + 2a_1 a_2 \rho \sigma_1 \sigma_2.$$

b. Let $X_1, \ldots, X_n$ be independent, with $X_i \sim N(\mu_i, \sigma_i^2)$. Let

$$U = \sum_i a_i X_i \quad \text{and} \quad V = \sum_i b_i X_i.$$

Then

$$(U, V) \sim N_2(\nu_1, \nu_2, \tau_1^2, \tau_2^2, \rho), \qquad \nu_1 = \sum_i a_i \mu_i, \qquad \nu_2 = \sum_i b_i \mu_i,$$

$$\tau_1^2 = \sum_i a_i^2 \sigma_i^2, \qquad \tau_2^2 = \sum_i b_i^2 \sigma_i^2, \qquad \rho = \frac{\sum_i a_i b_i \sigma_i^2}{\tau_1 \tau_2}.$$

Proof

a. See the exercises.

b. The joint moment-generating function of the X's is the product of the marginal moment-generating functions, since they are independent, and is therefore given by

$$M_X(s_1, \ldots, s_n) = \exp\left(\sum_i s_i \mu_i + \frac{\sum_i s_i^2 \sigma_i^2}{2}\right).$$

The joint moment-generating function of U and V is

$$M_{U,V}(t_1, t_2) = E\ \exp\left(t_1 \sum_i a_i X_i + t_2 \sum_i b_i X_i\right) = E\ \exp\left[\sum_i X_i(a_i t_1 + b_i t_2)\right]$$

$$= M_X(a_1 t_1 + b_1 t_2, \ldots, a_k t_1 + b_k t_2)$$

$$= \exp\left[\sum_i (a_i t_1 + b_i t_2)\mu_i + \frac{\sum_i (a_i t_1 + b_i t_2)^2 \sigma_i^2}{2}\right]$$

$$= \exp\left[t_1 \nu_1 + t_2 \nu_2 + \frac{(t_1^2 \tau_1^2 + t_2^2 \tau_2^2 + 2\rho \tau_1 \tau_2 t_1 t_2)}{2}\right],$$

which is the moment-generating function of a bivariate normal distribution with means ν_1 and ν_2, variances τ_1^2 and τ_2^2, and correlation coefficient ρ. □

Part a of Theorem 5–10 says that if we take a linear function of a bivariate normal random vector, we get a univariate normal random variable. Part b says that when we take two linear combinations of independent univariate random variables, we get a bivariate normal random vector.

The following corollary of the theorem is often useful.

Corollary 5–10. If $(X_1, X_2) \sim N_2(\mu_1, \mu_2, \sigma_1^2, \sigma_2^2, \rho)$, then

$$X_1 - X_2 \sim N(\mu_1 - \mu_2, \sigma_1^2 + \sigma_2^2 - 2\rho\sigma_1\sigma_2)$$

and

$$X_1 + X_2 \sim N(\mu_1 + \mu_2, \sigma_1^2 + \sigma_2^2 + 2\rho\sigma_1\sigma_2).$$

In Exercise C2, it is shown that the joint density function of (X_1, X_2) is

$$f(x_1, x_2) = (2\pi\sigma_1\sigma_2)^{-1}(1 - \rho)^{-1/2} \exp[-q(x_1, x_2)/2(1 - \rho^2)], \quad -\infty < x_i < \infty, \tag{5.2}$$

where

$$q(x_1, x_2) = \left[\frac{x_1 - \mu_1}{\sigma_1}\right]^2 - 2\rho\left[\frac{x_1 - \mu_1}{\sigma_1}\right]\left[\frac{x_2 - \mu_2}{\sigma_2}\right] + \left[\frac{x_2 - \mu_2}{\sigma_2}\right]^2,$$

Fortunately, we shall not need this complicated density function.

Exercises—A

1. Suppose that verbal SAT score X_1 and math SAT score X_2 have a bivariate normal distribution with means 480 and 520, standard deviations 100 and 120, and correlation coefficient .6. Suppose a randomly chosen student takes these tests.

(a) Find the probability that the student gets higher than 500 on the verbal test; on the math test.

(b) Suppose the student gets 560 on the math test. Find the probability that he or she gets above 600 on the verbal test.

(c) Suppose the student gets 560 on the verbal test. Find the probability that he or she gets above 600 on the math test.

(d) Find the probability that the student gets a higher score on the math test than on the verbal test. (See Corollary 5–10.)

(e) Find the probability that the student's total score is over 1,100. (See Corollary 5–10 again.)

2. Suppose we make a study of heights of fathers and sons in a particular generation, letting X_1 be the father's height and X_2 the son's. Suppose that these heights have a bivariate normal distribution with both means 68 inches, both standard deviations 4 inches, and correlation coefficient .6.

   (a) Find the probability that a randomly chosen son's height is between 67 and 74 inches.

   (b) What is the probability that the son is taller than the father?

   (c) What is the probability that the son is at least 4 inches taller than the father?

   (d) Given that the father is 72 inches tall, find the probability that the son is taller than the father.

   (e) Given that the father is 64 inches tall, what is the probability that the son is taller than the father?

Exercises—B

1. Write out the density function and the moment-generating function for a bivariate normal distribution with means 0 and 3, variances 4 and 9, and correlation coefficient .8.

2. Let X_1 and X_2 have a bivariate normal distribution.

   (a) What is $q(b, c) = E(X_1 - bX_2 + c)^2$?

   (b) What choices for b and c minimize $q(b, c)$?

3. Differentiate $\psi(t_1, t_2) = \log[M(t_1, t_2)]$ to establish the means, variances, and covariance for the bivariate normal distribution.

4. (a) Let (X_1, X_2) have moment-generating function $M_X(t_1, t_2)$. Let $W = a_1 X_1 + a_2 X_2$. Show that W has moment-generating function $M_W(t) = M_X(a_1 t, a_2 t)$.

   (b) Prove part a of Theorem 5–10.

   (c) Prove Corollary 5–10.

Exercises—C

1. Let $Z \sim N(0, 1)$. Let U be 1 or -1 with probability $\frac{1}{2}$. Let $Y = ZU$.

   (a) Show that $Y \sim N(0, 1)$. (Show first that $Y|U \sim N(0, 1)$.)

   (b) Show that $EYZ|U = U$. (If $U = 1$, then $YZ = Z^2$, and if $U = -1$, then $YZ = -Z^2$; note that U and Z are independent.)

   (c) Show that Y and Z are uncorrelated.

   (d) Show that Y and Z are not independent. (What is $P(-1 < X < 1, -1 < Y < 1)$?)

2. Let X_1, X_2, Z_1, and Z_2 be as in the proof of Theorem 5–9.

   (a) Show that $Z_1 = ((X_1 - \mu_1)/\sigma_1 + \rho(X_2 - \mu_2)/\sigma_2)/(1 - \rho^2)^{1/2}$ and $Z_2 = (X_2 - \mu_2)/\sigma$.

   (b) Show that the joint density function of (Z_1, Z_2) is $f_Z(z) = (2\pi)^{-1} \exp[-(z_1^2 + z_2^2)/2]$.

   (c) Show that the Jacobian of the transformation from (Z_1, Z_2) to (X_1, X_2) is $J = (\sigma_1 \sigma_2)^{-1}(1 - \rho^2)^{1/2}$.

   (d) Verify Equation (5–2) for the joint density (X_1, X_2).

5.3 THE MULTIVARIATE NORMAL DISTRIBUTION

5.3.1 The Mean Vector and the Covariance Matrix

In presenting the multivariate normal distribution here, and in several later sections of the text, it is convenient to use vector and matrix algebra. Accordingly, we shall mark all sections with a # where vector and matrix algebra are used. Before studying the material in this section, it may be helpful to review the concepts and properties discussed in Appendix A.3.1.

We write both vectors and matrices in boldface. In the sections marked with a #, we follow convention and write vectors as columns. In the remaining sections, we use the more convenient notation of rows. Let R^k be the set of all k-dimensional (real) vectors. If a matrix $\mathbf{A}$ has p rows and q columns, we say that $\mathbf{A}$ is $p \times q$. We write $\mathbf{A}', \mathbf{A}^{-1}$, and $\det(\mathbf{A})$ for the transpose, inverse, and determinant of $\mathbf{A}$, respectively. (Note that $\mathbf{A}^{-1}$ and $\det(\mathbf{A})$ exist only if $\mathbf{A}$ is square ($p \times p$) and that $\mathbf{A}^{-1}$ may not exist even then.) If $\mathbf{A}$ and $\mathbf{B}$ are $p \times q$ and $r \times s$ matrices, then $\mathbf{A} + \mathbf{B}$, and $\mathbf{AB}$ are the sum and product, respectively, of $\mathbf{A}$ and $\mathbf{B}$. (The sum exists only if $p = r$ and $q = s$; the product exists only if $q = r$.) If $\mathbf{A}$ is a matrix and c is a real number, then $c\mathbf{A}$ is the scalar product of c and $\mathbf{A}$. If $\mathbf{a}$ is a $p \times 1$ vector, then $\|\mathbf{a}\| = (\mathbf{a}'\mathbf{a})^{1/2}$ is the length of $\mathbf{a}$. Finally, $\mathbf{I}_p$ or $\mathbf{I}$ is the $p \times p$ identity matrix, and $\mathbf{0}$ is a zero matrix or vector of any dimension. Note that in using matrix algebra, it is especially important to be careful about multiplication, since

$$\mathbf{AB} \neq \mathbf{BA}.$$

A quite useful way exists of summarizing the mean structure and the variance-covariance structure for random vectors. Let

$$\mathbf{X} = \begin{pmatrix} X_1 \\ \cdot \\ \cdot \\ \cdot \\ X_k \end{pmatrix}$$

be a k-dimensional random vector, and let

$$\mu_i = EX_i, \qquad \sigma_{ii} = \mathrm{var}(X_i) = \mathrm{cov}(X_i, X_i), \qquad \sigma_{ij} = \mathrm{cov}(X_i, X_j).$$

We define the *mean vector* $\boldsymbol{\mu}$ and *covariance matrix* $\boldsymbol{\Sigma}$ for $\mathbf{X}$ by

$$\boldsymbol{\mu} = E\mathbf{X} = \begin{pmatrix} \mu_1 \\ \cdot \\ \cdot \\ \cdot \\ \mu_k \end{pmatrix} \quad \text{and} \quad \boldsymbol{\Sigma} = \mathrm{cov}(\mathbf{X}) = \begin{pmatrix} \sigma_{11} & \cdots & \sigma_{1k} \\ \cdot & & \cdot \\ \cdot & & \cdot \\ \cdot & & \cdot \\ \sigma_{k1} & \cdots & \sigma_{kk} \end{pmatrix}.$$

That is, $\boldsymbol{\mu}$ is a $k \times 1$ vector whose ith component is $\mu_i = EX_i$, and $\boldsymbol{\Sigma}$ is a $k \times k$ matrix whose (i, j)th component is the covariance between X_i and X_j. In particular, $\boldsymbol{\Sigma}$ has the variance of X_i as its ith diagonal element.

Note that

$$\sigma_{ij} = \text{cov}(X_i, X_j) = \text{cov}(X_j, X_i) = \sigma_{ji}.$$

Therefore, $\mathbf{\Sigma}$ is a symmetric matrix.

The following theorem gives the basic result about mean vectors and covariance matrices. The *trace* of the $k \times k$ matrix $\mathbf{C}$, written $\text{tr}(\mathbf{C})$, is defined to be the sum of the diagonal elements of $\mathbf{C}$. If $\mathbf{A}$ is a $p \times k$ matrix with elements a_{ij} and B is a $k \times p$ matrix with elements b_{jk}, then it is easily verified that

$$\text{tr}(\mathbf{AB}) = \sum_i \sum_j a_{ij} b_{ji} = \text{tr}(\mathbf{BA}).$$

Theorem 5–11

a. Let $\mathbf{X}$ be a k-dimensional random vector with mean $\boldsymbol{\mu} = E\mathbf{X}$ and $\mathbf{\Sigma} = \text{cov}(\mathbf{X})$. Let $\mathbf{A}$ be a $p \times k$ matrix, and let $\mathbf{b}$ be a $p \times 1$ vector. Finally, let

$$\mathbf{Y} = \mathbf{AX} + \mathbf{b},$$

so that $\mathbf{Y}$ is a $p \times 1$ random vector. Then

$$E\mathbf{Y} = \mathbf{A}\boldsymbol{\mu} + \mathbf{b} = \mathbf{A}E\mathbf{X} + \mathbf{b} \quad \text{and} \quad \text{cov}(\mathbf{Y}) = \mathbf{A\Sigma A}' = \mathbf{A}(\text{cov}(\mathbf{X}))\mathbf{A}'.$$

b. If $\mathbf{C}$ is a symmetric $k \times k$ matrix, then

$$E\mathbf{X}'\mathbf{CX} = \boldsymbol{\mu}'\mathbf{C}\boldsymbol{\mu} + \text{tr}(\mathbf{C\Sigma}).$$

Proof

a. Let a_{ij} be the (i,j)th component of $\mathbf{A}$, let Y_j and b_j be the jth components of $\mathbf{Y}$ and $\mathbf{b}$, respectively, and let $\nu_i = EY_i$ be the ith component of $\boldsymbol{\nu} = E\mathbf{Y}$. Then

$$Y_i = \sum_n a_{in} X_n + b_i.$$

Therefore,

$$\nu_i = EY_i = E\left(\sum_n a_{in} X_n + b_i\right) = \sum_n a_{in} EX_n + b_i = \sum_n a_{in}\mu_n + b_i,$$

which is the ith component of $\mathbf{A}\boldsymbol{\mu} + \mathbf{b}$. Now, let $\Psi_{ij} = \text{cov}(Y_i, Y_j)$ and $\sigma_{ij} = \text{cov}(X_i, X_j)$ be the (i,j)th components of $\mathbf{\Psi} = \text{cov}(\mathbf{Y})$ and $\mathbf{\Sigma} = \text{cov}(\mathbf{X})$, respectively. Then

$$\Psi_{ij} = \text{cov}(Y_i, Y_j) = \text{cov}\left(\sum_n a_{in} X_n + b_i, \sum_m a_{jm} X_m + b_j\right)$$

$$= \sum_n \sum_m a_{in} a_{jm} \text{cov}(X_n, X_m) = \sum_n \sum_m a_{in} \sigma_{nm} a_{jm},$$

which is the (i,j)th component of $\mathbf{A\Sigma A}'$.

b. Let c_{ij} be the (i,j)th component of $\mathbf{C}$. Then

$$\mathbf{X}'\mathbf{CX} = \sum_i \sum_j c_{ij} X_i X_j.$$

Therefore,

$$E\mathbf{X}'\mathbf{C}\mathbf{X} = \sum_i \sum_j c_{ij} EX_i X_j = \sum\sum c_{ij}(\sigma_{ij} + \mu_i \mu_j)$$

$$= \sum_i \sum_j c_{ij}\sigma_{ij} + \sum_i \sum_j c_{ij}\mu_i\mu_j = \mathrm{tr}(\mathbf{C}\boldsymbol{\Sigma}) + \boldsymbol{\mu}'\mathbf{C}\boldsymbol{\mu}. \quad \square$$

Observe that part a of Theorem 5–11 is a fairly natural generalization of the univariate results that $E(aX + b) = aEX + b$ and $\mathrm{var}(aX + b) = a^2\,\mathrm{var}(X)$. Part b is used in Chapters 12 and 13.

Example 5–8

Let X_1, X_2, and X_3 satisfy $EX_1 = 2$, $EX_2 = 4$, $EX_3 = 8$, $\mathrm{var}(X_1) = 4$, $\mathrm{var}(X_2) = 10$, $\mathrm{var}(X_3) = 16$, $\mathrm{cov}(X_1, X_2) = 2$, $\mathrm{cov}(X_1, X_3) = -3$, and $\mathrm{cov}(X_2, X_3) = -5$. Then $\mathbf{X} = (X_1, X_2, X_3)'$ has mean vector

$$\boldsymbol{\mu} = \begin{Bmatrix} 2 \\ 4 \\ 8 \end{Bmatrix}$$

and covariance matrix

$$\boldsymbol{\Sigma} = \begin{Bmatrix} 4 & 2 & -3 \\ 2 & 10 & -5 \\ -3 & -5 & 16 \end{Bmatrix}.$$

Now, let $Y_1 = X_1 - 2X_2 + 3X_3 - 4$ and $Y_2 = 2X_1 + X_2 - 3X_3 + 5$. Then

$$\mathbf{Y} = \begin{Bmatrix} Y_1 \\ Y_2 \end{Bmatrix} = \begin{Bmatrix} 1 & -2 & 3 \\ 2 & 1 & -3 \end{Bmatrix} \begin{Bmatrix} X_1 \\ X_2 \\ X_3 \end{Bmatrix} + \begin{Bmatrix} -4 \\ 5 \end{Bmatrix}.$$

Therefore $\mathbf{Y} = \mathbf{A}\mathbf{X} + \mathbf{b}$, where

$$A = \begin{Bmatrix} 1 & -2 & 3 \\ 2 & 1 & -3 \end{Bmatrix} \quad \text{and} \quad \mathbf{b} = \begin{Bmatrix} -4 \\ 5 \end{Bmatrix}.$$

Hence,

$$\boldsymbol{\nu} = E\mathbf{Y} = \begin{Bmatrix} 1 & -2 & 3 \\ 2 & 1 & -3 \end{Bmatrix} \begin{Bmatrix} 2 \\ 4 \\ 8 \end{Bmatrix} + \begin{Bmatrix} -4 \\ 5 \end{Bmatrix} = \begin{Bmatrix} 14 \\ -11 \end{Bmatrix}$$

and

$$\boldsymbol{\Psi} = \mathrm{cov}(\mathbf{Y}) = \begin{Bmatrix} 1 & -2 & 3 \\ 2 & 1 & -3 \end{Bmatrix} \begin{Bmatrix} 4 & 2 & -3 \\ 2 & 10 & -5 \\ -3 & -5 & 16 \end{Bmatrix} \begin{Bmatrix} 1 & 2 \\ -2 & 1 \\ 3 & -3 \end{Bmatrix} = \begin{Bmatrix} 222 & -216 \\ -216 & 244 \end{Bmatrix}.$$

Consequently,

$$EY_1 = 14, \qquad EY_2 = -11, \qquad \mathrm{var}(Y_1) = 222, \qquad \mathrm{var}(Y_2) = 244, \qquad \mathrm{cov}(Y_1, Y_2) = -216.$$

Using matrix notation does not really make the preceding calculations any easier, but it does make for nicer formulas which can be remembered more easily.

We say that a $k \times k$ matrix $\mathbf{Q}$ is *nonnegative definite* and write

$$\mathbf{Q} \geqq 0$$

if $\mathbf{Q}$ is symmetric and $\mathbf{s}'\mathbf{Q}\mathbf{s} \geqq 0$ for all $\mathbf{s} \in R^k$. (Note that $\mathbf{s}'\mathbf{Q}\mathbf{s}$ is just a number.) We say that $\mathbf{Q}$ is *positive definite* and write

$$\mathbf{Q} > 0$$

if $\mathbf{Q}$ is symmetric and $\mathbf{s}'\mathbf{Q}\mathbf{s} > 0$ for all $\mathbf{s} \neq \mathbf{0}, \mathbf{s} \in R^k$. (Note that a positive definite matrix is also nonnegative definite.)

Theorem 5–12

a. If $\mathbf{X}$ is a random vector with covariance matrix $\mathbf{\Sigma}$, then $\mathbf{\Sigma}$ is nonnegative definite ($\mathbf{\Sigma} \geqq 0$).
b. If $\mathbf{X}$ is a continuous random vector with covariance matrix $\mathbf{\Sigma}$, then $\mathbf{\Sigma}$ is positive definite ($\mathbf{\Sigma} > 0$).

Proof

a. We have already observed that $\mathbf{\Sigma}$ is symmetric. Let $\mathbf{s} \in R^k$, and let $Y = \mathbf{s}'\mathbf{X}$ (so that Y is a one-dimensional random vector, i.e., a random variable). Then the covariance matrix, $\text{cov}(Y)$, is a 1×1 matrix whose sole element is $\text{var}(Y)$. But then, by Theorem 5–11,

$$\mathbf{s}'\mathbf{\Sigma}\mathbf{s} = \text{cov}(\mathbf{s}'\mathbf{X}) = \text{var}(Y) \geqq 0,$$

since every random variable has nonnegative variance.

b. By part a, $\mathbf{s}'\mathbf{\Sigma}\mathbf{s} \geqq 0$ for all $\mathbf{s}$. Suppose, then, that $\mathbf{s}'\mathbf{\Sigma}\mathbf{s} = 0$ for some $\mathbf{s} \neq 0$. By the calculation in part a,

$$\text{var}(\mathbf{s}'\mathbf{X}) = 0,$$

and hence $\mathbf{s}'\mathbf{X}$ is degenerate at $E\mathbf{s}'\mathbf{X} = \mathbf{s}'\boldsymbol{\mu}$. However, if $\mathbf{X}$ has a continuous density function, then

$$P(\mathbf{s}'\mathbf{X} = c) = 0$$

for all $\mathbf{s} \neq \mathbf{0}$ and c. Hence, if $\mathbf{X}$ has a continuous distribution, $\mathbf{s}'\mathbf{\Sigma}\mathbf{s} \neq 0$ for any $\mathbf{s} \neq \mathbf{0}$. □

In the next section, it is shown that for any $\boldsymbol{\mu} \in R^k$ and any $k \times k$ matrix $\mathbf{\Sigma} \geqq 0$, there is a random vector with $E\mathbf{X} = \boldsymbol{\mu}$ and $\text{cov}(\mathbf{X}) = \mathbf{\Sigma}$. Similarly, it is shown that for any such $\boldsymbol{\mu}$ and $\mathbf{\Sigma} > 0$, there is a continuous random vector with mean $\boldsymbol{\mu}$ and covariance matrix $\mathbf{\Sigma}$. Therefore, $(\boldsymbol{\mu}, \mathbf{\Sigma})$ is a possible mean vector and covariance matrix if and only if $\mathbf{\Sigma} \geqq 0$, and $(\boldsymbol{\mu}, \mathbf{\Sigma})$ is a possible mean vector and covariance matrix for a continuous random vector if and only if $\mathbf{\Sigma} > 0$. (This is again a natural generalization of the univariate case, in which σ^2 must always be nonnegative, and must be positive if the random variable is continuous.)

It is often difficult to tell whether a matrix is nonnegative definite or positive definite. (In Exercise B2, an example is given in which $\mathbf{Q} > 0$ but some of the components of $\mathbf{Q}$ are negative, and another example is given in which $\mathbf{R}$ has all positive elements but $\mathbf{R}$ is not nonnegative definite.) In this book, we shall assume that particular covariance matrices are positive or nonnegative definite. However, we shall not verify that they are.

We next look at the mean vector and covariance matrix for a multinomial distribution.

Example 5–9

Let $\mathbf{X} \sim M_k(n, \mathbf{p})$, where $\mathbf{X} = (X_1, \ldots, X_k)'$ and $p = (p_1, \ldots, p_k)'$. From Section 5.2.1, we know that

$$EX_i = np_i, \qquad \operatorname{var}(X_i) = np_i(1-p_i), \qquad \operatorname{cov}(X_i, X_j) = -np_i p_j.$$

Therefore, $\mathbf{X}$ has mean vector

$$\boldsymbol{\mu} = E\mathbf{X} = \begin{pmatrix} np_1 \\ np_2 \\ \cdot \\ \cdot \\ \cdot \\ np_k \end{pmatrix}$$

and covariance matrix

$$\boldsymbol{\Sigma} = \operatorname{cov}(\mathbf{X}) = \begin{pmatrix} np_1(1-p_1) & -np_1p_2 & \cdots & -np_1p_k \\ -np_2p_1 & np_2(1-p_2) & \cdots & -np_2p_k \\ \cdot & \cdot & & \\ \cdot & \cdot & & \\ \cdot & \cdot & & \\ -np_kp_1 & -np_kp_2 & \cdots & np_k(1-p_k) \end{pmatrix}.$$

Since $\boldsymbol{\Sigma}$ is a covariance matrix, $\boldsymbol{\Sigma} \geqq 0$.

Now, let $\mathbf{1} = (1, \ldots, 1)'$ be a k-dimensional vector, all of whose elements are 1. Then $\mathbf{1}'$ times the first column of $\boldsymbol{\Sigma}$ is

$$np_1(1-p_1) - np_1p_2 - \cdots - np_1p_k = np_1(1 - p_1 - p_2 - \cdots - p_k) = 0,$$

since the p_i sum to one. In a similar way $\mathbf{1}'$ times any column of $\boldsymbol{\Sigma}$ is zero, and therefore,

$$\mathbf{1}'\boldsymbol{\Sigma}\mathbf{1} = 0.$$

Hence, $\boldsymbol{\Sigma}$ is not positive definite; in fact,

$$\mathbf{1}'\boldsymbol{\Sigma}\mathbf{1} = \operatorname{var}(\mathbf{1}'\mathbf{X}) = \operatorname{var}\left(\sum_i X_i\right).$$

However,

$$\sum_i X_i = n,$$

so that $\Sigma_i X_i$ is degenerate at n, and $\operatorname{var}(\Sigma_i X_i)$ must be zero.

In Appendix A.3.2, it is shown that if $\boldsymbol{\Sigma} > 0$, then $\boldsymbol{\Sigma}$ is invertible and $\boldsymbol{\Sigma}^{-1} > 0$. (If $\boldsymbol{\Sigma} \geqq 0$, but is not positive definite, then $\boldsymbol{\Sigma}$ is not invertible.) It is also shown that for any $\boldsymbol{\Sigma} \geqq 0$, there is a matrix $\boldsymbol{\Sigma}^{1/2}$ such that $\boldsymbol{\Sigma}^{1/2} \geqq 0$, and $\boldsymbol{\Sigma} = (\boldsymbol{\Sigma}^{1/2})(\boldsymbol{\Sigma}^{1/2})$. If $\boldsymbol{\Sigma} > 0$, then $\boldsymbol{\Sigma}^{1/2} > 0$ and, hence, is invertible. In that case, we define

$$\boldsymbol{\Sigma}^{-1/2} = (\boldsymbol{\Sigma}^{1/2})^{-1}.$$

We call $\boldsymbol{\Sigma}^{1/2}$ and $\boldsymbol{\Sigma}^{-1/2}$ the *square root* and *inverse square root* matrices of $\boldsymbol{\Sigma}$, respectively. In later sections, we often use the properties of $\boldsymbol{\Sigma}^{1/2}$ and $\boldsymbol{\Sigma}^{-1/2}$ given in the following theorem.

Theorem 5–13. If $\mathbf{\Sigma} > 0$, then $\mathbf{\Sigma}^{1/2}$ and $\mathbf{\Sigma}^{-1/2}$ are symmetric matrices, and

$$(\mathbf{\Sigma}^{1/2})^2 = \mathbf{\Sigma}, \qquad (\mathbf{\Sigma}^{-1/2})^2 = \mathbf{\Sigma}^{-1}, \qquad \mathbf{\Sigma}^{-1/2}\mathbf{\Sigma}\mathbf{\Sigma}^{-1/2} = \mathbf{I},$$

$$\det(\mathbf{\Sigma}^{1/2}) = (\det(\mathbf{\Sigma}))^{1/2}, \qquad \det(\mathbf{\Sigma}^{-1/2}) = (\det(\mathbf{\Sigma}))^{-1/2}.$$

Proof. See Appendix A.3.2. □

The equalities in Theorem 5–13 are not surprising. They merely indicate that we can apply the usual rules of exponents to square root matrices. Note again the similarity to numbers, where positive numbers have positive square roots and nonnegative numbers have nonnegative square roots.

The square root matrix is difficult to compute. However, for our purposes, all we shall need is existence. We shall never have to find a square root matrix. (It should be emphasized that only nonnegative definite matrices have square root matrices, and only positive definite matrices have inverse square root matrices.)

We now give a brief discussion of the joint moment-generating function. Let $\mathbf{X}$ be a k-dimensional random variable, and let $\mathbf{t}$ be a k-dimensional vector. Then $\mathbf{X}$ has moment-generating function

$$M(\mathbf{t}) = E\ \exp\left(\sum_i X_i t_i\right) = E\ \exp(\mathbf{X}'\mathbf{t}),$$

and the following theorem obtains.

Theorem 5–14. Let $\mathbf{X}$ be a k-dimensional random vector with moment-generating function $M_X(\mathbf{s})$. Let $\mathbf{A}$ be a $p \times k$ matrix, let $\mathbf{b}$ be a $p \times 1$ vector, and let

$$\mathbf{Y} = \mathbf{AX} + \mathbf{b}.$$

Then $\mathbf{Y}$ has moment-generating function

$$M_Y(\mathbf{t}) = \exp[\mathbf{b}'\mathbf{t}]M_X(\mathbf{A}'\mathbf{t}).$$

Proof

$$M_Y(\mathbf{t}) = E\ \exp[\mathbf{Y}'\mathbf{t}] = E\ \exp[(\mathbf{AX} + \mathbf{b})'\mathbf{t}]$$
$$= \exp(\mathbf{b}'\mathbf{t})E\ \exp[\mathbf{X}'(\mathbf{A}'\mathbf{t})] = \exp(\mathbf{b}'\mathbf{t})M_X(\mathbf{A}'\mathbf{t}). \quad \square$$

We note again the similarity of the preceding result to the univariate case.

Exercises—B

1. Let X_1, X_2, X_3, and X_4 have means $2, 5, 7, 9$, respectively, and variances $4, 6, 12, 10$, respectively. Suppose that $\text{cov}(X_1, X_2) = 2$, $\text{cov}(X_2, X_3) = 1$, $\text{cov}(X_3, X_4) = -1$, and the remaining three covariances are all zero.
 (a) What are the mean vector and covariance matrix of $\mathbf{X} = (X_1, X_2, X_3, X_4)'$?
 (b) Let $Y_1 = 3X_1 + 2X_3 - X_4 + 2$ and $Y_2 = X_1 + 2X_2 - 3X_3 - 4$. What are the mean vector and covariance matrix of $\mathbf{Y} = (Y_1, Y_2)'$?

2. Let $\mathbf{P} = \begin{Bmatrix} 1 & 1 \\ 1 & 1 \end{Bmatrix}$, $\mathbf{Q} = \begin{Bmatrix} 4 & -1 \\ -1 & 4 \end{Bmatrix}$, and $\mathbf{R} = \begin{Bmatrix} 1 & 4 \\ 4 & 2 \end{Bmatrix}$. Let $\mathbf{s} = (s_1, s_2)'$.

(a) Show that $\mathbf{s}'\mathbf{P}\mathbf{s} = (s_1 + s_2)^2$ and that $\mathbf{P} \geqq 0$, but $\mathbf{P}$ is not positive definite. (Note that $\mathbf{s}'\mathbf{P}\mathbf{s} = 0$ when $\mathbf{s} = (1, -1)'$.)
(b) Show that $\mathbf{s}'\mathbf{Q}\mathbf{s} = 3s_1^2 + (s_1 - s_2)^2 + 3s_3^2$ and $\mathbf{Q} > 0$.
(c) Let $\mathbf{s} = (1, -1)'$. Show that $\mathbf{s}'\mathbf{R}\mathbf{s} = -5$, so that $\mathbf{R}$ is not nonnegative definite.

5.3.2 The Multivariate Normal Distribution

In Chapter 4 and in Section 5.2.1, we defined the univariate and bivariate normal distributions. We now extend these definitions to define a general k-variate normal distribution.

Let $\mathbf{X}$ be a k-variate random vector. Let $\boldsymbol{\mu}$ be a k-dimensional vector, and let $\boldsymbol{\Sigma}$ be a $k \times k$ nonnegative definite matrix. Then $\mathbf{X}$ has a *k-variate normal distribution* with mean vector $\boldsymbol{\mu}$ and covariance matrix $\boldsymbol{\Sigma}$, written

$$\mathbf{X} \sim N_k(\boldsymbol{\mu}, \boldsymbol{\Sigma}),$$

if $\mathbf{X}$ has joint moment-generating function

$$M(\mathbf{t}) = \exp\left(\boldsymbol{\mu}'\mathbf{t} + \frac{\mathbf{t}'\boldsymbol{\Sigma}\mathbf{t}}{2}\right).$$

(Note that $\boldsymbol{\mu}$ and $\mathbf{t}$ are $k \times 1$, so that $\boldsymbol{\mu}'\mathbf{t}$ is 1×1, i.e., a number. Similarly, $\mathbf{t}'\boldsymbol{\Sigma}\mathbf{t}$ is a number.) This definition is a natural generalization of the univariate normal moment-generating function, with $\boldsymbol{\mu}$ and $\boldsymbol{\Sigma}$ replacing μ and σ^2, respectively. We shall continue to use the notation $X \sim N(\mu, \sigma^2)$ to mean that X has a univariate normal distribution with mean μ and variance σ^2.

Theorem 5–15

a. The multivariate normal moment-generating function

$$M(\mathbf{t}) = \exp\left(\boldsymbol{\mu}'\mathbf{t} + \frac{\mathbf{t}'\boldsymbol{\Sigma}\mathbf{t}}{2}\right)$$

is a moment-generating function for all k-dimensional vectors $\boldsymbol{\mu}$ and $k \times k$ nonnegative definite matrices $\boldsymbol{\Sigma}$.

b. If $\mathbf{X} \sim N_k(\boldsymbol{\mu}, \boldsymbol{\Sigma})$, then

$$E\mathbf{X} = \boldsymbol{\mu} \quad \text{and} \quad \text{cov}(\mathbf{X}) = \boldsymbol{\Sigma}.$$

Proof. Let $\mathbf{Z} = (Z_1, \ldots, Z_k)'$ be a k-dimensional random vector such that the Z_i are independent and $Z_i \sim N(0, 1)$. Then $EZ_i = 0$, $\text{var}(Z_i) = 1$, and $\text{cov}(Z_i, Z_j) = 0$. Therefore,

$$E\mathbf{Z} = \mathbf{0} \quad \text{and} \quad \text{cov}(\mathbf{Z}) = \mathbf{I}_k.$$

Since the Z_i are independent, $\mathbf{Z}$ has joint moment-generating function

$$M_Z(\mathbf{t}) = \prod_i \exp\left(\frac{t_i^2}{2}\right) = \exp\left(\frac{\sum_i t_i^2}{2}\right) = \exp\left(\frac{\mathbf{t}'\mathbf{t}}{2}\right).$$

Now, let $\boldsymbol{\Sigma}$ be a $k \times k$ nonnegative definite matrix, and let $\boldsymbol{\mu} \in R^k$. Finally, let

$$\mathbf{X} = \boldsymbol{\Sigma}^{1/2}\mathbf{Z} + \boldsymbol{\mu},$$

where $\boldsymbol{\Sigma}^{1/2}$ is the square root matrix of $\boldsymbol{\Sigma}$.

a. By Theorem 5–14,

$$M_X(\mathbf{t}) = \exp(\boldsymbol{\mu}'\mathbf{t})M_Z((\boldsymbol{\Sigma}^{1/2})'\mathbf{t}) = \exp\left[\boldsymbol{\mu}'\mathbf{t} + \frac{(\boldsymbol{\Sigma}^{1/2}\mathbf{t})'(\boldsymbol{\Sigma}^{1/2}\mathbf{t})}{2}\right]$$

$$= \exp\left(\boldsymbol{\mu}'\mathbf{t} + \frac{\mathbf{t}'\boldsymbol{\Sigma}^{1/2}\boldsymbol{\Sigma}^{1/2}\mathbf{t}}{2}\right) = \exp\left(\boldsymbol{\mu}'\mathbf{t} + \frac{\mathbf{t}'\boldsymbol{\Sigma}\mathbf{t}}{2}\right).$$

Therefore, $\mathbf{X} \sim N_k(\boldsymbol{\mu}, \boldsymbol{\Sigma})$, and $M(\mathbf{t})$ is a moment-generating function.

b. $E\mathbf{X} = \boldsymbol{\Sigma}^{1/2}E\mathbf{Z} + \boldsymbol{\mu} = \boldsymbol{\mu}$, and $\operatorname{cov}(\mathbf{X}) = \boldsymbol{\Sigma}^{1/2}\operatorname{cov}(\mathbf{Z})\boldsymbol{\Sigma}^{1/2} = \boldsymbol{\Sigma}^{1/2}\boldsymbol{\Sigma}^{1/2} = \boldsymbol{\Sigma}$. □

The next theorem presents what is probably the most commonly used fact about the multivariate normal distribution.

Theorem 5–16. Let $\mathbf{X} \sim N_k(\boldsymbol{\mu}, \boldsymbol{\Sigma})$. Let $\mathbf{A}$ be a $p \times k$ matrix, and let $\mathbf{b}$ be a $p \times 1$ vector. Finally, let $\mathbf{Y} = \mathbf{A}\mathbf{X} + \mathbf{b}$. Then

$$\mathbf{Y} \sim N_p(\boldsymbol{\nu}, \boldsymbol{\Psi}), \qquad \boldsymbol{\nu} = \mathbf{A}\boldsymbol{\mu} + \mathbf{b}, \qquad \boldsymbol{\Psi} = \mathbf{A}\boldsymbol{\Sigma}\mathbf{A}'.$$

Proof. Note first that $\mathbf{A}\boldsymbol{\Sigma}\mathbf{A}' \geqq 0$. Now, by Theorem 5–14,

$$M_Y(\mathbf{t}) = \exp(\mathbf{b}'\mathbf{t})M_X(\mathbf{A}'\mathbf{t}) = \exp(\mathbf{b}'\mathbf{t})\ \exp\left[\boldsymbol{\mu}'(\mathbf{A}'\mathbf{t}) + \frac{(\mathbf{A}'\mathbf{t})'\boldsymbol{\Sigma})(\mathbf{A}'\mathbf{t})}{2}\right]$$

$$= \exp\left[(\mathbf{A}\boldsymbol{\mu} + \mathbf{b})'\mathbf{t} + \frac{\mathbf{t}'(\mathbf{A}\boldsymbol{\Sigma}\mathbf{A}')t}{2}\right] = \exp\left(\boldsymbol{\nu}'\mathbf{t} + \frac{\mathbf{t}'\boldsymbol{\Psi}\mathbf{t}}{2}\right),$$

which is the moment-generating function of a p-variate normal distribution with mean vector $\boldsymbol{\nu}$ and covariance matrix $\boldsymbol{\Psi}$. The desired result then follows from the uniqueness theorem. □

Recall that the mean vector and covariance matrix for $\mathbf{Y}$ were derived in the previous section. The main point of the preceding proof derivation is to show that if $\mathbf{X}$ has a multivariate normal distribution, then $\mathbf{Y}$ does also.

We can use Theorem 5–16 to find the marginal distribution for any component of $\mathbf{X}$.

Corollary 5–16. Let $\mathbf{X}$ and $\boldsymbol{\mu}$ have ith components X_i and μ_i, respectively, and let $\boldsymbol{\Sigma}$ have (i, j)th component σ_{ij}. Then $X_i \sim N(\mu_i, \sigma_{ii}^2)$.

Proof. Let $\mathbf{a}_i$ be a $1 \times k$ matrix with a one in the ith place and zeros in the remaining places. Then

$$\mathbf{a}_i\mathbf{X} = X_i, \qquad \mathbf{a}_i\boldsymbol{\mu} = \mu_i, \qquad \mathbf{a}_i\boldsymbol{\Sigma}\mathbf{a}_i' = \sigma_{ii},$$

and the desired result follows from Theorem 5–16. □

The multivariate normal distribution is used to model multivariate measurements of many kinds. The multivariate central limit theorem, set forth in the next chapter, implies that any vector of measurements which are themselves sums of many small effects should have an approximate multivariate normal distribution.

Example 5–10

Suppose the prices of objects A, B, and C are jointly normally distributed with means \$1, \$3, and \$6 and variances 3, 5, and 9 dollars squared. Suppose further that the covariance between the prices of A and B is 2, that between the prices of A and C is 3, and that between the prices of B and C is 5 (dollars squared). Let W, X, and Y be the prices of A, B, and C, respectively, on a particular day. Also, let

$$\mathbf{V} = \begin{Bmatrix} W \\ X \\ Y \end{Bmatrix}, \qquad \boldsymbol{\mu} = E\mathbf{V} = \begin{Bmatrix} 1 \\ 3 \\ 6 \end{Bmatrix}, \qquad \boldsymbol{\Sigma} = \text{cov}(\mathbf{V}) = \begin{Bmatrix} 3 & 2 & 3 \\ 2 & 5 & 5 \\ 3 & 5 & 9 \end{Bmatrix}.$$

Then

$$\mathbf{V} \sim N_3(\boldsymbol{\mu}, \boldsymbol{\Sigma}).$$

Now, suppose a person wants to buy three of object A, four of object B, and one of object C. Then the fee will be $U = 3W + 4X + Y = \mathbf{aV}$ dollars, where $\mathbf{a} = (3, 4, 1)$. Now,

$$\mathbf{a}\boldsymbol{\mu} = 21 \quad \text{and} \quad \mathbf{a}\boldsymbol{\Sigma}\mathbf{a}' = 222.$$

Therefore, by Theorem 5–16,

$$U \sim N(21, 222).$$

Hence, the probability that the person spends more than \$30 is

$$P(U > 30) = P\left(Z > \frac{30 - 21}{(222)^{1/2}}\right) = P(Z > .60) = .2743.$$

We conclude the section with a theorem about the multivariate normal distribution which will be useful later.

Theorem 5–17

a. Let $\mathbf{X} = (X_1, \ldots, X_k)'$ have a multivariate normal distribution. If the X_i are uncorrelated, then they are independent.
b. If $\mathbf{Z} \sim N_k(\mathbf{0}, \mathbf{I})$, then $\mathbf{Z}'\mathbf{Z} \sim \chi^2_k$.
c. If $\mathbf{X} \sim N_k(\boldsymbol{\mu}, \boldsymbol{\Sigma})$ and $\boldsymbol{\Sigma} > 0$, then $(\mathbf{X} - \boldsymbol{\mu})'\boldsymbol{\Sigma}^{-1}(\mathbf{X} - \boldsymbol{\mu}) \sim \chi^2_k$.

Proof

a. Suppose that $\mathbf{X} \sim N(\boldsymbol{\mu}, \boldsymbol{\Sigma})$ and that the X_i are uncorrelated. Then $\boldsymbol{\Sigma}$ is a diagonal matrix. Let μ_i be the ith component of $\boldsymbol{\mu}$ and σ_{ii} be the ith diagonal component of $\boldsymbol{\Sigma}$. Then $\mathbf{X}$ has joint moment-generating function

$$M(\mathbf{t}) = \exp\left(\boldsymbol{\mu}'\mathbf{t} + \frac{\mathbf{t}'\boldsymbol{\Sigma}\mathbf{t}}{2}\right) = \exp\left(\sum_i \mu_i t_i + \frac{\sum_i t_i^2 \sigma_{ii}}{2}\right) = \prod_i \exp\left(\mu_i t_i + \frac{t_i^2 \sigma_{ii}}{2}\right),$$

which is the moment-generating function of a random vector $\mathbf{X}$ whose components are independently $N(\mu_i, \sigma_{ii})$ distributed.

b. Let $\mathbf{Z}$ have ith component Z_i. By part a, the Z_i are independent. Also, by Corollary 5–16, $Z_i \sim N(0, 1)$. Therefore, by the definition of the χ^2 distribution,

$$\mathbf{Z}'\mathbf{Z} = \sum_i Z_i^2 \sim \chi_k^2.$$

c. Let $\mathbf{Z} = \mathbf{\Sigma}^{-1/2}(\mathbf{X} - \boldsymbol{\mu})$. By Theorem 5–16,

$$\mathbf{Z} \sim N_k(\mathbf{\Sigma}^{-1/2}(\boldsymbol{\mu} - \boldsymbol{\mu}), \mathbf{\Sigma}^{-1/2}\mathbf{\Sigma}\mathbf{\Sigma}^{-1/2}) = N_k(\mathbf{0}, \mathbf{I}).$$

Therefore, by part b,

$$(\mathbf{X} - \boldsymbol{\mu})'\mathbf{\Sigma}^{-1}(\mathbf{X} - \boldsymbol{\mu}) = \mathbf{Z}'\mathbf{Z} \sim \chi_k^2. \quad \square$$

Part c of Theorem 5–17 is the theoretical basis for the distribution of Pearson's χ^2 tests discussed in Chapter 14.

Any random variables which are independent are pairwise independent, and any pairwise independent random variables are uncorrelated. Part a of Theorem 5–17 says that if the random variables are jointly normally distributed and uncorrelated, then they are independent. Therefore, for jointly normally distributed random variables, independence, pairwise independence, and absence of correlation are all equivalent.

In Exercise C1, it is shown that if $\mathbf{X} \sim N_k(\boldsymbol{\mu}, \mathbf{\Sigma})$ and $\mathbf{\Sigma}$ is positive definite, then $\mathbf{X}$ has the continuous density function

$$f(\mathbf{x}) = (2\pi)^{-k/2}(\det(\mathbf{\Sigma}))^{-1/2} \exp\left[\frac{-(\mathbf{x} - \boldsymbol{\mu})'\mathbf{\Sigma}^{-1}(\mathbf{x} - \boldsymbol{\mu})}{2}\right], \qquad \mathbf{x} \in R^k. \tag{5–3}$$

Although this density function seems a fairly natural generalization of the univariate normal density function, it is somewhat difficult to use because of the presence of $\mathbf{\Sigma}^{-1}$. Fortunately, we shall not need the multivariate normal density function in this text.

Let $\mathbf{X} \sim N_k(\boldsymbol{\mu}, \mathbf{\Sigma})$. If $\mathbf{\Sigma}$ is nonnegative definite but not positive definite, then $M(\mathbf{t})$ exists, by Theorem 5–15. However, by Theorem 5–12, $\mathbf{X}$ has no continuous density function. (In particular, note that $\mathbf{\Sigma}$ is not invertible in this case.) It is convenient to allow the possibility of such singular distributions. If we did not, then in Theorem 5–16 we would have to assume that the rows of $\mathbf{A}$ were linearly independent (in order to have $\mathbf{A\Sigma A}'$ positive definite). But this in turn would lead to further complications in the proof of Theorem 5–19 in the next section. It so happens that it is no inconvenience to allow the possibility that $\mathbf{\Sigma}$ is not positive definite, since we do not use the density function anywhere. (If $k = 1$ and $\Sigma = \sigma^2$ is nonnegative but not positive, then $\sigma^2 = 0$. If $X \sim N(\mu, 0)$, then X is degenerate at μ; see Exercise B5.)

Exercises—A

1. Suppose that a randomly chosen student's college grade point average (GPA) X_1, high school GPA X_2, and verbal SAT score X_3 are jointly normally distributed with means 2.5, 2.8, and 550 and variances .25, .36, and 10,000, and with $\text{cov}(X_1, X_2) = .10$, $\text{cov}(X_1, X_3) = 20$, and $\text{cov}(X_2, X_3) = 30$.

(a) What are the mean vector and covariance matrix for $\mathbf{X} = (X_1, X_2, X_3)$?
(b) What is the distribution of the high school GPA and the college GPA?
(c) What is the probability that the college GPA is higher than the high school GPA?

Exercises—B

1. Let $\mathbf{X} = (X_1, X_2, X_3)'$ have a trivariate normal distribution with means 6, 4, and 2 and variances 16, 25, and 64, and with $\text{cov}(X_1, X_2) = 6$ and $\text{cov}(X_1, X_3) = \text{cov}(X_2, X_3) = 0$.
   (a) What are the mean vector and covariance matrix for $\mathbf{X}$?
   (b) Find the density function of $\mathbf{X}$.
   (c) Find the moment-generating function of $\mathbf{X}$.
2. Let $Y_1 = 2X_1 + 3X_2 + X_3 + 2$ and $Y_2 = 4X_1 + X_3 + 2$, where X_1, X_2, and X_3 are as in the previous problem.
   (a) What is the joint distribution of $\mathbf{Y} = (Y_1, Y_2)'$?
   (b) What is the correlation coefficient between Y_1 and Y_2?
   (c) What is the moment-generating function of $\mathbf{Y}$?
   (d) What is the joint density function of $\mathbf{Y}$?
3. Show that the bivariate normal density defined in Section 5.2.1 is a special case of the multivariate normal distribution defined in this section.
4. Verify that the formula for the moment-generating function for the bivariate normal given in Section 5.2.1 is a special case of the formula given in this section.
5. Let $\mathbf{X} \sim N_k(\boldsymbol{\mu}, \mathbf{0})$.
   (a) What is the moment-generating function of $\mathbf{X}$?
   (b) Show that $\mathbf{X}$ is degenerate at $\boldsymbol{\mu}$.

Exercises—C

1. Let $\mathbf{X}$ and $\mathbf{Z}$ be defined as in the proof of Theorem 5–15, so that $\mathbf{X} \sim N_k(\boldsymbol{\mu}, \boldsymbol{\Sigma})$. Suppose that $\boldsymbol{\Sigma}$ is positive definite, so that $\boldsymbol{\Sigma}^{-1/2}$ exists.
   (a) Show that $\mathbf{Z}$ has joint density function $f_Z(\mathbf{z}) = (2\pi)^{-k/2} \exp(-\mathbf{z}'\mathbf{z}/2), \mathbf{z} \in R^k$.
   (b) Verify that $\mathbf{X}$ has the joint density function given in Equation (5–3). (Let $h(\mathbf{X}) = \boldsymbol{\Sigma}^{-1/2}(\mathbf{X} - \boldsymbol{\mu})$. Then $\mathbf{Z} = h(\mathbf{X})$ and $J_h = \det(\boldsymbol{\Sigma}^{-1/2}) = (\det(\boldsymbol{\Sigma}))^{-1/2}$.
2. Let $\mathbf{X}$ be a random vector with moment-generating function $M(\mathbf{t}) = \exp(\mathbf{b}'\mathbf{t} + \mathbf{t}'\mathbf{A}\mathbf{t}/2)$, where $\mathbf{a} \in R^k$ and $\mathbf{A}$ is a symmetric $k \times k$ matrix. (If $\mathbf{A} \geqq 0$, then $\mathbf{X}$ has a normal distribution.)
   (a) Show that $E\mathbf{X} = \mathbf{b}$ and $\text{cov}(\mathbf{X}) = \mathbf{A}$. (*Hint*: $\Psi(\mathbf{t}) = \log(M(\mathbf{t})) = \Sigma b_i t_i + \Sigma\Sigma_{i,j} a_{ij} t_i t_j/2$.)
   (b) Show that $M(\mathbf{t})$ is a moment-generating function only if $\mathbf{A} \geqq 0$.

5.3.3 Marginal and Conditional Distributions

In this section, we show that for multivariate normal distributions, the marginal and conditional distributions of any subvectors are normal.

Theorem 5–18. Let $\mathbf{X} \sim N_k(\boldsymbol{\mu}, \boldsymbol{\Sigma})$. Partition $\mathbf{X}$, $\boldsymbol{\mu}$, and $\boldsymbol{\Sigma}$ as

$$\mathbf{X} = \begin{Bmatrix} \mathbf{X}_1 \\ \mathbf{X}_2 \end{Bmatrix}, \qquad \boldsymbol{\mu} = \begin{Bmatrix} \boldsymbol{\mu}_1 \\ \boldsymbol{\mu}_2 \end{Bmatrix}, \qquad \boldsymbol{\Sigma} = \begin{Bmatrix} \boldsymbol{\Sigma}_{11} & \boldsymbol{\Sigma}_{12} \\ \boldsymbol{\Sigma}_{21} & \boldsymbol{\Sigma}_{22} \end{Bmatrix},$$

where $\mathbf{X}_1$ and $\boldsymbol{\mu}_1$ are $p \times 1$, $\mathbf{X}_2$ and $\boldsymbol{\mu}_2$ are $q \times 1$, $\boldsymbol{\Sigma}_{11}$ is $p \times p$, $\boldsymbol{\Sigma}_{12}$ is $p \times q$, $\boldsymbol{\Sigma}_{21}$ is $q \times p$, and $\boldsymbol{\Sigma}_{22}$ is $q \times q$ (so that $p + q = k$).

a. The marginal distribution of $\mathbf{X}_1$ is

$$\mathbf{X}_1 \sim N_p(\boldsymbol{\mu}_1, \boldsymbol{\Sigma}_{11}).$$

b. $\mathbf{X}_1$ and $\mathbf{X}_2$ are independent if and only if $\boldsymbol{\Sigma}_{12} = \mathbf{0}$.

Proof

a. Let $\mathbf{t}' = (\mathbf{t}_1', \mathbf{t}_2')$, where $\mathbf{t}_1$ is $p \times 1$ and $\mathbf{t}_2$ is $q \times 1$. Now,

$$\boldsymbol{\mu}'\mathbf{t} = \boldsymbol{\mu}_1'\mathbf{t}_1 + \boldsymbol{\mu}_2'\mathbf{t}_2 \quad \text{and} \quad \mathbf{t}'\boldsymbol{\Sigma}\mathbf{t} = \mathbf{t}_1'\boldsymbol{\Sigma}_{11}\mathbf{t}_1 + \mathbf{t}_2'\boldsymbol{\Sigma}_{22}\mathbf{t}_2 + 2\mathbf{t}_1'\boldsymbol{\Sigma}_{12}\mathbf{t}_2.$$

The joint moment-generating function of $\mathbf{X} = (\mathbf{X}_1', \mathbf{X}_2')'$ is

$$M_X(\mathbf{t}) = M_{X_1,X_2}(\mathbf{t}_1, \mathbf{t}_2) = \exp\left(\boldsymbol{\mu}'\mathbf{t} + \frac{\mathbf{t}'\boldsymbol{\Sigma}\mathbf{t}}{2}\right)$$

$$= \exp\left(\boldsymbol{\mu}_1'\mathbf{t}_1 + \frac{\mathbf{t}_1\boldsymbol{\Sigma}_{11}\mathbf{t}_1}{2}\right)\exp\left(\boldsymbol{\mu}_2'\mathbf{t}_2 + \frac{\mathbf{t}_2'\boldsymbol{\Sigma}_{22}\mathbf{t}_2}{2}\right)\exp(\mathbf{t}_1'\boldsymbol{\Sigma}_{12}\mathbf{t}_2).$$

Now, $\mathbf{X}_1$ has moment-generating function

$$M_{X_1}(\mathbf{t}_1) = M_{X_1,X_2}(\mathbf{t}_1, \mathbf{0}) = \exp\left(\boldsymbol{\mu}_1'\mathbf{t}_1 + \frac{\mathbf{t}_1'\boldsymbol{\Sigma}_{11}\mathbf{t}_1}{2}\right),$$

which is the moment-generating function of a p-variate normal distribution with mean vector $\boldsymbol{\mu}_1$ and covariance matrix $\boldsymbol{\Sigma}_{11}$. By the uniqueness theorem,

$$\mathbf{X}_1 \sim N_p(\boldsymbol{\mu}_1, \boldsymbol{\Sigma}_{11}).$$

b. Now, $\mathbf{X}_1$ and $\mathbf{X}_2$ are independent if and only if

$$M_{X_1,X_2}(\mathbf{t}_1, \mathbf{t}_2) = M_{X_1,X_2}(\mathbf{t}_1, \mathbf{0})M_{X_1,X_2}(\mathbf{0}, t_2)$$

if and only if

$$\mathbf{t}_1'\boldsymbol{\Sigma}_{12}\mathbf{t}_2 = 0$$

for all $\mathbf{t}_1$ and $\mathbf{t}_2$, which in turn occurs if and only if $\boldsymbol{\Sigma}_{12} = \mathbf{0}$. □

We now find the conditional distribution of $\mathbf{X}_1$ given $\mathbf{X}_2$.

Theorem 5–19. Let $\mathbf{X} \sim N_k(\boldsymbol{\mu}, \boldsymbol{\Sigma})$, where $\mathbf{X}$, $\boldsymbol{\mu}$, and $\boldsymbol{\Sigma}$ are partitioned as in Theorem 5–18. If $\boldsymbol{\Sigma}_{22}$ is invertible, then the conditional distribution of $\mathbf{X}_1$ given $\mathbf{X}_2$ is

$$\mathbf{X}_1|\mathbf{X}_2 \sim N_p(\boldsymbol{\mu}_1 + \boldsymbol{\Sigma}_{12}\boldsymbol{\Sigma}_{22}^{-1}(\mathbf{X}_2 - \boldsymbol{\mu}_2), \boldsymbol{\Sigma}_{11} - \boldsymbol{\Sigma}_{12}\boldsymbol{\Sigma}_{22}^{-1}\boldsymbol{\Sigma}_{21}).$$

Proof. Let $\mathbf{U} = \mathbf{X}_1 - \boldsymbol{\Sigma}_{12}\boldsymbol{\Sigma}_{22}^{-1}\mathbf{X}_2$ and $\mathbf{V} = \mathbf{X}_2$. By Theorem 5–16,

$$\begin{Bmatrix}\mathbf{U}\\ \mathbf{V}\end{Bmatrix} = \begin{Bmatrix}\mathbf{I} & -\boldsymbol{\Sigma}_{12}\boldsymbol{\Sigma}_{22}^{-1}\\ \mathbf{0} & \mathbf{I}\end{Bmatrix}\begin{Bmatrix}\mathbf{X}_1\\ \mathbf{X}_2\end{Bmatrix} \sim N\left(\begin{Bmatrix}\boldsymbol{\mu}_1 - \boldsymbol{\Sigma}_{12}\boldsymbol{\Sigma}_{22}^{-1}\boldsymbol{\mu}_2\\ \boldsymbol{\mu}_2\end{Bmatrix}, \begin{Bmatrix}\boldsymbol{\Sigma}_{11} - \boldsymbol{\Sigma}_{12}\boldsymbol{\Sigma}_{22}^{-1}\boldsymbol{\Sigma}_{21} & \mathbf{0}\\ \mathbf{0} & \boldsymbol{\Sigma}_{22}\end{Bmatrix}\right).$$

But by Theorem 5–18, $\mathbf{U}$ and $\mathbf{V}$ are independent. Therefore, the conditional distribution of $\mathbf{U}$ given $\mathbf{X}_2$ $(=\mathbf{V})$ is the same as the marginal distribution of $\mathbf{U}$. Using Theorem 5–18 again, we see that

$$\mathbf{U}|\mathbf{X}_2 \sim N_p(\boldsymbol{\mu}_1 - \boldsymbol{\Sigma}_{12}\boldsymbol{\Sigma}_{22}^{-1}\boldsymbol{\mu}_2, \boldsymbol{\Sigma}_{11} - \boldsymbol{\Sigma}_{12}\boldsymbol{\Sigma}_{22}^{-1}\boldsymbol{\Sigma}_{21}).$$

Now, $\mathbf{X}_1 = \mathbf{U} + \boldsymbol{\Sigma}_{12}\boldsymbol{\Sigma}_{22}^{-1}\mathbf{X}_2$, and $\mathbf{X}_2$ is a constant in the conditional distribution of $\mathbf{U}$ given $\mathbf{X}_2$. Therefore, by Theorem 5–16 with $\mathbf{A} = \mathbf{I}$ and $\mathbf{b} = \boldsymbol{\Sigma}_{12}\boldsymbol{\Sigma}_{22}^{-1}\mathbf{X}_2$,

$$\mathbf{X}_1|\mathbf{X}_2 \sim N_p(\boldsymbol{\mu}_1 + \boldsymbol{\Sigma}_{12}\boldsymbol{\Sigma}_{22}^{-1}(\mathbf{X}_2 - \boldsymbol{\mu}_2), \boldsymbol{\Sigma}_{11} - \boldsymbol{\Sigma}_{12}\boldsymbol{\Sigma}_{22}^{-1}\boldsymbol{\Sigma}_{21}). \quad \square$$

In a similar manner to the preceding proof, we can show that if $\boldsymbol{\Sigma}_{11}$ is invertible, then

$$\mathbf{X}_2 \sim N_q(\boldsymbol{\mu}_2, \boldsymbol{\Sigma}_{22}) \quad \text{and} \quad \mathbf{X}_2|\mathbf{X}_1 \sim N_q(\boldsymbol{\mu}_2 + \boldsymbol{\Sigma}_{21}\boldsymbol{\Sigma}_{11}^{-1}(\mathbf{X}_1 - \boldsymbol{\mu}_1), \boldsymbol{\Sigma}_{22} - \boldsymbol{\Sigma}_{21}\boldsymbol{\Sigma}_{11}^{-1}\boldsymbol{\Sigma}_{12}).$$

Note that the argument given for the conditional distribution of $\mathbf{X}_1$ given $\mathbf{X}_2$ is rather unintuitive. However, a straightforward approach to finding the conditional distribution leads to some difficult matrix calculations. In the exercises, it is shown that even if $\boldsymbol{\Sigma}_{22}$ is not invertible, the conditional distribution of $\mathbf{X}_1$ given $\mathbf{X}_2$ is still a multivariate normal distribution.

Actually, Theorems 5–18 and 5–19 can be used to find the marginal distribution of any set of components of $\mathbf{X}$. We first rearrange the components of $\mathbf{X}$ so that the components we want in the marginal distribution come first, making the appropriate adjustments in the mean vector and covariance matrix. The resulting permuted vector $\mathbf{X}$ still has a multivariate normal distribution by Theorem 5–16, because the permutation is a linear function. We can then use Theorem 5–18 to find the marginal distribution. Similarly, to find the conditional distribution of any subset of the components of $\mathbf{X}$ given a set of other components, we reorder $\mathbf{X}$ so that the variables whose conditional distribution we want are first, the conditioning variables are second, and the remaining variables are third, making the appropriate adjustments in the mean vector and the covariance matrix again. We then use Theorem 5–18 to eliminate the remaining variables and Theorem 5–19 to find the desired conditional distribution. (It is hard to describe this operation precisely; however, the method is clear from the following example.)

Example 5–11

Let $\mathbf{X} = (X_1, X_2, X_3, X_4)'$ have a multivariate normal distribution with mean vector

$$\boldsymbol{\mu} = E\mathbf{X} = \begin{pmatrix} 1 \\ 2 \\ 3 \\ 4 \end{pmatrix}$$

and covariance matrix

$$\boldsymbol{\Sigma} = \begin{pmatrix} 1 & 1 & 1 & 1 \\ 1 & 2 & 2 & 2 \\ 1 & 2 & 3 & 3 \\ 1 & 2 & 3 & 4 \end{pmatrix}.$$

Now, let $\mathbf{X}_1 = (X_1, X_2)'$ and $\mathbf{X}_2 = (X_3, X_4)'$. By Theorem 5–18,

$$\mathbf{X}_1 = \begin{Bmatrix} X_1 \\ X_2 \end{Bmatrix} \sim N_2\left(\begin{Bmatrix} 1 \\ 2 \end{Bmatrix}, \begin{Bmatrix} 1 & 1 \\ 1 & 2 \end{Bmatrix}\right).$$

That is, X_1 and X_2 have a bivariate normal distribution with means 1 and 2, variances 1 and 2, and covariance 1. Hence, the correlation coefficient is $1/(1 \times 2)^{1/2} = (\frac{1}{2})^{1/2}$.

We next find the conditional distribution of X_3 and X_4 given X_1 and X_2. We have

$$\mathbf{\Sigma}_{11}^{-1} = \begin{Bmatrix} 1 & 1 \\ 1 & 2 \end{Bmatrix}^{-1} = \begin{Bmatrix} 2 & -1 \\ -1 & 1 \end{Bmatrix}, \qquad \mathbf{\Sigma}_{21}\mathbf{\Sigma}_{11}^{-1} = \begin{Bmatrix} 1 & 2 \\ 1 & 2 \end{Bmatrix}\begin{Bmatrix} 2 & -1 \\ -1 & 1 \end{Bmatrix} = \begin{Bmatrix} 0 & 1 \\ 0 & 1 \end{Bmatrix},$$

$$\boldsymbol{\mu}_2 + \mathbf{\Sigma}_{21}\mathbf{\Sigma}_{11}^{-1}(\mathbf{X}_1 - \boldsymbol{\mu}_1) = \begin{Bmatrix} 3 \\ 4 \end{Bmatrix} + \begin{Bmatrix} 0 & 1 \\ 0 & 1 \end{Bmatrix}\begin{Bmatrix} X_1 - 1 \\ X_2 - 2 \end{Bmatrix} = \begin{Bmatrix} X_2 + 1 \\ X_2 + 2 \end{Bmatrix},$$

$$\mathbf{\Sigma}_{22} - \mathbf{\Sigma}_{21}\mathbf{\Sigma}_{11}^{-1}\mathbf{\Sigma}_{12} = \begin{Bmatrix} 3 & 3 \\ 3 & 4 \end{Bmatrix} - \begin{Bmatrix} 1 & 2 \\ 1 & 2 \end{Bmatrix}\begin{Bmatrix} 2 & -1 \\ -1 & 1 \end{Bmatrix}\begin{Bmatrix} 1 & 1 \\ 2 & 2 \end{Bmatrix} = \begin{Bmatrix} 1 & 1 \\ 1 & 2 \end{Bmatrix}.$$

Therefore, the conditional distribution of $(X_3, X_4)'$ given $(X_1, X_2)'$ is

$$\begin{Bmatrix} X_3 \\ X_4 \end{Bmatrix} \Bigg| \begin{Bmatrix} X_1 \\ X_2 \end{Bmatrix} \sim N_2\left(\begin{Bmatrix} X_2 + 1 \\ X_2 + 2 \end{Bmatrix}, \begin{Bmatrix} 1 & 1 \\ 1 & 2 \end{Bmatrix}\right).$$

That is, conditionally on X_1 and X_2, X_3 and X_4 have a bivariate normal distribution with means $X_2 + 1$ and $X_2 + 2$, variances 1 and 2, and covariance 1. Hence, the conditional correlation coefficient is $1/(1 \times 2)^{1/2} = 2^{-1/2}$.

Now suppose we want the marginal (joint) distribution of X_1 and X_3. By Theorem 5–16,

$$\begin{Bmatrix} X_1 \\ X_3 \\ X_2 \\ X_4 \end{Bmatrix} \sim N_4\left(\begin{Bmatrix} 1 \\ 3 \\ 2 \\ 4 \end{Bmatrix}, \begin{Bmatrix} 1 & 1 & 1 & 1 \\ 1 & 3 & 2 & 3 \\ 1 & 2 & 2 & 2 \\ 1 & 3 & 2 & 4 \end{Bmatrix}\right).$$

It is not actually necessary to use any matrix algebra to find the mean and covariance matrix of the vector. For example, we know that $EX_3 = 3$, so the second component of the new mean vector must be 3. Similarly, $\text{cov}(X_3, X_4) = 3$, so that the $(2, 4)$ component of the new covariance matrix must be 3. In fact, to find the new mean vector, we just interchange the second and third elements of $\boldsymbol{\mu}$, and to find the new covariance matrix, we just interchange the second and third rows of $\mathbf{\Sigma}$ and the second and third columns of $\mathbf{\Sigma}$. Therefore, by Theorem 5–18,

$$\begin{Bmatrix} X_1 \\ X_3 \end{Bmatrix} \sim N_2\left(\begin{Bmatrix} 1 \\ 3 \end{Bmatrix}, \begin{Bmatrix} 1 & 1 \\ 1 & 3 \end{Bmatrix}\right).$$

Next, suppose we want the conditional distribution of X_1 and X_3 given X_2. By Theorem 5–18 again,

$$\begin{Bmatrix} X_1 \\ X_3 \\ X_2 \end{Bmatrix} \sim N_3\left(\begin{Bmatrix} 1 \\ 3 \\ 2 \end{Bmatrix}, \begin{Bmatrix} 1 & 1 & 1 \\ 1 & 3 & 2 \\ 1 & 2 & 2 \end{Bmatrix}\right).$$

Therefore, by Theorem 5–19 again, X_1 and X_3 given X_2 have a bivariate normal distribution with mean vector

$$\begin{Bmatrix} 1 \\ 3 \end{Bmatrix} + \begin{Bmatrix} 1 \\ 2 \end{Bmatrix} 2^{-1}(X_2 - 2) = \begin{Bmatrix} X_2/2 \\ X_2 + 1 \end{Bmatrix}$$

and covariance matrix

$$\begin{Bmatrix} 1 & 1 \\ 1 & 3 \end{Bmatrix} - \begin{Bmatrix} 1 \\ 2 \end{Bmatrix} 2^{-1} \{1 \quad 2\} = \begin{Bmatrix} .5 & 0 \\ 0 & 1 \end{Bmatrix}.$$

Note that X_1 and X_3 are independent in the conditional distribution given X_2 but are not independent in their marginal distribution.

From the case of finding the marginal distribution of X_1 and X_3, we see that to find the marginal distribution of any subset of the components of $\mathbf{X}$, we take the mean $\boldsymbol{\mu} = E\mathbf{X}$ and cross out the components that we associated with the X's we do not want (μ_2 and μ_4 in Example 5–11) and take the covariance matrix $\boldsymbol{\Sigma}$ and cross out the rows and columns associated with the components we do not want (the second and fourth rows and columns in the example). However, to find conditional distributions of one set of components given another set, we must reorder $\mathbf{X}$.

Example 5–12

Suppose the weight X_1, height X_2, and age X_3 of a randomly chosen male have a multivariate normal distribution with means 170, 68, and 40 and variances 400, 16, and 256, and with $\text{cov}(X_1, X_2) = 64$, $\text{cov}(X_1, X_3) = 128$, and $\text{cov}(X_2, X_3) = 0$. That is, $\mathbf{X} = (X_1, X_2, X_3)' \sim N_3(\boldsymbol{\mu}, \boldsymbol{\Sigma})$, where

$$\boldsymbol{\mu} = \begin{Bmatrix} 170 \\ 68 \\ 40 \end{Bmatrix} \quad \text{and} \quad \boldsymbol{\Sigma} = \begin{Bmatrix} 400 & 64 & 128 \\ 64 & 16 & 0 \\ 128 & 0 & 256 \end{Bmatrix}.$$

Therefore,

$$\begin{Bmatrix} X_1 \\ X_2 \end{Bmatrix} \sim N_2\left(\begin{Bmatrix} 170 \\ 68 \end{Bmatrix}, \begin{Bmatrix} 400 & 64 \\ 64 & 16 \end{Bmatrix}\right),$$

which is the same distribution assumed in Example 5–7. In that example, we showed that

$$X_1|X_2 \sim N(4X_2 - 102, 144) \quad \text{and} \quad X_1|(X_2 = 72) \sim N(186, 144).$$

We now find the conditional distribution of X_1 given X_2 and X_3, that is, the conditional distribution of the weight given the height and the age. First, note that

$$170 - \{64 \quad 128\} \begin{Bmatrix} 16 & 0 \\ 0 & 256 \end{Bmatrix}^{-1} \begin{Bmatrix} X_2 - 68 \\ X_3 - 40 \end{Bmatrix} = 4X_2 + .5X_3 - 122,$$

$$400 - \{64 \quad 128\} \begin{Bmatrix} 16 & 0 \\ 0 & 256 \end{Bmatrix}^{-1} \begin{Bmatrix} 64 \\ 128 \end{Bmatrix} = 80,$$

$$X_1|(X_2, X_3) \sim N(4X_2 + .5X_3 - 122, 80).$$

Hence, if we have a male who is 6′ (=72″) tall and 24 years old, then his weight has the distribution

$$X_1|(X_2 = 72, X_3 = 24) \sim N(178, 80).$$

If we compare this result with the conditional distribution of weight given that the height is 72, we see that conditional mean has been reduced from 186 to 178, and the variance has been reduced from 144 to 80, because of the additional information about age. (It seems clear that the distribution of weight given both height and age should be tighter than that of weight given height alone.)

Exercises—A

1. Consider again Exercise A1 of the last section.
   (a) What is the conditional distribution of college GPA given high school GPA and SAT score?
   (b) What is the conditional distribution of SAT score given high school and college GPA?
   (c) What is the conditional distribution of the average of high school and college GPA given the SAT score? (First find the joint distribution of Y_1, the average of the GPA's, and Y_2, the SAT score.)

Exercises—B

1. Let $\mathbf{X} = (X_1, X_2, X_3, X_4)'$ have a four-variate normal distribution with $EX_1 = 2$, $EX_2 = 1$, $EX_3 = 3$, $EX_4 = 6$, $\text{var}(X_1) = 1$, $\text{var}(X_2) = 2$, $\text{var}(X_3) = 3$, $\text{var}(X_4) = 4$, and $\text{cov}(X_i, X_j) = 1$ for all $i \neq j$.
   (a) What are the mean vector and covariance matrix of $\mathbf{X}$?
   (b) What is the marginal distribution of (X_1, X_3)?
   (c) What is the conditional distribution of $(X_1, X_2)'$ given $(X_3, X_4)'$?
   (d) What is the conditional distribution of $(X_1, X_4)'$ given $(X_2, X_3)'$?
   (e) What is the conditional distribution of $(X_1, X_4)'$ given X_3?
   (f) What is the joint distribution of $Y_1 = X_1 + X_2 + X_3$ and $Y_2 = X_1 - 2X_4$?
   (g) What is the conditional distribution of Y_1 given Y_2?
2. Let $\mathbf{X} = (X_1, X_2, X_3)$ have a trivariate normal distribution with means 6, 4, and 2, variances 16, 25, and 64, and $\text{cov}(X_1, X_2) = 6$, $\text{cov}(X_1, X_3) = \text{cov}(X_2, X_3) = 0$.
   (a) Find the conditional distribution of X_1 given X_2 and X_3.
   (b) Find the conditional distribution of $(X_1, X_2)'$ given X_3.
   (c) Let $Y_1 = 2X_1 + 3X_2 + X_3 + 2$ and $Y_2 = 4X_1 + X_3 + 2$. Find the joint distribution of Y_1 and Y_2.
   (d) Find the conditional distribution of Y_2 given Y_1.
3. Verify that the marginal and conditional distributions given in Section 5.2.1 are special cases of the ones given in this section.
4. Use Theorem 5–16 to prove part a of Theorem 5–18.
5. Verify the mean vector and covariance matrix for $(\mathbf{U}', \mathbf{V}')'$ in the proof of Theorem 5–19.

Exercises—C

1. Suppose that $\mathbf{t}'\mathbf{A}\mathbf{s} = 0$ for all $\mathbf{t}$ and $\mathbf{s}$. Show that $\mathbf{A} = \mathbf{0}$. (*Hint*: Let $\mathbf{t}_i$ have a one in the ith place and zeros elsewhere, and let $\mathbf{s}_j$ have a one in the jth place and zeros elsewhere. Then $\mathbf{t}_i\mathbf{A}\mathbf{s}_j$ is the (ij)th element of $\mathbf{A}$.)
2. Let $\mathbf{X} \sim N_k(\boldsymbol{\mu}, \boldsymbol{\Sigma})$, where $\mathbf{X}$, $\boldsymbol{\mu}$, and $\boldsymbol{\Sigma}$ are partitioned as in Theorem 5–18. Let $\mathbf{A}$ be a matrix such that $\mathbf{A}\boldsymbol{\Sigma}_{22} = \boldsymbol{\Sigma}_{12}$.
   (a) Let $\mathbf{U} = \mathbf{X}_1 - \mathbf{A}\mathbf{X}_2$ and $\mathbf{V} = \mathbf{X}_2$. Find the joint distribution of $\mathbf{U}$ and $\mathbf{V}$.
   (b) Show that $\mathbf{X}_1|\mathbf{X}_2 \sim N_p(\boldsymbol{\mu}_1 + \mathbf{A}(\mathbf{X}_2 - \boldsymbol{\mu}_2), \boldsymbol{\Sigma}_{11} - \mathbf{A}\boldsymbol{\Sigma}_{22}\mathbf{A}')$.
   (c) Show that if $\boldsymbol{\Sigma}_{22}$ is invertible, then $\mathbf{A} = \boldsymbol{\Sigma}_{12}\boldsymbol{\Sigma}_{22}^{-1}$, and that this result reduces to the result given in Theorem 5–19.

   (For any possible $\boldsymbol{\Sigma}_{12}$ and $\boldsymbol{\Sigma}_{22}$, it can be shown that such an $\mathbf{A}$ exists. This problem therefore establishes the conditional distribution of $\mathbf{X}_1$ given $\mathbf{X}_2$ for any normal distribution.)

5.3.4 Linear Functions and Quadratic Forms

Many of the models we consider later in the book, including all the analysis-of-variance and regression models in Chapters 12 and 13, have the form wherein we observe X_i to be independent random variables such that

$$X_i \sim N(\mu_i, \sigma^2).$$

That is, the X_i are independently normally distributed with possibly unequal means but equal variances. In this section, we use matrix algebra to give the basic results for such a model. We first give a multivariate reformulation of the model.

Theorem 5–20. Let $\mathbf{X} = (X_1, \ldots, X_n)'$ and $\boldsymbol{\mu} = (\mu_1, \ldots, \mu_n)'$. If $\mathbf{X} \sim N_n(\boldsymbol{\mu}, \sigma^2 \mathbf{I})$, then the X_i are independent and $X_i \sim N(\mu_i, \sigma^2)$.

Proof. If $\mathbf{X} \sim N_n(\boldsymbol{\mu}, \sigma^2 \mathbf{I})$, then, by Theorem 5–17, the X_i are independent. But then, by Corollary 5–16, $X_i \sim N(\mu_i, \sigma^2)$. □

We say that a matrix $\mathbf{P}$ is *idempotent* if $\mathbf{P}$ is symmetric and

$$\mathbf{P}^2 = \mathbf{P}.$$

The next theorem presents the basic facts about linear functions and quadratic forms for the cases considered in this section.

Theorem 5–21. Let $\mathbf{X} \sim N_n(\boldsymbol{\mu}, \sigma^2 \mathbf{I})$, $\sigma^2 > 0$.

a. Let $\mathbf{A}$ be a $p \times n$ matrix. Then

$$\mathbf{AX} \sim N_p(\mathbf{A}\boldsymbol{\mu}, \sigma^2 \mathbf{AA}').$$

b. Let $\mathbf{B}$ be an $n \times n$ symmetric matrix. Then

$$\frac{\mathbf{X}'\mathbf{BX}}{\sigma^2} \sim \chi_k^2$$

if and only if $\mathbf{B}$ is an idempotent matrix such that $k = \text{tr}(\mathbf{B})$ and $\boldsymbol{\mu}'\mathbf{B}\boldsymbol{\mu} = 0$.

c. 1. Let $\mathbf{A}_1$ and $\mathbf{A}_2$ be $p \times n$ and $q \times n$ matrices. Then $\mathbf{A}_1\mathbf{X}$ and $\mathbf{A}_2\mathbf{X}$ are independent if and only if $\mathbf{A}_1\mathbf{A}_2' = \mathbf{0}$.

2. Let $\mathbf{B}_1$ and $\mathbf{B}_2$ be $n \times n$ symmetric matrices. Then $\mathbf{X}'\mathbf{B}_1\mathbf{X}$ and $\mathbf{X}'\mathbf{B}_2\mathbf{X}$ are independent if and only if $\mathbf{B}_1\mathbf{B}_2 = \mathbf{0}$.

3. Let $\mathbf{A}$ be a $p \times n$ matrix, and let $\mathbf{B}$ be an $n \times n$ symmetric matrix. Then $\mathbf{AX}$ and $\mathbf{X}'\mathbf{BX}$ are independent if and only if $\mathbf{AB} = \mathbf{0}$.

Proof

a. The desired result follows directly from Theorem 5–16.

b. Suppose that $\mathbf{B}$ is an idempotent matrix such that $\boldsymbol{\mu}'\mathbf{B}\boldsymbol{\mu} = 0$ and $\text{tr}(\mathbf{B}) = k$. In Appendix A.3.2, we show that there exists a $k \times n$ matrix $\mathbf{C}$ such that

$$\mathbf{B} = \mathbf{C}'\mathbf{C}, \qquad \mathbf{CC}' = \mathbf{I}_k,$$

where $\mathbf{I}_k$ is a k-dimensional identity matrix. Now,

$$0 = \boldsymbol{\mu}'\mathbf{B}\boldsymbol{\mu} = \boldsymbol{\mu}'\mathbf{C}'\mathbf{C}\boldsymbol{\mu} = (\mathbf{C}\boldsymbol{\mu})'(\mathbf{C}\boldsymbol{\mu}) = \|\mathbf{C}\boldsymbol{\mu}\|^2.$$

Therefore, $\mathbf{C}\boldsymbol{\mu} = \mathbf{0}$. By Theorem 5–16 again,

$$\mathbf{Z} = \sigma^{-1}\mathbf{C}\mathbf{X} \sim N_k(\sigma^{-1}\mathbf{C}\boldsymbol{\mu}, \sigma^{-1}\mathbf{C}(\sigma^2\mathbf{I})\mathbf{C}'\sigma^{-1}) = N_k(\mathbf{0}, \mathbf{I}).$$

Now, since

$$\frac{\mathbf{X}'\mathbf{B}\mathbf{X}}{\sigma^2} = (\sigma^{-1}\mathbf{C}\mathbf{X})'(\sigma^{-1}\mathbf{C}\mathbf{X}) = \mathbf{Z}'\mathbf{Z},$$

it follows, by Theorem 5–17, that

$$\frac{\mathbf{X}'\mathbf{B}\mathbf{X}}{\sigma^2} = \mathbf{Z}'\mathbf{Z} \sim \chi_k^2.$$

The proof of the converse is somewhat more difficult. (See Hogg and Craig (1978), pp. 410–413.) However, since we shall not need it in this text, we have not included it.

c. 1. By Theorem 5–16,

$$\begin{Bmatrix}\mathbf{A}_1\mathbf{X}\\ \mathbf{A}_2\mathbf{X}\end{Bmatrix} = \begin{Bmatrix}\mathbf{A}_1\\ \mathbf{A}_2\end{Bmatrix}\mathbf{X} \sim N_{p+q}\left(\begin{Bmatrix}\mathbf{A}_1\boldsymbol{\mu}\\ \mathbf{A}_2\boldsymbol{\mu}\end{Bmatrix}, \sigma^2\begin{Bmatrix}\mathbf{A}_1\mathbf{A}_1' & \mathbf{A}_1\mathbf{A}_2'\\ \mathbf{A}_2\mathbf{A}_1' & \mathbf{A}_2\mathbf{A}_2'\end{Bmatrix}\right).$$

Therefore, by part a of Theorem 5–18, $\mathbf{A}_1\mathbf{X}$ and $\mathbf{A}_2\mathbf{X}$ are independent if and only if $\mathbf{A}_1\mathbf{A}_2' = \mathbf{0}$.

2. A proof of this result in its full generality is beyond the level of this book. (See Hogg and Craig (1978), pp. 414–419 for such a proof.) Instead, we give a simple proof for the only case we use in the text, viz., that $\mathbf{B}_1\mathbf{B}_2 = \mathbf{0}$ implies independence when the $\mathbf{B}_i$ are idempotent. In this case,

$$\mathbf{X}'\mathbf{B}_i\mathbf{X} = \mathbf{X}'\mathbf{B}_i'\mathbf{B}_i\mathbf{X} = (\mathbf{B}_i\mathbf{X})'(\mathbf{B}_i\mathbf{X}) = \|\mathbf{B}_i\mathbf{X}\|^2.$$

Now, by part 1, if $\mathbf{B}_1\mathbf{B}_2 = \mathbf{B}_1\mathbf{B}_2' = \mathbf{0}$, then $\mathbf{B}_1\mathbf{X}$ and $\mathbf{B}_2\mathbf{X}$ are independent, and hence, so are $\|\mathbf{B}_1\mathbf{X}\|^2$ and $\|\mathbf{B}_2\mathbf{X}\|^2$.

3. This result is also beyond the level of the text. A proof for the only case we shall use is in Exercise B2. □

Let $\mathbf{X} \in R^n$ have components X_i, and let $\mathbf{B}$ be an $n \times n$ symmetric matrix with components b_{ij}. Then

$$\mathbf{X}'\mathbf{B}\mathbf{X} = \sum_{ij}\sum X_i X_j b_{ij} = \sum_i b_{ii}X_i^2 + \sum_{i<j}\sum 2b_{ij}X_iX_j.$$

Now, let

$$q(\mathbf{X}) = \sum_i a_i X_i^2 + \sum_{i<j}\sum a_{ij}X_iX_j.$$

($q(\mathbf{X})$ is called a *quadratic form* in $\mathbf{X}$.) We see that

$$q(\mathbf{X}) = \mathbf{X}'\mathbf{B}\mathbf{X},$$

where $\mathbf{B}$ has ith diagonal element a_{ii} and (ij)th off-diagonal element $a_{ij}/2$.

Example 5–13

Let X_1, X_2, and X_3 be independent, with $X_1 \sim N(1, \sigma^2)$, $X_2 \sim N(-1, \sigma^2)$, and $X_3 \sim N(0, \sigma^2)$. Let

$$q_1(\mathbf{X}) = \frac{X_1^2 + X_2^2 + 2X_3^2 + 2X_1X_2}{2},$$

$$q_2(\mathbf{X}) = \frac{X_1^2 + X_2^2 - 2X_1X_2}{2},$$

$$q_3(\mathbf{X}) = \frac{X_1^2 + X_2^2 + X_3^2 + 2X_1X_2 + 4X_1X_3 + 4X_2X_3}{6}.$$

Using the preceding rules, we see that $q_i(\mathbf{X}) = \mathbf{X}'\mathbf{B}_i\mathbf{X}$, where

$$\mathbf{B}_1 = 2^{-1}\begin{Bmatrix} 1 & 1 & 0 \\ 1 & 1 & 0 \\ 0 & 0 & 2 \end{Bmatrix}, \qquad \mathbf{B}_2 = 2^{-1}\begin{Bmatrix} 1 & -1 & 0 \\ -1 & 1 & 0 \\ 0 & 0 & 0 \end{Bmatrix}, \qquad \mathbf{B}_3 = 6^{-1}\begin{Bmatrix} 1 & 1 & 2 \\ 1 & 1 & 2 \\ 2 & 2 & 1 \end{Bmatrix}.$$

Now,

$$\mathbf{B}_1^2 = 4^{-1}\begin{Bmatrix} 1 & 1 & 0 \\ 1 & 1 & 0 \\ 0 & 0 & 2 \end{Bmatrix}\begin{Bmatrix} 1 & 1 & 0 \\ 1 & 1 & 0 \\ 0 & 0 & 2 \end{Bmatrix} = 4^{-1}\begin{Bmatrix} 2 & 2 & 0 \\ 2 & 2 & 0 \\ 0 & 0 & 4 \end{Bmatrix} = \mathbf{B}_1.$$

Moreover,

$$\boldsymbol{\mu}'\mathbf{B}_1\boldsymbol{\mu} = 2^{-1}\{1 \quad -1 \quad 0\}\begin{Bmatrix} 1 & 1 & 0 \\ 1 & 1 & 0 \\ 0 & 0 & 2 \end{Bmatrix}\begin{Bmatrix} 1 \\ -1 \\ 0 \end{Bmatrix} = 0 \quad \text{and} \quad \operatorname{tr}(\mathbf{B}_1) = 2^{-1}(1 + 1 + 2) = 2.$$

Therefore,

$$\frac{q_1(\mathbf{X})}{\sigma^2} \sim \chi_2^2.$$

Similarly, $\mathbf{B}_2^2 = \mathbf{B}_2$ and $\operatorname{tr}(\mathbf{B}_2) = 1$. However, $\boldsymbol{\mu}'\mathbf{B}_2\boldsymbol{\mu} = 2$, so that $\mathbf{X}'\mathbf{B}_2\mathbf{X}/\sigma^2$ does not have a χ^2 distribution. Further, $\mathbf{B}_1\mathbf{B}_2 = \mathbf{0}$, so that $q_1(\mathbf{X})$ and $q_2(\mathbf{X})$ are independent. On the other hand, $\mathbf{B}_3$ is not idempotent, and $\mathbf{B}_1\mathbf{B}_3 \neq \mathbf{0}$. Therefore, $q_3(\mathbf{X})/\sigma^2$ does not have a χ^2 distribution, and $q_1(\mathbf{X})$ and $q_3(\mathbf{X})$ are not independent.

Example 5–14

Let $X_1, \ldots, X_n$ be independent and $X_i \sim N(\mu, \sigma^2)$. Let

$$\overline{X} = n^{-1}\sum_i X_i \quad \text{and} \quad S^2 = (n-1)^{-1}\left(\sum_i X_i^2 - n\overline{X}^2\right)$$

be the sample mean and sample variance. We give an alternative derivation of the joint distribution of $\overline{X}$ and S^2. Let $\mathbf{X} = (X_1, \ldots, X_n)'$, $\boldsymbol{\mu} = (\mu, \ldots, \mu)'$, and $\mathbf{a} = (n^{-1}, \ldots, n^{-1})$. (That is, $\boldsymbol{\mu}$ is a $k \times 1$-dimensional vector, all of whose components are μ, and $\mathbf{a}$ is a $1 \times k$-row vector, all of whose elements are n^{-1}.) By Theorem 5–20,

$$\mathbf{X} \sim N_n(\boldsymbol{\mu}, \sigma^2\mathbf{I}).$$

Furthermore,

$$\overline{X} = \mathbf{aX}, \qquad \overline{X}^2 = (\mathbf{aX})^2 = (\mathbf{aX})'(\mathbf{aX}) = \mathbf{X}'\mathbf{a}'\mathbf{aX},$$

and hence,

$$(n-1)S^2 = \sum X_i^2 - n\overline{X}'\overline{X} = \mathbf{X}'\mathbf{IX} - \mathbf{X}'(n\mathbf{a}'\mathbf{a})\mathbf{X} = \mathbf{X}'\mathbf{BX},$$

where

$$\mathbf{B} = \mathbf{I} - n\mathbf{a}'\mathbf{a}.$$

(Note that $\overline{X}$ is a linear function of $\mathbf{X}$ and $(n-1)S^2$ is a quadratic form in $\mathbf{X}$.)

Now,

$$\mathbf{a}\mathbf{a}' = n^{-2} + \cdots + n^{-2} = n^{-1} \quad \text{and} \quad \mathbf{a}\boldsymbol{\mu} = n^{-1}\mu + \cdots + n^{-1}\mu = \mu,$$

so that

$$\overline{X} \sim N(\mathbf{a}\boldsymbol{\mu}, \sigma^2 \mathbf{a}\mathbf{a}') = N\left(\mu, \frac{\sigma^2}{n}\right),$$

by part a of Theorem 5–21. Also, $\mathbf{B}$ is a symmetric matrix, and

$$\mathbf{B}^2 = (\mathbf{I} - n\mathbf{a}'\mathbf{a})(\mathbf{I} - n\mathbf{a}'\mathbf{a}) = I - 2n\mathbf{a}'\mathbf{a} + n^2\mathbf{a}'(\mathbf{a}\mathbf{a}')\mathbf{a} = I - 2n\mathbf{a}'\mathbf{a} + n\mathbf{a}'\mathbf{a} = \mathbf{B},$$

so that $\mathbf{B}$ is idempotent. In addition,

$$\boldsymbol{\mu}'\mathbf{B}\boldsymbol{\mu} = \boldsymbol{\mu}'\boldsymbol{\mu} - n(\boldsymbol{\mu}'\mathbf{a})(\mathbf{a}'\boldsymbol{\mu}) = \boldsymbol{\mu}'\boldsymbol{\mu} - n(\mathbf{a}'\boldsymbol{\mu})^2 = n\mu^2 - n\mu^2 = 0.$$

Finally,

$$\text{tr}(\mathbf{B}) = \text{tr}(\mathbf{I}) - \text{tr}(n\mathbf{a}'\mathbf{a}) = \text{tr}(\mathbf{I}) - \text{tr}(n\mathbf{a}\mathbf{a}') = n - \text{tr}(1) = n - 1.$$

(Note that $\mathbf{I}$ is $n \times n$ and 1 is a number.) Therefore, by part b of Theorem 5–21,

$$\frac{(n-1)S^2}{\sigma^2} \sim \chi^2_{n-1},$$

In addition,

$$\mathbf{a}\mathbf{B} = \mathbf{a}(\mathbf{I} - n\mathbf{a}'\mathbf{a}) = \mathbf{a} - n(\mathbf{a}\mathbf{a}')\mathbf{a} = \mathbf{a} - \mathbf{a} = \mathbf{0},$$

so that $\overline{X}$ and S^2 are independent by part c of Theorem 5–21.

One advantage of the method used in Example 5–14 to derive the joint distribution of $\overline{X}$ and S^2 is that we can use it to derive joint distributions for other linear and quadratic functions of normal random variables.

In Exercise C2, Theorem 5–21 is extended to the case in which $\mathbf{X}$ has an arbitrary positive definite covariance matrix.

Exercises—B

1. In Example 5–13,
   (a) Verify that $\mathbf{B}_2^2 = \mathbf{B}_2$ and $\text{tr}(\mathbf{B}_2) = 1$.
   (b) Verify that $\mathbf{B}_3$ is not idempotent.
   (c) Verify that $\mathbf{B}_1\mathbf{B}_2 = \mathbf{0}$ and $\mathbf{B}_1\mathbf{B}_3 \neq \mathbf{0}$.
2. Let $\mathbf{X} \sim N_n(\boldsymbol{\mu}, \sigma^2\mathbf{I})$. Suppose that $\mathbf{A}$ is a $p \times n$ matrix and $\mathbf{B}$ is an $n \times n$ idempotent matrix. Show that if $\mathbf{AB} = \mathbf{0}$, then $\mathbf{AX}$ and $\mathbf{X}'\mathbf{BX}$ are independent.
3. Suppose $X_i \sim N(i\theta, \tau^2)$ are independent, $i = 1, \ldots, n$. Let $k = \Sigma_i i^2 = n(n+1)(2n+1)/6$, $U = \Sigma_i iX_i/k$, and $V^2 = \Sigma_i(X_i - iU)^2/(n-1)$.
   (a) Show that $(n-1)V^2 = \Sigma X_i^2 - kU^2$.
   (b) Let $\mathbf{X} = (X_1, \ldots, X_n)'$ and $\mathbf{b} = (1/k, 2/k, \ldots, n/k)$. Show that $\mathbf{X} \sim N_n(k\theta\mathbf{b}', \tau^2\mathbf{I})$.
   (c) Let $\mathbf{C} = \mathbf{I} - k\mathbf{b}'\mathbf{b}$. Show that $U = \mathbf{bX}$ and $(n-1)V^2 = \mathbf{X}'\mathbf{CX}$.
   (d) Show that $U \sim N(\theta, \tau^2/k)$ $(k\theta\mathbf{bb}' = \theta, \tau^2\mathbf{bIb}' = \tau^2/k)$.
   (e) Show that $(n-1)V^2/\tau^2 \sim \chi^2_{n-1}$ ($\mathbf{C}$ is idempotent, $(k\theta\mathbf{b}')'\mathbf{C}(k\theta\mathbf{b}') = 0$, $\text{tr}(\mathbf{C}) = n-1$).
   (f) Show that U and V^2 are independent ($\mathbf{bC} = \mathbf{0}$).
   (These results are used often in the exercises in Chapters 7–10.)

Exercises—C

1. Let $\mathbf{Y} \sim N_n(\mathbf{X}\boldsymbol{\beta}, \sigma^2\mathbf{I})$, where $\mathbf{Y}$ is an $n \times 1$ random vector, $\mathbf{X}$ is an $n \times p$ matrix whose columns are linearly independent (so that $\mathbf{X}'\mathbf{X}$ is invertible), $\boldsymbol{\beta}$ is a $p \times 1$ vector, and σ^2 is a number. Let $\mathbf{B} = (\mathbf{X}'\mathbf{X})^{-1}\mathbf{X}'\mathbf{Y}$ and $S^2 = (\mathbf{Y} - \mathbf{X}\mathbf{B})'(\mathbf{Y} - \mathbf{X}\mathbf{B})/(n - p)$. We seek the joint distribution of $\mathbf{B}$ and S^2.
   (a) Show that $(n - p)S^2 = \mathbf{Y}'(\mathbf{I} - \mathbf{X}(\mathbf{X}'\mathbf{X})^{-1}\mathbf{X}')\mathbf{Y}$.
   (b) Let $\mathbf{M} = (\mathbf{X}'\mathbf{X})^{-1}$, $\mathbf{A} = \mathbf{M}\mathbf{X}'$, and $\mathbf{C} = \mathbf{I} - \mathbf{X}'\mathbf{M}\mathbf{X}$. Show that $\mathbf{B} = \mathbf{A}\mathbf{Y}$ and $(n - p)S^2 = \mathbf{Y}'\mathbf{C}\mathbf{Y}$.
   (c) Show that $\mathbf{B} \sim N_p(\boldsymbol{\beta}, \sigma^2\mathbf{M})$ (i.e., $\mathbf{A}\mathbf{X} = \mathbf{I}, \mathbf{A}\mathbf{A}' = \mathbf{M}$).
   (d) Show that $(n - p)S^2/\sigma^2 \sim \chi^2_{n-p}$ (i.e., $\mathbf{C}^2 = \mathbf{C} = \mathbf{C}', \mathbf{X}'\mathbf{C}\mathbf{X} = \mathbf{0}, \operatorname{tr}(\mathbf{C}) = n - p$).
   (e) Show that $\mathbf{B}$ and S^2 are independent (i.e., that $\mathbf{A}\mathbf{C} = \mathbf{0}$).
   (This exercise gives the basic distribution theory for multiple regression.)
2. Suppose that $\mathbf{Y} \sim N_n(\boldsymbol{\mu}, \boldsymbol{\Sigma}), \boldsymbol{\Sigma} > 0$. Let $\mathbf{A}$ be a $p \times n$ matrix, and let $\mathbf{B}$ be a symmetric matrix.
   (a) Show that $\mathbf{A}\mathbf{Y} \sim N_p(\mathbf{A}\boldsymbol{\mu}, \mathbf{A}\boldsymbol{\Sigma}\mathbf{A}')$.
   (b) Show that if $\mathbf{B}\boldsymbol{\Sigma}\mathbf{B} = \mathbf{B}$, $\boldsymbol{\mu}'\mathbf{B}\boldsymbol{\mu} = 0$, and $\operatorname{tr}(\mathbf{B}\boldsymbol{\Sigma}) = k$, then $\mathbf{Y}'\mathbf{B}\mathbf{Y} \sim \chi^2_k$. (*Hint*: Let $\mathbf{X} = \boldsymbol{\Sigma}^{-1/2}\mathbf{Y}$ and $\boldsymbol{\nu} = \boldsymbol{\Sigma}^{-1/2}\boldsymbol{\mu}$. Then $\mathbf{X} \sim N_n(\boldsymbol{\nu}, \mathbf{I})$. (Why?) Let $\mathbf{C} = \boldsymbol{\Sigma}^{1/2}\mathbf{B}\boldsymbol{\Sigma}^{1/2}$. Then $\mathbf{Y}'\mathbf{B}\mathbf{Y} = \mathbf{X}'\mathbf{C}\mathbf{X}$. Show that $\mathbf{C}$ is idempotent iff $\mathbf{B}\boldsymbol{\Sigma}\mathbf{B} = \mathbf{B}$, that $\boldsymbol{\mu}'\mathbf{B}\boldsymbol{\mu} = \boldsymbol{\nu}'\mathbf{C}\boldsymbol{\nu}$ and $\operatorname{tr}(\mathbf{C}) = \operatorname{tr}(\mathbf{B}\boldsymbol{\Sigma})$.)
   (c) Show that if $\mathbf{B}\boldsymbol{\Sigma}\mathbf{B} = \mathbf{B}$ and $\mathbf{A}\boldsymbol{\Sigma}\mathbf{B} = \mathbf{0}$, then $\mathbf{A}\mathbf{Y}$ and $\mathbf{Y}'\mathbf{B}\mathbf{Y}$ are independent.
   (In this problem, we extend the results of Theorem 5–21 to the case of general $\boldsymbol{\Sigma} > 0$.)

6

ASYMPTOTIC DISTRIBUTIONS

As we have seen in previous chapters, it is often difficult to find explicit expressions for the distribution functions of certain random variables. Even when such expressions can be found, it is often convenient to approximate complicated distributions by somewhat simpler ones. In Chapter 4, for example, we use the central limit theorem to approximate probabilities for binomial, Poisson, gamma, and χ^2 distributions. For instance, for the binomial distribution, we theoretically would need a table of the distribution function for every possible n and p. Instead, the normal approximation allows us to approximate many binomial distributions using a single table, the distribution function for the standard normal distribution. In this chapter, we develop a theory of such approximations.

The results we derive are also quite useful in statistics. For most inference problems involving discrete distributions, there are no finite sample methods possible, and we must rely on asymptotic methods. In addition, for many models involving continuous random variables, we can use the asymptotic results to argue that the exact procedures we derive are not too sensitive to some of the assumptions used in deriving them.

6.1 BASIC DEFINITIONS AND THEOREMS

6.1.1 Convergence in Probability and Distribution

Let $X_1, X_2, X_3, \ldots$ be a sequence of random variables such that X_n has distribution function $F_n(x)$, and let X be a random variable with distribution function $F(x)$. Recall that X_n *converges in distribution* to X, written $X_n \xrightarrow{d} X$, if

$$\lim_{n\to\infty} F_n(x) = F(x)$$

at all points x where $F(x)$ is continuous. We call $F(x)$ the *asymptotic distribution* of X_n.

Let b be a number. Recall that a random variable X *is degenerate at* b, written $X \equiv b$, if $P(X = b) = 1$. If $X \equiv b$, then X has the distribution function

$$F(x) = \begin{cases} 0 & \text{if } x < b \\ 1 & \text{if } x \geqq b \end{cases}.$$

If $X_n \xrightarrow{d} X$, where $X \equiv b$, we say that X_n *converges in probability* to b and write $X_n \xrightarrow{P} b$. That is, $X_n \xrightarrow{P} b$ if

$$\lim_{n\to\infty} F_n(x) = \begin{cases} 0 & \text{if } x < b \\ 1 & \text{if } x > b \end{cases}.$$

(Note that b is a point of discontinuity of this distribution function, so that it is not necessary that $F_n(b)$ converge to $F(b) = 1$.)

Example 6–1

Let $X_1, X_2, \ldots$ be a sequence of independent random variables with X_i uniformly distributed on the interval $(0, 1)$. Let

$$U_n = \max_{i \leqq n} X_i.$$

That is, U_n is the maximum of the first n of the X_i. Then the distribution function of U_n is

$$F_n(u) = P(U_n \leqq u) = P(X_1 \leqq u, \ldots, X_n \leqq u) = (P(X_1 \leqq u))^n = \begin{cases} 0 & \text{if } u \leqq 0 \\ u^n & \text{if } 0 < u < 1 \\ 1 & \text{if } u \geqq 1 \end{cases}.$$

If $0 < u < 1$, then $u^n \to 0$ as $n \to \infty$. Therefore,

$$\lim_{n\to\infty} F_n(u) = F(u) = \begin{cases} 0 & \text{if } u < 1 \\ 1 & \text{if } u \geqq 1, \end{cases}$$

$F(u)$ is the distribution function of a random variable which is degenerate at 1; hence,

$$U_n \xrightarrow{P} 1$$

Now, let $W_n = n(1 - U_n)$. Then the distribution function of W_n is

$$F_n^*(w) = P(W_n \leqq w) = P(n(1 - U_n) \leqq w) = P\left(U_n \geqq 1 - \frac{w}{n}\right) = 1 - F_n\left(1 - \frac{w}{n}\right)$$

$$= \begin{cases} 0 & \text{if } w \leqq 0 \\ 1 - (1 - w/n)^n & \text{if } 0 \leqq w \leqq n \\ 1 & \text{if } w \geqq n \end{cases} \rightarrow \quad F^*(w) = \begin{cases} 0 & \text{if } w \leqq 0 \\ 1 - e^{-w} & \text{if } w \geqq 0 \end{cases}.$$

But $F^*(w)$ is the distribution function of an exponential distribution with mean one, i.e., $F^*(w)$ is a $\Gamma(1,1)$ distribution. Therefore,

$$W_n \xrightarrow{d} T \sim \Gamma(1,1).$$

Example 6–1 illustrates a common occurrence in asymptotic theory. We often have a sequence X_n of random variables such that $X_n \xrightarrow{P} a$. In order to find a nondegenerate asymptotic distribution, we look at $Q_n = n^b(X_n - a)$ (or in this case, $n^b(a - X_n)$). Typically, there is one choice for b such that this distribution is nondegenerate. In this case, $b = 1$. If we look at $n^{1/2}(1 - U_n)$ in the example, we find that it converges to zero, while if we look at $n^2(1 - U_n)$, we find that it goes to infinity. (See the exercises.) Note that approximating a distribution by a degenerate distribution is fairly uninteresting, so that we want to find an asymptotic distribution which is nondegenerate.

Example 6–2

Let $X_1, X_2, \ldots$ be a sequence of independent random variables with $X_i \sim N(0, 1)$. As usual, let $\overline{X}_n$ be the sample mean computed from the first n of the X_i. Then $\overline{X}_n \sim N(0, 1/n)$. Therefore, the distribution function of $\overline{X}_n$ is

$$F_n(x) = P(\overline{X}_n \leqq x) = P(n^{1/2}\overline{X}_n \leqq n^{1/2}x) = P(Z \leqq n^{1/2}x) = N(n^{1/2}x)$$

(where $N(z)$ is the distribution function of the standard normal distribution). Now, if $x > 0$, then $n^{1/2}x \to \infty$ and $N(n^{1/2}x) \to 1$; if $x < 0$, then $n^{1/2}x \to -\infty$ and $N(n^{1/2}x) \to 0$; and if $x = 0$, then $n^{1/2}x = 0$ and $N(n^{1/2}x) = .5$. Therefore,

$$\lim_{n\to\infty} F_n(x) = F(x) = \begin{cases} 0 & \text{if } x < 0 \\ .5 & \text{if } x = 0 \\ 1 & \text{if } x > 0 \end{cases}.$$

Now, $F(x)$ is equal to the distribution function $G(x)$ of a random variable which is degenerate at zero (except at the point 0, where G is discontinuous). Therefore, $\overline{X}_n$ converges in distribution to a random variable which is degenerate at zero, and hence,

$$\overline{X}_n \xrightarrow{P} 0.$$

Now, let $Y_n = n^{1/2}\overline{X}_n \sim N(0,1)$. Then, since the distribution of Y_n does not depend on n,

$$Y_n \xrightarrow{d} Y \sim N(0,1).$$

Let $X \equiv 0$. Then $\overline{X}_n \xrightarrow{d} X$. However,

$$P(\overline{X}_n < 0) = .5, \qquad P(X < 0) = 0, \qquad P(\overline{X}_n \leqq 0) = .5, \qquad P(X \leqq 0) = 1,$$

so that $P(\overline{X}_n < 0)$ and $P(\overline{X}_n \leqq 0)$ do not go to $P(X < 0)$ and $P(X \leqq 0)$ as $n \to \infty$.

The following theorem shows that complications such as the one preceding can only happen at points where $F(x)$ is discontinuous.

Theorem 6–1. Let $X_n \xrightarrow{d} X$. Then

$$P(X_n < a) \to P(X < a) \quad \text{and} \quad P(X_n \leqq a) \to P(X \leqq a) \qquad \text{as } n \to \infty \tag{6–1}$$

for all a where $F(x)$ is continuous.

Proof. Let a be a point where $F(x)$ is continuous, so that

$$P(X < a) = P(X \leqq a) = F(a).$$

By the definition of convergence in distribution,

$$P(X_n \leqq a) = F_n(a) \to F(a) = P(X_n \leqq a).$$

Now $F(x)$ is continuous at a and is nondecreasing. Therefore, for all $c > 0$, there exists $a^* < a$ such that $F(a^*) > F(a) - c$ and $F(x)$ is continuous at a^*. Now,

$$P(X_n \leqq a^*) \leqq P(X_n < a) \leqq P(X_n \leqq a),$$

$$P(X_n \leqq a^*) \to F(a^*) \geqq F(a) - c, \qquad P(X_n \leqq a) \to F(a).$$

Therefore, for all $c > 0$, $P(X_n < a)$ is trapped between a sequence which goes to a number at least $F(a) - c$ and one which goes to $F(a)$. But this implies that

$$P(X_n < a) \to F(a) = P(X < a). \quad \square$$

As we have seen in Chapter 4, if Equation (6–1) is satisfied for all a, we may use probabilities for the distribution of X_n to approximate those for X for any sets we consider.

Corollary 6–1

a. If $X_n \xrightarrow{d} X$ and X is a continuous random variable, then Equation (6–1) is satisfied for all a.

b. If $X_n \xrightarrow{d} X$ and X_n and X are nonnegative integer-valued random variables, then Equation (6–1) is satisfied for all a.

Proof

a. If X is continuous, then $F(x)$ is continuous everywhere, and the desired result follows from Theorem 6–1.

b. If X is integer-valued, then $F(x)$ is continuous for all x which are not integers, and hence, Equation (6–1) is satisfied for all a which are not integers. Also, since X_n and X are both integer valued, for any integer k,

$$P(X_n \leqq k) = P(X_n \leqq k + .5) \to P(X \leqq k + .5) = P(X \leqq k).$$

(Note that $F(x)$ is continuous at $k + .5$.) Similarly,

$$P(X_n < k) = P(X_n \leqq k - .5) \to P(X \leqq k - .5) = P(X < k). \quad \square$$

Corollary 6–1 was stated in Chapter 4 to justify using the various approximations dealt with there.

We next give a condition that is equivalent to convergence in probability.

Theorem 6–2. $X_n \xrightarrow{P} b$ if and only if for all $c > 0$,

$$P(|X_n - b| > c) \to 0 \quad \text{as } n \to \infty. \tag{6–2}$$

Proof. Let $X_n \xrightarrow{P} b$, and let $X \equiv b$, so that $X_n \xrightarrow{d} X$. Then the distribution function of X is continuous at $b - c$ and $b + c$. Therefore, by Theorem 6–1,

$$P(|X_n - b| \leqq c) = P(X_n < b + c) - P(X_n \leqq b - c) \to P(X < b + c) - P(X \leqq b - c)$$

$$= P(|X - b| \leqq c) = 1.$$

For the converse, suppose that Equation (6–2) is satisfied. Then

$$0 \leqq F_n(b - c) \leqq F_n(b + c) \leqq 1,$$

and

$$F_n(b + c) = P(X_n \leqq b + c) \geqq P(|X_n - b| < c) \to 1 = F(b + c),$$

$$F_n(b - c) = P(X_n \leqq b - c) \leqq P(|X_n - b| \geqq c) \leqq P\left(|X_n - b| > \frac{c}{2}\right) \to 0 = F(b - c).$$

Hence,

$$F_n(b + c) \to 1 = F(b + c) \quad \text{and} \quad F_n(b - c) \to 0 = F(b - c),$$

for all $c > 0$. Therefore, $F_n(x) \to F(x)$ at all $x \neq b$, and it follows that $X_n \xrightarrow{P} b$. $\square$

Equation (6–2) is often used as the definition of convergence in probability.

The following theorem relates moment-generating functions to convergence in distribution and is the basic theorem of asymptotic theory. We say that the sequence $X_1, X_2, \ldots$ of random variables is *uniformly bounded* if there exists $q < \infty$ such that $P(|X_n| > q) = 0$ for all n. (Note that q does not depend on n, so that q is a bound for $|X_n|$ uniformly in n.)

Theorem 6–3. (Continuity theorem) Let $X_1, X_2, \ldots$ be a sequence of random variables with X_n having moment-generating function $M_n(t)$, and let X be a random variable having moment-generating function $M(t)$.

a. Suppose that there exists $a > 0$ such that $\lim_{n\to\infty} M_n(t)$ for all $t \in (-a, a)$. Then $X_n \xrightarrow{d} X$.
b. Conversely, if $X_1, X_2, \ldots$ are uniformly bounded random variables and $X_n \xrightarrow{d} X$, then $\lim_{n\to\infty} M_n(t) = M(t)$.

Proof. The proof of this theorem is, in general, quite difficult and will not be given here. (See Billingsley (1979), pp. 302–304 for such a proof.) Instead, we shall give a proof of the special case in which both X_n and X are nonnegative integer-valued random variables.

a. Let $f_{nk} = P(X_n = k)$, $f_k = P(X = k)$, $P_n(s) = M_n(\log(s)) = E(s^{X_n}) = \Sigma_{k=0}^{\infty} f_{nk} s^k$, and $P(s) = M(\log(s)) = E(s^X) = \Sigma_{k=0}^{\infty} f_k s^k$. Then $P_n(s)$ and $P(s)$ are both power series, and $0 \leqq P_n(s) < 1$ and $0 \leqq P(s) < 1$ for $0 < s < 1$. Therefore, $P_n(s) \to P(s)$ only if $f_{nk} \to f_k$, which in turn occurs only if $X_n \xrightarrow{d} X$. (See Theorem 4–1.)
b. Conversely, if $X_n \xrightarrow{d} X$, then $f_{nk} \to f_k$. But if X is bounded by q (which we may assume to be an integer), then

$$M_n(t) = \sum_{k=1}^{q} f_{nk} e^{tk} \to \sum_{k=1}^{q} f_k e^{tk},$$

since the sum is only a finite sum. □

Note, in Theorem 6–3, that $M_n(0) = M(0) = 1$. The restriction that the condition be satisfied on $(-a, a)$ for some $a > 0$ implies that the condition be satisfied on an interval on both sides of zero. All the moment-generating functions encountered in this text are defined on such an interval, and we shall ignore this condition from now on.

If the X_n or X do not have moment-generating functions, then characteristic functions may be substituted in the theorem. Interestingly, if we substitute characteristic functions for moment-generating functions, then the converse of the theorem is also true without assuming uniform boundedness. (By contrast, in Exercises C1 and C2 we present examples which show that for moment-generating functions the converse may be false if the sequences are not uniformly bounded.)

Example 6–3

Let $X_n \sim B(n, a/n)$, where a is a fixed constant. Then X_n has moment-generating function

$$M_n(t) = 1 - \frac{a}{n} + \frac{ae^t}{n} = \left(1 + \frac{a(e^t - 1)}{n}\right)^n \to \exp(a(e^t - 1)),$$

which is the moment-generating function of a Poisson random variable with mean a.

Therefore, by the continuity theorem, $X_n \xrightarrow{d} X \sim P(a)$. This is the same result we derived in Chapter 4 using density functions.

By Equation (6–2), if $X_n \xrightarrow{P} b$, then X_n is getting quite close to b for large n. However, when $X_n \xrightarrow{d} X$, it is not necessary that X_n get close to X in any way, or that $X_n - X$ converge to anything (even in distribution). All that we know in that case is that the distribution function of X_n is getting close to the distribution function of X.

In Exercise C1, an example is given in which $X_n \xrightarrow{P} a$, but EX_n does not converge to a, so that we cannot conclude that if $X_n \xrightarrow{P} a$, then EX_n goes to a. In Exercise C2, another example is given in which $X_n \xrightarrow{d} Z \sim N(0, 1)$, but EX_n does not go to zero and $\text{var}(X_n)$ does not go to one. Therefore, it is not necessarily true that if $X_n \xrightarrow{d} X$, then EX_n goes to EX or $\text{var}(X_n)$ goes to $\text{var}(X)$.

Exercises—B

1. Let $X_1, X_2, \ldots$ be a sequence of independent random variables with X_i uniformly distributed on the interval $(0,1)$. Let $V_n = \min_{i \leqq n} X_i$ be the minimum of the first n of the X_i.
   (a) Find the distribution function of V_n.
   (b) Show that $V_n \xrightarrow{P} 0$.
   (c) Find the distribution function of $Q_n = nV_n$.
   (d) Show that $Q_n \xrightarrow{d} \Gamma(1,1)$.
2. Let $X_1, X_2, \ldots$ be a sequence of independent random variables which are exponentially distributed with a mean of 3. Let V_n be the minimum of the first n of the X_i.
   (a) Find the limiting distribution of V_n.
   (b) Find the limiting distribution of $W_n = nV_n$.
3. Let $X_n \sim \Gamma(n, 1/n)$.
   (a) What is the moment-generating function of X_n?
   (b) Show that $X_n \xrightarrow{P} 1$.
4. (a) Let $X_n \sim N(3, n^{-1})$. Use the continuity theorem to show that $X_n \xrightarrow{P} 3$.
   (b) Let $X_n \sim N(n^{-1}, 1 - n^{-1})$. Use the continuity theorem to show that $X_n \xrightarrow{d} Z \sim N(0,1)$.
5. In Example 6–1,
   (a) Find the distribution function of $Q_n = n^{1/2}(1 - U)$, and show that $Q_n \xrightarrow{P} 0$.
   (b) Find the distribution function of $R_n = n^2(1 - U)$, and show that this distribution function converges to a function which is zero everywhere (which implies that R_n has "drifted off" to infinity).
6. Let X_n be a discrete random variable with density function $f_n(x) = 1$ if $x = 1/n$, so that X_n is degenerate at $1/n$.
   (a) What is $\lim_{n\to\infty} f_n(x)$?
   (b) Show that $X_n \xrightarrow{P} 0$.
   (This is an example in which $X_n \xrightarrow{P} 0$, but the density function of X_n converges to zero, which is not a density function.)
7. Let X_n have a negative binomial distribution with parameters n and $p = 1 - a/n$. Use the continuity theorem to show that $X_n \xrightarrow{d} X \sim P(a)$.
8. In Theorem 6–3, show that if $0 < s < 1$, then $0 \leqq P_n(s) < 1$ and $0 \leqq P(s) < 1$.

Exercises—C

1. Let $X_1, X_2, \ldots$ be a sequence of discrete random variables such that X_n has density function f_n given by $f_n(0) = 1 - (1/n)$ and $f_n(n) = 1/n$.

(a) Show that $EX_n = 1$ and $\mathrm{var}(X_n) \to \infty$.
(b) Show that $X_n \xrightarrow{P} 0$.
(c) Show that $M_n(t) = E\ \exp(X_n t) \to \infty \neq \exp(0t) = 1$.
(This is an example in which $X_n \xrightarrow{P} 0$, but EX_n and $\mathrm{var}(X_n)$ do not go to zero and $M_n(t)$ does not go to one.)

2. Let X_n be defined as in the previous exercise, let $U \sim N(0, 1)$ be independent of X_n. Let $W_n = U + X_n$.
(a) Show that $EW_n = 1$ and $\mathrm{var}(W_n) \to \infty$.
(b) Show that $W_n \xrightarrow{d} Z \sim N(0, 1)$.
(*Hint*: $F_n(w) = (1 - (1/n))P(U \leqq w) + (1/n)P(U \leqq w - n)$. Why?)
(c) Show that $M_n(t) = E\ \exp(W_n t) \to \infty \neq M(t) = E\ \exp(Zt)$.
(This is an example in which $X_n \xrightarrow{d} Z$, but EX_n, $\mathrm{var}(X_n)$, and $M_n(t)$ do not go to EZ, $\mathrm{var}(Z)$, and $M(t)$, respectively.)

3. Suppose that $P(|X_n| > q) = 0$ for all n and that $X_n \xrightarrow{d} X$. Show that $P(|X| > q) = 0$.

4. Suppose that $P(|X| > q) = 0$. Show that X has a finite moment-generating function.

6.1.2 Taylor's Theorem

Several of the results of this chapter rely on a theorem of the calculus called *Taylor's theorem*. In this book, we use Young's form of Taylor's theorem, which is not usually given in basic calculus texts and whose proof is somewhat difficult to locate. We therefore include a proof.

Let $f(x)$ be a function such that $f^{(n)}(c)$ exists. Then $f^{(j)}(x)$ exists in an interval around c for $j < n$. In particular, $f^{(j)}(c)$ exists for $j \leqq n$. Let

$$P_n(x) = \sum_{j=0}^{n} \frac{(x - c)^j f^{(j)}(c)}{j!} = f(c) + (x - c)f'(c) + \frac{(x - c)^2 f''(c)}{2} + \cdots + \frac{(x - c)^n f^{(n)}(c)}{n!}.$$

In particular,

$$P_1(x) = f(c) + f'(c)(x - c)$$

and

$$P_2(x) = f(c) + f'(c)(x - c) + \frac{f''(c)(x - c)^2}{2}.$$

$P_n(x)$ is called the nth degree *Taylor polynomial* of $f(x)$ at c. Note that

$$f(c) - P_n(c) = 0.$$

Finally, let $R_n(c) = 0$, and for $x \neq c$, let

$$R_n(x) = \frac{n![f(x) - P_n(x)]}{(x - c)^n},$$

so that

$$\begin{aligned} f(x) &= P_n(x) + \frac{(x - c)^n R_n(x)}{n!} \\ &= f(c) + \cdots + \frac{(x - c)^{n-1} f^{(n-1)}(c)}{(n - 1)!} + \frac{(x - c)^n [f^{(n)}(c) + R_n(x)]}{n!}. \end{aligned}$$

Then $R_n(x)$ is the *remainder* for the nth degree Taylor polynomial at c. (Note that both $P_n(x)$ and $R_n(x)$ depend on c.)

Theorem 6–4. (Taylor's theorem) Let $f(x)$ be a function such that $f^{(n)}(c)$ exists for some real c. Let $R_n(x)$ be the remainder for the nth degree Taylor polynomial at c. Then $R_n(x)$ is continuous at c. That is,

$$\lim_{x \to c} R_n(x) = R_n(c) = 0.$$

Proof. When $n = 1$, for $x \neq c$,

$$P_1(x) = f(c) + (x - c)f'(c)$$

and

$$R_1(x) = \frac{f(x) - f(c)}{x - c} - f'(c),$$

which goes to zero as x approaches c, by the definition of $f'(c)$. When $n = 2$,

$$P_2(x) = f(c) + (x - c)f'(c) + \frac{(x - c)^2 f''(c)}{2}.$$

Furthermore,

$$P_2'(x) = f'(c) + (x - c)f''(c),$$

and $f'(x)$ exists in an interval about c because $f''(c)$ exists. Therefore, we can use L'Hospital's rule to find

$$\lim_{x \to c} R_2(x) = \lim_{x \to c} \left[\frac{2(f(x) - P_2(x))}{(x - c)^2}\right]$$

$$= \lim_{x \to c} \left[\frac{f'(x) - P_2'(x)}{x - c}\right] = \lim_{x \to c} \left[\frac{f'(x) - f'(c)}{x - c}\right] - f''(c),$$

which is zero by the definition of $f''(c)$. (We cannot use L'Hospital's rule to evaluate the last indeterminate form, because we don't know whether $f''(x)$ exists anywhere else but at c.) For general n, we apply L'Hospital's rule $n - 1$ times to find

$$\lim_{x \to c} \frac{R_n(x)}{(x - c)^n} = \lim_{x \to c} \left[\frac{n!(f(x) - P_n(x))}{(x - c)^n}\right] = \lim_{x \to c} \left[\frac{f^{(n-1)}(x) - f^{(n-1)}(c)}{x - c}\right] - f^{(n)}(c),$$

which is zero by the definition of $f^{(n)}(c)$. □

Example 6–4

Let $f(x) = \log(1 + x)$, and let $c = 0$. Then

$$f(0) = \log(1 + 0) = 0, \qquad f'(0) = (1 + 0)^{-1} = 1, \qquad f''(0) = -(1 + 0)^{-2} = -1.$$

Hence,

$$P_1(x) = 0 + (x - 0)(1) = x$$

and

$$P_2(x) = 0 + (x - 0)(1) + \frac{(x - 0)^2(-1)}{2} = x - \frac{x^2}{2}.$$

Therefore,

$$\log(x) = x(1 + R_1(x)) = x + \frac{x^2(-1 + R_2(x))}{2},$$

where $R_1(x)$ and $R_2(x)$ go to zero as x goes to zero.

Exercises—B

1. **a.** Find the second-order Taylor polynomial at $c = 0$ for $g(x) = \exp(2x)$.
   b. Find the second-order Taylor polynomial at $c = 0$ for $g(x) = \exp(x^2)$.
2. Let X be a random variable with mean μ, variance σ^2, and moment-generating function $M(t)$.
   a. Show that the first-order Taylor polynomial for $M(t)$ at $t = 0$ is $P_1(t) = 1 + \mu t$.
   b. Show that the second-order Taylor polynomial for $M(t)$ at $t = 0$ is $P_2(t) = 1 + \mu t + (\sigma^2 + \mu^2)t^2/2$.
   c. What are the first- and second-order Taylor polynomials at $t = 0$ for $\psi(t) = \log(M(t))$?

6.2 THE CENTRAL LIMIT THEOREM AND THE WEAK LAW

6.2.1 The Weak Law and the Central Limit Theorem

In this section, we derive the two most important results in limiting distribution theory. (In fact, it could be argued that the central limit theorem is the most important result in all of probability and statistics.)

We first need the following lemma.

Lemma. Let b_n be numbers such that $\lim_{n\to\infty} b_n = c$. Then

$$\lim_{n\to\infty}\left(1 + \frac{b_n}{n}\right)^n = \exp(c).$$

Proof. Taking the logarithm of both sides of the preceding equation, we obtain

$$\lim_{n\to\infty} n \log\left(1 + \frac{b_n}{n}\right) = c.$$

From Example 6–4,

$$\log(1 + t) = t(1 + R_1(t)) \quad \text{and} \quad \lim_{t\to 0} R_1(t) = 0.$$

Hence,

$$n \log\left(1 + \frac{b_n}{n}\right) = b_n\left(1 + R_1\left(\frac{b_n}{n}\right)\right),$$

and

$$\lim_{n\to\infty} R_1\left(\frac{b_n}{n}\right) = \lim_{t\to 0} R_1(t) = 0$$

(since $b_n/n \to 0$ as $n \to \infty$). Therefore,

$$\lim_{n\to\infty} n \log\left(1 + \frac{b_n}{n}\right) = \left(\lim_{n\to\infty} b_n\right)\left(\lim_{n\to\infty}\left[1 + R_1\left(\frac{b}{n}\right)\right]\right) = c. \quad \square$$

Now, let $X_1, X_2, X_3, \ldots$ be a sequence of random variables which are independently identically distributed. Also, let $\overline{X}_n$ be the average of the first n of these X_i, i.e.,

$$\overline{X}_n = \sum_{i=1}^{n} \frac{X_i}{n}.$$

Then we have the following theorem.

Theorem 6–5

a. (Weak law of large numbers) Suppose that $EX_i = \mu$ is finite. Then

$$\overline{X}_n \xrightarrow{P} \mu.$$

b. (Central limit theorem) Suppose, in addition to the preceding, that $\operatorname{var}(X_i) = \sigma^2 < \infty$. Then

$$\frac{n^{1/2}(\overline{X}_n - \mu)}{\sigma} \xrightarrow{d} Z \sim N(0, 1).$$

Proof. We use the continuity theorem and assume that the X_i have moment-generating functions. If not, characteristic functions (and a suitably generalized lemma) may be used.

a. Let $M(t)$ be the moment-generating function of the X_i. Then

$$M'(0) = EX_i = \mu.$$

By Taylor's theorem,

$$M(t) = M(0) + t(M'(0) + R(t)) = 1 + t(\mu + R(t))$$

and

$$\lim_{t\to 0} R(t) = 0.$$

Now, $\overline{X}_n$ has moment-generating function

$$M_n(t) = \left(M\left(\frac{t}{n}\right)\right)^n.$$

(See Exercise B1.) Therefore,

$$M_n(t) = \left(M\left(\frac{t}{n}\right)\right)^n = \left[1 + \frac{t(\mu + R(t/n))}{n}\right]^n,$$

$$\lim_{n\to\infty} t\left(\mu + R\left(\frac{t}{n}\right)\right) = t\mu + t\lim_{s\to 0} R(s) = \mu t$$

(since t/n goes to zero as n approaches infinity). But now, by the foregoing lemma,

$$\lim_{n\to\infty} M_n(t) = \lim_{n\to\infty}\left[1 + \frac{t(\mu + R(t/n))}{n}\right]^n = \exp(\mu t),$$

which is the moment-generating function of a random variable which is degenerate at μ. Hence, $\overline{X}_n \xrightarrow{P} \mu$.

b. Let $M^*(t)$ be the moment-generating function of $Y_i = (X_i - \mu)/\sigma$. Then

$$M^{*\prime}(0) = EY_i = 0 \quad \text{and} \quad M^{*\prime\prime}(0) = EY_i^2 = \text{var}(Y_i) = 1.$$

Therefore, by Taylor's theorem,

$$M^*(t) = 1 + \left(\frac{t^2}{2}\right)(1 + R^*(t)) \quad \text{and} \quad \lim_{t\to 0} R^*(t) = 0.$$

Also,

$$\left(\frac{t^2}{2}\right)(1 + R^*(n^{-1/2}t)) \to \left(\frac{t^2}{2}\right)(1 + R^*(0)) = \frac{t^2}{2} \quad \text{as } n \to \infty.$$

Now, the random variable

$$Z_n = \frac{n^{1/2}(\overline{X}_n - \mu)}{\sigma} = n^{1/2}\,\overline{Y}_n$$

has moment-generating function

$$M_n^*(t) = (M^*(n^{-1/2}t))^n = \left[1 + \frac{(t^2/2)(1 + R^*(n^{-1/2}t))}{n}\right]^n \to \exp\left(\frac{t^2}{2}\right)$$

(using the lemma again) which is the moment-generating function of a standard-normal random variable. The desired result then follows from the continuity theorem. □

We have seen in Chapter 4 how useful the central limit theorem is for finding probabilities. We shall also see in Chapters 8 and 10 that it is quite useful in statistics. For one thing, it implies that many procedures which are derived for normal models are not too sensitive to the normal assumption. For another, it implies that many continuous random variables (such as weight or IQ test score) which are averages of many small effects should be approximately normally distributed. In fact, the central limit theorem is the primary justification for using the normal distribution as often as we do in statistics. We should also mention that the central limit theorem is true in even more generality than given here: Under certain conditions, it can be extended to averages and sums of random variables which are neither independent nor identically distributed.

The weak law of large numbers implies that as the sample size n increases, the probability that the sample mean $\overline{X}$ is close to the true mean μ gets quite high.

Example 6–5

Consider an experiment, such as flipping a coin, which may be replicated many times under the same conditions. Suppose the probability of some particular outcome of the experiment (say, getting a head) is p. Suppose we replicate the experiment n independent times, and let U_n be the proportion of times the particular outcome occurs in the first n trials (e.g., the proportion of times a head turns up in the first 100 trials). Then

$$U_n \xrightarrow{P} p.$$

For let $V_i = 1$ if the outcome occurs on the ith trial and $V_i = 0$ if not. Then the V_i are i.i.d., and

$$EV_i = (1)p + (0)(1-p) = p.$$

Now, $U_n = \overline{V}_n$, the average of the first n of the V_i's. (Why?) Therefore, by the weak law,

$$U_n = \overline{V}_n \xrightarrow{P} p.$$

This fact is often taken as a justification for the frequentist approach to probability, discussed in Chapter 1, which interprets the probability of an event E as the limit, as n approaches infinity, of the proportion of the time E occurs in n trials of the experiment.

It is not surprising that where there is a weak law of large numbers there is also a strong law of large numbers. However, we shall not use that law in this book, and so shall not state it here. (See Billingsley (1979), pp. 247–258, for a discussion of the strong law.)

Exercises—B

1. Let $X_1, X_2, \ldots$ be a sequence of i.i.d. random variables with moment-generating function $M(t)$. Let $\overline{X}_n$ be the average of the first n of the X_i.
   (a) Show that $\overline{X}_n$ has moment-generating function $M_n(t) = (M(t/n))^n$.
   (b) Let $Y_n = n^{1/2}\overline{X}_n$. Show that Y_n has moment-generating function $M^*(t) = (M(t/n^{1/2}))^n$.
2. Let V_i be independent, $V_i \sim \chi_1^2$, and $U_n = \Sigma_{i=1}^n V_i$.
   (a) How is U_n distributed?
   (b) Show that $U_n/n \xrightarrow{P} 1$.
3. Let $W_i \sim B(1,p)$, with W_i independent. Let $X_n = \Sigma_{i=1}^n W_i$.
   (a) How is X_n distributed?
   (b) Show that $X_n/n \xrightarrow{P} p$.

Exercises—C

1. **(a)** Let $h(u)$ be a function such that $h(0) = 1$ and $h'(u)$ is continuous at zero. Use L'Hospital's rule to show that $\lim_{u\to 0} \log[h(u)]/u = h'(0)$.
   (b) Use the result of part a with $u = n^{-1}$ and $h(u) = M(tu)$ to show that $\lim_{n\to\infty} \log[M_n(t)] = \exp(\mu t)$ in the proof of the weak law of large numbers.
2. **(a)** Let $h(u)$ be a function such that $h(0) = 1$, $h'(0) = 0$, and $h''(u)$ is continuous at zero. Use L'Hospital's rule to show that $\lim_{u\to 0} \log[h(u)]/u^2 = h''(0)/2$.
   (b) Use the result of part a with $u = n^{-1/2}$ and $h(u) = M^*(tu)$ to prove that $\lim_{n\to\infty} \log[M_n^*(t)] = t^2/2$ in the proof of the central limit theorem.
3. Let $X_1, X_2, \ldots$ be a sequence of independently identically distributed random variables having the common density function $f(x) = \pi^{-1}(1+x^2)^{-1}$, $-\infty < x < \infty$.
   (a) Show that $f(x)$ is symmetric about the point zero.
   (b) Show that $\overline{X}_n$ has the same distribution as X_1. (*Hint*: Use induction and the fact than $\overline{X}_{n+1} = (n\overline{X}_n + X_{n+1})/(n+1)$.)
   (c) Show that $\overline{X}_n$ does not converge in probability to anything.
   (Note that EX_n does not exist for the Cauchy distribution.)

* 6.2.2 Asymptotic Distribution of the Sample Median

Let $F(x)$ be the distribution function of a continuous random variable X with density function $f(x)$. The *median* of F is the number θ which satisfies

$$F(\theta) = \frac{1}{2}.$$

We assume in this section that $F(\theta)$ is unique and that the density function $f(x)$ is continuous at θ.

Now, let $X_1, \ldots, X_n$ be independently distributed having the common distribution function $F(x)$. The *sample median* of the X_i is the $(n+1)/2$ largest of the X_i if n is odd and the average of the $n/2$ and $1 + n/2$ largest X_i if n is even. Thus, if $n = 6$, and the X_i are 4, 3, 1, 5, 6, and 7, then the sample median is $(4+5)/2$. In this section, we find the asymptotic distribution of the sample median. It is rather surprising that this distribution should be normal, since the sample median is the average of at most two random variables.

Theorem 6–6. Let $X_1, X_2, \ldots$ be a sequence of independently identically distributed continuous random variables having common distribution function $F(x)$ and density function $f(x)$. Let θ be the median of F, and let $\tilde{X}_n$ be the sample median computed from the first n of the X_i. Then

$$n^{1/2}(\tilde{X}_n - \theta) \xrightarrow{d} Y \sim N(0, (4f^2(\theta))^{-1}).$$

Proof. (Heuristic) A rigorous proof is beyond the scope of the text, but there is a fairly convincing heuristic argument which we now give. For simplicity, we assume that n is odd, so that $\tilde{X}_n$ is the $(n+1)/2$ largest of the first n X_i's. Let $U_{n,a}$ be the number of the first n X_i's which are less than or equal to a. Then

$$\tilde{X}_n \leqq a \Leftrightarrow U_{n,a} \geqq \frac{n}{2}.$$

(Why?) Therefore,

$$P(n^{1/2}(\tilde{X}_n - \theta) \leqq b) = P(\tilde{X}_n \leqq a_n) = P\left(U_{n,a_n} \geqq \frac{n}{2}\right), \qquad a_n = \theta + bn^{-1/2}.$$

Now, let $p_n = F(a_n)$, so that

$$U_{n,a_n} \sim B(n, p_n).$$

Then

$$P\left(U_{n,a_n} \geqq \frac{n}{2}\right) = P(Z_n \geqq c_n), \qquad Z_n = \frac{U_{n,a_n} - p_n}{(np_n(1-p_n))^{1/2}},$$

$$c_n = \frac{n^{1/2}(1 - 2p_n)}{2(p_n(1-p_n))^{1/2}},$$

It can be shown that

$$Z_n \xrightarrow{d} Z \sim N(0,1), \qquad c_n \to -2f(\theta)b.$$

The second of the foregoing limits is derived in the exercises. If p_n did not depend on n, then the first fact would just be the normal approximation to the binomial distribution. In this case, $p_n \to \frac{1}{2}$, and the limit given can be established using a generalization of the central limit theorem called the Lindeberg-Feller theorem. It can also be shown that

$$\frac{Z_n}{c_n} \xrightarrow{d} \frac{-Z}{2f(\theta)b}.$$

(This is an application of Slutzky's theorem of the next section.) Hence,

$$P(n^{1/2}(\tilde{X}_n - \theta) \leqq b) = P\left(\frac{Z_n}{c_n} \geqq 1\right) \to P\left(\frac{-Z}{2f(\theta)b} \geqq 1\right) = P\left(\frac{Z}{2f(\theta)} \leqq b\right)$$

and it follows by the definition of convergence in distribution that

$$n^{1/2}(\tilde{X}_n - \theta) \xrightarrow{d} \frac{Z}{2f(\theta)} \sim N(0, (4f^2(\theta))^{-1}). \quad \square$$

Corollary 6–6. $\tilde{X}_n \xrightarrow{P} \theta$.

Proof. $\tilde{X}_n = n^{-1/2}(n^{1/2}(\tilde{X}_n - \theta)) + \theta \xrightarrow{d} 0Y + \theta = \theta$, and hence, $\tilde{X}_n \xrightarrow{P} \theta$. $\square$

If the distribution of X is symmetric about θ (i.e., $f(\theta - x) = f(\theta + x)$), and if the distribution has a mean μ, then $\mu = \theta$. (See Exercise B4.) If, in addition, the distribution has a variance σ^2, then, by the central limit theorem,

$$n^{1/2}(\overline{X}_n - \theta) \xrightarrow{d} Y^* \sim N(0, \sigma^2).$$

A natural question to ask is whether the sample mean $\overline{X}_n$ or the sample median $\tilde{X}_n$ is a better estimator of θ. One sensible answer is to take the one which has smaller asymptotic variance. That is, take the sample median when $(4f^2(\theta))^{-1} < \sigma^2$ and the sample mean when $(4f^2(\theta))^{-1} > \sigma^2$. For example, if the X_i are normally distributed, then

$$(4f^2(\theta))^{-1} = \frac{\pi\sigma^2}{2} > \sigma^2,$$

so that the sample mean is better by this criterion for normal random variables.

It should be emphasized that $\tilde{X}_n \xrightarrow{P} \theta$ for any continuous distribution, but $\overline{X}_n \xrightarrow{P} \theta$ only if the distribution has a finite mean μ and $\mu = \theta$. Similarly, the asymptotic distribution derived for $\tilde{X}_n$ is correct for any continuous distribution, while that for $\overline{X}_n$ is correct only when the distribution has a finite variance. (See Exercise C3, Section 6.2.1.)

Exercises—B

1. In the context of the proof of Theorem 6–6,
   a. Show that $p_n \to 1/2$ as $n \to \infty$.
   b. Show that $n^{1/2}(1 - 2p_n) \to -2bF(\theta)$ as $n \to \infty$. (*Hint*: Let $h(x) = (1 - 2F(\theta + bx))|x$. Use L'Hospital's rule to show $\lim_{x \to 0} h(x) = -2bf(\theta)$.)
   c. Show that $c_n \to -2bF(\theta)$ as $n \to \infty$.

2. Suppose the variables X_i are uniformly distributed. Which has the smaller asymptotic variance, the sample mean or the sample median?
3. Suppose the variables X_i have a t-distribution on three degrees of freedom. Which has smaller asymptotic variance, the sample median or the sample mean?
4. Let X be a continuous random variable with density function $f(x)$ which is symmetric about θ (i.e., $f(\theta - t) = f(\theta + t)$ for all t).
   a. Show that if $EX < \infty$, then $EX = \theta$. (*Hint:* Show that $E(X - \theta) = 0$ by showing that $\int_{-\infty}^{\theta}(x - \theta)f(x)dx = -\int_{\theta}^{\infty}(x - \theta)f(x)dx$. Let $t = -(x - \theta)$ in the first integral and let $t = x - \theta$ in the second one.)
   b. Show that θ is a median of X. (*Hint:* Show that $P(X \leqq \theta) = P(X \geqq \theta)$.)

6.3 FURTHER RESULTS ON UNIVARIATE ASYMPTOTICS

6.3.1 Continuous Functions and Slutzky's Theorem

In this section, we present some additional results about convergence in probability and in distribution. We first give a simple method which can often be used to show that a sequence converges in probability.

Theorem 6–7. Let $X_1, X_2, \ldots$ be a sequence of random variables with $EX_n = \mu_n$ and $\text{var}(X_n) = \sigma_n^2$. If $\mu_n \to a$ and $\sigma_n^2 \to 0$, then $X_n \xrightarrow{P} a$.

Proof. Observe that

$$E(X_n - a)^2 = E(X_n - \mu_n + \mu_n - a)^2 = \sigma_n^2 + (\mu_n - a)^2.$$

Therefore, by Markov's inequality,

$$0 \leqq P(|X_n - a| > \epsilon) = P((X_n - a)^2 > \epsilon^2) \leqq \frac{E(X_n - a)^2}{\epsilon^2} = \frac{\sigma_n^2 + (\mu_n - a)^2}{\epsilon^2} \to 0$$

as n goes to infinity. □

In Exercise C1, Section 6.1.1, an example is given in which $X_n \xrightarrow{P} 0$, but $EX_n = 1$ and $\text{var}(X_n) = \infty$, so that the converse of Theorem 6–7 is not true.

Corollary 6–7. Let a_n be a sequence of constants such that $\lim_{n\to\infty} a_n = b$. Then

$$a_n \xrightarrow{P} b.$$

Proof. We treat the a_n as degenerate random variables. Then $EX_n = a_n \to b$ and $\text{var}(a_n) = 0$. The desired result then follows from Theorem 6–7. □

We next use Theorem 6–7 to give a second derivation of the weak law of large numbers.

Example 6–6

Let $X_1, X_2, \ldots$ be a sequence of i.i.d. random variables, with $EX_i = \mu$ and $\text{var}(X_i) = \sigma^2$. Let $\overline{X}_n$ be the sample mean computed from the first n of the X_i. In Chapter 4, we showed that

$$E\overline{X}_n = \mu \quad \text{and} \quad \text{var}(\overline{X}_n) = \frac{\sigma^2}{n} \to 0.$$

Therefore, $\overline{X}_n \xrightarrow{P} \mu$ by Theorem 6–7.

The next theorem gives some results which are often useful in deriving asymptotic distributions.

Theorem 6–8

a. If $X_n \xrightarrow{P} a$ and $g(x)$ is continuous at $x = a$, then

$$g(X_n) \xrightarrow{P} g(a).$$

b. If $X_n \xrightarrow{P} a$, $Y_n \xrightarrow{P} b$, and $g(x, y)$ is continuous at $(x, y) = (a, b)$, then

$$g(X_n, Y_n) \xrightarrow{P} g(a, b).$$

c. If $X_n \xrightarrow{d} X$ and $g(x)$ is continuous for all x in the range of X, then

$$g(X_n) \xrightarrow{d} g(X).$$

d. (Slutzky) If $X_n \xrightarrow{d} X$, $Y_n \xrightarrow{P} b$, and $g(x, y)$ is continuous at (x, b) for all x in the range of X, then

$$g(X_n, Y_n) \xrightarrow{d} g(X, b).$$

Proof

a. See Exercise C1.

b. Since g is continuous at (a, b), for all $\epsilon > 0$ there exists $\delta > 0$ such that if $|x - a| < \delta$ and $|y - b| < \delta$, then $|g(x, y) - g(a, b)| < \epsilon$. Therefore, if $|g(X_n, Y_n) - g(a, b)| > \epsilon$, then either $|X_n - a| > \delta$ or $|Y_n - b| > \delta$ (or both). Hence,

$$P(|g(X_n, Y_n) - g(a, b)| > \epsilon) \leqq P(|X_n - a| > \delta) + P(|Y_n - b| > \delta),$$

and both of these probabilities go to zero by the definition of convergence in probability.

c. and d. Proofs of these results are beyond the scope of this book. For such proofs, see Billingsley (1979), pp. 278–279. □

Note that for parts b and d of Theorem 6–8, it is not necessary to assume anything about the joint distribution of (X_n, Y_n). In particular, we do not need to assume that X_n and Y_n are independent. (Unfortunately, it is possible that $X_n \xrightarrow{d} X$, $Y_n \xrightarrow{d} Y$, and $g(x, y)$ is continuous, but $g(X_n, Y_n)$ does not converge in distribution to $g(X, Y)$. To guarantee that $g(X_n, Y_n) \xrightarrow{d} g(X, Y)$, it is necessary to assume that (X_n, Y_n) converges jointly to (X, Y). (See Section 6.4.1 for a discussion of joint convergence.))

Parts a and b of the theorem are fairly easy to interpret. For example,

$$X_n \xrightarrow{P} 3, \; Y_n \xrightarrow{P} 4, \; \Rightarrow X_n + Y_n \xrightarrow{P} 3 + 4 = 7$$

and

$$X_n \log(Y_n) \xrightarrow{P} 3 \log(4).$$

Parts c and d are a little more difficult to interpret. We first note that the statements

$$X_n \xrightarrow{d} Z \sim N(0, 1) \quad \text{and} \quad X_n \xrightarrow{d} Y \sim N(0, 1) \tag{6–3}$$

mean the same thing. They both imply that as n approaches infinity, the distribution of X_n goes to that of a standard normal distribution. We see, therefore, that it doesn't matter what letter we use to denote the limiting random variable; the interpretation is still the same. (For this reason, some authors use the notation $X_n \xrightarrow{d} N(0, 1)$ for Equations (6–3). However, with this notation, parts c and d of the theorem are much harder to state.) Now, suppose that Equations (6–3) are satisfied. Then part c of Theorem 6–8 implies that

$$X_n^2 \xrightarrow{d} Z^2, \qquad Z \sim N(0, 1). \tag{6–4}$$

That is, the asymptotic distribution of X_n^2 is the distribution of $U = Z^2$ when $Z \sim N(0, 1)$. Now, if $Z \sim N(0, 1)$, then $U = Z^2 \sim \chi_1^2$ (by the definition of the χ^2 distribution). Therefore, an equivalent version to Equations (6–4) is that

$$X_n^2 \xrightarrow{d} U \sim \chi_1^2.$$

Now, suppose, in addition, that $Y_n \xrightarrow{P} 5$. Then $Q = 5Z \sim N(0, 25)$ and $R = 5Z + 5 \sim N(5, 25)$. Therefore, by Slutzky's theorem,

$$Y_n X_n \xrightarrow{d} Q \sim N(0, 25) \quad \text{and} \quad Y_n X_n + Y_n \xrightarrow{d} R \sim N(5, 25).$$

We next derive some simple corollaries of Theorem 6–8.

Corollary 6–8a. If $n^{1/2}(\overline{X}_n - \mu)/\sigma \xrightarrow{d} Z \sim N(0, 1)$, then

$$n^{1/2}(\overline{X}_n - \mu) \xrightarrow{d} U \sim N(0, \sigma^2).$$

Proof. By part c of Theorem 6–8,

$$n^{1/2}(\overline{X}_n - \mu) = \sigma\left(\frac{n^{1/2}(\overline{X}_n - \mu)}{\sigma}\right) \xrightarrow{d} \sigma Z = U \sim N(0, \sigma^2). \quad \square$$

In using the theorems of this chapter, it is quite important that the limiting distribution does not depend on n. In particular, $n^{1/2}\overline{X}_n$ does *not* converge in distribution to an $N(n^{1/2}\mu, \sigma^2)$ distribution, and $\overline{X}_n$ does *not* converge in distribution to an $N(\mu, \sigma^2/n)$ distribution.

Corollary 6–8b. If $X_n \xrightarrow{P} a$ and $Y_n \xrightarrow{P} b$, then

$$X_n + Y_n \xrightarrow{P} a + b, \qquad X_n - Y_n \xrightarrow{P} a - b, \qquad X_n Y_n \xrightarrow{P} ab.$$

Also, if $b \neq 0$, then $X_n / Y_n \xrightarrow{P} a/b$.

Proof. These results follow from part b of Theorem 6–8, since $x + y$, $x - y$, and xy are continuous functions, as is x/y when $y \neq 0$. $\square$

Using Corollary 6–8b twice, we see that if $X_n \xrightarrow{P} a$, $Y_n \xrightarrow{P} b$, and $Z_n \xrightarrow{P} c$, then

$$X_n + Y_n + Z_n = (X_n + Y_n) + Z_n \xrightarrow{P} (a + b) + c = a + b + c.$$

Similarly, $X_n Y_n Z_n \xrightarrow{P} abc$.

Corollary 6–8c. If $X_n \xrightarrow{d} X$, $Y_n \xrightarrow{P} 1$, and $W_n \xrightarrow{P} 0$, then

$$X_n Y_n \xrightarrow{d} X, \qquad \frac{X_n}{Y_n} \xrightarrow{d} X, \qquad X_n + W_n \xrightarrow{d} X, \qquad X_n W_n \xrightarrow{P} 0.$$

Proof. The first three results follow directly from Slutzky's theorem and the continuity of the functions xy, x/y (when $y = 1$), and $x + w$. To prove the fourth result, note that

$$W_n X_n \xrightarrow{d} 0(X) = 0$$

by Slutzky's theorem, and therefore, $X_n W_n \xrightarrow{P} 0$. □

Using Corollaries 6–8b and c, we see that if $X_n \xrightarrow{d} X$, $V_n \xrightarrow{P} 0$, and $W_n \xrightarrow{P} 0$, then

$$X_n + V_n + W_n = X_n + (V_n + W_n) \xrightarrow{d} X.$$

Corollary 6–8c is what is often called Slutzky's theorem. However, we use the more general theorem often enough that is is nice to have a name for it. Hence, we use the term Slutzky's theorem for part d of Theorem 6–8, and also for Corollary 6–8c, which follows directly from that result.

Example 6–7

Let X_i be independent, with $X_i \sim U(0, 1)$ (i.e., the X_i are uniformly distributed on the interval $(0, 1)$). Let U_n be the maximum of the first n observations. From Section 6.1.1 we see that

$$U_n \xrightarrow{P} 1 \quad \text{and} \quad n(1 - U_n) \xrightarrow{d} T \sim \Gamma(1, 1).$$

Also, by part a of Theorem 6–8,

$$\exp(U_n) \xrightarrow{P} \exp(1) = e \quad \text{and} \quad (U_n + 1)^2 - U_n \xrightarrow{P} (1 + 1)^2 - 1 = 3.$$

By part c of Theorem 6–8,

$$\exp[-n(1 - U_n)] \xrightarrow{d} \exp(-T).$$

Now, if $T \sim \Gamma(1, 1)$, then $Q = \exp(-T) \sim U(0, 1)$. Therefore,

$$\exp[-n(1 - U_n)] \xrightarrow{d} Q \sim U(0, 1).$$

By part d of Theorem 6–8,

$$(U_n + 1)^2[n(1 - U_n)] \xrightarrow{d} (1 + 1)^2 T = 4T \sim \Gamma(1, 4).$$

From this example, we see that once we know that $n(1 - U_n) \xrightarrow{d} T$, it is quite easy to find the asymptotic distribution of many functions of U_n.

We present one further corollary to Theorem 6–8, which was mentioned in Chapter 4.

Corollary 6–8d. Let T_n have a t-distribution with n degrees of freedom.

Then

$$T_n \xrightarrow{d} Z \sim N(0,1).$$

Proof. See Exercise B6. □

Theorem 6–8 leads to some results which are quite important in statistics.

Theorem 6–9. Let $X_1, X_2, \ldots$ be a sequence of independently identically distributed random variables with $EX_i = \mu$. Let $\bar{X}_n$ and S_n^2 respectively be the sample mean and sample variance computed from the first n of the X_i. Let

$$t_n = \frac{n^{1/2}(\bar{X}_n - \mu)}{S_n}.$$

a. If $\sigma^2 = \text{var}(X_i) < \infty$, then

$$S_n^2 \xrightarrow{P} \sigma^2, \qquad S_n \xrightarrow{P} \sigma, \qquad t_n \xrightarrow{d} Z \sim N(0,1).$$

b. Suppose that $\gamma = E(X-\mu)^4/\sigma^4 < \infty$. Then

$$n(S_n^2 - \sigma^2) \xrightarrow{d} W \sim N(0, (\gamma - 1)\sigma^4).$$

Proof

a. Let $H_i = X_i - \mu$ and $U_i = H_i^2$. Then

$$\left(\frac{n-1}{n}\right)S_n^2 = \frac{\sum(X_i - \bar{X}_n)^2}{n} = \frac{\sum(H_i - \bar{H})^2}{n} = \frac{\sum H_i^2}{n} - \bar{H}_n^2 = \bar{U}_n - \bar{H}_n^2.$$

Now, $EH_i = 0$ and $EU_i = E(X_i - \mu)^2 = \sigma^2$. Therefore, by the weak law of large numbers, $\bar{U}_n \xrightarrow{P} \sigma^2$ and $\bar{H}_n \xrightarrow{P} 0$. In addition, $n/(n-1) \xrightarrow{P} 1$. Therefore, by two applications of Corollary 6–8b,

$$S_n^2 = \left(\frac{n}{n-1}\right)(\bar{U}_n - \bar{H}_n^2) \xrightarrow{P} (1)(\sigma^2 + 0) = \sigma^2.$$

By part a of Theorem 6–8,

$$S_n \xrightarrow{P} \sigma \quad \text{and} \quad R_n = \frac{S_n}{\sigma} \xrightarrow{P} 1$$

(using the continuity of the functions $x^{1/2}$ and x/σ). Also, by the central limit theorem,

$$Z_n = \frac{n^{1/2}(\bar{X}_n - \mu)}{\sigma} \xrightarrow{d} Z \sim N(0,1).$$

Therefore, by Slutzky's theorem,

$$t_n = \frac{Z_n}{R_n} \xrightarrow{d} \frac{Z}{1} = Z \sim N(0,1).$$

b. Note first that

$$S_n^2 = \left(\frac{n-1}{n}\right)S_n^2 + n^{-1}S_n^2 = \bar{U}_n - \bar{H}_n^2 + n^{-1}S_n^2,$$

(where U_i and H_i are as given in the proof of part a). Therefore,

$$n^{1/2}(S_n^2 - \sigma^2) = n^{1/2}(\overline{U}_n - \sigma^2) - n^{1/2}\overline{H}_n^2 + n^{-1/2}S_n^2.$$

Now, $EU_i = \sigma^2$ and $\text{var}(U_i) = (\gamma - 1)\sigma^4$. Therefore, by the central limit theorem, and Corollary 6–8a, we obtain

$$n^{1/2}(\overline{U}_n - \sigma^2) \xrightarrow{d} W \sim N(0, (\gamma - 1)\sigma^4).$$

Now, $EH_i = 0$ and $\text{var}(H_i) = \sigma^2$. Therefore, by the weak law, the central limit theorem, and Corollary 6–8c,

$$\overline{H}_n \xrightarrow{P} 0 \quad \text{and} \quad n^{1/2}\overline{H}_n \xrightarrow{d} Q \sim N(0, \sigma^2) \Rightarrow n^{1/2}\overline{H}_n^2 = \overline{H}_n(n^{1/2}\overline{H}_n) \xrightarrow{P} 0.$$

Finally,

$$n^{-1/2} \xrightarrow{P} 0 \quad \text{and} \quad S_n^2 \xrightarrow{P} \sigma^2 \Rightarrow n^{-1/2}S_n^2 \xrightarrow{P} 0.$$

Therefore, by two applications of Slutzky's theorem,

$$n^{1/2}(S_n^2 - \sigma^2) = n^{1/2}(\overline{U}_n - \sigma^2) - n^{1/2}\overline{H}_n + n^{1/2}S_n^2 \xrightarrow{d} W + 0 + 0 = W. \quad \square$$

We call $\gamma = E(X - \mu)^4/\sigma^2$ the *kurtosis* of X.

Exercises—B

1. Let $X_1, X_2, \ldots$ be a sequence of random variables with $EX_n = \mu_n$ and $\text{var}(X_n) = \sigma_n^2$. Show that $E(X_n - a)^2 = \sigma_n^2 + (\mu_n - a)^2$.
2. **(a)** Show that $(3n + 4)/(2n + 6) \to \frac{3}{2}$ and $(6n^2 + 4n)/(5n^2 + 4) \to \frac{6}{5}$ as $n \to \infty$. (*Hint*: $(3n + 4)/(2n + 6) = (3 + n^{-1}4)/(2 + n^{-1}6)$.)
   (b) Show that $(3n + 4)/(4n^2 + 6n)$ goes to zero as n approaches infinity.
3. Let $X_n \sim P(4n)$. Let $Y_n = X_n/n$.
   (a) Show that $Y_n \xrightarrow{P} 4$. (What are EY_n and $\text{var}(Y_n)$?)
   (b) Show that $Y_n^2 + Y_n^{1/2} \xrightarrow{P} 18$ and $(n^2 Y_n^2 + nY_n)/(nY_n + n^2) \xrightarrow{P} 16$.
4. Let $\overline{X}_n$ and S_n^2 be the sample mean and the sample variance computed from the first n of a sequence of i.i.d. random variables with mean μ and variance $\sigma^2 > 0$. To what do the following converge in probability?
   (a) $\overline{X}_n S_n$ and $\overline{X}_n^2/S_n^2$.
   (b) $n(\overline{X}_n + S_n)/(n + 1)$ and $n(\overline{X}_n + S_n)/(n^2 + 1)$.
5. Suppose that $X_n \xrightarrow{P} 3$ and $Y_n \xrightarrow{P} 6$. To what do the following converge in probability?
   (a) $X_n Y_n^2$, X_n/Y_n, and $X_n/(Y_n + 6)$.
   (b) $n(X_n + Y_n^2)/(n + 1)^2$ and $n^2(X_n^2 + Y_n)/(n^2 + n - 3)$.
6. **(a)** Let $U_n \sim \chi_n^2$. Show that $U_n/n \xrightarrow{P} 1$. (What are EU_n and $\text{var}(U_n)$?)
   (b) Prove Corollary 6–8d.
7. Let $X_n \xrightarrow{d} Z \sim N(0, 1)$ and $Y_n \xrightarrow{P} 4$.
   (a) Show that $4X_n/Y_n \xrightarrow{d} Z$, and that $16X_n^2/Y_n^2 \xrightarrow{d} Z^2 \sim \chi_1^2$.
   (b) Show that $(4n + Y_n)X_n/(nY_n + Y_n^2) \xrightarrow{d} Z$.
8. Let $X \sim N(\mu, \sigma^2)$. Show that X has kurtosis 3. (Let $Z = (X - \mu)/\sigma$. Then X and Z have the same kurtosis. Why?)

Exercises—C

1. Prove part a of Theorem 6–8. (*Hint*: If $g(x)$ is continuous at a, then for all $\epsilon > 0$, there exists $\delta > 0$ such that $|x - a| < \delta$ implies that $|g(x) - g(a)| < \epsilon$.)
2. Let $(X_1, Y_1), (X_2, Y_2), \ldots$ be a sequence of independent bivariate random vectors having the same joint distribution, with $EX_i = \mu$, $EY_i = \nu$, $\text{var}(X_i) = \sigma^2$, $\text{var}(Y_i) = \tau^2$, and $\text{corr}(X_i, Y_i) = \rho$. Let S_n^2 and T_n^2 be the sample variances of the X's and the Y's computed from the first n observations. Let $Q_n = \sum_{i=1}^n (X_i - \overline{X}_n)(Y_i - \overline{Y}_n)/(n-1)$ and $R_n = Q_n/S_n T_n$. Finally, let $H_i = X_i - \mu$, $G_i = Y_i - \nu$, and $W_i = H_i G_i$.
   (a) Show that $Q_n = (n/(n-1))(\overline{W}_n - \overline{H}_n \overline{G}_n)$.
   (b) Show that $Q_n \xrightarrow{P} \rho\sigma\tau$. (*Hint*: $EW_i = \rho\sigma\tau$. Why?)
   (c) Show that $R_n \xrightarrow{P} \rho$.
   (Q_n and R_n are called the *sample covariance* and *sample correlation coefficient*, respectively, between the X_i and the Y_i.)
3. Let $\delta = E(X_i - \mu)^2 (Y_i - \nu)^2/\sigma^2\tau^2$. Using the information and results of the previous problem,
   (a) Show that $Q_n = \overline{W}_n - \overline{H}_n\overline{G}_n + Q_n/n$.
   (b) Show that $EW_i = \rho\sigma\tau$ and $\text{var}(W_i) = (\delta - \rho^2)\sigma^2\tau^2$.
   (c) Show that $n^{1/2}(Q_n - \rho\sigma\tau) \xrightarrow{d} A \sim N(0, (\delta - \rho^2)\sigma^2\tau^2)$.
   (The asymptotic distribution of R_n is dealt with in the exercises in Section 6.4.2.)

* 6.3.2 The Cramer δ Theorem

We present one final theorem about univariate limiting distributions.

Theorem 6–10. (Cramer δ) Suppose that $n^b(X_n - a) \xrightarrow{d} X$ for some $b > 0$. Let $g(x)$ be a function which is differentiable at $x = a$. Then

$$n^b(g(X_n) - g(a)) \xrightarrow{d} g'(a)X.$$

Proof. By Taylor's theorem,

$$g(x) = g(a) + g'(a)(x - a) + (x - a)R(x),$$

where $R(x) \to R(a) = 0$ as $x \to a$ (and hence, $R(x)$ is continuous at a). Now,

$$X_n = n^{-b}(n^b(X_n - a)) + a \xrightarrow{P} 0 + a = a.$$

(Why?) Therefore, $R(X_n) \xrightarrow{P} R(a) = 0$. Now, since $n^b(X_n - a) \xrightarrow{d} X$,

$$n^b(X_n - a)R(X_n) \xrightarrow{P} 0,$$

by Slutzky's theorem. Also, because $n^b(X_n - a) \xrightarrow{d} X$,

$$n^b(X_n - a)g'(a) \xrightarrow{d} g'(a)X$$

(since $g'(a)x$ is a continuous function of x). Therefore,

$$n^b(g(X_n) - g(a)) = n^b(X_n - a)g'(a) + n^b(X_n - a)R(X_n) \xrightarrow{d} g'(a)X + 0,$$

by Slutzky's theorem. □

Theorem 6–10 is not necessarily true when $b = 0$. (See Exercise C1.)

Example 6–8

Suppose $n^{1/2}(X_n - 5) \xrightarrow{d} X \sim N(0,4)$. Let $g(x) = x^2$, so that $g'(5) = 10$. If $X \sim N(0,4)$, then $10X \sim N(0,400)$. Hence, by the Cramer δ theorem,

$$n^{1/2}(X_n^2 - 25) \xrightarrow{d} 10X \sim N(0,400).$$

If we choose $g(x) = e^{2x}$, then $g'(5) = 2e^{10}$, and it follows that

$$n^{1/2}(\exp(2X_n) - \exp(10)) \xrightarrow{d} 2e^{10}X \sim N(0, 16e^{20}).$$

Example 6–9

Let $X_n \sim \chi_n^2$. In Chapter 4, we showed that

$$W_n = \frac{X_n - n}{(2n)^{1/2}} \xrightarrow{d} Z \sim N(0,1),$$

or equivalently,

$$n^{1/2}\left(\frac{X_n}{n} - 1\right) \xrightarrow{d} 2^{1/2}Z.$$

Let $g(x) = (2x)^{1/2}$, so that $g'(x) = (2x)^{-1/2}$. Then $g(1) = 2^{1/2}$ and $g'(1) = 2^{-1/2}$. Now, let

$$V_n = n^{1/2}(g(X_n) - g(1)) = (2X_n)^{1/2} - (2n)^{1/2}.$$

By the Cramer δ theorem, then,

$$V_n \xrightarrow{d} Z \sim N(0,1).$$

It turns out that the distribution of V_n converges to a standard normal distribution faster than the distribution of W_n, so that the normal approximation based on V_n is the one which is often used for the χ^2 distribution.

We can use the Cramer δ theorem to find the asymptotic distribution of the sample standard deviation $S_n = (S_n^2)^{1/2}$.

Corollary 6–10. $n^{1/2}(S_n - \sigma) \xrightarrow{d} W \sim N(0, (\gamma - 1)\sigma^2/4)$.

Proof. From the previous section,

$$n^{1/2}(S_n^2 - \sigma^2) \xrightarrow{d} Q \sim N(0, (\gamma - 1)\sigma^4).$$

Let $g(x) = x^{1/2}$, so that $g'(x) = x^{-1/2}/2$. Then

$$S_n = g(S_n^2), \qquad \sigma = g(\sigma^2), \qquad g'(\sigma^2) = \frac{\sigma^{-1}}{2}.$$

Therefore, the Cramer δ theorem implies that

$$n^{1/2}(S_n - \sigma) \xrightarrow{d} \frac{\sigma^{-1}Q}{2} \sim N\left(0, (\gamma - 1)\frac{\sigma^2}{4}\right). \quad \square$$

Exercises—B

1. Suppose that $n^{1/2}(X_n - a) \xrightarrow{d} Z \sim N(0,1)$. Show that
   (a) $n^{1/2}(X_n^2 - a^2) \xrightarrow{d} 2aZ \sim N(0, 4a^2)$.
   (b) $n^{1/2}(\log(X_n) - \log(a)) \xrightarrow{d} Z/a \sim N(0, a^{-2})$.

(c) $X_n \xrightarrow{P} a$. (*Hint*: $(X_n - a) = n^{-1/2}(n^{1/2}(X_n - a))$; use Slutzky's theorem.)
(d) $n(X_n - a)^2 \xrightarrow{d} \chi_1^2$. (*Hint*: See Theorem 6–8, part c.)

2. Suppose that $n(X_n - 3) \xrightarrow{d} X$. Find the limiting distribution of
   (a) $n(X_n^3 - 27)$.
   (b) $h(\exp(X_n) - \exp(3))$.
   (c) $n^{1/2}(X_n^3 - 27)$.
   (d) $n^3(X_n - 3)^3$.
3. Let $\overline{X}_n$ be the sample mean computed from a sample of size n from a distribution with mean μ and variance σ^2. Find the asymptotic distribution of
   (a) $n^{1/2}(\overline{X}_n^2 - \mu^2)$.
   (b) $n(\overline{X}_n - \mu)^2$.
   (c) $n^{1/2}(\overline{X} - \mu)^2$.

Exercises—C

1. Let $X_n \sim N(0,1)$, so that $n^0(X_n - 0) = X_n \xrightarrow{d} Z \sim N(0,1)$. Let $g(x) = x^2$. Show that $n^0(g(X_n) - g(0))$ does not converge in distribution to $g'(0)Z$—in other words, the Cramer δ theorem need not be true when $b = 0$.

6.4 MULTIVARIATE ASYMPTOTICS

*# 6.4.1 Multivariate Asymptotic Distributions

In this section, we extend the results of previous sections to multivariate distributions. Let $\mathbf{X}_1, \mathbf{X}_2, \ldots$ be a sequence of p-dimensional random vectors such that $\mathbf{X}_n$ has distribution function $F_n(\mathbf{x})$, and let $\mathbf{X}$ be a p-dimensional random vector which has distribution function $F(\mathbf{x})$. We say that $\mathbf{X}_n$ *converges in distribution* to $\mathbf{X}$ and write $\mathbf{X}_n \xrightarrow{d} \mathbf{X}$ if

$$\lim_{n\to\infty} F_n(\mathbf{x}) = F(\mathbf{x})$$

at all points $\mathbf{x}$ where $F(\mathbf{x})$ is continuous. $\mathbf{X}_n$ *converges in probability* to a fixed vector $\mathbf{b}$, written $\mathbf{X}_n \xrightarrow{P} \mathbf{b}$, if

$$\lim_{n\to\infty} P(\|\mathbf{X}_n - \mathbf{b}\| > \epsilon) = 0$$

for all $\epsilon > 0$, (where, as usual, $\|\mathbf{a}\| = (\Sigma a_i^2)^{1/2}$). It can be shown that $\mathbf{X}_n \xrightarrow{P} \mathbf{b}$ if and only if $\mathbf{X}_n \xrightarrow{d} \mathbf{X}$, where $\mathbf{X}$ is degenerate at $\mathbf{b}$. However, we shall not use that result in this book.

We first give the multivariate version of the continuity theorem. A sequence $\mathbf{X}_1, \mathbf{X}_2, \ldots$ of random vectors is *uniformly bounded* if there exists a positive number $q < \infty$ such that

$$P(\|\mathbf{X}_n\|^2 \leq q) = 1$$

for all n.

Theorem 6–11. (Continuity theorem) Let $\mathbf{X}_1, \mathbf{X}_2, \ldots$ be a sequence of p-dimensional random vectors, and let $\mathbf{X}$ be another p-dimensional random vector. Let $\mathbf{X}_n$ and $\mathbf{X}$ have moment-generating functions $M_n(\mathbf{t})$ and $M(\mathbf{t})$, respectively.

a. Suppose that there exists $a > 0$ such that $M_n(\mathbf{t}) \to M(\mathbf{t})$ for all $\mathbf{t}$ such that $\|\mathbf{t}\| < a$. Then $\mathbf{X}_n \xrightarrow{d} \mathbf{X}$.
b. Conversely, if $\mathbf{X}_1, \mathbf{X}_2, \ldots$ is a uniformly bounded sequence of random vectors such that $\mathbf{X}_n \xrightarrow{d} \mathbf{X}$, then $M_n(\mathbf{t}) \to M(\mathbf{t})$ for all $\mathbf{t}$.

Proof. A general proof is beyond the level of the text. A proof in the discrete case can be constructed along the lines of that given in the univariate case for Theorem 6–1. □

We now present a corollary that was used in Chapter 5.

Corollary 6–11. Let $(X_n, Y_n)'$ have a trinomial distribution with parameters n, a/n, and b/n. Then $(X_n, Y_n) \xrightarrow{d} (U, V)$, where U and V are independent and have Poisson distributions with means a and b.

Proof. $(X_n, Y_n)'$ has moment-generating function

$$M_n(s, t) = \left(1 + \frac{ae^s + be^t - a - b}{n}\right)^n \to \exp[a(e^s - 1)]\exp[b(e^t - 1)],$$

which is the moment-generating function of (U, V) with U and V independent and having Poisson distributions with means a and b. □

The next theorem relates multivariate convergence in probability to univariate convergence in probability.

Theorem 6–12. Let $\mathbf{X}_n' = (X_{1n}, \ldots, X_{pn})$ and $\mathbf{a} = (a_1, \ldots, a_p)$. Then $\mathbf{X}_n \xrightarrow{P} \mathbf{a}$ if and only if $X_{in} \xrightarrow{P} a_i$ for $i = 1, \ldots, p$.

Proof. Suppose that $\mathbf{X}_n \xrightarrow{P} \mathbf{a}$. Then since $|X_{in} - a_i| \leqq \|\mathbf{X}_n - \mathbf{a}\|$,

$$P(|X_{in} - a_i| > \epsilon) \leqq P(\|\mathbf{X}_n - \mathbf{a}\| > \epsilon) \to 0 \quad \text{as } n \to \infty.$$

Therefore, $X_{in} \xrightarrow{P} a_i$ for all i.

Conversely, suppose that $X_{in} \xrightarrow{P} a_i$ for all i. If $\|\mathbf{X}_n - \mathbf{a}\| > \epsilon$, then $|X_{in} - a_i| > \epsilon/p$ for some i. Therefore,

$$P(\|\mathbf{X}_n - \mathbf{a}\| > \epsilon) \leqq \sum_i P(|X_{in} - a_i| > \epsilon/p) \to 0 \quad \text{as } n \to \infty,$$

and hence $\mathbf{X}_n \xrightarrow{P} \mathbf{a}$. (Note that the sum has a finite number of terms.) □

Theorem 6–12 says that $\mathbf{X}_n$ converges in probability to $\mathbf{a}$ if and only if each component of $\mathbf{X}_n$ converges to the appropriate component of $\mathbf{a}$. Thus, when $p = 2$, the theorem says that

$$(X_n, Y_n)' \xrightarrow{P} (a, b)' \Leftrightarrow X_n \xrightarrow{P} a \quad \text{and} \quad Y_n \xrightarrow{P} b.$$

We now show how to reduce certain problems involving multivariate convergence in distribution to problems involving univariate convergence in distribution. The result is not quite as simple as that for convergence in probability.

Theorem 6–13. (Cramer-Wold) $\mathbf{X}_n \xrightarrow{d} \mathbf{X}$ if and only if $\mathbf{t}'\mathbf{X}_n \xrightarrow{d} \mathbf{t}'\mathbf{X}$ for all $\mathbf{t} \in R^p$.

Proof. We give a proof for the case in which $\mathbf{X}_1, \mathbf{X}_2, \ldots$ are uniformly bounded random vectors, so that the converse of the continuity theorem holds. (The proof can be extended to nonbounded random vectors by replacing the moment-generating functions with characteristic functions.) Let $M_n(\mathbf{t})$ and $M(\mathbf{t})$ be the moment-generating functions of $\mathbf{X}_n$ and $\mathbf{X}$, respectively. Let

$$U_{n,t} = \mathbf{t}'\mathbf{X}_n \quad \text{and} \quad U_t = \mathbf{t}'\mathbf{X},$$

and let $N_{n,t}(s)$ and $N_t(s)$ be the moment-generating functions of $U_{n,t}$ and U_t. (Note that $U_{1,t}, U_{2,t}$ is a uniformly bounded set of random variables, so that the converse of the continuity theorem applies.) Then

$$N_{n,t}(s) = M_n(s\mathbf{t}) \quad \text{and} \quad N_t(s) = M(s\mathbf{t}).$$

Suppose first that

$$U_{n,t} \xrightarrow{d} U_t$$

for all $\mathbf{t}$. Then, by the converse of the univariate continuity theorem,

$$M_n(\mathbf{t}) = N_{n,t}(1) \to N_t(1) = M(\mathbf{t}).$$

Hence, by the multivariate continuity theorem,

$$\mathbf{X}_n \xrightarrow{d} \mathbf{X}.$$

Now suppose that $\mathbf{X}_n \xrightarrow{d} \mathbf{X}$. Then, by the converse of the multivariate continuity theorem,

$$N_{n,t}(s) = M_n(s\mathbf{t}) \to M(s\mathbf{t}) = N_t(s)$$

for all t, and hence, by the univariate continuity theorem,

$$U_{n,t} \xrightarrow{d} U_t. \quad \square$$

When $p = 2$, Theorem 6–13 says that

$$(X_n, Y_n)' \xrightarrow{d} (X, Y)' \Leftrightarrow (sX_n + tY_n) \xrightarrow{d} (sX + tY)$$

for all s and t.

Corollary 6–13. $\mathbf{X}_n \xrightarrow{d} \mathbf{X} \sim N_p(\boldsymbol{\mu}, \boldsymbol{\Sigma})$ if and only if $\mathbf{t}'\mathbf{X}_n \xrightarrow{d} \mathbf{t}'\mathbf{X} \sim N(\mathbf{t}'\boldsymbol{\mu}, \mathbf{t}'\boldsymbol{\Sigma}\mathbf{t})$ for all $\mathbf{t}$.

We next give the multivariate versions of the weak law and the central limit theorem.

Theorem 6–14. Let $\mathbf{X}_1, \mathbf{X}_2, \ldots$ be a sequence of i.i.d. p-dimensional random vectors, and let

$$\overline{\mathbf{X}}_n = n^{-1} \sum_{i=1}^{n} \mathbf{X}_i.$$

(Note that $\overline{\mathbf{X}}_n$ is a p-dimensional vector.)

a. (Weak law of large numbers) If $E\mathbf{X}_i = \boldsymbol{\mu}$ exists, then $\overline{\mathbf{X}}_n \xrightarrow{P} \boldsymbol{\mu}$.

b. (Central limit theorem) If, in addition, $\mathbf{X}_i$ has finite covariance matrix $\boldsymbol{\Sigma}$, then

$$n^{1/2}(\overline{\mathbf{X}}_n - \boldsymbol{\mu}) \xrightarrow{d} \mathbf{Y} \sim N(\mathbf{0}, \boldsymbol{\Sigma}).$$

Proof

a. Let $\mathbf{X}_n = (X_{1n}, \ldots, X_{pn})'$ and $\boldsymbol{\mu} = (\mu_1, \ldots, \mu_p)'$. Then $\overline{\mathbf{X}}_n = (\overline{X}_{1n}, \ldots, \overline{X}_{pn})'$. By the univariate weak law, $\overline{X}_{in} \xrightarrow{P} \mu_i$ as $n \to \infty$ for all i. Therefore, by Theorem 6–12, $\overline{\mathbf{X}}_n \xrightarrow{P} \boldsymbol{\mu}$.

b. Fix $\mathbf{t} \in R^p$, and let $U_i = \mathbf{t}'\mathbf{X}_i$. Then the U_i are i.i.d., and

$$EU_i = \mathbf{t}'\boldsymbol{\mu} \quad \text{and} \quad \text{var}(U_i) = \mathbf{t}'\boldsymbol{\Sigma}\mathbf{t}.$$

Therefore, by the univariate central limit theorem,

$$\mathbf{t}'(n^{1/2}(\overline{\mathbf{X}}_n - \boldsymbol{\mu})) = n^{1/2}(\overline{U}_n - \mathbf{t}'\boldsymbol{\mu}) \xrightarrow{d} \mathbf{t}'\mathbf{Y} \sim N(\mathbf{0}, \mathbf{t}'\boldsymbol{\Sigma}\mathbf{t}),$$

and the desired result follows from Corollary 6–13. □

We now use Theorem 6–14 to derive the normal approximation to the multinomial distribution.

Corollary 6–14. Let $\mathbf{Y}_n'$ have a k-dimensional multinomial distribution with parameters n and $\mathbf{p}' = (p_1, \ldots, p_k)$. Let $\mathbf{V}$ be a $k \times k$ matrix with (i,j)th element v_{ij}, where $v_{ii} = p_i(1 - p_i)$ and $v_{ij} = -p_i p_j$. Then

$$n^{-1/2}(\mathbf{Y}_n - n\mathbf{p}) \xrightarrow{d} \mathbf{X} \sim N_k(\mathbf{0}, \mathbf{V})$$

and

$$\frac{n^{-1/2}(\mathbf{t}'\mathbf{Y}_n - n\mathbf{t}'\mathbf{p})}{\mathbf{t}'\mathbf{V}t} \xrightarrow{d} Z \sim N(0, 1).$$

Proof. Let $\mathbf{X}_i$ be independent k-dimensional random vectors such that $\mathbf{X}_i \sim M_k(1, \mathbf{p})$. Then $E\mathbf{X}_i = \mathbf{p}$ and $\text{cov}(\mathbf{X}_i) = \mathbf{V}$. Therefore, by the multivariate central limit theorem,

$$n^{1/2}(\overline{\mathbf{X}}_n - \mathbf{p}) \xrightarrow{d} \mathbf{X} \sim N_k(\mathbf{0}, \mathbf{V}).$$

Now, let $\mathbf{Y}_n = n\overline{\mathbf{X}}_n$. Then $\mathbf{Y}_n \sim M_k(n, \mathbf{p})$. (See Exercise B1.) Moreover,

$$n^{1/2}(\overline{\mathbf{X}}_n - \mathbf{p}) = n^{-1/2}(\mathbf{Y}_n - n\mathbf{p}),$$

which establishes the first result. The second result follows from the Cramer-Wold theorem. □

Exercises—B

1. Let $\mathbf{X}_i$ be independent, with $\mathbf{X}_i \sim M_k(1, \mathbf{p})$. Let $Y_n = n\overline{X}_n$. Show that $Y_n \sim M_k(n, \mathbf{p})$.
2. In the notation of Theorem 6–14, what is the asymptotic distribution of the difference between the first and second components of $\overline{\mathbf{X}}_n$? Of the sum of those components?
3. Let $\mathbf{X}_n$, $\mathbf{Y}_n$, $\mathbf{X}$, and $\mathbf{Y}$ be random vectors such that $(\mathbf{X}_n', \mathbf{Y}_n')' \xrightarrow{d} (\mathbf{X}', \mathbf{Y}')'$. Use the Cramer-Wold theorem to show that $\mathbf{X}_n \xrightarrow{d} \mathbf{X}$.

4. Show that if $\mathbf{X}_1, \mathbf{X}_2, \ldots$ is a uniformly bounded set of random vectors, then $\mathbf{t}'\mathbf{X}_1, \mathbf{t}'\mathbf{X}_2, \ldots$ is a uniformly bounded set of random variables.
5. Let $\mathbf{X}$ and $\mathbf{Y}$ be p-dimensional random vectors with finite moment-generating functions.
   (a) Show that $\mathbf{X}$ and $\mathbf{Y}$ have the same distribution if and only if $\mathbf{t}'\mathbf{X}$ and $\mathbf{t}'\mathbf{Y}$ have the same distribution for all $t \in R^p$.
   (b) Show that $\mathbf{X} \sim N_p(\boldsymbol{\mu}, \boldsymbol{\Sigma})$ if and only if $\mathbf{t}'\mathbf{X} \sim N(\mathbf{t}'\boldsymbol{\mu}, \mathbf{t}'\boldsymbol{\Sigma}\mathbf{t})$ for all $\mathbf{t} \in R^p$.

Exercises—C

1. Let (X_n, Y_n) have a trinomial distribution with parameters n and (p_1, p_2).
   (a) If $n^{-1/2}((X_n, Y_n)' - n\mathbf{a}) \xrightarrow{d} (U, V)'$, what is $\mathbf{a}$? How is $(U, V)'$ distributed?
   (b) The conditional distribution of X_n given Y_n is a binomial distribution. What is the normal approximation to this distribution?
   (c) Show that the normal approximation to the conditional distribution of X_n given Y_n is different from the conditional distribution of U given V. (Exercise B2 above indicates that the limit of the marginal distribution is the same as the marginal distribution of the limit. This exercise indicates that conditional distributions are not as well behaved.)

6.4.2 *Multivariate Slutzky and Cramer δ Theorems*

In this section, we generalize the theorems of previous sections on continuous functions. The first theorem is proved in the exercises, while the other two proofs are somewhat beyond the level of the text and are omitted. (See Billingsley (1979), pp. 330–331, for the proofs.)

Theorem 6–15

a. Let $g(\mathbf{x})$ be a function which is continuous at $\mathbf{x} = \mathbf{a}$. Then

$$\mathbf{X}_n \xrightarrow{P} \mathbf{a} \Rightarrow g(\mathbf{X}_n) \xrightarrow{P} g(\mathbf{a}).$$

b. Let $g(\mathbf{x})$ be a function which is continuous in the range of $\mathbf{X}$. Then

$$\mathbf{X}_n \xrightarrow{d} \mathbf{X} \Rightarrow g(\mathbf{X}_n) \xrightarrow{d} g(\mathbf{X}).$$

c. (Slutzky's theorem) Let $g(\mathbf{x}, \mathbf{y})$ be a function which is continuous at $(\mathbf{x}, \mathbf{a})$ for all $\mathbf{x}$ in the range of $\mathbf{X}$. Then

$$\mathbf{X}_n \xrightarrow{d} \mathbf{X},\ \mathbf{Y}_n \xrightarrow{P} \mathbf{a} \Rightarrow g(\mathbf{X}_n, \mathbf{Y}_n) \xrightarrow{d} g(\mathbf{X}, \mathbf{a}).$$

Note that g may be a vector-valued function in Theorem 6–15, and that $\mathbf{X}_n$ and $\mathbf{Y}_n$ need not be independent in part c.

Example 6–10

Suppose that $(X_n, Y_n, Z_n) \xrightarrow{d} (X, Y, Z)$, $U_n \xrightarrow{P} a$, and $V_n \xrightarrow{P} b$. Then

$$Z_n + (V_n Y_n) \log(U_n X_n) \xrightarrow{d} Z + (bY) \log(aX).$$

We next give the multivariate version of the Cramer δ theorem. Recall from the calculus that if $y = g(\mathbf{x})$ is a function from R^k to R^1, then the *gradient*, $\nabla g(\mathbf{x})$, is the k-dimensional column vector whose ith component is the derivative of y with respect to x_i. That is,

$$\nabla g(\mathbf{x}) = \left(\frac{\partial}{\partial x_1} g(\mathbf{x}), \ldots, \frac{\partial}{\delta x_k} g(\mathbf{x})\right)'.$$

Theorem 6–16. (Cramer δ) Let $\mathbf{X}_n$ be a sequence of random vectors such that

$$n^b(\mathbf{X}_n - \mathbf{a}) \xrightarrow{d} \mathbf{X}$$

for somc $b > 0$. If $g(\mathbf{x})$ is a real-valued function with gradient $\nabla g(\mathbf{a})$ at $\mathbf{a}$, then

$$n^b(g(\mathbf{X}_n) - g(\mathbf{a})) \xrightarrow{d} (\nabla g(\mathbf{a}))'\mathbf{X}.$$

Proof. Note that $\mathbf{X}_n - \mathbf{a} = n^{-b}(n^b(\mathbf{X}_n - \mathbf{a})) \xrightarrow{d} 0\mathbf{X} = \mathbf{0}$, and hence, $\mathbf{X}_n \xrightarrow{P} \mathbf{a}$. By the multivariate version of Taylor's theorem,

$$g(\mathbf{X}_n) - g(\mathbf{a}) = (\nabla g(\mathbf{a}) + R(\mathbf{X}_n))'(\mathbf{X}_n - \mathbf{a}),$$

where $R(\mathbf{X}_n) \xrightarrow{P} R(\mathbf{a}) = \mathbf{0}$. Therefore, by Slutzky's theorem,

$$n^b(g(\mathbf{X}_n) - g(\mathbf{a})) = (\nabla g(\mathbf{a}) + R_n(\mathbf{X}_n))'(n^b(\mathbf{X}_n - \mathbf{a})) \xrightarrow{d} (\nabla g(\mathbf{a}))'\mathbf{X}. \quad \square$$

Corollary 6–16. If $n^b(\mathbf{X}_n - \mathbf{a}) \xrightarrow{d} \mathbf{X} \sim N_p(\mathbf{0}, \mathbf{\Sigma})$, and if $g(\mathbf{x})$ is a function from R^p to R^1 which is differentiable at $\mathbf{a}$, then

$$n^b(g(\mathbf{X}_n) - g(\mathbf{a})) \xrightarrow{d} [\nabla g(\mathbf{a})]'\mathbf{X} \sim N(0, [\nabla g(\mathbf{a})]'\mathbf{\Sigma}[\nabla g(\mathbf{a})]).$$

Example 6–11

Suppose

$$n\left(\begin{Bmatrix} X_n \\ Y_n \end{Bmatrix} - \begin{Bmatrix} 3 \\ 4 \end{Bmatrix}\right) \xrightarrow{d} \mathbf{X} \sim N_2\left(\begin{Bmatrix} 0 \\ 0 \end{Bmatrix}, \begin{Bmatrix} 3 & 1 \\ 1 & 2 \end{Bmatrix}\right).$$

Let $g(x, y) = xy^2$, so that $g(3, 4) = 48$, $\nabla g(x, y) = (y^2, 2xy)'$, and $\nabla g(3, 4) = (16, 24)'$. Then

$$(\nabla g(3, 4))'\mathbf{\Sigma}(\nabla g(3, 4)) = (16, 24)\begin{Bmatrix} 3 & 1 \\ 1 & 2 \end{Bmatrix}\begin{Bmatrix} 16 \\ 24 \end{Bmatrix} = 2{,}688,$$

so that

$$n(X_n Y_n^2 - 48) \xrightarrow{d} (16, 24)\mathbf{X} \sim N(0, 2{,}688).$$

Example 6–12

Suppose we observe $(X_1, Y_1), (X_2, Y_2), \ldots$, a sequence of i.i.d. random vectors with

$$E\begin{Bmatrix} X_i \\ Y_I \end{Bmatrix} = \begin{Bmatrix} \mu \\ \nu \end{Bmatrix} \quad \text{and} \quad \operatorname{cov}\begin{Bmatrix} X_i \\ Y_i \end{Bmatrix} = \begin{Bmatrix} \sigma^2 & \rho\sigma\tau \\ \rho\sigma\tau & \tau \end{Bmatrix},$$

and suppose we want the asymptotic distribution of $\overline{X}_n/\overline{Y}_n$. We assume that $\nu \neq 0$. By the multivariate central limit theorem,

$$n^{1/2}\left(\begin{Bmatrix} \overline{X}_n \\ \overline{Y}_n \end{Bmatrix} - \begin{Bmatrix} \mu \\ \nu \end{Bmatrix}\right) \xrightarrow{d} \mathbf{X} \sim N_2(\mathbf{0}, \begin{Bmatrix} \sigma^2 & \rho\sigma\tau \\ \rho\sigma\tau & \tau^2 \end{Bmatrix}).$$

Let $g(x, y) = x/y$. Then

$$g(\overline{X}_n, \overline{Y}_n) = \frac{\overline{X}_n}{\overline{Y}_n}, \qquad g(\mu, \nu) = \frac{\mu}{\nu},$$

$$\nabla g(x, y) = (y^{-1}, -xy^{-2})', \qquad \nabla g(\mu, \nu) = (\nu^{-1}, -\mu\nu^{-2})'.$$

Therefore,

$$n^{1/2}(\overline{X}_n\overline{Y}_n^{-1} - \mu\nu^{-1}) \xrightarrow{d} U \sim N\left(0, \frac{\sigma^2}{\nu^2} - \frac{2\mu\rho\sigma\tau}{\nu^3} + \frac{\mu^2\tau^2}{\nu^4}\right). \tag{6–5}$$

Note that if the X_i and Y_i are independent, then $\rho = 0$ and

$$U \sim N\left(0, \frac{\sigma^2}{\nu^2} + \frac{\mu^2\tau^2}{\nu^4}\right).$$

We now find the asymptotic joint distribution of $\overline{X}_n$ and S_n^2. Let $X_1, X_2, \ldots$ be a sequence of i.i.d. random variables with

$$EX_i = \mu, \qquad \text{var}(X_i) = \sigma^2, \qquad \frac{E(X_i - \mu)^3}{\sigma^3} = \delta, \qquad \frac{E(X_i - \mu)^4}{\sigma^4} = \gamma.$$

(δ and γ are respectively called the *skewness* and *kurtosis* of the distribution of X_i.)

Theorem 6–17. Let $\overline{X}_n$ be the sample mean and S_n^2 be the sample variance computed from the first n X_i's. Then

$$n^{1/2}\left(\begin{Bmatrix}\overline{X}_n\\ S_n^2\end{Bmatrix} - \begin{Bmatrix}\mu\\ \sigma^2\end{Bmatrix}\right) \xrightarrow{d} \mathbf{T} \sim N_2\left(\begin{Bmatrix}0\\0\end{Bmatrix}, \begin{Bmatrix}\sigma^2 & \delta\sigma^3\\ \delta\sigma^3 & (\gamma-1)\sigma^4\end{Bmatrix}\right).$$

Proof. Let $H_i = X_i - \mu$ and $U_i = H_i^2$. Then $(H_i, U_i)'$ are i.i.d. two-dimensional random vectors with

$$E\begin{Bmatrix}H_i\\ U_i\end{Bmatrix} = \begin{Bmatrix}0\\ \sigma^2\end{Bmatrix} \quad \text{and} \quad \text{cov}\begin{Bmatrix}H_i\\ U_i\end{Bmatrix} = \mathbf{\Sigma} = \begin{Bmatrix}\sigma^2 & \delta\sigma^3\\ \delta\sigma^3 & (\gamma-1)\sigma^4\end{Bmatrix}.$$

Therefore, by the multivariate central limit theorem,

$$n^{1/2}\begin{Bmatrix}\overline{H}_n\\ \overline{U}_n - \sigma^2\end{Bmatrix} \xrightarrow{d} \mathbf{T} \sim N_2(\mathbf{0}, \mathbf{\Sigma}).$$

Now, $\overline{H}_n = \overline{X}_n - \mu$. Also, in Section 6.3.1 we have shown that

$$h_n = n^{1/2}(S_n^2 - \overline{U}_n) \xrightarrow{P} 0.$$

Therefore, by Slutzky's theorem,

$$n^{1/2}\begin{Bmatrix}\overline{X}_n - \mu\\ S_n^2 - \sigma^2\end{Bmatrix} = n^{1/2}\begin{Bmatrix}\overline{H}_n\\ \overline{U}_n - \sigma^2\end{Bmatrix} + \begin{Bmatrix}0\\ h_n\end{Bmatrix} \xrightarrow{d} \mathbf{T}. \quad \square$$

From Theorem 6–17, we see that $(\overline{X}_n, S_n^2)$ is asymptotically jointly normally distributed when sampled from any distribution with a finite fourth moment. We also see that if the skewness is zero, then $\overline{X}_n$ and S_n^2 are asymptotically uncorrelated.

We next apply the Cramer δ theorem to find the asymptotic distribution of $\overline{X}_n/S_n$. Let $g(x, y) = xy^{-1/2}$. Then

$$g(\overline{X}_n, S_n^2) = \frac{\overline{X}_n}{S_n}, \qquad g(\mu, \sigma^2) = \frac{\mu}{\sigma},$$

$$\nabla g(x, y) = \left(y^{-1/2}, \frac{-xy^{-3/2}}{2}\right)', \qquad \nabla g(\mu, \sigma^2) = \left(\sigma^{-1}, \frac{-\mu\sigma^{-3}}{2}\right)'.$$

Finally,

$$(\nabla g(\mu,\sigma^2))'\begin{Bmatrix}\sigma^2 & \delta\sigma^3\\ \delta\sigma^3 & (\gamma-1)\sigma^4\end{Bmatrix}(\nabla g(\mu,\sigma^2)) = 1 - \frac{\mu\delta}{\sigma} + \frac{(\gamma-1)\mu^2}{4\sigma^2}.$$

Therefore,

$$n^{1/2}\left(\frac{\overline{X}_n}{S_n}-\frac{\mu}{\sigma}\right)\xrightarrow{d} N\left(0, 1-\frac{\mu\delta}{\sigma}+\frac{(\gamma-1)\mu^2}{4\sigma^2}\right).$$

Exercises—B

1. Suppose that $(X_n, Y_n)' \xrightarrow{d} (X, Y)' \sim N_2(\mathbf{0}, \mathbf{I})$, $U_n \xrightarrow{P} 2$, and $V_n \xrightarrow{P} 4$. What are the limiting distributions of the following?
   (a) $U_n X_n + V_n Y_n$ and $U_n + V_n X_n + Y_n$.
   (b) $U_n X_n + V_n Y_n/n$ and $U_n X_n + (n+2)V_n Y_n/n$.
2. Suppose that $n^{1/2}((X_n, Y_n)' - (2,3)') \xrightarrow{d} \mathbf{Z} \sim N_2(\mathbf{0}, \mathbf{I})$. What are the limiting distributions of
   (a) $U_n = n^{1/2}(X_n Y_n^2 - 18)$?
   (b) $V_n = n^{1/2}(\exp(X_n Y_n) - \exp(6))$?
3. In the notation of Theorem 6–17, verify the formula for the covariance matrix of (H_i, U_i).
4. Finish the derivation of Equation (6–5).
5. Find the asymptotic distributions of $n^{1/2}(\overline{X}_n S_n - \mu\sigma)$ and $n^{1/2}(\overline{X}_n^2 S_n^4 - \mu^2\sigma^4)$.
6. Suppose that $n^{1/2}(\mathbf{X}_n - \mathbf{a}) \xrightarrow{d} \mathbf{U} \sim N_k(\mathbf{0}, \mathbf{V})$, $\mathbf{V} > 0$.
   (a) Show that $n(\mathbf{X}_n - \mathbf{a})'\mathbf{V}^{-1}(\mathbf{X}_n - \mathbf{a}) \xrightarrow{d} \mathbf{U}'\mathbf{V}^{-1}\mathbf{U} \sim \chi_k^2$.
   (b) Suppose that $\mathbf{V}_n \xrightarrow{P} \mathbf{V}$. Show that $n(\mathbf{X}_n - \mathbf{a})'\mathbf{V}_n^{-1}(\mathbf{X}_n - \mathbf{a}) \xrightarrow{d} \mathbf{U}'\mathbf{V}^{-1}\mathbf{U} \sim \chi_k^2$.
7. Let X be a random variable with finite mean μ and skewness δ. Show that if $X - \mu$ has the same distribution as $-(X - \mu)$, then $\delta = 0$.

Exercises—C

1. Prove part a of Theorem 6–15. (*Hint*: If $g(\mathbf{x})$ is continuous at $\mathbf{b}$, then for all $\epsilon > 0$, there is a $\delta > 0$ such that if $\|\mathbf{x} - \mathbf{b}\| < \delta$ then $\|g(\mathbf{x}) - g(\mathbf{b})\| < \epsilon$.)
2. Using the notation of Exercise C2, Section 6.3.1, let $\delta_{ij} = E(X_k - \mu)^i (Y_k - \nu)^j/\sigma^i\tau^j$. Let $G_i = X_i - \mu$ and $H_i = Y_i - \nu$, and let $U_i = G_i^2$, $V_i = H_i^2$, and $W_i = G_i H_i$. Then $S_n^2 = \overline{U}_n - \overline{H}_n^2 + S_n^2/n$, $T_n^2 = \overline{V}_n - \overline{G}_n^2 + T_n^2/n$, and $Q_n = \overline{W}_n - \overline{H}_n\overline{G}_n + Q_n/n$.
   (a) Show that
   $$\mathbf{m} = E\begin{Bmatrix} U_i \\ V_i \\ W_i \end{Bmatrix} = \begin{Bmatrix} \sigma^2 \\ \tau^2 \\ \rho\sigma\tau \end{Bmatrix}$$
   and
   $$\mathbf{K} = \operatorname{cov}\left(\begin{Bmatrix} U_i \\ V_i \\ W_i \end{Bmatrix}\right) = \begin{Bmatrix} \delta_{40}\sigma^4 & \delta_{22}\sigma^2\tau^2 & \delta_{31}\sigma^3\tau \\ \delta_{22}\sigma^2\tau^2 & \delta_{04}\tau^4 & \delta_{13}\sigma\tau^3 \\ \delta_{31}\sigma^3\tau & \delta_{13}\sigma\tau^3 & \delta_{22}\sigma^2\tau^2 \end{Bmatrix} - \mathbf{mm}'.$$
   (b) Show that $n^{1/2}((\overline{U}_n, \overline{V}_n, \overline{W}_n)' - \mathbf{m}) \xrightarrow{d} \mathbf{A} \sim N_3(\mathbf{0}, \mathbf{K})$.
   (c) Show that $n^{1/2}((S_n^2, T_n^2, Q_n)' - \mathbf{m}) \xrightarrow{d} \mathbf{A} \sim N_3(\mathbf{0}, \mathbf{K})$.
   (d) Show that $n^{1/2}(R_n - \rho) \xrightarrow{d} B \sim N(0, \omega^2)$, where $\omega^2 = \rho^2(\delta_{40} + \delta_{04} + 2\delta_{22})/4 - \rho(\delta_{31} + \delta_{13}) + \delta_{22}$. (*Hint*: Let $g(x, y, z) = z/(xy)^{1/2}$. Then $R_n = g(S_n^2, T_n^2, Q_n)$.)
3. In the notation of the previous problem, suppose that (X_i, Y_i) have a bivariate normal distribution.
   (a) Use the joint moment-generating function of (G_i, H_i) to show that $\delta_{40} = \delta_{04} = 3$, $\delta_{13} = \delta_{31} = 3\rho$, and $\delta_{22} = 2\rho^2 + 1$.
   (b) Show that $n^{1/2}(R_n - \rho) \xrightarrow{d} B \sim N(0, (1 - \rho^2)^2)$.
   (c) Let $h(x) = (\frac{1}{2})\log[(1+x)/(1-x)]$. Show that $n^{1/2}(h(R_n) - h(\rho)) \xrightarrow{d} Z \sim N(0, 1)$.
   (d) Show that $(n-3)^{1/2}(h(R_n) - h(\rho) - \rho/2(n-1)) \xrightarrow{d} Z \sim N(0, 1)$. ($(n-3)^{1/2}(h(R_n) - h(\rho) - \rho/2(n-1))$ is called *Fisher's Z-transformation*.)

Chapter 7

ESTIMATION

Consider the following experiment in statistics. A researcher wants to determine the effects of jogging on a persons' resting pulse rate. She finds 30 male joggers and determines each of their resting pulse rates. She then finds the sample mean for these 30 numbers to be 55 and the sample variance to be 225. On the basis of these observations, she wants to answer several questions about μ, the average resting pulse rate for all joggers. (Note that 55 is the average *only* for the 30 joggers observed.) The first question she might entertain is how to estimate μ. This is an example of an *estimation* problem, which is the main topic of this chapter. A second question is how to find an interval in which she is confident that μ lies. This is an example of a *confidence interval* problem, which is discussed in the next chapter. Yet another question is whether μ is less than 72, the average pulse rate for all men. This is an example of a *testing problem,* which is also discussed in the next chapter. A fourth question is whether knowing just the sample mean and sample variance is enough to answer all questions about the data. This is an example of a *data reduction* problem and is discussed in Chapter 10. In Chapters 7–10, all of these problems are treated abstractly. Practical examples are then discussed in Chapters 11–14.

7.1 THE LIKELIHOOD FUNCTION AND MAXIMUM LIKELIHOOD ESTIMATORS

7.1.1 Statistical Models and the Likelihood Function

A *statistical model* consists of

1. An observed random vector $\mathbf{X} \in \chi$;
2. An unknown constant parameter vector $\boldsymbol{\theta} \in \Omega$; and
3. A function $f(\mathbf{x}; \boldsymbol{\theta})$ which represents the density function of $\mathbf{X}$ for each $\boldsymbol{\theta}$.

We call χ the *sample space* and Ω the *parameter space*. Throughout Chapters 7–10, we use n for the dimension of $\mathbf{X}$ and p for the dimension of $\boldsymbol{\theta}$.

A function $\boldsymbol{\tau} = \tau(\boldsymbol{\theta})$ of the unknown parameter vector $\boldsymbol{\theta}$ is a *parameter*, and any function $\mathbf{T} = T(\mathbf{X})$ of the observed random vector $\mathbf{X}$ is a *statistic*. Note that both $\boldsymbol{\tau}$ and $\mathbf{T}$ may be vector valued.

For the reaminder of the book, we employ the following conventions:

1. We use Greek letters for unknown parameters.
2. We use Roman letters for statistics and observed constants (such as the number n of observations).

That is, we use Roman letters for quantities that we can compute from the data and Greek letters for the unknown constants about which we want to make inferences.

Let $X_1, \ldots, X_{30}$ be the observed pulse rates for the 30 joggers discussed previously. As a model for this investigation, we might assume that the X_i are independent, with $X_i \sim N(\mu, \sigma^2)$, where μ and σ^2 are unknown parameters. Then $\mathbf{X} = (X_1, \ldots, X_{30})$, $\boldsymbol{\theta} = (\mu, \sigma)$, $p = 2$, and

$$f(\mathbf{x};\boldsymbol{\theta}) = (2\pi)^{-15}\sigma^{-30}\exp\left(-\sum_{i=1}^{30}\frac{(x_i-\mu)^2}{2\sigma^2}\right).$$

The sample space for this model is R^{30}, and the parameter space is the half-plane of all $\boldsymbol{\theta} = (\mu, \sigma)$, where $\mu \in R^1$ and $\sigma > 0$.

In this chapter and following chapters, it is very important to keep the statistics separate from the parameters. The goal of statistical analysis is to use the observed statistics to draw inferences about the unknown parameters. In the example just cited, the sample mean $\overline{X}$ and the sample variance S^2 are computed from the observations X_i and are therefore statistics, while the mean μ and the variance σ^2 are parameters. We wish to use the statistics $\overline{X}$ and S^2 to draw conclusions about the unknown parameters μ and σ^2.

The *likelihood function* for a statistical model is

$$L_X(\boldsymbol{\theta}) = f(\mathbf{X};\boldsymbol{\theta}).$$

That is, the likelihood function is the function formed when we substitute the data $\mathbf{X}$ for the dummy variable $\mathbf{x}$ in the density function. Therefore, the likelihood is a random function. We have written $\mathbf{X}$ as a subscript to indicate that we are considering $L_X(\boldsymbol{\theta})$ a function of $\boldsymbol{\theta}$ for each fixed $\mathbf{X}$. (We consider the density $f(\mathbf{x};\boldsymbol{\theta})$ a function of $\mathbf{x}$ for each fixed $\boldsymbol{\theta}$.) If $\boldsymbol{\theta}_0$ and $\boldsymbol{\theta}_1$ are two possible values of $\boldsymbol{\theta}$, we say that $\boldsymbol{\theta}_0$ is *more likely* than $\boldsymbol{\theta}_1$ (for a given observation vector $\mathbf{X}$) if

$$L_X(\boldsymbol{\theta}_0) > L_X(\boldsymbol{\theta}_1).$$

The main topic of this chapter is estimation, in which we want to estimate the k-dimensional parameter $\boldsymbol{\tau} = \tau(\boldsymbol{\theta})$. We often take $\tau(\boldsymbol{\theta}) = \boldsymbol{\theta}$. An *estimator* of $\boldsymbol{\tau}$ is a k-dimensional statistic $\mathbf{T} = T(\mathbf{X})$. Suppose we want to estimate $\tau(\mu, \sigma) = \mu$ in the joggers' pulse example. Both $\overline{X}$ and S^2 are estimators of μ. However, S^2 is obviously a rather silly estimator. In this chapter, we discuss methods for finding "sensible" estimators for functions $\tau(\boldsymbol{\theta})$.

Let us consider some examples which will be used in this chapter, as well as the three following ones.

Example A

This example consists of three separate cases in which we observe X_i independent, $i = 1, \ldots, n$, such that

$$EX_i = i\theta,$$

where θ is the unknown parameter. That is,

$$EX_1 = \theta, EX_2 = 2\theta, EX_3 = 3\theta, \ldots .$$

In the first model, X_i has a Poisson distribution with mean $i\theta$; in the second, X_i has a normal distribution with mean $i\theta$ and variance one; in the third, X_i has an exponential distribution with mean $i\theta$. Now, let

$$a_n = \sum_{i=1}^{n} i = \frac{n(n+1)}{2} \quad \text{and} \quad b_n = \sum_{i=1}^{n} i^2 = \frac{n(n+1)(2n+1)}{6}.$$

We consider three different estimators for θ:

$$P \doteq \sum_{i=1}^{n} \left(\frac{X_i}{a_n}\right), \qquad Q = \sum_{i=1}^{n} \left(\frac{iX_i}{b_n}\right), \qquad R = \sum_{i=1}^{n} \left(\frac{X_i}{ni}\right).$$

As long as $EX_i = i\theta$,

$$EP = \sum_{i=1}^{n} \frac{EX_i}{a_n} = \sum_{i=1}^{n} \left(\frac{i\theta}{a_n}\right) = \frac{\theta a_n}{a_n} = \theta.$$

In a similar manner,

$$EQ = ER = \theta,$$

so that all three estimators would seem sensible estimators of θ for all three models. In later sections of this chapter, we shall see that P is the "correct" estimator to use for the first model, Q is "correct" for the second, and R is "correct" for the third.

As an example, suppose that $n = 5$ (so that $a_n = 15$ and $b_n = 55$) and that we observe $X_1 = 10$, $X_2 = 11$, $X_3 = 13$, $X_4 = 15$, and $X_5 = 16$. Then

$$P = \frac{10 + 11 + 13 + 15 + 16}{15} = 4.33,$$

$$Q = \frac{10 + 22 + 39 + 60 + 80}{55} = 3.84,$$

and

$$R = \frac{10 + (11/2) + (13/3) + (15/4) + (16/5)}{5} = 5.36,$$

so that the three estimators can yield different results.

We now find the likelihood $L_X(\theta)$ for all three models.

Example A1

The X_i are independent and $X_i \sim P(i\theta), \theta > 0$. Therefore,

$$f(\mathbf{x};\theta) = \prod_{i=1}^{n} \frac{(i\theta)^{i} e^{-i\theta}}{x_i!} = h_1(\mathbf{x}) \exp\left[\sum_{i=1}^{n} x_i \log(\theta) - \theta a_n\right],$$

and hence, the likelihood function for this model is

$$L_X(\theta) = h_1(\mathbf{X}) \exp\left[\sum_{i=1}^{n} X_i \log(\theta) - \theta a_n\right].$$

Example A2

The X_i are independent and $X_i \sim N(i\theta, 1), \theta \in R^1$. Therefore,

$$f(\mathbf{x};\theta) = \prod_{i=1}^{n} (2\pi)^{-1/2} \exp\left[\frac{-(x_i - i\theta)^2}{2}\right] = h_2(\mathbf{x}) \exp\left(\theta \sum_{i=1}^{n} ix_i - \frac{b_n \theta^2}{2}\right),$$

and hence, the likelihood function for this model is

$$L_X(\theta) = h_2(\mathbf{X}) \exp\left(\theta \sum_{i=1}^{n} iX_i - \frac{b_n \theta^2}{2}\right).$$

Example A3

The X_i are independent and $X_i \sim E(i\theta), \theta > 0$. Therefore,

$$f(\mathbf{x};\theta) = \prod_{i=1}^{n} (i\theta)^{-1} \exp\left(\frac{-x_i}{i\theta}\right) = h_3 \theta^{-n} \exp\left[-\theta^{-1} \sum_{i=1}^{n} \left(\frac{x_i}{i}\right)\right].$$

and hence, the likelihood function for this model is

$$L_X(\theta) = h_3 \theta^{-n} \exp\left[-\theta^{-1} \sum_{i=1}^{n} \left(\frac{X_i}{\mathrm{i}}\right)\right].$$

Note that in all three of the preceding examples, we have pulled out functions $h_i(\mathbf{X})$ which depend only on $\mathbf{X}$, and we did not worry about their form. As we shall see in later sections, any multiplicative function involving only $\mathbf{X}$ is irrelevant for statistical inference, and its form can be ignored.

The second example to be used in Chapters 7–10 has two parameters and is just an extension of the joggers' pulse rate model to a sample of size n.

Example B

In this example, we observe X_i independent, with $X_i \sim N(\mu, \sigma^2)$ for $i = 1, \ldots, n$, $\mu \in R^1, \sigma > 0$. Therefore, $\boldsymbol{\theta} = (\mu, \sigma)$, and

$$f(\mathbf{x}; (\mu, \sigma)) = (2\pi)^{-n/2} \sigma^{-n} \exp\left[-\sum_{i=1}^{n} \frac{(x_i - \mu)^2}{2\sigma^2}\right],$$

and hence, the likelihood function for this model is

$$L_X(\mu, \sigma) = (2\pi)^{-n/2} \sigma^{-n} \exp\left[-\sum_{i=1}^{n} \frac{(X_i - \mu)^2}{2\sigma^2}\right].$$

If $\overline{X}$ and S^2 are the sample mean and sample variance of the X_i's, then we have shown that $\overline{X}$ and S^2 are independent, and

$$\overline{X} \sim N\left(\mu, \frac{\sigma^2}{n}\right) \quad \text{and} \quad S^2 \sim \Gamma\left(\frac{n-1}{2}, \frac{2\sigma^2}{n-1}\right).$$

The next example is often used for homework problems.

Example C

Let $X_1, \ldots, X_n$ be independent and $X_i \sim N(i\theta, \sigma^2)$, where θ and $\sigma^2 > 0$ are unknown parameters. Let

$$Q = \frac{\sum iX_i}{b_n} \quad \text{and} \quad T^2 = \frac{\sum (X_i - iQ)^2}{(n-1)} = \frac{\sum X_i^2 - b_n Q^2}{(n-1)}.$$

In Exercise B3, Section 5.3.4, we have shown that Q and T^2 are independent,

$$Q \sim N\left(\theta, \frac{\sigma^2}{b_n}\right) \quad \text{and} \quad T^2 \sim \Gamma\left(\frac{n-1}{2}, \frac{2\sigma^2}{n-1}\right).$$

These results are, of course, similar to those for Example B, and can be used in a similar manner.

Throughout Chapters 7–10, Examples A1, A2, A3, B, and C will be used to refer to the examples given above.

Exercises—B

1. Verify that as long as $EX_i = i\theta$, $EQ = \theta$ and $ER = \theta$.
2. Verify the formula for $f(\mathbf{x}; (\mu,\sigma))$ in Example B.
3. Find the likelihood function for Example C.
4. Let $X_1, \ldots, X_n$ be independent, with $X_i \sim N(\theta, \theta^2)$. Find the likelihood function for this model.
5. Let $X_1, \ldots, X_n$ be independent, with $X_i \sim \Gamma(i, \theta)$. Find the likelihood function for this model.

7.1.2 Maximum Likelihood Estimators

We now present a method which often leads to sensible estimators. We first look at estimating the whole p-dimensional parameter vector $\boldsymbol{\theta}$. A p-dimensional statistic $\mathbf{T} = T(\mathbf{X})$ of $\boldsymbol{\theta}$ is a *maximum likelihood estimator* (MLE) of $\boldsymbol{\theta}$ if for each $\mathbf{X}$, $T(\mathbf{X})$ is more likely than any other possible value of $\boldsymbol{\theta}$. That is, $\mathbf{T} = T(\mathbf{X})$ is a maximum lieklihood estimator of $\boldsymbol{\theta}$ if for all $\mathbf{X} \in \chi, \boldsymbol{\theta} \in \Omega$,

1. $T(\mathbf{X}) \in \Omega$, and
2. $L_X(T(\mathbf{X}) \geqq L_X(\boldsymbol{\theta})$.

If for each $\mathbf{X}$, the maximum $T(\mathbf{X})$ is unique, then $\mathbf{T} = T(\mathbf{X})$ is the *unique* maximum likelihood estimator of $\boldsymbol{\theta}$.

Now, suppose that we want to estimate some function $\boldsymbol{\tau} = \tau(\boldsymbol{\theta})$. We define the *maximum likelihood estimator* of $\tau(\boldsymbol{\theta})$ to be $\tau(\mathbf{T})$, where $\mathbf{T}$ is the MLE of $\boldsymbol{\theta}$. This definition is called the *invariance principle* for MLEs. It takes some work to show that the definition is consistent. The invariance principle is discussed further in the next section.

We often use the notation $\hat{\tau}$ for an estimator of $\tau(\boldsymbol{\theta})$. (Note that $\hat{\tau}$ is a statistic; that is, $\hat{\tau} = T(\mathbf{X})$ for some function $T(\mathbf{X})$.) If $\hat{\boldsymbol{\theta}}$ is the MLE of $\boldsymbol{\theta}$ and $\hat{\tau}$ the MLE of $\tau(\boldsymbol{\theta})$, then the invariance principle says that

$$\hat{\tau} = \tau(\hat{\theta}).$$

In finding MLEs, it is often helpful to use the fact that $\log u$ is an increasing function of u, and therefore, maximizing $L_X(\boldsymbol{\theta})$ is the same as maximizing $\log[L_X(\boldsymbol{\theta})]$. We often do the latter by setting first derivatives equal to zero.

Let us look again at Example A.

Example A1

Let $X_i \sim P(i\theta)$, independent. For this model,

$$L_X(\theta) = h_1(\mathbf{X}) \exp[\textstyle\sum X_i \log(\theta) - \theta a_n], \qquad \log(L_X(\theta)) = \log(h_1(\mathbf{X})) + \textstyle\sum X_i \log(\theta) - \theta a_n,$$

$$\frac{\partial}{\partial\theta} \log(L_X(\theta)) = \frac{\sum X_i}{\theta} - a_n .$$

Setting the last equal to zero, we get $\hat{\theta} = \Sigma X_i / a_n$. Therefore, $P = \Sigma X_i / a_n$ is the unique MLE of θ for this model. (To see that the critical point $\hat{\theta}$ is actually a maximum, note that $(\partial^2/\partial\theta^2)\log[L_X(\theta)] = -\Sigma X_i/\theta^2 < 0$. In subsequent examples, we shall not check the second derivative, leaving that exercise to the interested reader.)

Example A2

Let $X_i \sim N(i\theta, 1)$, independent. Then the MLE is $Q = \Sigma i X_i / b_n$.

Example A3

Let $X_i \sim E(i\theta)$, independent. Then the MLE is $R = \Sigma(X_i/ni)$.

The last two MLEs are left as exercises. Observe that the MLEs are different for the three models and that for each model the MLE is unique.

We next look at Example B.

Example B

Let $X_1, \ldots, X_n$ be independent, $n \geqq 2$, and $X_i \sim N(\mu, \sigma^2)$, where $\boldsymbol{\theta} = (\mu, \sigma)'$ is the unknown parameter. Then

$$\log[L_X(\mu, \sigma)] = -n \log\sigma - \frac{\sum (X_i - \mu)^2}{2\sigma^2} - \frac{n(\log(2\pi))}{2}.$$

Also, let

$$0 = \left(\frac{\partial}{\partial\mu}\right) \log[L_X(\mu, \sigma)] = \frac{\sum (X_i - \mu)}{\sigma^2} = \frac{\sum X_i - n\mu}{\sigma^2},$$

$$0 = \left(\frac{\partial}{\partial\sigma}\right) \log[L_X(\mu, \sigma)] = \frac{-n}{\sigma} + \frac{\sum (X_i - \mu)^2}{\sigma^3}.$$

Solving the first equation, we obtain

$$\hat{\mu} = \frac{\sum X_i}{n} = \overline{X}.$$

Substituting this in for μ in the second yields

$$\hat{\sigma} = \left(\frac{\sum (X_i - \overline{X})^2}{n}\right)^{1/2} = \left[\frac{(n-1)S^2}{n}\right]^{1/2}$$

(where S^2 is the sample variance of the X's). Therefore, $(\hat{\mu}, \hat{\sigma})$ is the MLE for (μ, σ).

By the invariance principle for MLEs, it follows that the MLE for σ^2 is

$$\hat{\sigma}^2 = \frac{(n-1)S^2}{n} = \frac{\sum(X_i - \overline{X})^2}{n}.$$

(A direct derivation of this equation, without the invariance principle, is given in the exercises.) By the invariance principle again, the MLEs of μ/σ and $\mu\sigma$ are $\hat{\mu}/\hat{\sigma}$ and $\hat{\mu}\hat{\sigma}$, respectively.

We now present an example in which there is no MLE.

Example B

Now let $n = 1$, and let $X \sim N(\mu, \sigma^2)$ (that is, X is a sample of size 1). Then

$$\log[L_X(\mu, \sigma)] = -\log \sigma - \frac{(X-\mu)^2}{2\sigma^2} - \frac{\log(2\pi)}{2}.$$

If we differentiate this with respect to μ, see that $\hat{\mu} = X$. Now,

$$\log(L_X(X, \sigma)) = -\log \sigma - \frac{\log(2\pi)}{2},$$

which can be made arbitrarily large by letting σ go to zero. Therefore, $\log(f(X; (\mu, \sigma))$ has no maximum and, hence, no MLE. (Actually, X is not even the MLE for μ in this case, since there is no maximum to the likelihood.)

In Sections 7.2.1–7.2.3, we take a different approach to estimation, defining a property called unbiasedness that we would like an estimator to have. In Sections 7.2.3 and 10.3.1, we describe two methods for finding the best estimators having this property. The method of maximum likelihood is somewhat different from this approach, in that it merely gives a method for constructing estimators which often turn out to be sensible. It can happen that MLEs can be quite poor as we shall see in multiple regression. However, in Section 7.3.1, we derive a result which shows that, asymptotically, MLEs are doing as well as possible. One other advantage to MLEs is the invariance property, which is not shared by other estimators (such as best unbiased estimators). It seems intuitively clear that if $\hat{\theta}$ is a sensible estimator of θ, then $\tau(\hat{\theta})$ should be a sensible estimator of $\tau(\theta)$. For example, if S^2 is a sensible estimator of σ^2, then S should be a sensible estimator of σ.

Exercises—B

1. For Example A2, show that Q is the MLE for θ. What is the MLE for θ^3?
2. For Example A3, show that R is the MLE for θ. What is the MLE for θ^2?
3. Let $X_1, \ldots, X_n$ be independent, with $X_i \sim P(\theta)$. What is the MLE of θ? of e^{θ}?
4. Let $X_1, \ldots, X_n$ be independent, with $X_i \sim \Gamma(i, \theta)$. What are the MLEs of θ and θ^{-1}?
5. Let $X_1, \ldots, X_n$ be independent, with X_i geometrically distributed with parameter θ. That is, X_i has density $f(x; \theta) = \theta(1-\theta)^x, x = 0, 1, \ldots, 0 < \theta < 1$. Find the MLEs of θ and $(1-\theta)^{-1}$.
6. Let $X_1, \ldots, X_n$ be independently identically distributed with X_i having density $f(x; \theta) = \theta x^{\theta - 1}, 0 < x < 1, \theta > 0$. Find the MLE of θ.
7. For Example C in Section 7.1.1, find the MLE of (θ, σ). What is the MLE of $\theta\sigma$? of $\theta\sigma^2$?

8. Let (X,Y) have a trinomial distribution with parameters n, θ^2, $2\theta(1-\theta)$, and $(1-\theta)^2$. That is, (X,Y) has joint density function

$$f(x,y) = k(x,y)(\theta^2)^x(2\theta(1-\theta))^y((1-\theta)^2)^{n-x-y}, \quad x=0,1,\ldots,y=0,1,\ldots,x+y \leqq n.$$

Suppose n is known. Show that the MLE of θ is $(2X+Y)/2n$.

9. Let (X,Y) have a trinomial distribution with parameters n, θ_1, and θ_2. Show that the MLE of (θ_1,θ_2) is $(X/n, Y/n)$.

Exercises—C

1. Let $X_1,\ldots,X_n$ be independent, with $X_i \sim N(\theta,\theta^2)$. Find the MLE of θ. (Be sure to verify which root of the quadratic leads to the maximum.)
2. Let $X_1,\ldots,X_n$ be independent, with $X_i \sim N(\mu,\delta)$. That is, X_i is normally distributed with mean μ and variance δ. Hence, $\boldsymbol{\theta} = (\mu,\delta)$ for this model. Without using the invariance principle, show that the MLE for (μ,δ) is given by $\hat{\mu} = \overline{X}, \hat{\delta} = (n-1)S^2/n$.

* 7.1.3 More on the Invariance Principle

One of the most important properties of MLEs is the invariance principle, which says that if $\hat{\boldsymbol{\theta}}$ is the MLE of $\boldsymbol{\theta}$, then $\tau(\hat{\boldsymbol{\theta}})$ is the MLE of $\tau(\boldsymbol{\theta})$. This property is often quite helpful in finding MLEs, but it also seems like a property that a "sensible" estimation principle would satisfy.

The next two examples show that the invariance property is a strange mixture, being part theorem and part definition.

Example A1

Suppose we observe $X_i \sim P(i\theta)$, independent. Let $\delta = \theta^2$. We can find the MLE of δ directly from the definition of an MLE, without using the invariance principle. We have $\theta > 0$, so that $\theta = \delta^{1/2}$. Therefore, $X_i \sim P(i\delta^{1/2})$. Hence, the likelihood in terms of δ is

$$L_X^*(\delta) = h(\mathbf{X}) \exp\left[\frac{\sum X_i(\log\delta)}{2} - \delta^{1/2} a_n\right],$$

and

$$\log L_X^*(\delta) = \log(h(\mathbf{X})) + \frac{\sum X_i(\log\delta)}{2} - a_n\delta^{1/2}.$$

If we set the derivative with respect to δ of the log likelihood to zero, we see that $\hat{\delta} = (\Sigma X_i/a_n)^2 = P^2$, where P is the MLE for θ. For this example, therefore, the invariance principle is a true fact.

Example A2

Suppose now that we observe $X_i \sim N(i\theta, 1)$, independent, and want to estimate $\delta = \theta^2$. In this case, θ can be either $\pm\delta^{1/2}$, so that θ is not a function of δ. Thus, there is no way to use the definition of an MLE to find the MLE for δ, and the invariance principle is a definition here.

We now return to a general model with parameter $\boldsymbol{\theta}$. We say that $\boldsymbol{\delta} = h(\boldsymbol{\theta})$ is a *reparametrization* if h is an invertible function (i.e., if there exists k such that $\boldsymbol{\theta} = k(\boldsymbol{\delta})$.) The following theorem is the theoretical justification for the invariance principle. We do not assume that principle in its statement or proof.

Theorem 7–1. Consider a model with parameter $\boldsymbol{\theta}$, and let $\boldsymbol{\delta} = h(\boldsymbol{\theta})$ be a reparametrization. Let $\hat{\boldsymbol{\theta}}$ be an MLE of $\boldsymbol{\theta}$. Then $\hat{\boldsymbol{\delta}} = h(\hat{\boldsymbol{\theta}})$ is an MLE of $\boldsymbol{\delta}$. Also, if $\hat{\boldsymbol{\theta}}$ is the unique MLE of $\boldsymbol{\theta}$, then $\hat{\boldsymbol{\delta}}$ is the unique MLE of $\boldsymbol{\delta}$.

Proof. Let $\hat{\boldsymbol{\theta}} = T(\mathbf{X})$ be an MLE of $\boldsymbol{\theta}$. Let Ω be the set of possible values of $\boldsymbol{\theta}$, and let Ω^* be the set of possible values of $\boldsymbol{\delta}$. (That is, Ω^* is the set of all $h(\boldsymbol{\theta}), \boldsymbol{\theta} \in \Omega$.) Then $T(\mathbf{X}) \in \Omega$ for all $\mathbf{x} \in \chi$, and therefore,

$$h(T(\mathbf{X})) \in \Omega^* \quad \text{for all } \mathbf{X}.$$

Let $k(\boldsymbol{\delta}) = \boldsymbol{\theta}$, let $f(\mathbf{x};\boldsymbol{\theta})$ be the density of $\mathbf{X}$ for a given $\boldsymbol{\theta} \in \Omega$, and let $f^*(\mathbf{x};\boldsymbol{\delta})$ be the density of $\mathbf{X}$ for a given $\boldsymbol{\delta} \in \Omega^*$. Then $f(\mathbf{x};\mathbf{a})$ is the density of $\mathbf{X}$ when $\boldsymbol{\theta} = \mathbf{a}$, and $f^*(\mathbf{x};\mathbf{b})$ is the density of $\mathbf{X}$ when $\boldsymbol{\delta} = \mathbf{b}$, or equivalently, $\boldsymbol{\theta} = k(\mathbf{b})$. Therefore, $f(\mathbf{x};k(\mathbf{b})) = f^*(\mathbf{x};\mathbf{b})$, or equivalently, $f(\mathbf{x};\mathbf{a}) = f^*(\mathbf{x};h(\mathbf{a}))$. Hence,

$$f^*(\mathbf{x};h(\boldsymbol{\theta})) = f(\mathbf{x};\boldsymbol{\theta}) \quad \text{and} \quad L_X^*(h(\boldsymbol{\theta})) = L_X(\boldsymbol{\theta}).$$

(Note that we do not multiply by the Jacobian, even in the continuous case.) Therefore, if $T(\mathbf{X})$ is the MLE of $\boldsymbol{\theta}$, then

$$L_X^*(h[T(\mathbf{X})]) = L_X(T(\mathbf{X})) \geqq L_X(\boldsymbol{\theta}) = L_X^*(h(\boldsymbol{\theta})) = L_X^*(\boldsymbol{\delta}),$$

for all $\boldsymbol{\delta}$. Consequently, $\hat{\boldsymbol{\delta}} = h(T(\mathbf{X}))$ is the MLE of $\boldsymbol{\delta}$. Also, if $T(\mathbf{X})$ is a unique maximum of L_X for all $\mathbf{X}$, then $h(T(\mathbf{X}))$ is a unique maximum for L_X^* for all $\mathbf{X}$, and hence, $\hat{\boldsymbol{\delta}}$ is unique if $\hat{\boldsymbol{\theta}}$ is. □

Theorem 7–1 implies that the invariance principle is a true theorem for invertible functions $h(\boldsymbol{\theta})$.

Now, let $p(\boldsymbol{\theta})$ be a noninvertible function of $\boldsymbol{\theta}$, and let $p^*(\boldsymbol{\delta}) = p(k(\boldsymbol{\delta}))$, so that $p(\boldsymbol{\theta})$ and $p^*(\boldsymbol{\delta})$ are really the same parameter. If $\hat{\boldsymbol{\theta}}$ is an MLE of $\boldsymbol{\theta}$, then $\hat{\boldsymbol{\delta}} = h(\hat{\boldsymbol{\theta}})$ is an MLE of $\boldsymbol{\theta}$, by Theorem 7–1. Also,

$$p^*(\hat{\boldsymbol{\delta}}) = p(k(\hat{\boldsymbol{\delta}})) = p(k(h(\hat{\boldsymbol{\theta}}))) = p(\hat{\boldsymbol{\theta}}).$$

This result implies that the invariance principle gives consistent answers when we reparametrize the problem, so that the invariance principle is logically consistent even for noninvertible functions.

Exercises—B

1. In Example A3, let $\delta = \theta^{1/2}$, so that $\theta = \delta^2$.
   (a) What is the likelihood function in terms of δ?
   (b) Without using the invariance principle, find the MLE $\hat{\delta}$ for δ, and verify that $\hat{\delta} = R^{1/2}$.
2. Let $X_1, \ldots, X_n$ be independently geometrically distributed with parameter θ. Let $\mu = EX_i$.
   (a) Show that $\mu = h(\theta)$ for some invertible function θ.
   (b) Find the MLE $\hat{\mu}$ of μ using the definition of an MLE, and verify that $\hat{\mu} = h(\hat{\theta})$.

7.2 PROPERTIES OF ESTIMATORS

7.2.1 Unbiased Estimators

Let $\tau(\boldsymbol{\theta})$ be a one-dimensional parameter. Let $T = T(\mathbf{X})$ be an estimator of $\tau(\boldsymbol{\theta})$. The *bias* of T is defined as

$$\text{bias}(T) = E_\theta T(\mathbf{X}) - \tau(\boldsymbol{\theta}).$$

(We write $E_\theta T(\mathbf{X})$ for $E(T(\mathbf{X}))$ to indicate the dependence of E on the parameter $\boldsymbol{\theta}$.) An estimator is called *unbiased* if it has bias zero. Otherwise, it is called *biased.* It seems that the property of unbiasedness is a reasonable property for an estimator to have.

Typically, there are many unbiased estimators of a given function $\tau(\boldsymbol{\theta})$, as will be seen in the examples that follow. Therefore, it is useful to have some criterion for comparing unbiased estimators. One criterion which is often used is the variance of the unbiased estimator T (denoted here by $\text{var}_\theta(T)$ to emphasize its dependence on $\boldsymbol{\theta}$). If $T_1 = T_1(\mathbf{X})$ and $T_2 = T_2(\mathbf{X})$ are both unbiased estimators of $\tau(\boldsymbol{\theta})$, and if

$$\text{var}_\theta(T_1) = E(T_1 - \tau(\theta))^2 < \text{var}_\theta(T_2) = E(T_2 - \tau(\theta))^2,$$

then $(T_1 - \tau(\boldsymbol{\theta}))^2$ is, on the average, less than $(T_2 - \tau(\boldsymbol{\theta}))^2$, which indicates that T_1 is "nearer" $\tau(\boldsymbol{\theta})$ than T_2 is. For that reason, we say that the unbiased estimator T_1 is *as good as* the unbiased estimator T_2 if

$$\text{var}_\theta(T_1) \leqq \text{var}_\theta(T_2)$$

for all $\boldsymbol{\theta}$, and T_1 is *better than* T_2 if, in addition,

$$\text{var}_\theta(T_1) < \text{var}_\theta(T_2)$$

for at least one $\boldsymbol{\theta}$. We say that the unbiased estimator T is the *best unbiased estimator* of $\tau(\boldsymbol{\theta})$ if it is better than any other unbiased estimator of $\tau(\boldsymbol{\theta})$. In Section 7.2.3, we give one method which can occasionally be used for finding best unbiased estimators. A more general method is given in Chapter 10.

In this section and the next, we often compute means and variances of statistics. Recall here the following simple properties of means and variances:

$$E(aX + b) = aEX + b, \qquad \text{var}(aX + b) = a^2 \,\text{var}(X), \qquad E\sum a_i X_i = \sum a_i EX_i.$$

If the X_i are independent, then

$$\text{var}(\sum a_i X_i) = \sum a_i^2 \,\text{var}(X_i).$$

Example A

In this example, there are three different models, all with $X_1, \ldots, X_n$ independent and $EX_i = i\theta$. As before, let $P = \Sigma X_i/a_n$, $Q = \Sigma iX_i/b_n$, $R = \Sigma(X_i/ni)$. In Section 7.1.1, it was shown that as long as $EX_i = i\theta$,

$$EP = EQ = ER = \theta,$$

so that all three of these estimators are unbiased for all three models.

To see which estimator is better, we must find the variances of the estimators and compare them. The variances, of course, depend on the model chosen.

Example A1

If $X_i \sim P(i\theta)$, then $\text{var}(X_i) = EX_i = i\theta$. Therefore,

$$\text{var}(P) = \sum_{i=1}^{n} \frac{\text{var}(X_i)}{a_n^2} = \frac{\theta}{a_n},$$

$$\text{var}(Q) = \sum_{i=1}^{n} \frac{i^2\,\text{var}(X_i)}{b_n^2} = \theta \sum_{i=1}^{n} \frac{i^3}{b_n^2},$$

$$\text{var}(R) = \sum_{i=1}^{n} (in)^{-2}\,\text{var}(X_i) = \sum_{i=1}^{n} (in)^{-2} i\theta = \theta \sum_{i=1}^{n} \frac{i^{-1}}{n^2}.$$

In the exercises, it is shown that $\text{var}(P) < \text{var}(Q)$ and $\text{var}(P) < \text{var}(R)$. Therefore, P is better than Q or R for this model. Note that P is the MLE for the model.

Example A2

If $X_i \sim N(i\theta, 1)$, then $\text{var}(X_i) = 1$. Using this information, we can show that for this model, Q is better than P and R. Q is the MLE for the model.

Example A3

If $X_i \sim E(i\theta)$, then $\text{var}(X_i) = i^2\theta^2$. Using this information, we can show that for this model, R, the MLE, is better than Q or P.

We thus see that although all three estimators are unbiased for all three models, the MLE picks out the best one for each model. Unfortunately, MLEs need not be unbiased, as we now show.

We consider estimating θ^2 for the preceding models.

Example A1

If $X_i \sim P(i\theta)$, then, from the invariance property of MLEs, P^2 is the MLE for θ^2. However,

$$EP^2 = (EP)^2 + \text{var}(P) = \theta^2 + \frac{\theta}{a_n},$$

so that the MLE P^2 is not unbiased. However,

$$T = P^2 - \frac{P}{a_n}$$

is unbiased for the model, since

$$ET = EP^2 - \frac{EP}{a_n} = \theta^2 + \frac{\theta}{a_n} - \frac{\theta}{a_n} = \theta^2.$$

Examples A2, A3

The MLEs Q^2 and R^2 are not unbiased estimators for these models. Unbiased estimators based on the MLEs are asked for in the exercises.

Estimating θ^2 in Example A makes it clear that there is no invariance principle for unbiased estimators. If T is an unbiased estimator for θ, $\tau(T)$ is rarely an unbiased estimator of $\tau(\theta)$. (See Exercises B6 and C1.)

Example B

Let $X_1, \ldots, X_n$ be independent, with $X_i \sim N(\mu, \sigma^2)$, where $\theta = (\mu, \sigma)'$ is unknown. First consider estimating μ. Note that $\hat{\mu} = \bar{X} \sim N(\mu, \sigma^2/n)$. Therefore, the MLE is unbiased for μ. Now, $X_1 \sim N(\mu, \sigma^2)$. Therefore, X_1 is also an unbiased estimator of μ. However, X_1 has variance $\sigma^2 > \sigma^2/n$, so that $\bar{X}$ is better than X_1.

Now consider estimating σ^2. We have shown that

$$ES^2 = \sigma^2,$$

so that the sample variance S^2 is an unbiased estimator of σ^2. Now the MLE $\hat{\sigma}^2 = (n-1)S^2/n$. Therefore, $E\hat{\sigma}^2 = (n-1)\sigma^2/n$, so that the MLE is not an unbiased estimator of σ^2 for this model.

Finally, consider estimating σ. In Exercise B11, it is shown that

$$ES = \frac{\Gamma(n/2)2^{1/2}}{\Gamma((n-1)/2)(n-1)^{1/2}}\sigma.$$

From this equation, we see that

$$\frac{\Gamma((n-1)/2)(n-1)^{1/2}}{\Gamma(n/2)2^{1/2}}S$$

is an unbiased estimator of σ and that neither S nor $\hat{\sigma}$ is unbiased. We would rarely use the unbiased estimator of σ: Its complicated formula negates any benefit that might be derived from its unbiasedness.

In a general statistical model, let $\hat{\boldsymbol{\theta}}$ be the MLE of $\boldsymbol{\theta}$, and suppose that $T(\mathbf{X}) = h(\hat{\boldsymbol{\theta}})$ is an unbiased estimator of $\tau(\boldsymbol{\theta})$. Then we say that $T(\mathbf{X})$ is an unbiased estimator *based on* the MLE. In Chapter 10, we show that under fairly general conditions, there is only one unbiased estimator based on the MLE, and that estimator is the best unbiased estimator. Therefore, in finding a sensible unbiased estimator for a parameter τ, we typically start with the MLE of τ. If it is unbiased, then we have found the estimator we want. If it is not unbiased, we decide what we need to do to modify the MLE to make it unbiased. (Often, there is no obvious modification which works, and in those situations, we shall typically not be able to find a sensible unbiased estimator.)

It is often useful to have a criterion for comparing estimators that are not necessarily unbiased. We define the *mean squared error* (MSE) of a (possibly biased) estimator $T(\mathbf{X})$ of $\tau(\boldsymbol{\theta})$ by

$$MSE_T(\boldsymbol{\theta}) = E(T(\mathbf{X}) - \tau(\boldsymbol{\theta}))^2 = \text{var}(T(\mathbf{X})) + (\text{bias}(T(\mathbf{X})))^2.$$

An estimator with a small MSE is one whose bias and variance are both small. We say that $T(\mathbf{X})$ is *as good as* $S(\mathbf{X})$ if $MSE_T(\boldsymbol{\theta}) \leqq MSE_S(\boldsymbol{\theta})$ for all $\boldsymbol{\theta}$, and that T is *better than* S if, in addition, $MSE_T(\boldsymbol{\theta}) < MSE_S(\boldsymbol{\theta})$ for some $\boldsymbol{\theta}$. If T and S are unbiased, then the MSE is the same as the variance, so this definition reduces to the one given in the case of unbiased estimators.

Example B

Let us continue with the problem of estimating σ^2. We have two possible estimators, the sample variance S^2, which is unbiased, and the MLE $\hat{\sigma}^2$. In Exercise B10, it is shown that the MSE for S^2 is $2(n-1)^{-1}$ and that the MSE for $\hat{\sigma}^2$ is $(2n-1)/n^2 < 2/(n-1)$. Therefore, $\hat{\sigma}^2$ is better than S^2. However, it is also shown in the exercises that the estimator $(n-1)S^2/(n+1)$ is better than either S^2 or $\hat{\sigma}^2$ and is better than any other estimator in the class kS^2. The problem of estimating σ is also considered and the best estimator σ of the form mS is found. Unfortunately, it is quite messy again and is unlikely to be used.

Estimating σ in Example B illustrates that the properties of an estimator cannot be completely summarized by its MSE. Two important considerations that are ignored by the MSE are computational simplicity and the explanation of the procedure to the experimenter who wants to interpret the data. It seems likely that statisticians will continue to use either the sample standard deviation S or the MLE $\hat{\sigma}$ to estimate σ, rather than the "better" estimator $\tilde{\sigma}$, particularly since the gain in MSE from using $\tilde{\sigma}$ instead of $\hat{\sigma}$ or S is quite small.

Example A3

We consider this example for the case where $n = 1$, so that we observe X_1 having an exponential distribution with mean μ. We look at estimators of μ of the form $T_c(X_1) = cX_1$. The mean squared error for this estimator is

$$MSE_c(\mu) = E(cX_1 - \mu)^2 = c^2 EX_1^2 - 2c\mu EX_1 + \mu^2 = \mu^2(2c^2 - 2c + 1).$$

The choice of c which minimizes the MSE is $c = .5$. In fact, the MSE for the estimator $T_{.5}(X_1) = .5X_1$ is $.5\mu^2$, and the MSE for the unbiased estimator $T_1(X_1) = X_1$ is μ^2. In Section 7.2.3 and Chapter 10, we shall see that X_1 is the best unbiased estimator of μ. This example points out that it is occasionally possible to find a biased estimator which has considerably smaller mean squared error than any unbiased estimator. In this case, the biased estimator would be preferred to the unbiased estimator. (In the exercises, it is shown that for larger samples, the unbiased estimator can always be improved, but the improvement goes away as n goes to infinity.)

In assessing the quality of an estimator, we often think of the bias as a measure of the "accuracy" of an estimator: An unbiased estimator is one whose expectation is correct, so it is accurate on the average. We think of the variance of the estimator as measuring the "precision" of the estimator: An estimator with low variance is one which is quite precise. However, if it is not accurate, it is precisely wrong. For example, in Example B, we could ignore the data and estimate μ always to be zero. (That is, we could use the estimator $T(\mathbf{X})$ identically equal to zero.) This estimator has variance zero and is quite precise. However, it is a rather silly estimator, being very precise but not at all accurate. In finding estimators with low MSE, we are finding estimators which are both accurate (have low bias) and precise (have low variance).

Exercises—B

1. For Example 7A1, show that $\text{var}(P) < \text{var}(Q)$ and $\text{var}(P) < \text{var}(R)$. (*Hint*: If $a_i > 0$ and $b_i > 0$, then $\Sigma a_i b_i < \Sigma a_i \Sigma b_i$ and $(\Sigma a_i b_i)^2 < \Sigma a_i^2 \Sigma b_i^2$ unless $a_i = kb_i$. Why?)
2. Show that the MLEs of θ in Exercises B3 and B4, Section 7.1.2, are unbiased.
3. Find unbiased estimators of θ^2 in Examples A2 and A3.
4. For the model in Exercise B8, Section 7.1.2, show that the MLE of θ is unbiased.
5. For Example C in Section 7.1.1, find unbiased estimators of θ, σ^2, and $\theta\sigma^2$.

6. Let $\hat{\theta}$ be an unbiased estimator of θ. Show that $\hat{\theta}^2$ is not an unbiased estimator of θ^2 unless $\text{var}(\hat{\theta}) = 0$.
7. For Example A3 with $n = 1$,
   (a) Verify the calculations for the MSE of $T_c(X_1) = cX_1$.
   (b) Verify that the MSE is minimized when $c = .5$.
   (c) Verify the MSEs for $.5X_1$ and X_1.
8. Let $n > 1$ in Example A3.
   (a) Find the MSE for the estimator $T_c(R) = cR$.
   (b) Find the choice for c which minimizes the mean squared error.
   (c) Let V be the ratio of the MSE for the optimal c over the MSE for the unbiased estimator R. Show that $V \to 1$ as $n \to \infty$.
9. (a) For Example A3, show that $R \sim \Gamma(n, \theta/n)$. (Show first that $U_i = X_i/in \sim \Gamma(1, \theta/n)$.)
   (b) Find ER^{-1}. (See Theorem 4–26.)
   (c) Find an unbiased estimator of θ^{-1}.
   (d) Find the MSE of cR^{-1} as an estimator of θ^{-1}, and find the c that minimizes this MSE.
10. (a) For Example B, find the MSE of the estimator cS^2 as an estimator of σ^2. Verify the formulas given for the MSE for S^2 and $\hat{\sigma}^2 = (n-1)S^2/n$.
    (b) Show that the choice of c that minimizes the preceding MSE is $(n-1)/(n+1)$.
11. Let $U = S^2$ in Example B.
    (a) How is U distributed?
    (b) Verify the formula in the text for $ES = EU^{1/2}$. (See Theorem 4–26.)
    (c) Find the MSE for the estimator cS as an estimator of σ.
    (d) Find the value of c that minimizes the preceding MSE.
12. Verify that $MSE_T(\boldsymbol{\theta}) = \text{var}_{\boldsymbol{\theta}}(T) + (\text{bias}_{\boldsymbol{\theta}}(T))^2$.

Exercises—C

1. Use Jensen's inequality to show that if $h(u)$ is a strictly convex function and $\hat{\theta}$ is an unbiased estimator of θ which is not degenerate, then $h(\hat{\theta})$ is not an unbiased estimator of $h(\theta)$.

7.2.2 Consistent Estimators

We continue studying the problem of estimating the one-dimensional function $\tau(\boldsymbol{\theta})$. We consider the problem as n, the number of observations, goes to infinity. Suppose that for each n, we have an estimator T_n of $\tau(\boldsymbol{\theta})$. We would hope that as n goes to infinity, the estimator T_n would get close to $\tau(\boldsymbol{\theta})$. We say that a sequence of estimators T_n is a *consistent* sequence of estimators for $\tau(\boldsymbol{\theta})$ if T_n converges in probability to $\tau(\boldsymbol{\theta})$.

The following theorem gives an easy way to establish consistency for many models.

Theorem 7–2

a. If $\text{bias}(T_n)$ and $\text{var}(T_n)$ go to zero, then T_n is a consistent sequence of estimators.
b. If T_n is a consistent sequence for $\tau(\boldsymbol{\theta})$ and $h(t)$ is a continuous function, then $h(T_n)$ is a consistent sequence for $h(\tau(\boldsymbol{\theta}))$.

Proof. These are just Theorems 6–7 and 6–8a. □

Since the MSE of an estimator is its variance plus its bias squared, an equivalent statement for part a of Theorem 7–2 is that if the MSE for T_n goes to zero, then T_n is a consistent sequence. Recall that the converse of part a of the theorem is not true.

In the following examples and the remainder of the book, we use the shorter expression "consistent estimator" for "consistent sequence of estimators." We also suppress the subscript n in order to simplify the notation.

Example A1

Suppose we observe $X_i \sim P(i\theta)$, independent. In the last section, it was shown that P is an unbiased estimator of θ and $\text{var}(P) = \theta/a_n$. Therefore, P is consistent, since $a_n \to \infty$. By part b of Theorem 7–2, P^2 is a consistent estimator of θ^2. Consider now the unbiased estimator $T = P^2 - P/a_n$ of θ^2. We have

$$T - P^2 = \frac{-P}{a_n}, \qquad P \xrightarrow{P} \theta, \qquad -a_n^{-1} \to 0.$$

By Corollaries 6–7 and 6–8c, $-P/a_n$ converges in probability to zero. Hence, T is also a consistent estimator of θ^2, by Theorem 6–8a.

Example A2

In the exercises it is stated that for this model, the MLE Q is a consistent estimator of θ and that both the MLE and the unbiased estimators of θ^2 are consistent.

Example A3

For this model, it is stated that the MLE R is a consistent estimator of θ and that both the MLE and the unbiased estimator of θ^2 are consistent.

Example B

For this model, $E\overline{X} = \mu$ and $\text{var}(\overline{X}) = \sigma^2/n \to 0$, so that $\overline{X}$ is a consistent estimator of μ. Since $(n-1)S^2/\sigma^2 \sim \chi^2_{n-1}$, $ES^2 = \sigma^2$ and $\text{var}(S^2) = 2\sigma^4/(n-1) \to 0$. Therefore, S^2 is a consistent estimator of σ^2. (In Chapter 6, it is shown that $\overline{X}$ and S^2 are consistent estimators of μ and σ^2, respectively, for samples from any distribution with finite mean μ and finite variance σ^2.) Now, the MLE of σ^2 is $\hat{\sigma}^2 = (n-1)S^2/n = S^2 - S^2/n$. But $S^2/n \xrightarrow{P} 0$; therefore, $\hat{\sigma}^2$ is also a consistent estimator of σ^2. By part b of Theorem 7–2, S and $\hat{\sigma}$ are both consistent estimators of σ. Finally, by Theorem 6–8b, $\overline{X}S$ and $\overline{X}/S$ are consistent estimators of $\mu\sigma$ and μ/σ.

The foregoing examples indicate an interesting fact about MLEs: Typically, they are consistent. It is not possible to make such a strong statement about unbiased estimators, since there are often many unbiased estimators for a particular function. In Example B, X_1, the first observation, is also an unbiased estimator of μ, but X_1 is not consistent. When we use the expression "the" unbiased estimator in these examples, we mean the unbiased estimator which is based on the MLE. In Chapter 10, we shall see that for many models there is only one such estimator. Typically, this estimator is also consistent. In fact, for most models, any reasonable estimator is at least consistent.

Exercises—B

1. For Example A2, show that Q is a consistent estimator of θ as n approaches infinity and that both the MLE and the unbiased estimators of θ^2 are consistent estimators of θ^2.
2. For Example A3, show that R is a consistent estimator of θ as n approaches infinity and that both the MLE and the unbiased estimators of θ^2 are consistent estimators of θ^2.
3. For the model in Exercise B4, Section 7.1.2, show that the MLE of θ is consistent.
4. For Example C of Section 7.1.1, show that the MLEs and unbiased estimators of θ, σ^2, and $\theta\sigma^2$ are consistent estimators.
5. For the model in Exercise B5, Section 7.1.2, show that the MLE of θ is consistent. (*Hint*: First show that $\overline{X}$ is a consistent estimator of $\mu = (1-\theta)/\theta$.)
6. For the model in Exercise B8, Section 7.1.2, show that the MLE of θ is consistent.

Exercises—C

1. For the model in Exercise C1, Section 7.1.2, show that the MLE of θ is consistent.
2. Let $X_1, \ldots, X_n$ be independently distributed with the common Cauchy density function $f(x;\theta) = \pi^{-1}(1+(x-\theta)^2)^{-1}$.
   (a) Show that this distribution is symmetric about θ.
   (b) Is the sample median a consistent estimator of θ? (See Section 6.4.)
   (c) Is the sample mean a consistent estimator of θ? (See Exercise C3 of Section 6.2.1.)

* 7.2.3 Fisher Information and Efficient Estimators

In this section, we return to the problem of estimating $\tau(\theta)$ with a fixed sample size n. We assume that both τ and θ are one-dimensional. (In previous sections we only assumed that τ was one-dimensional.) Extensions to vector $\boldsymbol{\theta}$ are given in Section 7.3.2.

Careful statements and derivations of the results in this and the following sections in this chapter require mathematics beyond the scope of the text. Therefore, the theorems tacitly begin with "if certain regularity conditions are satisfied," and the derivations are called arguments instead of proofs. Suitable regularity conditions and proofs are given in Lehmann (1983), pp. 115–130 and 409–435. In particular, as will be seen shortly, it is necessary to pull derivatives with respect to θ inside integrals. Therefore, the results are not valid for random variables whose ranges depend on θ, such as uniform random variables. As long as the range of the random variable does not depend on θ, we can and will assume without loss of generality that $f(\mathbf{x};\theta) > 0$ for all $\mathbf{x} \in \chi$.

Let $\mathbf{X}$ be a random vector with likelihood function $L_X(\theta)$. Define

$$S = S(\mathbf{X};\theta) = \frac{\partial}{\partial\theta}\log[L_X(\theta)] \quad \text{and} \quad I(\theta) = \text{var}[S(\mathbf{X};\theta)].$$

$S(\mathbf{X};\theta)$ is called the *score function*. $I(\theta)$ is called the *Fisher information* in $\mathbf{X}$ and is often used to measure the "information" in the observations.

Theorem 7–3

a. $ES(\mathbf{X};\theta) = 0$.

b. $I(\theta) = E(S(\mathbf{X};\theta))^2 = E\left[-\left(\frac{\partial}{\partial\theta}\right) S(\mathbf{X};\theta)\right] = E\left[-\left(\frac{\partial^2}{\partial\theta^2}\right) \log(L_X(\theta))\right].$

c. Let $\mathbf{X} = (X_1, \ldots, X_n)'$, where the X_i are independent. Let $I_i(\theta)$ be the information in X_i. Then

$$I(\theta) = \sum_{i=1}^{n} I_i(\theta).$$

If, in addition, the X_i are identically distributed, then

$$I(\theta) = nI_1(\theta).$$

Argument. The arguments are given in the continuous case. Arguments for the discrete case are similar. Note first that

$$\frac{\partial}{\partial\theta}\log[f(\mathbf{x};\theta)] = \frac{\frac{\partial}{\partial\theta} f(\mathbf{x};\theta)}{f(\mathbf{x};\theta)},$$

or equivalently,

$$\frac{\partial}{\partial\theta} f(\mathbf{x};\theta) = \left(\frac{\partial}{\partial\theta}\log[f(\mathbf{x};\theta)]\right) f(\mathbf{x};\theta).$$

a. Using the preceding result, we see that

$$ES(\mathbf{X};\theta) = \int_{\mathcal{X}} \left[\frac{\partial}{\partial\theta}\log(f(\mathbf{x};\theta))\right] f(\mathbf{x};\theta)d\mathbf{x} = \int_{\mathcal{X}} \frac{\partial}{\partial\theta} f(\mathbf{x};\theta)d\mathbf{x}$$
$$= \frac{\partial}{\partial\theta}\int_{\mathcal{X}} f(\mathbf{x};\theta)d\mathbf{x} = \frac{\partial}{\partial\theta}1 = 0.$$

b. The first equality in part b follows from part a and the last one from the definition of $S(\mathbf{X};\theta)$. To obtain the second equality, we note that by part a,

$$0 = ES(\mathbf{X};\theta) = \int_{\mathcal{X}} \left(\frac{\partial}{\partial\theta}\log[f(\mathbf{x};\theta)]\right) f(\mathbf{x};\theta)d\mathbf{x}.$$

Differentiating both sides with respect to θ, we obtain, from the product rule for derivatives,

$$0 = \int_{\mathcal{X}} \frac{\partial}{\partial\theta}\left[\frac{\partial}{\partial\theta}\log(f(\mathbf{x};\theta)) f(\mathbf{x};\theta)\right] d\mathbf{x}$$
$$= \int_{\mathcal{X}} \left[\frac{\partial^2}{\partial\theta^2}\log(f(\mathbf{x};\theta))\right] f(\mathbf{x};\theta)d\mathbf{x} + \int_{\mathcal{X}} \left[\frac{\partial}{\partial\theta}\log(f(\mathbf{x};\theta))\right]^2 f(\mathbf{x};\theta)d\mathbf{x}$$
$$= E\frac{\partial}{\partial\theta}S(\mathbf{X};\theta) + E(S(\mathbf{X};\theta))^2,$$

and the desired result follows.

c. This part is left as an exercise. □

The next theorem gives the main result of this section.

Theorem 7–4. (Information inequality) Let $T(\mathbf{X})$ be an unbiased estimator of $\tau(\theta)$. Then

$$\operatorname{var}(T(\mathbf{X})) \geqq \frac{(\tau'(\theta))^2}{I(\theta)},$$

with equality if and only if

$$L_X(\theta) \equiv \exp(h(\theta)T(\mathbf{X}) + k(\theta) + u(\mathbf{X}))$$

for some functions $h(\theta)$, $k(\theta)$, and $u(\mathbf{X})$.

Argument. Since $T(\mathbf{X})$ is unbiased,

$$\tau(\theta) = ET(\mathbf{X}) = \int_X T(\mathbf{X}) f(\mathbf{x};\theta)d\mathbf{x}.$$

Differentiating both sides of this expression, we obtain

$$\tau'(\theta) = \int_X T(\mathbf{x})\left[\frac{\partial}{\partial\theta} f(\mathbf{x};\theta)\right]d\mathbf{x} = \int_X T(\mathbf{x})\left(\frac{\partial}{\partial\theta}\log[f(\mathbf{x};\theta)]\right) f(\mathbf{x};\theta)d\mathbf{x}$$
$$= ET(\mathbf{X})S(\mathbf{X};\theta).$$

However, by part a of Theorem 7–3, $ES(\mathbf{X};\theta) = 0$, so that $\operatorname{cov}(T(\mathbf{X}), S(\mathbf{X};\theta)) = \tau'(\theta)$. Therefore, by the Cauchy-Schwarz inequality,

$$(\tau'(\theta))^2 \leqq \operatorname{var}(T(\mathbf{X}))\ \operatorname{var}(S(\mathbf{X};\theta)) = \operatorname{var}(T(\mathbf{X}))I(\theta),$$

and the desired inequality follows. The inequality is an equality if and only if $S(\mathbf{X};\theta) \equiv h'(\theta)T(\mathbf{X}) + k'(\theta)$ for some functions h' and k'. Now, $S(\mathbf{X};\theta) = (\partial/\partial\theta)\log[L_X(\theta)]$. Therefore, the equality holds if and only if

$$\log[L_X(\theta)] \equiv h(\theta)T(\mathbf{X}) + k(\theta) + u(\mathbf{X})$$

for some functions h, k, and u, and the desired result follows. □

The lower bound $(\tau'(\theta))^2/I(\theta)$ for the variance of an unbiased estimator of $\tau(\theta)$ is called the *information bound* for the variance. (It is also called the *Frechet-Cramer-Rao bound.*)

If an unbiased estimator $T(\mathbf{X})$ for $\tau(\theta)$ has variance equal to the information bound, we say that $T(\mathbf{X})$ is an *efficient estimator* of $\tau(\theta)$. By Theorem 7–4, there can be at most one efficient estimator of any parameter.

Corollary 7–4. If $T(\mathbf{X})$ is an efficient unbiased estimator of $\tau(\theta)$, then it is the best unbiased estimator of $\tau(\theta)$.

Unfortunately, Corollary 7–4 is useful for finding the best unbiased estimator only when $L_X(\theta)$ has the form stated in Theorem 7–4, and even when it has that form, efficient unbiased estimators exist only for functions $\tau(\theta) = c\tau_0(\theta)$ for a single function $\tau_0(\theta)$. Therefore, the corollary has limited usefulness for finding the best unbiased estimators. A broader method for finding such estimators is given in Chapter 10.

We now consider Example A again.

Example A1

Suppose we observe X_i independent, with $X_i \sim P(i\theta)$. From Section 7.1.2, we have

$$S(\mathbf{X};\theta) = -a_n + \frac{\sum X_i}{\theta}.$$

Therefore,

$$I(\theta) = \text{var}\left(-a_n + \frac{\sum x_i}{\theta}\right) = \theta^{-2}\sum \text{var}(X_i) = \frac{a_n}{\theta}.$$

Hence, the information bound for the variance of unbiased estimators of $\tau(\theta) = \theta$ is

$$\frac{1}{I(\theta)} = \frac{\theta}{a_n} = \text{var}(P).$$

Hence, P is an efficient unbiased estimator of θ and is the best unbiased estimator of θ. The information lower bound for the variance of an unbiased estimator of $\tau(\theta) = \theta^2$ is

$$\frac{(2\theta)^2}{I(\theta)} = \frac{4\theta^3}{a_n}.$$

In Chapter 10, we show that there is no efficient estimator for θ^2, but that there is a best unbiased estimator.

Examples A2 and A3

In the exercises, it is shown that Q is an efficient unbiased estimator for Example A2 and that R is an efficient unbiased estimator for Example A3.

In all three of the preceding examples, we see that the MLE is an efficient unbiased estimator of θ. This does not happen in all models. However, we can prove that if any estimator of $\tau(\theta)$ is efficient, then it must be the MLE.

Theorem 7–5. Let $T(\mathbf{X})$ be an efficient unbiased estimator of $\tau(\theta)$. Then $T(\mathbf{X})$ is the MLE of $\tau(\theta)$.

Argument. Suppose that $T(\mathbf{X})$ is an efficient unbiased estimator of $\tau(\theta)$. Then $S(\mathbf{X};\theta) = h'(\theta)T(\mathbf{X}) + k'(\theta)$, by Theorem 7–4. But by part a of Theorem 7–3,

$$0 = ES(\mathbf{X};\theta) = h'(\theta)ET(\mathbf{X}) + k'(\theta) = h'(\theta)\tau(\theta) + k'(\theta),$$

and hence,

$$\tau(\theta) = \frac{-k'(\theta)}{h'(\theta)}.$$

Now, as we have seen in Section 7.1.2, the MLE $\hat{\theta}$ can be found by setting $S(\mathbf{X};\theta) = 0$, and therefore, $\hat{\theta}$ satisfies

$$0 = S(\mathbf{X};\hat{\theta}) = h'(\hat{\theta})T(\mathbf{X}) + k'(\hat{\theta}).$$

By the invariance principle for MLEs, the MLE of $\tau(\theta) = -h'(\theta)/k'(\theta)$ is $-k'(\hat{\theta})/h'(\hat{\theta}) = T(\mathbf{X})$. □

Exercises—B

1. For Example A2,
   (a) Show that Q is an efficient unbiased estimator of θ.
   (b) Find the lower bound for an unbiased estimator of θ^2.
2. For Example A3,
   (a) Show that R is an efficient unbiased estimator of θ.
   (b) Find the lower bound for an unbiased estimator of θ^{-1}.
   (c) Find an unbiased estimator of θ^{-1}. ($R \sim \Gamma(n, \theta/n)$.)
   (d) Is this estimator efficient?
3. For the model in Exercise B8, Section 7.1.2, is the MLE an efficient unbiased estimator of θ?
4. (a) For the model in Exercise B5, Section 7.1.2, find the lower bound for the variance of an unbiased estimator of $\mu = (1 - \theta)/\theta$.
   (b) Is $\overline{X}$ an efficient unbiased estimator of μ?
5. Prove part c of Theorem 7–3.

Exercises—C

1. Consider a model with a univariate parameter θ. Let $\delta = h(\theta)$ be a reparametrization of the model with inverse $\theta = k(\delta)$. Suppose that h is differentiable. Let $I(\theta)$ be the Fisher information when θ is the parameter, and let $I^*(\delta)$ be the Fisher information when δ is the parameter.
   a. Let $S^*(\mathbf{X}, \delta)$ and $S(\mathbf{X}, \theta)$ be the score functions for δ and θ, respectively. Show that $S^*(\mathbf{X}, \delta) = S(\mathbf{X}, k(\delta))k'(\delta)$. (Recall that $L_X^*(\delta) = L_X(k(\delta))$, where $L_X^*(\delta)$ and $L_X(\theta)$ are the likelihoods of θ and δ, respectively.
   b. Show that $I^*(\delta) = (k'(\delta))^2 I(k(\delta))$.
   c. Let $\tau(\theta)$ be a function, and let $\tau^*(\delta) = \tau(k(\delta))$. Show that $(\tau'(k(\delta)))^2/I(k(\delta)) = (\tau^{*\prime}(\delta))^2/I^*(\delta)$.

   (This result implies that the information lower bound is unaffected by reparametrization.)

7.3 ASYMPTOTIC PROPERTIES OF MAXIMUM LIKELIHOOD ESTIMATORS

* 7.3.1 Asymptotic Properties of Maximum Likelihood Estimators

We continue studying the problem of estimating $\tau(\theta)$ where both τ and θ are one-dimensional and as n, the number of observations, goes to infinity. Let $\hat{\theta}_n$ and $\hat{\tau}_n = \tau(\hat{\theta}_n)$ be the MLEs of θ and $\tau = \tau(\theta)$ based on n observations, and let $I_n(\theta)$ be the Fisher information based on n observations. For many models, when $I_n(\theta)$ goes to infinity, then

$$(I_n(\theta))^{1/2}(\hat{\tau}_n - \tau) \xrightarrow{d} V \sim N(0, (\tau'(\theta))^2). \tag{7–1}$$

Equation (7–1) implies that $\hat{\tau}_n$ is a consistent estimator of τ. (See Exercise B3.) It also implies that, asymptotically,

$$\hat{\tau}_n \mathrel{\dot\sim} N\left(\tau, \frac{(\tau'(\theta))^2}{I_n(\theta)}\right)$$

(where $\dot\sim$ means "is approximately distributed as"). Since the mean of the asymptotic distribution is τ, we say that $\hat{\tau}_n$ is *asymptotically unbiased*, and since the variance of the asymptotic distribution is equal to the information lower bound, we say that $\hat{\tau}_n$ is *asymptotically efficient*. Since, in addition, the asymptotic distribution is normal, we say that τ_n is *best asymptotically normal*.. Equation (7–1) indicates that, asymptotically, the MLE does as well as an estimator can.

Before deriving Equation (7–1) in a particular case, we look again at Example A.

Example A1

Since the X_i are independent, with $X_i \sim P(i\theta)$, it follows from Theorem 4–22 that $Y_n = \Sigma X_i \sim P(\theta a_n)$. Also, since $a_n \to \infty$, by the normal approximation for the Poisson distribution, we have

$$(I_n(\theta))^{1/2}(P - \theta) = \frac{Y_n - \theta a_n}{(\theta a_n)^{1/2}} \xrightarrow{d} Z \sim N(0, 1),$$

and therefore, Equation (7–1) obtains.

Example A2

For this model, it is shown in the exercises that

$$(I_n(\theta))^{1/2}(Q - \theta) \sim N(0, 1)$$

exactly for all n.

Example A3

It is also shown in the exercises that this model satisfies Equation (7–1).

We see, therefore, that for all three models, Equation (7–1) is satisfied as long as $I_n(\theta) \to \infty$.

We now derive Equation (7–1) in the particular case where the observations are independently identically distributed (i.i.d.), and hence, $I_n(\theta) = nI(\theta)$, where $I(\theta) = I_1(\theta)$ is the Fisher information in one observation.

Theorem 7–6. Let $X_1, X_2, \ldots$ be i.i.d. observations with common Fisher information $I(\theta)$. Let $\hat{\theta}_n$ be the MLE of θ based only on $X_1, \ldots, X_n$. Under fairly general conditions of regularity, $\hat{\theta}_n$ is a consistent estimator of θ, and

$$(nI(\theta))^{1/2}(\hat{\theta}_n - \theta) \xrightarrow{d} Z \sim N(0, 1),$$

or equivalently,

$$n^{1/2}(\hat{\theta}_n - \theta) \xrightarrow{d} Y \sim N(0, [I(\theta)]^{-1}).$$

Argument. Let X_i have common density function $g(x; \theta)$, and let

$$h(x, \theta) = \frac{\partial}{\partial\theta} \log[g(x, \theta)] \quad \text{and} \quad k(x, \theta) = \frac{\partial}{\partial\theta} h(x, \theta).$$

In Section 7.1.2., we saw that the MLE is typically computed by setting the derivative with respect to θ of

$$\log[L_X(\theta)] = \log\left[\prod_{i=1}^{n} g(X_i, \theta)\right] = \sum_{i=1}^{n} \log[g(X_i, \theta)]$$

equal to zero. That is,

$$0 = \sum_{i=1}^{n} \frac{\partial}{\partial\theta} \log[g(X_i, \hat{\theta}_n)] = \sum_{i=1}^{n} h(X_i, \hat{\theta}_n).$$

By a Taylor series approximation of $h(X_i, \hat{\theta}_n)$ as a function of $\hat{\theta}_n$ about θ, (considering X_i as a constant), we have

$$h(X_i, \hat{\theta}_n) = h(X_i, \theta) + (\hat{\theta}_n - \theta)[k(X_i, \theta) + R(X_i, \hat{\theta}_n)],$$

where

$$\lim_{\hat{\theta}_n \to \theta} R(X_i, \hat{\theta}_n) = R(X_i, \theta) = 0. \tag{7–2}$$

Therefore,

$$0 = \sum_{i=1}^{n} \frac{h(X_i, \hat{\theta}_n)}{n} = \overline{U}_n + (\hat{\theta}_n - \theta)(\overline{V}_n + \overline{R}_n),$$

where

$$U_i = h(X_i; \theta), \qquad V_i = k(X_i; \theta), \qquad R_{ni} = R(X_i, \hat{\theta}_n).$$

($\overline{U}_n$ and $\overline{V}_n$ are the averages of the first n U_i's and V_i's, respectively, and $\overline{R}_n$ is the average of the R_{ni}.) Hence,

$$(\hat{\theta}_n - \theta) = \frac{-\overline{U}_n}{\overline{V}_n + \overline{R}_n}.$$

Now, the U_i are i.i.d., $EU_i = 0$ (see part a of Theorem 7–3), and $\text{var}(U_i) = I(\theta)$ (by the definition of Fisher information). Therefore, by the central limit theorem,

$$n^{1/2}\,\overline{U}_n \xrightarrow{d} W \sim N(0, I(\theta)).$$

Now, the V_i are also i.i.d., and $EV_i = -I(\theta)$ (by part b of Theorem 7–3). Therefore, by the weak law of large numbers,

$$\overline{V}_n \xrightarrow{P} -I(\theta).$$

It can be shown that $\overline{R}_n \xrightarrow{P} 0$. (This is not surprising in view of Equation (7–2).) Therefore, $\overline{V}_n + \overline{R}_n \xrightarrow{P} -I(\theta)$. By Slutzky's theorem,

$$n^{1/2}(\hat{\theta}_n - \theta) = \frac{-n^{1/2}\,\overline{U}_n}{\overline{V}_n + \overline{R}_n} \xrightarrow{d} Y = (I(\theta))^{-1} W \sim N(0, [I(\theta)]^{-1}).$$

The consistency of $\hat{\theta}_n$ then follows from Slutzky's theorem because

$$(\hat{\theta}_n - \theta) = n^{-1/2}(n^{1/2}(\hat{\theta}_n - \theta)) \xrightarrow{d} 0U = 0.$$

Hence, $(\hat{\theta}_n - \theta) \xrightarrow{P} 0$. □

We now return to the more general problem of estimating $\tau(\theta)$.

Corollary 7–6. Let $\hat{\tau}_n = \tau(\hat{\theta}_n)$ be the MLE of $\tau(\theta)$ based on $X_1, \ldots, X_n$. Under the conditions of Theorem 7–6, if τ is a differentiable function, then $\hat{\tau}_n$ is a consistent estimator of τ, and

$$n^{1/2}(\hat{\tau}_n - \tau) \xrightarrow{d} W \sim N\left(0, \frac{(\tau'(\theta))^2}{I(\theta)}\right).$$

Proof. See Exercise B4. □

In the preceding theorems, we have assumed that the X_i are random variables. However, a careful reading of the proof shows that it works just as well if the $\mathbf{X}_i$ are i.i.d. random vectors.

Exercises—B

1. For Example A2, verify that $(I_n(\theta))^{1/2}(Q - \theta) \sim N(0, 1)$.
2. For Example A3, verify that $(I_n(\theta))^{1/2}(R - \theta) \xrightarrow{d} Z \sim N(0, 1)$. (*Hint*: $nR \sim \Gamma(n, \theta)$. Why? Use the normal approximation to the gamma distribution.)
3. Show that Equation (7–1) implies that $\hat{\tau}_n$ is a consistent estimator of τ.
4. Verify Corollary 7–6. (Use the Cramer δ theorem.)

*# 7.3.2 Extension to the Multiparameter Case

Let us now consider the case in which $\boldsymbol{\theta}$ is a vector, as in Example B. To state the results in this section, it is necessary to use matrices.

Let $\mathbf{X}$ be a random vector with likelihood $L_X(\boldsymbol{\theta})$, where $\boldsymbol{\theta}$ is a p-dimensional vector. The *score function* $S(\mathbf{X};\boldsymbol{\theta})$ is the p-dimensonal vector defined by

$$S(\mathbf{X};\boldsymbol{\theta}) = \nabla(\log[L_X(\boldsymbol{\theta})]).$$

That is, $S(\mathbf{X},\boldsymbol{\theta})$ has ith component

$$S_i(\mathbf{X};\boldsymbol{\theta}) = \frac{\partial}{\partial\theta_i}\log[L_X(\boldsymbol{\theta})].$$

The *Fisher information matrix* $I(\boldsymbol{\theta})$ is the covariance matrix of the score function, i.e.,

$$I(\boldsymbol{\theta}) = \text{cov}(S(\mathbf{X};\theta)).$$

We can show that

$$ES(\mathbf{X};\boldsymbol{\theta}) = 0 \quad \text{and} \quad I(\boldsymbol{\theta}) = -EH(\mathbf{X};\boldsymbol{\theta}),$$

where $H(\mathbf{X};\boldsymbol{\theta})$ is the Hessian matrix for $\log[L_X(\boldsymbol{\theta})]$. That is, $H(\mathbf{X};\boldsymbol{\theta})$ has (i,j)th element

$$h_{ij} = \frac{\partial^2}{\partial\theta_i\,\partial\theta_j}\log[L_X(\boldsymbol{\theta})].$$

(See the exercises.) These definitions and results are natural generalizations of results for the univariate case.

Now, let $\tau = \tau(\boldsymbol{\theta})$ be a univariate parameter, and let $\nabla\tau(\boldsymbol{\theta})$ be the gradient of $\tau(\boldsymbol{\theta})$, i.e.,

$$\nabla\tau(\boldsymbol{\theta}) = \left(\frac{\partial}{\partial\theta_1}\tau(\boldsymbol{\theta}), \ldots, \frac{\partial}{\partial\theta_p}\tau(\boldsymbol{\theta})\right)'.$$

Theorem 7–7. (Information inequality) Let $T(\mathbf{X})$ be an unbiased estimator of the univariate parameter $\tau(\boldsymbol{\theta})$. Under general, regularity conditions,

$$\operatorname{var}(T(\mathbf{X})) \geqq [\nabla\tau(\boldsymbol{\theta})]'[I(\boldsymbol{\theta})]^{-1}[\nabla\tau(\boldsymbol{\theta})].$$

Proof. See Exercise C1. □

If an unbiased estimator $T(\mathbf{X})$ has variance equal to this lower bound, then we say that it is an *efficient estimator*, and it must be the best unbiased estimator.

We next give the extension of Theorem 7–6 to the case of vector $\boldsymbol{\theta}$.

Theorem 7–8. Let $\mathbf{X}_1, \mathbf{X}_2, \ldots$ be i.i.d. observations, each with Fisher information matrix $I(\boldsymbol{\theta})$. Let $\hat{\boldsymbol{\theta}}_n$ be the MLE of the p-dimensional vector $\boldsymbol{\theta}$ based on $\mathbf{X}_1, \ldots, \mathbf{X}_n$, and let $\hat{\tau}_n = \tau(\hat{\boldsymbol{\theta}}_n)$ be the maximum likelihood estimator of the univariate parameter $\tau = \tau(\boldsymbol{\theta})$. Then under general, regularity conditions,

a. $\hat{\boldsymbol{\theta}}_n$ is a consistent estimator of $\boldsymbol{\theta}$.

b.

$$n^{1/2}(\hat{\boldsymbol{\theta}}_n - \boldsymbol{\theta}) \xrightarrow{d} \mathbf{Y} \sim N_p(\mathbf{0}, (I(\boldsymbol{\theta}))^{-1}).$$

c.

$$n^{1/2}(\hat{\tau} - \tau) \xrightarrow{d} W \sim N(0, \nabla\tau(\boldsymbol{\theta})'(I(\boldsymbol{\theta}))^{-1}\nabla\tau(\boldsymbol{\theta})).$$

Heuristic Proof. See Exercise C2. □

Thus, even for the more general situation, the MLE is consistent, asymptotically unbiased, asymptotically efficient, and asymptotically normal. Hence, asymptotically, MLEs are doing as well as is possible even for the case of vector $\boldsymbol{\theta}$.

Example B

Let $X_i \sim N(\mu, \sigma^2)$, independent, so that $\theta = (\mu, \sigma)'$. For this model,

$$\frac{\partial^2}{\partial\mu^2} L_X(\mu,\sigma) = \frac{-n}{\sigma^2}, \qquad -E\left(\frac{-n}{\sigma^2}\right) = \frac{n}{\sigma^2},$$

$$\frac{\partial^2}{\partial\mu\,\partial\sigma} L_X(\mu,\sigma) = \frac{-2\sum(X_i - \mu)}{\sigma^3}, \qquad -E\left(\frac{-2\sum(X_i - \mu)}{\sigma^3}\right) = 0,$$

$$\frac{\partial^2}{\partial\sigma^2} L_X(\mu,\sigma) = \frac{(n/\sigma^2) - 3\sum(X_i - \mu)^2}{\sigma^4}, \qquad -E\left(\frac{(n/\sigma^2 - 3\sum(X_i - \mu)^2}{\sigma^4}\right) = \frac{2n}{\sigma^2}.$$

Therefore, the Fisher information matrix for the model is

$$I(\mu,\sigma) = \left(\frac{n}{\sigma^2}\right)\begin{Bmatrix} 1 & 0 \\ 0 & 2 \end{Bmatrix}, \qquad (I(\mu,\sigma))^{-1} = \left(\frac{\sigma^2}{2n}\right)\begin{Bmatrix} 2 & 0 \\ 0 & 1 \end{Bmatrix}.$$

Now, let $\tau(\mu,\sigma) = \mu$. Then $\nabla\tau(\mu,\sigma) = (1,0)'$. Therefore, the information bound for an unbiased estimator of μ is $(\nabla\tau)'(\mathbf{I}^{-1})(\nabla\tau) = \sigma^2/n$. Now, $\overline{X}$ is an unbiased estimator of μ, and $\text{var}(\overline{X}) = \sigma^2/n$, so that $\overline{X}$ is an efficient estimator of μ and, hence, the best unbiased estimator of μ. Now, let $\tau^*(\mu,\sigma) = \sigma^2$. Then $\nabla\tau^*(\mu,\sigma) = (0,2\sigma)'$. Therefore, the lower bound for the variance of an unbiased estimator of σ^2 is $2\sigma^4/n$. Now,

$$S^2 \sim \Gamma\left(\frac{n-1}{2}, \frac{2\sigma^2}{n-1}\right), \qquad ES^2 = \sigma^2, \qquad \text{var}(S^2) = \frac{2\sigma^4}{n-1} > \frac{2\sigma^4}{n},$$

so that S^2 is not an efficient estimator of σ^2 (although we shall see in Chapter 10 that it is the best unbiased estimator of σ^2). Finally, let $\tilde{\tau}(\mu,\sigma) = \mu\sigma^2$. Then $\nabla\tilde{\tau}(\mu,\sigma) = (\sigma^2, 2\mu\sigma)'$. Therefore, the lower bound for the variance of an unbiased estimator of $\mu\sigma^2$ is $\sigma^4(\sigma^2 + 2\mu^2)/n$. Now, $\overline{X}S^2$ is an unbiased estimator of $\mu\sigma^2$, but, from the exercises

$$\text{var}(\overline{X}S^2) = \frac{2\sigma^4}{n(n-1)} + \frac{2\mu^2\sigma^2}{n-1} + \frac{\sigma^6}{n} > \frac{\sigma^4(\sigma^2+2\mu^2)}{n}.$$

Therefore, $\overline{X}S^2$ is not an efficient estimator of $\mu\sigma^2$.

Exercises—B

1. For Example B,
   (a) Verify the formula for the lower bound for unbiased estimators of $\mu\sigma^2$.
   (b) Verify the formula for $\text{var}(\overline{X}S^2)$. ($E(\overline{X}S^2)^2 = EX^2\,EU^2, U = S^2$.)
2. For Example B again,
   (a) Find an unbiased estimator of μ/σ^2. ($E(\overline{X}/S^2) = E\overline{X}EU^{-1}, U = S^2$.)
   (b) Find the variance of this estimator.
   (c) Find the lower bound for the variance of an unbiased estimator of μ/σ^2.
3. Consider the model of Example C of Section 7.1.1. Let $\boldsymbol{\theta} = (\theta,\sigma)'$.
   (a) Find the Fisher information matrix $I(\theta,\sigma)$.
   (b) Find the lower bound for the variance of unbiased estimators of θ. Is Q efficient?
   (c) Find the lower bound for the variance of an unbiased estimator of σ^2. Is T^2 efficient?
   (d) Find the lower bound for the variance of unbiased estimators of $\theta\sigma^2$. Is QT^2 efficient?

Exercises—C

1. Let $\boldsymbol{\theta}$ be a p-dimensional vector.
   (a) Show that $ES(\mathbf{X},\boldsymbol{\theta}) = \mathbf{0}$ (i.e., that $E(\partial/\partial\theta_i)\log[L_X(\boldsymbol{\theta})] = 0$).
   (b) Show that $I(\boldsymbol{\theta}) = -EH(\mathbf{X};\boldsymbol{\theta})$, where $H(\mathbf{X};\boldsymbol{\theta})$ is the Hessian matrix of $\log[L_X(\boldsymbol{\theta})]$.
   (c) Let $T(\mathbf{X})$ be an unbiased estimator of the univariate function $\tau(\boldsymbol{\theta})$. Show that $\nabla\tau(\boldsymbol{\theta}) = ET(\mathbf{X})S(\mathbf{X},\boldsymbol{\theta})$ (i.e., that $(\partial/\partial\theta_i)(\tau(\boldsymbol{\theta})) = ET(X)S_i(\mathbf{X};\boldsymbol{\theta})$).
   (d) Let $U(\mathbf{X},\boldsymbol{\theta}) = [\nabla\tau(\boldsymbol{\theta})]'[I(\boldsymbol{\theta})]^{-1}S(\mathbf{X},\boldsymbol{\theta})$. Note that U is univariate. Show that $\text{var}(U(\mathbf{X},\boldsymbol{\theta})) = [\nabla\tau(\boldsymbol{\theta})]'[I(\boldsymbol{\theta})]^{-1}[\nabla\tau(\boldsymbol{\theta})]$. (If $\boldsymbol{\Sigma} = \text{cov}(\mathbf{Y})$, then $\text{var}(\mathbf{a}'\mathbf{Y}) = \mathbf{a}'\boldsymbol{\Sigma}\mathbf{a}$.)
   (e) Show that $\text{cov}(U(\mathbf{X},\boldsymbol{\theta}), T(\mathbf{X})) = [\nabla\tau(\boldsymbol{\theta})]'[I(\boldsymbol{\theta})]^{-1}[\nabla\tau(\boldsymbol{\theta})]$. (*Hint*: $\text{cov}(U,T) = [\nabla\tau(\boldsymbol{\theta})]' \times [I(\boldsymbol{\theta})]^{-1}E[S(\mathbf{X},\boldsymbol{\theta})T(\mathbf{X})]$. Why?)
   (f) Show that $\text{var}(T(\mathbf{X})) \geqq [\nabla\tau(\boldsymbol{\theta})]'[I(\boldsymbol{\theta})]^{-1}[\nabla\tau(\boldsymbol{\theta})]$.
2. Using the notation of Theorem 7–8, let $f(\mathbf{x};\boldsymbol{\theta})$ be the common (marginal) density function of the X_i. Let $h(\mathbf{x};\boldsymbol{\theta})$ and $k(\mathbf{x};\boldsymbol{\theta})$ be the gradient and Hessian matrices, respectively, for $\log f(\mathbf{x};\boldsymbol{\theta})$. Let $\mathbf{U}_i = h(\mathbf{X}_i;\boldsymbol{\theta})$ and $\mathbf{V}_i = k(\mathbf{X}_i;\boldsymbol{\theta})$. (Note that $\mathbf{U}_i$ is $p \times 1$ and $\mathbf{V}_i$ is $p \times p$. Exercise C1 implies that $E\mathbf{U}_i = 0$, $\text{cov}(\mathbf{U}_i) = I(\boldsymbol{\theta})$, and $E\mathbf{V}_i = -I(\boldsymbol{\theta})$.)

(a) Show that $\overline{\mathbf{U}}_n = -(\overline{\mathbf{V}}_n + \overline{\mathbf{R}}_n)(\hat{\boldsymbol{\theta}}_n - \boldsymbol{\theta})$, where $R_{ni}(\mathbf{X},\hat{\boldsymbol{\theta}}_n) = R(X_i,\hat{\boldsymbol{\theta}}_n)$, and that $R(X_i,\hat{\boldsymbol{\theta}}_n) \rightarrow \mathbf{0}$ as $\hat{\boldsymbol{\theta}}_n \rightarrow \boldsymbol{\theta}$. (Note that R_{ni} is $p \times p$.)

(b) Show that $n^{1/2}\overline{\mathbf{U}}_n \xrightarrow{d} N_p(0, I(\boldsymbol{\theta}))$ and $\overline{\mathbf{V}}_n \xrightarrow{P} -I(\boldsymbol{\theta})$.

(c) Argue that $n^{1/2}(\hat{\boldsymbol{\theta}}_n - \boldsymbol{\theta}) \xrightarrow{d} \mathbf{Y} \sim N_p(0, (I(\boldsymbol{\theta}))^{-1})$. (Note that $\hat{\boldsymbol{\theta}}_n - \boldsymbol{\theta} = -(\overline{\mathbf{V}}_n + \mathbf{R}_n)^{-1}\overline{\mathbf{U}}_n$.)

(d) Show that $\hat{\boldsymbol{\theta}}_n$ is a consistent estimator of $\boldsymbol{\theta}$.

(e) Let $\tau(\boldsymbol{\theta})$ be a univariate differentiable function, and let $\nabla\tau(\boldsymbol{\theta})$ be the gradient of $\tau(\boldsymbol{\theta})$. Let $\hat{\tau}_n$ be the MLE of τ. Argue that $n^{1/2}(\hat{\tau}_n - \tau) \xrightarrow{d} W \sim N(0, [\nabla\tau(\boldsymbol{\theta})]'[I(\boldsymbol{\theta})]^{-1}[\nabla\tau(\boldsymbol{\theta})])$.

Chapter 8

CONFIDENCE INTERVALS AND TESTS

In this short chapter, we introduce two very important concepts: confidence intervals and tests.

In the last chapter, we developed methods for finding sensible point estimators for a parameter τ. However, in practice, we typically want an interval estimator for τ, that is, an interval

$$L(\mathbf{X}) \leqq \tau \leqq R(\mathbf{X})$$

in which we are fairly confident that τ lies. Since $L = L(\mathbf{X})$ and $R = R(\mathbf{X})$ are random variables, we often use the shorter notation

$$L \leqq \tau \leqq R.$$

(L stands for "left end," R for "right end.") If $P(L(\mathbf{X}) \leqq \tau \leqq R(\mathbf{X})) = 1 - \alpha$, we say that the interval $L(\mathbf{X}) \leqq \tau \leqq R(\mathbf{X})$ is a $(1 - \alpha)$ *confidence interval* for τ. Typically, α is chosen to be .05, so that we find a .95 confidence interval for τ. In that case, we often say that the interval is a 95% confidence interval for τ.

Another very important problem is *hypothesis testing*. The problem occurs when we have collected some data and want to know whether the data can be considered statistical disproof of a scientific hypothesis. For example, suppose we want to prove that the mean for joggers' pulses is lower than 72, the mean pulse for all men. In order to test this hypothesis, we collect some data $\mathbf{X}$, develop a statistical model (e.g., $X_i \sim N(\mu, \sigma^2)$, independent), and formulate a null hypothesis $\mu = 72$. On the basis of the data we must then decide whether to accept or reject the hypothesis that $\mu = 72$. Perhaps the most obvious approach is to reject the hypothesis $\mu = 72$ whenever $\bar{X} \neq 72$. However, even if $\mu = 72$, $P(\bar{X} \neq 72) = 1$, so that this approach does not work.

Instead, we want a rule which tells how far $\bar{X}$ must be from 72 in order to reject the null hypothesis. We call such a rule a *test*. For example, we might reject the null hypothesis whenever $|\bar{X} - 72| > 2S$, where S is the sample standard deviation of the X_i). If $P(|\bar{X} - 72| > 2S) = \alpha$ when $\mu = 72$, we say that this test is a size-α test. In general, a test has *size* α if it has probability α of rejecting the null hypothesis when the null hypothesis is true. Typically, α is chosen to be .05. If we reject the null hypothesis that $\mu = 72$ in this problem with a size-.05 test, we often say that μ is significantly different from 72 with a size-.05 test. In Section 8.1, we give a fairly simple way to find $1-\alpha$ confidence intervals and size-α tests for many models.

In Section 8.2, we present a method for assessing the sensitivity of confidence intervals and tests to the assumptions under which they were derived, at least asymptotically. We then use these results to indicate that the procedures designed for drawing inferences about means of normal distributions are not too sensitive to the normal assumption under which they were derived, but that procedures for variances are quite sensitive to that assumption. We give an example in which a confidence interval which has confidence coefficient .95 if the normal model is correct, but has asymptotic confidence coefficient .67 for a particular nonnormal model. Hence, the confidence intervals and test derived for σ^2 in a normal distribution should be used only when we believe that the observations are exactly normally distributed, which seems an unlikely occurrence.

Later on in the section, we show that procedures are even more sensitive to the independence assumption. In fact, we show that if we compute a 95% confidence interval for the mean for the normal model with independence, then this interval is an asymptotic 0% confidence interval as long as there is any positive correlation between the observations. Similarly, the supposed size-.05 tests are either size-1 tests or size-.5 tests asymptotically.

In this chapter, we shall only be concerned with finding procedures which have low probability α of rejecting a true hypothesis. Typically, however, there are many such procedures. In the next chapter, we discuss methods for choosing procedures which also have small probability of accepting a false hypothesis.

8.1 $(1-\alpha)$ CONFIDENCE INTERVALS AND SIZE-α TESTS

8.1.1 Confidence Intervals and Pivotal Quantities

Often, rather than finding a point estimator of a univariate parameter $\tau = \tau(\boldsymbol{\theta})$, we would like to find an interval in which we feel "confident" that τ lies. We say that the random interval

$$L(\mathbf{X}) \leqq \tau \leqq R(\mathbf{X})$$

is a $(1-\alpha)$ *confidence interval* for τ if

$$P(L(\mathbf{X}) \leqq \tau \leqq R(\mathbf{X})) = 1-\alpha.$$

(Note that the interval depends on the observations in $\mathbf{X}$ and is random. τ is a fixed, but unknown, constant.) We call $1-\alpha$ the *confidence coefficient* for the $1-\alpha$ confidence interval.

Example B

Let X_i be independent, with $X_i \sim N(\mu, \sigma^2)$. Suppose we want a confidence interval for $\tau(\mu, \sigma) = \mu$. We have shown that

$$t = \frac{n^{1/2}(\overline{X}-\mu)}{S} \sim t_{n-1}.$$

Recall that t_k^a is the number such that

$$P(t > t_{n-1}^a) = a$$

and that the distribution of t is symmetric, so that

$$t_k^{1-a} = -t_k^a.$$

(t_k^a for certain a can be found in Table 2 at the back of the book.) Therefore,

$$1 - \alpha = P\left(-t_{n-1}^{\alpha/2} \leqq \frac{n^{1/2}(\overline{X} - \mu)}{S} \leqq t_{n-1}^{\alpha/2}\right)$$

$$= P(\overline{X} - t_{n-1}^{\alpha/2} S n^{-1/2} \leqq \mu \leqq \overline{X} + t_{n-1}^{\alpha/2} S n^{-1/2}),$$

since the two events in the probabilities are the same. Hence, the interval

$$\overline{X} - t_{n-1}^{\alpha/2} S n^{-1/2} \leqq \mu \leqq \overline{X} + t_{n-1}^{\alpha/2} S n^{-1/2}$$

is a $(1 - \alpha)$ confidence interval for α. We often write this interval as

$$\mu \in \overline{X} \pm t_{n-1}^{\alpha/2} S n^{-1/2}.$$

Now, suppose that we want a $(1 - \alpha)$ confidence interval for $\tau(\mu, \sigma) = \sigma^2$. We have shown that

$$U = \frac{(n-1)S^2}{\sigma^2} \sim \chi_{n-1}^2.$$

Recall that $\chi_{n-1}^{2\ a}$ is the number such that

$$P(U > \chi_{n-1}^{2\ a}) = a,$$

and the χ^2 distribution is *not* symmetric. ($\chi_{n-1}^{2\ a}$ for certain a can be found in Table 3 at the back of the book.) Now,

$$1 - \alpha = P\left(\chi_{n-1}^{2\ 1-\alpha/2} \leqq \frac{(n-1)S^2}{\sigma^2} \leqq \chi_{n-1}^{2\ \alpha/2}\right)$$

$$= P\left(\frac{(n-1)S^2}{\chi_{n-1}^{2\ \alpha/2}} \leqq \sigma^2 \leqq \frac{(n-1)S^2}{\chi_{n-1}^{2\ 1-\alpha/2}}\right),$$

since the events are again the same. Therefore, the interval

$$\frac{(n-1)S^2}{\chi_{n-1}^{2\ \alpha/2}} \leqq \sigma^2 \leqq \frac{(n-1)S^2}{\chi_{n-1}^{2\ 1-\alpha/2}}$$

is a $(1 - \alpha)$ confidence interval for σ^2.

In most sciences and social sciences, α is chosen to be .05, and .95 confidence intervals or 95% confidence intervals are used. In this example, therefore,

$$\mu \in \overline{X} \pm t_{n-1}^{.025} S n^{-1/2}$$

and

$$\frac{(n-1)S^2}{\chi_{n-1}^{2\ .025}} \leqq \sigma^2 \leqq \frac{(n-1)S^2}{\chi_{n-1}^{2\ .975}}$$

are 95% confidence intervals for μ and σ^2. In the joggers' pulse example, $n = 30$, $\overline{X} = 55$, and $S^2 = 225$. Suppose we choose $\alpha = .05$. Then $t_{29}^{.025} = 2.045$. Therefore, a 95% confidence interval for μ is

$$\mu \in 55 \pm 2.045\left(\frac{225}{30}\right)^{1/2}, \quad \text{or} \quad \mu \in 55 \pm 5.6.$$

Also, $\chi_{29}^{2\ .975} = 16$ and $\chi_{29}^{2\ .025} = 45.7$. Therefore, a 95% confidence interval for σ^2 is

$$\frac{(29)(225)}{45.7} \leqq \sigma^2 \leqq \frac{(29)(225)}{16}, \quad \text{or} \quad 142.78 \leqq \sigma^2 \leqq 407.8.$$

A 95% confidence interval for σ is

$$(142.78)^{1/2} \leqq \sigma \leqq (407.8)^{1/2}, \quad \text{or} \quad 11.95 \leqq \sigma \leqq 20.19.$$

Let us consider the interpretation of confidence intervals. Either the parameter (for example, μ) is in the 95% confidence interval, or it is not. In a particular experiment, we, of course, do not know whether it is in the interval or out of it. However, if we compute 95% confidence intervals many times, about 95% of them will be correct.

As a practical example, we might imagine a chemistry class in which the professor has prepared a chemical for the class to analyze. He knows the true amount μ of some particular chemical in his samples, because he prepared them. He gives the samples to the students to analyze. Each student does the experiment and computes his or her own 95% confidence interval. In this situation, μ is fixed. The confidence intervals are random and are different for each student. However, about 95% of the intervals should contain the true amount μ, although the students will not know whose intervals are correct. (However, the teacher will.)

We now discuss a general method for finding $(1 - \alpha)$ confidence intervals for $\tau = \tau(\boldsymbol{\theta})$. A *pivotal quantity* for τ is is a function $h(\mathbf{X}; \tau)$ whose distribution does not depend on $\boldsymbol{\theta}$ (i.e., the distribution is completely known). The function h is allowed to depend on the observations $\mathbf{X}$, but depends on $\boldsymbol{\theta}$ only through $\tau(\boldsymbol{\theta})$. If τ takes on a continuum of values, then h must have a continuous distribution, which we henceforth assume. (In fact, pivotal quantities are not possible for most discrete models.)

In Example B, $\overline{X}$ and S^2 are both functions of $\mathbf{X}$, and the distribution of

$$t(\mathbf{X}, \mu) = \frac{n^{1/2}(\overline{X} - \mu)}{S}$$

is completely known. Therefore, $t(\mathbf{X}, \mu)$ is a pivotal quantity for μ. (Note that the distribution of $n^{1/2}(\overline{X} - \mu)/\sigma$ does not depend on any unknown parameters, but is not a pivotal quantity for μ because it depends on σ. Similarly, although $n^{1/2}(\overline{X} - \mu)$ does not depend on σ, its distribution does, so it is not a pivotal quantity either.) Also,

$$U(\mathbf{X}, \sigma) = \frac{(n-1)S^2}{\sigma^2}$$

is a pivotal quantity for σ^2.

The next theorem shows how to use a pivotal quantity for τ to find a confidence interval for τ. It simply formalizes the method used to find confidence intervals for μ and σ^2.

Theorem 8–1. Let $h(\mathbf{X}, \tau)$ be a pivotal quantity for τ, which is a continuous random variable. Let h^a be the number such that $P(h(\mathbf{X}, \tau) > h^a) = a$, and let $\beta + \gamma = \alpha$. Suppose that

$$h^{1-\gamma} \leqq h(\mathbf{X}, \tau) \leqq h^{\beta} \Leftrightarrow L(\mathbf{X}) \leqq \tau \leqq R(\mathbf{X})$$

for some statistics $L(\mathbf{X})$ and $R(\mathbf{X})$. Then $L(\mathbf{X}) \leqq \tau \leqq R(\mathbf{X})$ is a $(1-\alpha)$ confidence interval for τ.

Proof. Since $h(\mathbf{X}, \tau)$ has a continuous distribution, $P(h(\mathbf{X}, \tau) < h^{1-\gamma}) = \gamma$. Hence,

$$\begin{aligned} P(L(\mathbf{X}) \leqq \tau \leqq R(\mathbf{X})) &= P(h^{1-\gamma} \leqq h(\mathbf{X}, \tau) \leqq h^{\beta}) \\ &= P(h(\mathbf{X}, \tau) \leqq h^{\beta}) - P(h(\mathbf{X}, \tau) < h^{1-\gamma}) = 1 - \beta - \gamma = 1 - \alpha. \end{aligned}$$

□

In applying Theorem 8–1, we typically take $\beta = \gamma = \alpha/2$. We say that the $(1-\alpha)$ confidence interval constructed in this way is *based* on the pivotal quantity $h(\mathbf{X}, \tau)$. If $\beta = \gamma = \alpha/2$, we call the confidence interval the *equal tails* $(1-\alpha)$ confidence interval based on $h(\mathbf{X}, \tau)$.

Let us next consider Example A. No pivotal quantity exists for Example A1.

Example A2

Let $X_i \sim N(i\theta, 1)$, with X_i independent. Then $Q \sim N(\theta, b_n^{-1})$. Therefore,

$$Z = (Q - \theta)b_n^{1/2} \sim N(0, 1).$$

Since the distribution of Z is completely specified, Z is a pivotal quantity for θ for this model. Furthermore, because of the symmetry of Z's distribution,

$$z^{1-a} = -z^{a}.$$

(We can find z^a from Table 1 or the last row of Table 2 at the back of the book.) Also,

$$-z^{\alpha/2} \leq (Q - \theta)(b_n)^{1/2} \leq z^{\alpha/2} \Leftrightarrow \theta \in Q + z^{\alpha/2}(b_n)^{-1/2}.$$

Therefore, the interval

$$\theta \in Q \pm z^{\alpha/2}(b_n)^{-1/2}$$

is a $(1-\alpha)$ confidence interval for θ. For example, suppose that $n = 10$ (so that $b_n = 385$), $Q = 8$, and $\alpha = .10$ (so that $z^{.05} = 1.645$). Then we get the following 90% confidence interval for θ:

$$\theta \in 8 \pm 1.645(385)^{-1/2}, \quad \text{or} \quad \theta \in 8 \pm .084.$$

Example A3

Let $X_i \sim E(i\theta)$, with X_i independent. Then

$$U = \frac{2nR}{\theta} \sim \chi^2_{2n}.$$

(See Exercise B2.) Therefore, U is a pivotal quantity for θ, and a $(1-\alpha)$ confidence interval for θ is

$$\frac{2nR}{\chi^{2\ \alpha/2}_{2n}} \leqq \theta \leqq \frac{2nR}{\chi^{2\ 1-\alpha/2}_{2n}}.$$

For example, if $n = 10$, $\alpha = .05$ (so that $\chi^{2\ .025}_{20} = 34.2$ and $\chi^{2\ .975}_{20} = 9.59$), and $R = 12$, we get the 95% confidence interval

$$\frac{20(12)}{34.2} \leqq \theta \leqq \frac{20(12)}{9.59}, \quad \text{or} \quad 7.02 \leqq \theta \leqq 25.03.$$

For many estimation problems involving discrete random variables, and even for some involving continuous ones, no pivotal quantities exist, and we cannot find confidence intervals. However, we can often find asymptotic pivotal quantities and confidence intervals. Let $L_n(\mathbf{X})$ and $R_n(\mathbf{X})$ be sequences of statistics. Then

$$L_n(\mathbf{X}) \leqq \tau \leqq R_n(\mathbf{X})$$

is an *asymptotic* $(1-\alpha)$ *confidence interval* for τ if

$$\lim_{n\to\infty} P(L_n(\mathbf{X}) \leqq \tau \leqq R_n(\mathbf{X})) = 1-\alpha.$$

An *asymptotic pivotal quantity* for τ is a sequence of functions $h_n(\mathbf{X},\tau)$ of the observations and τ whose limiting distribution does not depend on $\boldsymbol{\theta}$. (To distinguish asymptotic pivotal quantities from just plain pivotal quantities, we sometimes use the term *exact* pivotal quantity for the latter. And similarly, we sometimes use the term *exact* $(1-\alpha)$ confidence interval for what we have been calling a $(1-\alpha)$ confidence interval.)

If we have an asymptotic pivotal quantity, we can find an asymptotic confidence interval by means of the next theorem.

Theorem 8–2. Let $h_n(\mathbf{X},\tau)$ be an asymptotic pivotal quantity for τ. Suppose that $h_n(\mathbf{X},\tau)\xrightarrow{d}H$, a continuous random variable. Let h^a be the number such that $P(H > h^a) = a$, and let $\beta+\gamma=\alpha$. Finally, suppose that

$$h^{1-\gamma} \leqq h_n(\mathbf{X},\tau) \leqq h^{\beta} \Leftrightarrow L_n(\mathbf{X}) \leqq \tau \leqq R_n(\mathbf{X}).$$

Then $L_n(\mathbf{X}) \leqq \tau \leqq R_n(\mathbf{X})$ is an asymptotic $(1-\alpha)$ confidence interval for τ.

Proof. Since $h_n(\mathbf{X},\tau)\xrightarrow{d}H$,

$$\lim_{n\to\infty} P(L_n(\mathbf{X}) \leqq \tau \leqq R_n(\mathbf{X})) = \lim_{n\to\infty} P(h^{1-\gamma} \leqq h_n(\mathbf{X},\tau) \leqq h^{\beta})$$

$$= P(h^{1-\gamma} \leqq H \leqq h^{\beta}) = 1-\beta-\gamma = 1-\alpha. \qquad \square$$

We now return to Example A1.

Example A1

Suppose we observe $X_i \sim P(i\theta)$, independent. Then

$$a_n P = \sum_{i=1}^{n} X_i \sim P(a_n\theta).$$

Since $a_n \to \infty$ as $n \to \infty$, it follows by the normal approximation to the Poisson distribution that

$$Z_n = (P-\theta)\left(\frac{a_n}{\theta}\right)^{1/2} \xrightarrow{d} Z \sim N(0,1),$$

so that Z_n is an asymptotic pivotal quantity for θ. Unfortunately, it is somewhat difficult to solve $-z^{\alpha/2} \leq Z_n \leq z^{\alpha/2}$ for θ. (See Exercise C3.) We accordingly find an asymptotic pivotal quantity which is easier to use. We have seen that P is a consistent estimator of θ, and, therefore,

$$\left(\frac{P}{\theta}\right)^{1/2} \xrightarrow{P} 1$$

(since $k(U) = (U/\theta)^{1/2}$ is a continuous function). By Slutzky's theorem,

$$Z_n^* = (P - \theta)\left(\frac{a_n}{P}\right)^{1/2} = \frac{(P-\theta)(a_n/\theta)^{1/2}}{(P/\theta)^{1/2}} \xrightarrow{d} Z \sim N(0,1),$$

so that Z_n^* is also an asymptotic pivotal quantity. Now,

$$-z^{\alpha/2} \le Z_n^* \le z^{\alpha/2} \Leftrightarrow \theta \in P \pm z^{\alpha/2}\left(\frac{P}{a_n}\right)^{1/2},$$

so that an asymptotic $(1-\alpha)$ confidence interval for θ is

$$\theta \in P \pm z^{\alpha/2}\left(\frac{P}{a_n}\right)^{1/2}.$$

For example, if $n = 20$ (so that $a_n = 210$), $\alpha = .05$ (so that $z^{.025} = 1.96$), and $P = 30$, we get the asymptotic 95% confidence interval

$$\theta \in 30 \pm 1.96\left(\frac{30}{210}\right)^{1/2}, \quad \text{or} \quad \theta \in 30 \pm .74.$$

Note that for Example A1, we have found an asymptotic confidence interval for θ which is centered at P; for Example A2, we have found a confidence interval for θ which is centered at Q; and for Example A3, we have used R to construct the confidence interval for θ (although the interval is not symmetric about R because the χ^2 distribution is not symmetric). That is, in all three cases, we have used the MLE for θ as the basis for the confidence interval. In most problems we start with the MLE in constructing the pivotal quantity and confidence interval, so that the confidence interval is compatible with the point estimator.

Exercises—B

1. **(a)** For Example A2, show that $b_n^{1/2}(Q - \theta) \sim N(0,1)$.
   (b) If $X_1 = 10$, $X_2 = 18$, $X_3 = 33$, $X_4 = 44$, and $X_5 = 40$, find a 95% confidence interval for θ.
2. **(a)** For Example A3, show that $2nR/\theta \sim \chi^2_{2n}$.
   (b) For the data in part b of Problem 1, find a 98% confidence interval for θ.
3. **(a)** For the example in Exercise B4, Section 7.1.2, show that $2a_n P/\theta \sim \chi^2_{2a_n}$.
   (b) Find a 90% confidence interval for θ with the data in part b of Problem 1.
4. Let $X_1, \ldots, X_n$ be independent, with $X_i \sim P(\theta)$.
   (a) Show that $n^{1/2}(\bar{X} - \theta)/\bar{X}^{1/2} \xrightarrow{d} Z \sim N(0,1)$. (*Hint*: By the central limit theorem, $n^{1/2}(\bar{X} - \theta)/\theta^{1/2} \xrightarrow{d} Z$, and by the weak law of large numbers, $\bar{X} \xrightarrow{P} \theta$. Use Slutzky's theorem.)
   (b) Find a $(1-\alpha)$ asymptotic confidence interval for θ. Evaluate this confidence interval when $1 - \alpha = .95$, $\bar{X} = 81$, and $n = 100$.
5. For the model in Example C, in Section 7.1.1,
   (a) Find a pivotal quantity and a $(1-\alpha)$ confidence interval for σ^2.
   (b) Show that $b_n(Q - \theta)/T \sim t_{n-1}$ is a pivotal quantity for θ.
   (c) Find a $(1-\alpha)$ confidence interval for θ.
   (d) Evaluate the preceding confidence intervals for the data given in part b of Exercise 1 when $1 - \alpha = .90$.
6. Consider the model for Exercise B8, Section 7.1.2.
   (a) Find an asymptotic pivotal quantity for θ. (See Theorem 5–4.)

(b) Find an asymptotic .95 confidence interval for θ when $\hat{\theta} = .6$, $n = 100$.

7. Consider the model in Exercise B6, Section 7.1.2.
   (a) Find the moment-generating function of the MLE for θ. (*Hint*: $E\exp(t\log(X_i)) = EX_i^t$.)
   (b) Show that $2n\hat{\theta}/\theta \sim \chi^2_{2n}$.
   (c) Find a .95 confidence interval for θ when $\hat{\theta} = 3$, $n = 10$.

Exercises—C

1. Let $X_1, \ldots, X_n$ be independent random variables having the Cauchy density function $f(x;\theta) = \pi^{-1}(1 + (x - \theta)^2)^{-1}$. Let $\tilde{X}_n$ be the sample median of the X_i.
   (a) Show that θ is the median of the X_i.
   (b) Use Theorem 6–6 to find an asymptotic pivotal quantity for θ.
   (c) Find a $(1 - \alpha)$ asymptotic confidence interval for θ.
2. Let $X_1, \ldots, X_n$ be independent, with X_i geometrically distributed with parameter θ. Let $\hat{\theta} = (1 + \overline{X})^{-1}$ be the MLE of θ.
   (a) Show that $n^{1/2}(\hat{\theta} - \theta)/(\theta(1-\theta)^{1/2}) \xrightarrow{d} Z \sim N(0, 1)$. (*Hint*: By CLT, $n^{1/2}(\overline{X} - (1-\theta)/\theta) \xrightarrow{d} (1-\theta)^{1/2}Z/\theta$. Use the Cramer δ theorem.)
   (b) Show that $\hat{\theta}$ is a consistent estimator of θ. (Show first that $\overline{X}$ is a consistent estimator of $(1 - \theta)/\theta$.)
   (c) Show that $n^{1/2}(\hat{\theta} - \theta)/(\hat{\theta}(1 - \hat{\theta})^{1/2}) \xrightarrow{d} Z \sim N(0, 1)$.
   (d) Find an asymptotic $(1 - \alpha)$ confidence interval for θ.
   (e) Find an asymptotic .95 confidence interval for θ when $\hat{\theta} = .2$ and $n = 100$.
3. For Example A1, find an asymptotic $(1 - \alpha)$ confidence interval based on $Z_n = (P - \theta)(a_n/\theta)^{1/2}$.

8.1.2 Testing Statistical Hypotheses and Test Statistics

In this section, we discuss what we mean when we say that some data provide "statistical proof" of some hypothesis (e.g., that heavy smoking is associated with a higher risk of cancer). In order to make such a statement, we must first give a statistical model for the data $\mathbf{X} \in \chi$, in which $\mathbf{X}$ has known density $f(\mathbf{x}; \boldsymbol{\theta})$, $\boldsymbol{\theta} \in \Omega$, where $\boldsymbol{\theta}$ is again an unknown parameter. A statistical *hypothesis* is a statement about $\boldsymbol{\theta}$. A testing problem consists of a statistical model together with two hypotheses, the *null hypothesis* H_0 that $\boldsymbol{\theta} \in N$, and the *alternative hypothesis* H_A that $\boldsymbol{\theta} \in A$. We call N and A the *null* and *alternative sets,* respectively, and assume that they are disjoint. Often, the null hypothesis H_0 has the form

$$H_0\colon \tau(\boldsymbol{\theta}) = c$$

for some one-dimensional function τ and some known c, and the alternative hypothesis has one of three forms:

$$H_A\colon \tau(\boldsymbol{\theta}) > c, \qquad H_A\colon \tau(\boldsymbol{\theta}) < c, \quad \text{or} \quad H_A\colon \tau(\boldsymbol{\theta}) \neq c.$$

The first two alternative hypotheses are called *one-sided* alternatives, and the third is called the *two-sided* alternative. In setting up a testing problem, we establish the null hypothesis as what we want to disprove. Typically, as we shall see, there is no way to prove the null hypothesis, because no amount of data could prove that $\tau(\boldsymbol{\theta})$ exactly equaled c. However, it is often possible to prove that $\tau(\boldsymbol{\theta})$ is not c.

A (nonrandomized) *test* is a rule which, for each $\mathbf{X} \in \chi$, decides whether to *accept* the null hypothesis H_0 or *reject* the null hypothesis H_0 in favor of the alternative hypothesis H_A. The set C of $\mathbf{X}$ for which the test rejects H_0 is called the *critical region* for the test. The *size* α of the test with critical region C is

$$\alpha = \sup_{\theta \in N} P_\theta(C).$$

That is, the size of a test is the maximum of the probability of falsely rejecting the null hypothesis. Typically, this probability is constant when $\boldsymbol{\theta} \in N$, so that it is rarely necessary to find a maximum. If we reject the hypothesis $\tau(\boldsymbol{\theta}) = c$ in favor of either a one-sided or two-sided alternative, we often say that $\tau(\boldsymbol{\theta})$ is *significantly different* from c.

In doing a statistical test, we set up a small α (often $\alpha = .05$) and find a size-α test for testing that $\boldsymbol{\theta} \in N$ against $\boldsymbol{\theta} \in A$. If we reject the null hypothesis, we consider it statistical proof that the null hypothesis is false, since the probability of our falsely rejecting it is less than or equal to α. However, if we accept the null hypothesis, we do not consider it proof that the null hypothesis is true, since we have not controlled the probability of falsely accepting the null hypothesis. For example, if we accept the null hypothesis with a test with $\alpha = .05$, but would have rejected it with $\alpha = .10$, we have data which indicate that the null hypothesis is false, but are not quite strong enough evidence to prove that it is. For this reason, many authors use the phrase "fail to reject" instead of the term "accept."

Let us return to the 30 male joggers whose pulse rates were measured, giving a sample mean of 55 and a sample standard deviation of 15. We wish to know whether the mean pulse rate for joggers is less than 72, the mean pulse rate for all males. For a statistical model, we assume again that the X_i, the observations on the 30 joggers, are independent, with $X_i \sim N(\mu, \sigma^2)$. As in the previous chapter, $\mathbf{X} = (X_1, \ldots, X_{30})$ and $\boldsymbol{\theta} = (\mu, \sigma)$. On the basis of these data, we wish to decide whether μ is less than 72 or not. That is, we want to test the null hypothesis H_0: $\mu = 72$ against the one-sided alternative H_A: $\mu < 72$.

It should be apparent that no amount of evidence could ever convince us that $\mu = 72$ (as opposed to $\mu = 71.99$ or 71.9999 or . . .). However, it is quite possible that evidence could convince us that $\mu < 72$. It is not enough, however, that $\overline{X} < 72$. Even if $\mu = 72$, $\overline{X}$ would be less than 72 half the time. (Note that $P(\overline{X} = 72) = 0$, since $\overline{X}$ is a continuous random variable.) The question is, How low must $\overline{X}$ be before we are convinced that $\mu < 72$?

Now, consider the test which rejects the null hypothesis if

$$t(\overline{X}, S) = \frac{n^{1/2}(\overline{X} - 72)}{S} < -1.699 \qquad \left(\text{i.e., } \overline{X} < 72 - \frac{1.699S}{n^{1/2}}\right).$$

When $\mu = 72$, $t(\overline{X}, S) \sim t_{29}$, and therefore, from Table 2 at the back of the text, we see that the size α of this procedure is

$$\alpha = P(t(\overline{X}, S) < -1.699 | \mu = 72) = .05.$$

We have, therefore, found a .05 test for this problem. Note that $t(55, 15) = -6.2$, so we would reject the null hypothesis that $\mu = 72$ in favor of the alternative hypothesis that $\mu < 72$ with these data. This would, therefore, be considered statistical proof that the mean pulse rate μ for these male joggers is less than 72, the mean for all men.

We now make several important statements about this testing problem and others.

1. Our conclusions are only as good as our model. If, for example, the data are not normally distributed, then the foregoing inference may be suspect. (However, in Section 8.2.1, we shall show, for reasonable samples, that the size of the preceding procedure is not too sensitive to the assumption of a normal distribution.)
2. Our inference is valid only for the population from which we have sampled. For example, if the men turned out to be weight lifters instead of joggers, then we would not have proved anything about joggers' pulse rates.
3. If we do size-.05 tests all our lives, we will incorrectly reject the null hypothesis about 5% of the time. If this seems too high an error rate, then we should do .01 or .001 tests.
4. The size of the test is computed before we look at the data. Therefore, the test must be selected before we look at the data. (For example, we cannot change from a one-sided test to a two-sided test after looking at the data.)

Note that the test which rejects the null hypothesis if $t(\overline{X}, S) > 1.699$ also has size .05 for the foregoing situation. However, it would be a rather silly test, since it would reject $\mu = 72$ in favor of $\mu < 72$ when $\overline{X}$ is much bigger than 72. For this reason, it is important to choose a "sensible" size-α test for problems.

We now return to the general problem of testing that $\boldsymbol{\theta} \in N$ against $\boldsymbol{\theta} \in A$. A statistic $T(\mathbf{X})$ is a *test statistic* if the distribution of $T(\mathbf{X})$ under the null hypothesis does not depend on any unknown parameters. In the previous example, the distribution of $t(\overline{X}, S)$ does not depend on any unknown parameters (σ) when $\mu = 72$, so $t(\overline{X}, S)$ is a test statistic. Let $T(\mathbf{X})$ be a test statistic, and let T^{α} be the upper α point of the null distribution of T. Then three size-α tests for testing that $\boldsymbol{\theta} \in N$ are:

1. Reject the null hypothesis if $T > T^{\alpha}$.
2. Reject the null hypothesis if $T < T^{1-\alpha}$.
3. Reject the null hypothesis if $T > T^{\alpha/2}$ or $T < T^{1-\alpha/2}$.

Typically, the first two tests are sensible for (different) one-sided alternatives, while the third one is sensible for two-sided alternatives. However, the shape of the critical region (i.e., which of the three tests to use) depends on the hypotheses and the choice of T.

A particularly nice situation occurs when we are testing $\tau(\boldsymbol{\theta}) = c$ against a one-sided or two-sided alternative and there is a pivotal quantity $h(\mathbf{X}, \tau)$ for τ.

Theorem 8–3. Let $h(\mathbf{X}, \tau)$ be a pivotal quantity for τ, and let $T(\mathbf{X}) = h(\mathbf{X}, c)$. Then $T(\mathbf{X})$ is a test statistic for testing $\tau(\boldsymbol{\theta}) = c$ against any alternative.

Proof. If $h(\mathbf{X}, \tau)$ is a pivotal quantity, then its distribution does not depend on any unknown parameters. Therefore, the distribution of $h(\mathbf{X}, c)$ does not depend on any unknown parameters under the null hypothesis that $\tau = c$. □

Although the pivotal quantity $h(\mathbf{X}, \tau)$ is not a statistic, because it depends on the unknown parameter τ, $h(\mathbf{X}, c)$ is a statistic, since the value c is known.

Typically, the pivotal quantity is a decreasing function of τ. In such a situation, if we want to test that $\tau(\boldsymbol{\theta}) = c$ against $\tau(\boldsymbol{\theta}) > c$, we would reject the null hypothesis when $T(\mathbf{X}) = h(\mathbf{X}, c)$ is too large. If $\tau(\boldsymbol{\theta}) > c$, then $h(\mathbf{X}, \tau) < h(\mathbf{X}, c) = T(\mathbf{X})$, and hence $T(\mathbf{X})$ would be larger than we would expect $h(\mathbf{X}, \tau)$ to be. A similar argument could be applied to designing sensible tests for the other one-sided alternative and the two-sided alternative. This discussion leads to the following method for designing size-α tests for problems in which we have a pivotal quantity of the correct form:

Heuristic Rule for Sensible Tests. Let $h(\mathbf{X}, \tau)$ be a pivotal quantity for τ which is decreasing in τ. Let $T(\mathbf{X}) = h(\mathbf{X}, c)$, and let T^{α} be the upper α point of the distribution of $h(\mathbf{X}, \tau)$.

a. A sensible size-α test for testing that $\tau(\boldsymbol{\theta}) = c$ against $\tau(\boldsymbol{\theta}) > c$ rejects the null hypothesis if $T(\mathbf{X}) > T^{\alpha}$.
b. A sensible size-α test for testing $\tau(\boldsymbol{\theta}) = c$ against $\tau(\boldsymbol{\theta}) < c$ rejects the null hypothesis if $T(\mathbf{X}) < T^{1-\alpha}$.
c. A sensible size-α test for testing $\tau(\boldsymbol{\theta}) = c$ against $\tau(\boldsymbol{\theta}) \neq c$ rejects the null hypothesis if $T(\mathbf{X}) > T^{\beta}$ or $T(\mathbf{X}) < T^{1-\gamma}$, where $\beta + \gamma = \alpha$.

We call the tests in a and b the one-sided tests *based on* the pivotal quantity $h(\mathbf{X}, \tau)$. The test in c is called a two-sided test *based on* $h(\mathbf{X}, \tau)$. If $\beta = \gamma = \alpha/2$, the two-sided test is an *equal tails* test based on $h(\mathbf{X}, \tau)$. We call the numbers T^{α} and $T^{1-\alpha}$ the *critical values* for the tests given in a and b. The *critical values* for the test given in c are T^{β} and T^{γ}.

We shall use the rule for sensible tests often in this and later sections for designing tests for models with pivotal quantities. In later chapters, we shall often show or state that tests designed in this way have certain attractive properties. However, it is possible to design models for which the rule leads to inappropriate procedures. (Occasionally, we shall find pivotal quantities which are increasing functions of τ. In that case, the inequalities in a and b are just reversed.)

We now apply this rule to design sensible size-α tests for the examples discussed earlier.

Example A2

In this example, we observe X_i independent, with $X_i \sim N(i\theta, 1)$. From Section 8.1.1,

$$(Q - \theta)b_n^{1/2} \sim N(0, 1)$$

is a pivotal quantity which is decreasing in θ. Therefore, a test statistic for testing $\theta = c$ against one-sided or two-sided alternatives is

$$Z = (Q - c)b_n^{1/2} \sim N(0, 1)$$

under the null hypothesis. To test that $\theta = c$ against $\theta > c$, we reject the null hypothesis if $Z > z^{\alpha}$; to test that $\theta = c$ against $\theta < c$, we reject the null hypothesis if $Z < -z^{\alpha}$ $(= z^{1-\alpha})$; and to test that $\theta = c$ against $\theta \neq c$, we reject the null hypothesis if $Z > z^{\alpha/2}$ or $Z < -z^{\alpha/2}$. For example, to test that $\theta = 2$ against $\theta < 2$ with a .05 test when $n = 5$ (so that $b_n = 55$), we reject the null hypothesis if

$$Z < -z^{.05} = -1.645.$$

If we observe $Q = 1$, then

$$Z = (1 - 2)(55)^{1/2} = -7.416,$$

so we reject the null hypothesis that $\theta = 2$ in favor of $\theta < 2$ with these data.

Example A3

Here, we observe X_i independently exponentially distributed with mean $i\theta$. In the previous section, it was shown that in this case $2nR/\theta \sim \chi^2_{2n}$ is a pivotal quantity for θ. Since this quantity is also a decreasing function of θ, let

$$U = \frac{2nR}{c}.$$

Then a sensible size-α test that $\theta = c$ against $\theta > c$ would reject the null hypothesis if $U > \chi^{2\ \alpha}_{2n}$. For example, to test that $\theta = 2$ against $\theta > 2$ if $\alpha = .05$ and $n = 5$, we reject the null hypothesis if

$$U > \chi^{2\ .05}_{10} = 18.31.$$

If $R = 3$, then,

$$U = \frac{10(3)}{2} = 15,$$

so we accept the null hypothesis that $\theta = 2$ with these data.

Example B

In this case, we observe $X_i \sim N(\mu, \sigma^2)$ with X_i independent. We first consider testing that $\mu = c$. From the previous section, $n^{1/2}(\overline{X} - \mu)/S \sim t_{n-1}$ is a pivotal quantity for, as well as a decreasing function of, μ. Therefore, let

$$t = \frac{n^{1/2}(\overline{X} - c)}{S}.$$

A sensible size-α test for testing that $\mu = c$ against $\mu < c$ would reject the null hypothesis if $t < -t^{\alpha}_{n-1}$ $(= t^{1-\alpha}_{n-1})$, and one for testing $\mu = c$ against $\mu \neq c$ would reject the null hypothesis if $t > t^{\alpha/2}_{n-1}$ or $t < -t^{\alpha/2}_{n-1}$. For example, in testing that $\mu = 3$ against $\mu \neq 3$ with a .05 test when $n = 10$, we reject the null hypothesis if

$$t > t_9^{.025} = 2.262, \quad \text{or} \quad t < -t_9^{.025} = -2.262.$$

If $\overline{X} = 5$ and $S = 6$, then

$$t = \frac{(10)^{1/2}(5 - 3)}{6} = 1.05,$$

so we accept the null hypothesis that $\mu = 3$ with these data.

Now, consider testing that $\sigma^2 = d$. From the previous section, $(n-1)S^2/\sigma^2 \sim \chi^2_{n-1}$ is a pivotal quantity for, and a decreasing function of, σ^2. Therefore, let

$$U = \frac{(n-1)S^2}{d}.$$

Then a sensible test for testing that $\sigma^2 = d$ against $\sigma^2 \neq d$ rejects the null hypothesis if

$$U > \chi^{2\ \alpha/2}, \quad \text{or} \quad U < \chi^{2\ 1-\alpha/2}.$$

For example, to test that $\sigma^2 = 50$ against $\sigma^2 \neq 50$ with a .05 test when $n = 10$, we reject the null hypothesis if

$$U > \chi_9^{2\ .025} = 19.02, \quad \text{or} \quad U < \chi_9^{2\ .975} = 2.700.$$

if $\overline{X} = 5$ and $S = 6$, then

$$U = \frac{9(36)}{50} = 6.48,$$

so we accept the hypothesis that $\sigma^2 = 50$ with these data.

A sequence $T_n(\mathbf{X})$ of statistics is an *asymptotic test statistic* if its asymptotic null distribution does not depend on any unknown parameters. A sequence of tests is *asymptotically size* α if the supremum of the probability of falsely rejecting the null hypothesis goes to α as n goes to infinity. (To emphasize the distinction between test statistics and asymptotic test statistics, we sometimes use the term *exact* test statistics for test statistics. Similarly, we sometimes use the term *exact* size-α tests for what we have called size-α tests.) We can use asymptotic test statistics to design asymptotic size-α tests in the same way as we have described for using exact test statistics to design exact size-α tests.

Theorem 8–4. Let $h_n(\mathbf{X}, \tau)$ be an asymptotic pivotal quantity for τ, and let $T_n(\mathbf{X}) = h_n(\mathbf{X}, c)$. Then $T_n(\mathbf{X})$ is an asymptotic test statistic for testing that $\tau(\boldsymbol{\theta}) = c$ against any alternatives.

Proof. See the exercises. □

If we have an asymptotic pivotal quantity for τ which is decreasing in τ, we can design sensible asymptotic tests for testing that $\tau = c$ against one-sided and two-sided alternatives in the same way as we have given for designing sensible exact size-α tests from exact pivotal quantities which are decreasing in τ.

Example A1

In the previous section, it was mentioned that there is no pivotal quantity for θ in the model where we observe $X_i \sim P(i\theta)$, with X_i independent, but that

$$\frac{a_n^{1/2}(P - \theta)}{\theta^{1/2}} \xrightarrow{d} Z \sim N(0, 1)$$

is an asymptotic pivotal quantity which is a decreasing function of θ. Therefore, let

$$Z_n = \frac{a_n^{1/2}(P - c)}{c^{1/2}}.$$

Then Z_n is an asymptotic test statistic for testing that $\theta = c$. That is, an asymptotic size-α test that $\theta = c$ against $\theta < c$ rejects the null hypothesis if $Z_n < -z^{\alpha}$. For example, to test that $\theta = 4$ against $\theta < 4$ when $n = 10$ (so that $a_n = 55$) with an asymptotic .05 test, we reject the null hypothesis if

$$Z_n < -z^{.05} = -1.645.$$

If $P = 2$, then

$$Z_n = \frac{(55)^{1/2}(2-4)}{(4)^{1/2}} = -7.416,$$

so we reject the null hypothesis that $\theta = 4$ in favor of $\theta < 4$ with these data with an asymptotic .05 test.

In the previous section, we also saw that

$$\frac{a_n^{1/2}(P-\theta)}{P^{1/2}} \xrightarrow{d} Z$$

is an asymptotic pivotal quantity for θ, so that $Z_n^* = a_n^{1/2}(P-4)/P^{1/2}$ is also an asymptotic test statistic for testing that $\theta = 4$. Using the given data, we see that

$$Z_n^* = -10.48 < -z^{.05} = -1.645,$$

so that we still reject the null hypothesis. Typically, in testing problems there is at most one sensible exact pivotal quantity to use, but several sensible asymptotic pivotal quantities. It is then often unclear which asymptotic pivotal quantity is best. The ambiguity usually occurs because there are several possible denominators to use. In this example we could use $4^{1/2}$, as in Z, which is appropriate only when the null hypothesis is true, or we could use $P^{1/2}$, as in Z^*, which is also appropriate under the alternative hypothesis. When finding a confidence interval, it is customary to "studentize" by an estimator of the standard deviation, such as $P^{1/2}$, since the computations are easier. However, when testing, it is customary to "studentize," when possible, by the actual standard deviation under the null hypothesis, such as $4^{1/2}$. It is not obvious that this approach leads to better procedures. (Nor is it even obvious what "better" means in asymptotic procedures.) In fact, the issue of optimal asymptotic "studentization" has not been resolved. Unfortunately, the choice of "studentization" can make a difference in the test statistic. In this book, we shall follow the traditional approach of "studentizing" the test statistics with the true value under the null hypothesis.

One difficulty with the approach to hypothesis testing discussed in this section is the necessity for choosing a size, α, for the test. An approach that is often used to avoid this problem is to give the p-value for the data. Suppose that for each α, $0 \leqq \alpha \leqq 1$, there is a size-α test we would use, and suppose that if $\alpha_1 < \alpha_2$, then the critical region for the size-α_1 test is contained in the critical region for the size-α_2 test (i.e., we are not doing one-sided tests for some α and two-sided tests for other α). Then the *p-value* for the data is defined as the infimum of α such that we would reject the null hypothesis with those data. For example, if a researcher reports a p-value of .012, it would mean that the null hypothesis could be rejected for any size greater than .012. It could be rejected, for example, with a size-.05 test, but not with a size-.01 test.

One appealing aspect of the p-value is that it is a more continuous response to the data than is the approach of deciding to accept or reject the null hypothesis. It will sometimes happen that two data points are quite close to each other, yet we would reject the null hypothesis for one and accept it for the other. Thus, in Example A2, if $Z = 1.65$, we would reject the null hypothesis with a one-sided size-.05 test, and if $Z = 1.64$, we would accept it since $z^{.05} = 1.645$. The p-value, however, is typically a continuous function of the data. The p-value associated with $Z = 1.65$ would be the infimum of α such that $1.65 > z^{\alpha}$. That is, the p-value is the α such that $z^{\alpha} = 1.65$, which is .045. Similarly, the p-value for 1.64 is .055.

Although p-values are often useful in applied statistics, it is difficult to give a theory of statistics based on p-values. For the remainder of this text, we shall assume that the researcher chooses a size for the procedure in question before looking at the data.

Exercises—B

In doing the following problems, be sure to list the critical regions, including the critical values, for all tests.

1. Use the data in Exercise B1, Section 8.1.1, to do a size-.05 test that $\theta = 11$ against $\theta \neq 11$ for the model in Example A2.
2. Use the data in Exercise B1, Section 8.1.1, to do a size-.01 test that $\theta = 10$ against $\theta > 10$ for the model in Example A3.
3. Use the data in Exercise B1, Section 8.1.1, to do a size-.05 test that $\theta = 10$ against $\theta \neq 10$ for the model in Exercise B3, Section 8.1.1.
4. For the data and model given in Exercise B4, Section 8.1.1, do an asymptotic .02 test that $\theta = 85$ against $\theta < 85$.
5. (a) For the data and model given in Exercise B5, Section 8.1.1, do a size-.01 test that $\theta = 10$ against $\theta < 10$.
   (b) Using the same data and model, do a size-.05 test that $\sigma^2 = 50$ against $\sigma^2 < 50$.
6. For the model of Exercise B6, Section 8.1.1, do an asymptotic-.01 test that $\theta = .4$ against $\theta \neq .4$.
7. For the model of Exercise B7, Section 8.1.1, do a size-.05 test that $\theta = 2$ against $\theta > 2$ when $n = 10$ and $\hat{\theta} = 3$.
8. Prove Theorem 8–4.

Exercises—C

1. Find an asymptotic size-α test for testing that $\theta = 0$ against $\theta > 0$ for the model of Exercise C1, Section 8.1.1.
2. Use the data and model of Exercise C2, Section 8.1.1, to test that $\theta = .4$ against $\theta < .4$.
3. Consider testing that $\tau(\boldsymbol{\theta}) = c$ against $\tau(\boldsymbol{\theta}) \neq c$. Let $h(X, \tau)$ be a pivotal quantity for τ which is decreasing in τ. Let $L(\mathbf{X}) \leqq \tau \leqq R(X)$ be a $(1 - \alpha)$ confidence interval for τ based on $h(\mathbf{X}, \tau)$. Consider the test which rejects $\tau(\boldsymbol{\theta}) = c$ if and only if c is not in this confidence interval.
   a. Show that this test has size α.
   b. Show that this test is the same as the "sensible" test for this problem suggested by the heuristic rule in this section.

8.2 SENSITIVITY TO ASSUMPTIONS

8.2.1 Asymptotic Insensitivity to Assumptions

In the last two sections, we have shown how to use pivotal quantities to construct $(1 - \alpha)$ confidence intervals and size-α tests for many models. In this section, we discuss the sensitivity of these procedures to certain assumptions. This aspect of a test or confidence interval is often quite important, since we rarely believe that the assumptions of a model are completely satisfied. For instance, in the jogging example, we probably do not believe that the observations are exactly normally distributed, but believe merely that they are nearly normally distributed (i.e., that the density function starts low on the left, increases to a single maximum, and then decreases on the right).

Suppose we have a test designed under certain assumptions and we want to discuss its sensitivity to a particular assumption. We say that the *nominal* size of the test is its size when the assumption is included in the model, and its *true* size is its size when the assumption is removed. Similarly, the *nominal* confidence coefficient of a confidence interval is the confidence coefficient of the interval when the assumption is satisfied, and its *true* confidence coefficient is its confidence coefficient when the assumption is removed. We define the asymptotic nominal and true size and confidence coefficient similarly.

It is unfortunately not possible to give a theory of sensitivity to assumptions for finite samples. Accordingly, we discuss asymptotic results which are helpful in assessing the sensitivity of procedures to particular assumptions.

Consider a sequence of tests that test $\theta \in N$ against $\theta \in A$ which have size α or asymptotic size α for a given model. We say that such a sequence is *asymptotically insensitive* to a particular assumption in the model if the tests are still asymptotically size-α tests when the assumption is removed (that is, if the asymptotic nominal and true sizes are the same). Similarly, a sequence of $(1-\alpha)$ or asymptotic $(1-\alpha)$ confidence intervals for $\tau(\theta)$ in a given model is *asymptotically insensitive* to an assumption in that model if the intervals are asymptotically $(1-\alpha)$ confidence intervals when the assumption is removed. If a sequence of tests or confidence intervals is asymptotically insensitive to a particular assumption, then for large n, the tests or intervals should have approximately size α or confidence coefficient $(1-\alpha)$, even if the assumption is false. On the other hand, if a sequence is *not* asymptotically insensitive to an assumption, then the size or confidence coefficient is not right, even asymptotically, unless the assumptions of the model are met exactly. Such procedures should not be used unless we are fairly certain that the assumptions are met.

The next theorem is the main result of this section. Although it seems rather obvious, a careful proof is surprisingly difficult. We shall therefore only outline an argument for the theorem.

Theorem 8–5. Let $U_n(\tau)$ be a sequence of (asymptotic) pivotal quantities for τ which are decreasing in τ. Suppose there exists $Q_n(\tau)$, an increasing function of $U_n(\tau)$, such that

$$Q_n(\tau) \xrightarrow{d} Z \sim N(0,1)$$

if the model is true. Suppose that when a particular assumption is removed,

$$Q_n(\tau) \xrightarrow{d} bZ \sim N(0,b^2)$$

for some number $b > 0$.

a. Consider a sequence of equal-tailed confidence intervals for τ based on $U_n(\tau)$

which have nominal (asymptotic) confidence coefficient $1-\alpha$. When the assumption is removed, these have true asymptotic confidence coefficient $1-\alpha^*$, where

$$\alpha^* = 2P\left(Z > \frac{z^{\alpha/2}}{b}\right).$$

b. Consider the sequence of equal-tailed two-sided tests that $\tau = c$ against $\tau \neq c$ based on $U_n(c)$ which have nominal (asymptotic) size α. When the particular assumption is removed, these tests have true asymptotic size α^* as just given.

c. Consider the sequence of one-sided tests for testing that $\tau = c$ against $\tau > c$ which have nominal (asymptotic) size α. When the assumption is removed, these tests have true asymptotic size

$$\alpha^{**} = P\left(Z > \frac{z^{\alpha}}{b}\right).$$

Argument. We outline an argument for the case where Q_n and U_n are exact pivotal quantities. Then U_n and Q_n must each have continuous distributions. Let q_n^α and u_n^α be the upper α points of the distributions of Q_n and U_n, respectively, when the model is true. Since $Q_n \xrightarrow{d} Z$ when the model is true,

$$q_n^\alpha \to z^\alpha.$$

(This seemingly obvious fact is surprisingly difficult to prove; see Randles and Wolf (1979), pp. 423–424.)

a. Let $L_n \leqq \tau \leqq R_n$ be the confidence interval based on $U_n(\tau)$. Then

$$L_n \leqq \tau \leqq R_n \Leftrightarrow u_n^{1-\alpha/2} \leqq U_n(\tau) \leqq u_n^{\alpha/2}$$
$$\Leftrightarrow q_n^{1-\alpha/2} \leqq Q_n(\tau) \leqq q_n^{\alpha/2}$$

(since Q_n is an increasing function of U_n). If the assumption is not satisfied, then $Q_n(\tau) \xrightarrow{d} bZ$. Therefore,

$$P(L_n \leqq \tau \leqq R_n) = P(q_n^{1-\alpha/2} \leqq Q_n(\tau) \leqq q_n^{\alpha/2}) \to$$
$$P(z^{1-\alpha/2} \leqq bZ \leqq z^{\alpha/2}) = 1 - 2P\left(Z > \frac{z^{\alpha/2}}{b}\right) = 1 - \alpha^*.$$

b. See the exercises.

c. The test based on the test statistic $U_n(c)$ rejects the null hypothesis when $U_n(c) > u_n^\alpha$. Since Q_n is an increasing function of U_n, the test also rejects the null hypothesis if $Q_n(c) > q_n^\alpha$. Under the null hypothesis, $Q_n(c) = Q_n(\tau)$ (since $\tau = c$). But since $Q_n(\tau) \xrightarrow{d} bZ$ when the assumption is removed, it follows that

$$\alpha^{**} = P(Q_n(\tau) > q_n^\alpha) \to P(bZ > z^\alpha) = P\left(Z > \frac{z^\alpha}{b}\right). \quad \square$$

Corollary 8–5. Under the assumptions of Theorem 8–5, confidence intervals and tests based on $U_n(\tau)$ are asymptotically insensitive to the assumption if and only if $b = 1$.

Part c of Theorem 8–5 could obviously be extended to the case of one-sided tests in the other direction.

Next, let us consider Example B again, in which we observe $X_i \sim N(\mu, \sigma^2)$, with X_i independent, $i = 1, \ldots, n$. We call this the *normal* model. We study the sensitivity of procedures for μ and σ^2 to the normal assumption in this model. By the model without the normal assumption, we mean the model in which we observe X_i independently and identically distributed with $EX_i = \mu$ and $\text{var}(X_i) = \sigma^2 < \infty$. Let $\overline{X}_n$ and S_n^2 respectively be the sample mean and sample variance computed from a sample of size n.

We first look at inferences about μ. Tests and confidence intervals for μ are based on the pivotal quantity

$$t_n = \frac{n^{1/2}(\overline{X}_n - \mu)}{S_n} \sim t_{n-1}$$

if the model is true. In Chapter 6, we showed that

$$t_n \xrightarrow{d} Z \sim N(0, 1)$$

for both the normal model and the model without the normal assumption. Therefore, the confidence intervals and tests about μ are asymptotically insensitive to the normal assumption used in deriving them.

Now let us consider inferences about σ^2. Here, confidence intervals and tests are based on the pivotal quantity

$$U_n = \frac{(n-1)S_n^2}{\sigma^2} \sim \chi^2_{n-1}$$

if the model is true. By the normal approximation to the χ^2 distribution,

$$Q_n = \frac{U_n - (n-1)}{[2(n-1)]^{1/2}} \xrightarrow{d} Z \sim N(0, 1)$$

when the model is true. Since Q_n is an increasing function of U_n, we are now in a position to use Theorem 8–5. However, we need to find the asymptotic distribution of Q_n for the model without the normal assumption. Accordingly, let $\delta = E(X_i - \mu)^4/\sigma^4$ be the kurtosis of the distribution of the X_i. (If X_i are normally distributed, then $\delta = 3$.) In Chapter 6, we showed that

$$W_n = \frac{n^{1/2}(S_n - \sigma^2)}{(\delta - 1)^{1/2}\sigma^2} \xrightarrow{d} Z \sim N(0, 1).$$

Therefore,

$$Q_n = \left[\frac{(n-1)(\delta - 1)}{2n}\right]^{1/2} W_n \xrightarrow{d} \left(\frac{\delta - 1}{2}\right)^{1/2} Z \sim N\left(0, \frac{\delta - 1}{2}\right).$$

Hence, the procedures for σ^2 are *not* asymptotically insensitive to the normal assumption used in deriving them unless $(\delta - 1)/2 = 1$ or $\delta = 3$. Since it is difficult to predict the kurtosis of a distribution, these procedures should be used only when we are fairly confident that the normal assumption is satisfied.

As an indication of how far off these procedures can be, consider the case where $\alpha = .05$ and $\delta = 9$. Using Theorem 8–5 with

$$b = \left(\frac{\delta - 1}{2}\right)^{1/2} = 2,$$

we get

$$\alpha^* \doteq 2P\left(Z > \frac{1.96}{2}\right) = 2P(Z > .98) = .33$$

and

$$\alpha^{**} = P\left(Z > \frac{1.645}{2}\right) = P(Z > .82) = .21$$

(from Table 1). Therefore, if we compute a nominal 95% confidence interval for σ^2, and the distribution has kurtosis 9, then the confidence interval is truly only a 67% confidence interval. Similarly, if we do a nominal size-.05 test, then the test will have true size .33 or .21, depending on whether it is a one-sided or two-sided test.

If $\alpha = .05$ and $\delta = \frac{17}{8}$, then

$$\alpha^* = .010 \quad \text{and} \quad \alpha^{**} = .014.$$

We see, then, that the error in the size or confidence coefficient can go in either direction. In fact, procedures based on U_n are essentially meaningless unless we are sampling from a distribution with kurtosis 3.

The reason that the procedures for drawing inferences about σ^2 are so sensitive to the normal assumption is that the statistic has been "studentized" by an invalid estimator of the variance of σ^2. The asymptotic calculations indicate that the variance of σ^2 (at least for large n) is about $(\delta - 1)\sigma^4$, so that for large samples, we should use a pivotal quantity of the form

$$Z_n = \frac{n^{1/2}(S_n^2 - \sigma^2)}{(\hat{\delta} - 1)^{1/2}\sigma^2},$$

where $\hat{\delta}$ is a consistent estimator of δ. Instead, when we use U_n, we are using

$$\frac{(n-1)^{1/2}(S_n^2 - \sigma^2)}{2^{1/2}\sigma^2}.$$

We are thus using the number 2 as an estimator of $\delta - 1$. Obviously, this estimator is sensible only when $\delta = 3$.

Although we should use the normal theory procedures for σ^2 only when we really believe the normal assumption, in practice this is not done because reasonable alternative procedures which are not as sensitive to the assumed normality have not been developed.

Exercises—B

1. Verify that $\alpha^* = .01$ and $\alpha^{**} = .014$ for procedures for σ^2 based on the usual χ^2 pivotal quantity when $\delta = \frac{17}{8}$ and $\alpha = .05$.
2. Let $X \sim N(\mu, \sigma^2)$. Verify that X has kurtosis 3. (*Hint*: Let $Z = (X - \mu)/\sigma \sim N(0,1)$. Then $\delta = EZ^4$.)

3. Let $X \sim E(\theta)$. Show that X has kurtosis 9.
4. Make an argument for Theorem 8–5, part b, when U_n and Q_n are exact pivotal quantities.

Exercises—C

1. Make an argument for Theorem 8–5, part a, when U_n and Q_n are asymptotic pivotal quantities.

* 8.2.2 Sensitivity to Independence

All of the statistical procedures discussed in this book make strong assumptions about independence of the observations. Unfortunately, however, they are all quite sensitive to that assumption. In this section, we illustrate that sensitivity using Example B. For this model, we can even discuss sensitivity for finite samples.

Recall that confidence intervals and tests for μ in Example B are based on the pivotal quantity

$$t(\mu) = \frac{n^{1/2}(\overline{X} - \mu)}{S} \sim t_{n-1}$$

if the model is correct. In order to discuss the sensitivity of these procedures to the assumption of independence, we consider the model in which the X_i are jointly normally distributed with

$$EX_i = \mu, \qquad \operatorname{var}(X_i) = \sigma^2, \qquad \operatorname{cov}(X_i, X_j) = \rho\sigma^2.$$

If $\rho = 0$, then the X_i are independent, and the model is exactly that of Example B.

Theorem 8–6. Let

$$\delta = \left(\frac{1 + (n-1)\rho}{1 - \rho}\right)^{1/2}, \qquad t^*(\mu) = \frac{t(\mu)}{\delta}.$$

Then

$$t^*(\mu) \sim t_{n-1}.$$

Proof. Note first that

$$E\overline{X} = n^{-1}\sum EX_i = \mu$$

and

$$\operatorname{var}(\overline{X}) = n^{-2}\left[\sum \operatorname{var}(X_i) + \sum_{i \neq j}\sum \operatorname{cov}(X_i, X_j)\right] = \sigma^2(1 + (r-1)\rho) = \frac{\sigma^2\delta^2(1-\rho)}{n}.$$

Since the X_i are jointly normally distributed, and since $\overline{X}$ is a linear function of the X_i,

$$\overline{X} \sim N\left(\mu, \frac{\sigma^2(1-\rho)\delta^2}{n}\right).$$

It can be shown that

$$\frac{(n-1)S^2}{\sigma^2(1-\rho)} \sim \chi^2_{n-1}$$

independently of $\bar{X}$. (See Arnold (1981), pp. 232–238). In Exercise B1, it is shown that $E(n-1)S^2 = (n-1)\sigma^2(1-\rho)$.) Therefore, by the definition of the t-distribution,

$$t^*(\mu) = \frac{n^{1/2}(\bar{X}-\mu)}{S\delta} \sim t_{n-1}. \quad \square$$

We can see from Theorem 8–6 why the procedures are sensitive to the independence assumption. The variance of $\bar{X}$ is $\sigma^2\delta^2(1-\rho)/n$. In using $t_n(\mu)$, we are "studentizing" with the square root of an estimator of $\sigma^2(1-\rho)/n$. Therefore, we have not "studentized properly" unless

$$\delta = 1 \Leftrightarrow \rho = 0.$$

To illustrate the effect of correlation between the X_i, let $n = 21$ and $\rho = .2$, so that $\delta = 2.5$ and $t^*(\mu) = t(\mu)/2.5$. A nominal 95% confidence interval for μ is the set of all μ such that

$$-t_{20}^{.975} \leqq t(\mu) \leqq t_{20}^{.975} \Leftrightarrow -2.09 \leqq t(\mu) \leqq 2.09.$$

Using the fact that $t^*(\mu) = t(\mu)/2.5 \sim t_{20}$, we see that the true confidence coefficient for this procedure when $\rho = .2$ is

$$P(-2.09 \leqq t(\mu) \leqq 2.09) = P(-.84 \leqq t^*(\mu) \leqq .84) \approx .6.$$

(Note that $P(-.861 \leqq t^*(\mu) \leqq .861) = .6$, so that the actual probability of the foregoing is a little less than .6.) Therefore, in this setting, we have computed what we think is a 95% confidence interval for μ, but it is actually only a 60% confidence interval.

Now, suppose we are testing the hypothesis that $\mu = 2$ against $\mu > 2$. If we assume independence, we reject the null hypothesis if

$$t(2) > t_{20}^{.05} = 1.72.$$

Now, suppose again that the observations are not really independent, but $\rho = .2$, so that $t^*(2) = t(2)/2.5 \sim t_{20}$ under the null hypothesis. Then the true probability of rejecting the null hypothesis is

$$P(t(2) > 1.72) = P(t^*(2) > .69) = .25.$$

Therefore, the nominal size for this procedure is .05, but its true size when $\rho = .2$ is .25.

We next look at the asymptotic sensitivity of the procedures to the assumption of independence.

Theorem 8–7

a. An equal-tailed confidence interval for μ which has nominal confidence coefficient $1-\alpha$ for the independence model has true asymptotic confidence coefficient zero for any $\alpha > 0$ and $\rho > 0$.

b. An equal-tailed two-sided test for μ which has nominal size α for the independence model has true asymptotic size one for $\alpha > 0$ and $\rho > 0$.

c. A one-sided test for μ which has nominal size α for the independence model has true asymptotic size .5 as long as $\rho > 0$ and $\alpha > 0$.

Proof. Note that

$$t^*(\mu) \xrightarrow{d} Z, \qquad t_{n-1}^{\alpha} \to z^{\alpha}, \qquad \delta^{-1} \to 0$$

as $n \to \infty$. Therefore, by Slutzky's theorem,

$$\frac{t(\mu) - t_{n-1}^{\alpha}}{\delta} = t^*(\mu) - t_{n-1}^{\alpha}\delta^{-1} \xrightarrow{d} Z - z^{\alpha}0 = Z \sim N(0,1),$$

and

$$P(t(\mu) > t_{n-1}^{\alpha}) = P\left(\frac{t(\mu) - t_{n-1}^{\alpha}}{\delta} > 0\right) \to P(Z > 0) = .5$$

for any α, as long as $\rho > 0$.

a. By the preceding,

$$P(-t_{n-1}^{\alpha/2} \leqq t(\mu) \leqq t_{n-1}^{\alpha/2}) = P(t(\mu) \leqq t_{n-1}^{\alpha/2}) - P(t(\mu) \leqq -t_{n-1}^{\alpha/2}) \to .5 - .5 = 0.$$

Parts b and c follow similarly. □

Theorem 8–7 implies that, for any $\alpha > 0$, the $(1 - \alpha)$ confidence interval derived for μ under the assumption of independence has a true asymptotic probability of zero of covering μ for any $\rho > 0$. Similarly, for any $\alpha > 0$, two-sided tests with nominal size α have an asymptotic probability of one of rejecting a true null hypothesis, and one-sided tests have asymptotic probability .5 of rejecting a true null hypothesis.

These calculations indicate that the procedures based on $t(\mu)$ for drawing inferences about μ have no validity if the observations are not independent. Therefore, in designing an experiment it is extremely important to design it with independent observations. (Many procedures have been developed for dealing with dependent observations. Some of these come under headings such as repeated measures models, profile analysis, and growth curves analysis; see Arnold (1981), pp. 209–242, 342, and 374–390.)

Next, consider inferences about σ^2. Since such inferences are based on the pivotal quantity

$$U(\sigma^2) = \frac{(n-1)S^2}{\sigma^2} \sim \chi_{n-1}^2.$$

when the observations are independent, for general ρ,

$$\frac{U(\sigma^2)}{1-\rho} \sim \chi_{n-1}^2$$

Therefore, the procedures for drawing inferences about σ^2 would also be inappropriate if $\rho > 0$.

In all the models we examine in this book, the observations are assumed independent. Procedures for drawing inferences for all these models are quite sensitive to this assumed independence. Therefore, the experiments should be designed to encourage independence of the observations.

Now, suppose ρ is an unknown parameter in the model. Then there is no MLE of (μ, σ^2, ρ) for this model, and neither are there any unbiased estimators for ρ or σ^2. Perhaps of most concern, there is no nonrandomized size-α test for testing that $\rho = 0$ against $\rho > 0$. Therefore, there is no way we can tell from the data whether or not $\rho = 0$. (See Arnold (1981), pp. 232–239 for details.)

Exercises—B

1. Show that $ES^2 = \sigma^2(1 - \rho)$.
2. Suppose $n = 40$ and $\rho = .1$. Find the true confidence coefficient of a confidence interval which has nominal confidence coefficient .95 for independent data. (Use the normal approximation to approximate the t-values.)
3. Show that the true size of a one-sided test which has nominal size .01 for independent data when $n = 10$ and $\rho = .15$ is more than .05.
4. Prove parts b and c of Theorem 8–7.

Exercises—C

1. Without using Theorem 8–6, under the assumptions of this section,
   (a) Show that $S^2 \xrightarrow{P} \sigma^2(1 - \rho)$.
   (b) Show that $t^*(\mu) \xrightarrow{d} Z \sim N(0, 1)$.

Chapter 9

OPTIMAL TESTS

In the previous chapter, we developed $(1-\alpha)$ confidence intervals and size-α tests using pivotal quantities. The method used there is often quite easy to implement. Typically, however, there are many pivotal quantities and test statistics (or asymptotic pivotal quantities and test statistics) in a problem.

Example B

Suppose we observe $X_1, \ldots, X_n$ independent, with $X_i \sim N(\mu, \sigma^2)$. In the last chapter, we used the pivotal quantity $t(\mu) = n^{1/2}(\bar{X}-\mu)/S$ for μ. However, let $\tilde{X}$ be the sample median of the X_i, and let R be the sample range. (That is, $\tilde{X}$ is the middle of the ordered observations if n is odd and is the average of the two middle ones if n is even, and R is the difference between the largest and the smallest observations). Then

$$Q(\mu) = \frac{\tilde{X}-\mu}{R}$$

is also a pivotal quantity for μ (see the exercises). Let Q^a be the upper a point for Q. Then a $1-\alpha$ confidence interval for μ based on $Q(\mu)$ is

$$-q^{\alpha/2} \leqq Q(\mu) \leqq q^{\alpha/2} \Leftrightarrow \mu \in \tilde{X} \pm q^{\alpha/2} R,$$

an interval different from the interval based on $t(\mu)$. Similarly, the heuristic rule suggests that another sensible size-α test that $\mu = a$ against $\mu > a$ would reject the null hypothesis if $Q(a) > q^{\alpha}$.

We could find many more pivotal quantities for μ for this model, each leading to its own confidence intervals and tests. In this chapter, we shall develop methods for characterizing which of the many size-α tests is the best.

In choosing which of many possible size-α tests to use, we try to find size-α procedures which have small probability of falsely accepting an untrue null hypothesis. (Note that we have already controlled the probability of falsely rejecting true hypotheses by choosing a small α.)

In Section 9.1, we extend the notion of a test to allow the possibility of randomized tests. Although the possibility of randomized tests makes the theory nicer, it seems doubtful that such tests would be used very often in practice. In Section 9.2 we define the power function of a test. As much as possible, we want to characterize the performance of a test in terms of the test's power function. In the rest of Section 9.2, we show how to find tests that have optimal power functions in some simple cases. Then in Section 9.3, we look at likelihood ratio tests. As with maximum likelihood estimators, likelihood ratio tests typically have good properties, especially for large data sets.

In general, there are many possible confidence intervals for a parameter. (See Example B.) Since we have controlled the probability of covering the true value of the parameter so that it is $1 - \alpha$, we look for confidence intervals which have small probability of covering incorrect values. (For example, it seems reasonable that the probability of covering an incorrect value should be no more than the probability of covering the true value.) There is also a theory of confidence intervals similar to the theory of testing developed in this chapter. We shall not include that theory here, however. Loosely speaking, it says that if the two-sided test based on a pivotal quantity is optimal, then the confidence interval is also. (See Lehmann (1986), pp. 213–216, for a discussion of optimal confidence intervals.)

Exercises—B

1. For Example B, let $Z_i = (X_i - \mu)/\sigma$. Let $\tilde{X}$ and $\tilde{Z}$ be the sample medians of the X_i and the Z_i, respectively, and let R and T be the sample ranges of the X_i and Z_i, respectively.
   (a) Show that $(\tilde{X} - \mu)/R = \tilde{Z}/T$.
   (b) Show that $Q(\mu) = (\tilde{X} - \mu)/R$ is a pivotal quantity for μ. (Note that distribution of Z_i does not depend on μ or σ.)

9.1 RANDOMIZED TESTS

9.1.1 Randomized Tests

In developing the theory of hypothesis testing, it is useful to enlarge the class of possible tests to include randomized tests. A *randomized test* is a rule which assigns to each $\mathbf{X}$ a probability $\Phi(\mathbf{X})$ of rejecting the null hypothesis. The function $\Phi(\mathbf{X})$ is called the *critical function* of the test. For example, we might have

$$\Phi(\mathbf{X}) = \begin{cases} 1 & \text{if } \mathbf{X} \in C \\ \frac{1}{2} & \text{if } \mathbf{X} \in D \\ 0 & \text{if } \mathbf{X} \in E \end{cases}$$

(where C, D, and E are disjoint sets such that $C \cup D \cup E = \chi$). In this case, if we observe $\mathbf{X} \in C$, we reject the null hypothesis; if we observe $\mathbf{X} \in E$, we accept the null hypothesis; and if we observe $\mathbf{X} \in D$, we flip a fair coin and we reject the null hypothesis if we get a head and accept it if we get a tail.

As defined in the previous chapter, nonrandomized tests are special cases of randomized tests, as defined here. A nonrandomized test with critical region C is the same as a randomized test with critical function

$$\Phi(\mathbf{X}) = \begin{cases} 1 & \text{if } \mathbf{X} \in C \\ 0 & \text{if } \mathbf{X} \notin C \end{cases}.$$

For testing $\boldsymbol{\theta} \in N$ against $\boldsymbol{\theta} \in A$, the *size* α of a randomized test is defined to be

$$\alpha = \sup_{\boldsymbol{\theta} \in N} E_{\boldsymbol{\theta}} \Phi(\mathbf{X}).$$

That is, the size is the supremum of the probability of rejecting a true null hypothesis, averaged over both the randomization and the distribution of $\mathbf{X}$. As in the last chapter, $E_{\boldsymbol{\theta}}\Phi(\mathbf{X})$ is typically constant for $\boldsymbol{\theta} \in N$, so that it is rarely necessary to take a supremum. Note that this definition reduces to the one in the last chapter when the test is nonrandomized.

Consider again Example A1, in which we observe $X_i \sim P(i\theta)$, independent. As mentioned in the previous chapter, there is no pivotal quantity for this model. However,

$$V = \sum X_i = a_n P \sim P(\theta a_n)$$

is a test statistic for testing that $\theta = c$ against one-sided or two-sided alternatives (since $V \sim P(ca_n)$ under the null hypothesis). We now show how to use V to find size-α tests for this model.

Suppose, for definiteness, that $n = 3$ ($a_n = 6$), $c = 1$, and $\alpha = .05$. Consider the problem of testing that $\theta = 1$ against $\theta > 1$. Since V is an increasing function of P, a sensible test would reject the null hypothesis if V is too large, that is, if $V > k$, where $P(V > k|\theta = 1) = .05$. Under the null hypothesis, $V \sim P(6)$, and from Table 6, we see that

$$P(V > 9|\theta = 1) = 1 - .916 = .084 \quad \text{and} \quad P(V > 10|\theta = 1) = 1 - .957 = .043.$$

However, since V takes on only integer values, it is not possible to find such a nonrandomized test.

Now, consider the randomized test

$$\Phi(V) = \begin{cases} 1 & \text{if } V > 10 \\ \frac{7}{41} & \text{if } V = 10 \\ 0 & \text{if } V < 10 \end{cases}.$$

Then the size of Φ is

$$E(\Phi(V)|\theta = 1) = (1)P(V > 10) + \left(\frac{7}{41}\right)P(V = 10) + (0)P(V < 10)$$

$$= .043 + \left(\frac{7}{41}\right)(.041) = .05.$$

Therefore, the test Φ has size .05.

From this example, we see that if we want a size-α one-sided test that $\theta = c$ based on a discrete random variable X, and if $P(X > m|\theta = c) = \alpha - a$ and $P(X \geqq m|\theta = c) = \alpha + b$, then the test

$$\Phi(X) = \begin{cases} 1 & \text{if } X > m \\ a/(a + b) & \text{if } X = m \\ 0 & \text{if } X < m \end{cases}$$

is a size-α test. We could design two-sided size-α tests for discrete distributions in a similar way. For most problems involving discrete distributions, it is not possible to find a nonrandomized test which has size exactly equal to a specified number, and it is necessary to use randomized tests to achieve that size exactly.

In practice, however, it seems unlikely that we would ever actually use a randomized test. Rather, we would probably just change α. In the example given above, we would probably just do a size-.043 test. We might also just report the p-value, which is a discrete random variable for this model. For example, the test which rejects the null hypothesis if $V > 10$ has size .043, and the test which rejects it if $V > 9$ has size .084. Therefore, if we observe that $V = 10$, we report a p-value of .084.

Although randomized tests would rarely be used in practice, allowing them makes the theory somewhat nicer, and we shall use them for the remainder of the chapter. We use the expression "the test Φ" for the expression "the test with critical function Φ."

Exercises—B

1. Let $X \sim B(10, \theta)$. Consider testing that $\theta = .3$ against $\theta \neq .3$ with the test

$$\Phi(X) = \begin{cases} 1 & \text{if } X > 6 \text{ or } X < 1 \\ .4 & \text{if } X = 6 \\ .2 & \text{if } X = 1 \\ 0 & \text{if } 1 < X < 6 \end{cases}$$

   What is the size of this test?

2. Let $X_1, \ldots, X_{10}$ be independent, with $X_i \sim B(1, \theta)$. Suppose we want to test the null hypothesis that $\theta = .2$.
   (a) Show that $U = \Sigma X_i \sim B(10, .2)$ under the null hypothesis.
   (b) Consider testing that $\theta = .2$ against $\theta > .2$. Find a size-.05 test for this situation.
   (c) Find a .05 test for testing that $\theta = .2$ against $\theta < .2$.
   (d) Find a .10 test for testing that $\theta = .2$ against $\theta \neq .2$.
   (e) Find a .05 test for testing that $\theta = .2$ against $\theta \neq .2$.

3. Let X have a geometric distribution with parameter θ.
   (a) Design a sensible size-.05 test of the null hypothesis that $\theta = .4$ against the alternative that $\theta \neq .4$. (Recall that $P(X \geq k) = (1 - \theta)^k$.)
   (b) Design a sensible size-.05 test that $\theta = .4$ against $\theta < .4$. (We should reject the null hypothesis when X is large. Why?)

4. Suppose $P(X < m) = \alpha - a$ and $P(X \leqq m) = \alpha + b$. Let $\Phi(X) = 1$ if $X < m$, $\Phi(X) = k$ if $X = m$, and $\Phi(X) = 0$ if $X > m$. If $\Phi(X)$ has size α, what is k?

5. Suppose that $P(X < m) = -a + \alpha/2$, $P(X \leqq m) = b + \alpha/2$, $P(X > M) = -c + \alpha/2$, and $P(X \geqq M) = d + \alpha/2$. Let $\Phi(X) = 1$ if $X < m$ or $X > m$, $\Phi(X) = k$ if $X = m$ or $X = M$, and $\Phi(X) = 0$ if $m < X < M$. If $\Phi(X)$ has size α, what is k?

9.2 POWERFUL TESTS

9.2.1 The Power Function

When we make a decision in a hypothesis-testing problem, there are two possible errors we may make. We make a *type I error* when we reject a true null hypothesis and a *type II error* when we accept a false null hypothesis. That is, a type I error is a false rejection, and a type II error is a false acceptance.

The size of a test is the supremum of the probability of a type I error. Thus, if we set the size of a procedure to .05, we are guaranteeing that the probability of a type I error is at most .05. However, we are not making any statement about the probability of a type II error. One approach to dealing with this difficulty is to control γ, the supremum of the probability of a type II error. Unfortunately, typically, $\gamma = 1 - \alpha$ for all possible procedures, so that this approach is not useful.

We define the *power function* $K_\Phi(\boldsymbol{\theta})$ to be the probability that we reject the null hypothesis with the procedure Φ when $\boldsymbol{\theta}$ is the true value. That is,

$$K_\Phi(\boldsymbol{\theta}) = E_{\boldsymbol{\theta}}\,\Phi(\mathbf{X}).$$

If the test is a nonrandomized test with critical region C, then

$$K_\Phi(\boldsymbol{\theta}) = P_{\boldsymbol{\theta}}(\mathbf{X} \in C).$$

The probability of a type I error for Φ is given by $K_\Phi(\boldsymbol{\theta})$ for $\boldsymbol{\theta}$ in the null set N, and the probability of a type II error is given by $1 - K_\Phi(\boldsymbol{\theta})$ for $\boldsymbol{\theta}$ in the alternative set A. The size of Φ is

$$\alpha = \sup_{\boldsymbol{\theta} \in N} K_\Phi(\boldsymbol{\theta}).$$

A good procedure for dealing with this problem is one whose power is low when $\boldsymbol{\theta} \in N$ and high when $\boldsymbol{\theta} \in A$.

Let Φ and Φ^* be two size-α tests. We say that Φ is *as powerful as* Φ^* if $K_\Phi(\boldsymbol{\theta}) \geqq K_{\Phi^*}(\boldsymbol{\theta})$ for all $\boldsymbol{\theta}$ in A and that Φ is *more powerful than* Φ^* if, in addition, $K_\Phi(\boldsymbol{\theta}) > K_{\Phi^*}(\boldsymbol{\theta})$ for some $\boldsymbol{\theta}$ in A. If Φ is more powerful than Φ^*, then we consider Φ to be a better test than Φ^* (because they both have the same maximum probability of type I error, but Φ has smaller probability of type II error).

The approach we take to finding good tests is to limit discussion to procedures of size α (often .05) and find (if possible) the size-α test which is more powerful than any other size-α test. In this way, we first guarantee that the probability of a type I error is small. Then, out of those procedures with small probability of a type I error, we find the procedure which has the smallest probability of a type II error. In the example that follows, we shall see that even this best procedure will have a high probability of a type II error for some $\boldsymbol{\theta}$. This is why accepting the null hypothesis is not considered proof that it is true, but merely lack of evidence that it is false.

Let us return to Example A2, in which we observe $X_i \sim N(i\theta, 1)$, independent. We assume for simplicity that $n = 1$. Then

$$X_1 = Q \sim N(\theta, 1).$$

The .05 one-sided test Φ^* for testing that $\theta = 0$ against $\theta > 0$ rejects the null hypothesis if

$$Q > 1.645.$$

To find the power function of this test, we calculate

$$\begin{aligned} K_{\Phi^*}(\theta) &= P_\theta(Q > 1.645) = P_\theta(Q - \theta > 1.645 - \theta) = P(Z > 1.645 - \theta) \\ &= 1 - N(1.645 - \theta), \end{aligned}$$

where $N(z)$ is the distribution function of a standard normal random variable. Therefore,

$$K_{\Phi^*}(1) = 1 - N(.645) = .26.$$

The interpretation of this statement is that the probability that we reject the null hypothesis with Φ^* when $\theta = 1$ is .26. Therefore, the probability that we falsely accept the null hypothesis $\theta = 0$ when in fact $\theta = 1$ with this procedure is $1 - .26 = .74$. Other values of the power function of Φ^* are shown in Table 9–1.

Now, let Φ_* be the .05 one-sided test on the other side which rejects the null hypothesis when

$$Q < -1.645.$$

Then

$$K_{\Phi_*}(\theta) = K_{\Phi^*}(-\theta).$$

(Why?) Values of this power function are given in Table 9–1 also.

Finally, let Φ be the two-sided size-.05 test which rejects the null hypothesis if

$$Q > 1.96 \quad \text{or} \quad Q < -1.96.$$

Then the power function of Φ is given by

$$K_{\Phi}(\theta) = 1 - P(-1.96 < Q < 1.96) = 1 - N(1.96 - \theta) + N(-1.96 - \theta).$$

For example,

$$K_{\Phi}(1) = 1 - N(.96) + N(-2.96) = .17.$$

Other values of the power function of Φ are given in Table 9–1. We have graphed these three power functions in Figure 9–1.

From the table, we see that $K(0) = .05$ for all three tests, because they were set up as .05 tests. Also, Φ^* has the highest power function when $\theta > 0$, which is why we use this test to test that $\theta = 0$ against $\theta > 0$. Furthermore, when $\theta < 0$, the power of Φ^* is $< .05$, so that Φ^* also has size .05 for testing that $\theta \leqq 0$ against $\theta > 0$. Similarly, the test Φ_* has the highest power when $\theta < 0$, so that we use this test for testing that $\theta = 0$ (or $\theta \geqq 0$) against $\theta < 0$. Finally, the two-sided test Φ does not have the highest power function anywhere. However, it is much higher than the power function of Φ^* when $\theta < 0$ and not too much lower when $\theta > 0$, and it is much higher than the power function of Φ_* when $\theta > 0$ and not much lower when $\theta < 0$. Accordingly, we use Φ for testing $\theta = 0$ against $\theta \neq 0$.

Note that as θ goes to zero, the power functions of all three tests go to .05. This

TABLE 9–1 VALUES OF POWER FUNCTIONS

θ	−2	−1.75	−1.5	−1.25	−1	−.75	−.5	−.25	0	.25	.5	.75	1	1.25	1.5	1.75	2
Φ^*	0	0	0	0	0	.01	.02	.03	.05	.08	.13	.19	.26	.35	.44	.54	.64
Φ_*	.64	.54	.44	.35	.26	.19	.13	.08	.05	.03	.02	.01	0	0	0	0	0
Φ	.52	.42	.32	.24	.17	.12	.08	.06	.05	.06	.08	.12	.17	.24	.32	.42	.52

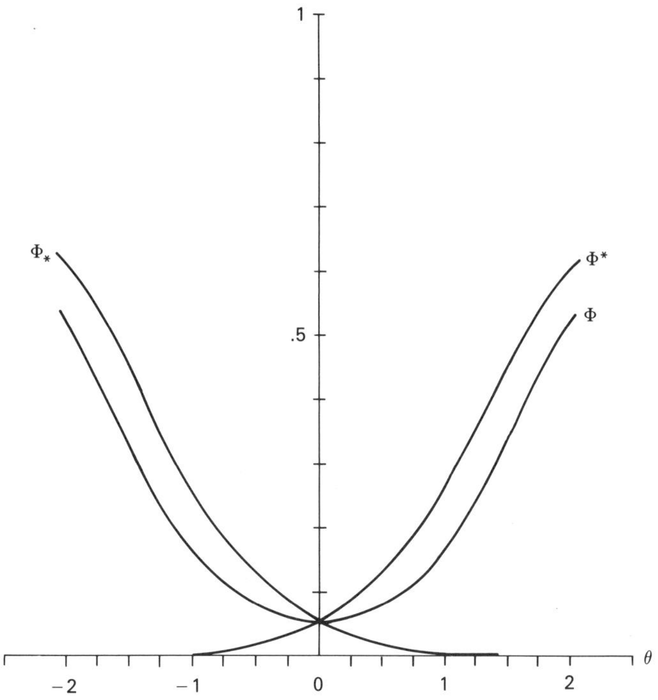

Figure 9–1 Power functions of Φ_*, Φ^*, and Φ.

is because the power function is a continuous function of θ. Therefore, the probability of a type II error can be made arbitrarily close to .95, and hence, the supremum of the probability of a type II error is .95. In fact, if $\theta = .25$, then Φ^* has power .08, which means that if $\theta = .25$, we would falsely accept the null hypothesis (and make a type II error) 92% of the time. Hence, when we accept the null hypothesis, it may well be that $\theta = .25$. Consequently, when we accept $\theta = 0$, we only consider it as lack of evidence that θ is not zero. On the other hand, the probability of a type I error is .05, so that when we reject the null hypothesis, we consider it evidence that the null hypothesis is false.

From this discussion, it is apparent that we consider type I errors to be more important than type II errors. One explanation for this approach is the following. If we reject a hypothesis, this implies that the null model (which is typically the model which had been used) is false. Therefore, if we falsely reject a true null hypothesis, then we shall be directing future researchers down a blind alley. On the other hand, if we incorrectly accept a false hypothesis, it should not be considered proof of anything. Therefore, falsely rejecting a true hypothesis often has a more serious effect on future research than does incorrectly accepting a false hypothesis.

Exercises—B

1. Verify the entries in Table 9–1 for the power of Φ^* and Φ for $\theta = -1.25, -.75, -.25, .25, .75$, and 1.25.

2. Show that $K_{\Phi_*}(\theta) = K_{\Phi^*}(-\theta)$.
3. Find the power of the test defined in Exercise B1, Section 9.1.1, when $\theta = .1, .3, .5, .7$, and .9.
4. Let $X_1, \ldots, X_5 \sim P(\theta)$.
   (a) Find a size-α test for testing that $\theta = .4$ against $\theta < .4$.
   (b) Find the power of the preceding test at $\theta = .2, .4, .6, .8$, and 1.

9.2.2 Simple Hypotheses and the Neyman-Pearson Theorem

We say that the hypothesis $\boldsymbol{\theta} \in B$ is a *simple hypothesis* if the set B contains only one point. In this section, we find the best test for testing the simple null hypothesis $\boldsymbol{\theta} = \mathbf{a}$ against the simple alternative $\boldsymbol{\theta} = \mathbf{b}$. Although this situation rarely occurs in practice, the Neyman-Pearson theorem describing the test is the basic theoretical result of hypothesis testing. The test Φ is the *most powerful size-α test* for this problem testing the simple null hypothesis $\boldsymbol{\theta} = \boldsymbol{a}$ against the simple alternative hypothesis $\boldsymbol{\theta} = \boldsymbol{b}$ if Φ is as powerful as any size-α test. That is, Φ is the most powerful size-α test if

1. Φ has size α (i.e., $K_\Phi(\mathbf{a}) = \alpha$).
2. If Φ^* is any other size-α test, then $K_\Phi(\mathbf{b}) \geqq K_{\Phi^*}(\mathbf{b})$.

Theorem 9–1. (Neyman-Pearson) Consider testing the null hypothesis that $\boldsymbol{\theta} = \mathbf{a}$ against the alternative hypothesis that $\boldsymbol{\theta} = \mathbf{b}$. Let

$$\lambda(\mathbf{X}) = \frac{L_X(\mathbf{b})}{L_X(\mathbf{a})} \quad \text{and} \quad \Phi(\mathbf{X}) = \begin{cases} 1 & \text{if } \lambda(\mathbf{X}) > k \\ d & \text{if } \lambda(\mathbf{X}) = k \\ 0 & \text{if } \lambda(\mathbf{X}) < k \end{cases},$$

where $k \geqq 0$ and $0 \leqq d \leqq 1$ are chosen so that Φ has size α. Let $\Phi^*(\mathbf{X})$ be a test of size $\gamma \leqq \alpha$. Then

$$K_\Phi(\mathbf{b}) \geqq K_{\Phi^*}(\mathbf{b}).$$

Proof. As usual, we assume that $\mathbf{X}$ has a continuous density; the proof for the discrete case is similar. Note that

$$(\Phi(\mathbf{x}) - \Phi^*(\mathbf{x}))(f(\mathbf{x};\mathbf{b}) - kf(\mathbf{x};\mathbf{a})) \geqq 0$$

(since if $f(\mathbf{x};\mathbf{b}) - kf(\mathbf{x};\mathbf{a}) > 0$, then $\Phi(\mathbf{x}) = 1 \geqq \Phi^*(\mathbf{x})$, and if $f(\mathbf{x};\mathbf{b}) - kf(\mathbf{x};\mathbf{a}) < 0$, then $\Phi(\mathbf{x}) = 0 \leqq \Phi^*(\mathbf{x})$). Therefore,

$$\begin{aligned} K_\Phi(\mathbf{b}) - K_{\Phi^*}(\mathbf{b}) - k(\alpha - \gamma) &= K_\Phi(\mathbf{b}) - K_{\Phi^*}(\mathbf{b}) - k(K_\Phi(\mathbf{a}) - K_{\Phi^*}(\mathbf{a})) \\ &= \int_{\mathcal{X}} (\Phi(\mathbf{x}) - \Phi^*(\mathbf{x}))(f(\mathbf{x};\mathbf{b}) - kf(\mathbf{x};\mathbf{a}))d\mathbf{x} \geqq 0. \end{aligned}$$

Hence,

$$K_\Phi(\mathbf{b}) - K_{\Phi^*}(\mathbf{b}) \geqq k(\alpha - \gamma) \geqq 0. \quad \square$$

Corollary 9–1. $\Phi(\mathbf{X})$ is the most powerful size-α test for testing the null hypothesis that $\boldsymbol{\theta} = \mathbf{a}$ against the alternative hypothesis that $\boldsymbol{\theta} = \mathbf{b}$.

The statistic $\lambda(\mathbf{X})$ is called the *Neyman-Pearson test statistic* for the situation given. Theorem 9–1 implies that the most powerful size-α test rejects the null hypothesis if $\lambda(\mathbf{X}) > k$, randomizes (if necessary) when $\lambda(\mathbf{X}) = k$, and accepts the null hypothesis when $\lambda(\mathbf{X}) < k$, where k and the randomization are chosen so that the test has size α.

In Section 9.1.1, we indicated how to find k and d so that Φ has size α for $0 < \alpha < 1$ when $\lambda(\mathbf{X})$ has a discrete distribution. When $\lambda(\mathbf{X})$ has a continuous distribution, $P(\lambda(\mathbf{X}) = k) = 0$, so that we can choose d to be anything. In this case, for the remainder of the book, we shall choose d to be zero, so that when $\lambda(\mathbf{X})$ has a continuous distribution, the most powerful size-α test is given by

$$\Phi(\mathbf{X}) = \begin{cases} 1 & \text{if } \lambda(\mathbf{X}) > k \\ 0 & \text{if } \lambda(\mathbf{X}) \leqq k \end{cases},$$

where k is chosen so that the test has size α. Note that this test is a nonrandomized test with critical region

$$C = \{\mathbf{X}: \lambda(\mathbf{X}) > k\}.$$

When $P(\lambda(\mathbf{X}) = k) = 0$, a slightly more delicate version of the proof of Theorem 9–1 can be used to show that Φ is more powerful than any other size-α test $\Phi^*(\mathbf{X}) \not\equiv \Phi(\mathbf{X})$, i.e., that $K_\Phi(\mathbf{b}) > K_{\Phi^*}(\mathbf{b})$.

Example 9–1

Let X be a discrete random variable with density $f(x;\theta)$ given in the following table:

x	0	1	2	3
$f(x;0)$	.05	.05	.10	.80
$f(x;1)$	.05	.15	.50	.30
$\lambda(x)$	1	3	5	$\frac{3}{8}$

($\lambda(x) = f(x;1)/f(x;0)$). Suppose it is desired that the null hypothesis $\theta = 0$ be tested against the alternative hypothesis $\theta = 1$ with a .05 test. Three possible tests are

$$\Phi_1(X) = \begin{cases} 1 & \text{if } X = 0 \\ 0 & \text{if } X \neq 0 \end{cases}, \quad \Phi_2(X) = \begin{cases} 1 & \text{if } X = 1 \\ 0 & \text{if } X \neq 1 \end{cases}, \quad \Phi_3(X) = \begin{cases} .5 & \text{if } X = 2 \\ 0 & \text{if } X \neq 2 \end{cases}.$$

The first two of these tests are nonrandomized with size $\alpha = .05$. The third test is randomized with

$$\alpha = .5P(X = 2 | \theta = 0) = .05.$$

Now, Φ_1 has probability .05 of rejecting the null hypothesis when $\theta = 1$, and Φ_2 has probability .15 of rejecting the null hypothesis when $\theta = 1$, so that Φ_2 is more powerful than Φ_1. However, Φ_3 has power

$$K_{\Phi_3}(b) = .5P(X = 2 | \theta = 1) = .25.$$

Therefore, the randomized test Φ_3 is more powerful than either of the .05 nonrandomized tests. In fact,

$$\Phi_3(X) = \begin{cases} 1 & \text{if } \lambda(X) > 5 \\ .5 & \text{if } \lambda(X) = 5 \\ 0 & \text{if } \lambda(X) < 5 \end{cases},$$

so that Φ_3 is the most powerful size-.05 test that $\theta = 0$ against $\theta = 1$, by the Neyman-Pearson theorem. In a similar manner, the most powerful size-.10 test is the non-randomized test which rejects the null hypothesis if $X = 2$, and the most powerful size-.125 test is the randomized test which rejects the null hypothesis if $X = 2$ and rejects it half the time when $X = 1$.

From this example, we see how the Neyman-Pearson theorem works in the discrete case. Note that $\lambda(x) = f(x;1)/f(x;0)$ represents the "units" of probability under the alternative hypothesis for each "unit" under the null. Therefore, when $\lambda(x)$ is highest, we are getting the maximum possible probability under the alternative for each "unit" of probability under the null hypothesis. The Neyman-Pearson theorem says that we first look for the point q which maximizes $\lambda(X)$. The optimal test rejects the null hypothesis at q with as high a probability as is compatible with $\alpha = E(\Phi(X)|\theta = 0)$. If we have not used all of α, we then look for the point r where $\lambda(X)$ is second highest and reject the null hypothesis there also, with as high a probability as possible. We continue in this fashion until we have used all of α.

Example A1

In this example, we observe $X_i \sim P(i\theta)$, independent. Therefore,

$$L_X(\theta) = h_1(\mathbf{X}) \exp[\textstyle\sum X_i \log(\theta) - a_n \theta].$$

Now, consider testing that $\theta = 2$ against $\theta = 3$. Then

$$\lambda(\mathbf{X}) = \frac{L_X(3)}{L_X(2)} = \exp\left[\sum X_i \log\left(\frac{3}{2}\right) - a_n\right].$$

Now,

$$\lambda(\mathbf{X}) > k \Leftrightarrow U(\mathbf{X}) \geqq k^*,$$

where

$$U(\mathbf{X}) = \sum X_i \quad \text{and} \quad k^* = \frac{\log(k) + a_n}{\log(\frac{3}{2})}.$$

Therefore, the most powerful size-α test can be written as

$$\Phi(\mathbf{X}) = \begin{cases} 1 & \text{if } U(\mathbf{X}) > k^* \\ d & \text{if } U(\mathbf{X}) = k^* \\ 0 & \text{if } U(\mathbf{X}) < k^* \end{cases},$$

where $U(\mathbf{X}) = \Sigma X_i$ and k^* and d are chosen so that Φ has size α. Now,

$$U(\mathbf{X}) = \sum X_i = Pa_n \sim P(\theta a_n).$$

Suppose we take $\alpha = .05$ and $n = 3$. Then $U(\mathbf{X}) \sim P(12)$ under the null hypothesis that $\theta = 2$. Using Table 6, we then see that $k^* = 18$, $d = \frac{1}{2}$, and

$$\Phi(\mathbf{X}) = \begin{cases} 1 & \text{if } U(\mathbf{X}) > 18 \\ \frac{1}{2} & \text{if } U(\mathbf{X}) = 18 \\ 0 & \text{if } U(\mathbf{X}) < 18 \end{cases}. \tag{9–1}$$

We could now solve for $k = (\frac{3}{2})^{k^*} \exp(-a_n) = (\frac{3}{2})^{18} \exp(-6)$ and rewrite the test in terms of $\lambda(\mathbf{X})$. However, the test is much more complicated in terms of $\lambda(\mathbf{X})$. The test

$\Phi(\mathbf{X})$ is the same, whether written in terms of $U(\mathbf{X})$ as in Equation (9–1) or in terms of $\lambda(\mathbf{X})$ as in Theorem 9–1. The form given in the equation is much simpler. In fact, in most problems, we start with the form of the test given in the theorem and manipulate it until it is in simpler form.

Example A2

Here, we observe $X_i \sim N(i\theta, 1)$, independent, so that

$$L_X(\theta) = h_2(\mathbf{X}) \exp\left(\theta \sum iX_i - \frac{b_n \theta^2}{2}\right).$$

Now, consider testing that $\theta = 2$ against $\theta = 4$. Then

$$\lambda(\mathbf{X}) = \frac{L_X(4)}{L_X(2)} = \exp(2\sum iX_i - 6b_n).$$

Now,

$$\lambda(\mathbf{X}) > k \Leftrightarrow V(\mathbf{X}) > k^*,$$

where

$$V(\mathbf{X}) = \sum iX_i \quad \text{and} \quad k^* = 3b_n + \frac{\log(k)}{2}.$$

Since $\lambda(\mathbf{X})$ has a continuous distribution, the most powerful size-α test is given by

$$\Phi(\mathbf{X}) = \begin{cases} 1 & \text{if } V(\mathbf{X}) > k^* \\ 0 & \text{if } V(\mathbf{X}) \leqq k^* \end{cases},$$

where k^* is chosen so that the test has size α.

Now,

$$V(\mathbf{X}) = Qb_n \sim N(\theta b_n, b_n).$$

Suppose that $n = 3$ ($b_n = 14$) and $\alpha = .05$. Then $V(\mathbf{X}) \sim N(28, 14)$ under the null hypothesis. Therefore,

$$k^* = (1.645)(14)^{1/2} + 28 = 34.16.$$

Hence, the most powerful test for this situation is

$$\Phi(\mathbf{X}) = \begin{cases} 1 & \text{if } V(\mathbf{X}) > 34.16 \\ 0 & \text{if } V(\mathbf{X}) \leqq 34.16 \end{cases}.$$

Now, consider testing that $\theta = 2$ against $\theta = 6$ in this same situation. Then

$$\lambda(\mathbf{X}) = \exp(4V(\mathbf{X}) - 16b_n) = \exp(4V(\mathbf{X}) - 224)$$

and

$$\lambda(\mathbf{X}) > k \Leftrightarrow V(\mathbf{X}) > k^{**}.$$

Hence, the most powerful size-α test rejects the null hypothesis if $V(\mathbf{X}) > k^{**}$, where k^{**} is chosen so that the test has size α. However, this testing problem has the same null hypothesis as the previous one ($\theta = 2$), and so $V(\mathbf{X})$ has the same null distribution. Consequently, $k^* = k^{**} = 34.16$. We see, therefore, that the most powerful test for testing that $\theta = 2$ against $\theta = 6$ is the same as the most powerful test for testing that $\theta = 2$ against $\theta = 4$. In the next section, we shall see that the most powerful size-α test for testing that $\theta = a$ against $\theta = b > a$ in this model does not depend on b.

Exercises—B

1. Let X be a discrete random variable having the following density function:

x	0	1	2	3	4	5
$f(x; 0)$	.05	.05	.10	.10	.20	.50
$f(x; 1)$	.10	.15	.25	.15	.25	.10

(a) Find the most powerful .05 test that $\theta = 0$ against $\theta = 1$.
(b) Find the most powerful .10 test that $\theta = 0$ against $\theta = 1$.
(c) Find the most powerful .15 test that $\theta = 0$ against $\theta = 1$.
(d) Find the most powerful .10 test that $\theta = 1$ against $\theta = 0$.
(e) Find the most powerful .15 test that $\theta = 1$ against $\theta = 0$.

2. In Example A2, find the most powerful size-.05 test for testing that $\theta = 1$ against $\theta = 0$ when $n = 3$. Do we accept or reject the null hypothesis when $Q = .4$?

3. **a.** For Example A3, with $n = 10$, find the most powerful size-.05 test for testing that $\theta = 1$ against $\theta = 2$. Do we accept or reject the null hypothesis when $R = 2.2$?
b. For Example A3, with $n = 10$, find the most powerful size-.05 test for testing that $\theta = 1$ against $\theta = 3$. Do we accept or reject the null hypothesis when $R = 2.2$?

4. For Example B, find the form of the most powerful size-α test that $\mu = 0$ and $\sigma^2 = 1$ against $\mu = 1$ and $\sigma^2 = 4$. (Do not attempt to simplify the test or find the critical value.)

5. Let $X_1, \ldots, X_n$ be independent, and let $f(x_i, \theta) = \theta x_i^{\theta - 1}$. Find the form of the most powerful size-α test that $\theta = 1$ against $\theta = 2$, and show that it rejects the null hypothesis if ΠX_i is too large.

6. Prove the Neyman-Pearson theorem in the discrete case.

7. Let (X, Y) have a trinomial distribution with $n = 2$ and parameters (θ_1, θ_2). Find the most powerful size-.16 test of the null hypothesis that $\theta_1 = \theta_2 = .4$ against $\theta_1 = .2$ and $\theta_2 = .6$. (*Hint*: List the six possible points.)

8. Let X have a uniform distribution on $(0, \theta)$. Show that a most powerful size-.05 test of the null hypothesis $\theta = 2$ against the alternative $\theta = 1$ is $\Phi(X) = .1$ if $X < 1$ and $\Phi(X) = 0$ if $X > 1$.

Exercises—C

1. Let $\Phi(\mathbf{X})$ be the most powerful size-α test for testing that $\theta = a$ against $\theta = b$. Show that the power at b is at least α. (*Hint*: Let $\Phi^*(\mathbf{X}) \equiv \alpha$. Then Φ is more powerful than Φ^*. Why?)

2. Let $\mathbf{X}$ have the Cauchy density $f(x; \theta) = \pi^{-1}(1 + (x - \theta)^2)^{-1}, -\infty < x < \infty$. Consider testing the null hypothesis that $\theta = -1$ against the alternative hypothesis $\theta = -1$.
(a) What is the Neyman-Pearson test statistic $\Lambda(X)$ for this situation?
(b) Show that $\Lambda(X) > 1$ if and only if $X > 0$. What is α for this test?
(c) Show that $\Lambda(X) > 5$ if and only if $X > -1$ or $X < -2$. What is α for this test?
(d) Show that $\Lambda(X) > .2$ if and only if $1 < X < 2$. What is α for this test?
(Therefore, for one choice of α, the most powerful size-α test for this problem rejects the null hypothesis for X too large (as we would expect), for another α, the test rejects the

null hypothesis if X is outside an interval, and for a third α, the test rejects the null hypothesis if X is inside an interval.)

9.2.3 Uniformly Most Powerful Tests

The hypothesis $\boldsymbol{\theta} \in B$ is a *composite hypothesis* if B consists of more than one point (i.e., if the hypothesis is not simple). In this section, we find optimal tests for certain testing problems involving composite hypotheses. A size-α test Φ for testing the null hypothesis that $\boldsymbol{\theta} \in N$ against the alternative hypothesis that $\boldsymbol{\theta} \in A$ is a *uniformly most powerful* (*UMP*) size-α test if it is as powerful as any other size-α test. That is, Φ is UMP size α if for any other size-α test Φ^*,

$$K_{\Phi}(\boldsymbol{\theta}) \geqq K_{\Phi^*}(\boldsymbol{\theta}) \quad \text{for all } \boldsymbol{\theta} \in A.$$

If a UMP size-α test exists for a testing problem, it is the correct size-α test to use for that problem. A UMP size-α test minimizes the probability of a type II error out of all possible size-α tests. (Note again that we have controlled the probability of the type I error by setting α to be small.)

UMP tests typically exist only when θ is a one-dimensional parameter, which we now assume. Let $T(\mathbf{X})$ be a univariate statistic. Then a model has *monotone likelihood ratio* (MLR) in $T(\mathbf{X})$ if, for all $\theta_1 > \theta_0$,

$$\frac{L_X(\theta_1)}{L_X(\theta_0)}$$

depends only on $T(\mathbf{X})$ and is a nondecreasing function of $T(\mathbf{X})$. That is,

$$\frac{L_X(\theta_1)}{L_X(\theta_0)} = h(T(\mathbf{X}); (\theta_0, \theta_1)),$$

where for all $\theta_1 > \theta_0$, $h(t; \theta_0, \theta_1)$ is a nondecreasing function of t.

Theorem 9–2. Let $\theta \in R^1$, and suppose the model has monotone likelihood ratio in $T(\mathbf{X})$.

a. Consider testing that $\theta = c$ against $\theta > c$. A UMP size-α test for this situation is given by

$$\Phi_1(\mathbf{X}) = \begin{cases} 1 & \text{if } T(\mathbf{X}) > k \\ d & \text{if } T(\mathbf{X}) = k \\ 0 & \text{if } T(\mathbf{X}) < k \end{cases},$$

where d and k are chosen so that Φ_1 has size α.

b. Consider testing that $\theta = c$ against $\theta < c$. A UMP size-α test for this situation is given by

$$\Phi_2(\mathbf{X}) = \begin{cases} 1 & \text{if } T(\mathbf{X}) < k^* \\ d^* & \text{if } T(\mathbf{X}) = k^* \\ 0 & \text{if } T(\mathbf{X}) > k^* \end{cases},$$

where d^* and k^* are chosen so that Φ_2 has size α.

Proof

a. Consider testing the null hypothesis $\theta = c$ against the alternative $\theta = b$, where

$b > c$. By the Neyman-Pearson theorem, the most powerful size-α test for this situation is given by

$$\Phi^*(\mathbf{X}) = \begin{cases} 1 & \text{if } \lambda(\mathbf{X}) > k^{**} \\ d & \text{if } \lambda(\mathbf{X}) = k^{**} \\ 0 & \text{if } \lambda(\mathbf{X}) < k^{**} \end{cases}, \qquad \lambda(\mathbf{X}) = \frac{L_X(b)}{L_X(c)},$$

where k^{**} and d are chosen so that this test has size α. Now, by the definition of a monotone likelihood ratio, $\lambda(\mathbf{X})$ is a nondecreasing function of $T(\mathbf{X})$. Therefore,

$$\lambda(\mathbf{X}) > k^{**} \Leftrightarrow T(\mathbf{X}) > k$$

for some k. Also, d and k^{**} (and hence, k) are chosen so that $K_{\Phi^*}(c) = \alpha$. That is, d and k are chosen so that the probability of rejecting the null hypothesis when $\theta = c$ is α (and hence, d and k do not depend on b). Therefore, the test Φ^* is the same as the test Φ_1 given in the statement of the theorem. Hence, Φ_1 is the most powerful size-α test for testing that $\theta = c$ against $\theta = b$ for all $b > c$ and is, therefore, a UMP size-α test for testing that $\theta = c$ against $\theta > c$.

b. This proof is left to the reader. □

What part a of Theorem 9–2 says is that if the model has monotone likelihood ratio in $T(\mathbf{X})$, then the UMP test for testing that $\theta = c$ against $\theta > c$ rejects the null hypothesis if $T(\mathbf{X})$ is too large (possibly randomizing on the boundary). Part b says that in this situation, the UMP test for testing that $\theta = c$ against $\theta < c$ rejects the null hypothesis if $T(\mathbf{X})$ is too small. How large or small $T(\mathbf{X})$ must be in order to reject the null hypothesis is determined by the size of the procedure.

If $T(\mathbf{X})$ is a discrete random variable, we find k and d by the methods of Section 9.1.1. If $T(\mathbf{X})$ is a continuous random variable, we can assume that $d = 0$ and get a nonrandomized UMP size-α test.

We now look at Example A again. (We cannot look at Example B because $\boldsymbol{\theta}$ has dimension two for that example.)

Example A1

We observe $X_i \sim P(i\theta)$, independent. Hence,

$$\frac{L_X(\theta_1)}{L_X(\theta_0)} = \left(\frac{\theta_1}{\theta_0}\right)^{\sum X_i} \exp(a_n(\theta_0 - \theta_1)),$$

which is an increasing function of $U(\mathbf{X}) = \Sigma X_i$ when $\theta_1 > \theta_0$. Therefore, the UMP size-α test that $\theta = 2$ against $\theta > 2$ is

$$\Phi_1(\mathbf{X}) = \begin{cases} 1 & \text{if } U(\mathbf{X}) > k \\ d & \text{if } U(\mathbf{X}) = k \\ 0 & \text{if } U(\mathbf{X}) < k \end{cases},$$

where k and d are chosen so that the probability of rejecting the null hypothesis when $\theta = 2$ is α. Now, $U(\mathbf{X}) \sim P(\theta a_n)$. So if $n = 3$ and $\alpha = .05$, we see from the previous section that $k = 18$ and $d = \frac{1}{2}$.

Example A2

Here, we observe $X_i \sim N(i\theta, 1)$, independent, so that

$$\frac{L_X(\theta_1)}{L_X(\theta_0)} = \exp\left((\theta_1 - \theta_0)\sum iX_i + \frac{(\theta_0^2 - \theta_1^2)b_n}{2}\right),$$

which is an increasing function of $V(\mathbf{X}) = \Sigma iX_i$ when $\theta_1 > \theta_0$, and hence, the density has monotone likelihood ratio in $V(\mathbf{X})$. Therefore, the UMP size-α test that $\theta = 2$ against $\theta > 2$ is

$$\Phi_1(\mathbf{X}) = \begin{cases} 1 & \text{if } V(\mathbf{X}) > k \\ 0 & \text{if } V(\mathbf{X}) \leqq k \end{cases}.$$

If $n = 3$ and $\alpha = .05$, we see from the last section that $k = 34.16$.

Now, consider testing that $\theta = 2$ against $\theta < 2$. Under the preceding conditions, the UMP size-.05 test is

$$\Phi_2(\mathbf{X}) = \begin{cases} 1 & \text{if } V(\mathbf{X}) < 21.84 \\ 0 & \text{if } V(\mathbf{X}) \geqq 21.84 \end{cases}$$

$(21.84 = 28 - (14)^{1/2}(1.645))$.

Example A3

In the exercises, it is shown that the model in this example has monotone likelihood ratio in $\Sigma X_i/i$, and a UMP size-.05 test for testing that $\theta = 2$ against $\theta > 2$ when $n = 3$ is requested.

We have now discussed UMP size-.05 tests that $\theta = 2$ against $\theta > 2$ for each of the foregoing models when $n = 3$. These optimal tests are all different, even though the models seem rather similar.

Students often feel that in families with monotone likelihood ratio in $T(\mathbf{X})$, the UMP test for testing that $\theta = c$ against $\theta \neq c$ should reject the null hypothesis if $T(\mathbf{X})$ is too large or too small. However, there is rarely a UMP test for two-sided testing problems.

Unfortunately, UMP tests basically exist only for the rather limited situation considered in this section: testing one-sided hypotheses for models which have only a single parameter. Typically, there are no UMP tests for two-sided problems, nor any for hypotheses for models like that in Example B, which have more than one parameter. For such problems, a more subtle notion of optimality is needed.

We now return to the more general model in which $\boldsymbol{\theta}$ may be a vector. A size-α test $\Phi(\mathbf{X})$ for testing the null hypothesis $\boldsymbol{\theta} \in N$ against the alternative hypothesis $\boldsymbol{\theta} \in A$ is *unbiased* if

$$K_\Phi(\boldsymbol{\theta}) \geqq \alpha \quad \text{for all } \boldsymbol{\theta} \in A.$$

That is, Φ is unbiased if the probability of rejecting the null hypothesis when it is false is at least as great as the probability of rejecting it when it is true. An unbiased size-α test Φ is a *uniformly most powerful* (UMP) *unbiased* size-α test if it is as powerful as any other unbiased size-α test. That is, Φ is a UMP unbiased size-α test if for any other unbiased size-α test Φ^*,

$$K_\Phi(\boldsymbol{\theta}) \geqq K_{\Phi^*}(\boldsymbol{\theta}) \quad \text{for all } \boldsymbol{\theta} \in A.$$

It seems that any sensible size-α test should at least be unbiased, and therefore, a test which is UMP among all unbiased size-α tests should be quite good.

If there is a UMP size-α test for a problem, it must be unbiased (see Exercise C1). Therefore, the criterion of UMP unbiased tests is compatible with that of UMP tests.

In Section 9.2.1, we derived the power functions of three tests, Φ^*, Φ_*, and Φ, for testing the null hypothesis that $\theta = 0$ in Example A2 for the case $n = 1$. Φ^* is the

UMP size-.05 test for testing that $\theta = 0$ against $\theta > 0$, and Φ_* is the UMP size-.05 test for testing that $\theta = 0$ against $\theta < 0$. (See the exercises.) Now consider testing that $\theta = 0$ against $\theta \neq 0$. There is no UMP test for this situation, because Φ^* has the highest possible power among size-.05 tests when $\theta > 0$, while Φ_* has the highest possible power when $\theta < 0$. Therefore, there cannot be a single test which has the highest power on the whole alternative set. However, Φ^* and Φ_* are designed for testing one-sided hypotheses and are not really sensible for testing against the two-sided alternative that $\theta \neq 0$. One way to eliminate such tests is to look only at unbiased tests. Notc that Φ^* is not an unbiased test for the two-sided problem, since its power is less than .05 when $\theta < 0$. Similarly, Φ_* is not an unbiased test, because its power is less than .05 when $\theta > 0$. However, it appears from Table 9–1 that Φ is an unbiased size-.05 test for the two-sided problem. In fact, it can be shown that Φ is the UMP unbiased size-.05 test for this situation.

A rather extensive theory of UMP unbiased size-α tests has been developed which allows us to find UMP unbiased tests for testing that $\tau(\boldsymbol{\theta}) = a$ against one-sided or two-sided alternatives for many distributions, as long as $\tau(\boldsymbol{\theta})$ has dimension one. ($\boldsymbol{\theta}$ can have higher dimension.) This thoery is somewhat complicated and will not be presented here. (For treatments of the theory, see Lehmann (1986), pp. 134–281, or Ferguson (1967), pp. 224–238.) In later sections of the text, we shall often state (without proof) that certain tests are UMP unbiased.

We next present additional properties of the UMP tests just defined. These results are a digression and may be skipped without loss of continuity.

Theorem 9–3. Suppose that a model has monotone likelihood ratio in $T(\mathbf{X})$. Let $\Phi(\mathbf{X})$ be the test defined in part (a) of Theorem 9–2. Then

a. Φ is the uniformly most powerful size-α test for testing the null hypothesis that $\theta \leqq c$ against the alternative hypothesis that $\theta > c$.
b. If $\Phi^*(\mathbf{X})$ is another test such that $K_{\Phi^*}(c) = \alpha$, then

$$K_\Phi(\theta) \leqq K_{\Phi^*}(\theta) \text{ for all } \theta < c \quad \text{and} \quad K_\Phi(\theta) \geqq K_{\theta^*}(\theta) \text{ for all } \theta > c.$$

Proof

a. Let $a < c$, and let $\delta = K_\Phi(a)$. By part a of Theorem 9–2, Φ is the UMP size-δ test for testing that $\theta = a$ against $\theta > a$. Now, $\Phi^{**}(\mathbf{X}) \equiv \delta$ is a size-δ test for testing that $\theta = a$ against $\theta > a$, and hence, Φ is more powerful than Φ^{**} for all $\theta > d$, and it follows that

$$\alpha = K_\Phi(c) \geqq K_{\Phi^{**}}(c) = \delta.$$

Therefore, $K_\Phi(\theta) \leqq \alpha$ for all $\theta \leqq c$, and hence, Φ is a size-α test for testing that $\theta \leqq c$ against $\theta > c$. Now, let $\Phi^*(\mathbf{X})$ be any other size-α test for testing $\theta \leqq c$ against $\theta > c$. Then

$$K_{\Phi^*}(c) \leqq \sup_{\theta \leqq c} K_{\Phi^*}(\theta) = \alpha,$$

so that Φ^* has size $\gamma \leqq \alpha$ for testing that $\theta = c$ against $\theta > c$. Therefore,

$$K_\Phi(\theta) \geqq K_{\Phi^*}(\theta)$$

(see Exercise C1), and it follows that Φ is the UMP size-α test for this situation.

b. By Theorem 9–2, $K_\Phi(\theta) \geqq K_{\Phi^*}(\theta)$ for all $\theta > c$. To see the other inequality, let

$$\hat{\Phi}(\mathbf{X}) = 1 - \Phi(\mathbf{X}) = \begin{cases} 1 & \text{if } T(\mathbf{X}) < k \\ 1-d & \text{if } T(\mathbf{X}) = k \\ 0 & \text{if } T(\mathbf{X}) > k \end{cases}.$$

Then $K_{\hat{\Phi}}(c) = 1 - \alpha$. Now, by part b of Theorem 9–2, $\hat{\Phi}$ is the UMP size-$(1-\alpha)$ test for testing that $\theta = c$ against $\theta < c$. Let $\Phi^*(\mathbf{X})$ be any other test such that $K_{\Phi^*}(c) = \alpha$. Then $\hat{\Phi}^*(\mathbf{X}) = 1 - \hat{\Phi}(\mathbf{X})$ is a size-$(1-\alpha)$ test for testing that $\theta = c$ against $\theta < c$. Hence, for all $\theta < c$,

$$1 - K_\Phi(\theta) = K_{\hat{\Phi}}(\theta) \geqq K_{\hat{\Phi}^*}(\theta) = 1 - K_{\Phi^*}(\theta).$$

Therefore, for all $\theta < c$, $K_\Phi(\theta) \leqq K_{\Phi^*}(\theta)$. □

Part a of Theorem 9–2 shows that Φ_1 is the UMP size-α test for testing that $\theta = c$ against $\theta > c$. Part a of Theorem 9–3 shows that Φ_1 is also the UMP size-α test for testing $\theta \leqq c$ against $\theta > c$. Part b of this theorem is an especially strong result. It says that out of all tests whose power at c is α, Φ maximizes the power for $\theta > c$ and minimizes it for $\theta < c$. That is, out of this class of tests, Φ maximizes the probability of rejecting when the alternative hypothesis is true and maximizes the probability of accepting when the null hypothesis is true.

Theorem 9–3 could be extended in the obvious way to tests of the null hypothesis that $\theta \geqq c$ against $\theta < c$.

Exercises—B

1. For Example A3, find a UMP size-.05 test that $\theta = 2$ against $\theta > 2$ when $n = 3$. (*Hint*: $W = \Sigma X_i/i \sim \chi^2_6$ under the null hypothesis. Why?)
2. For Example A2, find a UMP size-.05 test that $\theta = 2$ against $\theta < 2$ when $n = 3$.
3. Let $X_1, \ldots, X_n$ be independent, with $X_i \sim N(0, \theta^2)$. Find the UMP size-α test for testing the null hypothesis that $\theta = 1$ against $\theta > 1$. If $\alpha = .05$, $n = 4$, $X_1 = 2$, $X_2 = 1$, $X_3 = -3$, and $X_4 = -2$, do we accept or reject the null hypothesis with this test?
4. Let X_1 and X_2 have independent geometric distributions with common parameter θ.
   (a) Show that this model has MLR in $-X_1 - X_2$.
   (b) Find the UMP size-.104 test that $\theta = .2$ against $\theta > .2$.
5. Let $X_1, \ldots, X_n$ be independent, with $X_i \sim B(1, \theta)$. Let $U = \Sigma X_i \sim B(n, \theta)$.
   (a) Show that this model has MLR in U.
   (b) Find a UMP size-.05 test that $\theta = .4$ against $\theta > .4$ when $n = 10$. (Note that this is a randomized test.)
   (c) Find a UMP size-.10 test that $\theta = .4$ against $\theta < .4$ when $n = 10$.
   (d) Find a UMP size-.05 test that $\theta = .2$ against $\theta > .2$ when $n = 100$. Use the normal approximation to approximate the critical value.
   (e) Find a UMP size-.05 test that $\theta = .01$ against $\theta > .01$ when $n = 100$. Use the Poisson approximation to approximate the critical value and the randomization.
6. Prove Theorem 9–2, part b.
7. Let $\mathbf{X}$ have density function $f(\mathbf{x}; \theta) = h(\mathbf{x})k(\theta)\exp[T(\mathbf{x})c(\theta)]$, where $c(\theta)$ is an increasing function of the one-dimensional parameter θ. Show that this model has MLR in $T(\mathbf{X})$.

8. Suppose that for $\theta_1 > \theta_0$, $L_X(\theta_1)/L_X(\theta_0)$ is a decreasing function of $T(\mathbf{X})$.
   (a) Show that this model has MLR in $T^*(\mathbf{X}) = -T(\mathbf{X})$.
   (b) Show that the UMP size-α test that $\theta = c$ against $\theta > c$ rejects the null hypothesis if $T^*(\mathbf{X})$ is too small.
9. For Example A2 with $n = 1$, let Φ^* and Φ_* be as defined in Section 9.2.1.
   (a) Show that Φ^* is the UMP size-.05 test that $\theta = 0$ against $\theta > 0$.
   (b) Show that Φ_* is the UMP size-.05 test that $\theta = 0$ against $\theta < 0$.

Exercises—C

1. Let $\Phi_1(\mathbf{X})$ be as given in part a of Theorem 9–2. Let $\Phi^*(\mathbf{X})$ have size $\gamma \leqq \alpha$ for testing that $\theta = c$ against $\theta > c$. Show that $K_\Phi(\theta) \geqq K_{\Phi^*}(\theta)$ for all $\theta > c$. (See Theorem 9–1.)
2. State and prove the extension of Theorem 9–3 to the problem of testing that $\theta \geqq b$ against $\theta < b$.
3. Let $\Phi_1(\mathbf{X})$ be the UMP size-α test given in Theorems 9–2 and 9–3. Show that $K_\Phi(\theta)$ is a nondecreasing function of θ. (*Hint*: Let $a < b$. Then Φ is the most powerful size-$\delta = K_\Phi(a)$ test for testing that $\theta = a$ against $\theta = b$. See Exercise C1 of the last section.)
4. Show that a UMP size-α test is unbiased and, hence, is UMP unbiased. (*Hint*: See Exercise C1, Section 9.2.2.)

9.3 LIKELIHOOD RATIO TESTS

9.3.1 Likelihood Ratio Tests

Consider the problem of testing the null hypothesis that $\boldsymbol{\theta} \in N$ against the alternative that $\boldsymbol{\theta} \in A$. Let $\hat{\boldsymbol{\theta}}$ be the MLE of $\boldsymbol{\theta}$ when $\boldsymbol{\theta} \in N \cup A$, and let $\hat{\hat{\boldsymbol{\theta}}}$ be the MLE of $\boldsymbol{\theta}$ when $\boldsymbol{\theta} \in N$. That is, $\hat{\boldsymbol{\theta}}$ is what we have been calling the MLE of $\boldsymbol{\theta}$, and $\hat{\hat{\boldsymbol{\theta}}}$ is the MLE of $\boldsymbol{\theta}$ for the restricted model that occurs when $\boldsymbol{\theta}$ is assumed to be in the null set N. We call $\hat{\boldsymbol{\theta}}$ (somewhat imprecisely) the MLE of $\boldsymbol{\theta}$ "under the alternative hypothesis" and $\hat{\hat{\boldsymbol{\theta}}}$ the MLE of $\boldsymbol{\theta}$ "under the null hypothesis."

Now let

$$\Lambda(\mathbf{X}) = \frac{L_X(\hat{\boldsymbol{\theta}})}{L_X(\hat{\hat{\boldsymbol{\theta}}})} \quad \text{and} \quad \Phi(\mathbf{X}) = \begin{cases} 1 & \text{if } \Lambda(\mathbf{X}) > k \\ d & \text{if } \Lambda(\mathbf{X}) = k \\ 0 & \text{if } \Lambda(\mathbf{X}) < k \end{cases},$$

where k and d are chosen so that Φ has size α. (If $\Lambda(\mathbf{X})$ is a continuous random variable, we take $d = 0$.) Then $\Phi(\mathbf{X})$ is the size-α *likelihood ratio test* (LRT), and $\Lambda(\mathbf{X})$ is the *likelihood ratio test statistic*. The likelihood ratio test rejects the null hypothesis if the likelihood ratio test statistic is too large. (Many books define the likelihood ratio test statistic as $1/\Lambda(\mathbf{X})$, and reject the null hypothesis when the statistic is small. Note that the test is the same in either case.)

$\Lambda(\mathbf{X})$ is a fairly natural generalization of the Neyman-Pearson test statistic given in Section 9.2.2, and hence, the likelihood ratio test is a natural generalization of the most powerful test given in that section. (Actually, a more natural generalization would define $\hat{\boldsymbol{\theta}}$ to be the MLE under A instead of $N \cup A$. However, for the models we are interested in, it can be shown that the MLE under $N \cup A$ is the same as the MLE under A with probability one. (See also Exercise C1.)

We first consider Example A2, in which we observe $X_i \sim N(i\theta, 1)$, independent. We propose testing the null hypothesis that $\theta = a$ against $\theta \neq a$. We have seen that the MLE under the alternative is $\hat{\theta} = Q = \Sigma i X_i / b_n$. Hence,

$$L_X(\hat{\theta}) = L_X(Q) = h_2(\mathbf{X}) \exp\left(\frac{-\sum(X_i - iQ)^2}{2}\right).$$

Since the null set consists of the single point a, the MLE of θ under the null hypothesis is $\hat{\hat{\theta}} = a$. Hence,

$$L_X(\hat{\hat{\theta}}) = L_X(a) = h_2(\mathbf{X}) \exp\left(\frac{-\sum(X_i - ia)^2}{2}\right).$$

Using the identity

$$\sum(X_i - ia)^2 = \sum(X_i - iQ)^2 + b_n(Q - a)^2,$$

we obtain

$$\Lambda(\mathbf{X}) = \frac{L_X(Q)}{L_X(a)} = \exp\left[\frac{b_n(Q-a)^2}{2}\right] = \exp\left(\frac{Z^2}{2}\right),$$

where

$$Z = b_n^{1/2}(Q - a)$$

is the test statistic for this problem derived in Chapter 8 using pivotal quantities. Therefore, the likelihood ratio test rejects the null hypothesis if

$$\Lambda(\mathbf{X}) > k \Leftrightarrow Z^2 > k^{*2} \Leftrightarrow Z > k^* \quad \text{or} \quad Z < -k^*,$$

for some k^*. In order to make this test a size-α test, we choose $k^* = z^{\alpha/2}$. The test is the same as the "sensible" test derived in Chapter 8.

We now look at Example B, in which we observe $X_i \sim N(\mu, \sigma^2)$, independent. First, consider testing that $\mu = a$ against $\mu \neq a$. From Chapter 7, the MLE of (μ, σ^2) under the alternative hypothesis is $(\overline{X}, V^2)$, where $V^2 = (n-1)S^2/n = \Sigma(X_i - \overline{X})^2/n$. Therefore,

$$L_X(\hat{\mu}, \hat{\sigma}^2) = L_X(\overline{X}, V^2) = (2\pi)^{-n/2}(V^2)^{-n/2} \exp\left(\frac{-\sum(X_i - \overline{X})^2}{2V^2}\right)$$

$$= (2\pi)^{-n/2}(V^2)^{-n/2} \exp\left(\frac{-n}{2}\right).$$

Let us now find the MLE of (μ, σ^2) under the null hypothesis that $\mu = a$. Clearly, the MLE of μ must be a, while the likelihood function under the null hypothesis is

$$L_X(a, \sigma^2) = (2\pi)^{-n/2}(\sigma^2)^{-n/2} \exp\left[\frac{-\sum(X_i - a)^2}{2\sigma^2}\right].$$

If we set the derivative of the log of this likelihood to zero, we find that the MLE for σ^2 under the null hypothesis is $W^2 = \Sigma(X_i - a)^2/n$. Therefore,

$$L_X(\hat{\mu}, \hat{\sigma}^2) = L_X(a, W^2) = (2\pi)^{-n/2}(W^2)^{-n/2} \exp\left(\frac{-\sum(X_i - a)^2}{2W^2}\right)$$

$$= (2\pi)^{-n/2}(W^2)^{-n/2} \exp\left(\frac{-n}{2}\right).$$

Hence,

$$\Lambda(\mathbf{X}) = \frac{L_X(\overline{X}, V^2)}{L_X(a, W^2)} = \left(\frac{W^2}{V^2}\right)^{n/2}.$$

Now,

$$W^2 = \frac{\sum(X_i - a)^2}{n} = V^2 + (\overline{X} - a)^2.$$

(See Exercise B4.) Therefore,

$$\Lambda(\mathbf{X}) = \left(1 + \frac{(\overline{X} - a)^2}{V^2}\right)^{n/2} = \left(1 + \frac{t^2}{n-1}\right)^{n/2},$$

where

$$t = \frac{n^{1/2}(\overline{X} - a)}{S} = \frac{(n-1)^{1/2}(\overline{X} - a)}{V}$$

is the test statistic suggested in Chapter 8 for this problem. Finally,

$$\Lambda(\mathbf{X}) > k \Leftrightarrow t^2 > (n-1)(k^{2/n} - 1) = (k^*)^2.$$

Hence, the LRT for this problem rejects the null hypothesis if

$$t > k^* \quad \text{or} \quad t < -k^*,$$

where k^* is chosen so that the test has size α, i.e., $k^* = t_{n-1}^{\alpha/2}$. Consequently, the LRT for this problem is the equal-tailed two-sided t-test suggested in Chapter 8.

Now, consider testing that $\sigma^2 = b^2$ against $\sigma^2 \neq b^2$. The MLE under the alternative is again given by $(\overline{X}, V^2)$. Under the null hypothesis, it is easily seen that the MLE is $(\overline{X}, b^2)$, and therefore,

$$\Lambda(\mathbf{X}) = \frac{(2\pi)^{-n/2}(V^2)^{-n/2} \exp[-\sum(X_i - \overline{X})^2/2V^2]}{(2\pi)^{-n/2}(b^2)^{-n/2} \exp[-\sum(X_i - \overline{X})^2/2b^2]}$$

$$= \left(\frac{V^2}{b^2}\right)^{-n/2} \exp\left[\frac{-n(1 - V^2/b^2)}{2}\right]$$

$$= \left(\frac{U}{n}\right)^{-n/2} \exp\left(\frac{U - n}{2}\right) = h(U),$$

where $U = nV^2/b^2$. Hence, the LRT is given by

$$\Phi(U) = \begin{cases} 1 & \text{if } h(U) > k \\ 0 & \text{if } h(U) \leqq k \end{cases}.$$

Unfortunately, it is not possible to solve the inequality $h(U) > k$ explicitly. However, we can describe the critical region for this test. In the exercises, it is shown that there exist c and d such that $h(U) > k$ if and only if $U > c$ or $U < d$. Therefore, the LRT rejects the null hypothesis if U is too small or too large. However, the LRT is not the equal-tails test, and it is difficult to find c and d such that the LRT has size α. Therefore, in applications, the equal-tails test is the one which is typically used.

We now state an important theorem which gives a simple formula for the asymptotic null distribution of $-2 \log \Lambda(\mathbf{X})$.

Theorem 9–4. Let $\mathbf{X}_1, \ldots, X_n$ be a sample from a distribution with parameter $\boldsymbol{\theta}$. Consider testing that $\tau(\boldsymbol{\theta}) = \mathbf{a}$ against $\tau(\boldsymbol{\theta}) \neq \mathbf{a}$, where $\boldsymbol{\tau}$ is r-dimensional. Let $\Lambda(\mathbf{X})$ be the LRT statistic for this problem. Then, under certain regularity conditions,

$$2 \log \Lambda(\mathbf{X}) \xrightarrow{d} Y \sim \chi_r^2$$

under the null hypothesis.

The proof of this theorem is beyond the scope of the text. (See Rao (1965), pp. 347–352, for a proof.) The regularity conditions include those for the asymptotic properties of MLEs.

Now suppose, in Example B, that we wish to test the null hypothesis that $\mu = 0$ and $\sigma^2 = 1$ against the alternative that $\mu \neq 0$ or $\sigma^2 \neq 1$ or both. The MLE under the alternative is $(\overline{X}, V^2)$, and we have seen previously that

$$L_X(\overline{X}, V^2) = (2\pi)^{-n/2}(V^2)^{-n/2} \exp\left(\frac{-n}{2}\right).$$

The MLE under the null hypothesis is $(0, 1)$, and

$$L_X(0, 1) = (2\pi)^{-n/2} \exp\left(\frac{-\sum X_i^2}{2}\right).$$

Therefore, the likelihood ratio test statistic for this problem is

$$\Lambda(\mathbf{X}) = (V^2)^{-n/2} \exp\left(\frac{n - \sum X_i^2}{2}\right).$$

Under the null hypothesis, there are no unknown parameters, so that the distribution of $\Lambda(\mathbf{X})$ could be tabled. However, it has not been. Therefore, we shall use Theorem 9–4 to find an asymptotic size-α test for this problem. We are testing that $\boldsymbol{\tau} = (\mu, \sigma^2) = (0, 1)$, so that $\boldsymbol{\tau}$ is a two-dimensional parameter. Thus,

$$2 \log[\Lambda(\mathbf{X})] \xrightarrow{d} Y \sim \chi_2^2$$

under the null hypothesis. Hence, the asymptotic size-α LRT rejects the null hypothesis if

$$2\log[(\Lambda(\mathbf{X})] > \chi_2^{2\,\alpha}.$$

Since $\boldsymbol{\tau}$ is a two-dimensional parameter, there is no way we could have used the method of pivotal quantities to design a test for this model.

There is no obvious reason why the LRT must be a good test for all problems. In fact, in Exercise C2, an example is given in which the probability of rejecting under the alternative hypothesis (when we should reject it) is less than the probability of rejecting under the null hypothesis (when we should not reject it). In this case, using the LRT is worse than ignoring the data. (Note that the UMP test can never be this bad; see Exercise C4, Section 9.2.3.) However, examples like this are quite artificial. Practically speaking, the LRT is usually a sensible test. Nonetheless, because the LRT can be bad in some contrived examples, there is always the chance that it could be bad in a practical situation. Therefore, it is still important to study the power function of an LRT to see if it is reasonable (e.g., to see if the test is unbiased or UMP unbiased). Such investigations are beyond the scope of this text, however.

Exercises—B

1. For Example A2, show that $\Sigma(X_i - ia)^2 = \Sigma(X_i - iQ)^2 + b_n(Q - a)^2$.
2. For Example A1, show that the likelihood ratio test statistic for testing $\theta = c$ against $\theta \neq c$ is given by $\Lambda(\mathbf{X}) = \exp[-a_n P\log(c/P) - a_n(P - c)]$.
3. For Example A3, show that the likelihood ratio test statistic for testing that $\theta = c$ against $\theta \neq c$ is $\Lambda(\mathbf{X}) = (W/2n)^{-n}\exp[-n + W/2]$, where $W = 2nR/c$ is the test statistic suggested in Chapter 8 for this model. What is the form of the LRT for this model?
4. For Example B, show that $W^2 = V^2 + (\overline{X} - a)^2$. (*Hint*: $X_i - a = X_i - \overline{X} + \overline{X} - a$, and $\Sigma(X_i - \overline{X}) = 0$. Why?)
5. Let $h(U)$ be defined as in the derivation of the LRT for testing that $\sigma^2 = b^2$. (See page 324.)
   (a) Show that $h(U) \to \infty$ as $u \to 0$ or $u \to \infty$.
   (b) Show that $h'(U) = 0$ if and only if $U = n$.
   (c) Sketch a graph of h, and show that there exist c and d such that $h(U) > k$ if and only if $U > c$ or $U < d$.
6. Let $X_1, \ldots, X_n$ be independent, with $X_i \sim N(\mu, 1)$. Find the size-α LRT for testing that $\mu = a$ against $\mu \neq a$, and show that it rejects the null hypothesis if $n^{1/2}(\overline{X} - a) > z^{\alpha/2}$ or $n^{1/2}(\overline{X} - a) < -z^{\alpha/2}$.
7. Consider the model in Example C.
   (a) Show that the MLE under the alternative hypothesis is $(\hat{\theta}, \hat{\sigma}^2)$, where $\hat{\theta} = \Sigma iX_i/b_n$ and $\hat{\sigma}^2 = \Sigma(X_i - i\hat{\theta})^2/n$.
   (b) Consider testing that $\theta = a$ against $\theta \neq a$. Show that the MLE under the null hypothesis is $(a, \hat{\sigma}^2 + b_n(\hat{\theta} - a)^2/n)$.
   (c) Show that the LRT statistic is $\Lambda = (1 + t^2/(n-1))^{n/2}$, where $t = b_n^{1/2}(\hat{\theta} - a)/T$, and that the size-$\alpha$ LRT rejects the null hypothesis if $t > t_{n-1}^{\alpha/2}$ or $t < -t_{n-1}^{\alpha/2}$.
8. For the same model as in Exercise 7, consider testing that $\sigma^2 = c^2$ against $\sigma^2 \neq c^2$.
   (a) Show that the MLE under the null hypothesis is $(\hat{\theta}, c^2)$.

(b) Show that the LRT statistic for this problem is $\Lambda = (W/n)^{-n/2} \exp[(W - n)/2]$, where $W = (n-1)T^2/c^2 = n\hat{\sigma}^2/c^2$. How is W distributed under the null hypothesis?
(c) Show that the LRT rejects the null hypothesis if W is too large or too small.

9. Find the LRT for testing that $\theta = 1$ and $\sigma^2 = 4$ in the model considered in the last two problems. How many degrees of freedom are there in the asymptotic χ^2 distribution?

10. Let (X, Y) have a trinomial distribution with parameters n and (θ_1, θ_2). Consider testing that $\theta_1 = \theta_2$ against $\theta_1 \neq \theta_2$.
(a) What is the MLE of (θ_1, θ_2) under the alternative hypothesis?
(b) Show that the MLE under the null hypothesis is $\hat{\theta}_1 = \hat{\theta}_2 = (X + Y)/n$.
(c) What is the LRT statistic for this problem?

Exercises—C

1. In the general problem of testing that $\boldsymbol{\theta} \in N$ against $\boldsymbol{\theta} \in A$, let $\hat{\boldsymbol{\theta}}^*$ be the MLE of $\boldsymbol{\theta}$ over the set A. Let $\Lambda^*(\mathbf{X}) = L_X(\hat{\boldsymbol{\theta}}^*)/L_X(\hat{\hat{\boldsymbol{\theta}}})$.
(a) Show that $L_X(\hat{\boldsymbol{\theta}}) = \max(L_X(\hat{\boldsymbol{\theta}}^*), L_X(\hat{\hat{\boldsymbol{\theta}}}))$.
(b) Show that $\Lambda(\mathbf{X}) = \max(\Lambda^*(\mathbf{X}), 1)$.
(c) Show that if $k > 1$, then $\Lambda(\mathbf{X}) > k$ if and only if $\Lambda^*(\mathbf{X}) > k$.

2. Suppose X is a discrete random variable taking on the values 1, 2, 3, and 4, and θ takes on the values -1, 0, and 1. Suppose also that the density of X is given in the following table:

x	1	2	3	4
$f(x; -1)$	.53	.30	.00	.17
$f(x; 0)$	.60	.20	.10	.10
$f(x; 1)$	.60	.22	.18	0

Consider testing that $\theta = 0$ against $\theta \neq 0$.
(a) Show that the .2 LRT for this problem rejects the null hypothesis if X is 3 or 4. (*Hint*: $\Lambda(1) = 1$, $\Lambda(2) = 1.5$, $\Lambda(3) = 1.8$, and $\Lambda(4) = 1.7$. Why?)
(b) Show that the probability of rejecting under the alternative hypothesis with this test is less than .2.
(c) Show that the test which rejects the null hypothesis if $X = 2$ is a size-.2 test which is more powerful than the LRT.

* 9.3.2 One-Sided Likelihood Ratio Tests

In this section, we derive LRTs for some one-sided problems. We first note that the LRT statistic $\Lambda(\mathbf{X})$ always satisfies

$$\Lambda(\mathbf{X}) \geqq 1$$

(since the numerator is the maximum of the likelihood over a larger set than the denominator is). The LRT rejects the null hypothesis if $\Lambda(\mathbf{X}) > k$. If $k < 1$, then the LRT always rejects the null hypothesis. Therefore, we assume that $k \geqq 1$.

Consider, first, Example A2, in which we observe $X_i \sim N(i\theta, 1)$, independent. We find the LRT for testing that $\theta = a$ against $\theta > a$. Under the null hypothesis, the

MLE is

$$\hat{\hat{\theta}} = a, \qquad L_X(\hat{\hat{\theta}}) = (2\pi)^{-n/2} \exp\left[\frac{-\sum(X_i - a)^2}{2}\right].$$

To find the MLE under the alternative hypothesis, we must maximize the likelihood over the set in which $\theta \geqq a$. Since Q is the MLE for θ over the space of all θ, Q maximizes the likelihood over the set $\theta \geqq a$ as long as $Q \geqq a$. If $Q < a$, then the likelihood looks like the graph in Figure 9–2. Therefore, in this case, the maximum for the likelihood occurs at the point $\theta_r = a$. Hence, the MLE over the set $\theta \geqq a$ is given by

$$\hat{\hat{\theta}}^* = \begin{cases} Q & \text{if } Q \geqq a \\ a & \text{if } Q < a \end{cases},$$

and the LRT statistic for this model is

$$\Lambda(\mathbf{X}) = \frac{L_X(\hat{\theta})}{L_X(\hat{\hat{\theta}}^*)} = \begin{cases} 1 & \text{if } Q < a \\ \exp[b_n(Q-a)^2/2] & \text{if } Q \geqq a \end{cases}.$$

The LRT rejects the null hypothesis if $\Lambda(\mathbf{X}) > k$ for some k. If $k \geqq 1$, the test rejects the null hypothesis if

$$\exp\left[\frac{b_n(Q-a)^2}{2}\right] > k \quad \text{and} \quad Q \geqq a,$$

that is, if

$$Z = b_n^{1/2}(Q - a) > k^* = (2\log(k))^{1/2}.$$

For a size-α test, we take $k^* = z^{\alpha}$. The LRT for this problem is then the same as the test derived earlier using pivotal quantities.

Now consider Example B, in which we observe $X_i \sim N(\mu, \sigma^2)$, independent. We find the LRT for testing that $\mu = a$ against $\mu < a$. From the previous section, we see that the MLE for (μ, σ^2) under the null hypothesis is

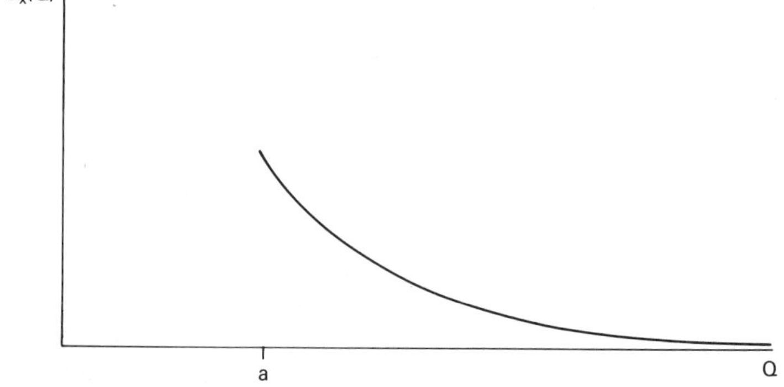

Figure 9–2 $L_x(Q) = (2\pi)^{-n/2} \exp\left[-\frac{1}{2}\sum_{i=1}^{n}(X_i - Q)^2\right].$

$$\hat{\mu} = a, \qquad \hat{\sigma}^2 = W^2 = \frac{\sum(X_i - a)^2}{n}, \qquad L_X(\hat{\mu}, \hat{\sigma}^2) = (2\pi)^{-n/2}(W^2)^{-n/2} \exp\left(\frac{-n}{2}\right).$$

To find the MLE under the alternative hypothesis, we must maximize the likelihood over the set $\mu \leqq a, \sigma^2 > 0$. If $\overline{X} \leqq a$, then this maximum occurs at $(\overline{X}, V^2)$, where $V^2 = \Sigma(X_i - \overline{X})^2/n$. If $\overline{X} > a$, then there is no critical point in the set $\mu \leqq a$, so that the maximum occurs on the boundary, i.e., the set where $\mu = a$. Hence, when $Q > a$, the maximum occurs at (a, W^2). Therefore, the MLE under the alternative hypothesis is

$$(\hat{\mu}^*, \hat{\sigma}^{*2}) = \begin{cases} (\overline{X}, V^2) & \text{if } \overline{X} \leqq a \\ (a, W^2) & \text{if } \overline{X} > a \end{cases},$$

and it follows that the LRT statistic $\Lambda^*(\mathbf{X}) = 1$ if $\overline{X} > a$, and is the same as the LRT statistic $\Lambda(\mathbf{X})$ for the two-sided testing problem if $\overline{X} \leqq a$ (because if $\overline{X} \leqq a$, the MLEs under the null and alternative hypotheses are the same as they are for the two-sided problem). That is,

$$\Lambda^*(\mathbf{X}) = \begin{cases} 1 & \text{if } \overline{X} > a \\ (1 + t^2/(n-1))^{n/2} & \text{if } \overline{X} \leqq a \end{cases},$$

where $t = n^{1/2}(\overline{X} - a)/S$. Note that $\overline{X} \leqq a$ if and only if $t < 0$. Therefore, $\Lambda^*(\mathbf{X}) > k \geqq 1$ if and only if $\Lambda(\mathbf{X}) > k$, and $t < 0$ if and only if

$$t < k^* = -(n-1)(k^{2/n} - 1).$$

For a size-α test, we take $k^* = -t_{n-1}^{\alpha}$. Therefore, the LRT for this model is the same as the test derived earlier using pivotal quantities.

From these examples, we see that the one-sided LRT statistic $\Lambda^*(\mathbf{X})$ is often equal to the two-sided LRT statistic $\Lambda(\mathbf{X})$ on a set A, and $\Lambda^*(\mathbf{X}) = 1$ otherwise. Therefore, if $k > 1$, then

$$\Lambda^*(\mathbf{X}) > k \Leftrightarrow \Lambda(\mathbf{X}) > k \quad \text{and} \quad \mathbf{X} \in A.$$

(Note that the critical value k (or k^*) necessary to make the one-sided LRT have size α is different from the critical value necessary for the two-sided test to do the same.)

Exercises—B

1. For Example A3, show tht the LRT for testing that $\theta = c$ against $\theta < c$ rejects the null hypothesis if $U = 2nR/c$ is too small.
2. For Example A1, show that the LRT for testing that $\theta = c$ against $\theta < c$ rejects the null hypothesis if P is too small.
3. In Example B, consider testing that $\mu = a$ against $\mu > a$.
   (a) What is the MLE under the null hypothesis?
   (b) Argue that the MLE under the alternative hypothesis is $(\overline{X}, V^2)$ if $\overline{X} \geqq a$ and is (a, W^2) if $\overline{X} < a$.
   (c) Show that the LRT statistic $\Lambda(\mathbf{X})$ for this model is $(1 + t^2/(n-1))^{n/2}$ if $t \geqq 0$ and is one if $t < 0$ (where $t = (n-1)^{1/2}(\overline{X} - a)/W$).
   (d) Let $0 < k < 1$. Show that $\Lambda(\mathbf{X}) > k$ if and only if $t > k^*$ for some k^*.

4. In Example B, consider testing that $\mu \geqq a$ against $\mu < a$. (Note that the null hypothesis for this problem is the same as the union of the null and alternative hypotheses for the previous problem.)
   (a) Show that the LRT statistic $\Lambda(\mathbf{X})$ in this case is $(1 + t^2/(n-1))^{n/2}$ if $t \leqq 0$ and is one if $t > 0$.
   (b) Show that the LRT for this model rejects the null hypothesis if t is too small.
5. In Example B, find the size-α LRT for testing that $\sigma^2 = c^2$ against $\sigma^2 > c^2$, and show that it rejects the null hypothesis if $U > \chi^{2\ \alpha}_{n-1}$, where $U = (n-1)S^2/c^2$.

Chapter 10

SUFFICIENT STATISTICS

Sufficient statistics contain "all the information" in the data about the unknown parameters. Often, sufficient statistics have much lower dimension than the original data, so we can reduce a problem fairly dramatically with no loss. For instance, in the joggers' pulse example, the sample mean and sample variance are sufficient so that we can reduce the data from 30 numbers to 2 numbers with no loss of information.

A careful discussion of sufficient statistics in the case of continuous random variables involves measure theory. Therefore, many of the proofs are given only in the discrete case. Recall that if $\mathbf{X}$ and $\mathbf{Y}$ are random variables or vectors, then

$$\mathbf{X} \equiv \mathbf{Y} \Leftrightarrow P_{\boldsymbol{\theta}}(\mathbf{X} \neq \mathbf{Y}) = 0$$

for all $\boldsymbol{\theta}$.

10.1 DEFINITION AND BASIC PROPERTIES

10.1.1 Definition

Consider the general statistical model in which we observe the random vector $\mathbf{X}$ having density $f(\mathbf{x}; \boldsymbol{\theta})$. Let $\mathbf{T} = T(\mathbf{X})$ be a (possibly vector-valued) statistic. Then $\mathbf{T}$ is a *sufficient statistic* for this model if the conditional distribution of $\mathbf{X}$ given $\mathbf{T}$ does not depend on the unknown parameter $\boldsymbol{\theta}$.

In the next paragraph, we argue that for any procedure based on the original data $\mathbf{X}$, there is a procedure based on $\mathbf{T}$ which is just as good. Therefore, a person who knows $\mathbf{T}$ really knows just as much about $\boldsymbol{\theta}$ as someone who knows $\mathbf{X}$, and hence, we need not keep track of $\mathbf{X}$, but need only know $\mathbf{T}$. (In the jogging example, we need only keep $(\overline{X}, S^2)$ and can throw away the original data.) The *principle of sufficiency*, which most statisticians accept, says that any procedure we use for drawing inferences about $\boldsymbol{\theta}$ should depend only on the sufficient statistic $\mathbf{T}$.

We now present a heuristic argument which shows how to find a procedure that depends only on the sufficient statistic $\mathbf{T}$ which is as good as a given procedure based on the original sample $\mathbf{X}$. To keep the ideas straight, suppose we have one statistician who knows the original data vector $\mathbf{X}$ and another statistician who only knows the value of the sufficient statistic **T**. (In the jogging example, the first statistician knows the pulse rates of all 30 joggers, while the second one only knows the sample mean and sample variance of their pulse rates.) Obviously, any procedure which the second statistician can use, the first one can use also, since she can compute $\mathbf{T} = T(\mathbf{X})$. On the other hand, there are procedures which the first statistician can use which cannot be used by the second one, since he is unable to compute **X** from **T**. (There is no way to compute the actual pulse rates from the sample mean and sample variance.) However, since the conditional distribution of **X** given **T** does not depend on any unknown parameters, the second statistician knows that conditional distribution completely. He can therefore (at least in principle) artificially generate a new random vector $\mathbf{X}^*$ such that $\mathbf{X}^*$ given **T** has the same conditional distribution as **X** given **T**. Then the distribution of $\mathbf{X}^*$ is the same as the distribution of **X**. (Why?) Now, let $h(\mathbf{X})$ be a procedure (estimator, confidence interval, or test) which the first statistician can use for drawing inference about $\boldsymbol{\theta}$. Then, since **X** and $\mathbf{X}^*$ have the same distribution, $h(\mathbf{X}^*)$ has the same properties as $h(\mathbf{X})$. Hence, for any procedure based on the original observations **X** which the first statistician can use, the second statistician can apply the same procedure to $\mathbf{X}^*$ and get a procedure that is just as good. (If **T** were not a sufficient statistic, then the conditional distribution of **X** given **T** would depend on unknown parameters and, therefore, could not be artificially generated.)

To make these ideas more precise, consider the following simple example. We observe X_1 and X_2, independent Bernoulli random variables with the same probability of success θ, which is the unknown parameter. Let

$$T = T(X_1, X_2) = X_1 + X_2.$$

The conditional distribution of (X_1, X_2) given T is fairly simple. If $T = 0$, then (X_1, X_2) must be $(0,0)$. Similarly, if $T = 2$, then (X_1, X_2) must be $(1,1)$. If $T = 1$, then (X_1, X_2) could be either $(0,1)$ or $(1,0)$. Now, $T \sim B(2, \theta)$. Therefore,

$$P((X_1, X_2) = (1,0) | T = 1) = \frac{P(X_1 = 1, X_2 = 0)}{P(T = 1)}$$

$$= \frac{P(X_1 = 1)P(X_2 = 0)}{P(T = 1)} = \frac{\theta(1-\theta)}{2\theta(1-\theta)} = \frac{1}{2}.$$

Similarly, $P((X_1, X_2) = (0,1) | T = 1) = \frac{1}{2}$. This completes the conditional distribution of (X_1, X_2) given T. Note that this conditional distribution does not depend on θ, so that T is a sufficient statistic.

Now, we can generate (X_1^*, X_2^*) in the following fashion. If $T = 0$, let $(X_1^*, X_2^*) = (0,0)$, and if $T = 2$, let $(X_1^*, X_2^*) = (1,1)$. If $T = 1$, then flip a fair coin. If it comes up heads, let $(X_1^*, X_2^*) = (1,0)$, and if it comes up tails, let $(X_1^*, X_2^*) = (0,1)$. Now, let $h(X_1, X_2) = X_1$ be an estimator of θ. Then $h(X_1^*, X_2^*) = X_1^*$ would be just as good. Note that if $T = 0$, then $h = 0$, and if $T = 2$, then $h = 1$. If $T = 1$, then $h = 1$ with probability $\frac{1}{2}$ and $h = 0$ with probability $\frac{1}{2}$, depending on whether the coin comes up heads or tails. Note that although $h(X_1^*, X_2^*)$ and $h(X_1, X_2)$ would have the same properties, they are different procedures, because (X_1^*, X_2^*) does not necessarily equal (X_1, X_2); it just has the same distribution.

As mentioned, the preceding argument is heuristic in nature. In particular, it is not always obvious how to generate the vector $\mathbf{X}^*$ from the complicated conditional distribution of $\mathbf{X}$ given $\mathbf{T}$, even with the aid of a computer. However, in Section 10.1.3, we prove a series of results which make more specific the sense in which procedures based on a sufficient statistic are as good as those based on the original observations.

Exercises—B

1. Let X_1 and X_2 be independent, with $X_i \sim P(\theta)$. Let $T = X_1 + X_2$.
 (a) Find $P((X_1 = 2, X_2 = 1)|T = 3) = P(X_1 = 2, X_2 = 1)/P(T = 3)$. ($T \sim P(2\theta)$. Why?)
 (b) Find the conditional distribution of (X_1, X_2) given T, and show that T is a sufficient statistic for this model.

Exercises—C

1. Consider again Example A1. Let $T = a_n P = \Sigma X_i$.
 (a) How is T distributed?
 (b) Find $P(X_1 = x_1, \ldots, X_n = x_n|T = t) = P(X_1 = x_1, \ldots, X_n = x_n)/P(T = t)$.
 (c) Show that T is a sufficient statistic for this model.

10.1.2 The Factorization and Fisher-Neyman Criteria

For most models, it is quite difficult to find the conditional distribution of the original observations $\mathbf{X}$ given a statistic $\mathbf{T}$. Hence, for such models, it is difficult to use the definition to check whether a statistic is sufficient. Fortunately, there are two criteria for sufficiency which are easy to check. These criteria are true in both the discrete and continuous cases, and we shall use them in both cases in this book. However, in the continuous case, their proofs need measure theory. We therefore only give proofs for the case in which $\mathbf{X}$ (and hence $\mathbf{T}$) has a discrete distribution. (For a general proof, see Lehmann (1986), pp. 53–56.)

Let T be a statistic with density function $g(\mathbf{t}; \boldsymbol{\theta})$. Then T has likelihood function

$$L_T^*(\boldsymbol{\theta}) = g(\mathbf{T}; \boldsymbol{\theta}).$$

Theorem 10–1. Let $\mathbf{X}$ have likelihood $L_X(\boldsymbol{\theta})$, and let $\mathbf{T} = T(\mathbf{X})$ be a statistic with likelihood $L_T^*(\boldsymbol{\theta})$. Then $\mathbf{T}$ is a sufficient statistic if and only if either of the following criteria hold:

a. (Fisher-Neyman) $L_X(\boldsymbol{\theta}) = h(\mathbf{X}) L_{T(X)}^*(\boldsymbol{\theta})$ for some function $h(\mathbf{X}) \geqq 0$;
b. (Factorization) $L_X(\boldsymbol{\theta}) = h(\mathbf{X}) k(T(\mathbf{X}); \boldsymbol{\theta})$ for some functions $h(\mathbf{X})$ and $k(\mathbf{T}; \boldsymbol{\theta})$.

Proof. (Discrete case only) Let $g(\mathbf{t}; \boldsymbol{\theta})$ be the density of T. First, suppose that $\mathbf{T}$ is sufficient. Then $\mathbf{X} = \mathbf{x}$ if and only if $\mathbf{X} = \mathbf{x}$ and $T(\mathbf{X}) = T(\mathbf{x})$. Therefore,

$$\begin{aligned} f(\mathbf{x}; \boldsymbol{\theta}) &= P_\theta(\mathbf{X} = \mathbf{x}) = P_\theta(\mathbf{X} = \mathbf{x}, \mathbf{T} = T(\mathbf{x})) \\ &= P_\theta(\mathbf{X} = \mathbf{x} | \mathbf{T} = T(\mathbf{x})) P_\theta(\mathbf{T} = T(\mathbf{x})) = h(\mathbf{x}) g(T(\mathbf{x}); \theta) \end{aligned}$$

(since the conditional distribution of **X** given **T** does not depend on **θ**). Therefore, sufficiency implies the Fisher-Neyman criterion.

Now, clearly, the Fisher-Neyman criterion implies the factorization criterion. We shall therefore be finished when we show that the factorization criterion implies sufficiency. Accordingly, for any **t**, let

$$A_t = \{\mathbf{x}: T(\mathbf{x}) = \mathbf{t}\}.$$

(That is, A_t is the set of **x**'s such that T maps **x** into **t**.) Then the conditional density of **X** given **T** is

$$P_\theta(\mathbf{X} = \mathbf{x} | \mathbf{T} = \mathbf{t}) = \frac{P_\theta(\mathbf{X} = \mathbf{x}, \mathbf{T} = \mathbf{t})}{P_\theta(\mathbf{T} = \mathbf{t})}$$
$$= \begin{cases} P_\theta(\mathbf{X} = \mathbf{x})/P_\theta(\mathbf{T} = \mathbf{t}) & \text{if } \mathbf{x} \in A_t \\ 0 & \text{if } \mathbf{x} \notin A_t \end{cases}.$$

Now, A_t does not depend on **θ**. Also, if $\mathbf{x} \in A_t$, and if the factorization criterion holds, then

$$\frac{P_\theta(\mathbf{X} = \mathbf{x})}{P_\theta(\mathbf{T} = \mathbf{t})} = \frac{h(\mathbf{x})k(T(\mathbf{x}); \theta)}{\sum_{\mathbf{u} \in A_t} h(\mathbf{u})k(T(\mathbf{u}); \theta)} = \frac{h(\mathbf{x})k(\mathbf{t}; \theta)}{k(\mathbf{t}; \theta) \sum_{\mathbf{u} \in A_t} h(\mathbf{u})} = \frac{h(\mathbf{x})}{\sum_{\mathbf{u} \in A_t} h(\mathbf{u})}.$$

(Note that $k(T(\mathbf{x}); \theta) = k(\mathbf{t}; \theta)$ for all $\mathbf{x} \in A_t$, and hence, it can be pulled outside the sum.) Therefore, if the factorization criterion holds, then the conditional distribution of **X** given **T** does not depend on **θ**, and **T** is sufficient. □

Although we have proved the two criteria using density functions, we have stated them using likelihood functions. This is because the proof is more natural using densities. However, applications are more natural using **X**, indicating that $T(\mathbf{X})$ is a statistic. In order for T to be a sufficient statistic, it is necessary that these criteria hold for all **X** in the sample space. The Fisher-Neyman criterion merely says that we may take $k(\mathbf{T}; \boldsymbol{\theta})$ to be the likelihood of **T** and $h(\mathbf{x}) \geqq 0$ in the factorization criterion. Typically, we use the factorization criterion for showing that a statistic is sufficient. However, we often use the Fisher-Neyman criterion for deriving facts about sufficient statistics.

Before looking at examples of sufficient statistics, we prove a useful theorem.

Theorem 10–2. Let $\mathbf{T} = T(\mathbf{X})$ be a sufficient statistic, and let $\mathbf{S} = S(\mathbf{X})$ be an invertible function of **T**. Then **S** is also a sufficient statistic.

Proof. Since $S(\mathbf{X})$ is an invertible function of $T(\mathbf{X})$, there exists a function $R(\mathbf{S})$ such that $T(\mathbf{X}) = R(S(\mathbf{X}))$. But then, by the factorization criterion,

$$L_X(\boldsymbol{\theta}) = h(\mathbf{X})k(T(\mathbf{X}); \boldsymbol{\theta}) = h(\mathbf{X})k(R(S(\mathbf{X})); \boldsymbol{\theta}) = h(\mathbf{X})k^*(S(\mathbf{X}); \boldsymbol{\theta}).$$

And by the factorization criterion again, $S(\mathbf{X})$ is sufficient. □

Example A1

Let $X_i \sim P(i\theta)$, independent. Then

$$L_X(\theta) = h_1(\mathbf{X})\theta^{\Sigma X_i} \exp(-\theta a_n).$$

By the factorization criterion, ΣX_i is a sufficient statistic. Now, the MLE for θ for this model is $\Sigma X_i/a_n = P$, which is an invertible function of ΣX_i, so that P is also a sufficient statistic for this model.

Example A2

Let X_i be independent, with $X_i \sim N(i\theta, 1)$. Then

$$L_X(\theta) = h_2(\mathbf{X}) \exp\left(\theta \sum iX_i - \frac{\theta^2 b_n}{2}\right).$$

Therefore, by the factorization criterion, ΣiX_i is a sufficient statistic. Now, the MLE of θ for this model is $Q = \Sigma iX_i/b_n$, which is an invertible function of ΣiX_i and is therefore also a sufficient statistic.

Example A3

Let X_i be independent, with $X_i \sim E(i\theta)$. Then $R = \Sigma X_i/in$ is a sufficient statistic. (See Exercise B1.)

In the previous two chapters, we have derived many procedures for the three models in Example A. For Example A1 the procedures have all depended only on P, for Example A2 they have depended only on Q, and for Example A3, they have depended only on R. The reason we have designed procedures in this manner is that P is a sufficient statistic for model A1, Q is a sufficient statistic for model A2, and R is a sufficient statistic for model A3.

We now return to Example B.

Example B

Let $X_i \sim N(\mu, \sigma^2)$, independent. Let $\theta = (\mu, \sigma^2)$. Then

$$L_X(\mu, \sigma^2) = (2\pi)^{-n/2} \sigma^{-n} \exp\left(\frac{-(\sum X_i^2 - 2\mu \sum X_i + n\mu^2)}{2\sigma^2}\right),$$

and therefore, by the factorization criterion,

$$\mathbf{T} = (T_1, T_2) = \left(\sum X_i^2, \sum X_i\right)$$

is a sufficient statistic for this model. Now,

$$\overline{X} = \frac{T_2}{n}, \qquad S^2 = \frac{T_1 - T_2^2/n}{n-1},$$

$$T_1 = (n-1)S^2 + n\overline{X}^2, \qquad T_2 = n\overline{X}.$$

(See Exercise B2.) Therefore, $(\overline{X}, S^2)$ is an invertible function of $\mathbf{T}$ and, consequently, also a sufficient statistic. The MLE of (μ, σ^2) is an invertible function of $(\overline{X}, S^2)$ and is therefore also sufficient.

In this example, it is often tempting to think of $\overline{X}$ as sufficient for μ and S^2 as sufficient for σ^2. However, this is not a correct way of looking at the situation. Rather, $\overline{X}$ and S^2 are jointly sufficient for μ and σ^2. (In fact, if we know that $\sigma^2 = a$, then $\overline{X}$ is a sufficient statistic. However, if we know that $\mu = b$, then S^2 is not a sufficient statistic.)

From the preceding examples, we see that it is often quite easy to use the factorization criterion to find a sufficient statistic. First, we write down the likelihood function $L_X(\boldsymbol{\theta})$. Then we factor out anything which just depends on $\mathbf{X}$ $(h(\mathbf{X}))$. Then we look at what is left $(k(T(\mathbf{X}); \boldsymbol{\theta}))$ and list the functions of $\mathbf{X}$ which occur in it $(T_1(\mathbf{X}), \ldots, T_a(\mathbf{X}))$. Finally,

$$\mathbf{T} = (T_1(\mathbf{X}), \ldots, T_a(\mathbf{X}))$$

is a sufficient statistic.

Exercises—B

1. Show that R is a sufficient statistic for Example A3.
2. Show that, for Example B, (T_1, T_2) is an invertible function of $(\overline{X}, S^2)$.
3. Let X_i be independent, with $X_i \sim B(1, \theta)$. Show that $\overline{X}$ is a sufficient statistic.
4. Let X_i be independent, with $X_i \sim \Gamma(i, \theta)$. Show that $P = \Sigma X_i/a_n$ is a sufficient statistic.
5. Let (X, Y) have a trinomial distribution with parameters n, θ^2, and $2\theta(1-\theta)$, where n is known and θ is the unknown parameter. Show that $(2X + Y)/2n$ is a sufficient statistic.
6. For Example C of Section 7.1, show that the MLE of (θ, σ^2) is a sufficient statistic.
7. Let X_i be independent, with marginal density $f(x; \theta) = \theta x^{\theta - 1}, 0 < x < 1$. Show that ΠX_i is a sufficient statistic.

Exercises—C

1. Let θ be a univariate parameter, and suppose that the joint density $f(\mathbf{x}; \theta)$ of $\mathbf{X}$ has monotone likelihood ratio in $T(\mathbf{X})$. Show that $T(\mathbf{X})$ is a sufficient statistic. (*Hint*: Let a be a possible value for the parameter θ. Then $L_X(\theta)/L_X(a) = h(T(\mathbf{X}); \theta)$ for some function h, whether $\theta > a$, $\theta < a$, or $\theta = a$. Why?)
2. Consider a model in which we observe $\mathbf{X}$ having likelihood $L_X(\boldsymbol{\theta})$, where the parameter space consists only of the two points $\boldsymbol{\theta} = \mathbf{a}$ and $\boldsymbol{\theta} = \mathbf{b}$. Show that the Neyman-Pearson test statistic $\Lambda(\mathbf{X}) = L_X(\mathbf{a})/L_X(\mathbf{b})$ is a sufficient statistic. (*Hint*: Let $h(\Lambda, \theta) = \Lambda$ if $\boldsymbol{\theta} = \mathbf{a}$ and $h(\Lambda, \theta) = 1$ if $\boldsymbol{\theta} = \mathbf{b}$. Then $L_X(\boldsymbol{\theta}) = L_X(\mathbf{b}) h(\Lambda(\mathbf{X}), \boldsymbol{\theta})$.)

10.1.3 The Rao-Blackwell and Other Theorems

In Section 10.1.1, we gave a heuristic argument that for any rule based on the original observations, we could find a rule based on the sufficient statistic which is just as good. In this section, we give some more careful proofs of results of this type.

We first look at the problem of estimating $\tau = \tau(\boldsymbol{\theta})$, where τ is one dimensional. Let $d(\mathbf{X})$ be an estimator of τ based on the original data $\mathbf{X}$. The following theorem shows how to construct an estimator $d^*(\mathbf{T})$ based on the sufficient statistic $\mathbf{T}$ which is as good as $d(\mathbf{X})$.

Theorem 10–3. (Rao-Blackwell) Let $\mathbf{T} = T(\mathbf{X})$ be a sufficient statistic, and let $d(\mathbf{X})$ be an estimator of the one-dimensional function $\tau(\boldsymbol{\theta})$. Let

$$d^*(\mathbf{T}) = E(d(\mathbf{X})|\mathbf{T}).$$

Then

a. $d^*(\mathbf{T})$ is an estimator of τ.

b. $Ed^*(\mathbf{T}) = Ed(\mathbf{X})$.

c. $\text{Var}(d^*(\mathbf{T})) \leqq \text{var}(d(\mathbf{X}))$, with equality only if $d(\mathbf{X}) \equiv d^*(\mathbf{T})$.

d. $MSE_{d^*}(\boldsymbol{\theta}) \leqq MSE_d(\boldsymbol{\theta})$, with equality only if $d(\mathbf{X}) \equiv d^*(\mathbf{T})$.

Proof

a. By the definition of sufficiency, $d^*(\mathbf{T})$ does not depend on $\boldsymbol{\theta}$ and is therefore an estimator. (This is the only place in the proof in which we use the sufficiency of $\mathbf{T}$.)

b. By Theorem 3–22,

$$Ed^*(\mathbf{T}) = E(Ed(\mathbf{X})|\mathbf{T}) = Ed(\mathbf{X}).$$

c. By Theorem 3–23,

$$\text{var}(d(\mathbf{X})) = \text{var}(d^*(\mathbf{T})) + E(d(\mathbf{X}) - d^*(\mathbf{T}))^2 \geqq \text{var}(d^*(\mathbf{T})).$$

Now, $(d(\mathbf{X}) - d^*(\mathbf{T}))^2 \geqq 0$. Therefore, $E(d(\mathbf{X}) - d^*(\mathbf{T}))^2 \geqq 0$, with equality only if $(d(\mathbf{X}) - d^*(\mathbf{T}))^2 \equiv 0$, which occurs only if $d(\mathbf{X}) \equiv d^*(\mathbf{T})$ (see Theorem 3–5).

d. The mean squared error of an estimator is the bias squared plus the variance, so the desired result follows from parts b and c. □

Part b of Theorem 10–3 implies that $d^*(\mathbf{T})$ is unbiased if $d(\mathbf{X})$ is. Parts b and c imply that a best unbiased estimator (if it exists) must be a function of the sufficient statistic. Part d implies that $d^*(\mathbf{T})$ is better than $d(\mathbf{X})$ (in terms of MSE), unless $d(\mathbf{X})$ is already a function of $\mathbf{T}$. Therefore, for any estimator based on the original data which is not already a function of the sufficient statistic $\mathbf{T}$, there is an estimator based on the sufficient statistic $\mathbf{T}$ which is better. The theorem even tells how to find such an estimator. The process of finding $d^*(\mathbf{T})$ by finding $E(d(\mathbf{X})|\mathbf{T})$ is often called "Rao-Blackwellizing" $d(\mathbf{X})$. Exercises C3 and C4 give examples of Rao-Blackwellizing an unbiased estimator to get an unbiased estimator based on a sufficient statistic.

The next theorem is similar to the Rao-Blackwell theorem for testing.

Theorem 10–4. Let $\mathbf{T} = T(\mathbf{X})$ be a sufficient statistic, and let $\Phi(\mathbf{X})$ be a critical function for testing that $\boldsymbol{\theta} \in N$ against $\boldsymbol{\theta} \in A$. Let

$$\Phi^*(\mathbf{T}) = E(\Phi(\mathbf{X})|\mathbf{T}).$$

Then

a. $\Phi^*(\mathbf{T})$ is a critical function for testing that $\boldsymbol{\theta} \in N$ against $\boldsymbol{\theta} \in A$.

b. $\Phi^*(\mathbf{T})$ has the same power function as $\Phi(\mathbf{X})$.

Proof. See the exercises. □

The size of a critical function is determined by its power function. Therefore, Φ and Φ^* have the same size. Unbiasedness in testing also depends only on the power function, so that Φ^* is unbiased if Φ is. Similarly, UMP or UMP unbiasedness depends only on the power function, so that Φ^* is UMP size α or UMP unbiased size α if Φ is. Therefore, if there is a unique UMP or UMP unbiased test, it must depend only on the sufficient statistic $\mathbf{T}$.

Note that the result for testing is weaker than the result for estimation. In the first place, the induced test Φ^* has the same power function as Φ, while the induced estimator d^* has a better MSE than d. Also, it can happen in discrete cases that the original test is a nonrandomized test, but the induced one is not. (See Exercise C2.)

The last theorem in this section relates sufficiency to MLEs and LRTs. Let $\mathbf{T} = T(\mathbf{X})$ be a sufficient statistic with density $g(\mathbf{t}; \boldsymbol{\theta})$. By the *original model*, we mean the model in which we observe $\mathbf{X}$ having density $f(\mathbf{x}; \boldsymbol{\theta})$, and by the *reduced model*, we mean the model in which we observe $\mathbf{T}$ having density $g(\mathbf{t}; \boldsymbol{\theta})$.

Theorem 10–5

a. If $\hat{\boldsymbol{\theta}}$ is an MLE of $\boldsymbol{\theta}$ for the reduced model, then $\hat{\boldsymbol{\theta}}$ is an MLE of $\boldsymbol{\theta}$ for the original model.
b. If $\Lambda(\mathbf{T})$ is the LRT statistic for testing that $\boldsymbol{\theta} \in N$ against $\boldsymbol{\theta} \in A$ for the reduced model, then $\Lambda(T(\mathbf{X}))$ is the LRT statistic for the original model.

Proof

a. Let $L_X(\boldsymbol{\theta})$ and $L_T^*(\boldsymbol{\theta})$ be the likelihood functions for $\mathbf{X}$ and $\mathbf{T}$, respectively. By the Fisher-Neyman criterion,

$$L_X(\boldsymbol{\theta}) = h(\mathbf{X})L_{T(X)}^*(\boldsymbol{\theta}), \qquad h(\mathbf{X}) \geqq 0.$$

Since $\hat{\boldsymbol{\theta}}$ is the MLE of $\boldsymbol{\theta}$ for the reduced model,

$$L_T^*(\hat{\boldsymbol{\theta}}) \geqq L_T^*(\boldsymbol{\theta})$$

for all $\boldsymbol{\theta}$. Therefore,

$$L_X(\hat{\boldsymbol{\theta}}) = h(\mathbf{X})L_{T(X)}^*(\hat{\boldsymbol{\theta}}) \geqq h(\mathbf{X})L_{T(X)}^*(\boldsymbol{\theta}) = L_X(\boldsymbol{\theta})$$

for all $\boldsymbol{\theta}$. Hence, $\hat{\boldsymbol{\theta}}$ is an MLE of $\boldsymbol{\theta}$ for the original model.

b. Let $\hat{\hat{\boldsymbol{\theta}}}$ and $\hat{\boldsymbol{\theta}}$ be MLEs of $\boldsymbol{\theta}$ under the null and alternative hypotheses, respectively, for the reduced model. By part a, $\hat{\hat{\boldsymbol{\theta}}}$ and $\hat{\boldsymbol{\theta}}$ are also MLEs for the original model. Therefore, the LRT statistic for the original model is

$$\Lambda^*(\mathbf{X}) = \frac{L_X(\hat{\boldsymbol{\theta}})}{L_X(\hat{\hat{\boldsymbol{\theta}}})} = \frac{h(\mathbf{X})L_{T(X)}^*(\hat{\boldsymbol{\theta}})}{h(\mathbf{X})L_{T(X)}^*(\hat{\hat{\boldsymbol{\theta}}})} = \frac{L_{T(X)}^*(\hat{\boldsymbol{\theta}})}{L_{T(X)}^*(\hat{\hat{\boldsymbol{\theta}}})} = \Lambda(T(\mathbf{X})). \quad \square$$

Corollary 10–5. The LRT for a model depends only on the sufficient statistic $\mathbf{T}$. Moreover, if the MLE for the model is unique, then the MLE also depends only on the sufficient statistic.

Proof. The corollary follows directly from Theorem 10–5 and the fact that the LRT for a given problem is unique, even if the MLEs under the null and alternative hypotheses are not. (Why?) $\square$

Exercises—B

1. Prove Theorem 10–4.

Exercises—C

1. Show that the Fisher information in the reduced model is the same as the Fisher information in the original model, so that there is no Fisher information lost when we reduce to a sufficient statistic. (It can be shown that the information in a statistic is the same as the information in the original sample if and only if the statistic is sufficient; see Lindgren (1976), pp. 251–252.)
2. Let X_1 and X_2 be independent, with $X_i \sim B(1,\theta)$. Consider testing that $\theta = .4$ against $\theta > .4$. Let $\Phi(X_1, X_2)$ be the nonrandomized test which rejects the null hypothesis if $X_1 = 1$.
   (a) Show that $T = X_1 + X_2$ is a sufficient statistic.
   (b) Show that $\Phi^*(T) = E(\Phi(X_1, X_2)|T) = 0$ if $T = 0$, $\Phi^*(T) = \frac{1}{2}$ if $T = 1$, and $\Phi^*(T) = 1$ if $T = 2$. (*Hint*: See Section 10.1.1 for the conditional distribution of (X_1, X_2) given T.)
   (Therefore, the "Rao-Blackwellized" test Φ^* is a randomized test even though the original test Φ is nonrandomized.)
3. For model A1, let $h(X_1)$ be 1 if $X_1 = 0$ and 0 if $X_1 > 0$. Let $T = a_n P = \Sigma X_i$.
   (a) Show that $Eh(X_1) = \exp(-\theta)$.
   (b) Show that X_1 given T has the $B(T, a_n^{-1})$ distribution.
   (c) Use the Rao-Blackwell theorem to show that $H(T) = (1 - a_n^{-1})^T$ is an unbiased estimator of $\exp(-\theta)$.
4. Let $X_1, \ldots, X_n$ be independently geometrically distributed with parameter θ.
   (a) Show that $T = \Sigma X_i$ is a sufficient statistic for this problem. How is T distributed?
   (b) Let $h(X_1)$ be 0 if $X_1 = 0$ and be 1 if $X_1 \geqq 1$. Show that $Eh(X_1) = \theta$.
   (c) Show that if $n > 1$, then $U(T) = T/(n + T - 1)$ is an unbiased estimator of θ. (Hint: Rao-Blackwellize $h(X_1)$. Let $W = T - X_1$. Then $P(X_1 \geqq 1|T = t) = 1 - P(X_1 = 0)\,P(W = t)/P(T = t)$.)

10.2 MINIMAL AND COMPLETE SUFFICIENT STATISTICS

10.2.1 Minimal and Complete Sufficient Statistics

Typically, there are many different sufficient statistics for a given model. For example, the original data vector **X** is always sufficient. In this section, we shall be concerned with finding the "smallest" sufficient statistic, that is, the one that reduces the data the most.

A careful reading of the proof of Theorem 10–4 shows that what was proved is that if **T** is a sufficient statistic and **S** is another statistic from which **T** can be computed (i.e., **T** is a function of **S**), then **S** is also a sufficient statistic. In the example of the joggers' pulses, let $\tilde{X}$ be the sample median of the observations (the average of the 15th and 16th largest pulses), and let R be the range of the observations (the difference between the largest and smallest observation). Then $(\bar{X}, S^2)$ can be computed from $(\bar{X}, \tilde{X}, S^2, R)$, and $(\bar{X}, S^2)$ is sufficient. Therefore, $(\bar{X}, \tilde{X}, S^2, R)$ is also sufficient. However, in terms of reducing the data, $(\bar{X}, S^2)$ is better. Since $(\bar{X}, S^2)$ is sufficient, the additional information from $\tilde{X}$ and R is not useful for drawing inferences about (μ, σ^2) for this problem. Now, let $\mathbf{T} = T(\mathbf{X}) = (\Sigma X_i^2, \Sigma X_i)$. Since $(\bar{X}, S^2)$ can be computed from **T**, and **T** can be computed from $(\overline{X}, S^2)$, a statistician who knows only $(\overline{X}, S^2)$ can find **T**, and one who knows **T** can find $(\overline{X}, S^2)$. Therefore, **T** has the same information as $(\overline{X}, S^2)$.

With this example in mind, we set forth the following definitions. Let $\mathbf{R} = R(\mathbf{X})$ and $\mathbf{T} = T(\mathbf{X})$ be statistics. Then **R** and **T** are *equivalent* if they are invertible functions of each other, and **R** is a *reduction* of **T** if $\mathbf{R} = h(\mathbf{T})$ for some function h. Finally, a sufficient statistic **T** is *minimal sufficient* if it is a reduction of every other sufficient statistic.

The next theorem gives some basic properties of minimal sufficient statistics.

Theorem 10–6

a. For any model, there is a minimal sufficient statistic.
b. If **T** is a minimal sufficient statistic and **R** is equivalent to **T**, then **R** is minimal sufficient.
c. If **R** and **T** are both minimal sufficient statistics, then **R** and **T** are equivalent.
d. Suppose that $T(\mathbf{X})$ is a statistic such that

$$T(\mathbf{X}) = T(\mathbf{Y}) \Leftrightarrow L_X(\boldsymbol{\theta}) = k(\mathbf{X}, \mathbf{Y})L_Y(\boldsymbol{\theta}) \tag{10–1}$$

for some function $k(\mathbf{X}, \mathbf{Y}) > 0$. Then $T(\mathbf{X})$ is a minimal sufficient statistic.

Proof. We prove part a of the theorem last.

b. By Theorem 10–2, **R** is sufficient. Let **Q** be any other sufficient statistic. Since **R** is equivalent to **T** and **T** is a reduction of **Q**, we see that **R** is a reduction of **Q**.
c. By the definition of a minimal sufficient statistic, **R** is a reduction of **T** and **T** is a reduction of **R**. Therefore, **R** and **T** are equivalent.
d. Suppose $T(\mathbf{X})$ satisfies Equation (10–1). For all **t** in the range of T, let $A_t = \{\mathbf{x}: T(\mathbf{x}) = \mathbf{t}\}$, and let $\mathbf{u}_t = h(\mathbf{t})$ be a particular point in A_t. Now, if $\mathbf{x} \in A_t$, then

$$T(\mathbf{x}) = T(\mathbf{u}_t) = T(h(\mathbf{t})) \Rightarrow L_x(\boldsymbol{\theta}) = k(\mathbf{x}, h(\mathbf{t}))L_{h(t)}(\boldsymbol{\theta}).$$

Therefore, for any X,

$$L_X(\boldsymbol{\theta}) = k(\mathbf{X}, h(T(\mathbf{X})))L_{h(T(X))}(\boldsymbol{\theta}) = k^*(\mathbf{X})g(T(\mathbf{X}), \boldsymbol{\theta}).$$

By the factorization criterion, $T(\mathbf{X})$ is a sufficient statistic.

Now, let $S(\mathbf{X})$ be any other sufficient statistic. By the factorization criterion,

$$L_X(\boldsymbol{\theta}) = r(\mathbf{X})q(S(\mathbf{X}), \boldsymbol{\theta})$$

for some functions r and q. If $S(\mathbf{X}) = S(\mathbf{Y})$, then $q(S(\mathbf{X}), \boldsymbol{\theta}) = q(S(\mathbf{Y}), \boldsymbol{\theta})$, and

$$L_X(\boldsymbol{\theta}) = \left(\frac{r(\mathbf{X})}{r(\mathbf{Y})}\right)L_Y(\boldsymbol{\theta}).$$

(Note that if $r(\mathbf{y}) = 0$, then **y** is not in the range of **Y** for any $\boldsymbol{\theta}$, and it can be eliminated from the sample space.) Next, using Equation (10–1), we see that

$$S(\mathbf{X}) = S(\mathbf{Y}) \Rightarrow T(\mathbf{X}) = T(\mathbf{Y}),$$

which in turn implies that $T(\mathbf{X})$ is a function of $S(\mathbf{X})$. (See the exercises.) Hence, $T(\mathbf{X})$ is minimal sufficient.

a. Define the relation $\approx$ by

$$\mathbf{X} \approx \mathbf{Y} \Leftrightarrow L_X(\boldsymbol{\theta}) = k(\mathbf{X}, \mathbf{Y}) L_Y(\boldsymbol{\theta})$$

for some $k(\mathbf{X}, \mathbf{Y})$. Then $\approx$ is an equivalence relation. (See Exercise C2.) Therefore, there exists a partition of the sample space into disjoint sets A_α such that $\bigcup A_\alpha$ is the sample space and $\mathbf{X} \approx \mathbf{Y}$ if and only if $\mathbf{X}$ and $\mathbf{Y}$ are in the same A_α. Now, let $T(\mathbf{X})$ be any function which is constant on each of the sets A_α but is different for different A_α. Then

$$T(\mathbf{X}) = T(\mathbf{Y}) \Leftrightarrow X \approx Y \Leftrightarrow L_X(\boldsymbol{\theta}) = k(\mathbf{X}, \mathbf{Y}) L_Y(\boldsymbol{\theta}),$$

and hence, $T(\mathbf{X})$ is a minimal sufficient statistic. □

Parts b and c of Theorem 10–6 imply that the minimal sufficient statistic is not unique, but that it is unique up to equivalence.

The partition in the proof of part a is called the *minimal sufficient partition*. From that proof, we see that it is the partition which is really minimal sufficient. The minimal sufficient statistic is just a bookkeeping device for keeping track of the set of the partition in which the observation is located. However, the minimal sufficient partition is harder to use than a minimal sufficient statistic, so that we shall not use the minimal sufficient partition again in this book.

Corollary 10–6. Suppose that the range of $\mathbf{X}$ does not depend on $\boldsymbol{\theta}$. If

$$T(\mathbf{X}) = T(\mathbf{Y}) \Leftrightarrow \frac{L_X(\boldsymbol{\theta})}{L_Y(\boldsymbol{\theta})} \text{ does not depend on } \boldsymbol{\theta},$$

then $T(\mathbf{X})$ is a minimal sufficient statistic.

Example 10–1

Let $X_1, \ldots, X_n$ be independent, with $X_i \sim N(\mu, 1)$. Then

$$\frac{L_X(\mu)}{L_Y(\mu)} = \frac{(2\pi)^{-n/2} \exp[-\sum(X_i - \mu)^2]}{(2\pi)^{-n/2} \exp[-\sum(Y_i - \mu)^2]}$$

$$= \exp\left[\frac{-(\sum X_i^2 - \sum Y_i^2)}{2}\right] \exp[(\sum X_i - \sum Y_i)\mu].$$

This function does not depend on μ if and only if $\Sigma X_i = \Sigma Y_i$. Therefore, ΣX_i is a minimal sufficient statistic for this model. Since $\bar{X}$ is an invertible function of ΣX_i, $\bar{X}$ is also a minimal sufficient statistic.

In Section 10.3, we shall see that for estimation problems it is often helpful to have a complete sufficient statistic. Accordingly, we say that a statistic $\mathbf{T}$ has a *complete family of distributions* if $E_\theta h(\mathbf{T}) = 0$ for all 0 implies that $h(\mathbf{T}) \equiv 0$. Then, a *complete sufficient statistic* is a sufficient statistic with a complete family of distributions.

Theorem 10–7 gives some basic properties of complete sufficient statistics.

Theorem 10–7

a. A complete sufficient statistic is minimal sufficient.

b. If **T** is a complete sufficient statistic and **R** is equivalent to **T**, then **R** is also a complete sufficient statistic.

c. If **R** and **T** are both complete sufficient statistics, then **R** and **T** are equivalent.

Proof

a. Let **S** be a complete sufficient statistic. Let **T** be a minimal sufficient statistic (which exists by Theorem 10–6). Then, by the definition of minimal sufficiency, **T** is a function of **S**, say, $\mathbf{T} = h(\mathbf{S})$. If we now show that **S** is a function of **T**, we shall be done, by Theorem 10–6, part b. Accordingly, let

$$k(\mathbf{S}, \mathbf{T}) = \mathbf{S} - E(\mathbf{S}|\mathbf{T}) \quad \text{and} \quad q(\mathbf{S}) = k(\mathbf{S}, h(\mathbf{S})).$$

Note that k and q do not depend on $\boldsymbol{\theta}$, since **T** is sufficient. Now, $h(\mathbf{S}) = \mathbf{T}$, Therefore,

$$E_\theta q(\mathbf{S}) = E_\theta(k(\mathbf{S}, h(\mathbf{S}))) = E_\theta k(\mathbf{S}, \mathbf{T}) = E_\theta \mathbf{S} - E_\theta(E\mathbf{S}|\mathbf{T}) = E\mathbf{S} - E\mathbf{S} = 0$$

for all $\boldsymbol{\theta}$. But by the definition of completeness, $q(\mathbf{S}) \equiv 0$, and hence, $\mathbf{S} \equiv E\mathbf{S}|\mathbf{T}$. This fact can be shown (with measure theory) to imply minimal sufficiency.

b. By Theorem 10–2, **R** is sufficient. Then, since **R** and **T** are equivalent, $\mathbf{R} = g(\mathbf{T})$ for some g. Now, suppose that

$$0 = E_\theta h(\mathbf{R}) = E_\theta h(g(\mathbf{T}))$$

for all $\boldsymbol{\theta}$. Then, since **T** has a complete family,

$$h(g(\mathbf{T})) \equiv 0 \Rightarrow h(\mathbf{R}) \equiv 0$$

for all $\boldsymbol{\theta}$. Hence, **R** has a complete family.

c. By part a, **R** and **T** are both minimal sufficient, and so by part c of Theorem 10–6, **R** and **T** are equivalent. □

From Theorem 10–7, we see that a complete sufficient statistic is also unique up to equivalence.

From Theorem 10–6, there always exists a minimal sufficient statistic. A complete sufficient statistic exists only when the minimal sufficient statistic is complete. Theorems 10–6 and 10–7 imply that either all the minimal sufficient statistics are complete sufficient statistics, or none are. Typically, models for which a minimal sufficient statistic is complete are fairly "clean" to analyze statistically, while those models for which a minimal sufficient statistic is not complete are quite difficult to analyze.

Example 10–2

Let $X_1, \ldots, X_n$ be independent, with $X_i \sim N(\theta, 1)$, where a^2 is a known constant. In Example 10–1, we saw that $\overline{X}$ is a minimal sufficient statistic for this model. We now show that $\overline{X}$ is a complete sufficient statistic. Let $g(\overline{X})$ be a function such that

$$Eg(\overline{X}) = 0,$$

and let

$$g_+(x) = \max(g(x), 0) \quad \text{and} \quad g_-(x) = -\min(g(x), 0).$$

That is, $g_+(x) = g(x)$ if $g(x) > 0$ and $g_+(x) = 0$ if $g(x) \leqq 0$. Similarly, $g_-(x) = -g(x)$ if $g(x) < 0$ and $g_-(x) = 0$ if $g(x) \geqq 0$. Note that

$$g(x) = g_+(x) - g_-(x).$$

Now, $V = \bar{X} \sim N(\mu, n^{-1})$, and hence V has density function

$$f(v; \mu) = (n/2\pi)^{n/2} \exp(-n(v - \mu)^2/2) = k(\mu)h(v)\exp(nv\mu), \qquad -\infty < v < \infty.$$

Furthermore,

$$0 = Eg(\bar{X}) = Eg_+(\bar{X}) - Eg_-(\bar{X}) \Rightarrow Eg_+(\bar{X}) = Eg_-(\bar{X})$$

for all μ. Hence,

$$\int_{-\infty}^{\infty} g_+(v)h(v) \exp(nv\mu) = \int_{-\infty}^{\infty} g_-(v)h(v) \exp(nv\mu)dv \tag{10–2}$$

for all μ. Setting $\mu = 0$, we see that

$$\int_{-\infty}^{\infty} g_+(v)h(v)dv = \int_{-\infty}^{\infty} g_-(v)h(v)dv = K.$$

Now, let

$$j_+(v) = \frac{g_+(v)h(v)}{K} \quad \text{and} \quad j_-(v) = \frac{g_-(v)h(v)}{\mathrm{K}}.$$

Then, since $g_+(v)h(v) \geqq 0$ and $g_-(v)h(v) \geqq 0$, $j_+(v)$ and $j_-(v)$ are density functions. Let $M_+(t)$ and $M_-(t)$ be the moment-generating functions associated with $j_+(v)$ and $j_-(v)$, respectively. Equation (10–2) implies that

$$M_+(n\mu) = M_-(n\mu)$$

for all μ, and hence, $M_+(t) = M_-(t)$ for all t. Therefore, $j_+(v)$ and $j_-(v)$ are the same, by the uniqueness theorem for moment-generating functions. But then, this implies that $g_+(v) = g_-(v)$, which in turn implies that $g(v) = 0$. (Note that for any v, either $g_+(v) = 0$ or $g_-(v) = 0$.) We have now shown that if $Eg(\bar{X}) = 0$ for all μ, then $g(\bar{X}) = 0$. Hence, $\bar{X}$ has a complete family of distributions and is a complete sufficient statistic.

As this example indicates, it is often quite difficult to use the definition to show that a statistic is a complete sufficient statistic. In the next section, however, we shall give two criteria which make it quite simple to find complete sufficient statistics for many models.

Exercises—B

1. For model A1, show that $P = \Sigma X_i/a_n$ is a minimal sufficient statistic.
2. For model A2, show that $Q = \Sigma iX_i/b_n$ is a minimal sufficient statistic.
3. For model A3, show that $R = \Sigma X_i/in$ is a minimal sufficient statistic.
4. For model B, show that $(\Sigma X_i, \Sigma X_i^2)$ is a minimal sufficient statistic.
5. For model A2, show that $Q = \Sigma iX_i/b_n$ is a complete sufficient statistic.

Exercises—C

1. Let $S(\mathbf{x})$ and $T(\mathbf{x})$ be functions such that $S(\mathbf{x}) = S(\mathbf{y}) \Rightarrow T(\mathbf{x}) = T(\mathbf{y})$. For each $\mathbf{s}$ in the range of $S(\mathbf{x})$, let $A_s = \{\mathbf{x}: S(\mathbf{x}) = \mathbf{s}\}$. Let $\mathbf{y}_s = h(\mathbf{s})$ be a particular point in A_s, and let $g(\mathbf{s}) = T(h(\mathbf{s}))$.
   (a) Show that $T(\mathbf{x}) = g(\mathbf{s})$ for all $\mathbf{x} \in A_s$. (Note that $S(\mathbf{x}) = S(\mathbf{y}_s)$ for $x \in A_s$. Why?)
   (b) Show that $T(\mathbf{x}) = g(S(\mathbf{x}))$ for all $\mathbf{x}$.
2. Show that the relation $\approx$ in the proof of Theorem 10–6, part a, is an equivalence relation. That is, show that
   (a) $\mathbf{X} \approx \mathbf{X}$ (reflexivity).
   (b) $\mathbf{X} \approx \mathbf{Y} \Leftrightarrow \mathbf{Y} \approx \mathbf{X}$ (symmetry).
   (c) $\mathbf{X} \approx \mathbf{Y}, \mathbf{Y} \approx \mathbf{Z} \Rightarrow \mathbf{X} \approx \mathbf{Z}$ (transitivity).
3. Let θ be a univariate parameter, and suppose that the model has monotone likelihood ratio in $T(\mathbf{X})$. Show that $T(\mathbf{X})$ is a minimal sufficient statistic.
4. Suppose that $\boldsymbol{\theta}$ takes on only two values, $\boldsymbol{\theta} = \mathbf{a}$ or $\boldsymbol{\theta} = \mathbf{b}$. Let $\lambda(\mathbf{X}) = L_X(\mathbf{b})/L_X(\mathbf{a})$ be the Neyman-Pearson test statistic. Show that $\lambda(\mathbf{X})$ is a minimal sufficient statistic.

10.2.2 Exponential Families

It is often difficult to use the definitions to find minimal sufficient or complete sufficient statistics. However, most statistical models are examples of exponential families for which fairly simple criteria exist. We say that a random vector $\mathbf{X}$ with density function $f(\mathbf{x}; \boldsymbol{\theta})$ has a *p-dimensional exponential family* if

$$f(\mathbf{x}; \boldsymbol{\theta}) = h(\mathbf{x})k(\boldsymbol{\theta}) \exp[d_1(\boldsymbol{\theta})T_1(\mathbf{x}) + \cdots + d_p(\boldsymbol{\theta})T_p(\mathbf{x})], \qquad \mathbf{x} \in A,$$

or equivalently, if

$$L_X(\boldsymbol{\theta}) = h(\mathbf{X})k(\boldsymbol{\theta}) \exp[d_1(\boldsymbol{\theta})T_1(\mathbf{X}) + \cdots + d_p(\boldsymbol{\theta})T_p(\mathbf{X})], \qquad \mathbf{X} \in A, \tag{10–3}$$

where $h(\mathbf{X})$ and $k(\boldsymbol{\theta})$ are one-dimensional functions and

$$T(\mathbf{X}) = (T_1(\mathbf{X}), \ldots, T_p(\mathbf{X})) \quad \text{and} \quad d(\boldsymbol{\theta}) = (d_1(\boldsymbol{\theta}), \ldots, d_p(\boldsymbol{\theta}))$$

are p-dimensional functions. Note that the dimension of the family (or of $d(\boldsymbol{\theta})$) need not be the same as the dimension of $\boldsymbol{\theta}$. (See Example 10–3.) It is quite important that the set A of possible values of $\mathbf{X}$ does not depend on $\boldsymbol{\theta}$.

By the factorization criterion, $\mathbf{T} = T(\mathbf{X})$ is a sufficient statistic for this exponential family. In this section, we give conditions for $\mathbf{T}$ to be a minimal sufficient or complete sufficient statistic. We assume throughout that the components of $d(\boldsymbol{\theta})$ take values on an interval. For simplicity, we look first at one-dimensional exponential families, so that $d(\boldsymbol{\theta})$ and $T(\mathbf{X})$ are one dimensional.

Theorem 10–8. (Exponential criterion I) Under fairly general conditions, if $\mathbf{X}$ has a one-dimensional exponential family given in Equation (10–3), and if $d(\boldsymbol{\theta})$ takes values on an interval, then $T = T(\mathbf{X})$ is a complete sufficient statistic (and is therefore also minimal sufficient).

Proof. A proof of the theorem in its full generality is beyond the level of this book, but can be found in Lehmann (1986), pp. 142–143. We give a proof for the case in which T is a nonnegative integer-valued random variable. In that case, let A_t be the set of all $\mathbf{x}$'s such that $T(\mathbf{x}) = t$. Then

$$P(T = t) = P(\mathbf{X} \in A_t) = \sum_{A_t} h(\mathbf{x})k(\boldsymbol{\theta}) \exp[T(\mathbf{x})d(\boldsymbol{\theta})]$$

$$= k(\boldsymbol{\theta}) \exp(td(\boldsymbol{\theta})) \sum_{A_t} h(\mathbf{x}) = h^*(t)k(\boldsymbol{\theta}) \exp[td(\boldsymbol{\theta})].$$

(Note that $T(\mathbf{x}) = t$ for all $\mathbf{x} \in A_t$.) Therefore, if $E_\theta g(T) = 0$, then

$$0 = \sum_{t=0}^{\infty} g(t)h^*(t)k(\boldsymbol{\theta})(\exp(d(\boldsymbol{\theta})))^t = k(\boldsymbol{\theta}) \sum_{t=0}^{\infty} g(t)h^*(t) \exp(d(\boldsymbol{\theta}))^t.$$

Now, let $\delta = \exp(d(\boldsymbol{\theta}))$. Then

$$\sum_{t=0}^{\infty} g(t)h^*(t)\delta^t = 0 = \sum_{t=0}^{\infty} 0\delta^t.$$

Since δ takes on values in an interval, it follows by the uniqueness of power series that

$$g(t)h^*(t) = 0$$

for all integers t, and hence,

$$P(g(T) = 0) = 1.$$

Therefore, T is a complete sufficient statistic. □

Example A1

Suppose we observe $X_i \sim P(i\theta)$, independent. Then

$$L_X(\theta) = h_1(\mathbf{X}) \exp[-\theta a_n] \exp[(\log\theta)\sum X_i],$$

which is in the form of Theorem 10–8, with $T(\mathbf{X}) = \Sigma X_i$. Note that the space of possible values for θ is the set of all positive numbers, so that the space of possible values for $\delta = \log(\theta)$ is all numbers, which contains an interval. Therefore, $T = T(\mathbf{X})$ is a complete sufficient statistic for this model. Also, the MLE P is an invertible function of T and is a complete sufficient statistic for the model.

Example A2

Suppose we observe $X_i \sim N(i\theta, 1)$, independent. Then

$$L_X(\theta) = h_2(\mathbf{X}) \exp\left(\frac{-\theta^2 b_n}{2}\right) \exp(\theta(\sum iX_i)),$$

which is in exponential form with $T(\mathbf{X}) = \Sigma iX_i$. Therefore, $T = T(\mathbf{X})$ is a complete sufficient statistic for this model. Also, the MLE Q is an invertible function of T and is thus complete and sufficient.

Example A3

Suppose we observe X_i independent, with $X_i \sim E(i\theta)$. Then R is a complete sufficient statistic. (See the exercises.)

In each of the preceding examples, the MLE is a complete sufficient statistic for the model. Note, however, that the MLE P for Example A1 is not even sufficient for Example A2, nor is Q sufficient for Example A1.

We next seek to extend Theorem 10–8 to multiparameter families. Let $d(\boldsymbol{\theta}) = (d_1(\boldsymbol{\theta}), \ldots, d_p(\boldsymbol{\theta}))$ be a vector-valued function. Then the components of $d(\boldsymbol{\theta})$ are *linearly independent* if, whenever $\Sigma c_i d(\boldsymbol{\theta})$ does not depend on $\boldsymbol{\theta}$, the c_i are all zero. The components of $d(\boldsymbol{\theta})$ are *functionally independent* if the only functions g such that $g(d(\boldsymbol{\theta}))$ does not depend on $\boldsymbol{\theta}$ are constant functions. In other words, the components of $d(\boldsymbol{\theta})$ are linearly independent if no one of them can be written as a linear function of the others, and they are functionally independent if no one of them can be written as a (possibly nonlinear) function of the others. If the components are functionally independent, they are also linearly independent.

Theorem 10–9. (Exponential criterion II) Suppose that $\mathbf{X}$ has a p-dimensional exponential family as given in Equation (10–3). Then under fairly general regularity conditions,

a. If the components of $d(\boldsymbol{\theta})$ are linearly independent, then $\mathbf{T} = T(\mathbf{X})$ is a minimal sufficient statistic.
b. If the components of $d(\boldsymbol{\theta})$ are functionally independent, then $\mathbf{T} = T(\mathbf{X})$ is a complete sufficient statistic.

Proof

a. If $L_X(\boldsymbol{\theta})$ satisfies Equation (10–3), then

$$\frac{L_X(\boldsymbol{\theta})}{L_Y(\boldsymbol{\theta})} = \left[\frac{h(\mathbf{X})}{h(\mathbf{Y})}\right] \exp[\sum_i d_i(\boldsymbol{\theta})(T_i(\mathbf{X}) - T_i(\mathbf{Y}))].$$

Since the components of $d(\boldsymbol{\theta})$ are linearly independent, $L_X(\boldsymbol{\theta})/L_Y(\boldsymbol{\theta})$ does not depend on $\boldsymbol{\theta}$ if and only if $T_1(\mathbf{X}) = T_1(\mathbf{Y}), \ldots, T_p(\mathbf{X}) = T_p(\mathbf{Y})$. The desired result then follows from part d of Theorem 10–6.

b. See Lehmann (1986), pp. 142–143. A proof for the case in which the components of $\mathbf{T}$ are nonnegative and integer valued can be constructed along the lines of that for Theorem 10–8. □

If the components of $d(\boldsymbol{\theta})$ are not linearly independent, we can solve for one component and reduce the problem to one involving an exponential family for which the components of $d(\boldsymbol{\theta})$ are linearly independent. So we can assume without loss of generality that the components are linearly independent. However, if the components of $d(\boldsymbol{\theta})$ are linearly independent but not functionally independent, then the minimal sufficient statistic is typically not complete. Such families are called curved exponential families.

It is often somewhat difficult to tell whether the components of $d(\boldsymbol{\theta})$ are functionally independent. We present a heuristic rule which works for all practical models.

Heuristic Rule for Determining Functional Independence:

As long as the number of components of $d(\boldsymbol{\theta})$ is the same as the number of components of $\boldsymbol{\theta}$, then the components of $d(\boldsymbol{\theta})$ are functionally independent.

Although it is possible to construct counterexamples to this rule, such counterexamples are rather bizarre. We shall accordingly take the rule to be true in this and subsequent chapters.

Example B

Let $X_i \sim N(\mu, \sigma^2)$, independent. Then

$$L_X(\mu, \sigma^2) = (2\pi)^{-n/2}\left(\sigma^{-n} \exp\left[\frac{-n\mu^2}{2\sigma^2}\right]\right) \exp\left[\sum X_i^2\left(\frac{-1}{2\sigma^2}\right) + \sum X_i\left(\frac{\mu}{\sigma^2}\right)\right],$$

which is in the exponential family with

$$T(\mathbf{X}) = \left(\sum X_i^2, \sum X_i\right) \quad \text{and} \quad d(\mu, \sigma^2) = \left(\frac{-1}{2\sigma^2}, \frac{\mu}{\sigma^2}\right).$$

Note that $\boldsymbol{\theta} = (\mu, \sigma^2)$ has two components, as does $d(\mu, \sigma^2)$. By the foregoing rule, the components of $d(\mu, \sigma^2)$ are functionally independent, and therefore, $\mathbf{T} = T(\mathbf{X})$ is a complete sufficient statistic. Also, $(\bar{X}, S^2)$ is an invertible function of $\mathbf{T}$ and is consequently a complete sufficient statistic (and hence minimal sufficient).

From this example, we see that the exponential criterion is often easy to use. We first factor out any terms which involve just the random variables $(h(\mathbf{X}))$ and then factor out any terms which involve just the parameters $(k(\boldsymbol{\theta}))$. We then manipulate what is left and put it into the form

$$\exp[d_1(\boldsymbol{\theta})T_1(\mathbf{X}) + \cdots + d_p(\boldsymbol{\theta})T_p(\mathbf{X})],$$

if possible. Then we look at the components of

$$d(\boldsymbol{\theta}) = (d_1(\boldsymbol{\theta}), \ldots, d_p(\boldsymbol{\theta})).$$

If they are not linearly independent, we manipulate further to get the correct form with linearly independent components. If they are linearly independent but not functionally independent, then

$$\mathbf{T} = T(\mathbf{X}) = (T_1(\mathbf{X}), \ldots, T_p(\mathbf{X}))$$

is a minimal sufficient statistic but may not be a complete sufficient statistic. If the components of $d(\boldsymbol{\theta})$ are functionally independent, then $\mathbf{T}$ is a complete sufficient statistic.

We now consider an example in which the components of d are not functionally independent.

Example 10–3

Let $X_i \sim N(\theta, \theta^2)$, independent. (Note that there is only one parameter, θ, for this model.) Then

$$L_X(\theta) = (2\pi)^{-n/2}\theta^{-n}e^{-n/2} \exp\left(\sum X_i^2\left(\frac{-1}{2\theta^2}\right) + \sum X_i\left(\frac{1}{\theta}\right)\right),$$

which is an exponential family with

$$T(\mathbf{X}) = \left(\sum X_i^2, \sum X_i\right) \quad \text{and} \quad d(\theta) = \left(\frac{-1}{2\theta^2}, \frac{1}{\theta}\right).$$

The components of $d(\theta)$ are linearly independent. However, $d(\theta)$ has two components and θ only has one, so we would expect that the components of $d(\theta)$ might be functionally dependent, and in fact,

$$d_1(\theta) + \frac{(d_2(\theta))^2}{2} = 0.$$

$\mathbf{T} = T(\mathbf{X})$ is therefore a minimal sufficient statistic for this model, but it may not be complete. Therefore, $(\overline{X}, S^2)$ is minimal sufficient. Furthermore, $(\overline{X}, S^2)$ is a complete sufficient statistic if and only if $\mathbf{T}$ is. (Why?) Now,

$$ES^2 = \theta^2 \quad \text{and} \quad E\overline{X}^2 = \theta^2 + \frac{\theta^2}{n} = \frac{(n+1)\theta^2}{n}.$$

Hence,

$$E((n+1)S^2 - n\overline{X}^2) = 0.$$

However,

$$P((n+1)S^2 - n\overline{X}^2 = 0) = 0$$

(because $\overline{X}$ and S^2 are independent continuous random variables). Therefore, $(\overline{X}, S^2)$ is not a complete sufficient statistic for this model. Consequently, the model has no complete sufficient statistic.

The next example indicates why the naive approach of defining the minimal sufficient statistic as the sufficient statistic with the fewest components will not work.

Example 10–4

Let X be a single observation from $N(0, \theta^2)$. Then X is sufficient and is only one dimensional, so we might suspect that X would have to be minimal sufficient. However, X^2 is both sufficient and "smaller" than X, since a person who only knows X^2 does not know the sign of X (i.e., X^2 is a function of X, but X is not a function of X^2). Since X^2 is sufficient, the sign of X is irrelevant for making inferences about θ for this model. In fact, X^2 is the minimal sufficient statistic for the model (see the exercises) and X is not a minimal sufficient statistic. Furthermore, (X^2, X^4) is an invertible function of X^2, so that (X^2, X^4) is also a minimal sufficient statistic. (Note that someone who knows both X^2 and X^4 does not know any more than someone who only knows X^2.) Therefore, the two-dimensional sufficient statistic (X^2, X^4) is minimal sufficient, but the one-dimensional sufficient statistic X is not.

Exercises—B

1. Show that R is a complete sufficient statistic for model A3.
2. Let X_i be independent, with $X_i \sim \Gamma(i, \theta)$. Show that P is a complete sufficient statistic for this model.
3. Let (X, Y) have a trinomial distribution with parameters n, θ^2, and $2\theta(1-\theta)$. Show that $(2X + Y)/2n$ is a complete sufficient statistic.
4. Let X_i be independent with marginal density function $f(x; \theta) = \theta x^{\theta - 1}, 0 < x < 1$. Show that ΠX_i is a complete sufficient statistic.
5. For Example C, show that the MLE of (θ, σ^2) is a complete sufficient statistic.
6. Let $X \sim N(0, \sigma^2)$.
   a. Show that X^2 is a complete sufficient statistic.
   b. Show that $|X|$ is a minimal sufficient statistic.
7. Let $X_i, \ldots, X_n$ be independent, with $X_i \sim \Gamma(\alpha, \beta)$, where α and β are both unknown parameters. Show that $(\Sigma X_i, \Pi X_i)$ is a complete sufficient statistic.

8. Let θ be a univariate parameter, and let $\mathbf{X}$ have the one-dimensional exponential family $L_X(\theta) = h(\mathbf{X})k(\theta)\exp(d(\theta)T(\mathbf{X}))$, in which $d(\theta)$ is an increasing function of θ. Show that f has monotone likelihood ratio in $T(\mathbf{X})$.
9. Let $X_1, \ldots, X_n$ be independent, with $X_i \sim N(\mu, 1)$, where a^2 is a known constant. Using Theorem 10–8, show that $\bar{X}$ is a complete sufficient statistic for this model. Compare your argument with the argument for the same result presented in Example 10–2.

Exercises—C

1. Let θ be a univariate parameter, and let $\mathbf{X}$ have a univariate exponential family.
   (a) Show that $ET(\mathbf{X}) = -k'(\theta)/k(\theta)d'(\theta)$. (*Hint*: Differentiate the formula $1 = \int_{-\infty}^{\infty} f(\mathbf{x}; \theta)d\mathbf{x}$ with respect to θ.)
   (b) Show that the Fisher information is

$$\left(\frac{k'(\theta)}{k(\theta)}\right)^2 - \left(\frac{k''(\theta)}{k(\theta)}\right) + \left(\frac{d''(\theta)}{d'(\theta)}\right)\left(\frac{k'(\theta)}{k(\theta)}\right).$$

10.3 UNBIASED ESTIMATION

10.3.1 Best Unbiased Estimators and the Lehmann-Scheffé Theorem

Recall from Chapter 7 that an estimator $U(\mathbf{X})$ of the univariate function $\tau(\boldsymbol{\theta})$ is *unbiased* if

$$E_\theta U(\mathbf{X}) = \tau(\boldsymbol{\theta}),$$

and it is the *best unbiased estimator* of $\tau(\boldsymbol{\theta})$ if, in addition, its variance is as small as that of any unbiased estimator. In this section, we develop a general method which can be used to find best unbiased estimators in any model which has a complete sufficient statistic.

Theorem 10–10. (Lehmann-Scheffé) Consider a model with a complete sufficient statistic $\mathbf{T} = T(\mathbf{X})$.

a. If $h(\mathbf{T})$ is an unbiased estimator of $\tau(\boldsymbol{\theta})$, it is the best unbiased estimator of $\tau(\boldsymbol{\theta})$.
b. The best unbiased estimator is unique.
c. If there is any unbiased estimator of $\tau(\boldsymbol{\theta})$, there is a best unbiased estimator of $\tau(\boldsymbol{\theta})$.

Proof

a. Let $U(\mathbf{X})$ be another unbiased estimator of $\tau(\boldsymbol{\theta})$, and let $g(\mathbf{T}) = E(U(\mathbf{X})|\mathbf{T})$. By the Rao-Blackwell theorem, $g(\mathbf{T})$ is an unbiased estimator of $\tau(\boldsymbol{\theta})$, and

$$\text{var}_\theta(g(\mathbf{T})) \leqq \text{var}_\theta(U(\mathbf{X})).$$

Now,

$$E_\theta(g(\mathbf{T}) - h(\mathbf{T})) = \tau(\boldsymbol{\theta}) - \tau(\boldsymbol{\theta}) = 0.$$

By the definition of completeness, $g(\mathbf{T}) \equiv h(\mathbf{T})$, and hence,

$$\text{var}_\theta(h(\mathbf{T})) = \text{var}_\theta(g(\mathbf{T})) \leqq \text{var}_\theta(U(\mathbf{X})).$$

Hence, $h(\mathbf{T})$ is the best unbiased estimator of $\tau(\theta)$.

b. By the Rao-Blackwell theorem, $\text{var}_\theta(U(\mathbf{X})) = \text{var}_\theta(g(\mathbf{T}))$ only if

$$U(\mathbf{X}) \equiv g(\mathbf{T}) \equiv h(\mathbf{T}),$$

and thus, $h(\mathbf{T})$ is unique.

c. If $U(\mathbf{X})$ is an unbiased estimator of $\tau(\boldsymbol{\theta})$, then $h(\mathbf{T}) = E(U(\mathbf{X})|\mathbf{T})$ is a best unbiased estimator of $\tau(\boldsymbol{\theta})$. □

Theorem 10–10 implies that as long as we have a complete sufficient statistic, there is a best unbiased estimator for any function $\tau(\boldsymbol{\theta})$ that has an unbiased estimator. To find the best unbiased estimator, we have to find a function of the complete sufficient statistic which is unbiased. The definition of completeness implies that there is only one such function.

Example A1

Here we observe $X_i \sim P(i\theta)$, independent. We showed in earlier sections that P is a complete sufficient statistic and is an unbiased estimator for this model. It is therefore the best unbiased estimator of θ.

Now, from Chapter 7, we also know that

$$P^2 - \frac{P}{a_n}$$

is an unbiased estimator of θ^2, and therefore, it is the best unbiased estimator of θ^2 (since it is a function of P).

Example A2

In this case we observe $X_i \sim N(i\theta, 1)$, independent. In previous sections, we showed that Q is a complete sufficient statistic and an unbiased estimator of θ for this model, so that Q is the best unbiased estimator of θ. Now,

$$Q^2 - b_n^{-1}$$

is an unbiased estimator of θ^2 for the model and is hence the best unbiased estimator of θ^2.

Example B

Here we observe $X_i \sim N(\mu, \sigma^2)$, independent. In the last section, we showed that $(\overline{X}, S^2)$ is a complete sufficient statistic for this model. Now, $\overline{X}$ is an unbiased estimator of μ and is therefore the best unbiased estimator of μ. Similarly, S^2 is an unbiased estimator of σ^2 and is therefore the best unbiased estimator of σ^2. Since $\overline{X}$ and S^2 are independent,

$$E\overline{X}S^2 = E\overline{X}ES^2 = \mu\sigma^2,$$

so that $\overline{X}S^2$ is the best unbiased estimator of $\mu\sigma^2$. In Chapter 7, we showed that

$$\frac{\Gamma((n-1)/2)(n-1)^{1/2}}{\Gamma(n/2)2^{1/2}}S$$

is an unbiased estimator of σ and is therefore the best unbiased estimator of σ.

Note that it is not enough to find an unbiased estimator for the preceding models; in order for an unbiased estimator to be a best unbiased estimator, it must depend on the data only through the complete sufficient statistic.

The next theorem was first mentioned in Chapter 7.

Theorem 10–11. Let $\hat{\boldsymbol{\theta}}$ be the unique MLE of $\boldsymbol{\theta}$ in a model with a complete sufficient statistic. Suppose that $h(\hat{\boldsymbol{\theta}})$ is an unbiased estimator of $\tau(\boldsymbol{\theta})$. Then $h(\hat{\boldsymbol{\theta}})$ is the best unbiased estimator of $\tau(\boldsymbol{\theta})$.

Proof. Let $\mathbf{T}$ be a complete sufficient statistic. By Theorem 10–5, $\hat{\boldsymbol{\theta}}$ is a function of $\mathbf{T}$, say, $\hat{\boldsymbol{\theta}} = k(\mathbf{T})$. Therefore, $h(\hat{\boldsymbol{\theta}}) = h(k(\mathbf{T}))$ is a function of the complete sufficient statistic. The desired result then follows from the Lehmann-Scheffé theorem. □

Theorem 10–11 implies that if we manipulate the MLE to find an unbiased estimator of $\tau(\boldsymbol{\theta})$, then that unbiased estimator must be the best unbiased estimator. Note that we do not actually need to know the complete sufficient statistic $\mathbf{T}$ to use the theorem; we only need to know that a complete sufficient statistic exists.

The next example shows that the best unbiased estimator can sometimes be rather silly.

Example A1

First, consider this model when $n = 1$, so that $P \sim P(\theta)$. We have

$$E(-1)^P = \frac{\sum_p (-1)^p \theta^p e^{-\theta}}{p!} = \frac{e^{-\theta} \sum_p (-\theta)^p}{p!} = e^{-2\theta}.$$

Since P is a complete sufficient statistic for the model,

$$U(P) = (-1)^P$$

is the best unbiased estimator of $e^{-2\theta}$. But observe that $(-1)^P = 1$ if P is even and $(-1)^P = -1$ if P is odd, so that this is a rather silly estimator of a function $e^{-2\theta}$ which is always positive.

Now consider the case of general n. In the exercises, it is shown that the best unbiased estimator for $e^{-2\theta}$ is

$$U(P) = \left[\frac{(a_n - 2)}{a_n}\right]^{Pa_n}.$$

Note that $U_n(P)$ is always positive when $n > 1$, so that the silly estimation occurs only when $n = 1$.

If we believe in the principle of sufficiency given earlier in this chapter, the only estimators we should consider should be based on the complete sufficient statistic $\mathbf{T}$. By the definition of completeness, there is only one such estimator which is unbiased. Therefore, in some sense, the best unbiased estimator is best in a class of one. Consequently, perhaps we should call this estimator "the unbiased estimator" instead of "the best unbiased estimator."

The invariance principle for MLEs presents a very appealing property for a class of estimators to have. It seems reasonable that if $T(\mathbf{X})$ is a "sensible" estimator of σ, then $(T(\mathbf{X}))^2$ should be a "sensible" estimator of σ^2. Unfortunately, unbiased estimators do not have this property.

Exercises—B

1. For model A3, find the best unbiased estimators of θ, θ^2, and θ^{-1}.
2. Let X_i be independent, with $X_i \sim \Gamma(i, \theta)$. Find the best unbiased estimators of θ, $\theta^{1/2}$, and θ^2.
3. Let (X, Y) have a trinomial distribution with parameters n, θ^2, and $2\theta(1-\theta)$. Find the best unbiased estimator of θ.
4. For model C, find the best unbiased estimators of θ, σ^2, σ, $\theta\sigma^2$, and $\theta\sigma$.
5. Let $X_1, \ldots, X_n$ be independent, with $X_i \sim B(1, \theta)$. Find the best unbiased estimators of θ and $\theta(1-\theta)$.
6. Let $X_1, \ldots, X_n$ be independent, with $X_i \sim \Gamma(\alpha, \beta)$, where α and β are both unknown.
   (a) Find the best unbiased estimator of $\alpha\beta$.
   (b) Find the best unbiased estimator of $\alpha^n \beta^n$.
7. Let $W \sim P(m)$. Show that $E(d^W) = \exp[m(d-1)]$.
8. (a) For model A1, show that $U(P) = [(a_n - 2)/a_n]^{Pa_n}$ is the best unbiased estimator of $\exp(-2\theta)$.
   (b) What is the MLE of $\exp(-2\theta)$ for model A1?
9. Consider model A1 when $n = 1$. Let $\tau = e^{-2\theta}$.
   (a) Find the mean squared error for $U(P) = (-1)^P$ for estimating τ.
   (b) What is the MLE of τ?
   (c) Find the mean squared error for the MLE of τ.

Exercises—C

1. Show that the best unbiased estimators of θ^{-1} and θ^2 in Example A3 are not efficient.
2. Let $X_1, \ldots, X_n$ be independent, having a geometric distribution with parameter θ. Find the best unbiased estimator of θ. (See Exercise C2, Section 10.1.3.)
3. For model A1, consider estimating $\tau = \exp(-\theta)$.
   (a) Find the best unbiased estimator of τ using Exercise C4, Section 10.1.2.
   (b) Find the best unbiased estimator of τ using Exercise A7.
4. Use the converse of the information inequality to show that if there is an efficient unbiased estimator of $\tau(\theta)$, then there is also a complete sufficient statistic. (This fact implies that the Lehmann-Scheffé approach to finding best unbiased estimators is more general than the efficiency approach given in Chapter 7.)

* 10.3.2 Convex Loss Functions

In previous sections, we characterized the performance of an estimator $d(\mathbf{X})$ of $\tau(\boldsymbol{\theta})$ by its mean squared error,

$$MSE_d(\boldsymbol{\theta}) = E_\theta(d(\mathbf{X}) - \tau(\boldsymbol{\theta}))^2.$$

This is certainly not the only criterion we can consider, however. We might want to look at the mean absolute error

$$E_\theta|d(\mathbf{x}) - \tau(\boldsymbol{\theta})|$$

or even measures of the form

$$E_\theta(d(\mathbf{X}) - \tau(\boldsymbol{\theta}))^{2k} \quad \text{or} \quad E_\theta|d(\mathbf{X}) - \tau(\boldsymbol{\theta})|^k,$$

where k is an arbitrary positive integer. In this section, we extend the Rao-Blackwell and Lehmann-Scheffé theorems to measures of the form

$$E_\theta L(d(\mathbf{X}), \boldsymbol{\theta})$$

where $L(d, \boldsymbol{\theta})$ is a convex function of d which we call the loss function. (Note that $|d - \tau(\boldsymbol{\theta})|^k$ and $(d - \tau(\boldsymbol{\theta}))^{2k}$ are convex functions of d; see Chapter 3 for the definition of a convex function.)

We need first the following simple extension of Jensen's inequality.

Theorem 10–12. If h is convex on I, U is a random variable such that $U \in I$, and $\mathbf{V}$ is a random variable or vector, then

$$E(h(U)|\mathbf{V}) \geqq h(E(U|\mathbf{V})).$$

If h is strictly convex and U is not a function of $\mathbf{V}$, then

$$E(h(U)|\mathbf{V}) > h(E(U|\mathbf{V})).$$

Proof. The theorem follows from Jensen's inequality, because conditional expectation is ordinary expectation with respect to the conditional distribution of U given $\mathbf{V}$, and the conditional distribution of U given V is degenerate if and only if U is a function of V. □

We now return to a statistical model in which we observe the random vector $\mathbf{X}$ having density $f(\mathbf{x}; \boldsymbol{\theta})$, where $\boldsymbol{\theta}$ is an unknown parameter. Let $L(d, \boldsymbol{\theta})$ be a convex function of d. For any estimator $d(\mathbf{X})$, we define the *risk function* $R(d, \boldsymbol{\theta})$ associated with the loss function $L(d, \boldsymbol{\theta})$ by

$$R(d, \boldsymbol{\theta}) = E_\theta(L(d(\mathbf{X}), \boldsymbol{\theta})).$$

Now, let $d_1(\mathbf{X})$ and $d_2(\mathbf{X})$ be estimators of $\tau(\boldsymbol{\theta})$. Then $d_1(\mathbf{X})$ is *as good as* $d_2(\mathbf{X})$ for $L(d, \boldsymbol{\theta})$ if

$$R(d_1, \boldsymbol{\theta}) \leqq R(d_2, \boldsymbol{\theta})$$

for all $\boldsymbol{\theta}$, and $d_1(\mathbf{X})$ is *better than* $d_2(\mathbf{X})$ for $L(d, \boldsymbol{\theta})$ if, in addition, the inequality is strong for some $\boldsymbol{\theta}$. (If $L(d, \boldsymbol{\theta}) = (d - \tau(\theta))^2$, then this definition reduces to the definition given earlier in terms of MSE.)

Theorem 10–13

a. (Generalized Rao-Blackwell) Let $\mathbf{T} = T(\mathbf{X})$ be a sufficient statistic, and let $d(\mathbf{X})$ be an estimator of $\tau(\boldsymbol{\theta})$. Let

$$d^*(\mathbf{T}) = Ed(\mathbf{X})|\mathbf{T}.$$

Then $d^*(\mathbf{T})$ is as good as $d(\mathbf{X})$ for any convex loss function $L(d, \boldsymbol{\theta})$ and is better than $d(\mathbf{X})$ for any strictly convex loss function, unless $d^*(\mathbf{T}) \equiv d(\mathbf{X})$.

b. (Generalized Lehmann-Scheffé) Let $\mathbf{T} = T(\mathbf{X})$ be a complete sufficient statistic, and let $d^*(\mathbf{T})$ be an unbiased estimator of $\tau(\boldsymbol{\theta})$. Then $d^*(\mathbf{T})$ is as good as any unbiased estimator of $\tau(\boldsymbol{\theta})$ for any convex loss function and is better than any other unbiased estimator $d(\mathbf{X}) \neq d^*(\mathbf{T})$ for any strictly convex loss function.

Proof. See the exercises. □

Part a of Theorem 10–13 implies that the Rao-Blackwellized estimator $d^*(\mathbf{T})$ improves on $d(\mathbf{X})$ for any strictly convex loss function $L(d, \boldsymbol{\theta})$, unless $d(\mathbf{X})$ is already a function of $\mathbf{T}$. If $d(\mathbf{X})$ is not already a function of $\mathbf{T}$, then

$$E(d^*(\mathbf{T}) - \tau(\boldsymbol{\theta}))^{2k} < E(d(\mathbf{X}) - \tau(\boldsymbol{\theta}))^{2k}$$

for $k = 1, 2, \ldots$, and

$$E|d^*(\mathbf{T}) - \tau(\boldsymbol{\theta})|^k \leqq E|d(\mathbf{X}) - \tau(\boldsymbol{\theta})|^k$$

for $k = 1, 2, \ldots$, with strong equality for $k = 2, 3, \ldots$. Therefore, the Rao-Blackwellized estimator $d^*(\mathbf{T})$ is better than the original estimator $d(\mathbf{X})$ in a very strong way.

In a similar manner, part b implies that an unbiased estimator $d^*(\mathbf{T})$ based on a complete sufficient statistic $\mathbf{T}$ is the best unbiased estimator for any convex loss function. (As mentioned in the previous section, this is not too surprising, since in some sense, it is "best" in a class with only one estimator.)

Exercises—C

1. Prove part a of Theorem 10–13.
2. Prove part b of Theorem 10–13.

Chapter 11

SOME ONE-SAMPLE, TWO-SAMPLE, AND PAIRED MODELS

In this chapter, we apply the results of the last three chapters to some one-sample, two-sample, and paired models. A *sample* of size n from a distribution function $F(\mathbf{x};\boldsymbol{\theta})$ is a set of independently identically distributed (i.i.d.) random variables or vectors $\mathbf{X}_1, \ldots, \mathbf{X}_n$, each having the (marginal) distribution function $F(\mathbf{x};\boldsymbol{\theta})$. A *one-sample model* is a model in which we observe a single sample from a univariate distribution function. Here, we are interested primarily in inferences about the mean μ of the distribution. A *two-sample model* is one in which we observe two *independent* samples $X_1, \ldots, X_m$ and $Y_1, \ldots, Y_n$ from different univariate distribution functions. In this case, we are interested primarily in inference about $\delta = EX_i - EY_j$, the difference in the means of the two distributions. A *paired model* is one in which we observe a sample $(X_1, Y_1), \ldots, (X_n, Y_n)$ from a bivariate distribution function. For paired models, we are also interested primarily in inferences about δ, the difference in the means.

We shall discuss the asymptotic sensitivity of the procedures considered to the assumptions under which they were derived. All these procedures are very sensitive to the assumption of independently identically distributed observations. Although we shall not mention this sensitivity any more throughout the chapter, it should be kept in mind.

Let $T = T(\mathbf{X})$ be a test statistic for testing that $\tau(\boldsymbol{\theta}) = c$ for some statistical model. Let t^{α} be the upper α point of the null distribution of T (which does not depend on any unknown parameters, since T is a test statistic). By "the one-sided size-α test based on T for testing $\tau(\boldsymbol{\theta}) = c$ against $\tau(\boldsymbol{\theta}) > c$," we mean the test which rejects the null hypothesis when $T > t^{\alpha}$. Similarly, the one-sided size-α test based on T for testing that $\tau(\boldsymbol{\theta}) = c$ against $\tau(\boldsymbol{\theta}) < c$ is the test which rejects the null hypothesis if $T < t^{1-\alpha}$. When we say that "the one-sided tests based on T" have certain properties for the "one-sided testing problems" involving $\tau(\boldsymbol{\theta})$, we mean that for each of the one-sided problems, the appropriate test just defined has those properties. By "a two-sided size-α test for testing that $\tau(\boldsymbol{\theta}) = c$ against $\tau(\boldsymbol{\theta}) \neq c$," we mean a test which rejects the null hypothesis if $T > t^{\beta}$ or $T < t^{1-\gamma}$, where $\beta + \gamma = \alpha$. If $\beta = \gamma = \alpha/2$, we call the test the equal-tailed size-α test for the hypothesis in question.

Now, let $Q = Q(\mathbf{X}, \tau)$ be a pivotal quantity for τ. Let q^{α} be the upper α point for the distribution of Q (which is completely known, since Q is a pivotal quantity).

Then a $(1-\alpha)$ confidence interval based on Q is the interval determined by

$$L(\mathbf{X}) \leqq \tau \leqq R(\mathbf{X}) \Leftrightarrow q^{1-\gamma} \leqq Q(\mathbf{X}, \tau) \leqq q^{\beta},$$

where $\beta + \gamma = \alpha$. Again, if $\beta = \gamma = \alpha/2$, we say the interval is the equal-tailed confidence interval based on Q.

We make the same conventions as those described when $T(\mathbf{X})$ or $Q(\mathbf{X}, \tau)$ is only an asymptotic test statistic or asymptotic pivotal quantity, and hence the tests and confidence intervals are only asymptotic size-α tests or asymptotic $(1-\alpha)$ confidence intervals.

In a one-sample model, we say that μ is significantly different from c at size .05 if we reject the hypothesis $\mu = c$ in favor of the alternative $\mu \neq c$ with a size-α test. We say that μ is significantly greater than c if we reject $\mu = c$ in favor of $\mu > c$. If we accept the null hypothesis $\mu = c$, we say that μ is not significantly different from c. In a two-sample or paired model, we say μ and ν are significantly different if we reject the hypothesis $\delta\,(=\mu - \nu) = 0$ against the alternative $\delta \neq 0$. We say that μ is significantly greater than ν if we reject $\delta = 0$ in favor of $\delta > 0$. If we accept $\delta = 0$, we say that there is no significant difference between μ and ν.

Some further notational conventions are as follows. We suppress the $\mathbf{X}$ in pivotal quantities and test statistics, writing $Q(\tau)$ for $Q(\mathbf{X}, \tau)$ and T for $T(\mathbf{X})$. If $Q(\mathbf{X}, \tau)$ is a pivotal quantity for τ, we write $Q(c)$ for the test statistic for testing that $\tau(\boldsymbol{\theta}) = c$ given by $Q(\mathbf{X}, c)$. Finally, when it causes no confusion, we suppress the subscripts n and m from asymptotic pivotal quantities and test statistics.

Most of the models considered in this chapter have at least two parameters. Since there are no UMP tests for such models, we derive likelihood ratio tests. All the hypotheses considered in this chapter involve only a single function of the parameters. For such problems, there is typically a UMP unbiased test which we also state. Derivations of UMP unbiased tests are beyond the scope of the text, however. They are given in Lehmann (1986), Chapters 4–5.

11.1 LARGE-SAMPLE INFERENCE FOR MEANS

11.1.1 Basic One-Sample and Two-Sample Asymptotic Results

We now derive the basic asymptotic results. We use these results in three ways: to derive asymptotic tests and confidence regions for a very general model (Section 11.1.2), to establish asymptotic sensitivity or insensitivity to various assumptions of the exact procedures presented (Section 11.2), and to derive asymptotic pivotal quantities and test statistics for discrete models for which no exact pivotal quantities or test statistics exist (Sections 11.3 and 11.4).

We first give the basic results for one-sample asymptotic models.

Theorem 11–1. Let $X_1, \ldots, X_n$ be a sample from a distribution function with mean μ and variance σ^2. Let $\overline{X}_n$ and S_n^2 be the sample mean and sample variance from this sample.

a. $\overline{X}_n$ and S_n^2 are consistent, unbiased estimators of μ and σ^2.
b. $n^{1/2}(\overline{X}_n - \mu)/\sigma \xrightarrow{d} Z \sim N(0,1)$ as $n \to \infty$.
c. If $\hat{\sigma}_n^2$ is a consistent estimator of σ^2, then $n^{1/2}(\overline{X}_n - \mu)/\hat{\sigma}_n \xrightarrow{d} Z \sim N(0,1)$ as $n \to \infty$.

Proof. See Chapter 6. □

We now give the basic results for two-sample asymptotics. Consider first a heuristic argument which is useful in remembering the results. Let $\overline{X}_m$ and $\overline{Y}_n$ be the sample means from independent samples of size m and n from distributions with means μ and ν and variances σ^2 and τ^2. By the central limit theorem,

$$\frac{m^{1/2}(\overline{X}_m - \mu)}{\sigma} \xrightarrow{d} Z_1 \sim N(0,1) \quad \text{and} \quad \frac{n^{1/2}(\overline{Y}_n - \nu)}{\tau} \xrightarrow{d} Z_2 \sim N(0,1).$$

Therefore, for large m and n,

$$\overline{X}_m \mathrel{\dot{\sim}} N\left(\mu, \frac{\sigma^2}{m}\right), \qquad \overline{Y}_n \mathrel{\dot{\sim}} N\left(\nu, \frac{\tau^2}{n}\right),$$

where $\approx$ means "is approximately distributed as." Hence,

$$\overline{X}_m - \overline{Y}_n \mathrel{\dot{\sim}} N(\mu - \nu, m^{-1}\sigma^2 + n^{-1}\tau^2),$$

or equivalently,

$$\frac{(\overline{X}_m - \overline{Y}_n) - (\mu - \nu)}{(m^{-1}\sigma^2 + n^{-1}\tau^2)^{1/2}} \approx N(0,1)$$

(since $\overline{X}_m$ and $\overline{Y}_n$ are independent).

The next theorem makes the preceding result more precise and also implies that we may substitute consistent estimators for σ^2 and τ^2 without affecting the asymptotic distribution.

Theorem 11–2. Let $\overline{X}_m$ and $\overline{Y}_n$ be sample means computed from independent samples of size m and n, with means μ and ν and variances σ^2 and τ^2. Let $m = aq$ and $n = bq$, where a and b are fixed and $q \to \infty$.

a. Let

$$Z_q = \frac{(\overline{X}_m - \overline{Y}_n) - (\mu - \nu)}{\left(\dfrac{\sigma^2}{m} + \dfrac{\tau^2}{n}\right)^{1/2}}.$$

Then, as q approaches infinity,

$$Z_q \xrightarrow{d} Z \sim N(0,1).$$

b. Let $\hat{\sigma}_m^2$ and $\hat{\tau}_n^2$ be consistent estimators of σ^2 and τ^2, and let

$$Z_q^* = \frac{(\overline{X}_m - \overline{Y}_n) - (\mu - \nu)}{\left(\dfrac{\hat{\sigma}_m^2}{m} + \dfrac{\hat{\tau}_n^2}{n}\right)^{1/2}}.$$

Then, as q approaches infinity,

$$Z_q^* \xrightarrow{d} Z \sim N(0,1).$$

c. If Q_m and R_n are consistent estimators of a parameter ϵ, then $(mQ_m + nR_n)/(n+m)$ is also a consistent estimator of ϵ.

Proof

a. We assume, for simplicity, that a and b are integers. Let

$$U_i = \frac{X_{a(i-1)+1} + \cdots + X_{ai}}{a} \quad \text{and} \quad V_i = \frac{Y_{b(i-1)+1} + \cdots + Y_{bi}}{b}.$$

For example, if $a = 4$ and $b = 7$, then U_1 is the average of the first four X's, U_2 is the average of the next four, etc. Similarly, V_1 is the average of the first seven Y's, V_2 is the average of the next seven, etc. Let

$$W_i = U_i - V_i.$$

Then, because $m = aq$ and $n = bq$,

$$\overline{X}_m - \overline{Y}_n = \overline{U}_q - \overline{V}_q = \overline{W}_q.$$

Now, the W_i are independently identically distributed, with

$$E(W_i) = E(U_i - V_i) = \mu - \nu \quad \text{and} \quad \operatorname{var}(W_i) = \operatorname{var}(U_i) + \operatorname{var}(V_i) = a^{-1}\sigma^2 + b^{-1}\tau^2.$$

Therefore, by the central limit theorem,

$$Z_q = \frac{q^{1/2}(\overline{W}_q - (\mu - \nu))}{(a^{-1}\sigma^2 + b^{-1}\tau^2)^{1/2}} \xrightarrow{d} Z \sim N(0,1).$$

b. Let

$$W_q = \left(\frac{\dfrac{\hat{\sigma}_m^2}{a} + \dfrac{\hat{\tau}_n}{b}}{\dfrac{\sigma^2}{a} + \dfrac{\tau^2}{b}} \right)^{1/2}.$$

Then $Z_q^* = Z_q/W_q$. But since $\hat{\sigma}_m^2 \xrightarrow{P} \sigma^2$ and $\hat{\tau}_n^2 \xrightarrow{P} \tau^2$, and since a and b are fixed, $W_q \xrightarrow{P} 1$, and the desired result follows from Slutzky's theorem.

c. See the exercises. □

In later sections of this chapter, we shall use these results somewhat less precisely. For example, if we let $Z_{m,n}$ be Z_q of part a of Theorem 11–2, then we shall say that

$$Z_{m,n} \xrightarrow{d} Z \sim N(0,1)$$

as m and n both approach infinity. That is, we shall not formally state that m/n is held constant as n approaches infinity.

Exercises—B

1. Prove part c of Theorem 11–2.
2. Suppose that Z_1 and Z_2 are independent, with $Z_i \sim N(0,1)$. Show that $(cZ_1 + dZ_2)/(c^2 + d^2)^{1/2} \sim N(0,1)$.

11.1.2 Large-Sample Inferences About Means

Consider a model in which we observe a sample $X_1, \ldots, X_n$ from a (univariate) distribution with mean μ and variance σ^2 (but not necessarily normally distributed). We could use this model for the example of the joggers' pulses discussed in earlier chapters. Alternatively, suppose we find the cholesterol levels of 40 smokers and want to decide whether the mean cholesterol level for smokers is significantly higher than the mean cholesterol level for nonsmokers, which we assume is known (for now). Finally, we might try a new fertilizer on 45 fields and measure the yields for each field. We could then test whether the mean yield for the new fertilizer is higher than the mean yield for the old fertilizer (which we must also assume is known to use this model).

Let $\overline{X}$ and S^2 respectively be the sample mean and variance computed from the X's.

Theorem 11–3

a. $\overline{X}$ is a consistent, unbiased estimator of μ.

b. An asymptotic pivotal quantity for μ is

$$Z_1(\mu) = \frac{n^{1/2}(\overline{X} - \mu)}{S} \xrightarrow{d} Z \sim N(0,1).$$

Proof. The desired results follow directly from Theorem 11–1. □

From Theorem 11–3, an asymptotic $(1 - \alpha)$ confidence interval for μ is

$$\mu \in \overline{X} \pm \frac{z^{\alpha/2} S}{n^{1/2}},$$

and an asymptotic size-α test that $\mu = d$ against $\mu < d$ rejects the null hypothesis if

$$Z_1(d) < -z^{\alpha} \; (= z^{1-\alpha}).$$

We call tests based on $Z_1(d)$ *one-sample Z-tests*.

An important question at this time is how large n must be in order to use the preceding asymptotic results. A common rule of thumb is that the results may be used as long as $n \geqq 30$.

The procedures suggested in the foregoing discussion are not optimal (UMP, best unbiased, etc.) for the model considered. In fact, the model is so general that it is not possible to find optimal procedures. However, the weak assumptions under which the procedures are derived makes them quite appealing in practice.

We now look at the model in which we observe two *independent* samples—$X_1, \ldots, X_m$ from a distribution with mean μ and variance σ^2, and $Y_1, \ldots, Y_n$ from a distribution with mean ν and variance τ^2. Our main interest is inferences about $\delta = \mu - \nu$.

One possible application of this model is to joggers' pulses. In previous sections, we assumed that the mean pulse rate of nonjoggers is known. If we did not know this pulse rate, then we could get two samples: $X_1, \ldots, X_m$ of pulse rates of joggers, and $Y_1, \ldots, Y_n$ of pulse rates of nonjoggers. In this case, δ is the difference between the mean pulse rates of joggers and nonjoggers. Similarly, if we wanted to compare the mean cholesterol levels of smokers and nonsmokers and did not know either of these, we could get two samples, one from smokers and one from nonsmokers, and compare their means. In this case, δ would be the difference between the mean cholesterol levels of smokers and nonsmokers.

Accordingly, let $\overline{X}$ and S^2 respectively be the sample mean and variance computed from the X's, and let $\overline{Y}$ and T^2 respectively be the sample mean and sample variance computed from the Y's. Let $\hat{\delta} = \overline{X} - \overline{Y}$.

Theorem 11–4

a. $\hat{\delta}$ is a consistent, unbiased estimator of δ.

b. An asymptotic pivotal quantity for δ is

$$Z_2(\delta) = \frac{\hat{\delta} - \delta}{\left(\dfrac{S^2}{m} + \dfrac{T^2}{n}\right)^{1/2}} \xrightarrow{d} Z \sim N(0, 1).$$

Proof

a. $\overline{X}$ and $\overline{Y}$ are consistent, unbiased estimators of μ and ν. Therefore, $\hat{\delta}$ is a consistent, unbiased estimator of δ.

b. Since S^2 and T^2 are consistent estimators of σ^2 and τ^2, the desired result follows from Theorem 11–2. □

From Theorem 11–4, an asymptotic $(1 - \alpha)$ confidence interval for δ is

$$\delta \in \hat{\delta} \pm z^{\alpha/2} \left(\frac{S^2}{m} + \frac{T^2}{n}\right)^{1/2},$$

and an asymptotic size-α test that $\delta = d$ against $\delta \neq d$ rejects the null hypothesis if

$$Z_2(d) > z^{\alpha/2} \quad \text{or} \quad Z_2(d) < -z^{\alpha/2}.$$

A common rule of thumb is that these asymptotic procedures can be used as long as $n \geqq 30$ and $m \geqq 30$. Tests based on $Z_2(d)$ are called *two-sample Z-tests*.

Again, the foregoing procedures are not optimal for the model in question, since no optimal procedures exist. However, the weak assumptions under which they are derived often make them appealing in practice.

Now consider the paired model in which we observe a sample $(X_1, Y_1), \ldots, (X_n, Y_n)$ from a bivariate distribution with

$$EX_i = \mu, \qquad EY_i = \nu, \qquad \text{var}(X_i) = \sigma^2, \qquad \text{var}(Y_i) = \tau^2, \qquad \text{cov}(X_i, Y_i) = \rho\sigma\tau.$$

Note that the X_i and Y_i are *not* independent in this model. Our main interest is again inferences about $\delta = \mu - \nu$.

Another experiment to compare the pulse rates of joggers and nonjoggers is to take a sample of 50 nonjoggers, record their pulse rates, and then have them take up jogging for a year and record their new pulse rates. If we used this design for our experiment, the data would be paired, and it would be incorrect to use the preceding two-sample procedure. Similarly, to test for the difference between the mean yields of two fertilizers, we could take our fields and divide each one in half. Then we would use the new fertilizer on one half and the old fertilizer on the other half and measure the yield on each half-field. The data would then be paired, since the yields on a common field would be dependent. Again, it would be incorrect to use the two-sample model discussed for this kind of experiment.

Let $U_i = X_i - Y_i$. Let $\overline{U}$ and V^2 be the sample mean and sample variance of the U's. Let $\hat{\delta} = \overline{U} = \overline{X} - \overline{Y}$.

Theorem 11–5

a. $\hat{\delta}$ is a consistent, unbiased estimator of δ.

b. An asymptotic pivotal quantity for δ is

$$Z_3(\delta) = \frac{n^{1/2}(\hat{\delta} - \delta)}{V} \xrightarrow{d} Z \sim N(0, 1).$$

Proof. See Exercise B1. □

From Theorem 11–5, an asymptotic confidence interval for δ for the model in question is

$$\delta \in \hat{\delta} \pm \frac{z^{\alpha/2}V}{n^{1/2}},$$

and an asymptotic size-α test that $\delta = d$ against $\delta > d$ rejects the null hypothesis if

$$Z_3(d) > z^{\alpha}.$$

Again, the standard rule of thumb says that the foregoing procedures may be used as long as $n \geqq 30$. Tests based on $Z_3(d)$ are called *paired Z-tests*.

It is important to distinguish the paired model discussed here from the two-sample model discussed previously. Since the observations X_i and Y_i are not independent for paired data, it is incorrect to use the two-sample model for such data.

In designing an experiment, we can often design it as either a two-sample experiment or a paired experiment, as in the joggers' pulse and fertilizer examples. When it is possible to design the experiment either way, it is almost always better to use a paired experiment. In a paired experiment, each individual acts as its own control, which eliminates some of the variation in the data. Suppose, for example, we have a choice of a two-sample experiment with n observations in each sample or a paired experiment with n pairs of observations. We note that $\hat{\delta} = \overline{X} - \overline{Y}$ is the same for both experiments and is consistent and unbiased for both. For the two-sample experiment,

$$\operatorname{var}(\hat{\delta}) = \frac{\sigma^2 + \tau^2}{n}, \tag{11–1}$$

while for the paired experiment,

$$\operatorname{var}(\hat{\delta}) = \frac{\sigma^2 + \tau^2 - 2\rho\sigma\tau}{n}. \tag{11–2}$$

Therefore, as long as $\rho > 0$, the variance of $\hat{\delta}$ is smaller for the paired experiment. For both experiments, the asymptotic confidence interval for δ has the same midpoint $\hat{\delta}$ and expected squared length equal to

$$4(z^{\alpha/2})^2 \operatorname{var}(\hat{\delta}). \tag{11–3}$$

Therefore, when $\rho > 0$, the expected squared length of the asymptotic confidence interval is shorter for the paired experiment than for the two-sample experiment. In a similar way, we could show that the tests have higher asymptotic power for the paired experiment than for the two-sample experiment. In sum, as long as X_i and Y_i are positively correlated, we get better asymptotic results using a paired experiment than a two-sample one.

It should be emphasized that in the preceding paragraph we are comparing two different methods of designing the experiment. Once data collection is complete, the data should be analyzed as either a two-sample experiment or a paired one (but not both), depending on how the experiment was designed.

A certain difficulty in interpretation often arises when we do tests on very large samples. Suppose we want to test the equality of two means in a two-sample model in which we have sample sizes $m = n = 1{,}000{,}000$. Suppose we observe $\overline{X} = 10.12$, $\overline{Y} = 10.11$, $S^2 = 10$, and $T^2 = 6$. Then

$$Z_2(0) = \frac{10.12 - 10.11}{\left(\dfrac{10}{1{,}000{,}000} + \dfrac{6}{1{,}000{,}000}\right)^{1/2}} = 2.5 > 1.96.$$

Therefore, we can reject the hypothesis that $\mu = \nu$ against $\mu \neq \nu$ with an asymptotic .05 test. Yet it seems clear that, in most situations, there is no practical difference between 10.12 and 10.11. Thus, the situation under discussion is an example in which there is a statistically significant difference between the two sample means which has little practical significance. In fact, it is hard to imagine very many practical situations in which the means of two different populations would be identical. Therefore, in most practical situations, if we take large enough samples, we will be able to reject the hypothesis of equal means, even though there is no important difference between the two means. For this reason, with very large sample sizes, it often makes more sense to find a confidence interval for $\delta = \mu - \nu$. In this example, we find that an asymptotic 95% confidence interval for δ is

$$\delta \in .01 \pm .0078. = (.0022, .0178).$$

Altough this interval does not cover the value zero, it is so close to zero as to indicate that there is little practical significance between the two means. The phenomenon of results that are statistically significant without being practically significant occurs in any model when we have huge data sets. In particular, it occurs in one-sample and paired models.

Exercises—A

For all testing problems from now on, show the critical region for the test and tell whether the hypothesis is accepted or rejected. Also, for story problems, write down the assumptions you are making.

1. Pure rain has a pH of 5.7, while acid rain has a lower pH. Suppose we take 64 rain samples, measure the pH of each, and get a sample mean of 3.9 and a sample standard deviation of 4. Can we conclude that μ, the mean pH for the 64 samples, is significantly lower than 5.7 with an asymptotic .01 test? Also, find an asymptotic 99% confidence interval for μ.
2. A light bulb manufacturer claims that his light bulbs have a mean lifetime of 900 hours. To test this claim, a consumer group bought 100 light bulbs and tested them to see how long they worked. The group found a sample mean of 750 hours and a sample standard deviation of 900 hours. Can the group conclude that the mean lifetime μ of the bulbs is significantly less than 800 with an asymptotic .01 test? Also, find an asymptotic 95% confidence interval for μ.
3. An experiment was performed to see whether smokers have higher blood pressure than nonsmokers. One hundred smokers had a sample mean of 120 and a sample standard deviation of 30, while 80 nonsmokers had a sample mean of 90 with a sample standard deviation of 40. Can we conclude that smokers had significantly higher blood pressure than nonsmokers? Use an asymptotic .05 test. Also, find an asymptotic 98% confidence interval for δ, the difference between the mean blood pressures of smokers and nonsmokers.
4. A company ran a test to compare miles per gallon (MPG) for two different small cars. In a sample of 80 cars of brand A, the sample mean of the MPG was 28 and the sample standard deviation was 5, while for brand B, the sample mean was 30 and the sample standard deviation was 4 in a sample of 70. Is there any significant difference between the mean MPGs of the two brands with an asymptotic .01 test? Find an asymptotic 95% confidence interval for δ, the difference between the MPGs of the two brands.
5. An experiment was performed to see whether large doses of vitamin C are helpful in controlling cholesterol. One hundred men with high cholesterol levels were given 600 mg of vitamin C per day. After two months, their cholesterol levels were measured and compared to their cholesterol levels at the start of the experiment. The sample mean of the difference between the cholesterol levels at the start and at the end was found to be -50, and the sample standard deviation of the differences was 200. Can we reject the hypothesis that the vitamin C had no effect on the cholesterol level with an asymptotic .05 test? Also, find an asymptotic 99% confidence interval for δ, the mean of the differences between the before and after measurements.

Exercises—B

1. Prove Theorem 11–5. (Note that the U_i are a sample from a distribution with mean δ and finite variance γ.)
2. **(a)** Verify Equations (11–1) and (11–2).
   (b) Verify Equation (11–3).

11.2 NORMAL MODELS

11.2.1 The One-Sample Normal Model

Let us consider the finite sample model in which we have a sample of size n from a normal distribution with mean μ and variance σ^2. We have already used this model for the example of the joggers' pulses. We could also use the model for the other one-sample examples presented in Section 11.1.2. Let $\overline{X}$ and S^2 be the sample mean and sample variance, respectively, computed from this sample.

Theorem 11–6

a. $(\overline{X}, S^2)$ is a complete sufficient statistic for the one-sample normal model.

b. $\overline{X}$ and S^2 are independent, with $\overline{X} \sim N(\mu, \sigma^2/n)$ and $(n-1)S^2/\sigma^2 \sim \chi^2_{n-1}$.

c. $\overline{X}$ and S^2 are the best unbiased estimators of μ and σ^2 and are consistent.

d. $(\overline{X}, \hat{\sigma}^2)$ is the MLE of (μ, σ^2), where $\hat{\sigma}^2 = (n-1)S^2/n$.

e. A pivotal quantity for μ is

$$t(\mu) = \frac{n^{1/2}(\overline{X} - \mu)}{S} \sim t_{n-1},$$

1. Confidence intervals and tests based on $t(\mu)$ are asymptotically insensitive to the normal assumption used in deriving them.
2. The one-sided tests and the equal-tailed two-sided test based on $t(c)$ are likelihood ratio tests for testing that $\mu = c$ against the appropriate alternative.

f. A pivotal quantity for σ^2 is

$$U(\sigma^2) = \frac{(n-1)S^2}{\sigma^2} \sim \chi^2_{n-1}.$$

1. Confidence intervals and tests based on $U(\sigma^2)$ are *not* asymptotically insensitive to the normal assumption used in deriving them.
2. The one-sided tests based on $U(d)$ are likelihood ratio tests for testing that $\sigma^2 = d$ against the appropriate hypotheses. The likelihood ratio test for testing that $\sigma^2 = d$ against $\sigma^2 \neq d$ is a two-sided test based on $U(d)$, but is not the equal-tailed test.

Proof. These results have been derived in the last three chapters. □

Tests based on $t(c)$ are called *one sample t-tests*. The power functions of these t-tests are also asymptotically insensitive to the normal assumption used in deriving them. (See Exercise C1.)

The one-sided tests and the equal-tailed two-sided test based on $t(c)$ are UMP unbiased for the appropriate hypotheses, as are the one-sided tests based on $U(d)$. For testing $\sigma^2 = d$ against $\sigma^2 \neq d$, the UMP unbiased test is a two-sided test based on $U(d)$, but is not the equal-tailed test or the likelihood ratio test. In fact, the likelihood ratio test and the UMP unbiased tests for two-sided hypotheses about σ^2 are rarely used, due to the lack of appropriate tables. Fortunately, the equal-tailed test is a reasonable approximation to these tests.

Tests and confidence intervals for σ^2 based on U are quite sensitive to the normal assumption, as we have seen in Chapter 8. Therefore, these procedures should be used only when we are relatively certain that the normal assumption is satisfied (exactly), which seems somewhat unlikely in practice. Fortunately, we are not often seriously interested in inferences about σ^2; rather, we compute S^2 primarily to help with inferences about μ. Unfortunately, when we are interested in inferences about σ^2, there are no alternative procedures which are less sensitive to the normal assumption. (We would often rather have a suboptimal procedure which is not too sensitive to the normal assumption than an optimal procedure which is sensitive to that assumption.)

In Sections 11.2.2 through 11.2.4, we consider other models in which the observations are assumed normally distributed. In Chapters 12 and 13, we consider some more complicated models in which the observations are assumed normally distributed. There are at least four reasons for the heavy use of normal models in statistics:

1. It is easier to find optimal procedures for normal models than for many other models.
2. The data from many experiments look fairly normally distributed.
3. The central limit theorem implies that many observations (such as weight or score on an intelligence test) should be nearly normally distributed, since they are averages of many small effects.
4. The central limit theorem implies that many procedures derived under the assumption of normality are asymptotically insensitive to the normal assumption.

Exercises—A

1. SAT scores in a state have an average value of 580. A particular town wants to know whether its mean SAT score is significantly higher than the average. So it takes a sample of 25 such SAT scores and finds a sample mean of 600 and a sample standard deviation of 90. Can the town conclude that its mean SAT score μ is greater than 580 with a .05 test? Also, find a 98% confidence interval for μ.
2. A soft drink company has a can-filling machine which it suspects is filling the cans too unevenly. So it takes a sample of 20 cans and measures the quantity of drink in each can. It finds a sample mean of 10 oz and a sample standard deviation of 3 oz. Can the company conclude that the standard deviation for the machine is significantly greater than 2 oz? Use $\alpha = .05$. (*Hint*: Test that $\sigma^2 = 4$.)

Exercises—B

1. In a sample of size 3 we observe 4, 6, and 8. Can we reject the hypothesis that $\mu = 3$ against $\mu \neq 3$ with these data? Use $\alpha = .05$.
2. Let $\hat{\sigma}^2$ be the MLE for σ^2.
   (a) Show that $t^*(\mu) = (n-1)^{1/2}(\overline{X} - \mu)/\hat{\sigma}$ is a pivotal quantity for μ.
   (b) Show that the equal-tailed $(1-\alpha)$ confidence interval for μ based on $t^*(\mu)$ is the same as the equal-tailed $(1-\alpha)$ confidence interval for μ based on $t(\mu) = n^{1/2}(\overline{X} - \mu)/S$.

(c) Show that $U^*(\sigma^2) = n\hat{\sigma}^2/\sigma^2$ is a pivotal quantity for σ^2.

(d) Show that the equal-tailed $(1-\alpha)$ confidence interval for σ^2 based on $U^*(\sigma^2)$ is the same as the equal-tailed $(1-\alpha)$ confidence interval for σ^2 based on $U(\sigma^2) = (n-1)S^2/\sigma^2$.

Exercises—C

1. Let $X_1, \ldots, X_n$ be independently identically distributed with $EX_i = \mu$ and $\text{var}(X_i) = \sigma^2$. Let $t_n(\mu) = n^{1/2}(\overline{X}_n - \mu)/S_n$.
   (a) Show that for all μ, $U_n = \overline{X} - c - t_{n-1}^{\alpha} S/n^{1/2} \xrightarrow{P} \mu - c$.
   (b) Show that $P_{\mu,\sigma}(t_n(c) > t_{n-1}^{\alpha}) \to 1$ for all $\mu > c$. (*Hint*: $t_n(c) > t_{n-1}^{\alpha}$ iff $U_n > 0$. Why?)
   (c) Now let (μ_n, σ_n) satisfy $n^{1/2}(\mu_n - c)/\sigma_n \to \delta$, a fixed constant. Show that $t_n(c) \xrightarrow{d} Y \sim N(\delta, 1)$. (*Hint*: Let $W_n = t_n(\mu) - t_n(c)$. Then $t_n(\mu_n) \xrightarrow{d} Z$, $W_n \xrightarrow{P} \delta$. Why?)

   (Part b of this problem shows that for any fixed $\mu > c$, the asymptotic power of the t-test is one, even without the normal assumption. In part c, we let μ converge to c to get a nontrivial asymptotic power function. That result shows that the power function of the t-test is also asymptotically insensitive to the normal assumption under this condition.)

11.2.2 The Two-Sample Normal Model with Equal Variances

Consider the model in which we observe a sample $X_1, \ldots, X_m$ from a normal distribution with mean μ and variance σ^2, and a second *independent* sample $Y_1, \ldots, Y_n$ from a normal distribution with mean ν and variance σ^2. We are primarily interested in inferences about $\delta = \mu - \nu$. Examples of two-sample experiments were presented in Section 11.1.2.

The most unfortunate assumption about this model is that the two populations have equal variances. It often seems unreasonable to assume that the variances of the two populations are the same, when the means may be different. At the end of the section, we show that the procedures we derive are *not* asymptotically insensitive to the equal-variance assumption unless $n = m$. However, we present some other calculations which indicate that the procedures are not too sensitive to this assumption, unless the sample sizes and variances are quite different. In the next section, we consider the two-sample normal model without the equal-variance assumption. In that case optimal procedures do not exist, nor are there reasonable exact pivotal quantities or test statistics, so that most confidence intervals and tests which have been suggested are only approximate.

Let $\overline{X}$ and S^2 be the sample mean and sample variance computed from the X's, and let $\overline{Y}$ and T^2 be the sample mean and sample variance computed from the Y's. Let

$$\hat{\delta} = \overline{X} - \overline{Y} \quad \text{and} \quad S_p^2 = \frac{(m-1)S^2 + (n-1)T^2}{m+n-2}.$$

S_p^2 is called the *pooled variance estimator*.

We first find a sufficient statistic and its joint distribution.

Theorem 11–7

a. $(\overline{X}, \overline{Y}, S_p^2)$ is a complete sufficient statistic for the two-sample normal model with equal variances.

b. $\overline{X}$, $\overline{Y}$, and S_p^2 are independent, with $\overline{X} \sim N(\mu, \sigma^2/m)$, $\overline{Y} \sim N(\nu, \sigma^2/n)$, and $(m+n-2)S_p^2/\sigma^2 \sim \chi^2_{m+n-2}$.

Proof

a. The joint density of the X_i and the Y_j is given by

$$f(\mathbf{x}, \mathbf{y}; \mu, \nu, \sigma^2) = (2\pi)^{-(m+n)/2}(\sigma^2)^{-(m+n)/2} \exp\left[\frac{-(\sum(x_i-\mu)^2 + \sum(y_j-\nu)^2)}{2\sigma^2}\right],$$

and therefore, the likelihood function for this model is

$$L_{X,Y}(\mu,\nu,\sigma) = L(\mu,\nu,\sigma^2) = c(\sigma^2)^{-(m+n)/2} \exp\left[\frac{-(\sum(X_i-\mu)^2 + \sum(Y_j-\nu)^2)}{2\sigma^2}\right],$$

$$= c(\sigma^2)^{-(m+n)/2} \exp\frac{(-m\mu^2 - n\nu^2)}{2\sigma^2} \exp\left(-\frac{(\sum X_i^2 + \sum Y_j^2)}{2\sigma^2} + \frac{\sum X_i \mu}{\sigma^2} + \frac{\sum Y_j \nu}{\sigma^2}\right),$$

which is in exponential form, with

$$\mathbf{T} = T(\mathbf{X},\mathbf{Y}) = (\textstyle\sum X_i^2 + \sum Y_j^2, \sum X_i, \sum Y_j) \quad \text{and} \quad \mathbf{d} = d(\mu,\nu,\sigma^2) = \left(\frac{-1}{2\sigma^2}, \frac{\mu}{\sigma^2}, \frac{\nu}{\sigma^2}\right).$$

Now, $\mathbf{d}$ has three components and there are three parameters for this family, so that $\mathbf{T} = T(\mathbf{X},\mathbf{Y})$ is a complete sufficient statistic for the model. But $(\overline{X}, \overline{Y}, S_p^2)$ is an invertible function of $\mathbf{T}$ and is therefore also complete and sufficient.

b. $(\overline{X}, S^2)$ and $(\overline{Y}, T^2)$ are independent because they come from independent samples. By known results about the one-sample model, $\overline{X}$ and S^2 are independent, as are $\overline{Y}$ and T^2. Therefore, $\overline{X}$, S^2, $\overline{Y}$, and T^2 are all independent. Hence, $\overline{X}$, $\overline{Y}$, and S_p^2 are independent. The distributions of $\overline{X}$ and $\overline{Y}$ follow from known results about the one-sample model. From these results, we also know that

$$\frac{(m-1)S^2}{\sigma^2} \sim \chi^2_{m-1} \quad \text{and} \quad \frac{(n-1)T^2}{\sigma^2} \sim \chi^2_{n-1}.$$

Since these variables are independent,

$$\frac{(m+n-2)S_p^2}{\sigma^2} = \frac{(m-1)S^2}{\sigma^2} + \frac{(n-1)T^2}{\sigma^2} \sim \chi^2_{m+n-2}. \quad \square$$

We next find optimal estimators for the model under examination.

Theorem 11–8

a. $\overline{X}$, $\overline{Y}$, S_p^2, and $\hat{\delta}$ are the best unbiased estimators of μ, ν, σ^2, and δ and are consistent.

b. $(\overline{X}, \overline{Y}, \hat{\sigma}^2)$ is the MLE of (μ, ν, σ^2), where $\hat{\sigma}^2 = (m+n-2)S_p^2/(m+n)$. $\hat{\delta}$ is the MLE of δ.

Proof

a. $E\overline{X} = \mu$, $E\overline{Y} = \nu$, $E\hat{\delta} = \delta$, and $ES_p^2 = ((m-1)ES^2 + (n-1)ET^2)/(m+n-2) = \sigma^2$. Since these estimators are both unbiased and functions of the complete sufficient statistic, they are the best unbiased estimators by the Lehmann-Scheffé theorem. Now, $\overline{X}$ and $\overline{Y}$ are consistent estimators of μ and ν, and hence, $\hat{\delta}$ is a consistent estimator of δ. Using the distribution of S_p^2, we see that $\text{var}(S_p^2) = 2\sigma^4/(m+n-2) \to 0$. Since S_p^2 is also unbiased, S_p^2 is a consistent estimator of σ^2.

b. Differentiating $\log[L(\mu, \nu, \sigma)]$, we get the following three equations:

$$0 = \frac{\sum(X_i - \mu)}{\sigma^2} = \frac{m(\overline{X} - \mu)}{\sigma^2}, \qquad 0 = \frac{\sum(Y_j - \nu)}{\sigma^2} = \frac{n(\overline{Y} - \nu)}{\sigma^2},$$

$$0 = -(m+n)\sigma^{-1} + \sigma^{-3}\left(\sum(X_i - \mu)^2 + \sum(Y_j - \nu)^2\right).$$

Solving the first two of these, we obtain $\hat{\mu} = \overline{X}$ and $\hat{\nu} = \overline{Y}$. Putting these in for μ and ν in the third then yields

$$\hat{\sigma}^2 = \frac{\left(\sum(X_i - \overline{X})^2 + \sum(Y_j - \overline{Y})^2\right)}{m+n}.$$

Therefore, the MLE of (μ, ν, σ) is $(\overline{X}, \overline{Y}, \hat{\sigma})$. By the invariance principle, it then follows that the MLEs of σ^2 and δ are $\hat{\sigma}^2$ and $\overline{X} - \overline{Y}$. □

The next theorem deals with inferences about δ.

Theorem 11–9

a. A pivotal quantity for δ is

$$t(\delta) = \frac{\hat{\delta} - \delta}{S_p(m^{-1} + n^{-1})^{1/2}} \sim t_{m+n-2}.$$

b. Tests and confidence intervals based on $t(\delta)$ are asymptotically insensitive to the normal assumption.

c. Tests and confidence intervals based on $t(\delta)$ are *not* asymptotically insensitive to the assumption of equal variances, unless $\lim(m/n) = 1$.

d. One-sided tests and the equal-tailed two-sided test based on $t(d)$ are likelihood ratio tests for testing that $\delta = d$ against the appropriate hypothesis.

Proof

a. By Theorem 11–7, $\hat{\delta} \sim N(\delta, \sigma^2(m^{-1} + n^{-1}))$. Therefore,

$$Z = \frac{\hat{\delta} - \delta}{\sigma(m^{-1} + n^{-1})^{1/2}} \sim N(0, 1).$$

By Theorem 11–7 again, $U = (m+n-2)S_p^2/\sigma^2 \sim \chi^2_{m+n-2}$, independently of Z. But then, by the definition of the t-distribution,

$$t(\delta) = \frac{Z}{(U/(m+n-2))^{1/2}} \sim t_{m+n-2}.$$

b. See Exercises C1.

c. See Theorem 11.11.

d. The derivations of the indicated likelihood ratio tests are straightforward, but messy. We illustrate with the likelihood ratio test for testing that $\delta = 0$ against $\delta \neq 0$. From Theorem 11–8 the MLE of (μ, ν, σ) under the alternative hypothesis is given by $\hat{\mu} = \bar{X}$, $\hat{\nu} = \bar{Y}$, $\hat{\sigma}^2 = V^2 = (m+n-2)S_p^2/(m+n)$. Therefore,

$$\begin{aligned} L(\hat{\mu}, \hat{\nu}, \hat{\sigma}) &= L(\bar{X}, \bar{Y}, V) \\ &= (2\pi)^{-(m+n)/2} V^{-(m+n)} \exp\left[\frac{-(\sum(X_i - \bar{X})^2 + \sum(Y_j - \bar{Y})^2)}{2V^2}\right] \\ &= (2\pi)^{-(m+n)/2}(V^2)^{-(m+n)/2} \exp\left[\frac{-(m+n)}{2}\right]. \end{aligned}$$

Under the null hypothesis, we can think of the X_i and Y_j as a sample of size $m+n$ from a normal distribution with mean μ and variance σ^2. Therefore, the MLE of (μ, ν, σ) under the null hypothesis is given by (Q, Q, W), where

$$Q = \frac{\sum X_i + \sum Y_j}{m+n} = \frac{m\bar{X} + n\bar{Y}}{m+n}$$

and

$$W^2 = \frac{\sum(X_i - Q)^2 + \sum(Y_j - Q)^2}{m+n} = \frac{V^2 + mn(\bar{X} - \bar{Y})^2}{(m+n)^2}.$$

(See Exercise B2.) Hence,

$$L(\hat{\hat{\mu}}, \hat{\hat{\nu}}, \hat{\hat{\sigma}}) = L(Q, Q, W) = (2\pi)^{-(m+n)/2}(W^2)^{-(m+n)/2} \exp\left[\frac{-(m+n)}{2}\right].$$

Finally,

$$\begin{aligned} \Lambda(\mathbf{X}, \mathbf{Y}) &= \frac{L(\bar{X}, \bar{Y}, V)}{L(Q, Q, W)} = \left(\frac{W^2}{V^2}\right)^{(m+n)/2} \\ &= \left(1 + \frac{nm(\bar{X} - \bar{Y})^2}{(m+n)^2 V^2}\right)^{(m+n)/2} = \left(1 + \frac{t^2}{m+n-2}\right)^{(m+n)/2}, \end{aligned}$$

where

$$t = t(0) = \frac{\bar{X} - \bar{Y}}{S_p(m^{-1} + n^{-1})^{1/2}}.$$

Therefore, the likelihood ratio test rejects the null hypothesis when t^2 is large, or equivalently, when $t > a$ or $t < -a$ for some a. For a size-α test, $a = t_{m+n-2}^{\alpha/2}$. □

Tests based on the statistic $t(d)$ are called *two-sample t-tests*. In Exercise C2, it is shown that the power functions of the two-sample tests are also asymptotically insensitive to the normal assumption. Lehmann (1986) shows that the one-sided tests and equal-tailed two-sided tests based on $t(d)$ are UMP unbiased for the appropriate hypotheses.

Theorem 11–10

a. A pivotal quantity for σ^2 is

$$U(\sigma^2) = \frac{(m+n-2)S_p^2}{\sigma^2} \sim \chi^2_{m+n-2}.$$

b. Tests and confidence intervals based on $U(\sigma^2)$ are *not* asymptotically insensitive to the normal assumption.

c. The one-sided tests based on $U(c)$ are likelihood ratio tests for testing that $\sigma^2 = c$ against the appropriate one-sided hypothesis. For testing that $\sigma^2 = c$ against $\sigma^2 \neq c$, the likelihood ratio test is a two-sided test based on $U(c)$, but is not the equal-tailed test.

Proof. The proof of this theorem is essentially the same as that of part f of Theorem 11–5 and is left to the reader. □

Due to their sensitivity to the normal distribution, tests and confidence intervals based on $U(\sigma^2)$ should be used only when we are relatively certain that the observations are normally distributed.

We now turn to sensitivity to the assumption of equal variances. In Section 11.2.3, it is shown that $\overline{X}$, $\overline{Y}$, and $\hat{\delta}$ are the optimal estimators of μ, ν, and δ, even without the assumption of equal variances. Inferences about σ^2 are not meaningful without the assumption of equal variances. We therefore focus on inferences about δ based on the pivotal quantity $t(\delta)$ (or the test statistic $t(d)$ for testing that $\delta = d$). Suppose that the X_i and Y_j are still independent samples and that $EX_i = \mu$ and $EY_j = \nu$, but that $\text{var}(X_i) = \sigma^2$ and $\text{var}(Y_j) = \tau^2$. Then

$$\text{var}(\hat{\delta}) = \text{var}(\overline{X}) + \text{var}(\overline{Y}) = \frac{\sigma^2}{m} + \frac{\tau^2}{n}.$$

When we use $t(\delta)$, we are estimating this variance by

$$S_p^2(m^{-1} + n^{-1}).$$

Now,

$$S_p^2 = \frac{(m-1)S^2 + (n-1)T^2}{m+n-2} \approx \frac{mS^2 + nT^2}{m+n}$$

is essentially estimating $(m\sigma^2 + n\tau^2)/(m+n)$. Hence, $S_p^2(m^{-1} + n^{-1})$ is essentially estimating

$$\frac{(m^{-1} + n^{-1})(m\sigma^2 + n\tau^2)}{m+n} = \frac{\sigma^2}{n} + \frac{\tau^2}{m}.$$

(Note that the denominators of the formula for $\text{var}(\hat{\delta})$ have been interchanged in this result.) Therefore, $t(\delta)$ and $t(d)$ have been improperly studentized unless $m = n$ or $\sigma^2 = \tau^2$. This simple calculation shows why procedures based on $t(\delta)$ or $t(d)$ are sensitive to the assumption of equal variances unless $m = n$.

We now give a more formal result regarding this dependence on equal variances. Note that

$$t(\delta) \xrightarrow{d} Z \sim N(0, 1)$$

if the model is true. (This result is true even without the normal assumption, as long as the variances are equal; see the exercises.)

Theorem 11–11. Suppose that $\xi = \sigma^2/\tau^2$ and $m/n \to r$. Then

$$t(\delta) \xrightarrow{d} Q \sim N\left(0, \frac{\xi + r}{\xi r + 1}\right).$$

Proof. To keep the proof simple, we again assume that $m = aq$ and $n = bq$, where a and b are fixed and $q \to \infty$, so that $r = a/b$. Now, $\hat{\sigma}^2 = (m - 1)S^2/m$ and $\hat{\tau}^2 = (n - 1)T^2/n$ are consistent estimators of σ^2 and τ^2, respectively. Therefore,

$$S_p^2 = \frac{m + n}{m + n - 2} \frac{a\hat{\sigma}^2 + b\hat{\tau}^2}{a + b} \xrightarrow{P} \frac{a\sigma^2 + b\tau^2}{a + b}.$$

By Theorem 11–2,

$$Z_{mn} = \frac{\overline{X} - \overline{Y} - \delta}{\left(\dfrac{\sigma^2}{m} + \dfrac{\tau^2}{n}\right)^{1/2}} \xrightarrow{d} Z \sim N(0, 1).$$

Now, $t(\delta) = Z_{mn} W_{mn}$, where

$$W_{mn} = \frac{\left(\dfrac{\sigma^2}{m} + \dfrac{\tau^2}{n}\right)^{1/2}}{S_p(m^{-1} + n^{-1})^{1/2}} = \frac{(b\sigma^2 + a\tau^2)^{1/2}}{S_p(a + b)^{1/2}}$$

$$\xrightarrow{P} \frac{(b\sigma^2 + a\tau^2)^{1/2}}{(a\sigma^2 + b\tau^2)^{1/2}} = \left(\frac{\xi + r}{\xi r + 1}\right)^{1/2}.$$

The desired result then follows from Slutzky's theorem in Chapter 6. □

Thus, by Theorem 11–11, procedures based on $t(\delta)$ or $t(d)$ are asymptotically insensitive to the assumption of equal variances only if $(\xi + r)/(\xi r + 1) = 1$, which happens only if $\xi = 1$ or $r = 1$. Therefore, these procedures are asymptotically insensitive to that assumption only when the ratio of the sample sizes approaches one.

Let us make some elementary calculations using Theorem 11–11 to determine how sensitive the foregoing procedures are, asymptotically, to the assumption of equal variances. As an example, let $r = 1.5$ (so that m is one-and-a-half times n), and let $\xi = 4$ (so that $\text{var}(X_i) = 4\ \text{var}(Y_i)$). Then

$$\theta = \frac{\xi + r}{\xi r + 1} = \frac{5.5}{7} = .79 \quad \text{and} \quad \theta^{1/2} = .89.$$

Now, suppose we are using a one-sided test based on $t(d)$ which has nominal size .05. Then, by Theorem 8–5, the true asymptotic size for this case of unequal variances is

$$\alpha^{**} = P\left(Z > \frac{z^{.05}}{.89}\right) = P(Z > 1.85) = .0322.$$

Similarly, if we are using a two-sided test with nominal size .05, then for this case, the test has the true asymptotic size of

$$\alpha^{*} = 2P\left(Z > \frac{z^{.025}}{.89}\right) = 2P(Z > 2.20) = .0278.$$

Finally, a 95% confidence interval for the equal variances model is a 97% confidence interval for this situation ($.97 = 1 - .03$).

Now suppose that $r = 1.5$ still, but $\xi = .25$ (so that 4 $\text{var}(X_i) = \text{var}(Y_j)$). Then

$$\theta = \frac{1.75}{1.375} = 1.27 \quad \text{and} \quad \theta^{1/2} = 1.13.$$

Therefore, in this situation, if $\alpha = .05$ still, then

$$\alpha^{*} = 2P\left(Z > \frac{1.96}{1.13}\right) = 2P(Z > 1.75) = .0802$$

and

$$\alpha^{**} = P\left(Z > \frac{1.645}{1.13}\right) = P(Z > 1.46) = .0722.$$

From this example, we see that even when one sample is one-and-a-half times as large as the other and one variance is four times the other, the sizes and confidence coefficients are not too dramatically affected. We also see that when the large sample is from the distribution with large variance, the true sizes are smaller than the nominal size, .05, and when the large sample is from the distribution with small variance, the true sizes are larger than the nominal size.

Exercises—A

1. A certain hormone in a cow's food is supposed to increase the cow's milk production. To test this hypothesis, 16 cows were given the hormone and 15 were not. The sample mean of the milk production for the 16 cows that received the hormone was 25, with a sample standard deviation of 8, while the sample mean for the 15 cows that were not given the hormone was 20, with a sample standard deviation of 6. Use a .01 test to test the hypothesis that there is no significant increase in milk production due to the hormone. Find a 95% confidence interval for the difference in mean milk production for the two treatments. Find a 95% confidence interval for the variance with these data.

Exercises—B

1. Show that $(\overline{X}, \overline{Y}, S_p^2)$ is an invertible function of $(\Sigma X_i^2 + \Sigma Y_j^2, \Sigma X_i, \Sigma Y_j)$.
2. Verify that $W^2 = V^2 + mn(\overline{X} - \overline{Y})^2/(m + n)^2$.

3. Show that the likelihood ratio test for testing that $\sigma^2 = c$ against $\sigma^2 \neq c$ rejects the null hypothesis if $U(c)$ is too large or too small.
4. Suppose that $\text{var}(X_i) = (.10)\ \text{var}(Y_i)$ and $m = 8n$. Find the true asymptotic size of a one-sided test which has nominal size .05 when the variances are assumed equal. Also, find the true confidence coefficient for a confidence interval which has nominal confidence coefficient .95.

Exercises—C

1. Consider the model described in this section without the normal assumption. That is, let $X_1, \ldots, X_m$ and $Y_1, \ldots, Y_n$ be independent samples from distributions such that $EX_i = \mu$, $EY_i = \nu$, and $\text{var}(X_i) = \text{var}(Y_i) = \sigma^2$.
   (a) Show that S_p^2 is a consistent estimator of σ^2. (*Hint*: $S_p^2 = ((m+n)/(m+n-2)) \times (m\hat{\sigma}^2 + n\hat{\tau}^2)/(m+n)$, and both $\hat{\sigma}^2$ and $\hat{\tau}^2$ are consistent estimators of σ^2. Use Theorem 11–2, part c.)
   (b) Show that $t(\delta) \xrightarrow{d} Z \sim N(0, 1)$, even without the normal assumption. (Use Theorem 11–2, part b.)
   (c) Show that inferences based on $t(\delta)$ are asymptotically insensitive to the normal assumption.
2. **(a)** Show that if $\mu - \nu \neq d$, then $P(t(d) > t^{\alpha}_{m+n-2}) \to 1$ as $m + n \to \infty$.
   (b) Show that if μ_m, ν_n, and σ^2 satisfy $(\mu_m - \nu_n - d)/\sigma(m^{-1} + n^{-1}) \to \theta$, then $t(d) \xrightarrow{d} W \sim N(\theta, 1)$.
   (See Exercise C1, Section 11.2.1. The answer to this problem implies that the power function of the two-sample t-test is also asymptotically insensitive to the normal assumption.)
3. Show that tests and confidence intervals based on $U(\sigma^2)$ are not asymptotically insensitive to the normal assumption used in their derivation.
4. Show that the likelihood ratio test for testing that $\sigma^2 = c$ against $\sigma^2 > c$ rejects the null hypothesis if $U(\sigma^2)$ is too large.
5. Show that $\hat{\delta}$ is an efficient estimator of δ. (See Section 7.2.3.)
6. Consider testing the null hypothesis that $\mu = \nu$ against the alternative that $\mu > \nu$. Let $V^2 = (m+n-2)S_p^2/(m+n)$, $W^2 = V^2 + mn(\bar{X} - \bar{Y})^2/(m+n)$, $U = (m\bar{X} + n\bar{Y})/(m+n)$, $t = (\bar{X} - \bar{Y})/S_p(m^{-1} + n^{-1})^{1/2}$.
   (a) Show that the MLE under the alternative hypothesis is
   $$(\hat{\mu}, \hat{\nu}, \hat{\sigma}^2) = \begin{cases} (\bar{X}, \bar{Y}, V^2) & \text{if } \bar{X} > \bar{Y} \\ (U, U, W^2) & \text{if } \bar{X} \leqq \bar{Y} \end{cases}.$$
   (b) Show that the likelihood ratio test statistic for this problem is
   $$\Lambda = \begin{cases} (1 + t^2/(m+n))^{(m+n)/2} & \text{if } t > 0 \\ 1 & \text{if } t \leqq 0 \end{cases}.$$
   (c) Show that the size α likelihood ratio test for this model rejects if $t > t^{\alpha}$.

11.2.3 The Two-Sample Normal Model with Unequal Variances

Consider the model in which we observe independent samples, $X_1, \ldots, X_m$ and $Y_1, \ldots, Y_n$, where the X_i are normally distributed with mean μ and variance σ^2 and the Y_j are normally distributed with mean ν and variance τ^2. We examine inferences about $\delta = \mu - \nu$ as well as $\gamma = \sigma^2/\tau^2$. Let $\overline{X}$ and S^2 be the sample mean and sample variance computed from the X's, and let $\overline{Y}$ and T^2 be the sample mean and sample variance computed from the Y's. Let $\hat{\delta} = \overline{X} - \overline{Y}$.

Theorem 11–12

a. $(\overline{X}, \overline{Y}, S^2, T^2)$ is a complete sufficient statistic for the two-sample normal model with unequal variances.
b. $\overline{X}$, $\overline{Y}$, S^2 and T^2 are independent, with $\overline{X} \sim N(\mu, \sigma^2/m)$, $Y \sim N(\nu, \tau^2/n)$, $(m-1)S^2/\sigma^2 \sim \chi^2_{m-1}$, and $(n-1)T^2/\tau^2 \sim \chi^2_{n-1}$.
c. X, Y, $\hat{\delta}$, S^2, and T^2 are the best unbiased estimators of μ, ν, δ, σ^2, and τ^2, respectively, and are consistent.
d. $(\overline{X}, \overline{Y}, \hat{\sigma}^2, \hat{\tau}^2)$ is the MLE of $(\mu, \nu, \sigma^2, \tau^2)$, where $\hat{\sigma}^2 = (m-1)S^2/m$ and $\hat{\tau}^2 = (n-1)T^2/n$. $\hat{\delta}$ is the MLE of δ.

Proof. See the exercises. □

Theorem 11–13. An asymptotic pivotal quantity for δ is

$$Z(\delta) = \frac{\hat{\delta} - \delta}{\left(\dfrac{S^2}{m} + \dfrac{T^2}{n}\right)^{1/2}} \xrightarrow{d} Z \sim N(0, 1).$$

Asymptotic confidence intervals and tests based on $Z(\delta)$ are asymptotically insensitive to the normal assumption.

Proof. This theorem follows directly from Theorem 11–2. □

In regard to inferences about δ, note that $\hat{\delta}$ is the optimal estimator of δ, whether we assume that the variances are equal or not. However, unless $m = n$, asymptotic tests and confidence intervals based on $Z(\delta)$ just defined are different (even asymptotically) from the tests and confidence intervals based on $t(\delta)$ defined in the last section. Even though the tests based on $Z(\delta)$ are not optimal for any model, while those based on $t(\delta)$ are optimal for the model with equal variances, the procedures based on $Z(\delta)$ seem preferable for large samples, unless we are quite confident that the variances are equal (which is rather unlikely). $Z(\delta)$ has been studentized by the square root of an unbiased, consistent estimator of the variance of $\hat{\delta}$, whether the variances are equal or not, while $t(\delta)$ has been studentized by the square root of an inconsistent, biased estimator, unless $m = n$ or $\sigma^2 = \tau^2$. (If $n = m$, then $Z(\delta) = t(\delta)$; see Exercise B4.) Often, we would prefer a suboptimal procedure which is not too sensitive to the assumption of equal variances to an optimal procedure which is sensitive to that assumption.

If at least one of the samples is not large, then we cannot use the asymptotic procedures based on $Z(\delta)$. The problem of testing $\delta = 0$ for finite samples for unequal variances is called the *Behrens-Fisher problem*. It has been studied extensively, but without much success.

An exact pivotal quantity for δ is given in Exercise B6 when $m = n$. It can be generalized to an exact pivotal quantity for any n and m. Tests based on this pivotal quantity can be shown to be unbiased, and confidence intervals are centered at $\hat{\delta} = \bar{X} - \bar{Y}$. The pivotal quantity is called the *Scheffé solution to the Behrens-Fisher problem*. Unfortunately, tests and confidence intervals based on $t^*(\delta)$ are not functions of the complete sufficient statistic and can be changed by reordering the observations. For these reasons, procedures based on $t^*(\delta)$ are seldom used in practice. There does not seem to be any sensible exact pivotal quantity for this problem.

In practice, we seldom know before looking at our data whether $\sigma^2 = \tau^2$ or not. It is therefore usually difficult to decide whether to use procedures based on $t(\delta)$, which are optimal when $\sigma^2 = \tau^2$, but are not even studentized properly when $\sigma^2 \neq \tau^2$, or those based on $Z(\delta)$, which are not optimal (or even exact) for any model, but have at least been studentized properly. In the previous section, however, we indicated that $t(\delta)$ is not too badly studentized, as long as m and n are not too unequal, nor are σ^2 and τ^2. In practice, the following rules of thumb are often given:

1. If $m \geqq 30$ and $n \geqq 30$, use the asymptotic procedures based on $Z(\delta)$.
2. If either $m < 30$ or $n < 30$, assume that the variances are equal and use procedures based on $t(\delta)$, unless there is reason to believe that the variances are dramatically unequal. In this case, get a larger sample.
3. When possible, design the experiment with m approximately equal to n, so that it will not matter much whether $t(\delta)$ or $Z(\delta)$ is used.

MISPRINT

We next find an exact pivotal quantity for $\gamma = \sigma^2/\tau^2$. Let $\hat{\gamma} = S^2T^2$. Recall that if U and V are independent, with $U \sim \chi_j^2$ and $V \sim \chi_k^2$, then $kU/jV \sim F_{j,k}$. Upper α values, $F_{j,k}^{\alpha}$, for the F-distribution are given in Table 4 at the back of the book. We often use the fact that

$$F_{j,k}^{1-\alpha} = (F_{k,j}^{\alpha})^{-1}.$$

(See the exercises; note that the degrees of freedom j and k have been interchanged on the right side of the preceding equation.)

Theorem 11–14

a. A pivotal quantity for γ is

$$F(\gamma) = \frac{\hat{\gamma}}{\gamma} \sim F_{m-1,n-1}.$$

b. Confidence intervals and tests based on $F(\gamma)$ are *not* asymptotically insensitive to the normal assumption.

c. One-sided tests based on $F(c)$ for testing that $\gamma = c$ against the appropriate one-sided alternatives are likelihood ratio tests. For testing that $\gamma = c$ against $\gamma \neq c$, the likelihood ratio test is a two-sided test based on $F(c)$, but is not the equal-tailed test unless $m = n$.

Proof

a. We have

$$U = \frac{(m-1)S^2}{\sigma^2} \sim \chi_{m-1}^2 \quad \text{and} \quad V = \frac{(n-1)T^2}{\tau^2} \sim \chi_{n-1}^2,$$

with U and V independent. Therefore, by the definition of the F-distribution,

$$F(\gamma) = \frac{(n-1)U}{(m-1)V} \sim F_{m-1,n-1}.$$

The proofs of parts b and c are left as exercises. □

Using the pivotal quantity $F(\gamma) = \hat{\gamma}/\gamma \sim F_{m-1,n-1}$, we see that a $(1-\alpha)$ confidence interval for γ is

$$\hat{\gamma} F^{\alpha/2}_{n-1,m-1} \geqq \gamma \geqq \frac{\hat{\gamma}}{F^{\alpha/2}_{m-1,n-1}}.$$

Similarly, to test that $\gamma = c$ against $\gamma < c$ with a size-α test, we would reject the null hypothesis if

$$F(c) = \frac{\hat{\gamma}}{c} < (F^{\alpha}_{n-1,m-1})^{-1}.$$

For example, suppose $m = 21$, $n = 16$, $S^2 = 30$, and $T^2 = 40$. Then from Table 4,

$$F^{.01}_{20,15} = 3.37, \qquad F^{.01}_{15,20} = 3.09, \qquad \hat{\gamma} = \frac{30}{40} = .75.$$

Therefore, a 98% confidence interval for $\gamma = \sigma^2/\tau^2$ is

$$.75(3.09) > \gamma > \frac{.75}{3.37}, \qquad 2.32 > \gamma > .22.$$

To test that $\gamma = 1$ against $\gamma < 1$ with a .05 test, we reject the null hypothesis if

$$\frac{S^2}{T^2} < \frac{1}{F^{.05}_{15,20}} = \frac{1}{2.2} = .45.$$

Since $S^2/T^2 = .75$, we accept the null hypothesis that $\gamma = 1$ (or equivalently, that $\sigma^2 = \tau^2$) with these data.

In addition to being likelihood ratio tests, the one-sided F-tests are UMP unbiased for the appropriate hypotheses. For the two-sided problem, the UMP unbiased test is a two-sided test based on F, but is not the equal-tailed test (or the likelihood ratio test) unless $m = n$. However, the equal-tailed test is typically used in practice.

In the exercises, it is shown that

$$\frac{\log(F(\gamma))}{2^{1/2}(m^{-1} + n^{-1})^{1/2}} \xrightarrow{d} Z \sim N(0,1),$$

as long as the assumptions of the model are satisfied. We can use this normal approximation to find asymptotic critical points for confidence intervals and tests based on $F(\gamma)$. It is also shown that if the normal assumption is removed, then

$$\frac{\log(F(\gamma))}{(m^{-1}(\theta_1 - 1) + n^{-1}(\theta_2 - 1))^{1/2}} \xrightarrow{d} Z \sim N(0,1),$$

where θ_1 and θ_2 are the kurtoses of the two distributions. Therefore, procedures based on $F(\gamma)$ are quite sensitive to the normal assumption and should be used only when we are confident that that assumption is satisfied.

One common application of tests based on $F(\gamma)$ is as a pretest before testing hypotheses about $\delta = \mu - \nu$. We first test that $\gamma = 1$ (or $\sigma^2 = \tau^2$). If we reject the null hypothesis, we use the procedure based on $Z(\delta)$, which does not assume equal variances. If we accept that $\gamma = 1$, we use procedures based on $t(\delta)$. There are certain problems associated with this use of the F-test. First, we might reject the null hypothesis because the data are not normally distributed (but the variances are equal). The procedures based on $t(\delta)$ are not too sensitive to the normal assumption, and we would then not use those procedures when they are valid. Second, if we accept the hypothesis that $\sigma^2 = \tau^2$, it is not proof that that hypothesis is true. (Recall that accepting any hypothesis is not proof that it is true.) If we accept it with a .05 test but would have rejected it with a .10 test, we hardly have proof that the variances are equal. We merely have not quite enough proof to conclude that they are different. For this reason, doing a preliminary F-test is not a very satisfactory resolution of the difficult problem of making inferences about δ in the presence of possibly unequal variances.

Exercises—A

1. Two diets are to be compared. Suppose that 35 men on the first diet had a sample mean weight loss of 20 pounds with a sample standard deviation of 4 pounds, while 60 men on the second diet had a sample mean weight loss of 17.5 pounds with a sample standard deviation of 8 pounds.
   (a) Do an asymptotic two-sample Z-test to test the null hypothesis that there was no significant difference in the mean weight losses for the men on the two diets. Use a .05 test.
   (b) Do a two sample t-test to test the same hypothesis with $\alpha = .05$.
2. An experiment was performed to compare two mixers. Five measurements are made of the amount of a particular chemical from the first mixer, giving observations of 2, 3, 7, 4, and 6. Seven measurements are made from the second mixer, giving 3, 6, 8, 9, 11, 7, and 12. Can we conclude that the measurements from the second mixer have significantly higher variance than those from the first mixer at $\alpha = .05$? Find a 98% confidence interval for the ratio between the two variances.

Exercises—B

1. Prove parts a and b of Theorem 11–12.
2. Prove parts c and d of Theorem 11–12.
3. **(a)** Show that the likelihood ratio test statistic for testing that $\sigma^2 = \tau^2$ against $\sigma^2 \neq \tau^2$ is $\Lambda = (1 + (m-1)F/(n-1))^{(m+n)/2}/hF^{m/2}$, where h is a constant and $F = S^2/T^2$.
   (b) Show that the likelihood ratio test rejects the null hypothesis if F is too large or too small. (Show that $\Lambda \to \infty$ as $F \to 0$ or $F \to \infty$ and that $\Lambda(F)$ has only one critical point.)
4. Show that if $m = n$, then $Z(\delta) = t(\delta)$.
5. Show that $F^{1-\alpha/2}_{m-1,n-1} = (F^{\alpha/2}_{n-1,m-1})^{-1}$. (*Hint*: If $Q \sim F_{m-1,n-1}$, then $Q^{-1} \sim F_{n-1,m-1}$. Why?)
6. **(a)** Suppose that $m = n$ in the Behrens-Fisher problem. Let $U_i = X_i - Y_i$, and let $\overline{U}$ and

V^2 be the sample mean and sample variance, respectively, of the U_i. Show that $t^*(\delta) = n^{1/2}(\overline{U} - \delta)/V \sim t_{n-1}$ is an exact pivotal quantity for δ.

(b) Show that confidence intervals based on $t^*(\delta)$ are centered at $\overline{X} - \overline{Y}$.

(c) Suppose we observe $m = n = 4$, $X_i = (1, 2, 3, 4)$, and $Y_j = (3, 4, 6, 7)$. Find a 95% confidence interval for δ using $t^*(\delta)$.

(d) Suppose we observe $m = n = 4$, $X_i = (1, 2, 3, 4)$, and $Y_j = (7, 6, 4, 3)$. Find a 95% confidence interval for δ using $t^*(\delta)$.

(Note that the data in part d is just a reordering of the data in part c.)

7. Consider the model in which we observe independent samples X_i and Y_j from normal distributions with the same mean but possibly different variances.
   (a) Show that a minimal sufficient statistic is $(\overline{X}, S^2, \overline{Y}, T^2)$, but that this statistic is not a complete sufficient statistic. $(E(\overline{X} - \overline{Y}) = 0.)$
   (b) Write down and try to solve the equations for the MLE for this model.

Exercises—C

1. Consider the two-sample model without the normal assumption. Let θ_1 and θ_2 be the kurtoses of the two samples.
   (a) Show that $m^{1/2}(\log(S^2/\sigma^2))/(\theta_1 - 1)^{1/2} \xrightarrow{d} Z \sim N(0, 1)$. (Use Cramer δ.)
   (b) Show that $\log(F(\gamma))/(m^{-1}(\theta_1 - 1) + n^{-1}(\theta_2 - 1))^{1/2} \xrightarrow{d} Z \sim N(0, 1)$ when $m = aq$ and $n = bq$, where a and b are fixed and $q \to \infty$. (Mimic the proof of Theorem 11–2.)
   (c) Show that if the observations are normally distributed, then $\log(F(\gamma))/2^{1/2} \times (m^{-1} + n^{-1})^{1/2} \xrightarrow{d} Z \sim N(0, 1)$.
   (d) Suppose that $\theta_1 = 9$, $\theta_2 = 7$, and $m/n = 8$. Find the true asymptotic size of a one-sided test which has nominal size .05 for the normal model.
2. Show that when $m = n$, the likelihood ratio test that $\sigma^2 = \tau^2$ against $\sigma^2 \neq \tau^2$ is an equal-tails test based on F. (*Hint*: When $m = n$, $\Lambda(1/F) = \Lambda(F)$ and $1/F$ has the same distribution as F.)
3. Show that the likelihood ratio test for testing that $\sigma^2 = \tau^2$ against $\sigma^2 > \tau^2$ rejects the null hypothesis if F is too large.

11.2.4 The Paired Normal Model

Consider the model in which we observe a sample $(X_1, Y_1), \ldots, (X_n, Y_n)$ from a bivariate normal distribution with $EX_i = \mu$, $\text{var}(X_i) = \sigma^2$, $EY_i = \nu$, $\text{var}(Y_i) = \tau^2$, and $\text{cov}(X_i, Y_i) = \rho\sigma\tau$. Note that although both the X_i and the Y_i are samples from a normal distribution, these samples are not independent unless $\rho = 0$. Our goal is to draw inferences about $\delta = \mu - \nu$.

Some examples of paired models are given in Section 11.1.2. As in that section, let $U_i = X_i - Y_i$, and let $\overline{U}$ and V^2 respectively be the sample mean and sample variance computed from the U_i. Then the U_i are a sample from a univariate normal distribution with mean δ and variance $\psi^2 = \sigma^2 + \tau^2 - 2\rho\sigma\tau$. By results for the one-sample normal model, a pivotal quantity for δ is

$$t(\delta) = \frac{n^{1/2}(\overline{U} - \delta)}{V} \sim t_{n-1}.$$

Therefore, a size-α test that $\delta = d$ against $\delta < d$ rejects the null hypothesis if

$$t(d) < t_{n-1}^{\alpha},$$

and a $(1-\alpha)$ confidence interval for δ is given by

$$\delta \in \bar{U} \pm \frac{t_{n-1}^{\alpha/2} V}{n^{1/2}}.$$

Notice that in reducing to the U_i, we have taken a bivariate model and reduced it to a univariate one. However, $(U_1, \ldots, U_n)$ is not a sufficient statistic for this model, so we have lost some information in the reduction. The following theorem implies that nonetheless, we have not lost any information about δ. The proof of the theorem is not really much harder than the proofs of the other theorems in this chapter, but it is much messier and thus will not be given here. Let $\bar{X}$, $\bar{Y}$, S^2, and T^2 be defined as in previous sections, and let

$$r = \frac{\sum_{i=1}^{n} (X_i - \bar{X})(Y_i - \bar{Y})}{(n-1)ST}$$

be the *sample correlation coefficient* between X and Y. As before, let

$$\hat{\delta} = \bar{X} - \bar{Y} = \bar{U}.$$

Theorem 11–15

a. $(\bar{X}, \bar{Y}, S^2, T^2, r)$ is a complete sufficient statistic for the paired normal model.
b. $\bar{X}$, $\bar{Y}$, S^2, T^2, and $\hat{\delta}$ are the best unbiased estimators of μ, ν, σ^2, τ^2, and δ, respectively, and are consistent; r is a biased but consistent estimator of ρ.
c. $(\bar{X}, \bar{Y}, \hat{\sigma}^2, \hat{\tau}^2, r)$ is the MLE of $(\mu, \nu, \sigma^2, \tau^2, \rho)$, where $\hat{\sigma}^2 = (n-1)S^2/n$ and $\hat{\tau}^2 = (n-1)T^2/n$. $\hat{\delta}$ is the MLE of δ.
d. A pivotal quantity for δ is

$$t(\delta) = \frac{n^{1/2}(\hat{\delta} - \delta)}{V} \sim t_{n-1}.$$

1. Confidence intervals and tests based on $t(\delta)$ are asymptotically insensitive to the normal assumption.
2. One-sided tests based on $t(d)$ and the equal-tailed two-sided test based on $t(d)$ are likelihood ratio tests for testing that $\delta = d$ against the appropriate hypotheses.

Tests based on $t(d)$ are called *paired t-tests* and are also UMP unbiased.

Clearly, procedures based on $t(\delta)$ are actually valid for more general models than those considered here. For example, we could have $EX_i = \mu + \alpha_i$ and $EY_i = \nu + \alpha_i$, where α_i represents an additive effect due to the individual.

It is again very important to distinguish the paired model in this section from the two-sample models considered in previous sections. Using two-sample procedures for paired data is wrong, because those models assume that the two samples are independent.

As mentioned in Section 11.1.2, when we have a choice of designing an experiment as a two-sample experiment or a paired experiment, it is nearly always better, asymptotically, to use a paired experiment. In the finite sample case, it is not quite so clear. If we collect two independent samples of size n, then S_p^2, the pooled variance estimator, has $2(n-1)$ degrees of freedom, as does the pivotal quantity t. In the paired model, both the variance estimator and the pivotal quantity have only $n-1$ degrees of freedom. However, the calculations of $\text{var}(\hat{\delta})$ given in Section 11.1.2 are still valid for the two models, so that $\text{var}(\hat{\delta})$ is smaller for the paired model as long as $\rho > 0$. Consequently, the paired model has a lower variance for $\hat{\delta}$, which leads to more powerful procedures than the two-sample model, but also has fewer degrees of freedom, which in turn leads to less powerful procedures. Hence, it is not clear which alternative is better in a particular case. Note again that this situation applies only to the design of the experiment: Once the data are collected, the model is either paired or independent (or neither), but not both, and should be analyzed using the appropriate procedures for the particular design.

A major advantage of the paired experiment over the two-sample experiment is that we do not need to worry about whether the variances are equal for the paired experiment.

Exercises—A

1. A taxi company wants to compare two brands of tires for wear. It chooses 10 cabs and puts a brand A tire on one rear wheel of each cab and a brand B tire on the other rear wheel. It then measures the tread depth after 30,000 miles of driving. It finds the following data:

A	5	7	6	8	9	10	6	4	11	7
B	7	9	7	8	8	13	8	6	14	6

   Can the company conclude that there is a significant difference between the mean tread depths of the two brands? Use $\alpha = .05$. Find a 90% confidence interval for the difference δ between the mean tread depths.

2. A researcher wants to know whether students' verbal SAT scores improve the second time they take the exam. So she samples nine students who took the SAT twice and finds their verbal scores each time, getting the following data:

1st	580	490	520	680	710	480	390	570	610
2nd	620	460	550	720	710	500	410	600	670

   Can she conclude that the mean scores are significantly higher the second time? Use $\alpha = .02$. Find a 98% confidence interval for the mean improvement.

Exercises—B

1. Show that $t(\mu) \sim t_{n-1}$.
2. (a) Show that $(\Sigma X_i, \Sigma Y_i, \Sigma X_i^2, \Sigma Y_i^2, \Sigma X_i Y_i)$ is a complete sufficient statistic for the paired model.
   (b) Show that $(\bar{X}, \bar{Y}, S^2, T^2, r)$ is a complete sufficient statistic.
3. Show that confidence intervals and tests based on $t(\delta)$ are asymptotically insensitive to the normal assumption.
4. Show that $\bar{U} = \bar{X} - \bar{Y}$ and $V^2 = S^2 + T^2 - 2rST$ and hence that $t(d)$ is a function of the complete sufficient statistic.

5. (a) Verify that $\overline{X}$ and S^2 are the best unbiased estimators of μ and σ^2 and are consistent.
   (b) Verify that $\hat{\delta}$ is the best unbiased estimator of δ and is consistent.
6. Find the best unbiased estimator and the MLE of δ^2.

Exercises—C

1. Find the equations that the MLEs must satisfy for the paired normal model and verify that the estimators given in part c of Theorem 11–15 satisfy those equations.

11.3 BERNOULLI MODELS

11.3.1 The One-Sample Bernoulli Model

Consider the model in which we observe $X_1, \ldots, X_n$, a sample from a Bernoulli distribution with mean μ. (Recall that a Bernoulli random variable X is a random variable which can only be zero or one, and $P(X = 1) = \mu$ and $P(X = 0) = 1 - \mu$.) We are interested in inferences about μ. Let $\overline{X}$ be the sample mean of the X_i. If we associate the number one with a success, then $\overline{X}$ is the proportion of the sample that are successes, while μ is the probability that a randomly chosen individual is a success. $\overline{X}$ is often called the *sample proportion* of successes.

As an example, we might do an experiment to determine the proportion of smokers who had heart attacks. We would take a sample of n smokers and find how many had heart attacks. Let $X_i = 1$ if the ith smoker had a heart attack, and $X_i = 0$ if not. Then $\overline{X}$ is the proportion of smokers in the sample who had heart attacks, while μ is the probability that a smoker chosen at random had a heart attack.

As another example, we might take a sample of women and find out the proportion in the sample who support a particular position. This proportion would be $\overline{X}$, while the true proportion of all women who support the position would be μ.

Theorem 11–16

a. $\overline{X}$ is a complete sufficient statistic for the one-sample Bernoulli model.
b. $\overline{X}$ is the best unbiased estimator and the MLE of μ and is consistent.

Proof

a. The joint density of the X_i is

$$f(\mathbf{x}; \mu) = \prod f(x_i; \mu) = \prod \mu^{x_i}(1-\mu)^{1-x_i} = \mu^{\Sigma x_i}(1-\mu)^{n-\Sigma x_i},$$

and therefore, the likelihood function for this model is

$$L_X(\mu) = L(\mu) = \exp[\textstyle\sum X_i \log(\mu) + (n - \sum X_i) \log(1-\mu)].$$

We now put this likelihood into exponential form. We have

$$L(\mu) = \exp[n \log(1-\mu)] \exp\left[\sum X_i \log\left(\frac{\mu}{1-\mu}\right)\right],$$

which is a one-dimensional exponential family with $T(\mathbf{X}) = \Sigma X_i$. Therefore,

$T = T(\mathbf{X}) = \Sigma X_i$ is a complete sufficient statistic. But then, $\overline{X} = T/n$ is an invertible function of T and is therefore also a complete sufficient statistic.

b. In Chapter 4, we showed that $EX_i = \mu$ and $\text{var}(X_i) = \mu(1-\mu)$. Therefore,

$$E\overline{X} = \mu \quad \text{and} \quad \text{var}(\overline{X}) = \frac{\mu(1-\mu)}{n}.$$

Hence, $\overline{X}$ is an unbiased estimator of μ. But since it is also a complete sufficient statistic, $\overline{X}$ is the best unbiased estimator, by the Lehmann-Scheffé theorem. Furthermore, since the bias is zero and the variance goes to zero, $\overline{X}$ is a consistent estimator of μ. (This also follows from the weak law of large numbers.) Now,

$$\frac{\partial}{\partial\mu}\log[L(\mu)] = \frac{\sum X_i}{\mu} - \frac{n - \sum X_i}{1-\mu}.$$

If we set the right side equal to zero, we see that $\hat{\mu} = \Sigma X_i/n = \overline{X}$. □

We now find two asymptotic pivotal quantities for μ.

Theorem 11–17. Two asymptotic pivotal quantities for μ are

$$Z_1(\mu) = \frac{n^{1/2}(\overline{X} - \mu)}{[\mu(1-\mu)]^{1/2}} \xrightarrow{d} Z \sim N(0,1)$$

and

$$Z_2(\mu) = \frac{n^{1/2}(\overline{X} - \mu)}{[\overline{X}(1-\overline{X})]^{1/2}} \xrightarrow{d} Z \sim N(0,1).$$

Proof. By the central limit theorem (Theorem 11–1, part b),

$$Z_1(\mu) = \frac{n^{1/2}(\overline{X} - \mu)}{[\mu(1-\mu)]^{1/2}} \xrightarrow{d} Z \sim N(0,1).$$

Now, $\overline{X}$ is a consistent estimator of μ, and $h(u) = u(1-u)$ is a continuous function. Therefore, $\overline{X}(1-\overline{X})$ is a consistent estimator of $\mu(1-\mu) = \text{var}(X_i)$. By Theorem 11–1,

$$Z_2(\mu) = \frac{n^{1/2}(\overline{X} - \mu)}{[\overline{X}(1-\overline{X})]^{1/2}} \xrightarrow{d} Z \sim N(0,1). \quad \square$$

The asymptotic pivotal quantity $Z_1(\mu)$ is rarely used for constructing confidence intervals, because they are quite complicated when based on $Z_1(\mu)$. (In Exercise B2, it is shown that $Z_1(\mu) - Z_2(\mu) \xrightarrow{P} 0$, so that the confidence intervals should be very close for large n.) Using $Z_2(\mu)$, we see that an asymptotic $(1-\alpha)$ confidence interval for μ is

$$\mu \in \overline{X} + \frac{z^{\alpha/2}(\overline{X}(1-\overline{X}))^{1/2}}{n^{1/2}}.$$

A common rule of thumb is that this interval may be used as long as

$$n\overline{X} \geqq 5 \quad \text{and} \quad n(1-\overline{X}) \geqq 5.$$

(This corresponds to the rule given earlier for the normal approximation to the binomial distribution, which may be used as long as $n\mu \geqq 5$ and $n(1-\mu) \geqq 5$.) Because the Bernoulli distribution is a discrete distribution, no exact pivotal quantity for μ is possible.

For testing that $\mu = c$, the asymptotic test statistic $Z_1(c)$ is often used. In order to do so, the usual rule of thumb says that we must have

$$nc \geqq 5 \quad \text{and} \quad n(1-c) \geqq 5.$$

We could also use the asymptotic test statistic $Z_2(c)$, but $Z_1(c)$ is the asymptotic test statistic which is typically used in practice.

Although there is no exact pivotal quantity for μ in this model there is an exact test statistic, which we now derive.

Theorem 11–18

a. An exact test statistic for testing that $\mu = c$ is

$$U = n\overline{X} \sim B(n, c)$$

under the null hypothesis.

b. One-sided tests based on U are UMP for the appropriate alternatives and are likelihood ratio tests.

c. The likelihood ratio test for testing $\mu = c$ against $\mu \neq c$ is a two-sided test based on U and is the equal-tailed two-sided test when $c = \frac{1}{2}$.

Proof

a. In Chapter 4, we showed that $U = n\overline{X} = \Sigma X_i \sim B(n, \mu)$, so that U is an exact test statistic and has the indicated null distribution.

b. Let $\mu_1 > \mu_0$, $U = \Sigma X_i$. Then

$$\frac{L(\mu_1)}{L(\mu_0)} = \left(\frac{\mu_1}{\mu_0}\right)^U \left(\frac{1-\mu_1}{1-\mu_0}\right)^{n-U}$$

$$= \left(\frac{1-\mu_1}{1-\mu_0}\right)^n \left(\frac{\mu_1(1-\mu_0)}{\mu_0(1-\mu_1)}\right)^U.$$

Since $\mu_1 > \mu_0$,

$$\frac{\mu_1(1-\mu_0)}{\mu_0(1-\mu_1)} > 1,$$

so that $L(\mu_1)/L(\mu_0)$ is an increasing function of U, and hence, the model has monotone likelihood ratio in U. It then follows that the UMP size-α test for testing that $\mu = c$ against $\mu > c$ rejects the null hypothesis if U is too large, and the UMP size-α test that $\mu = c$ against $\mu < c$ rejects the null hypothesis if U is too small. Hence, the one-sided size-α tests are UMP size-α tests for the appropriate one-sided hypotheses. (Note that U has a discrete distribution, so these tests would typically be randomized tests.) Proof that these one-sided tests are likelihood ratio tests is left as an exercise.

c. Now consider testing that $\mu = c$ against $\mu \neq c$. The MLE under the alternative hypothesis is $\overline{X}$, and the MLE under the null hypothesis is c. Therefore,

$$\Lambda(\mathbf{X}) = \frac{L(\overline{X})}{L(c)} = \left(\frac{\overline{X}}{c}\right)^{U}\left(\frac{1-\overline{X}}{1-c}\right)^{n-U}$$
$$= \left(\frac{U}{nc}\right)^{U}\left(\frac{n-U}{n-nc}\right)^{n-U} = h(U).$$

The likelihood ratio test rejects the null hypothesis if $h(U) > k$ and randomizes if $h(U) = k$. Now $h(0) = h(n) = n\log(n)$ and $h(u)$ has a minimum at $u = n/2$. Therefore, $h(U) > k$ if and only if $U < a$ or $U > b$ for some a and b. Hence the likelihood ratio test is a two-sided test based on U. If $c = \frac{1}{2}$, then $h(U) = h(n - U)$, so that the test rejects the null hypothesis if $U > k$ or $U < n - k$ (possibly randomizing on the boundary). When $\mu = \frac{1}{2}$, $P(U > k) = P(U < n - k)$, so that the likelihood ratio test is the equal-tails test when $c = \frac{1}{2}$. □

For small samples, the (exact) test statistic U may be used in the obvious way to construct tests. Note that $U > k$ if and only if $Z_1(c) > k^*$ for some k^*, so that tests based on U and $Z_1(c)$ are essentially the same, with $Z_1(c)$ using the asymptotic distribution to find the critical value for the test and U using the exact distribution.

Another approximation which is useful to remember is that if n is large and c is small in such manner that $nc < 5$, then U has approximately a Poisson distribution with mean nc under the null hypothesis. (If n is large and $1 - c$ is small, then $n - U$ has approximately a Poisson distribution with mean $n(1 - c)$.)

Although the procedures discussed in this section are quite sensitive to the assumption that the X_i have a Bernoulli distribution, this is really not an assumption. If X_i only takes on the values zero and one, it must have a Bernoulli distribution for some μ.

Exercises—A

1. A seed manufacturer guarantees that 99% of its seeds germinate. To test this hypothesis, a researcher takes 100 seeds and finds that only 95 germinate. Can we conclude that the true proportion of seeds which germinate is significantly less than .99 with a .08 test?
2. A detergent company claims that 40% of all households use its detergent. A sample is made of 100 households, and only 32 use the detergent. Can we conclude that the true proportion of households which use the detergent is significantly less than .40 with an asymptotic .05 test? Also, find an asymptotic 95% confidence interval for the true proportion of households which use the detergent.
3. A bomb maker claims that 90% of its bombs will go off. To test this hypothesis, 10 bombs are detonated, and only 6 go off. Is this statistical evidence that the true proportion of bombs which go off is less than .9 with $\alpha = .033$? (Use the exact test, which is randomized.)

Exercises—B

1. **(a)** Show that $\overline{X}(1 - \overline{X}) = (\Sigma X_i^2/n) - \overline{X}^2 = (n-1)S^2/n$. (*Hint*: Since $X_i = 0$ or $X_i = 1$, $X_i^2 = X_i$.)

(b) Use the result of part a to show that the asymptotic confidence interval for μ given in this section is nearly the same as the asymptotic confidence interval given in Section 11.1.2, as long as the X_i are only zero or one.

2. Find the best unbiased estimator and MLE for $\mu(1-\mu)$ for the model given in this section.

3. (a) Show that $Z_1(\mu) - Z_2(\mu) = Z_1(\mu)[1 - \bar{X}(1-\bar{X})/\mu(1-\mu)]^{1/2}$.
(b) Show that $Z_1(\mu) - Z_2(\mu) \xrightarrow{P} 0$. (See Slutzky's theorem.)

Exercises—C

1. (a) Show that $-a \leqq Z_1(\mu)/n^{1/2} \leqq a$ if and only if

$$\mu \in \frac{2\bar{X} + a^2}{2(1+a^2)} \pm \frac{(a^4 + 4a^2\bar{X}(1-\bar{X}))^{1/2}}{2(1+a^2)}.$$

(b) Construct a $(1-\alpha)$ confidence interval for μ based on $Z_1(\mu)$.
(c) Suppose $n = 100$ and $\bar{X} = .8$. Compute the 95% confidence intervals for μ based on $Z_1(\mu)$ and $Z_2(\mu)$.

2. Show that $\bar{X}$ is an efficient estimator of μ.

3. Show that the likelihood ratio test for testing that $\mu = c$ against $\mu > c$ rejects the null hypothesis if U is too large.

4. (a) Show that if $|U_n| \leqq |V_n|$ and $V_n \xrightarrow{P} 0$, then $U_n \xrightarrow{P} 0$.
(b) Show that $W_n = \max(Z_1(\mu), Z_2(\mu)) - Z_1(\mu) \xrightarrow{P} 0$. (*Hint*: $|W_n| \leqq |Z_2(\mu) - Z_1(\mu)|$; see Exercise B3 above.)
(c) Show that the test which rejects $\mu = c$ in favor of $\mu > c$ if either $Z_1(c) > z^\alpha$ or $Z_2(c) > z^\alpha$ is an asymptotic size-α test.

11.3.2 The Two-Sample Bernoulli Model

We next consider the model in which we observe two *independent* samples $X_1, \ldots, X_m$ and $Y_1, \ldots, Y_n$ from Bernoulli distributions, with

$$\mu = EX_i = P(X_i = 1) = 1 - P(X_i = 0) \quad \text{and} \quad \nu = EY_j = P(Y_j = 1) = 1 - P(Y_j = 0).$$

We seek to draw inferences about $\delta = \mu - \nu$.

As an example, we might want to determine whether a smoker is more likely to have a heart attack than a nonsmoker is. So we take samples of m smokers and n nonsmokers and find out whether they had heart attacks. Let $X_i = 1$ if the ith smoker had a heart attack and $X_i = 0$ otherwise, and let $Y_j = 1$ if the jth nonsmoker had a heart attack and $Y_j = 0$ if he did not. Then the X_i are a sample from a Bernoulli distribution with mean μ equal to the probability that a smoker has a heart attack, and the Y_j are a sample from a Bernoulli distribution with mean ν equal to the probability that a nonsmoker has a heart attack. For this example, δ is the difference between the probability that a smoker has a heart attack and the probability that a nonsmoker has a heart attack. If we could reject the hypothesis that $\delta = 0$ in favor of $\delta > 0$, this would be statistical proof that smokers are more likely to have heart attacks than nonsmokers.

As before, let $\bar{X}$ and $\bar{Y}$ be the sample means of the X_i and Y_i. (Then $\bar{X}$ is the proportion of the first sample that are ones, and $\bar{Y}$ is the proportion of the second sample that are ones.) Let $\hat{\delta} = \bar{X} - \bar{Y}$.

Theorem 11–19

a. $(\overline{X}, \overline{Y})$ is a complete sufficient statistic for the two-sample Bernoulli model.

b. $\overline{X}$, $\overline{Y}$, and $\hat{\delta}$ are the best unbiased estimators and MLEs of μ, ν, and δ, respectively, and are consistent.

c. An asymptotic pivotal quantity for δ is

$$Z(\delta) = \frac{\hat{\delta} - \delta}{\left[\dfrac{\overline{X}(1-\overline{X})}{m} + \dfrac{\overline{Y}(1-\overline{Y})}{n}\right]^{1/2}} \xrightarrow{d} Z \sim N(0,1).$$

d. An asymptotic test statistic for testing that $\delta = 0$ is

$$Z^* = \frac{\hat{\delta} - \delta}{[Q(1-Q)(m^{-1} + n^{-1})]^{1/2}} \xrightarrow{d} Z \sim N(0,1),$$

under the null hypothesis, where $Q = (m\overline{X} + n\overline{Y})/(m+n)$.

Proof. See Exercises B2 and B3. □

From part c of Theorem 11–19, an asymptotic confidence interval for δ is

$$\delta \in \hat{\delta} \pm z^{\alpha/2}\left[\frac{\overline{X}(1-\overline{X})}{m} + \frac{\overline{Y}(1-\overline{Y})}{n}\right]^{1/2}.$$

The usual rule of thumb says that this confidence interval may be used as long as

$$m\overline{X} \geqq 5, \quad m(1-\overline{X}) \geqq 5, \quad n\overline{Y} \geqq 5, \quad \text{and} \quad n(1-\overline{Y}) \geqq 5.$$

Unfortunately, due to the discreteness of the Bernoulli distribution, there is no exact pivotal quantity for δ.

Asymptotic tests that $\delta = d$ can also be designed using $Z(\delta)$. For the most common problem of testing that $\delta = 0$, tests are typically based on the asymptotic test statistic Z^*. (Z^* has been studentized with the square root of an estimator of $\text{var}(\hat{\delta})$ which is valid only under the null hypothesis, while $Z(\delta)$ has been studentized with the square root of an estimator which is valid all the time.) As expected, the rule of thumb is that tests based on Z^* are valid as long as

$$mQ \geqq 5, \quad m(1-Q) \geqq 5, \quad nQ \geqq 5, \quad \text{and} \quad n(1-Q) \geqq 5.$$

(Note that although Z^* is an asymptotic test statistic, it is not an asymptotic pivotal quantity and cannot be used to construct asymptotic confidence intervals.)

For small samples, there is no test statistic for testing that $\delta = 0$ against $\delta > 0$ (because of the discreteness of the X_i). However, there is an exact size-α test. Let

$$V = m\overline{X} \quad \text{and} \quad W = m\overline{X} + n\overline{Y}.$$

Now, if $\delta = 0$, then the conditional density of V given W has the hypergeometric distribution

$$f(v|w) = \frac{\begin{bmatrix} m \\ v \end{bmatrix}\begin{bmatrix} n \\ w-v \end{bmatrix}}{\begin{bmatrix} m+n \\ w \end{bmatrix}}.$$

(See Exercise B3.) This conditional distribution does not depend on any unknown parameters. Therefore, for any w, we can find $d(w)$ and $k(w)$ such that

$$P(V > d(W)|W) + k(W)P(V = d(W)|W) = \alpha$$

when $\delta = 0$. Let

$$\Phi(V, W) = \begin{cases} 1 & \text{if } V > d(W) \\ k\,(W) & \text{if } V = d(W). \\ 0 & \text{if } V < d(W) \end{cases}$$

Then Φ is designed so that

$$E\Phi(V, W)|W = \alpha,$$

and hence,

$$E\Phi(V, W) = E(E\Phi(V, W)|W) = E\alpha = \alpha.$$

Hence, Φ is an exact size-α test. This test is an example of what is called a *conditional test*, since the conditional size of the test given W is α. To see why Φ is a sensible test, note that it rejects the null hypothesis when the number of successes in the first sample is larger than would be expected from knowing the total number of successes when $\delta = 0$. This should only happen when the probability of success is higher in the first sample than in the second one.

To use the test Φ, it is only necessary to find $d(W)$ and $k(W)$ for the value of W actually observed. For example, suppose we observe 3 successes in 5 trials for the first sample and 1 success in 10 trials for the second sample. Then for these data, $V = 3$ and $W = 4$. Now, from the formula for $f(v|w)$,

$$P(V = 4|W = 4) = \frac{5}{1{,}365} = .004 \quad \text{and} \quad P(V = 3|W = 4) = \frac{100}{1{,}365} = .073.$$

Therefore, if we use a .05 test, we would use

$$\Phi(V, 4) = \begin{cases} 1 & \text{if } V = 4 \\ (.05 - .004)/.073 & \text{if } V = 3. \\ 0 & \text{if } V < 3 \end{cases}$$

Since we observe $V = 3$, we would use a randomized test and reject the null hypothesis with probability .63. On the other hand, if we used a .10 test, we could just reject the null hypothesis outright. In practice, we would probably just give the conditional p-value,

$$P(V \geqq 3|W = 4) = .004 + .073 = .077.$$

Similar tests could be given for testing that $\delta = 0$ against $\delta < 0$ or against $\delta \neq 0$. The one-sided exact tests are UMP unbiased, and the UMP unbiased test for the two-sided problem is a two-sided conditional test, but is not the equal-tailed test unless $m = n$. The one-sided and two-sided tests suggested here are called *Fisher's exact tests*.

As in the previous section, it should be mentioned that there is no distributional assumption made in this section. If a random variable takes on only the values zero and one, it must be a Bernoulli random variable.

In the exercises, procedures are developed for a paired Bernoulli model. This model would be appropriate, for example, in an experiment in which we have n pairs of sisters. In each pair, we give one sister treatment A and the other treatment B. We then see which women recover. We let $X_i = 1$ if the ith sister given treatment A recovers and $X_i = 0$ if she does not. Similarly, we let $Y_i = 1$ if the ith sister given treatment B recovers and $Y_i = 0$ if she does not. Then X_i and Y_i both have Bernoulli distributions, but are not independent. Accordingly, let

$$\delta = E(X_i - Y_i) = P(X_i \text{ recovers}) - P(Y_i \text{ recovers}).$$

If $\delta > 0$, then patients getting treatment A are more likely to recover than those getting treatment B. As with normal models, a paired Bernoulli design typically leads to more powerful procedures than a two-sample Bernoulli model. Unfortunately, it is often difficult to find sensible pairs in a Bernoulli experiment.

Exercises—A

1. We take a sample of 100 men and find 30 who smoke and a sample of 150 women and find 60 who smoke. Is there any significant difference between the probabilities that men and women smoke? Use $\alpha = .05$. Find an asymptotic 95% confidence interval for the difference in the probabilities.
2. Two bug sprays are to be compared as to their effectiveness in killing flies. In a controlled experiment, one spray kills 300 out of 500 bugs, while the other spray kills 400 out of 600. Is there any significant difference between the effectiveness of these two sprays with an asymptotic .01 test? Find an asymptotic 97% confidence interval for the difference between the probabilities of killing.
3. A new drug is to be compared to an old drug in treating cancer. Suppose that out of 6 cancer patients given the new drug, 5 stayed in remission for at least a year, while 7 of the 12 patients given the old drug stayed in remission for at least a year. Can we conclude that the new drug is significantly better than the old drug with a .05 test? Use Fisher's exact test.
4. Suppose we have a paired experiment involving sisters as described in the text, with 50 pairs. Suppose for 15 of these pairs both sisters recovered, for 10 pairs only those with treatment A recovered, for 20 pairs those with treatment B recovered, and for the remaining 5 pairs neither sister recovered. Can we conclude that the second treatment is significantly better than the first treatment with an asymptotic .04 test? Find an asymptotic 95% confidence interval for δ, the difference between the probabilities of recovery for treatments A and B. Use the results derived in Exercise C1.

Exercises—B

1. Prove parts a and b of Theorem 11–19.
2. Prove parts c and d of Theorem 11–19. (Use Theorem 11–2; for Z^*, note that under the null hypothesis Q is a consistent estimator of $\mu = \nu$ by part c of that theorem, and therefore, $Q(1 - Q)$ is a consistent estimator of $\text{var}(X_i) = \text{var}(Y_i)$.)
3. Let $V = m\overline{X}$, $U = n\overline{Y}$, and $W = U + V$, and assume that $\delta = 0$. Verify the formula for $P(V = v|W = w)$. (*Hint*: $P(V = v|W = w) = P(V = v)P(U = w - v)/P(W = w)$. Note that U and V are independent, $V \sim B(m, \mu)$, $U \sim B(n, \mu)$, and $W \sim B(m + n, \mu)$. Why?)

Exercises—C

1. We say that (X, Y) has a *bivariate Bernoulli distribution* if X and Y both have Bernoulli distributions. Let $\Pi_{ij} = P(X = i, Y = j)$, $i = 0, 1$, $j = 0, 1$. Then the joint density of (X, Y) is

$$f(x, y = \Pi_{00}^{(1-x)(1-y)}\Pi_{01}^{(1-x)y}\Pi_{10}^{x(1-y)}\Pi_{11}^{xy}, \qquad x = 0, 1, y = 0, 1.$$

In the paired Bernoulli model, we observe $(X_1, Y_1), \ldots, (X_n, Y_n)$, a sample from a bivariate Bernoulli distribution with parameters Π_{11}, Π_{10}, Π_{01}, and $\Pi_{00} = 1 - \Pi_{11} - \Pi_{01} - \Pi_{10}$. Let $U_{11} = \Sigma X_i Y_i$, $U_{01} = \Sigma(1 - X_i)Y_i$, $U_{10} = \Sigma X_i(1 - Y_i)$, and $U_{00} = \Sigma(1 - X_i)(1 - Y_i) = n - U_{11} - U_{10} - U_{01}$. That is, U_{10} is the number of pairs such that $X = 1$ and $Y = 0$.

(a) Show that $\mu = EX_i = \Pi_{11} + \Pi_{10}$, $\nu = EY_i = \Pi_{11} + \Pi_{01}$, and $\delta = \mu - \nu = \Pi_{10} - \Pi_{01}$.

(b) Show that $\overline{X} = (U_{11} + U_{10})/n$, $\overline{Y} = (U_{11} + U_{01})/n$, and $\hat{\delta} = \overline{X} - \overline{Y} = (U_{10} - U_{01})/n$.

(c) Show that (U_{11}, U_{10}, U_{01}) is a complete sufficient statistic.

(d) Show that U_{ij}/n is the MLE of Π_{ij} and that $\overline{X}$, $\overline{Y}$, and $\hat{\delta}$ are the MLEs of μ, ν, and δ, respectively.

(e) Show that $\overline{X}$, $\overline{Y}$, and $\hat{\delta}$ are the best unbiased estimators of μ, ν, and δ and are consistent.

2. In the paired Bernoulli model of Exercise C1, let $R = (U_{10} + U_{01})/n$, $Q = R - \hat{\delta}^2$.

(a) Show that $Z(\delta) = n^{1/2}(\hat{\delta} - \delta)/Q^{1/2} \xrightarrow{d} Z \sim N(0, 1)$, is an asymptotic pivotal quantity for δ.

(b) Show $Z^* = n^{1/2}\hat{\delta}/R^{1/2} \xrightarrow{d} Z \sim N(0, 1)$, when $\delta = 0$ is an asymptotic test statistic for testing that $\delta = 0$.

3. In the paired Bernoulli model of Exercises C1 and C2:

(a) Let $W = nR = U_{10} + U_{01}$. Show that if $\delta = 0$, then $U_{10}|W \sim B(W, \frac{1}{2})$.

(b) Use the latter fact to design an exact size-α test that $\delta = 0$ against $\delta > 0$.

11.4 FURTHER TOPICS

* 11.4.1 Simpson's Paradox

Consider an experiment to compare the effectiveness of two diets in preventing heart attacks. Suppose we get two samples of 100 patients and give one sample diet A and the other sample diet B. We find that in the first three years 23 of the patients who had diet A had heart attacks, and 37 of those on diet B had heart attacks. We use the standard Z-test to test for the equality of the proportions of heart attacks for the two diets. We note that $\overline{X}_A = \frac{23}{100} = .23$, $\overline{X}_B = \frac{37}{100} = .37$, and $Q = \frac{60}{200} = .30$. Therefore, the Z-statistic for this problem is

$$\frac{.23 - .37}{[.30(.70)(2/100)]^{1/2}} = -2.16 < -1.96.$$

Hence, we reject the hypothesis that there is no difference in effect between the two diets with a .05 two-sided test, and we would say that diet A is significantly better than diet B at preventing heart attacks.

Now suppose that a student checks the records and finds out that the sample for diet A had 25 smokers, of which 13 had heart attacks, and had 75 nonsmokers, of whom 10 had heart attacks. Suppose also that she finds that the sample for diet B had 75 smokers, of whom 35 had heart attacks, and had 25 nonsmokers, of whom 2 had heart attacks. We first consider the smokers. The respective proportions of smokers on diets A and B who had heart attacks are

$$\overline{X}_{S,A} = \frac{13}{25} = .52 \quad \text{and} \quad \overline{X}_{S,B} = \frac{35}{75} = .47.$$

That is, 52% of the smokers on diet A had a heart attack, while only 47% of the smokers on diet B did. Hence, it appears that diet B is better than diet A for the smokers. Similarly, if we look at the nonsmokers, we see that

$$\overline{X}_{N,A} = \frac{10}{75} = .13 \qquad \overline{X}_{N,B} = \frac{2}{25} = .08,$$

so that 13% of the nonsmokers on diet A had a heart attack, and only 8% of the nonsmokers on diet B did. It appears, therefore, that diet B is better than diet A for both smokers and nonsmokers. However, previously, we concluded that diet A was significantly better than diet B!

What is the problem with this example? In fact, it seems fairly clear what is happening. Smokers are much more likely to have heart attacks than nonsmokers. Since diet A has fewer smokers than diet B, its proportion of deaths is lower than the proportion of deaths for diet B, even though diet B had smaller proportions for both its smokers and its nonsmokers than did diet A. Therefore, for the given data, diet A's apparent superiority to diet B is really due to the fact that, by chance, diet A had more nonsmokers than diet B did, not because diet A is really any better than diet B.

In doing a .05 statistical test, there is a .05 chance that we reject the null hypothesis when it is true. The data set given probably represents one of those cases of false rejection. All the statistical inferences defined in this chapter are based on getting a random sample. However, sometimes the random sample may be poor, just by chance. Random sampling only guarantees that "on the average" random samples are balanced. That is, if we do several experiments involving diets A and B using random samples, it is unlikely that we would get good samples for A and poor samples for B every time, or even very often.

In fact, the researcher was very unlucky to get a division this poor. If we assume that the proportion of smokers in the whole population is .50, then U, the number of smokers in the first sample, and V the number of smokers in the second sample should have binomial distributions with $n = 100$ and $p = .50$. Using the normal approximation to the binomial distribution, we see that the probability of having at most 35 smokers in the first sample and at least 65 smokers in the second is less than .000004.

The phenomenon illustrated in this example, wherein the mean of sample A can be lower than the mean of sample B for each of two subpopulations and still be higher for the whole population is called *Simpson's paradox*. Simpson's paradox can occur only when the two samples contain unequal proportions of the sub-populations.

A natural question is what to do about the problem. Since it seems clear that diet A is not better than diet B, the sensible approach is to analyze the smokers and nonsmokers separately. In this case, we would find that there is no significant difference between diets A and B in preventing heart attacks, but there is some indication that diet B may be better than diet A for both smokers and nonsmokers.

Simpson's paradox can occur for continuous data as well as discrete data. An example is given in Exercise A3, in which the blood pressure of those individuals on diet A is significantly lower than that for those on diet B for both smokers and nonsmokers, but is significantly higher when the smokers and nonsmokers are pooled.

The main lesson to be learned from these examples is that if an important variable such as smoking is ignored in the analysis of an experiment, the results can be misleading. Simpson's paradox occurs when we are interested in the effects of variable Q (diet) on another variable R (heart attacks or blood pressure) in the presence of a third important variable S (smoking) when the samples for variable Q are not balanced in terms of S. One approach for dealing with such situations is to do the analysis separately for each of the levels of variable S. This is the approach we have just suggested. Another approach is to use a more complicated model which includes all three variables Q, R, and S. If R is a continuous random variable (such as blood pressure), one such model is the two-way analysis-of-variance model given in Chapter 13, in which the participants in the survey are classified by the two variables Q and S, and methods are used to determine the effects on R of both Q and S, as well as the interaction between these effects.

Exercises—A

1. Using the data given in the text, show that, for the smokers, there is no significant difference between the proportion of heart attacks suffered under the two diets. Do the same for the nonsmokers. Use $\alpha = .05$.
2. **(a)** Consider a sample which has n_S smokers and n_N nonsmokers. Let $\overline{X}_S$ and $\overline{X}_N$ be the sample means for the smokers and nonsmokers in the sample, and let $\overline{X}$ be the sample mean for the whole population. Show that $\overline{X} = p_S\overline{X}_S + (1 - p_S)\overline{X}_N$, where $p_S = n_S/(n_S + n_N)$ is the proportion of smokers in the population.
   (b) Consider samples from two different populations each with the same proportion of smokers. Show that if the sample mean for smokers in sample A is greater than the sample mean for smokers in sample B, and the sample mean for nonsmokers in A is greater than the sample mean for nonsmokers in B, then the sample mean for sample A must be greater than the sample mean for sample B.
3. Suppose an experiment is run to compare the effects of diet on blood pressure. We put 400 people on each of two diets. For simplicity, we assume that σ^2 is 100 for all populations.
   (a) Suppose the sample mean for the first diet is 140 and for the second diet is 135. Show that the first diet leads to significantly *higher* blood pressure than the second diet with a .05 test.
   (b) Now suppose that the first diet had 320 smokers and 80 nonsmokers and that the sample means for the smokers and nonsmokers on the first diet were 144 and 124, respectively. Suppose that the second diet had 80 smokers and 320 nonsmokers with sample means 151 and 131, respectively. Show that for smokers, the mean for the first diet is significantly *lower* than that for the second diet with $\alpha = .05$, and similarly for nonsmokers. Use $\alpha = .05$.
   (c) Comment on the preceding anomaly.

Exercises—B

1. Let X and Y be independent, with $X \sim B(100, .5)$ and $Y \sim B(100, .5)$. Find $P(X \leqq 35, Y \geqq 65)$.

* 11.4.2 Sampling from Finite Populations

In this section, we indicate the modifications that must be made to the theory developed in the previous sections when we sample without replacement from a finite population. We indicate that as long as the population size is much larger than the sample size, we may use the asymptotic procedures derived in earlier sections under the assumption that the observations are independent, or equivalently, that the sampling is done with replacement. (This is fortunate, because it is hard to imagine a practical situation where we would choose an individual for our sample and then allow the possibility of choosing him or her again.) We also indicate the modification that is necessary to earlier procedures when the sample size is an appreciable part of the population size.

Consider first the case of Bernoulli random variables. Suppose we are taking a sample of size n without replacement from a population of N individuals, M of which are labeled "successes." We call n the *sample size* and N the *population size*. We wish to make inferences about the proportion M/N of the population that are successes. Let $X_i = 1$ if the ith individual is a success and zero otherwise. Then the X_i are Bernoulli random variables with mean $\mu = M/N$. However, the X_i are not independent, since we are sampling without replacement. Therefore, we cannot immediately use the results derived in Section 11.3.1 for the one-sample Bernoulli model. Those results apply directly only when we sample with replacement or when $N = \infty$ (in which case it does not matter whether we replace the individual, since the chance of drawing him or her again is zero). Let $\overline{X}$ be the sample mean of the X_i. As in the Bernoulli model, $\overline{X}$ is called the *sample proportion of successes*.

As a simple example of sampling from a finite population, consider taking a sample of size 1,000 from a state with 3,000,000 voters, 1,000,000 of whom support a particular piece of legislation. Further, suppose that in this sample, 325 support the legislation. Then $n = 1{,}000$, $N = 3{,}000{,}000$, $M = 1{,}000{,}000$, $\mu = 1{,}000{,}000/3{,}000{,}000 = \frac{1}{3}$, and $\overline{X} = 325/1{,}000 = .325$. Note that the pollster would not know M or μ: If she did, she would not need to take the poll. However, we are assuming that she knows N, the population of the state.

Returning to the general model, let $U = n\overline{X}$ be the number of individuals in the sample who are successes. Then U has a hypergeometric distribution. (Why?) Thus, from results derived earlier for that distribution,

$$E\overline{X} = n^{-1} EU = \mu, \quad \text{and} \quad \operatorname{var}(\overline{X}) = n^{-2} \operatorname{var}(U) = \frac{(h_{n,N})^2 \mu(1-\mu)}{n} \leqq \frac{\mu(1-\mu)}{n},$$

where

$$h_{n,N} = \left(\frac{N-n}{N-1}\right)^{1/2}.$$

Therefore, $\overline{X}$ is a consistent unbiased estimator of μ (as $n \to \infty$), even when sampling without replacement. We call $h_{n,N}$ the *finite population correction*.

It can also be shown that if n and N go to infinity, then

$$Z(\mu) = \frac{n^{1/2}(\overline{X} - \mu)}{h_{n,N}(\mu(1-\mu))^{1/2}} \xrightarrow{d} Z \sim N(0,1)$$

is an asymptotic pivotal quantity for μ. Typically, $Z(c)$ is used as a test statistic for testing that $\mu = c$.

Since $\overline{X}$ is a consistent estimator of μ,

$$Z^*(\mu) = \frac{n^{1/2}(\overline{X} - \mu)}{(\overline{X}(1-\overline{X}))^{1/2}\, h_{n,N}} \xrightarrow{d} Z \sim N(0,1)$$

is an asymptotic pivotal quantity for μ. Hence, an asymptotic confidence interval for μ is

$$\mu \in \overline{X} \pm z^{\alpha/2}\left(\frac{\overline{X}(1-\overline{X})}{n}\right)^{1/2} h_{n,N}.$$

From Section 11.3.1, the formula for sampling with replacement is

$$\mu \in \overline{X} \pm z^{\alpha/2}\left(\frac{\overline{X}(1-\overline{X})}{n}\right)^{1/2},$$

so that the confidence interval for sampling without replacement can be computed from the confidence interval with replacement by multiplying the plus-or-minus term by $h_{n,N}$.

In the previous numerical example, $n = 1{,}000$, $N = 3{,}000{,}000$, and

$$h_{n,N} = \left(\frac{2{,}999{,}000}{2{,}999{,}999}\right)^{1/2} = .9998.$$

This implies that if we use the formula derived in Section 11.3.1 and assume that the trials are independent, we will have an interval which is $(.9998)^{-1} = 1.0002$ times too long, and the test statistic would only be .9998 as large as it should be.

From this example, we see that the inference for μ in the model for sampling without replacement is almost identical to that in the model for sampling with replacement, as long as N is enough larger than n so that the finite population correction is near one.

Now consider an example in which the finite population correction is important. Suppose we get a sample of 6,000 from a population of 10,000 and observe $\overline{X} = .6$. The finite population correction in this situation is

$$h_{6{,}000,\,10{,}000} = \left(\frac{4{,}000}{9{,}999}\right)^{1/2} = .63.$$

Therefore, a 95% asymptotic confidence interval for μ for these data is

$$\mu \in .6 \pm (1.96)\left(\frac{.24}{6{,}000}\right)^{1/2}(.63) = .6 \pm .008.$$

(Without the finite population correction, we would get the interval $\mu \in .6 \pm .012$.)

To test that $\mu = .59$ against $\mu > .59$ with a .05 test, we observe that

$$Z(.59) = \frac{(.60 - .59)}{(.59(.41)/6{,}000)^{1/2}(.63)} = 2.50 > 1.645,$$

so we reject the null hypothesis with a .05 test. (Note that we get 1.58 without the finite population correction, so that we would have accepted this hypothesis for these data if we had not used the correction.)

Next, consider sampling without replacement for more complicated models. Suppose we have a population of N individuals, on each of whom we can measure some quantity a. Let a_j be the measurement on the jth individual, and let μ and σ^2 be the mean and variance, respectively, for the whole population, i.e.,

$$\mu = \frac{\sum_{j=1}^{N} a_j}{N} \quad \text{and} \quad \sigma^2 = \frac{\sum_{j=1}^{N} (a_j - \mu)^2}{N}.$$

(Note that we can think of putting a discrete distribution on the a_j's, with each having probability N^{-1}.) Suppose we then take a sample of size n without replacement from these individuals (so that all possible subsets of size n are equally likely). Let X_i be the value of a for the ith individual in the sample, and let $\overline{X}$ and S^2 respectively be the sample mean and variance of the X_i. We seek to use the X_i to draw inferences about μ. Note that $EX_i = \mu$ and $\text{var}(X_i) = \sigma^2$, but the X_i are not independent, so that we cannot immediately use the results of Section 11.1.2 or 11.2.1 to make inferences about μ.

To illustrate the model in question, suppose a university has 20,000 students whose SAT scores we want to study. We get a sample of 500 students and record each of their scores. Then μ and σ^2 are the mean and variance for the SAT scores for all the students at the university, but $\overline{X}$ and S^2 are the mean and variance of the sample. We seek to use $\overline{X}$ and S^2 to draw inferences about μ.

Lemma. $E\overline{X} = \mu$ and $\text{var}(\overline{X}) = (h_{n,N})^2\sigma^2/n$, where $h_{n,N}$ is the finite population correction given previously.

Proof. We know that $EX_i = \mu$ and $\text{var}(X_i) = \sigma^2$. Therefore,

$$E\overline{X} = n^{-1}\sum_{i=1}^{n} EX_i = n^{-1}n\mu = \mu.$$

We now need to find $\text{cov}(X_i, X_j)$. Since all possible subsets are equally likely, $\text{cov}(X_i, X_j) = \text{cov}(X_1, X_2)$ for $i \neq j$. Now, suppose we continued sampling and found $X_1, X_2, \ldots, X_n, X_{n+1}, \ldots, X_N$. That is, suppose we sampled the whole population and let $X_1, \ldots, X_n$ be the first n sampled. Then

$$\sum_{i=1}^{N} X_i = N\mu$$

is constant. Therefore,

$$0 = \text{var}\left(\sum_{i=1}^{N} X_i\right) = \sum_{i=1}^{N} \text{var}(X_i) + \sum_{i=1}^{N}\sum_{\substack{j=1\\ i\neq j}}^{N} \text{cov}(X_i X_j)$$

$$= N\sigma^2 + N(N-1)\,\text{cov}(X_1, X_2).$$

Hence, $\operatorname{cov}(X_i, X_j) = -\sigma^2/(N-1)$, and it follows that

$$\operatorname{var}(\overline{X}) = n^{-2} \operatorname{var}\left(\sum_{i=1}^{n} X_i\right) = n^{-2}\left(\sum_{i=1}^{n} \operatorname{var}(X_i) + \sum_{i=1}^{n}\sum_{\substack{j=1 \\ i \neq j}}^{n} \operatorname{cov}(X_i, X_j)\right)$$

$$= n^{-2}\left(\frac{n\sigma^2 - n(n-1)\sigma^2}{N-1}\right) = \frac{\sigma^2(N-n)}{n(N-1)}. \quad \square$$

Note that $E\overline{X}$ is the same for sampling with or without replacement, but that $\operatorname{var}(\overline{X})$ for sampling without replacement is $h_{n,N}$ times $\operatorname{var}(\overline{X})$ for sampling with replacement, just as it is in the Bernoulli case. Let S^2 be the sample variance of the X_i. It can be shown that S^2 is a consistent estimator of σ^2 as $n \to \infty$ and

$$Z(\mu) = \frac{n^{1/2}(\overline{X} - \mu)}{h_{n,N} S} \xrightarrow{d} Z \sim N(0,1)$$

is an asymptotic pivotal quantity for μ, which we can use to find asymptotic tests and confidence intervals for μ. In particular, an asymptotic $(1-\alpha)$ confidence interval for μ is

$$\mu \in \overline{X} + z^{\alpha/2} S n^{-1/2} h_{n,N},$$

and an asymptotic test statistic for testing that $\mu = c$ is $Z(c)$.

Here again, the confidence interval and test statistic for sampling without replacement are the same as those for sampling with replacement up to the finite population correction $h_{n,N}$, so that if the finite population correction is near one, we may use the formulas derived under the assumption of independence even when we are sampling without replacement.

The results of this section imply that although we typically feel that bigger samples are better than smaller ones, statistical inference becomes more difficult when the sample size approaches the population size.

We now discuss a misconception that is common among people who are not statistically trained. Many such people feel that a sample of size 500 is enough from a population of 10,000 individuals, but is inadequate from a population of 100,000,000. However, the population size enters inferences only in the finite population correction. Thus, as long as the population size N is much larger than the sample size n, the population size is irrelevant to any inferences.

Exercises—A

1. Suppose a television station wants to declare a winner in an election in which 5,000 people voted. It takes a sample of 2,000 individuals and finds that 1,100 supported candidate A. Can the station conclude that the proportion of the population who supported candidate A is greater than one-half with an asymptotic .01 test (and hence declare candidate A the winner)? Find an asymptotic 98% confidence interval for the proportion of the voting population that supported candidate A.
2. A magazine wants to know the average income of its 10,000 subscribers. It takes a sample of 2,000 of those readers and finds their incomes. It finds a sample mean of \$25,000 with a sample standard deviation of \$8,000. Is this finding statistical proof that the average income of the magazine's subscribers is at least \$22,000 with an asymptotic .01 test? Find an asymptotic 99% confidence interval for the average income of all the subscribers.

3. A television survey company gets records from 5,000 homes in a city with 100,000 homes. It finds that 20% of the homes in its sample watched a particular program. Find an asymptotic 95% confidence interval for the proportion of homes in the city that watched the program.

Exercises—B

1. Show that $\overline{X}$ is a consistent estimator of μ for the Bernoulli finite sample model.
2. In the notation of the lemma given in the text, find ES^2 and show that it goes to σ^2 as n goes to ∞.

* 11.4.3 Poisson and Exponential Models: Some Exercises

This section deals solely with exercises that indicate how to make inferences with regard to one- and two-sample Poisson and exponential distributions. The Poisson distribution is useful for modeling phenomena such as accidents on a stretch of highway and telephone calls coming into a switchboard. The exponential distribution is useful for modeling the remaining lifetimes of cancer patients after treatment or time to failure of light bulbs.

Exercises—A

1. A study is made of typos in a manuscript. In a sample of 50 pages, the average number of typos per page is 4. Assuming that the typos per page are Poisson distributed, and using the results of Exercise B1, find an asymptotic 95% confidence interval for the mean number of typos per page in the whole book.
2. A project was funded to compare the accident rates on two stretches of highway. On the first stretch, there were 50 accidents in 100 days, while on the second stretch, there were 30 accidents in 100 days. Can we conclude that there is any significant difference between the mean numbers of accidents for these two roads with an asymptotic .01 test? (Exercise B3).
3. Do Exercise A2 of Section 11.1.2 assuming that the lifetime of light bulbs is exponentially distributed. (Exercise B5.)
4. Two drugs are given to cancer patients. The mean lifetime of the 10 patients given treatment A is 3.5 years and the mean lifetime for the 8 patients given treatment B is 2.5 years. Using the results of Exercise B8, can we say that the lifetime of patients receiving the first drug is significantly longer than that of patients receiving the second drug? Assume that the lifetimes are exponentially distributed.

Exercises—B

1. Let $X_1, \ldots, X_n$ be a sample from a Poisson distribution with mean μ.
   (a) Show that $\overline{X}$ is the MLE, best unbiased estimator, and a consistent estimator, of μ.
   (b) Show that $Z(\mu) = n^{1/2}(\overline{X} - \mu)/\overline{X}^{1/2} \xrightarrow{d} Z \sim N(0, 1)$. Find an asymptotic $(1 - \alpha)$ confidence interval for μ.
   (c) Show that $Z^*(\mu) = n^{1/2}(\overline{X} - \mu)/\mu^{1/2} \xrightarrow{d} Z \sim N(0, 1)$. Find an asymptotic size-α test that $\mu = c$ against $\mu > c$.

2. Using the model of Exercise B1,
   (a) Show that $U = n\overline{X} \sim P(n\mu)$. Describe an exact size-α test that $\mu = c$ against $\mu > c$.
   (b) Show that the test in part a is UMP size α.
   (c) Show that the size-α likelihood ratio test that $\mu = c$ against $\mu \neq c$ rejects the null hypothesis if U is too large or too small.
3. Suppose we observe two independent samples $X_1, \ldots, X_m$ and $Y_1, \ldots, Y_n$ from Poisson distributions with respective means μ and ν.
   (a) Show that $\hat{\delta} = \overline{X} - \overline{Y}$ is the MLE, the best unbiased estimator, and a consistent estimator, of $\delta = \mu - \nu$.
   (b) Show that $(\hat{\delta} - \delta)/(m^{-1}\overline{X} + n^{-1}\overline{Y}) \xrightarrow{d} Z \sim N(0, 1)$. Find an asymptotic $(1 - \alpha)$ confidence interval for δ.
   (c) Let $Q = (m\overline{X} + n\overline{Y})/(m + n)$. Show that if $\delta = 0$, then $(\hat{\delta} - \delta)/Q^{1/2}(m^{-1} + n^{-1}) \xrightarrow{d} Z \sim N(0, 1)$. Find an asymptotic size-α test that $\delta = 0$ against $\delta \neq 0$.
4. Let $X_1, \ldots, X_n$ be a sample from a distribution with mean μ and variance σ^2. Let $Z(\mu)$ be defined as in Exercise B1.
   (a) Show that $Z(\mu) \xrightarrow{d} (\sigma/\mu^{1/2})Z$, where $Z \sim N(0, 1)$, and hence, that procedures based on $Z(\mu)$ are not asymptotically insensitive to the Poisson assumption.
   (b) Suppose that $\mu = 10$ and $\sigma = 30$. Find the true asymptotic confidence coefficient of a confidence interval based on $Z(\mu)$ with nominal asymptotic confidence coefficient .95.
5. Let $X_1, \ldots, X_n$ be a sample from an exponential distribution with mean μ.
   (a) Show that $\overline{X}$ is the MLE, the best unbiased estimator, and a consistent estimator, of μ.
   (b) Show that $R(\mu) = 2n\overline{X}/\mu \sim \chi^2_{2n}$. Find a $(1 - \alpha)$ confidence interval for μ and a size-α test for testing that $\mu = c$ against $\mu < c$.
   (c) Show that the test given in part b is UMP size α.
6. Let $X_1, \ldots, X_m$ and $Y_1, \ldots, Y_n$ be independent samples from exponential distributions with means μ and ν, respectively.
   (a) Show that $\hat{\delta} = \overline{X} - \overline{Y}$ is the MLE, the best unbiased estimator, and a consistent estimator, of $\delta = \mu - \nu$.
   (b) Show that $(\hat{\delta} - \delta)/(m^{-1}\overline{X}^2 + n^{-1}\overline{Y}^2)^{1/2} \xrightarrow{d} Z \sim N(0, 1)$. Find an asymptotic $(1 - \alpha)$ confidence interval for δ.
   (c) Show that if $\delta = 0$, then $T = \overline{X}/\overline{Y} \sim F_{2m,2n}$. Find an exact size-$\alpha$ test that $\delta = 0$ against $\delta > 0$.
7. For the model of Exercise B6, show that the likelihood ratio test that $\delta = 0$ against $\delta \neq 0$ rejects the null hypothesis if T is too large or too small. Show that the test is an equal-tails test when $m = n$.
8. For the model of Exercise B6, let $\xi = \mu/\nu$ and $\hat{\xi} = \overline{X}/\overline{Y}$.
   (a) Show that $\hat{\xi}$ is the MLE and a consistent estimator of ξ.
   (b) Show that $\hat{\xi}/\xi \sim F_{2m,2n}$. Find an exact $(1 - \alpha)$ confidence interval for ξ.
   (c) Find the best unbiased estimator of ξ. ($E\hat{\xi} = E\overline{X}\,E(\overline{Y}^{-1})$.)
9. Let $X_1, \ldots, X_n$ be a sample from a distribution with mean μ and variance σ^2. Let $R(\mu) = 2n\overline{X}/\mu$.
   (a) Show that if the exponential model is true, then $Z(\mu) = (R(\mu) - 2n)/2n^{1/2} \xrightarrow{d} Z \sim N(0, 1)$.
   (b) Show that for general models, $Z(\mu) \xrightarrow{d} (\sigma/\mu)Z$ and hence procedures based on $R(\mu)$ are not asymptotically insensitive to the exponential assumption.
   (c) Find the true asymptotic size of a one-sided test based on $R(\mu)$ which has nominal size .05 for the exponential family when in fact $\mu = 10$ and $\sigma = 30$.

Exercises—C

1. For the two-sample Poisson model given in Exercise B3, let $U = m\overline{X}$, $V = n\overline{Y}$, and $W = U + V$.
   (a) Show that $U|W \sim B(W, \theta)$, where $\theta = m\mu/(m\mu + n\nu)$.
   (b) Find an exact size-α test that $\mu = \nu$ against $\mu > \nu$. ($\mu = \nu$ iff $\theta = m/(m+n)$.)

11.5 ONE-SAMPLE UNIFORM MODELS

* 11.5.1 The Uniform (0, θ) Model

In all the previous examples and exercises we have examined, the range of the random variables has not depended on the parameters. In this section and the next, we consider two models in which the range of the observations does depend on the parameters. For the model in this section, we get fairly nice results, much different, however, from those we have found in previous sections for exponential families.

The model we consider is one in which we observe $X_1, \ldots, X_n$, where the X_i are independently, uniformly distributed on the interval $(0, \theta)$. Therefore the joint density of the X_i is

$$f(\mathbf{x}, \theta) = \theta^{-n}, \qquad 0 \leqq x_i \leqq \theta.$$

Hence, the likelihood function for this model is

$$L_X(\theta) = L(\theta) = \begin{cases} \theta^{-n} & \text{if } \theta \geqq X_i \text{ for all } i \\ 0 & \text{if } \theta < X_i \text{ for some } i \end{cases}.$$

Now, let $T = \max(X_i)$. The first result we show is that T is the MLE for θ. If we differentiate $\log L(\theta)$ and set it equal to zero, we get $0 = n/\theta$, which has no solution. So we must attack the problem differently from the way we have dealt with earlier problems. We wish to maximize θ^{-n} or equivalently, minimize θ^n, subject to the constraint that $\theta \geqq X_i$ for all i. (If we choose $\theta < X_i$ for some i, then $L(\theta) = 0$, and we have clearly not maximized it.) We therefore choose θ as small as possible, subject to $\theta \geqq X_i$ for all i. That is, we choose $\hat{\theta} = \max(X_i)$. Therefore, $T = \max(X_i)$ is the MLE of θ.

We have graphed the likelihood in Figure 11–1. From this picture, it is also apparent that T is the MLE of θ.

T is a complete sufficient statistic for this model, as we now show. (We cannot use the exponential criterion for this model. Why?) We first use the factorization criterion to show that T is a sufficient statistic. The sample space S for this model is the set of all $\mathbf{x} = (x_1, \ldots, x_n)$ such that $x_1 \geqq 0, \ldots, x_n \geqq 0$. (Note that the sample space cannot depend on θ.) Let $h(T, \theta)$ be 1 if $0 \leqq T \leqq \theta$ and 0 if $T > \theta$. Then

$$L_X(\theta) = (1)\theta^{-n} h(T(\mathbf{X}), \theta)$$

all $\mathbf{X} \in S$. Therefore, by the factorization criterion, T is a sufficient statistic.

We now use the definition of completeness to show that T has a complete family of distributions. The density of T is

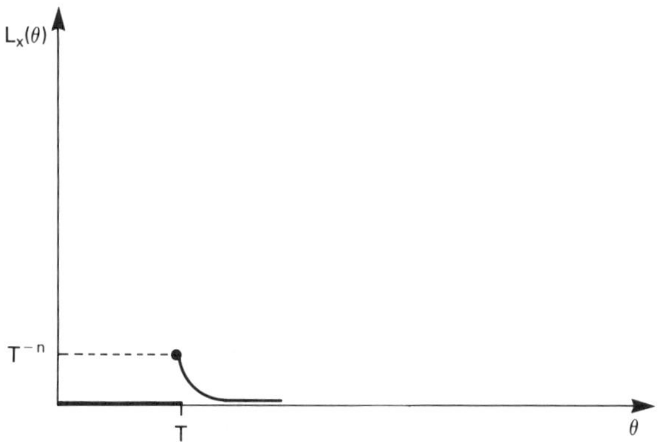

Figure 11-1 Graph of the likelihood of θ for fixed T.

$$h(t) = \frac{nt^{n-1}}{\theta^n}, \qquad 0 \leqq t \leqq \theta.$$

(See Exercise B1.) Suppose that $E_\theta g(T) = 0$ for all $\theta > 0$. In order to prove completeness, we must show that $g(T) \equiv 0$. It is not possible to establish this result rigorously at the level of the text. So instead, we establish it under the assumption that g is well behaved enough so that the following argument holds:

$$0 = \frac{\theta^n Eg(T)}{n} = \int_0^\theta g(t)t^{n-1}\,dt.$$

Differentiating both sides of this expression, and using the fundamental theorem of the calculus, we obtain

$$0 = \frac{d}{d\theta}\int_0^\theta g(t)t^{n-1}\,dt = g(\theta)\theta^{n-1},$$

and hence, $g(\theta) = 0$ for all $\theta > 0$, which finishes the proof. (Note that g must satisfy certain regularity conditions in order for us to use the fundamental theorem.)

We now find the best unbiased estimator of θ. Since

$$ET = \int_0^\theta \frac{nt^n}{\theta^n}dt = \frac{n\theta}{n+1},$$

$E(n+1)T/n = \theta$, and $(n+1)T/n$ is the best unbiased estimator of θ, by the Lehmann-Scheffé theorem. There is a nice interpretation for this estimator. Let the first spacing be the distance between zero and the smallest observation, let the second spacing be the distance between the smallest observation and the next smallest, . . . , and let the nth spacing be the distance between the next to the largest observation and the largest. Then T/n is the average of these n spacings. (Why?) Now, the best unbiased estimator of θ is

$$\frac{(n+1)T}{n} = T + \frac{T}{n}.$$

Therefore, the best unbiased estimator is estimating the distance between T and θ to be the average of the previous spacings. In fact, the best unbiased estimator seems preferable to the MLE, since the MLE estimates the distance between T and θ to be zero, when we know it must be positive. In Exercise B1, it is shown that both the MLE and the best unbiased estimator are consistent estimators of θ.

Now consider estimating $\mu = EX_i = \theta/2$. By the invariance principle, the MLE of μ is $T/2$. In addition, $E(n+1)T/2n = \theta/2 = \mu$, so that $(n+1)T/2n$ is the best unbiased estimator of μ. Another unbiased estimator of μ is $\overline{X}$, which is not a function of the sufficient statistic for this model. In Exercise B1, it is shown that

$$\operatorname{var}\left(\frac{(n+1)T}{2n}\right) = \frac{\theta^2}{4n(n+2)} < \operatorname{var}(\overline{X}) = \frac{\theta^2}{12n}.$$

For example, if $n = 20$, then $\operatorname{var}((n+1)T/2n) = \theta^2/1{,}760$, while $\operatorname{var}(\overline{X}) = \theta^2/240$. In this case, the variance of $\overline{X}$ is more than seven times the variance of the best unbiased estimator. Indeed, the variance of the best unbiased estimator even goes to zero at a faster rate (n^{-2}) than the variance of $\overline{X}$ does. This example shows that the sample mean $\overline{X}$ can sometimes be a poor estimator of the true mean of the distribution.

Now, let $U = T/\theta$. Then U has density function

$$k(u) = nu^{n-1}, \qquad 0 \leqq u \leqq 1$$

(see the exercises). Since this density does not depend on θ, U is a pivotal quantity for θ. (As usual, U is a decreasing function of θ.) Now, the upper α percentile of u is

$$u^{\alpha} = (1-\alpha)^{1/n}$$

(see the exercises). We can therefore use U to find confidence intervals and tests for θ in the usual way. For example, a $(1-\alpha)$ confidence interval for θ is

$$\frac{T}{u^{\alpha/2}} < \theta < \frac{T}{u^{1-\alpha/2}}.$$

In Exercise C1, it is shown that one-sided tests based on U are UMP for the appropriate hypotheses. It can also be shown that a particular two-sided test based on U is actually UMP for testing $\theta = c$ against $\theta \neq c$. This is essentially the only situation in which it is possible to find a UMP test for a two-sided problem. (See Lehmann (1986), p. 111.)

Exercises—B

1. In the model in which we observe a sample from $U(0, \theta)$,
   (a) Show that T has density function $f_T(t) = nt^{n-1}/\theta^n, 0 \leqq t \leqq \theta$. (*Hint*: $F_T(t) = (P(X_1 \leqq t))^n$. Why?)
   (b) Find the mean and variance of T.
   (c) Find the variance of $\overline{X}$ and of $(n-1)T/n$.
   (d) Show that T and $(n-1)T/n$ are both consistent estimators of θ.
2. Verify that $U = T/\theta$ has density $f_U(u) = nu^{n-1}, 0 \leqq u \leqq 1$, and that $u^{\alpha} = (1-\alpha)^{1/n}$.
3. Show that if $n > 1$, then $\theta^2/4n(n+2) < \theta^2/12n$.

4. Let $X_1, \ldots, X_n$ be a sample from the density function $f(x; \theta) = \exp(-(x - \theta)), x \geqq \theta$. Let S be the minimum of the X_i.
   (a) Show that S is the MLE of θ.
   (b) Show that S is a sufficient statistic for θ. (*Hint*: Let $k(S, \theta) = 1$ if $S > \theta$ and 0 if $S < \theta$. Show that $L_X(\theta) = g(\mathbf{X})k(S, \theta)$ for some function $g(\mathbf{X})$.
   (c) Show that S has density function $f_S(s) = n \exp(-n(s - \theta)), s \geqq \theta$. (*Hint*: $P(S \leqq s) = 1 - (P(X_1 \geqq s))^n$. Why?)
   (d) Show that S is a complete sufficient statistic for θ.
   (e) Find the best unbiased estimator of θ.
5. (a) Let $P = S - \theta$. For the model in Exercise B4, show that P is a pivotal quantity for θ and find a formula for p^α.
   (b) Construct a $(1 - \alpha)$ confidence interval for θ.
6. Let $X_1, \ldots, X_n$ be a sample from the discrete uniform distribution $f(x) = 1/\theta$, $x = 1, \ldots, \theta$, where θ is an unknown positive integer. Let T be the maximum of the X_i.
   (a) Show that T is the MLE of θ.
   (b) Show that T is a sufficient statistic for θ.
   (c) Show that T has density function $f_T(t) = (t^n - (t - 1)^n)/\theta^n, t = 1, \ldots, \theta$. (*Hint*: $F_T(t) = (P(X_1 \leqq t))^n, f_T(t) = F_T(t) - F_T(t - 1)$. Why?)
   (d) Show that T is a complete sufficient statistic. (*Hint*: Look at $E_\theta g(T) - E_{\theta - 1} g(T)$.)
   (e) Find the best unbiased estimator of θ when $n = 2$.

Exercises—C

1. (a) Show that the model discussed in this section has monotone likelihood ratio in T.
   (b) Show that the UMP size-α test that $\theta = c$ against $\theta > c$ rejects the null hypothesis if $U(c) = T/c$ is too large.
2. Let $X_1, \ldots, X_n$ be a sample from $U(\delta, \theta)$, where δ and θ are both unknown parameters. Let S and T be the minimum and maximum, respectively, of the X_i.
   (a) Show that (S, T) is the MLE of (δ, θ).
   (b) Show that (S, T) is a sufficient statistic for this model.
   (c) Show that the joint density of S and T is $f_{S,T}(s, t) = n(n - 1)(t - s)^{n-2}/(\theta - \delta)^n, \delta \leqq s \leqq t \leqq \theta$. ($F(s, t) = (P(X_1 \leqq t))^n - (P(s < X_1 \leqq t))^n$. Why?)
   (d) Show that (S, T) is a complete sufficient statistic for this model.
   (e) Find the best unbiased estimators of $\delta, \theta, \mu = (\delta + \theta)/2$, and $\beta = \theta - \delta$.
3. For the model of Exercise C2, let $W(\mu) = (M - \mu)/(T - S)$.
   (a) Show that W has density function $f(w) = (n - 1)(1 + 2|w|)^{n-1}$.
   (b) Show that $W(\mu)$ is a pivotal quantity for μ, find W^α, and find a $(1 - \alpha)$ confidence interval for μ.
4. For the model of Exercises C2 and C3, let $Q(\beta) = (T - S)/\beta$. Show that $Q(\beta)$ is a pivotal quantity for β, find q^α, and find a $(1 - \alpha)$ confidence interval for β.
5. Let $X_1, \ldots, X_n$ be a sample from the density function $f(x; \theta) = h(x)/H(\theta), 0 \leqq x \leqq \theta$, where $h(x) > 0$ for $x > 0$. Let $T = \max(X_i)$.
   (a) Show that $\int_0^\theta h(x)dx = H(\theta)$ and, hence, that $H'(x) = h(x)$.
   (b) Show that $F(x; \theta) = H(x)/H(\theta), 0 < x < \theta$ and, hence, that H is an increasing function.
   (c) Show that T is the MLE of θ.
   (d) Show that T has density $f_T(t) = nh(t)(H(t))^{n-1}/(H(\theta))^n, 0 \leqq t \leqq \theta$.
   (e) Show that T is a complete sufficient statistic.
   (f) Show that $U = H(T)/H(\theta)$ is a pivotal quantity for θ. What is the density of U?

* 11.5.2 The Uniform $(\mu - \frac{1}{2}, \mu + \frac{1}{2})$ Model

We now examine a model that has some rather surprising properties. In this model, we observe $X_1, \ldots, X_n$, where the X_i are independently distributed uniformly on the interval $(\mu - \frac{1}{2}, \mu + \frac{1}{2})$. The joint density of the X_i is

$$f(\mathbf{x}, \mu) = 1, \qquad \mu - \tfrac{1}{2} \leqq x_i \leqq \mu + \tfrac{1}{2}.$$

Let $S = \min(X_i)$ and $T = \max(X_i)$. Now, $\mu - \frac{1}{2} \leqq X_i \leqq \mu + \frac{1}{2}$ for all i if and only if $\mu - \frac{1}{2} \leqq S$ and $T \leqq \mu + \frac{1}{2}$. Hence, the likelihood function for the model is

$$L_X(\mu) = L(\mu) = \begin{cases} 1 & \text{if } \mu - \frac{1}{2} \leqq S \text{ and } T \leqq \mu + \frac{1}{2} \\ 0 & \text{otherwise} \end{cases}.$$

Therefore (S, T) is a sufficient statistic, by the factorization criterion. In fact, (S, T) is the minimal sufficient statistic. On the other hand, if $n \geqq 2$, then (S, T) is not a complete sufficient statistic, since

$$E\left(T - S - \frac{(n-1)}{n+1}\right) = 0$$

(see Exercise B1). If $n = 1$, then $S = T = X_1$, and even in this situation, X_1 is not a complete sufficient statistic, as we shall see.

Now consider the problem of estimating μ. The likelihood function is one as long as $T - \frac{1}{2} \leqq \mu$ and $S + \frac{1}{2} \geqq \mu$. Therefore, the likelihood is maximized for any number between $T - \frac{1}{2}$ and $S + \frac{1}{2}$. Accordingly, let $Q = Q(S, T)$ be any number such that

$$T - \tfrac{1}{2} \leqq Q(S, T) \leqq S + \tfrac{1}{2}.$$

Then Q is an MLE of μ. Note that the interval $[T - \frac{1}{2}, S + \frac{1}{2}]$ is not empty, so that an MLE exists, but this MLE is not unique.

The estimator that we usually choose is $M = (S + T)/2$, which lies in the preceding interval and is therefore an MLE. M is called the *midrange* of the X_i and is an unbiased consistent estimator of μ.

M is not the only possible unbiased MLE, however. To illustrate the possibilities, suppose that $n = 1$ and $S = T = X$. Then $M = X$. Let $Q = [X + \frac{1}{2}]$ (where $[y]$ is the largest integer less than or equal to y). Then

$$Q = [\mu] \Leftrightarrow \mu \leqq X + \tfrac{1}{2} < [\mu] + 1 \quad \text{and} \quad Q = [\mu] + 1 \Leftrightarrow [\mu] + 1 \leqq X + \tfrac{1}{2} \leqq \mu + 1.$$

Therefore,

$$P(Q = [\mu]) = [\mu] + 1 - \mu \quad \text{and} \quad P(Q = [\mu] + 1) = \mu - [\mu].$$

Hence,

$$EQ = \mu,$$

and Q is an unbiased estimator of μ.

Now, suppose that μ is an integer. Then

$$P(Q = \mu) = P(Q = [\mu]) = [\mu] + 1 - \mu = 1,$$

so that Q is certain to be correct when μ is an integer. Also,

$$X - \tfrac{1}{2} \leqq Q \leqq X + \tfrac{1}{2},$$

so that Q is an MLE of θ. Now, consider another estimator $Q_p = [X + \frac{1}{2} + p] - p$. Then, by an argument similar to that just given, Q_p is an unbiased estimator, and Q_p is exactly correct when μ is $N \div p$, where N is an integer. Therefore, for any μ, we can find an unbiased estimator which has zero variance for that μ. Hence, if there were a best unbiased estimator, it would have to have zero variance for all μ. Note also that Q and Q_p are functions of the sufficient statistic X, but

$$E(Q - Q_p) = \mu - \mu = 0,$$

so that X is not a complete sufficient statistic for the model under discussion.

We now return to the case where $n \geqq 1$. As before, let M be the midrange of the X_i, and let $V = M - \mu$. Then V has density

$$g(v) = n(1 - 2|v|)^{n-1}, \qquad -\tfrac{1}{2} \leqq v \leqq \tfrac{1}{2}$$

(see Exercise B4), so that V is a pivotal quantity for μ (and is, as usual, decreasing in μ). Furthermore, if $\alpha \leqq \frac{1}{2}$, then

$$v^{\alpha} = \frac{1 - (2\alpha)^{1/n}}{2} \quad \text{and} \quad v^{1-\alpha} = -v^{\alpha}.$$

Therefore, a $(1 - \alpha)$ confidence interval for μ is

$$\mu \in M \pm v^{\alpha/2}.$$

This confidence interval has the unfortunate property that for some observations, it is sure to cover the true parameter. For example, suppose that $\alpha = .05$ and $n = 8$. Then

$$v^{.025} = \frac{1 - (.05)^{1/8}}{2} = \frac{(1 - .688)}{2} = .156.$$

Suppose, in addition, that $T = .6$ and $S = -.3$. Then $M = .15$, and the confidence interval is

$$\mu \in .15 \pm .156 \Leftrightarrow -.006 \leqq \mu \leqq .306.$$

However, we know from before that $T - \frac{1}{2} \leqq \mu \leqq S + \frac{1}{2}$, so we know that

$$.1 \leqq \mu \leqq .2.$$

Therefore, we are saying that we are 95% certain that $-.006 \leqq \mu \leqq .306$ when we know that $.1 \leqq \mu \leqq .2$. (We might imagine that an engineer has brought a consulting statistician these data, and the statistician has found the 95% confidence interval to be $-.006 \leqq \mu \leqq .306$. The engineer would be a little disappointed if he already knew that $.1 \leqq \mu \leqq .2$.)

At this point, it seems that the confidence interval just derived must be incorrect. However, it is not. The confidence coefficient of a confidence interval is computed before the data are collected. Therefore, before we look at our observations, it is true that

$$P(\mu \in M + v^{\alpha/2}) = P(-v^{\alpha/2} \leqq M - \mu \leqq v^{\alpha/2}) = 1 - \alpha.$$

After we have collected the data, some intervals are longer than they should be, while others are shorter. The one in question ended up longer than it should be (because $T - S$ is large).

We can recast this example to make an interesting testing example. Suppose we are testing that $\mu = 0$ against $\mu \neq 0$. Using the heuristic rule given earlier, we should reject the null hypothesis if

$$|M| > v^{\alpha/2} = .156.$$

However, $M = .15$, so we would accept the hypothesis $\mu = 0$ with these data. Nonetheless, we know that $.1 \leqq \mu \leqq .2$, so we know that $\mu \neq 0$. Therefore, we accept the hypothesis that $\mu = 0$ with a size-.05 test, even when we know with certainty that $\mu \neq 0$.

The foregoing example has led many statisticians to suggest that any procedure which averages over the space of X-values should be viewed with some suspicion. We are only interested in inferences regarding the particular data set observed and, therefore, should not average over the values we did not see. This nonaveraging principle implies that we should never consider $(1 - \alpha)$ confidence intervals, size-α tests, power functions, unbiased estimators, or mean squared error. So the principle would eliminate most of the procedures discussed in this book. (It does not eliminate MLEs, however.) For an alternative approach to statistics which is compatible with the nonaveraging principle, see Chapter 16.

Exercises—B

1. For the model discussed in this section,
   (a) Show that the joint density of S and T is $f_{S,T}(s,t) = n(n-1)(t-s)^{n-2}$, $\mu - \frac{1}{2} \leqq s \leqq t \leqq \mu + \frac{1}{2}$. (*Hint*: $F_{S,T}(s,t) = (P(X_i \leqq t))^n - (P(s < X_i \leqq t))^n$. Why?)
   (b) Show that $E(T - S) = (n-1)/(n+1)$.
   (c) Show that M is an unbiased estimator of μ.
2. Show that $P(T - \frac{1}{2} < S + \frac{1}{2}) = 1$.
3. Show that $T - \frac{1}{2} \leqq M \leqq S + \frac{1}{2}$.
4. Let $V(\mu) = M - \mu$. Show that V has density $f_V(v) = n(1 - 2|v|)^{n-1}$, $-\frac{1}{2} \leqq v \leqq \frac{1}{2}$, and that $v^{\alpha} = (1 - (2\alpha)^{1/n})/2$ and $v^{1-\alpha} = -v^{\alpha}$.
5. Show that M is a consistent estimator of μ. (Note that $\text{var}(M) = \text{var}(M - \mu)$; use integration by parts.)

Exercises—C

1. For the model discussed in this section,
   (a) Show that the estimator $U = [X + \frac{1}{2} + p] - p$, is an unbiased estimator of μ (where the brackets denote "the greatest integer in").
   (b) Show that U is sure to be correct if $\mu = N - p$ for some integer N.
2. Show that (S, T) is the minimal sufficient statistic for the model of this section. (*Hint*: Use the necessary and sufficient condition given in Section 10.2.1.)

Chapter 12

REGRESSION ANALYSIS

In this chapter, we develop a statistical theory for fitting equations to points. As an example, suppose that a college administrator wants to predict a student's college grade point average (GPA) y from his high school GPA u and his combined SAT score v. Suppose that the administrator believes that the college GPA is best predicted by the function

$$y = \beta_1 + \beta_2 u + \beta_3 v + \beta_4 u^2 + \beta_5 v^2 + \beta_6 uv.$$

To study this relationship, she could check the records of 1,000 graduates to find their college GPAs, their high school GPAs, and their combined SAT scores. From these data, the administrator would hope to do several things. First, she would want to estimate the coefficients β_i. She then could use the preceding equation to predict the college GPA of an incoming freshman. However, she might want to have an interval estimate for the prediction she made, so she can tell how confident she should be in that prediction. She also might want to know if the SAT score is contributing to the prediction, or whether the college should stop asking for SAT scores. The regression model studied in this chapter gives a theory which allows us to give statistical answers to these and other questions.

The *multiple linear regression* model is one in which we observe $Y_1, \ldots, Y_n$ independent random variables such that

$$Y_i = x_{i1}\beta_1 + \cdots + x_{ip}\beta_p + e_i,$$

where the x_{ij} are known constants, the β_j are unknown parameters, and the e_i are unobserved random variables representing the variation in the measurements. We assume that the e_i are independent, with $e_i \sim N(0, \sigma^2)$, where $\sigma^2 > 0$ is another unknown parameter. We also write this model in the form

$$Y_i \sim N(\sum_j x_{ij}\beta_j, \sigma^2),$$

independent. We use capital letters for the Y_i's since we are thinking of them as observed random variables and small letters for the x's since we are thinking of them as observed constants.

The multiple linear regression model is an extremely flexible model for fitting equations to data. We need not assume that EY_i is a linear function of the x's; we only need to assume that it is a linear function of the unknown coefficients β_i.

We call the equation

$$y = \sum_j x_j \beta_j$$

the *regression function*; y is the *dependent variable*, and the x's are the *independent variables*.

In the previous example involving the predicted GPA of an incoming college freshman, let Y_i be the ith student's college GPA, and let

$$x_{i1} = 1, \qquad x_{i2} = u_i, \qquad x_{i3} = v_i, \qquad x_{i4} = u_i^2, \qquad x_{i5} = v_i^2, \qquad x_{i6} = u_i v_i$$

(where u_i and v_i are the ith student's high school GPA and SAT score, respectively). Then we have the model

$$Y_i = \beta_1 + \beta_2 u_i + \beta_3 v_i + \beta_4 u_i^2 + \beta_5 v_i^2 + \beta_6 u_i v_i + e_i,$$

where e_i is a random variable indicating the randomness of the students. The regression function for this model is that given earlier, viz.,

$$y = \beta_1 + \beta_2 u + \beta_3 v + \beta_4 u^2 + \beta_5 v^2 + \beta_6 uv.$$

Note that this function is a nonlinear function of u and v. In order to use the methods derived in this chapter, we only need the regression function to be linear in the coefficients β_i.

Now, suppose we have a multiple regression model in which $p = 2$, $x_{i1} = 1$, and $x_{i2} = x_i$. Then the model and the regression function reduce to

$$Y_i = \beta_1 + \beta_2 x_i + e_i.$$

We call this model the *simple linear regression model*. If $p > 2$ and $x_{ij} = x_i^{j-1}$, then this model becomes

$$Y_i = \beta_1 + \beta_2 x_i + \cdots + \beta_p x_i^{p-1} + e_i.$$

If $p = 3$, the model is called a *quadratic regression model*. For $p > 3$, the model is called a $(p-1)$*st-degree polynomial regression model*.

The multiple linear regression model involves the following assumptions about the joint distribution of the Y_i's:

1. *EY_i is a linear function of the x_{ij}* ($EY_i = \Sigma_j x_{ij} \beta_j$). The procedures developed in this chapter are quite sensitive to this assumption. If it is not satisfied, we can often remedy the problem by adding some more x's to the model. (For example, if a quadratic model is not good enough, we add a cubic term.) However, if we add too many terms to the model, the model becomes difficult to interpret. In practice, we often are faced with a situation in which we have too many possible x's for inclusion in the model. We wish to choose some subset of those x's in a sensible way and so are led to a collection of techniques called *stepwise regression*. Discussion of these techniques would take us far beyond the scope of this book and will not be given here. (See, for example, Draper and Smith (1981) for one such discussion.)

2. *Var*(Y_i) *is constant* (*Var* $Y_i = \sigma^2$ *for some* $\sigma^2 > 0$). This assumption is probably the most bothersome assumption in regression analysis. Unfortunately, the procedures involved are quite sensitive to it, but there does not seem to be any alternative. Although procedures have been derived for a few situations in which the assumption is not met, in practice, we typically have to hope the assumption is satisfied.
3. *The* Y_i *are independent*. The procedures are also quite sensitive to this assumption. Fortunately, in practice, it is often possible to design the experiment so that the assumption is satisfied.
4. *The* Y_i *are normally distributed*. This is probably the least bothersome assumption. One reason is that the procedures derived in this section regarding inferences about the β_i are typically asymptotically insensitive to the normal assumption. (See Arnold (1981), Chapter 10, for a discussion and derivation of conditions under which such procedures are insensitive to this assumption.) A second reason the assumption is not too bothersome is that there are sensible alternative procedures to use when the assumption is not satisfied. (See Huber (1981) and Hettmansperger (1984) for discussions of such robust procedures.) As usual, inferences about σ^2 are quite sensitive to the assumption of normality.

It should be noted that although we have been considering the x's as known constants, many of the procedures can be extended to the case in which the x's are themselves random variables. (See Sections 12.1.2 and 12.3.2.)

It is not feasible to deal with general multiple regression models without matrices. For that reason, in Sections 12.1 and 12.2, we deal only with the simple linear regression model and models related to it. In Section 12.3, we use matrices to look at the general multiple regression model. Even for simple linear regression, the derivations are more complicated without matrices, and the reader may choose to skip the derivations in Sections 12.1, doing them instead as special cases of those in Section 12.3. (The reader should, however, study the statements of the results in Section 12.1 and the examples of their application.)

In this chapter and the next, we have models in which we observe

$$Y_i \sim N(m_i(\boldsymbol{\theta}), \sigma^2),$$

independent, where the $m_i(\boldsymbol{\theta})$ are known functions of the unknown parameter $\boldsymbol{\theta}$. The *ordinary least squares estimator* of $\boldsymbol{\theta}$ is the estimator $\hat{\boldsymbol{\theta}}$ which minimizes

$$q(\boldsymbol{\theta}) = \sum (Y_i - m_i(\boldsymbol{\theta}))^2.$$

That is,

$$\sum (Y_i - m_i(\hat{\boldsymbol{\theta}}))^2 \leqq \sum (Y_i - m_i(\boldsymbol{\theta}))^2$$

for all $\boldsymbol{\theta}$. The reason that the ordinary least squares estimator is so important for these models is the following. The logarithm of the likelihood for these models is

$$\log[L_Y(\theta, \sigma^2)] = -\left(\frac{n}{2}\right) \log(2\pi) - n \log(\sigma) - \frac{q(\boldsymbol{\theta})}{2\sigma^2}.$$

Therefore, the $\boldsymbol{\theta}$ which maximizes the likelihood is the $\boldsymbol{\theta}$ which minimizes $q(\boldsymbol{\theta})$. Hence, the MLE of $\boldsymbol{\theta}$ is the same as the ordinary least squares estimator of $\boldsymbol{\theta}$. (Actually, least squares estimators were used for fitting curves to data long before MLEs were considered. However, ordinary least squares estimators would not be appropriate if the Y_i had unequal variances or were not independent.) Direct differentiation of the logarithm of the likelihood with respect to σ shows that the MLE of σ^2 is $\hat{\sigma}^2 = q(\hat{\boldsymbol{\theta}})/n$ for the models under consideration.

12.1 SIMPLE LINEAR REGRESSION

12.1.1 Estimation in Simple Linear Regression

In this section, we study the model in which we observe $Y_1, \ldots, Y_n$ independent, with

$$Y_i \sim N(\theta + \gamma x_i, \sigma^2),$$

where the x_i are known constants which are not all equal, and θ, γ, and σ^2 are unknown parameters. We call this model the *simple linear regression model* and call the line

$$y = \theta + \gamma x$$

the *regression line* of y on x. We call θ the *intercept* and γ the *slope* of this line.

We first consider the estimation of θ and γ for the model. Let $\hat{\theta}$ and $\hat{\gamma}$ be estimators of θ and γ, respectively. Then $(\hat{\theta}, \hat{\gamma})$ is the *ordinary least squares estimator* of (θ, γ) if $(\hat{\theta}, \hat{\gamma})$ minimizes

$$q(\theta, \gamma) = \sum_i (Y_i - (\theta + \gamma x_i))^2,$$

i.e., if the Y_i are closer to $\hat{\theta} + \hat{\gamma} x_i$ than they are to $\theta + \gamma x_i$ for any other θ and γ. Let

$$S_{xx} = \sum_i (x_i - \bar{x})^2 = \sum_i x_i^2 - n\bar{x}^2, \qquad S_{YY} = \sum_i (Y_i - \bar{Y})^2 = \sum_i Y_i^2 - n\bar{Y}^2,$$

$$S_{xY} = \sum_i (x_i - \bar{x})(Y_i - \bar{Y}) = \sum_i x_i Y_i - n\bar{x}\bar{Y}.$$

(Since the x_i are not identical, $S_{xx} > 0$.) Also, let

$$\hat{\gamma} = \frac{S_{xY}}{S_{xx}}, \qquad \hat{\theta} = \bar{Y} - \hat{\gamma}\bar{x}.$$

Theorem 12–1. $(\hat{\theta}, \hat{\gamma})$ is the ordinary least squares estimator of (θ, γ).

Proof. If we set the derivatives of $q(\theta, \gamma)$ with respect to θ and γ equal to zero, we get the two equations

$$0 = \sum (Y_i - \hat{\theta} - \hat{\gamma} x_i) \quad \text{and} \quad 0 = \sum x_i (Y_i - \hat{\theta} - \hat{\gamma} x_i).$$

From the first equation, we see that $\overline{Y} = \hat{\theta} + \hat{\gamma}\overline{x}$, or $\hat{\theta} = \overline{Y} - \hat{\gamma}\overline{x}$. Substituting this result into the second equation, we obtain

$$0 = \sum x_i(Y_i - \overline{Y} - \hat{\gamma}(x_i - \overline{x})) = \sum x_i(Y_i - \overline{Y}) - \hat{\gamma}\sum x_i(x_i - \overline{x}),$$

or, equivalently,

$$\hat{\gamma} = \frac{\sum x_i(Y_i - \overline{Y})}{\sum x_i(x_i - \overline{x})} = \frac{S_{xY}}{S_{xx}}. \quad \square$$

We call the line

$$y = \hat{\theta} + \hat{\gamma}x$$

the *sample regression line* of y on x.

We next find the joint distribution of $(\hat{\theta}, \hat{\gamma})$.

Theorem 12–2. $(\hat{\theta}, \hat{\gamma})$ has a bivariate normal distribution with

$$E\hat{\theta} = \theta, \qquad E\hat{\gamma} = \gamma, \qquad \operatorname{var}(\hat{\theta}) = \frac{\sigma^2(S_{xx} + n\overline{X}^2)}{nS_{xx}},$$

$$\operatorname{var}(\hat{\gamma}) = \frac{\sigma^2}{S_{xx}}, \qquad \operatorname{cov}(\hat{\theta}, \hat{\gamma}) = \frac{-\sigma^2\overline{x}}{S_{xx}}.$$

Proof. Let $b_i = (x_i - \overline{x})/S_{xx}$ and $a_i = n^{-1} - \overline{x}b_i$. Then

$$\hat{\gamma} = \frac{\sum (x_i - \overline{x})(Y_i - \overline{Y})}{S_{xx}} = \frac{\sum (x_i - \overline{x})Y_i}{S_{xx}} = \sum b_i Y_i,$$

$$\hat{\theta} = n^{-1}\sum Y_i - \overline{x}\hat{\gamma} = \sum a_i Y_i.$$

The Y_i are independently normally distributed. Therefore, by Theorem 5–10, $(\hat{\theta}, \hat{\gamma})$ has a bivariate normal distribution with

$$E\hat{\theta} = \sum a_i EY_i = \theta, \qquad E\hat{\gamma} = \sum b_i EY_i = \gamma,$$

$$\operatorname{var}(\hat{\gamma}) = \sigma^2\sum b_i^2 = \frac{\sigma^2}{S_{xx}}, \qquad \operatorname{var}(\hat{\theta}) = \sigma^2\sum a_i^2 = \frac{\sigma^2(S_{xx} + n\overline{x}^2)}{nS_{xx}},$$

$$\operatorname{cov}(\hat{\theta}, \hat{\gamma}) = \sigma^2\sum a_i b_i = \frac{-\sigma^2\overline{x}}{S_{xx}}$$

(see the exercises). $\square$

We now define the estimator for σ^2 for simple linear regression. Define S^2 by

$$(n-2)S^2 = \sum (Y_i - \hat{\theta} - \hat{\gamma}x_i)^2 = (S_{YY} - S_{xx}^{-1}S_{xY}^2) = S_{YY} - \hat{\gamma}^2 S_{xx}.$$

(S^2 is *not* the sample variance in this section.) Note that if the Y_i are close to $\hat{Y}_i = \hat{\theta} + x_i\hat{\gamma}$, then S^2 is small, and if they are far away, then S^2 is large.

Theorem 12–3. S^2 is an unbiased estimator of σ^2.

Proof. From Theorem 12–2,

$$E\hat{\gamma}^2 = (E\hat{\gamma})^2 + \text{var}(\hat{\gamma}) = \gamma^2 + \frac{\sigma^2}{S_{xx}}.$$

In addition,

$$ES_{YY} = (n-1)\sigma^2 + \gamma^2 S_{xx}.$$

(see the exercises). Therefore,

$$(n-2)ES^2 = ES_{YY} - S_{xx}E\hat{\gamma}^2 = (n-1)\sigma^2 + \gamma^2 S_{xx} - S_{xx}\left(\gamma^2 + \frac{\sigma^2}{S_{xx}}\right) = (n-2)\sigma^2,$$

so that S^2 is an unbiased estimator of σ^2. □

Example 12–1

Suppose we have $n = 7$, and we observe the following data:

x	1	2	3	4	5	6	7
Y	3	5	8	10	10	12	15

Then $\bar{x} = 4$, $\bar{Y} = 9$, $S_{xx} = 28$, $S_{YY} = 100$, and $S_{xY} = 52$. Hence, for these data, $\hat{\gamma} = \frac{52}{28} = 1.857$, $\hat{\theta} = 9 - 4(\frac{52}{28}) = 1.571$, $S^2 = (100 - (52)^2/28)/(7-2) = .686$.

Theorem 12–4 states the basic estimation results for simple linear regression.

Theorem 12–4

a. $(\hat{\theta}, \hat{\gamma}, S^2)$ is a complete sufficient statistic for the simple linear regression model.
b. $\hat{\theta}$, $\hat{\gamma}$, and S^2 are the best unbiased estimators of θ, γ, and σ^2.
c. The MLE of $(\theta, \gamma, \sigma^2)$ is $(\hat{\theta}, \hat{\gamma}, \hat{\sigma}^2)$, where $\hat{\sigma}^2 = (n-2)S^2/n$.

Proof

a. The joint density of the Y_i is

$$f(\mathbf{y}; \theta, \gamma, \sigma^2) = \prod\left((2\pi)^{-1/2}\sigma^{-1}\exp\left[\frac{-(y_i - \theta - \gamma x_i)^2}{2\sigma^2}\right]\right)$$

$$= (2\pi)^{-n/2}\sigma^{-n}\exp\left[\frac{-\sum(y_i - \theta - \gamma x_i)^2}{2\sigma^2}\right].$$

Therefore, the likelihood function for the model is

$$L_Y(\theta, \gamma, \sigma^2) = L(\theta, \gamma, \sigma^2) = (2\pi)^{-n/2}\sigma^{-n}\exp\left[\frac{-\sum(Y_i - \theta - \gamma x_i)^2}{2\sigma^2}\right]$$

$$= h(\theta, \gamma, \sigma^2) \exp\left(\frac{-\sum Y_i^2 + 2\theta \sum Y_i + 2\gamma \sum x_i Y_i}{2\sigma^2}\right)$$

which is in exponential form with

$$T(\mathbf{Y}) = (\sum_i Y_i^2, \sum_i Y_i, \sum_i x_i Y_i) \quad \text{and} \quad c(\theta, \gamma, \sigma^2) = (2\sigma^2)^{-1}(-1, 2\theta, 2\gamma).$$

(Since the x_i are observed constants, any function of x_i and Y_i can be considered a function of the random variables, and any function of x_i and $(\theta, \gamma, \sigma^2)$ can be considered a function of the parameters.) Because $T(\mathbf{Y})$ has dimension three and there are three parameters, $T(\mathbf{Y})$ is a complete sufficient statistic for the simple linear regression model.

We now show that $(\hat{\theta}, \hat{\gamma}, S^2)$ is an invertible function of $\mathbf{T} = (T_1, T_2, T_3)' = (\Sigma_i Y_i^2, \Sigma_i Y_i, \Sigma_i x_i Y_i)$. Note first that since the x_i are known constants and not random variables, when we say that $(\hat{\theta}, \hat{\gamma}, S^2)$ is a function of $\mathbf{T}$, we mean that it is a function of $(\mathbf{T}, \mathbf{x})$ (i.e., if someone knows $\mathbf{T}$ and $\mathbf{x}$, he or she can compute $(\hat{\theta}, \hat{\gamma}, S^2)$). A similar comment holds if we show that $\mathbf{T}$ is a function of $(\hat{\theta}, \hat{\gamma}, S^2)$. Now,

$$S_{YY} = T_1 - \frac{T_2^2}{n}, \qquad S_{xY} = T_3 - \bar{x} T_2, \qquad \bar{Y} = \frac{T_2}{n}.$$

Therefore,

$$\hat{\gamma} = \frac{T_3 - \bar{x} T_2}{S_{xx}}, \qquad \hat{\theta} = \frac{T_2}{n} - \frac{\bar{x}(T_3 - \bar{x} T_2)}{S_{xx}},$$

$$S^2 = \frac{T_1 - T_2^2/n - (T_3 - \bar{x}T_2)^2/S_{xx}}{n-2}$$

and $(\hat{\theta}, \hat{\gamma}, S^2)$ is a function of $\mathbf{T}$ (and $\mathbf{x}$). Conversely,

$$T_2 = n\bar{Y} = n(\hat{\theta} + \hat{\gamma}\bar{x}), \qquad T_3 = S_{xY} + \bar{x}T_2 = S_{xx}\hat{\gamma} + \bar{x}n(\hat{\theta} + \hat{\gamma}\bar{x}),$$

$$T_1 = S_{YY} + \frac{T_2^2}{n} = (n-2)S^2 + \hat{\gamma}^2 S_{xx} + n(\hat{\theta} + \hat{\gamma}\bar{x})^2.$$

Hence, someone who knows $(\hat{\theta}, \hat{\gamma}, S^2)$ (and $\mathbf{x}$) also knows $\mathbf{T}$. Consequently, $(\hat{\theta}, \hat{\gamma}, S^2)$ is an invertible function of $\mathbf{T}$ and is therefore also a complete sufficient statistic.

b. From Theorems 12–2 and 12–3, $\hat{\theta}$, $\hat{\gamma}$, and S^2 are unbiased estimators. Since they are functions of the complete sufficient statistic, they are the best unbiased estimators, by the Lehmann-Scheffé theorem.

c. The logarithm of the likelihood function for the simple linear regression model is

$$\log[L(\theta, \gamma, \sigma^2)] = \frac{-n \log(2\pi)}{2} - n \log \sigma - \frac{\sum_i (Y_i - \theta - \gamma x_i)^2}{2\sigma^2}.$$

To maximize this quantity for θ and γ, we must minimize

$$q(\theta, \gamma) = \sum_i (Y_i - \theta - \gamma x_i)^2,$$

so that the MLE for (θ, γ) is the least squares estimator $(\hat{\theta}, \hat{\gamma})$. If we set to zero the derivative of $\log[L(\hat{\theta}, \hat{\gamma}, \sigma^2)]$ with respect to σ we see that the MLE for σ^2 is

$$\hat{\sigma}^2 = \frac{\sum_i (Y_i - \hat{\theta} - \hat{\gamma} x_i)^2}{n} = \frac{(n-2)S^2}{n}. \quad \square$$

Exercises—A

1. An experiment was conducted to determine how the yield of wheat (Y) is influenced by rainfall (x). The following data were obtained:

rainfall (x)	9	10	13	15	18	13
yield (Y)	36	44	48	63	70	45

   (a) Find the sample regression line of y on x. Plot this line on a graph that includes the foregoing data points.
   (b) Find the unbiased estimator of the variance.
2. Suppose the data in problem 1 come from a model for regression through the origin model (Exercise B5 below). Answer questions a and b given in A1 for those data for this model.

Exercises—B

1. In the notation of the proof of Theorem 12–2, show that $\Sigma a_i EY_i = \theta$, $\Sigma b_i EY_i = \gamma$, $\Sigma b_i^2 = (S_{xx})^{-1}$, $\Sigma a_i^2 = (S_{xx} + n\bar{x}^2)/nS_{xx}$, $\Sigma a_i b_i = -\bar{x}/S_{xx}$.
2. Verify that $\Sigma(Y_i - \hat{\theta} - \hat{\gamma}x_i)^2 = S_{YY} - S_{xY}^2/S_{xx} = S_{YY} - \hat{\gamma}^2 S_{xx}$.
3. (a) Show that $EY_i^2 = \sigma^2 + (\theta + \gamma x_i)^2$.
   (b) Show that $\bar{Y} \sim N(\theta + \bar{x}\gamma, \sigma^2/n)$.
   (c) What is $E\bar{Y}^2$?
   (d) Verify that $ES_{YY} = (n-1)\sigma^2 + \gamma^2 S_{xx}$.
4. Suppose that $EY_i = \theta + x_i\gamma$, but that the Y_i are not necessarily independent or normally distributed and do not necessarily have equal variances. Show that $\hat{\theta}$ and $\hat{\gamma}$ are still unbiased estimators of θ and γ.
5. (Regression through the origin) Suppose we observe $Y_i \sim N(\delta x_i, \sigma^2)$, independent, where x_i are observed constants and δ and τ^2 are unknown parameters. Let $w = \Sigma x_i^2$, $\hat{\delta} = \Sigma x_i Y_i/w$, and $T^2 = \Sigma(Y_i - x_i\hat{\delta})^2/(n-1)$. We call the line $y = \delta x$ the *regression line* and the line $y = \hat{\delta}x$ the *sample regression line* for this model.
   (a) Show that $\hat{\delta}$ is the ordinary least squares estimator of δ, i.e., that $\hat{\delta}$ minimizes $q^*(\delta) = \Sigma_i(Y_i - \delta x_i)^2$.
   (b) Show that $\hat{\delta} \sim N(\delta, \tau^2/w)$.
   (c) Show that $(\hat{\delta}, T^2)$ is a complete sufficient statistic.
   (d) Show that $\hat{\delta}$ and T^2 are the best unbiased estimators of δ and σ^2.
   (e) Show that $(\hat{\delta}, (n-1)T^2/n)$ is the MLE for (δ, σ^2).

12.1.2 Inferences in Simple Linear Regression

We now consider confidence intervals and tests for the simple linear regression model. We continue with the notation from the previous section.

The proof of the following theorem is rather messy without matrices and is not given here. A more general version is proved in Section 12.3.2.

Theorem 12–5. $(n-2)S^2/\sigma^2 \sim \chi^2_{n-2}$ independently of $(\hat{\theta}, \hat{\gamma})$.

The following theorem gives the basic results regarding inferences for simple linear regression.

Theorem 12–6. Pivotal quantities for θ, γ, and σ^2 are given by

$$t_1(\theta) = \frac{(\hat{\theta} - \theta)(nS_{xx})^{1/2}}{(S_{xx} + n\overline{x}^2)^{1/2}\, S} \sim t_{n-2},$$

$$t_2(\gamma) = \frac{(\hat{\gamma} - \gamma)(S_{xx})^{1/2}}{S} \sim t_{n-2},$$

and

$$U(\sigma^2) = \frac{(n-2)S^2}{\sigma^2} \sim \chi^2_{n-2}.$$

Proof. From Theorem 12–2,

$$\hat{\theta} \sim N\left(\theta, \frac{\sigma^2(S_{xx} + n\overline{x}^2)}{nS_{xx}}\right)$$

or equivalently,

$$Z = \frac{(\hat{\theta} - \theta)(nS_{xx})^{1/2}}{\sigma(S_{xx} + n\overline{x}^2)^{1/2}} \sim N(0, 1)$$

and is independent of $U = (n-2)S^2/\sigma^2 \sim \chi^2_{n-2}$ by Theorem 12–5. By the definition of the t-distribution,

$$t_1(\theta) = \frac{Z}{[U/(n-2)]^{1/2}} \sim t_{n-2}.$$

The result for $t_2(\gamma)$ follows similarly, and the result for U follows directly from Theorem 12–5. □

We can use the pivotal quantities defined in Theorem 12–6 to find confidence intervals and do tests in the obvious way. For example, a $(1-\alpha)$ confidence interval for θ is

$$\theta \in \hat{\theta} \pm t^{\alpha/2}_{n-2} S \left(\frac{S_{xx} + n\overline{x}^2}{nS_{xx}}\right)^{1/2},$$

and a size-α test that $\gamma = 2$ against $\gamma > 2$ would reject the null hypothesis if

$$t_2(2) > t_{n-2}^{\alpha}.$$

The most important testing problem for the simple linear regression model is testing the null hypothesis that $\gamma = 0$ against one-sided or two-sided alternatives. When we test that $\gamma = 0$, we are testing that the distribution of Y is independent of X. This hypothesis is tested using $t_2(0)$.

By arguments similar to those in previous chapters, we could show that one-sided and equal-tailed two-sided tests based on $t_1(d)$ are likelihood ratio tests for testing that $\theta = d$ against the appropriate alternatives, as are those based on $t_2(c)$ for testing that $\gamma = c$ against appropriate alternatives. These tests are also UMP unbiased. Similarly, one-sided tests based on $U(b)$ are likelihood ratio tests and UMP unbiased for testing that $\sigma^2 = b$ against the appropriate one-sided alternatives. For the two-sided problem, the likelihood ratio test and the UMP unbiased test are (different) two-sided tests based on $U(b)$.

Example 12–2

Consider again Example 12–1, for which we found that

$$n = 7, \qquad \bar{x} = 4, \qquad S_{xx} = 28, \qquad \hat{\theta} = 1.57, \qquad \hat{\gamma} = 1.86, \qquad S^2 = .686.$$

Therefore, a 95% confidence interval for θ is given by

$$\theta \in 1.57 \pm \frac{2.57(.686)^{1/2}(28 + 7(16))^{1/2}}{(7(28))^{1/2}} = 1.57 \pm 1.80.$$

Similarly, to test that $\gamma = 0$ against $\gamma > 0$ with a .05 test, we reject the null hypothesis if $t_2(0) > 2.01$. Since, in fact,

$$t_2(0) = \frac{1.86(28)^{1/2}}{(.686)^{1/2}} = 11.9 > 2.01,$$

we reject the hypothesis that $\gamma = 0$.

In simple linear regression models, we often want a confidence interval for

$$\mu(t_0) = \theta + \gamma t_0,$$

for some number t_0. $\mu(t_0)$ represents the mean of the Y-value for an individual whose x-value is t_0. Now, let

$$T \sim N(\mu(t_0), \sigma^2),$$

independently of the Y_i. It is often useful to find a random interval

$$L(t_0) \leqq T \leqq R(t_0)$$

such that

$$P(L(t_0) \leqq T \leqq R(t_0)) = 1 - \alpha.$$

(Note that L and R also depend on $\mathbf{Y}$ and $\mathbf{x}$.) Since T is not a parameter, we do not call such an interval a confidence interval, but rather call it a *prediction interval* for T.

A $(1 - \alpha)$ *confidence band* for the regression line consists of two random functions $L(t)$ and $R(t)$ such that

$$P(L(t) \leqq \mu(t) \leqq R(t) \text{ for all } t) = 1 - \alpha.$$

A 95% confidence band for $\mu(x)$ is true, simultaneously for all t, 95% of the time. That is, the probability that it is false for at least one t is .05. (It is somewhat surprising that a confidence band is possible, since it must be true at infinitely many points simultaneously.)

In the previous example in which the x's are SAT scores and the Y's are GPAs, $\mu(t_0)$ represents the expected GPA of a person whose SAT score is t_0. Now, let T be the actual GPA of an entering student whose SAT score is t_0. Then T is a random variable, and if the model is correct, then $T \sim N(\mu(t_0), \sigma^2)$. A .95 prediction interval for T is a random interval in which T should lie 95% of the time. Suppose we have 20 new students with identical SAT scores, so that $\mu(t_0)$ is the same for all 20 students. Of course, the actual college GPAs would most likely be different for the 20. When we find a 95% confidence interval for $\mu(t_0)$, we are finding an interval in which we are 95% confident that the students' common mean lies. When we find a 95% prediction interval for the college GPA of a particular student, we are finding an interval in which we are 95% confident his or her actual GPA lies. In a 95% prediction interval, we have accounted for both the randomness in the estimator of the common mean and the randomness due to the variable performance of different students with the same entering SAT scores. If we compute such an interval for all 20 students, about 19 of them should be correct. (In four years, of course, we can check which of the prediction intervals are correct; by contrast we rarely ever find the true value of a parameter, so that we can rarely check whether a confidence interval is correct.) A .95 confidence band for $\mu(t)$ is a 95% band for the whole function $\mu(t)$.

Now, for all t, let

$$\hat{\mu}(t) = \hat{\theta} + t\hat{\gamma} \quad \text{and} \quad V(t) = [n^{-1} + S_{xx}^{-1}(t - \bar{x})^2]^{1/2}.$$

Theorem 12–7

a. $\hat{\mu}(t)$ is the MLE and best unbiased estimator of $\mu(t)$.

b. A $(1-\alpha)$ confidence interval for $\mu(t_0)$ is

$$\mu(t_0) \in \hat{\mu}(t_0) \pm t_{n-2}^{\alpha/2} SV(t_0).$$

c. A $(1-\alpha)$ prediction interval for a new observation U with mean $\mu(t_0)$ is

$$T \in \hat{\mu}(t_0) \pm t_{n-2}^{\alpha/2}(1 + V^2(t_0))^{1/2} S.$$

d. A $(1-\alpha)$ confidence band for $\mu(t)$ is given by

$$\mu(t) \in \hat{\mu}(t) \pm [2F_{2,n-2}^{\alpha}]^{1/2} SV(t).$$

Proof

a. $\hat{\mu}(t)$ is an unbiased estimator of $\mu(t)$ and a function of the complete sufficient statistic. It is therefore the best unbiased estimator. Also, since $\mu(t) = \theta + \gamma t$, $\hat{\mu}(t) = \hat{\theta} + \hat{\gamma} t$ is the MLE of $\mu(t)$, by the invariance principle for MLEs.

b. Since $\hat{\mu}(t_0)$ is a linear function of $(\hat{\theta}, \hat{\gamma})$ which has a bivariate normal distribution, $\hat{\mu}(t_0)$ has a normal distribution with mean

$$E\,\hat{\mu}(t_0) = E\hat{\theta} + t_0 E\hat{\gamma} = \theta + t_0\gamma = \mu(t_0)$$

and variance

$$\text{var}(\hat{\mu}(t_0)) = \text{var}(\hat{\theta}) + t_0^2\,\text{var}(\hat{\gamma}) + 2t_0\,\text{cov}(\hat{\theta},\hat{\gamma}) = \sigma^2 V^2(t_0).$$

Hence,

$$Z = \frac{\hat{\mu}(t_0) - \mu(t_0)}{\sigma V(t_0)} \sim N(0,1).$$

By Theorem 12–5 and the definition of a t-distribution,

$$t(\mu(t_0)) = \frac{\hat{\mu}(t_0) - \mu(t_0)}{V(t_0)S} = \frac{Z}{(U/(n-2))^{1/2}} \sim t_{n-2}.$$

The confidence interval for $\mu(x_0)$ follows from this result in the usual way.

c. Let $T \sim N(\mu(t_0), \sigma^2)$, independently of the Y_i. Then

$$T - \hat{\mu}(t_0) \sim N(\mu(t_0) - \mu(t_0), \sigma^2 + \sigma^2 V^2(t_0))$$

and hence,

$$Z^* = \frac{T - \hat{\mu}(t_0)}{\sigma(1 + V^2(t_0))^{1/2}} \sim N(0,1)$$

and is independent of $U = (n-2)S^2/\sigma^2$. Therefore,

$$t^* = \frac{T - \hat{\mu}(t_0)}{(1 + V^2(t_0))^{1/2}\,S} \sim t_{n-2},$$

by the definition of the t-distribution. Hence,

$$P(T \in \hat{\mu}(t_0) \pm t_{n-2}^{\alpha/2}(1 + V^2(t_0))^{1/2}\,S) = P(-t_{n-2}^{\alpha/2} \leq t^* \leq t_{n-2}^{\alpha/2}) = 1 - \alpha.$$

d. A proof of this part of the theorem is quite messy without matrices. A proof of a more general version is given in Section 12.3.2. □

Observe that the confidence interval for $\mu(t_0)$ and the prediction interval for T have the same center $\hat{\mu}(t_0)$. To get the prediction interval for T from the confidence interval for $\mu(t_0)$, we replace $V(t_0)$ with $(1 + V^2(t_0))^{1/2}$, so that the prediction interval is longer. The prediction interval should be longer, because it is accounting for the randomness in the estimator $\hat{\mu}(t_0)$ and the randomness in T. Note that the confidence band for $\mu(t)$ is centered at $\hat{\mu}(t)$. To get the confidence band from the confidence interval for $\mu(t)$, we replace $t_{n-2}^{\alpha/2}$ with $[2F_{2,n-2}^{\alpha}]^{1/2}$.

The width of both the confidence and prediction intervals is determined primarily by $V(t_0)$ (or $V(t)$). Now $V(t_0)$ is small when t_0 is close to $\bar{x}$—just how close is measured in terms of $(S_{xx})^{1/2}$. This causes the intervals to be small when t_0 is near the center of the data and large when t_0 is far from the center of the data.

Example 12–3

Consider again Example 12–1, and suppose we are interested in what happens when $t_0 = 8$. By part a of Theorem 12–7, the optimal estimator of $\mu(8) = \theta + 8\gamma$ is $1.57 +$

$8(1.86) = 16.45$. Also, $V^2(8) = (7^{-1} + (8-4)^2/28) = .7143$. Therefore, a 95% confidence interval for $\mu(8)$ is given by

$$\mu(8) \in 16.45 \pm 2.57(.7143)^{1/2}(.686)^{1/2} = 16.45 \pm 1.80.$$

A 95% prediction interval for a new observation T whose mean is $\mu(8)$ is

$$T \in 16.45 \pm 2.57(1.7143)^{1/2}(.686)^{1/2} = 16.45 \pm 2.79.$$

This interval is longer than the confidence interval because it is accounting for both the randomness in the estimator for $\mu(8)$ and the randomness in the distribution of T. The confidence band for $\mu(t)$ is

$$\mu(t) \in 1.57 + 1.86t \pm [2(5.79)]^{1/2}(.686)^{1/2}\left(\frac{1}{7} + \frac{(t-4)^2}{28}\right)^{1/2}$$
$$= 1.57 + 1.86t \pm 2.82(.143 + .036(t-4)^2)^{1/2}.$$

Figure 12–1 presents a graph of the preceding functions together with the function $\hat{\mu}(t) = 1.57 + 1.86t$. From the graph, the confidence band is smallest when t is near $4 = \bar{x}$ and gets larger as t gets further from 4. When $t = 8$, the confidence band is

$$16.45 \pm 2.82(.7143)^{1/2} = 16.45 \pm 2.38,$$

which is bigger than the confidence interval for $\mu(8)$, since the 95% confidence band is

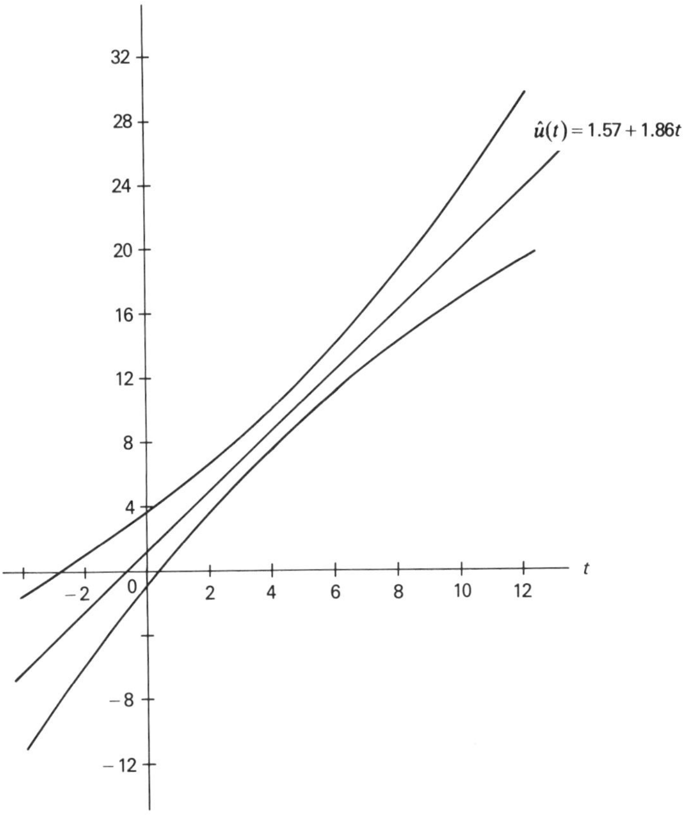

Figure 12–1 Graphs of $\hat{u}(t)$ and confidence band for $u(t)$.

true simultaneously for all t, while the confidence interval for $\mu(8)$ is true only when $t = 8$.

Until now, we have been assuming that the x_i are observed constants. Suppose, however, that we replace the observed constants x_i with observed random variables X_i (and replace the x_i by X_i in the preceding formulas). Suppose further that, conditionally on the X_i, the Y_i are independent and that

$$Y_i|X_i \sim N(\theta + X_i\gamma, \sigma^2).$$

In the conditional distribution of the $Y_i|X_i$, we treat the X_i as if they were known constants. Therefore, we can treat many of the results derived earlier as conditional results given the X_i. For example, we have

$$E\hat{\theta} = E[E\hat{\theta}|(X_1, \ldots, X_n)] = E\theta = \theta,$$

so that $\hat{\theta}$ is still an unbiased estimator of θ when we assume that the X_i are random variables. Similarly, $\hat{\gamma}$ and $\hat{\sigma}^2$ are unbiased estimators for this new model. Also,

$$t_1(\theta)|(X_1, \ldots, X_n) \sim t_{n-2}.$$

Since this conditional distribution does not depend on $(X_1, \ldots, X_n)$, $t_1(\theta)$ is independent of $(X_1, \ldots, X_n)$, and hence,

$$t_1(\theta) \sim t_{n-2}.$$

Therefore, $t_1(\theta)$ is still a pivotal quantity for θ for the model in which the X_i are random variables. Similarly, $t_2(\gamma)$ and $U(\sigma^2)$ are also pivotal quantities for the model with random X's.

Unfortunately, it is not necessarily true that the estimators are best unbiased estimators or MLEs for the model with random X's, or that the tests based on the pivotal quantities above are likelihood ratio tests or UMP unbiased tests for this model. However, the estimators are still unbiased, and the tests still have size α, so they should be sensible procedures for the model with random X's. (If we assume that the X_i are a sample from a normal distribution, then many of the optimality properties can be proved for the case of random X's; see Arnold (1981), Chapter 16.)

Exercises—A

1. For the data in Exercise A1, Section 12.1.1,
   (a) Find 95% confidence intervals for the slope, the intercept, and the variance.
   (b) Test the hypothesis that the slope is zero with a two-sided .05 test.
   (c) Find a 95% confidence interval for the mean yield when the rainfall is 12 and a 95% prediction interval for the actual yield when the rainfall is 12.
   (d) Find and graph the 95% confidence band for the mean yield.
2. Consider the data in Exercise A1 as coming from a regression through the origin model (Exercise B2 below). Answer questions a–c given in Exercise A1 for those data for this model (ignoring, of course, any questions about the intercept).

Exercises—B

1. Consider testing that $\gamma = 0$ against $\gamma \neq 0$.
   (a) Show that $L(\hat{\theta}, \hat{\gamma}, \hat{\sigma}^2) = (2\pi)^{-n/2}(\hat{\sigma}^2)^{-n/2} \exp(-n/2)$.
   (b) Show that the MLEs under the null hypothesis are $\hat{\hat{\theta}} = \overline{Y}$, $\hat{\hat{\sigma}}^2 = \Sigma(Y_i - \overline{Y})^2/n = \hat{\sigma}^2 + \hat{\gamma}^2 S_{xx}/n$, and $L(\hat{\hat{\theta}}, 0, \hat{\hat{\sigma}}^2) = (2\pi)^{-n/2}(\hat{\hat{\sigma}}^2)^{-n/2} \exp(-n/2)$.
   (c) Show that the likelihood ratio test statistic is $\Lambda(\mathbf{Y}) = (1 + t_2^2(0)/(n-2))^{n/2}$, where $t_2(0) = \hat{\gamma}(S_{xx})^{1/2}/\hat{\sigma}$, and hence, that the likelihood ratio test rejects the null hypothesis if $t_2(0) > k$ or $t_2(0) < -k$ for some k.
2. Consider the model of regression through the origin first presented in Exercise B1, Section 12.1.1. Assume that $(n-1)T^2/\tau^2 \sim \chi^2_{n-1}$ independently of $\hat{\delta}$.
   (a) Show that a pivotal quantity for δ is $(\hat{\delta} - \delta)W^{1/2}/T \sim t_{n-1}$.
   (b) Let $\nu(t) = \delta t$ and $\hat{\nu}(t) = \hat{\delta}t$. Show that a $(1-\alpha)$ confidence interval for $\nu(t_0)$ is $\nu(t_0) \in \hat{\nu}(t_0) \pm t^{\alpha/2}_{n-1} T|t_0|/W^{1/2}$.
   (c) Let $Q \sim N(\nu(t_0), \tau^2)$ independently of the Y_i. Show that a $(1-\alpha)$ prediction interval for Q is $Q \in \hat{\nu}(t_0) \pm t^{\alpha/2}_{n-1} T(1 + t_0^2/W)^{1/2}$.

12.2 CORRELATION ANALYSIS

12.2.1 Correlation Analysis

Consider the model in which we observe a sample $(X_1, Y_1), \ldots, (X_n, Y_n)$ from a bivariate normal distribution with

$$EX_i = \nu_1, \qquad EY_i = \nu_2, \qquad \operatorname{var}(X_i) = \tau_1^2, \qquad \operatorname{var}(Y_i) = \tau_2^2, \qquad \operatorname{cov}(X_i, Y_i) = \rho\tau_1\tau_2.$$

We desire to test that $\rho = 0$ against one-sided and two-sided alternatives. Since the (X_i, Y_i) have a bivariate normal distribution, $\rho = 0$ if and only if X_i and Y_i are independent.

As a simple example of where this model might be used, suppose we have a sample of students who have taken both the SAT test and an intelligence test. If we let X_i be the score on the SAT test and Y_i be the score on the intelligence test for the ith student, the model might be appropriate, and we could use the procedures discussed in this section to see whether there was any significant correlation between the performances on the SAT test and the intelligence test.

Continuing with the notation of the previous section, let

$$S_{YY} = \sum_i (Y_i - \overline{Y})^2, \qquad S_{XX} = \sum_i (X_i - \overline{X})^2, \qquad S_{XY} = \sum_i (X_i - \overline{X})(Y_i - \overline{Y}).$$

(We use capital letters for both Y and X, since they are both random variables.) From Chapter 5, the conditional distribution of Y_i given X_i is

$$Y_i|X_i \sim N\left(\nu_2 + \frac{\rho\tau_2(X_i - \nu_1)}{\tau_1}, \tau_2^2(1 - \rho^2)\right).$$

Also, given the X_i, the Y_i are independent, since the (X_i, Y_i) are a sample. This conditional model is a simple linear regression model with

$$\theta = \nu_2 - \frac{\rho\tau_2\nu_1}{\tau_1}, \qquad \gamma = \frac{\rho\tau_2}{\tau_1}, \qquad \sigma^2 = \tau_2^2(1-\rho^2).$$

Note that $\rho = 0$ for the paired model if and only if $\gamma = 0$ for the conditional model and that $\rho > 0$ if and only if $\gamma > 0$.

Now, let

$$r = \frac{S_{XY}}{(S_{XX}S_{YY})^{1/2}}$$

be the *sample correlation coefficient* between the X_i and the Y_i. Then r is the MLE of ρ (although the derivation of this fact is not trivial).

Theorem 12–8. A test statistic for testing that $\rho = 0$ in the bivariate normal model is

$$t^* = \frac{(n-2)^{1/2}r}{(1-r^2)^{1/2}} \sim t_{n-2}$$

when $\rho = 0$.

Proof. Let $t_2(0)$ be the test statistic defined in the previous section for testing that $\gamma = 0$ in the simple linear regression model. Then $t^* = t_2(0)$ (see the exercises). When $\rho = 0$, $\gamma = 0$; hence, when $\rho = 0$,

$$t_2(0)|(X_1, \ldots, X_n) \sim t_{n-2},$$

by the results of the previous section. Therefore, when $\rho = 0$, t^* is independent of $(X_1, \ldots, X_n)$, and $t^* \sim t_{n-2}$. □

Note that t^* is an increasing function of r, so that a sensible test that $\rho = 0$ against $\rho > 0$ is to reject the null hypothesis if $t^* > t_{n-2}^{\alpha}$. If we do reject this hypothesis, we say there is a significant positive correlation between the X's and the Y's. Similarly, to test that $\rho = 0$ against $\rho < 0$, we reject the null hypothesis if $t^* < -t^{\alpha}$, and to test that $\rho = 0$ against $\rho \neq 0$, we reject the null hypothesis if $t^* > t^{\alpha/2}$ or $t^* < -t^{\alpha/2}$.

One-sided and equal-tailed two-sided tests based on t^* are likelihood ratio tests for testing that $\rho = 0$ against the appropriate hypotheses. They are also UMP unbiased (see Lehmann (1986), Section 5.15) and are, under fairly general assumptions, asymptotically insensitive to the normal assumption (see Arnold (1981), Section 16.6).

Unfortunately, there is no pivotal quantity for ρ in this model, nor are there any test statistics for testing other hypotheses about ρ. However, if

$$h(x) = \left(\frac{1}{2}\right)\log\left(\frac{1+x}{1-x}\right). \tag{12–1}$$

then

$$Z(\rho) = n^{1/2}(h(r) - h(\rho)) \xrightarrow{d} Z \sim N(0,1)$$

is an asymptotic pivotal quantity for ρ. (See Exercise C3, Section 6.4.2.) Note that $h(\rho)$ is an increasing function of ρ, and hence, $Z(\rho)$ is a decreasing function of ρ. Therefore, we can design asymptotic $(1-\alpha)$ confidence intervals and do sensible asymptotic size-α one-sided and two-sided tests for ρ. For example, an asymptotic size-α test that $\rho = .5$ against $\rho > .5$ rejects the null hypothesis if

$$Z(.5) > z^{\alpha},$$

and an asymptotic $(1-\alpha)$ confidence interval for ρ is given by

$$-z^{\alpha} \leqq Z(\rho) \leqq z^{\alpha} \Leftrightarrow h^{-1}\left(h(r) - \frac{z^{\alpha/2}}{n^{1/2}}\right) \leqq \rho \leqq h^{-1}\left(h(r) + \frac{z^{\alpha/2}}{n^{1/2}}\right).$$

(Note that since $h(x)$ is increasing, the inverse function $h^{-1}(y)$ exists.) $Z(\rho)$ is often called *Fisher's Z-transformation*.

Actually, it is somewhat hard to interpret what the hypothesis $\rho = .6$ means, except to observe that it means that the observations are more correlated than $\rho = .5$. Similarly, it is not easy to interpret a confidence interval for ρ, except to see whether it covers zero or not. For this reason, there is not a great need for a pivotal quantity for ρ.

When analyzing paired data of the type discussed here, we can use either the bivariate model discussed in this section or the conditional regression model which occurs when we condition on the X's. We have seen that for testing that $\rho = 0$ (or equivalently, that $\gamma = 0$), it does not matter which approach we take: We get the same test based on $t^* = t_2(0)$. However, for most other purposes, the conditional regression model is preferable. It is difficult to know what the statement $\rho = .15$ means (except that there is a stronger linear relationship than when $\rho = .10$). When we do a regression analysis, we get a regression line which we can plot and which is fairly easy to interpret. In addition, we can find prediction intervals, a confidence band for $\mu(x)$, and the like. For that reason, the regression approach is usually preferable, at least as long as there is a fairly clear choice for the independent variable X and the dependent variable Y. Correlation analysis should be used only when the variables X and Y are symmetric, in the sense that it is not clear which one should be the independent variable and which should be the dependent one. An example of a pair like this might be mathematical and verbal SAT scores.

There is an interesting interpretation of r^2 for regression models. Note that $r^2 = (S_{YY} - (n-2)S^2)/S_{YY}$ (see Exercise B3). We think of S_{YY} as the variation in the Y's if we ignore their dependence on the X's, and $(n-2)S^2$ as the amount of variation left after accounting for the dependence of the Y's on the X's. Therefore, $S_{YY} - (n-2)S^2$ is the amount of variation in the Y's that is explained by their dependence on the X's, and it follows that r^2 is the proportion of the variation in the Y's that is explained by their dependence on the X's.

It is often felt that r^2 somewhat overestimates the amount of variation explained by the X's. Let $T^2 = S_{YY}/(n-1)$ be the sample variance of the Y_i's. Another statistic which is often preferred for measuring the proportion of the variance of the Y's explained by the X's is

$$r^{*2} = \frac{T^2 - S^2}{T^2} = 1 - \frac{(n-1)(1-r^2)}{n-2}.$$

(See the exercises.) The quantity r^{*2} is called the *adjusted* r^2 for this model. T^2 is an unbiased estimator of the variance of the Y's if we ignore the X's, and S^2 is an unbiased estimator of the conditional variance of the Y_i given the X_i, so that r^{*2} is the proportion of the variance of the Y's that is explained by their dependence on the X's. (Note that r^2 is the proportion of the variation explained, where the variation is the numerator of the variance.)

Both r^2 and adjusted r^2 are used in regression analysis as measures of how well the model fits. If either of them is close to one, then nearly all the variance in the Y's has been explained by the X's, and the model is considered to fit well, while if either is close to zero, then very little of the variance of the Y's has been explained by the X's, and the fit is considered poor.

Example 12–4

For the data from Example 12–1, we see that

$$r = \frac{52}{(28(100))^{1/2}} = .983.$$

To test that $\rho = 0$ against $\rho > 0$, we reject the null hypothesis if $t^* > 2.01$. Since

$$t^* = \frac{(5)^{1/2}.983}{(1-(.983)^2)^{1/2}} = 11.971,$$

we indeed reject the null hypothesis and say that we have a significant positive correlation between the X's and the Y's. (Note that t^* computed here is the same as $t_2(0)$ computed in the previous section.) Furthermore, for these data, $r^2 = .966$, so we have explained 96.6% of the variation in the Y's by their dependence on the X's. Finally for the data in question, adjusted r^2 is given by

$$r^{*2} = 1 - \frac{6(1-(.983)^2)}{5} = .96,$$

so that we have explained 96% of the variance if we use r^{*2} as the measure.

Exercises—A

1. An experiment was performed to see whether SAT scores and intelligence are independent of each other. The following data were obtained:

IQ	120	140	90	130	110
SAT	550	780	480	650	540

   Compute the sample correlation coefficient for these data. Use a .05 one-sided test to test whether there is any significant positive correlation between SAT scores and IQ scores.

2. Compute R^2 and adjusted R^2 for the data in Exercise A1, Section 12.1.1.
3. Suppose we repeated the experiment in Exercise 1 with sample size 64 and found that $r = .45$.
   (a) Use Fisher's Z-transformation to find an asymptotic .95 confidence interval for ρ.
   (b) Use Fisher's Z-transformation to do a .05 asymptotic test that $\rho = .3$ against $\rho > .3$. (See Exercise B5 below.)

Exercises—B

1. Let $U_i = aX_i + b$ and $V_i = cY_i + d$. Show that if $ac > 0$, then the sample correlation coefficient computed from the U_i and V_i is the same as that computed from the X_i and Y_i. What happens when $ac < 0$?
2. (a) Verify that $t_2(0)$ as defined in the previous section is the same as $t^* = (n-2)^{1/2}r/(1-r^2)^{1/2}$.
   (b) Show that t^* is an increasing function of r.
3. Show that $r^2 = (S_{YY} - (n-2)S^2)/S_{YY}$ and that $[(n-1)^{-1}S_{YY} - S^2]/[(n-1)^{-1}S_{YY}] = 1 - (n-1)(1-r^2)/(n-2)$.
4. Consider testing that $\rho = 0$ against $\rho \neq 0$. It can be shown that the MLE for $(\nu_1, \nu_2, \tau_1^2, \tau_2^2, \rho)$ is $(\overline{X}, \overline{Y}, T_1^2, T_2^2, r)$, where $T_1^2 = S_{XX}/n$ and $T_2^2 = S_{YY}/n$.
   (a) Show, for this model, that $L(\overline{X}, \overline{Y}, T_1^2, T_2^2, r) = (2\pi)^{-n}[T_1^2T_2^2(1-r^2)]^{-n/2}\exp(-n)$.
   (b) Show that the MLE under the null hypothesis that $\rho = 0$ is $(\overline{X}, \overline{Y}, T_1^2, T_2^2, 0)$ and that $L(\overline{X}, \overline{Y}, T_1^2, T_2^2, 0) = (2\pi)^{-n}(T_1^2T_2^2)^{-n/2}\exp(-n)$.
   (c) Show that the likelihood ratio test rejects the null hypothesis if $r^2 > k$ for some k.
   (d) Show that t^{*2} is an increasing function of r^2 and hence that the likelihood ratio test rejects the null hypothesis if $t^* > k^*$ or $t^* < -k^*$ for some k^*.
5. Show that $u = .5\log((1+x)/(1-x)) \Leftrightarrow x = (e^{2u}-1)/(e^{2u}+1))$.

* 12.2.2 Regression to the Mean

One of the early studies using least squares procedures studied the relationship between the height of a father and the height of his son. When the least squares line for predicting the son's height y from the father's height x was computed, it was noticed that the tallest fathers had sons who were tall, but not quite as tall as their fathers, while the shortest fathers had sons who were short, but not quite as short as their fathers. One explanation for this outcome is that the sons' heights had "regressed" from the fathers' heights, which in fact is the source of the word *regression* for the model considered in this chapter. The data were thought to indicate that there is some natural control mechanism affecting heights which keeps them from becoming too extreme.

It turns out that this regression phenomenon can be explained by the randomness in the data. Let X_i and Y_i be the heights of the ith father and son, respectively. We assume that (Y_i, X_i) are a sample from a bivariate normal distribution. It also seems natural to assume that

$$EY_i = EX_i = \mu \quad \text{and} \quad \text{var}(Y_i) = \text{var}(X_i) = \sigma^2.$$

From the previous section, the expected value of Y given X is

$$EY|X = \mu + \rho(X - \mu),$$

where ρ is the correlation coefficient between the Y_i and the X_i. Therefore, the true regression line is

$$y = \mu + \rho(x - \mu).$$

Now, suppose that $0 < \rho < 1$. Then, for any value of $x \neq \mu$,

$$|y - \mu| = \rho|x - \mu| < |x - \mu|.$$

This implies that the y-values should be closer to μ than the associated x-values, i.e., that the sons' heights should be "pulled inwards" from the fathers' heights. Therefore, when $\mu_X = \mu_Y$ and $\sigma_Y = \sigma_X$, the true regression line of Y on X always pulls the Y-value closer to the mean than the associated X-value, so that it is not surprising that the sample regression line does so also. In sum, the effect can be explained by the randomness in the data. We often call this pulling-inwards phenomenon *regression to the mean*.

If the preceding analysis is correct, we should find a similar phenomenon if we regress father's heights on son's heights. That is, we should find that the fathers of especially tall sons are taller than average, but not quite as tall as their sons and indeed, this is the case. It is hard to explain such an occurrence in terms of some natural control mechanism.

The phenomenon of regression to the mean can be given an intuitive explanation. Suppose that a person's height is partly determined by genes and is partly random, due to such environmental factors as the diet at certain periods of one's life, posture, accidents that happened when the person was young, etc. Then if we let v_i represent the part of the person's height which is determined genetically, let e_i represent the part which is determined randomly, and let Y_i be the person's actual height, we are assuming that

$$Y_i = v_i + e_i.$$

Now, if we choose the tallest father, he is likely to have v_i large, but also e_i large (since his height Y_i is large). If his son has the same genes as his father, then the son's height is $v_i + e_i^*$, where e_i^* represents the random component of the son's height, which we assume is independent of the random component e_i of the father's height. Now, e_i^* is unlikely to be as large as e_i because the father was essentially chosen so as to have maximum e_i. Therefore, the son of the tallest father is likely to be taller than other sons (because v_i is large), but is unlikely to be as tall as his father (because $e_i^* < e_i$). In a similar manner, the shortest father is likely to be one whose random component is very negative, so his son is likely to have a random component that is not so negative. Consequently, the son is likely to be taller than the father (but still short because his genetic part is small). (Obviously, this reasoning could bc reversed to argue that fathers of especially tall sons are not likely to be as tall as their sons.)

Regression to the mean can occur in situations that have nothing to do with regression models. For example, consider batting averages in baseball. It often happens that the batters who are hitting the best in June are still hitting well in September, but their averages are not as high in September as they were in June. This can again be explained by regression to the mean. The batter who has the highest average in June is likely to have had a few more lucky hits than the other batters (so that his random error is positive). However, as the season progresses, the lucky hits average out, and his average for the season regresses to the mean. Similarly, the hitters who are hitting the worst in June have probably had worse luck than other batters, and their luck is likely to get better as the year progresses. In like fashion, we would expect the batting average of the batter who had the highest batting average last year to be somewhat lower this year (but still quite high) because there is likely to have been some luck which helped him to have the highest average last year.

Exercises—A

1. Give an example of regression to the mean different from those in the text.

* 12.2.3 Correlation and Causation

Suppose we regress a variable y on another variable x and reject the hypothesis that the slope is zero in favor of the alternative that the slope is positive. It is then often tempting to conclude that high x-values cause high y-values. However, in most experiments this conclusion cannot be justified. As long as both the x's and the y's are random, there is no way to tell, statistically, whether the x-values are causing the y-values, or the y-values are causing the x-values, or both are being caused by some other variable that has not been measured. All that we can infer in this case is that high x-values and high y-values tend to go together, that is, that x and y are positively correlated. We cannot infer causation. By a similar argument, if the slope is negative, we cannot infer that high x's cause low y's, but only that high x's go with low y's.

For example, in the experiment involving sons' and fathers' heights discussed in the previous section, if we regressed the fathers' heights y on the sons' heights x, we would find that the slope is significant. However, it seems quite unlikely that tall sons cause tall fathers. All we can really conclude from the regression analysis is that tall fathers tend to have tall sons and vice versa. There is no way we can use statistics to decide whether it is the fathers' heights that are causing the sons', or the other way around. However, it seems likely from the biology of the problem that it is the fathers' heights that are causing the sons' heights.

The problem of correlation versus causation can occur in settings other than correlation and regression models. For example, many studies have shown that smokers are more likely to get lung cancer than nonsmokers. However, the studies do not prove that smoking causes lung cancer. Thus, the tobacco companies claim that it is possible that people who are likely to smoke are also (independently) likely to get lung cancer. That is, if they have the personality traits (or genes) which make them likely to smoke, then they have the personality traits (or genes) which make them likely to get lung cancer and would therefore be likely to get lung cancer even if they abstained from smoking. Although those who hold to this kind of argument may seem wrongheaded, there is in fact no way we can use the data that have been collected to reject such an interpretation. All we can conclude from the data is that people who smoke are more likely to get lung cancer, not that smoking causes lung cancer. That is, we can conclude that there is a correlation between smoking and lung cancer, but we cannot conclude that the one causes the other.

We can often get at the issue of causation by designing the study so that the variable which we think is causing the correlation is assigned by the experimenter. For example, in the case of smoking and cancer, we could collect some people and randomly assign some to smoke and others not to smoke. If we do this and find that the smokers still are more likely to get lung cancer than the nonsmokers, we can infer that the smoking is causing the cancer. The primary difference between this experiment and the one described before is that in this experiment we have not let the people choose whether or not to smoke, but have assigned them to do one or the other. Fortunately, no such experiment has been done with people, since it would be immoral. However, experiments of this sort have been done with other animals, and from those experiments, we can infer that smoking causes cancer in certain lower animals. Despite this, it is still a large step to conclude that smoking causes cancer in people from the fact that smoking causes cancer in lower animals. (However, in light of all the evidence we have, it is hard to believe that any reasonable person would doubt that smoking does cause cancer in people.)

Exercises—A

1. Give an example of an experiment, different from the one in the text, in which the issue of correlation versus causation arises.

12.3 MULTIPLE LINEAR REGRESSION

12.3.1 Estimation in Multiple Linear Regression

In this section, we consider the model in which we observe Y_i independent, with

$$Y_i \sim N(\mathbf{x}_i \boldsymbol{\beta}, \sigma^2), \qquad i = 1, \ldots, n,$$

where the $\mathbf{x}_i$ are known constant $1 \times p$ row vectors, $\boldsymbol{\beta}$ is an unobserved $p \times 1$ vector parameter, and σ^2 is an unobserved positive parameter. We assume that $n > p$. Such a model is called a *multiple linear regression model*, and the function $\mathbf{y} = \mathbf{x}\boldsymbol{\beta}$ is the *regression function*. Our primary goal for the model is to estimate and draw inferences about $\boldsymbol{\beta}$. An example of a multiple regression model is given at the beginning of the chapter.

The simple linear regression model treated in Section 12.1 is a special case of the multiple linear regression model with $p = 2$, and

$$\mathbf{x}_i = (1, x_i) \quad \text{and} \quad \boldsymbol{\beta}' = (\theta, \gamma).$$

In order to analyze a model as general as the multiple linear regression model, it is necessary to use matrix notation. Let

$$\mathbf{Y}' = (Y_1, \ldots, Y_n) \quad \text{and} \quad \mathbf{x}' = (\mathbf{x}_1', \ldots, \mathbf{x}_n').$$

That is, $\mathbf{Y}$ is an $n \times 1$ vector whose ith element is Y_i, and $\mathbf{x}$ is an $n \times p$ matrix whose ith row is $\mathbf{x}_i$. Since the Y_i are independent and have constant variance σ^2,

$$\mathbf{Y} \sim N_n(\mathbf{x}\boldsymbol{\beta}, \sigma^2 \mathbf{I}),$$

where $\mathbf{I}$ is the $n \times n$ identity matrix. This is the spherical normal distribution discussed in Chapter 5, and we use the results derived there.

In order to examine the multiple linear regression model, it is necessary to make one further assumption, namely, that the columns of $\mathbf{x}$ are linearly independent, or equivalently, that $\mathbf{x}$ is an $n \times p$ matrix of rank p. We make this assumption for the remainder of the section; it implies that $\mathbf{x}'\mathbf{x}$ is positive definite and, hence, is invertible.

Let $\tilde{\boldsymbol{\beta}}$ be an estimator of $\boldsymbol{\beta}$; that is, $\tilde{\boldsymbol{\beta}}$ is a $p \times 1$ function of $\mathbf{x}$ and $\mathbf{Y}$. Then $\tilde{\boldsymbol{\beta}}$ is the *ordinary least squares* estimator of $\boldsymbol{\beta}$ if it minimizes

$$q(\boldsymbol{\beta}) = (\mathbf{Y} - \mathbf{x}\boldsymbol{\beta})'(\mathbf{Y} - \mathbf{x}\boldsymbol{\beta})$$

(i.e., $\tilde{\boldsymbol{\beta}}$ is the ordinary least squares estimator of $\boldsymbol{\beta}$ if it minimizes the distance between $\mathbf{Y}$ and $\mathbf{x}\boldsymbol{\beta}$). Also, let

$$\hat{\boldsymbol{\beta}} = (\mathbf{x}'\mathbf{x})^{-1}\mathbf{x}'\mathbf{Y}.$$

Theorem 12–9. $\hat{\boldsymbol{\beta}}$ is the ordinary least squares estimator of $\boldsymbol{\beta}$.

Proof. Note that

$$q(\boldsymbol{\beta}) = q(\hat{\boldsymbol{\beta}}) + (\mathbf{x}\hat{\boldsymbol{\beta}} - \mathbf{x}\boldsymbol{\beta})'(\mathbf{x}\hat{\boldsymbol{\beta}} - \mathbf{x}\boldsymbol{\beta}) > q(\hat{\boldsymbol{\beta}}),$$

unless $\hat{\boldsymbol{\beta}} = \boldsymbol{\beta}$ (see the exercises). Therefore, $\hat{\boldsymbol{\beta}}$ is the (unique) ordinary least squares estimator of $\boldsymbol{\beta}$. □

We next define an estimator S^2 for σ^2. Let S^2 be defined by

$$(n - p)S^2 = (\mathbf{Y} - \mathbf{x}\hat{\boldsymbol{\beta}})'(\mathbf{Y} - \mathbf{x}\hat{\boldsymbol{\beta}}) = (\mathbf{Y}'\mathbf{Y} - \hat{\boldsymbol{\beta}}'\mathbf{x}'\mathbf{Y}) = \mathbf{Y}'(\mathbf{I} - \mathbf{x}(\mathbf{x}'\mathbf{x})^{-1}\mathbf{x}')\mathbf{Y}.$$

(Note that S^2 is a real number.) The next theorem gives the joint distribution of $\hat{\boldsymbol{\beta}}$ and S^2.

Theorem 12–10. $\hat{\boldsymbol{\beta}}$ and S^2 are independent,

$$\hat{\boldsymbol{\beta}} \sim N_p(\boldsymbol{\beta}, \sigma^2(\mathbf{x}'\mathbf{x})^{-1}) \quad \text{and} \quad \frac{(n-p)S^2}{\sigma^2} \sim \chi^2_{n-p}.$$

Proof. We merely outline the proof here; the details are left as an exercise. We use the basic results derived in Chapter 5 for the spherical normal distribution. Let

$$\mathbf{A} = (\mathbf{x}'\mathbf{x})^{-1}\mathbf{x}' \quad \text{and} \quad \mathbf{C} = \mathbf{I} - \mathbf{x}(\mathbf{x}'\mathbf{x})^{-1}\mathbf{x}'.$$

Then

$$\hat{\boldsymbol{\beta}} = \mathbf{A}\mathbf{Y} \quad \text{and} \quad (n-p)S^2 = \mathbf{Y}'\mathbf{C}\mathbf{Y}.$$

Now, $\mathbf{AC} = \mathbf{0}$, so by the results on the spherical normal distribution, $\hat{\boldsymbol{\beta}}$ and $(n-p)S^2$ are independent, and hence, $\hat{\boldsymbol{\beta}}$ and S^2 are. Also,

$$\mathbf{A}\mathbf{x}\boldsymbol{\beta} = \boldsymbol{\beta} \quad \text{and} \quad \mathbf{A}(\sigma^2\mathbf{I})\mathbf{A}' = \sigma^2(\mathbf{x}'\mathbf{x})^{-1},$$

so that $\hat{\boldsymbol{\beta}} = \mathbf{A}\mathbf{Y} \sim N_p(\boldsymbol{\beta}, \sigma^2(\mathbf{x}'\mathbf{x})^{-1})$. Furthermore,

$$\mathbf{C}' = \mathbf{C} \quad \text{and} \quad \mathbf{C}^2 = \mathbf{C},$$

so that $\mathbf{C}$ is an idempotent matrix. Also,

$$(\mathbf{x}\boldsymbol{\beta})'\mathbf{C}(\mathbf{x}\boldsymbol{\beta}) = 0 \quad \text{and} \quad \operatorname{tr}(\mathbf{C}) = n - p$$

(where $\operatorname{tr}(\mathbf{C})$ is the trace of $\mathbf{C}$). Therefore, $(n-p)S^2/\sigma^2 = \mathbf{Y}'\mathbf{C}\mathbf{Y}/\sigma^2 \sim \chi^2_{n-p}$. □

The next theorem gives the main result regarding estimation in multiple regression models.

Theorem 12–11

a. $(\hat{\boldsymbol{\beta}}, S^2)$ is a complete sufficient statistic for the multiple regression model.

b. $\hat{\boldsymbol{\beta}}$ and S^2 are the best unbiased estimators for $\boldsymbol{\beta}$ and σ^2.

c. $(\hat{\boldsymbol{\beta}}, \hat{\sigma}^2)$ is the MLE of $(\boldsymbol{\beta}, \sigma^2)$, where $\hat{\sigma}^2 = (n - p)S^2/n$.

Proof

a. The joint density of $\mathbf{Y}$ is

$$f(\mathbf{y}; \boldsymbol{\beta}, \sigma^2) = (2\pi)^{-n/2} \sigma^{-n} \exp\left[\frac{-(\mathbf{y} - \mathbf{x}\boldsymbol{\beta})'(\mathbf{y} - \mathbf{x}\boldsymbol{\beta})}{2\sigma^2}\right].$$

Therefore, the likelihood function for this model is

$$L_Y(\boldsymbol{\beta}, \sigma^2) = L(\boldsymbol{\beta}, \sigma^2) = (2\pi)^{-n/2} (\sigma^2)^{-n/2} \exp\left[\frac{-(\mathbf{Y} - \mathbf{x}\boldsymbol{\beta})'(\mathbf{Y} - \mathbf{x}\boldsymbol{\beta})}{2\sigma^2}\right],$$

or, in exponential form,

$$L(\boldsymbol{\beta}, \sigma^2) = h(\boldsymbol{\beta}, \sigma^2) \exp\left(\frac{-\mathbf{Y}'\mathbf{Y} + 2\mathbf{Y}'\mathbf{x}\boldsymbol{\beta}}{2\sigma^2}\right) = h(\boldsymbol{\beta}, \sigma^2) \exp[T(\mathbf{Y})'c(\boldsymbol{\beta}, \sigma^2)],$$

where

$$\mathbf{T} = T(\mathbf{Y}) = (\mathbf{Y}'\mathbf{Y}, \mathbf{Y}'\mathbf{x}) \quad \text{and} \quad c(\boldsymbol{\beta}, \sigma^2) = \sigma^{-2}(-\tfrac{1}{2}, \boldsymbol{\beta}')'.$$

Now, since $\mathbf{T}$ has $p + 1$ components and there are $p + 1$ unknown parameters, $\mathbf{T} = T(\mathbf{Y})$ is a complete sufficient statistic. But then, $(\hat{\boldsymbol{\beta}}, S^2)$ is an invertible function of $\mathbf{T}$ (see the exercises) and is therefore also a complete sufficient statistic.

b. By Theorem 12–10, $\hat{\boldsymbol{\beta}}$ and S^2 are unbiased estimators of $\boldsymbol{\beta}$ and σ^2, respectively. Since they are functions of the complete sufficient statistic, they are the best unbiased estimators of $\boldsymbol{\beta}$ and σ^2, by the Lehmann-Scheffé theorem.

c. The logarithm of the likelihood for this model is

$$\log[L(\boldsymbol{\beta}, \sigma^2)] = -(n/2) \log 2\pi + n \log \sigma - \frac{(\mathbf{Y} - \mathbf{x}\boldsymbol{\beta})'(\mathbf{Y} - \mathbf{x}\boldsymbol{\beta})}{2\sigma^2}.$$

Therefore, to find the MLE for $\boldsymbol{\beta}$, we must minimize $(\mathbf{Y} - \mathbf{x}\boldsymbol{\beta})'(\mathbf{Y} - \mathbf{x}\boldsymbol{\beta})$. Hence, the MLE for $\boldsymbol{\beta}$ is just the ordinary least squares estimator $\hat{\boldsymbol{\beta}}$. If we then set the derivative of $\log[L(\hat{\boldsymbol{\beta}}, \sigma^2)]$ with respect to σ to zero, we see that the MLE for σ^2 is

$$\frac{(\mathbf{Y} - \mathbf{x}\hat{\boldsymbol{\beta}})'(\mathbf{Y} - \mathbf{x}\hat{\boldsymbol{\beta}})}{n} = \frac{(n - p)S^2}{n}. \quad \square$$

If p/n is large, then the MLE for σ^2 is silly. For example, if $n = 100$ and $p = 50$, then the estimator of the form $c\hat{\sigma}^2$ which has the smallest mean squared error is

$$\frac{(n - p)S^2}{n - p + 2} = \frac{50S^2}{52}.$$

(see the exercises). However, the MLE is

$$\frac{(n-p)S^2}{n} = \frac{S^2}{2},$$

which is clearly much too small. In fact, if

$$\frac{p}{n} \to c > 0 \quad \text{as } n \to \infty,$$

then the MLE is drastically inconsistent. This example is an important one in which using MLEs can cause difficulty. Fortunately, for most practical models, the MLEs are quite reliable. In particular, the MLE is typically a very sensible estimator as long as the number of parameters is small compared to the number of observations. In the multiple regression model, if n is much larger than p, then it makes little difference whether we divide by n or by $n-p$.

Example 12–5

Suppose we have a regression model in which we observe Y_i independent, with

$$Y_i \sim N(\beta_1 + \beta_2 u + \beta_3 u^2 + \beta_4 v, \sigma^2).$$

Suppose also that we observe the following seven data points:

u	-3	-2	-1	0	1	2	3
v	-1	2	1	-4	1	2	-1
Y	-8	2	1	10	5	10	3

For these data,

$$\mathbf{x} = \begin{pmatrix} 1 & -3 & 9 & -1 \\ 1 & -2 & 4 & 2 \\ 1 & -1 & 1 & 1 \\ 1 & 0 & 0 & -4 \\ 1 & 1 & 1 & 1 \\ 1 & 2 & 4 & 2 \\ 1 & 3 & 9 & -1 \end{pmatrix} \quad \text{and} \quad \mathbf{Y} = \begin{pmatrix} -8 \\ 2 \\ 1 \\ 10 \\ 5 \\ 10 \\ 3 \end{pmatrix}.$$

The first column of $\mathbf{x}$ is always one, because β_1 is an intercept term. The second column contains the u_i's, the third column contains the squares of the u_i's, and the fourth column contains the v_i's, while $\mathbf{Y}$ contains the Y_i's. For this model, $n = 7$ and $p = 4$. Thus, for the data given,

$$\mathbf{x}'\mathbf{x} = \begin{pmatrix} 7 & 0 & 28 & 0 \\ 0 & 28 & 0 & 0 \\ 28 & 0 & 196 & 0 \\ 0 & 0 & 0 & 28 \end{pmatrix}, \quad \mathbf{M} = (\mathbf{x}'\mathbf{x})^{-1} = (84)^{-1} \begin{pmatrix} 28 & 0 & -4 & 0 \\ 0 & 3 & 0 & 0 \\ -4 & 0 & 1 & 0 \\ 0 & 0 & 0 & 3 \end{pmatrix}, \quad \mathbf{x}'\mathbf{Y} = \begin{pmatrix} 23 \\ 53 \\ 9 \\ -5 \end{pmatrix}.$$

Therefore, $\hat{\boldsymbol{\beta}} = \mathbf{M}\mathbf{x}'\mathbf{Y} = (84)^{-1}(608, 159, -83, -15)' = (7.24, 1.89, -.99, -.18)'$, so that the estimated regression function is

$$Y = 7.24 + 1.89U - .99U^2 - .18V.$$

Furthermore,

$$\mathbf{Y}'\mathbf{Y} = 303 \quad \text{and} \quad S^2 = \frac{\mathbf{Y}'\mathbf{Y} - \hat{\boldsymbol{\beta}}'\mathbf{x}'\mathbf{Y}}{n-p} = \frac{303 - 258.7}{3} = 14.8.$$

There is an interesting geometrical interpretation for the estimators in this section. Let $\hat{\mathbf{Y}} = \mathbf{x}\hat{\boldsymbol{\beta}}$. Then

$$\mathbf{x}'\hat{\mathbf{Y}} = \mathbf{x}'\mathbf{x}\hat{\boldsymbol{\beta}} = \mathbf{x}'\mathbf{Y}.$$

Now, let V be the set of all vectors $\mathbf{v}$ in R^n such that $\mathbf{v} = \mathbf{x}\mathbf{b}$ for some $\mathbf{b}$ in R^p. Then $\hat{\mathbf{Y}} = \mathbf{x}\hat{\boldsymbol{\beta}} \in V$. If we now let $\mathbf{R} = \mathbf{Y} - \hat{\mathbf{Y}}$ (R is called the *vector of residuals*) and $\mathbf{v} = \mathbf{x}\mathbf{b}$ be any other vector in V, then

$$\mathbf{v}'\mathbf{R} = \mathbf{b}'\mathbf{x}'(\mathbf{Y} - \hat{\mathbf{Y}}) = 0.$$

(Recall that $\mathbf{x}'\mathbf{Y} = \mathbf{x}'\hat{\mathbf{Y}}$.) Hence, $\mathbf{R}$ is orthogonal to any vector in v. In such case, we say that $\hat{\mathbf{Y}}$ is the (orthogonal) projection of $\mathbf{Y}$ onto V. In particular, $\mathbf{R}$ is orthogonal to $\hat{\mathbf{Y}}$, which implies that $\hat{\mathbf{Y}}$ and $\mathbf{R}$ are independent, and hence that

$$S^2 = \frac{\|\mathbf{R}\|^2}{n-p} \quad \text{and} \quad \hat{\boldsymbol{\beta}} = (\mathbf{x}'\mathbf{x})^{-1}\mathbf{x}'\hat{\mathbf{Y}}$$

are independent. In addition, if $\mathbf{v} \in V$, then $\hat{\mathbf{Y}} - \mathbf{v} \in V$, and hence, $\mathbf{R} = \mathbf{Y} - \hat{\mathbf{Y}}$ is orthogonal to $\hat{\mathbf{Y}} - \mathbf{v}$. Therefore, by the Pythagorean theorem,

$$\|\mathbf{Y} - \mathbf{v}\|^2 = \|(\mathbf{Y} - \hat{\mathbf{Y}}) + (\hat{\mathbf{Y}} - \mathbf{v})\|^2 = \|\mathbf{Y} - \hat{\mathbf{Y}}\|^2 + \|\hat{\mathbf{Y}} - \mathbf{v}\|^2 \geqq \|\mathbf{Y} - \hat{\mathbf{Y}}\|^2,$$

with equality only if $\hat{\mathbf{Y}} = \mathbf{v}$. This implies that $\hat{\mathbf{Y}}$ is the vector in V that is closest to $\mathbf{Y}$. (That is to say, $\mathbf{x}\hat{\boldsymbol{\beta}}$ is closer to $\mathbf{Y}$ than is $\mathbf{x}\boldsymbol{\beta}$ for any $\boldsymbol{\beta}$ in R^p, and hence $\hat{\boldsymbol{\beta}}$ is the ordinary least squares estimator of $\boldsymbol{\beta}$.)

To make the preceding ideas more precise, suppose that $n = 2$, $p = 1$, $\mathbf{Y}' = (3, 5)$, and $\mathbf{x}' = (1, 1)$. The space V of possible $\mathbf{x}b$ is the set of all points in R^2, both of whose components are the same. For the given data,

$$\hat{\beta} = [(1, 1)(1, 1)']^{-1}[(1, 1)(3, 5)'] = 4.$$

Therefore,

$$\hat{\mathbf{Y}} = 4(1, 1)' = (4, 4)' \quad \text{and} \quad \mathbf{Y} - \hat{\mathbf{Y}} = (3, 5)' - (4, 4)' = (-1, 1)'.$$

Since $\hat{\mathbf{Y}}'(\mathbf{Y} - \hat{\mathbf{Y}}) = 0$, $\hat{\mathbf{Y}}$ is the orthogonal projection of the point $(3, 5)'$ on that line, as shown in Figure 12–2. (For a more detailed development of regression (and analysis of variance) from a geometric standpoint, see Arnold (1981), Chapters 2–11.)

Exercises—A

1. Suppose we make a study on the causes of high blood pressure. We find four males (m) and seven females (f) and obtain their blood pressures y, their weights w, their ages a, and their sexes s. The data are as follows:

Sex	m	m	m	m	f	f	f	f	f	f	f
Weight	160	180	200	150	130	120	110	120	100	90	110
Age	50	60	40	45	35	40	50	45	40	30	55
Pressure	110	120	140	90	85	90	80	100	110	100	120

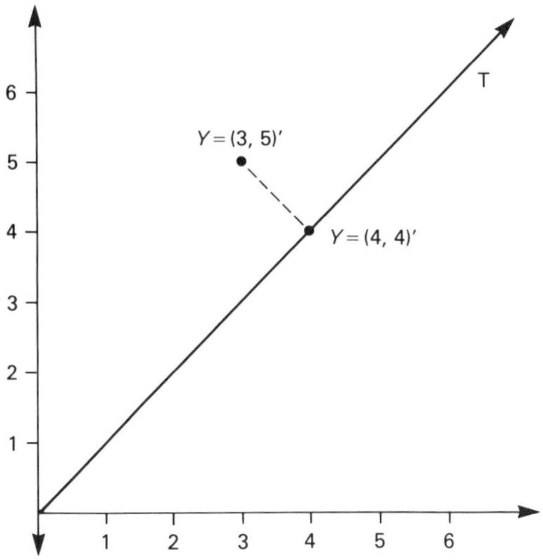

Figure 12-6

We wish to fit the regression equation $y = \beta_1 + \beta_2 s + \beta_3 w + \beta_4 w^2 + \beta_5 a + \beta_6 aw$ to the data.

(a) What are the **x** matrix and the **Y** vector for these data? (*Hint*: Code $m = 1, f = 0$).

(b) What is the mean blood pressure for a 40-year-old female who weighs 120 lb? For a 30-year-old male who weighs 150 lb? (Do *not* calculate the estimators unless you use a computer.)

Exercises—B

1. Consider the multiple regression model $Y_i \sim N(\beta_1 + \beta_2 x + \beta_3 z, \sigma^2)$, independent. Suppose we observe the following data:

x	−4	−2	−1	0	1	2	4
z	1	2	3	4	−3	−2	−5
Y	−10	−4	0	1	0	4	11

Find the ordinary least squares estimator of the β_i and the unbiased estimator of σ^2.

2. Consider a regression model in which we observe $Y_i \sim N(\beta_1 + \beta_2 x^2 + \beta_3 x, \sigma^2)$. Suppose we observe the following data:

x	−2	−1	0	1	2
Y	2	0	0	4	10

(Use fractions for the following problems, do not use decimals.)

(a) Find the $\hat{\beta}_i$ and S^2 for these data.

(b) Graph the curve $y = \hat{\beta}_1 + \hat{\beta}_2 x^2 + \hat{\beta}_3 x$ on a plot of the points (x, Y).

3. **(a)** Show that $q(\boldsymbol{\beta}) = q(\hat{\boldsymbol{\beta}}) + (\mathbf{x}\hat{\boldsymbol{\beta}} - \mathbf{x}\boldsymbol{\beta})'(\mathbf{x}\hat{\boldsymbol{\beta}} - \mathbf{x}\boldsymbol{\beta})$. (*Hint*: $\mathbf{Y} - \mathbf{x}\boldsymbol{\beta} = (\mathbf{Y} - \mathbf{x}\hat{\boldsymbol{\beta}}) + (\mathbf{x}\hat{\boldsymbol{\beta}} - \mathbf{x}\boldsymbol{\beta})$.)

(b) Show that $(\mathbf{x}\hat{\boldsymbol{\beta}} - \mathbf{x}\boldsymbol{\beta})'(\mathbf{x}\hat{\boldsymbol{\beta}} - \mathbf{x}\boldsymbol{\beta}) > 0$ unless $\hat{\boldsymbol{\beta}} = \boldsymbol{\beta}$. (*Hint*: $\hat{\boldsymbol{\beta}} - \boldsymbol{\beta} = (\mathbf{x}'\mathbf{x})^{-1}\mathbf{x}'(\mathbf{x}\hat{\boldsymbol{\beta}} - \mathbf{x}\boldsymbol{\beta})$.)

4. Show that $(\mathbf{Y} - \mathbf{x}\hat{\boldsymbol{\beta}})'(\mathbf{Y} - \mathbf{x}\hat{\boldsymbol{\beta}}) = \mathbf{Y}'\mathbf{Y} - \hat{\boldsymbol{\beta}}'\mathbf{x}'\mathbf{Y} = \mathbf{Y}'(\mathbf{I} - \mathbf{x}(\mathbf{x}'\mathbf{x})^{-1}\mathbf{x}')\mathbf{Y}$.
5. In the proof of Theorem 12–10, show that $\mathbf{AC} = \mathbf{0}$, $\mathbf{Ax\boldsymbol{\beta}} = \boldsymbol{\beta}$, $\mathbf{A}(\sigma^2\mathbf{I})\mathbf{A}' = \sigma^2(\mathbf{x}'\mathbf{x})^{-1}$, $(\mathbf{x}\boldsymbol{\beta})'\mathbf{C}(\mathbf{x}\boldsymbol{\beta}) = 0$, $\text{tr}(\mathbf{C}) = n - p$, $\mathbf{C}' = \mathbf{C}$, and $\mathbf{C}^2 = \mathbf{C}$. (*Hint:* $\text{tr}(\mathbf{x}[(\mathbf{x}'\mathbf{x})^{-1}\mathbf{x}']) = \text{tr}([(\mathbf{x}'\mathbf{x})^{-1}\mathbf{x}']\mathbf{x}) = \text{tr}(\mathbf{I})$. What is the dimension of $\mathbf{I}$?)
6. Let $\mathbf{T} = (T_1, \mathbf{T}_2')' = (\mathbf{Y}'\mathbf{Y}, \mathbf{Y}'\mathbf{x})'$. Let $\mathbf{M} = (\mathbf{x}'\mathbf{x})^{-1}$.
   (a) Show that $\hat{\boldsymbol{\beta}} = \mathbf{M}\mathbf{T}_2$ and $S^2 = (T_1 - \mathbf{T}_2'\mathbf{M}\mathbf{T}_2)/(n - p)$.
   (b) Solve for T_1 and $\mathbf{T}_2$ in terms of $\hat{\boldsymbol{\beta}}$ and S^2 (and $\mathbf{M}$).
7. Find the mean squared error of cS^2 as an estimator of σ^2, and show that it is minimized when $c = (n - p)/(n - p + 2)$.
8. (a) Show that S^2 is a consistent estimator of σ^2.
   (b) Show that if the diagonal elements of $\mathbf{M} = (\mathbf{x}'\mathbf{x})^{-1}$ go to zero, then $\hat{\beta}_i$ is a consistent estimator of β_i.
9. Suppose $p/n \to d$ as $n \to \infty$. Show that the MLE $\hat{\sigma}^2 \xrightarrow{P} (1 - d)\sigma^2$, so that the MLE is not even a consistent estimator of σ^2 unless $d = 0$.
10. (Simple linear regression) Let $p = 2$, $\mathbf{x}_i = (1, x_i)$, $\boldsymbol{\beta}' = (\theta, \gamma)$. Use the results in this section to do the following.
    (a) Verify the formulas given in Section 12.1 for $\hat{\theta}$, $\hat{\gamma}$, and S^2.
    (b) Verify the joint distribution of $\hat{\boldsymbol{\beta}}' = (\hat{\theta}, \hat{\gamma})$ given in Section 12.
11. (Regression through the origin) Let $p = 1$ (so that the x_i and β are univariate).
    (a) What are $\hat{\beta}$ and S^2 for this model?
    (b) How are $\hat{\beta}$ and S^2 distributed?
12. Consider a regression model in which we observe $Y_i \sim N(\gamma x_i + \delta z_i, \sigma^2)$, independent, where the x_i and z_i are observed constants. Let $T_{xx} = \Sigma x_i^2$, $T_{zz} = \Sigma z_i^2$, $T_{YY} = \Sigma Y_i^2$, $T_{xY} = \Sigma x_i Y_i$, $T_{zY} = \Sigma z_i Y_i$, $T_{xz} = \Sigma x_i z_i$, and $Q = T_{xx} T_{zz} - (T_{xz})^2$.
    (a) Show that the estimators of γ and δ are respectively given by $\hat{\gamma} = (T_{zz} T_{xY} - T_{xz} T_{zY})/Q$ and $\hat{\delta} = (T_{xx} T_{zY} - T_{xz} T_{xY})/Q$.
    (b) Show that $S^2 = (T_{YY} - \hat{\gamma} T_{xY} - \hat{\delta} T_{zY})/(n - 2)$.
    (c) Compute $\hat{\gamma}$, $\hat{\delta}$, and S^2 using the data in Exercise B1 above.

Exercises—C

1. Suppose we observe $\mathbf{Y} \sim N_n(\mathbf{x}\boldsymbol{\beta}, \sigma^2 \mathbf{v})$, where $\mathbf{x}$ is a known $n \times p$ matrix whose columns are independent, $\mathbf{v} > 0$ is a known $n \times n$ matrix, $\boldsymbol{\beta}$ is an unknown p-dimensional parameter, and $\sigma^2 > 0$ is an unknown one-dimensional parameter. Let $\mathbf{Y}^* = \mathbf{v}^{-1/2}\mathbf{Y}$ and $\mathbf{x}^* = \mathbf{v}^{-1/2}\mathbf{x}$.
   (a) Show that $\mathbf{Y}^* \sim N_n(\mathbf{x}^*\boldsymbol{\beta}, \sigma^2\mathbf{I})$ (so that the model for $\mathbf{Y}^*$ is just a multiple regression model).
   (b) Let $\hat{\boldsymbol{\beta}}^* = (\mathbf{x}^{*\prime}\mathbf{x}^*)^{-1}\mathbf{x}^{*\prime}\mathbf{Y}$ and $S^{*2} = (\mathbf{Y}^* - \mathbf{x}^*\hat{\boldsymbol{\beta}}^*)'(\mathbf{Y}^* - \mathbf{x}^*\hat{\boldsymbol{\beta}}^*)/(n - p)$ be the unbiased estimators for $\boldsymbol{\beta}$ and σ^2, respectively, for the multiple regression model in $\mathbf{Y}^*$. Show that $\hat{\boldsymbol{\beta}}^* = (\mathbf{x}'\mathbf{v}^{-1}\mathbf{x})^{-1}\mathbf{x}'\mathbf{v}^{-1}\mathbf{Y}$ and $S^{*2} = (\mathbf{Y} - \mathbf{x}\hat{\boldsymbol{\beta}}^*)'\mathbf{v}^{-1}(\mathbf{Y} - \mathbf{x}\hat{\boldsymbol{\beta}}^*)/(n - p)$.

   (This model is often called a *generalized multiple linear regression* model, and $\hat{\boldsymbol{\beta}}^*$ is called a *generalized least squares estimator* of $\boldsymbol{\beta}$. If $\mathbf{v}$ is a diagonal matrix, then $\hat{\boldsymbol{\beta}}^*$ is called a *weighted least squares estimator* of $\boldsymbol{\beta}$.)
2. It can be shown that for any $n \times p$ matrix $\mathbf{x}$ whose columns are independent, there exist matrices $\mathbf{Q}$ and $\mathbf{R}$ such that $\mathbf{x} = \mathbf{QR}$, $\mathbf{Q}$ is $n \times p$, $\mathbf{Q}'\mathbf{Q} = \mathbf{I}$, and $\mathbf{R}$ is a $p \times p$ lower triangular matrix. Let $\mathbf{Y}^* = \mathbf{Q}'\mathbf{Y}$.
   (a) Show that $\hat{\boldsymbol{\beta}} = \mathbf{R}^{-1}\mathbf{Y}^*$.
   (b) Show that $S^2 = (\mathbf{Y}'\mathbf{Y} - \mathbf{Y}^{*\prime}\mathbf{Y}^*)/(n - p)$.

   (The representation $\mathbf{x} = \mathbf{QR}$ is called the *QR decomposition* of $\mathbf{x}$. It is often useful for computing, since triangular matrices are easier to invert than general square matrices. The QR decomposition is found using the Gram-Schmidt process on the columns of $\mathbf{x}$.)

3. Consider the model in which we observe the n-dimensional random vector $\mathbf{Y}$ such that $E\mathbf{Y} = \mathbf{x}\boldsymbol{\beta}$, $\text{cov}(\mathbf{Y}) = \sigma^2\mathbf{I}$ where $\mathbf{x}$ is a known $n \times p$ matrix whose columns are independent and $\boldsymbol{\beta}$ and σ^2 are unknown parameters. (We are not assuming that $\mathbf{Y}$ is normally distributed.) Let $\hat{\beta} = (\mathbf{x}'\mathbf{x})^{-1}\mathbf{x}'\mathbf{Y}$, and let $\mathbf{L}$ be any $p \times n$ matrix such that $\mathbf{U} = \mathbf{L}\mathbf{Y}$ is an unbiased estimator of $\boldsymbol{\beta}$ and $\mathbf{U} \neq \hat{\boldsymbol{\beta}}$.
   (a) Show that $\text{cov}(\mathbf{U}) = \sigma^2\mathbf{L}'\mathbf{L}$.
   (b) Show that for any p-dimensional constant vector $\mathbf{a}$, $\text{var}(\mathbf{a}'\mathbf{U}) > \text{var}(\mathbf{a}'\hat{\beta})$. (*Hint*: The $\text{var}(\mathbf{a}'\mathbf{U})$ is the same whether $\mathbf{Y}$ is normally distributed or not. By Lehmann-Scheffé, $\mathbf{a}'\hat{\boldsymbol{\beta}}$ is the best unbiased estimator for the normal model.)
   (c) Show that $\text{cov}(U) - \text{cov}(\hat{\boldsymbol{\beta}})$ is positive definite. ($\mathbf{a}'(\text{cov}(\mathbf{U}) - \text{cov}(\hat{\boldsymbol{\beta}}))\mathbf{a} = \text{var}(\mathbf{a}'\mathbf{U}) - \text{var}(\mathbf{a}'\hat{\boldsymbol{\beta}})$. Why?)

   (This exercise shows that $\hat{\boldsymbol{\beta}}$ is the best estimator for $\boldsymbol{\beta}$ in the class of unbiased estimators of $\boldsymbol{\beta}$ which are linear functions. For this reason, it is often said that $\hat{\boldsymbol{\beta}}$ is the best linear unbiased (BLUE) estimator of $\boldsymbol{\beta}$. Parts b and c of this problem are often called the Gauss-Markov theorem.)

12.3.2 Inferences in Multiple Regression

In this section, we derive the basic theorems for finding confidence intervals and testing hypotheses in the model of the previous section. Let β_j and $\hat{\beta}_j$ be the jth components of $\boldsymbol{\beta}$ and $\hat{\boldsymbol{\beta}}$, respectively, and let M_{ij} be the ijth component of $\mathbf{M} = (\mathbf{x}'\mathbf{x})^{-1}$.

Theorem 12–12

a. A pivotal quantity for β_j is

$$t_1(\beta_j) = \frac{\hat{\beta}_j - \beta_j}{S(M_{jj})^{1/2}} \sim t_{n-p}.$$

b. Let $\mathbf{a} \neq \mathbf{0}$ be a p-dimensional fixed vector, and let $\xi = \mathbf{a}'\boldsymbol{\beta}$ and $\hat{\xi} = \mathbf{a}'\hat{\boldsymbol{\beta}}$. Then a pivotal quantity for ξ is

$$t_2(\xi) = \frac{\hat{\xi} - \xi}{S(\mathbf{a}'\mathbf{M}\mathbf{a})^{1/2}} \sim t_{n-p}.$$

c. Let $\mathbf{A}$ be a $k \times p$ matrix whose rows are linearly independent. Let $\boldsymbol{\delta} = \mathbf{A}\boldsymbol{\beta}$ and $\hat{\boldsymbol{\delta}} = \mathbf{A}\hat{\boldsymbol{\beta}}$. Then a pivotal quantity for $\boldsymbol{\delta}$ is

$$F(\boldsymbol{\delta}) = \frac{(\hat{\boldsymbol{\delta}} - \boldsymbol{\delta})'(\mathbf{A}\mathbf{M}\mathbf{A}')^{-1}(\hat{\boldsymbol{\delta}} - \boldsymbol{\delta})}{kS^2} \sim F_{k,n-p}.$$

d. A pivotal quantity for σ^2 is

$$U(\sigma^2) = \frac{(n-p)S^2}{\sigma^2} \sim \chi^2_{n-p}.$$

Proof

a. From Theorem 12–10, $\hat{\beta}_j \sim N(\beta_j, \sigma^2 M_{jj})$, and hence,

$$Z_1 = \frac{\hat{\beta}_j - \beta_j}{\sigma(M_{jj})^{1/2}} \sim N(0, 1),$$

independently of $U(\sigma^2) \sim \chi^2_{n-p}$. Therefore,

$$t_1(\beta_j) = \frac{Z_1}{[U/(n-p)]^{1/2}} \sim t_{n-p}.$$

b. The proof of this part is left as an exercise.

c. From Theorem 12–10 and results in Chapter 5,

$$\hat{\boldsymbol{\delta}} = \mathbf{A}\hat{\boldsymbol{\beta}} \sim N_k(\boldsymbol{\delta}, \sigma^2 \mathbf{AMA}').$$

Since the rows of $\mathbf{A}$ are linearly independent, $\mathbf{AMA}' > 0$ and is hence invertible. Therefore, by another result from Chapter 5,

$$T = (\hat{\boldsymbol{\delta}} - \boldsymbol{\delta})'(\sigma^2 \mathbf{AMA}')^{-1}(\hat{\boldsymbol{\delta}} - \boldsymbol{\delta}) \sim \chi^2_k,$$

independently of U, and hence,

$$F(\boldsymbol{\delta}) = \frac{(n-p)T}{kU} \sim F_{k,n-p}.$$

d. This result follows directly from Theorem 12–10. □

We can use the pivotal quantities $t_1(\beta_j)$ and $t_2(\xi)$ in the obvious way to construct confidence intervals and test hypotheses about β_j and ξ. For example, a $(1-\alpha)$ confidence interval for β_j is

$$\beta_j \in \hat{\beta}_j \pm t^{\alpha/2}_{n-p} S(M_{jj})^{1/2}.$$

A size-α test that $\xi = c$ against $\xi \neq c$ would reject the null hypothesis if

$$t_2(c) < -t^{\alpha/2}_{n-p} \quad \text{or} \quad t_2(c) > t^{\alpha/2}_{n-p}.$$

One-sided tests and the equal-tailed two-sided tests based on $t_1(b)$ or $t_2(c)$ are likelihood ratio tests and are UMP unbiased for testing that $\beta_j = b$ or $\xi = c$ against appropriate hypotheses. Similarly, $U(\sigma^2)$ can be used in the obvious way to construct confidence intervals and tests for σ^2. As usual, the one-sided tests are UMP unbiased and likelihood ratio tests, while the likelihood ratio test and the UMP unbiased test for the two-sided problem are (different) two-sided tests based on U, but are not equal-tailed tests.

Now consider inferences about the k-dimensional vector $\boldsymbol{\delta}$. We can use the pivotal quantity $F(\boldsymbol{\delta})$ to construct a size-α test that $\boldsymbol{\delta} = \mathbf{d}$ against $\boldsymbol{\delta} \neq \mathbf{d}$, where $\mathbf{d}$ is a known k-dimensional vector. Note that if $\boldsymbol{\delta} = \mathbf{d}$, then

$$F(\mathbf{d}) = \frac{(\hat{\boldsymbol{\delta}} - \mathbf{d})'(\mathbf{AMA}')^{-1}(\hat{\boldsymbol{\delta}} - \mathbf{d})}{kS^2}$$

should be small, and if $\boldsymbol{\delta} \neq \mathbf{d}$, $F(\mathbf{d})$ should be large. Therefore, a sensible test would reject the null hypothesis if F is too large, that is, if

$$F(\mathbf{d}) > F^{\alpha}_{k,n-p}.$$

(Observe that for testing problems involving a vector of parameters, there is no natural definition for a one-sided hypothesis.) This test is the likelihood ratio test for the problem. It is unbiased, but not UMP unbiased, since there is no UMP unbiased test for the problem. (There is rarely a UMP unbiased test for a hypothesis involving a vector parameter.) For a derivation of the unbiasedness of the test, together with other optimality properties, see Arnold (1981), Chapter 7. Since $\boldsymbol{\delta}$ is a vector, confidence intervals for $\boldsymbol{\delta}$ are not possible.

Now, let $\mathbf{t}$ be a p-dimensional column vector, and let

$$\mu(\mathbf{t}) = \mathbf{t}'\boldsymbol{\beta}, \qquad \hat{\mu}(\mathbf{t}) = \mathbf{t}'\hat{\boldsymbol{\beta}}, \qquad V(\mathbf{t}) = (\mathbf{t}'\mathbf{M}\mathbf{t})^{1/2}.$$

Then if $\mathbf{t}_0$ is a particular p-dimensional column vector, we are often interested in a confidence interval for $\mu(\mathbf{t}_0)$. Also, we are often interested in a $(1-\alpha)$ *confidence band* for $\mu(\mathbf{t})$, i.e., two functions $L(\mathbf{t})$ and $R(\mathbf{t})$ such that

$$P(L(\mathbf{t}) \leqq \mu(\mathbf{t}) \leqq R(\mathbf{t}) \text{ for all } \mathbf{t}) = 1 - \alpha.$$

(In other words, the probability that $\mu(\mathbf{t})$ is not in the band for at least one $\mathbf{t}$ is α.) Let T be a new observation such that

$$T \sim N(\mu(\mathbf{t}_0), \sigma^2)$$

(independently of $\mathbf{Y}$). Then a $(1-\alpha)$ *prediction interval* for T is a random interval $L \leqq T \leqq R$ such that

$$P(L \leqq T \leqq R) = 1 - \alpha.$$

(Note that T is a random variable, not a vector.)

Theorem 12–13

a. A $(1-\alpha)$ confidence interval for $\mu(\mathbf{t}_0)$ is

$$\mu(\mathbf{t}_0) \in \hat{\mu}(\mathbf{t}_0) \pm t_{n-p}^{\alpha/2} SV(\mathbf{t}_0).$$

b. A $(1-\alpha)$ prediction interval for T is

$$T \in \hat{\mu}(\mathbf{t}_0) \pm t_{n-p}^{\alpha/2} S(1 + V^2(\mathbf{t}_0))^{1/2}.$$

c. A confidence band for the function $\mu(\mathbf{t})$ is

$$\mu(\mathbf{t}) \in \hat{\mu}(\mathbf{t}) \pm (pF_{p,n-p}^{\alpha})^{1/2} SV(\mathbf{t}).$$

Proof

a. This result follows directly from part b of Theorem 12–12.

b. Since T and $\hat{\mu}(t_0)$ are independent,

$$T \sim N(\mu(\mathbf{t}_0), \sigma^2) \quad \text{and} \quad \hat{\mu}(\mathbf{t}_0) \sim N(\mu(\mathbf{t}_0), \sigma^2 V^2(t_0)),$$

we see that

$$T - \hat{\mu}(\mathbf{t}_0) \sim N(0, \sigma^2(1 + V^2(\mathbf{t}_0))) \quad \text{and} \quad t = \frac{(T - \hat{\mu}(\mathbf{t}_0))}{S(1 + V^2(\mathbf{t}_0))^{1/2}} \sim t_{n-p}.$$

Therefore

$$P(T \in \hat{\mu}(\mathbf{t}_0) \pm t_{n-p}^{\alpha/2} S(1 + V^2(\mathbf{t}_0))^{1/2}) = P(-t_{n-p}^{\alpha/2} \leqq t \leqq t_{n-p}^{\alpha/2}) = 1 - \alpha.$$

c. Let

$$F = \frac{(\hat{\boldsymbol{\beta}} - \boldsymbol{\beta})'\mathbf{x}'\mathbf{x}(\hat{\boldsymbol{\beta}} - \boldsymbol{\beta})}{pS^2}.$$

Then $F \sim F_{p,n-p}$, by part c of Theorem 12–12. Now,

$$(\hat{\boldsymbol{\beta}} - \boldsymbol{\beta})'\mathbf{x}'\mathbf{x}(\hat{\boldsymbol{\beta}} - \boldsymbol{\beta}) = \frac{\sup_{\mathbf{t} \neq \mathbf{0}}(\hat{\mu}(\mathbf{t}) - \mu(\mathbf{t}))^2}{V^2(\mathbf{t})}. \tag{12–2}$$

(see Exercise C3). The interval associated with $\mathbf{t} = \mathbf{0}$ is satisfied trivially. Therefore,

$$\begin{aligned} &P(\mu(\mathbf{t}) \in \hat{\mu}(\mathbf{t}) \pm (pF^{\alpha}_{p,n-p})^{1/2} SV(\mathbf{t}) \text{ for all } \mathbf{t}) \\ &= P\left(\frac{(\hat{\mu}(\mathbf{t}) - \mu(\mathbf{t}))^2}{V^2(\mathbf{t})} \leqq pS^2 F^{\alpha} \text{ for all } \mathbf{t} \neq 0\right) \\ &= P\left(\frac{(\sup_{\mathbf{t} \neq \mathbf{0}}(\hat{\mu}(\mathbf{t}) - \mu(\mathbf{t}))^2}{V^2(\mathbf{t})} \leqq pS^2 F^{\alpha}\right) = P(F \leqq F^{\alpha}) = 1 - \alpha. \quad \square \end{aligned}$$

Suppose, now, that the first component of $\boldsymbol{\beta}$ is an intercept, so that the first component of $\mathbf{x}_i$ is always one. Let

$$\mathbf{x}_i = (1, \mathbf{u}_i) \quad \text{and} \quad \boldsymbol{\beta}' = (\theta, \boldsymbol{\gamma}'),$$

where $\mathbf{u}_i$ is $1 \times (p - 1)$ and $\boldsymbol{\gamma}$ is $(p - 1) \times 1$. A measure of fit of the model under discussion is

$$R^2 = \frac{S_{YY} - (n - p) S^2}{S_{YY}},$$

where $S_{YY} = \Sigma(Y_i - \overline{Y})^2$. As in simple regression, S_{YY} is a measure of how much variation there is in the Y's when we ignore their dependence on the $\mathbf{u}_i$'s, while $(n - p)S^2$ is a measure of how much variation there is left after accounting for the dependence of the Y's on the $\mathbf{u}_i$'s. Therefore, R^2 represents the proportion of the variation in the Y's which has been explained by their dependence on the $\mathbf{u}_i$'s. If $R^2 \approx 1$, our model is a fairly good fit, but if $R^2 \approx 0$, the fit is not so good.

Let $\hat{\mathbf{Y}} = \mathbf{X}\hat{\boldsymbol{\beta}}$, and let r be the sample correlation coefficient between the components of $\mathbf{Y}$ and those of $\hat{\mathbf{Y}}$. Then $r \geqq 0$, and $r^2 = R^2$. This provides another interpretation of R^2, namely, as the square of the sample correlation coefficient between the observations (the Y_i) and their estimated means (the $\hat{Y}_i$). If R^2 is large, then the components of $\mathbf{Y}$ are highly correlated with those of $\hat{\mathbf{Y}}$, which again implies that the model fits well.

We often use R^2 to test the hypothesis that $\boldsymbol{\gamma} = \mathbf{0}$, or equivalently, that there is no dependence between the Y_i and the $\mathbf{u}_i$. Let $\mathbf{A} = (\mathbf{0}, \mathbf{I})$, where $\mathbf{0}$ is a $(p - 1)$-dimensional column vector of zeros and $\mathbf{I}$ is a $(p - 1) \times (p - 1)$-dimensional identity matrix. Then $\mathbf{A}\boldsymbol{\beta} = \boldsymbol{\gamma}$. Let $\boldsymbol{\delta} = \mathbf{A}\boldsymbol{\beta}$, and let F^* be the F-statistic for testing that $\boldsymbol{\delta} = 0$ suggested in Theorem 12–12. Then

$$F^* = \frac{(n - p)R^2}{(p - 1)(1 - R^2)} \tag{12–3}$$

(see Arnold (1981), Section 16.9), and hence, the optimal test for testing that $\boldsymbol{\gamma} = \mathbf{0}$ rejects the null hypothesis if

$$F^* > F^{\alpha}_{p-1,n-p}.$$

It should be mentioned that if we reject the hypothesis $\boldsymbol{\gamma} = 0$ with this test, we do not have statistical proof that the model fits adequately. Rather, we have statistical proof that there is significant improvement in fit when the $\mathbf{u}_i$ are included in the model.

It is often felt that we can get a better measure of the fit of the model by replacing S_{YY} with $T^2 = S_{YY}/(n-1)$ and replacing $(n-p)S^2$ with S^2, getting

$$R^{*2} = \frac{T^2 - S^2}{T^2} = 1 - \frac{(n-1)(1-R^2)}{n-p}.$$

Note that T^2 is the unbiased estimator of the variance of the Y_i if they are treated as a sample, ignoring their relation to the $\mathbf{x}_i$, and S^2 is the unbiased estimator of the variance of the Y_i after accounting for their dependence on the u_i. Therefore, R^{*2} is the proportion of the estimated variance in the Y_i that is explained by their dependence on the u_i. R^{*2} is called *adjusted* R^2.

Example 12–6

In Example 12–5, we found that

$$\mathbf{X'X} = \begin{Bmatrix} 7 & 0 & 28 & 0 \\ 0 & 28 & 0 & 0 \\ 28 & 0 & 196 & 0 \\ 0 & 0 & 0 & 28 \end{Bmatrix}, \qquad \mathbf{M} = (\mathbf{X'X})^{-1} = (84)^{-1}\begin{Bmatrix} 28 & 0 & -4 & 0 \\ 0 & 3 & 0 & 0 \\ -4 & 0 & 1 & 0 \\ 0 & 0 & 0 & 3 \end{Bmatrix},$$

$S^2 = 14.8$, and $\hat{\boldsymbol{\beta}} = \mathbf{MX'Y} = (84)^{-1}(608, 159, -83, -15)' = (7.24, 1.89, -.99, -.18)'$, so that the estimated regression function is

$$Y = 7.24 + 1.89U - .99U^2 - .18V.$$

Suppose we want to test that $\beta_4 = 0$, i.e., that there is no contribution from the V_i's. Then $M_{44} = \frac{3}{84} = .036$, and the appropriate test statistic is therefore

$$t = \frac{\hat{\beta}_4}{SM_{44}^{1/2}} = \frac{(-.18)}{(14.8(.036))^{1/2}} = -.247.$$

Since $t > -t_3^{.025} = -3.18$, we accept the hypothesis that $\beta_4 = 0$ with these data with a two-sided size-.05 test.

Suppose now we want to test that both β_2 and β_3 are zero, i.e., that the u_i's are not contributing to the prediction of the Y_i's. Let

$$\mathbf{A} = \begin{Bmatrix} 0 & 1 & 0 & 0 \\ 0 & 0 & 1 & 0 \end{Bmatrix}.$$

Then $\mathbf{A}\boldsymbol{\beta} = (\beta_2, \beta_3)'$. Therefore, this problem is equivalent to testing that $\mathbf{A}\boldsymbol{\beta} = \mathbf{0}$. Now,

$$\mathbf{AMA'} = 84^{-1}\begin{Bmatrix} 3 & 0 \\ 0 & 1 \end{Bmatrix}, \quad (\mathbf{AMA'})^{-1} = \begin{Bmatrix} 28 & 0 \\ 0 & 84 \end{Bmatrix}, \quad \mathbf{A}\hat{\boldsymbol{\beta}} = \begin{Bmatrix} 1.89 \\ -.99 \end{Bmatrix}.$$

Hence, the F-statistic for this test is

$$F = \frac{(\mathbf{A}\hat{\boldsymbol{\beta}})'(\mathbf{AMA'})^{-1}(\mathbf{A}\hat{\boldsymbol{\beta}})}{2S^2} = \frac{28(1.89)^2 + 84(-.99)^2}{29.6} = 6.16.$$

Since $6.16 < F_{2,3}^{.05} = 9.55$, we accept the hypothesis that $\beta_2 = \beta_3 = 0$ with a .05 test.

Now suppose we want to make inferences about the mean when $u = 2$ and $v = 3$. That is, we want to make inferences about

$$\beta_1 + 2\beta_2 + 2^2\beta_3 + 3\beta_3 = (1, 2, 4, 3)\boldsymbol{\beta} = \mathbf{t}_0\boldsymbol{\beta} = \mu(\mathbf{t}_0),$$

where $\mathbf{t}_0 = (1, 2, 4, 3)$. Then

$$\hat{\mu}(\mathbf{t}_0) = \mathbf{t}_0 \hat{\boldsymbol{\beta}} = 6.52 \quad \text{and} \quad V^2(\mathbf{t}_0) = \mathbf{t}_0 M \mathbf{t}_0' = .61.$$

Therefore, a 95% confidence interval for $\mu(\mathbf{t}_0)$ is

$$\mu(\mathbf{t}_0) \in 6.52 \pm (3.18)(14.8(.61))^{1/2} = 6.52 \pm 9.55,$$

a 95% prediction interval for a new observation T with mean $\mu(\mathbf{t}_0)$ is

$$T \in 6.52 \pm (3.18)(14.8(1.61))^{1/2} = 6.52 \pm 15.52,$$

and the 95% confidence band for the regression function, computed at $\mathbf{t}_0$ is

$$\mu(\mathbf{t}_0) \in 6.52 \pm [4(9.12)14.8(.61)]^{1/2} = 6.52 \pm 18.15.$$

Now $\overline{Y} = \frac{23}{7} = 3.29$ for the given data, so that

$$S_{YY} = \mathbf{Y}'\mathbf{Y} - 7\overline{Y}^2 = 227.2 \quad \text{and} \quad R^2 = \frac{227.2 - 3(14.8)}{227.2} = .80.$$

To test for the fit of the model, we use

$$F = \frac{3(.8)}{3(1 - .8)} = 4.$$

Since $F < F_{3,3}^{.05} = 9.28$, R^2 is not significant with this model. Note that F is the same as the F we would have obtained for testing that $\beta_2 = \beta_3 = \beta_4 = 0$. (See Exercise B10.) That is, if R^2 is not significant, then all the coefficients except the intercept are not significant. If we use adjusted R^2 instead of R^2, we get

$$R^{*2} = 1 - \frac{6(1 - .8)}{3} = .6,$$

so that R^2 and R^{*2} can be quite different, especially when the model does not fit very well.

As discussed in Section 12.1.2 for simple regression, the procedures derived in Sections 12.3.1 and 12.3.2 can also be used when the $\mathbf{x}_i$ are random vectors, as long as the conditional distribution of the Y_i satisfies the previously given conditions. The procedures derived are not necessarily optimal for the model with random $\mathbf{X}_i$ (although they are conditionally optimal), but they are still sensible. They can be shown to be optimal if the $\mathbf{X}_i'$ are a sample from a multivariate normal distribution (see Arnold (1981), Chapter 16).

Exercises—A

1. For the model in Exercise A1 of the previous section,
   (a) Write the hypothesis that age has no effect on blood pressure in the form $\mathbf{A}\boldsymbol{\beta} = \mathbf{0}$.
   (b) Write the hypothesis that weight has no effect on blood pressure in the form $\mathbf{C}\boldsymbol{\beta} = \mathbf{0}$.

Exercises—B

1. For the data in Exercise B1 of the previous section,
   (a) Test that $\beta_3 = 0$ against $\beta_3 \neq 0$ with $\alpha = .05$.
   (b) Find a 98% confidence interval for β_2.

(c) Test that $\beta_2 = \beta_3 = 0$. Use $\alpha = .05$.

(d) Let $\mathbf{t}_0 = (1, 1, 2)$. Find a 95% confidence interval for $\mu(t_0)$ and a 95% prediction interval for a new observation T with mean $\mu(t_0)$. Evaluate the 95% confidence band for the response surface at $\mu(t_0)$.

(e) Find R^2 for the given data. Is it significant with a .05 test? Find adjusted R^2.

2. Consider the model and data of Exercise B2, Section 12.3.1.

(a) Test the null hypothesis $\beta_2 = 0$ against the alternative $\beta_2 > 0$ with $\alpha = .05$.

(b) Test the null hypothesis that $\beta_1 = 0$, $\beta_3 = 0$ against the general alternative with $\alpha = .05$.

(c) Find a .95 prediction interval for a new observation with $x = 1$.

(d) Find a confidence band for the response surface $\mu(x) = \beta_1 + \beta_2 x^2 + \beta_3 x$, and graph this band on a plot of the points.

(e) What is R^2 for these data? Is it significant? What is adjusted R^2 for these data?

3. Derive Theorem 12–12, part b.

4. Consider testing that $\boldsymbol{\beta} = 0$. Show that the F-test for this hypothesis rejects the null hypothesis if $F = (\mathbf{x}\hat{\boldsymbol{\beta}})'(\mathbf{x}\hat{\boldsymbol{\beta}})/pS^2 > F^{\alpha}_{p,n-p}$.

5. Let $T^2 = S_{YY}/(n-1)$. Show that $(T^2 - S^2)/T^2 = 1 - (n-1)(1-R^2)/(n-p)$.

6. (Simple linear regression) Let $p = 2$, $\mathbf{x}_i = (1, x_i)$, $\boldsymbol{\beta}' = (\theta, \gamma)$. Use the results in this section to do the following.

(a) Verify the formulas for pivotal quantities for θ and γ given in Section 12.1.

(b) Let $\mathbf{t} = (1, t)$, $\mu(t) = \mathbf{t}\boldsymbol{\beta}$. Verify the formulas given in Section 12.1 for the $(1-\alpha)$ confidence interval for $\mu(t_0)$, the $(1-\alpha)$ prediction interval for an independent observation $T \sim N(\mu(t_0), \sigma^2)$ and the $(1-\alpha)$ confidence band for the response surface $\mu(t)$.

7. (Regression through the origin) Let $p = 1$ (so that x_i and β are univariate).

(a) Find a $(1-\alpha)$ confidence interval for β.

(b) Let $\mu(t) = t\beta$. Find a $(1-\alpha)$ confidence interval for $\mu(t_0)$, the $(1-\alpha)$ prediction interval for an independent observation $T \sim N(\mu(t_0), \sigma^2)$, and the $(1-\alpha)$ confidence band for the response surface $\mu(t)$.

8. Consider the regression model of Exercise B12, Section 12.3.1.

(a) Find a pivotal quantity for γ.

(b) Let $T \sim N(\gamma p_0 + \delta q_0, \sigma^2)$, independent of the Y_i. Find a $(1-\alpha)$ prediction interval for T.

(c) Using the data in Exercise B1, Section 12.3.1, compute the 95% confidence interval for γ and the 95% prediction interval for a new observation with mean $\gamma + \delta$.

9. Consider the multiple regression model with an intercept, in which $\mathbf{x}_i = (1, \mathbf{u}_i)$ and $\boldsymbol{\beta}' = (\theta, \boldsymbol{\gamma}')$. Consider testing that $\boldsymbol{\gamma} = \mathbf{0}$ against $\boldsymbol{\gamma} \neq \mathbf{0}$.

(a) Show that the likelihood under the alternative hypothesis is $L_Y(\hat{\boldsymbol{\beta}}, \hat{\sigma}^2) = (2\pi)^{-n/2}(\hat{\sigma}^2)^{-n/2}\exp(-n/2)$, where $\hat{\sigma}^2 = (n-p)S^2/n$.

(b) Show that the MLE under the null hypothesis is given by $\hat{\hat{\theta}} = \overline{Y}$, $\hat{\hat{\gamma}} = 0$, $\hat{\hat{\sigma}}^2 = S_{YY}/n$. (Note that when $\boldsymbol{\gamma} = \mathbf{0}$, this model is just a one-sample normal model as discussed in Section 11.2.1.)

(c) Show that $L_Y(\hat{\hat{\theta}}, \hat{\hat{\gamma}}, \hat{\hat{\sigma}}^2) = (2\pi)^{-n/2}(\hat{\hat{\sigma}}^2)^{-n/2}\exp(-n/2)$.

(d) Let F^* be as in Equation (12–3). Show that the likelihood ratio test statistic for this model is $\Lambda = (1 + (p-1)F^*/(n-p))^{n/2}$ and hence, that the likelihood ratio test for this problem rejects the null hypothesis if F^* is too large.

10. For the data in Example 12–6, compute the F-statistic for testing that $\beta_2 = \beta_3 = \beta_4 = 0$, and show that it is equal to $(n-p)R^2/p(1-R^2)$.

Exercises—C

1. In the notation of Exercise C1 of the last section,
   (a) Using the multiple regression model involving $\mathbf{Y}^*$, find the F-test for testing that $\mathbf{A\beta} = \mathbf{0}$. What is the F-statistic in terms of $\mathbf{Y}$?
   (b) Find the $(1-\alpha)$ confidence band for $\mu(\mathbf{t})$ in terms of $\mathbf{Y}^*$ and then in terms of $\mathbf{Y}$.
2. In the notation of Exercise C2 of the last section,
   (a) Show that the F-statistic for testing that $\mathbf{A\beta} = \mathbf{0}$ can be written as $(\mathbf{A}^*\mathbf{Y}^*)'(\mathbf{A}^*\mathbf{A}^{*\prime})^{-1}(\mathbf{A}^*\mathbf{Y}^*)/kS^2$, where $\mathbf{A}^* = \mathbf{AR}^{-1}$.
   (b) Show that the confidence band for the response surface can be written as $\mu(\mathbf{t}) = \mathbf{t}^{*\prime}\mathbf{Y}^* \pm S(pF^\alpha \mathbf{t}^{*\prime}\mathbf{t}^*)^{1/2}$, where $\mathbf{t}^* = (\mathbf{R}^{-1})'\mathbf{t}$.
3. Let $\mathbf{b}$ be a p-dimensional vector, and let $\mathbf{x}$ be an $n \times p$ matrix whose columns are independent. (Note that $\mathbf{x}'\mathbf{x} > 0$, so that $(\mathbf{x}'\mathbf{x})^{1/2}$ and $(\mathbf{x}'\mathbf{x})^{-1/2}$ exist.)
   (a) Show that for all $\mathbf{t} \neq \mathbf{0}$, $[\mathbf{t}'\mathbf{b}]^2/\mathbf{t}'(\mathbf{x}'\mathbf{x})^{-1}\mathbf{t} \leqq \mathbf{b}'\mathbf{x}'\mathbf{x}\mathbf{b}$. (Let $\mathbf{a} = (\mathbf{x}'\mathbf{x})^{-1/2}\mathbf{t}$ and $\mathbf{c} = (\mathbf{x}'\mathbf{x})^{1/2}\mathbf{b}$. Then by the Cauchy-Schwarz inequality for vectors, $(\mathbf{a}'\mathbf{c})^2 \leqq (\mathbf{a}'\mathbf{a})(\mathbf{c}'\mathbf{c})$.)
   (b) Let $\mathbf{t}^* = (\mathbf{x}'\mathbf{x})\mathbf{b}$. Show that $[\mathbf{t}^{*\prime}\mathbf{b}]^2/\mathbf{t}^{*\prime}(\mathbf{x}'\mathbf{x})^{-1}\mathbf{t}^* = \mathbf{b}'\mathbf{x}'\mathbf{x}\mathbf{b}$.
   (c) Show that $\mathbf{b}'\mathbf{x}'\mathbf{x}\mathbf{b} = \sup_{\mathbf{t}\neq\mathbf{0}}(\mathbf{t}'\mathbf{b})^2/\mathbf{t}'(\mathbf{x}'\mathbf{x})^{-1}\mathbf{t}$.
   (d) Verify Equation (12–2).

*# 12.3.3 Multicollinearity

In the multiple regression model, we must assume that the columns of $\mathbf{x}$ are linearly independent. Otherwise, the matrix $\mathbf{x}'\mathbf{x}$ is not invertible and the β_i are not well defined. For example, suppose that $p = 3$ and $x_{i3} = x_{i1} + 2x_{i2}$ for all i. Then $2x_{i1} + 2x_{i2} + x_{i3} = 3x_{i1} + 4x_{i1} + 0x_{i3}$, and there is no difference between $\boldsymbol{\beta} = (2, 2, 1)'$ and $\boldsymbol{\beta} = (3, 4, 0)'$.

Although it rarely happens that the columns of $\mathbf{x}$ are linearly dependent, it often happens they are "nearly" linearly dependent. In this case, the estimator $\hat{\boldsymbol{\beta}}$ is quite unstable, in the sense that small changes in the data can make large changes in the estimators.

Example 12–7

Suppose we have a multiple regression model with

$$\mathbf{x}' = \begin{Bmatrix} 1 & 1 & 1 & 1 & 1 \\ -2 & -1 & 0 & 1 & 2 \\ -.01 & 1.01 & 2 & 2.99 & 3.99 \end{Bmatrix}.$$

Note that the third column of $\mathbf{x}$ is approximately equal to twice the first column plus the second column. If $\mathbf{Y} = \mathbf{x}'\mathbf{b}$ for some vector $\mathbf{b}$, then $\hat{\boldsymbol{\beta}} = \mathbf{b}$. Suppose we choose $\mathbf{b} = (1, 2, 0)'$, so that

$$\mathbf{Y}' = (\mathbf{x}\mathbf{b})' = \mathbf{b}'\mathbf{x}' = (-3, -1, 1, 3, 5).$$

Therefore, if we observe $\mathbf{Y} = (-3, -1, 1, 3, 5)'$, we estimate $\boldsymbol{\beta}$ by $\hat{\boldsymbol{\beta}}' = (1, 2, 0)$. Now, let $\mathbf{b}^{*\prime} = (-3, 0, 2)$, so that

$$\mathbf{Y}^{*\prime} = \mathbf{b}^{*\prime}\mathbf{x}' = (-3.02, -.98, 1, 2.98, 4.98).$$

In this case, if we observe $\mathbf{Y}^* = (-3.02, -.98, 1, 2.98, 4.98)'$, we estimate $\boldsymbol{\beta}$ by $\hat{\boldsymbol{\beta}}' = (-3, 0, 2)$. Finally, if we take $\mathbf{b}' = (-1, 1, 1)$, then

$$\mathbf{Y}' = \mathbf{b}'\mathbf{x}' = (-3.01, -.99, 1, 2.99, 4.99).$$

Hence, if we observe $\mathbf{Y} = (-3.01, -.99, 1, 2.99, 4.99)'$, we estimate $\boldsymbol{\beta}$ by $\hat{\boldsymbol{\beta}}' = (-1, 1, 1)$. Now, plainly, $\mathbf{Y}$, $\mathbf{Y}^*$, and $\mathbf{Y}'$ are all nearly equal, but the estimators are quite different. Thus, with the $\mathbf{x}$ matrix given, there is no way to find out whether or not $\beta_3 = 0$. Of course, the test derived in the previous section can be used, but it is not very powerful. (Note that if $x_{31} \doteq 2x_{1i} + x_{2i}$, then

$$x_{1i} + 2x_{2i} + 0x_{3i} \doteq -3x_{1i} + 0x_{2i} + 2x_{3i} \doteq -x_{1i} + x_{2i} + x_{3i},$$

which explains why $\mathbf{b}$, $\mathbf{b}^*$, and $\mathbf{b}'$ lead to about the same values for $\mathbf{Y}$.)

To illustrate further the difficulties with multicollinearity, suppose we have a multiple regression experiment to predict blood pressure in which two of the independent variables are height and weight. Suppose also that in the particular data set we observe, weight and height are nearly collinear, and assume that weight is the primary cause of high blood pressure. Then when we fit the multiple regression model to the data, it is impossible to separate the effects of height and weight on blood pressure, because of the collinearity. Therefore, the effect of weight may be split between height and weight, and neither variable will appear important. It may also happen that the sample regression function puts most of the effect of weight onto height, and height appears important, but weight does not. Consequently, it is not possible to determine from the data whether it is height or weight which is causing the increased blood pressure. In fact, there is no way to use statistics to separate the effects of variables which are collinear.

One approach to dealing with multicollinearity is to increase the size of the data set so that the variables are no longer collinear. (In the example in the last paragraph, we might get some lightweight tall people and some heavyweight short people.) However, doing so may not always be feasible. Another approach is to use our understanding of the data to eliminate one of the variables from the model. If height and weight are collinear, then we might feel that height is not a major cause of high blood pressure and remove it from the model. It should be emphasized that we can only choose which variables to eliminate by using our knowledge of the data. There is no way we can use statistics to choose which variable to eliminate: Because of the instability of the estimators, the variable which best predicts the blood pressure may be determined primarily by the noise in the data.

Exercises—B

1. For Example 12–7, find $\mathbf{Y} = \mathbf{xb}$, where $\mathbf{b} = (0, 1.5, .5)$, and show that $\mathbf{Y}$ is quite near $(-3, -1, 1, 3, 5)$.

Chapter 13

ANALYSIS-OF-VARIANCE MODELS

In this chapter, we consider experiments in which we have classified individuals in several ways and measured some continuous random variable on each individual. Such models are called *analysis-of-variance* (ANOVA) models.

As an example, in studying the impact of diet on blood cholesterol, we might design an experiment in which we give each of a large number of people one of 12 different diets for a year. At the end of the year, we measure the cholesterol levels of the group. There are several ways we could analyze the data obtained. The simplest way is to divide the people only by diet. Let Y_{ij} be the cholesterol level of the jth person on the ith diet. We could assume the model

$$Y_{ij} \sim N(\theta + \alpha_i, \sigma^2),$$

where θ is an overall mean cholesterol level and α_i represents the effect of the ith diet. We call this model the *one-way analysis-of-variance* model, since we have classified the individuals in just one way. In testing that the diets have no effect on cholesterol level, we are testing that the α_i are zero. The basic results for the one-way model are given in Section 13.1.

We might suspect, however, that cholesterol level is affected by smoking. In that case, we might divide the people into three groups: nonsmokers, light smokers, and heavy smokers. We would give five nonsmokers each of the 12 diets, five light smokers each of the 12 diets, and five heavy smokers each of the 12 diets (so that we would have a total of $12 \times 3 \times 5 = 180$ people in the study). We could then let Y_{ijk} be the cholesterol level of the kth person in the jth smoking level who had the ith diet. Thus, Y_{132} would be the cholesterol level of the second person of the five who got the first diet and had the third smoking level. We might assume a model in which

$$Y_{ijk} \sim N(\theta + \alpha_i + \beta_j + \gamma_{ij}, \sigma^2),$$

where θ represents the overall mean cholesterol level, α_i represents the effect of the ith diet, β_j represents the effect of the jth smoking level, and γ_{ij} represents the interaction between the ith diet and the jth smoking level. Then, testing that diet has no overall effect on cholesterol is testing that the α_i are zero, testing that smoking has no overall effect on cholesterol is testing that the β_j are zero, and testing that smoking and diet have no interactive effect on cholesterol is testing that the γ_{ij} are zero. This model is called the *crossed two-way analysis-of-variance* model, since we have divided the people in two ways—by their smoking habits and by their diets—and since we have used all the diets on all smoking levels (crossed). When we make the simplifying assumption that the γ_{ij} are zero (i.e., that there is no interaction), we call the model an *additive* two-way model. Crossed two-way models are discussed in Section 13.2.

Suppose we also feel that sex has an important effect on cholesterol level. Then we could also classify the people by their sex and get a crossed three-way ANOVA model. If we had enough people, we could add additional classifications to get four-way or even higher crossed ANOVA models. The derivations for three-way models are given in exercises in Section 13.2. An elegant model which can occasionally be used to analyze three additive factors in a two-way design is the *Latin square* model, discussed in Section 13.3.

Suppose now that all the people in the hypothetical study are nonsmokers, so that there is no smoking effect. Suppose also that the 12 diets are of four different types, the first three low-salt diets, the second three low-cholesterol diets, the next three low-fat diets, and the last three "normal" diets. In this case, we could let Y_{ijk} be the cholesterol level of the kth person who had the jth diet of the ith type, so that Y_{123} would be the cholesterol level of the third person having the second type of low-salt diet. If we then suppose that there is no relationship between the jth diet in class i and the jth diet in class i' (e.g., there is no relationship between the first low-salt diet and the first low-fat diet), we could use the model

$$Y_{ijk} \sim N(\theta + \alpha_i + \delta_{ij}, \sigma^2),$$

where θ represents the overall mean cholesterol level, α_i represents the effect due to the type of diet, and δ_{ij} represents the effect of the different diets of a particular type. For example, α_1 should be low if the low-salt diets lead to lower cholesterol than the other types of diets, and δ_{11} should be low if the first low-salt diet leads to lower cholesterol than the other low-salt diets. In effect, α_1 is measuring the effect of the low-salt diet compared to the other types of diets, and δ_{11} is measuring the effect of the first low-salt diet relative to the other low-salt diets. We call this model the *two-fold nested analysis-of-variance* model because the diet is "nested" in the diet type and there are two levels of nesting. Such a model, considered in Section 13.3, could be extended directly to a k-fold nested model.

We could also have models with both crossed and nested factors. Suppose, in the cholesterol example, that we have twelve diets of four types as mentioned in the last paragraph and that we also classify people by sex. Let Y_{ijkm} be the cholesterol level of the mth person on the kth diet of the jth diet type and the ith sex. Then a model for this experiment is

$$Y_{ijkm} \sim N(\theta + \alpha_i + \beta_j + \gamma_{ij} + \delta_{jk}, \sigma^2).$$

In this situation, we say that diet is nested in type of diet, but that diet type is crossed with sex.

We say that an analysis-of-variance model is *balanced* if each possible cell has the same number of observations. For example, the two-way model discussed earlier is balanced if there is the same number of people in every possible combination of diet and smoking. There are sound philosophical reasons why we should try to design balanced experiments. However, often observations are lost (e.g., people on various diets may drop out of the study). Therefore, we must often analyze unbalanced models. Since it is much more difficult to present the theory for unbalanced models, in all the models we consider in this chapter, except the one-way model, we shall assume that the model is balanced. In Section 13.5, we present the regression approach to analysis of variance, which can be used for unbalanced models.

All the models we consider in this chapter are examples of what we call *orthogonal designs*. In Section 13.1 we state, without proof, the results for the one-way model, and then we state the basic results for general orthogonal designs and show that the results for the one-way model are special cases of the more general results. In Sections 13.2 and 13.3, we apply these more general results to various balanced higher way models. Section 13.4 gives the basic result about Scheffé simultaneous confidence intervals and indicates how they can be used as a follow-up to the tests derived in earlier sections. In Section 13.5, we derive the results for orthogonal designs. Although these results are fairly easy to state and apply without matrices, their derivations use matrices. Finally, we present the regression approach to analysis-of-variance models, which can be used to analyze unbalanced analysis-of-variance models.

Once the basic results on orthogonal designs are understood, the analysis-of-variance table and the Scheffé simultaneous confidence intervals can be written down for most balanced ANOVA models nearly immediately. (In fact, the basic results are just a formalization of a heuristic method that is sometimes used to remember many of the formulas pertaining to ANOVA models.)

In using the analysis-of-variance models described in this chapter, we make the following assumptions:

1. *The means have the correct structure*. In the one-way model described earlier, in which we are classifying the individuals only by their diet, we must assume that smoking and sex have no effect on cholesterol level. If this assumption is not satisfied, then the procedures we derive are not correct. In principle, if we are not sure whether the assumption is satisfied, we should add more terms. In practice, however, this is many times not possible. In this chapter, when we have not included enough terms in the model, the procedures derived are conservative, in that the test statistics are smaller than they should be. In this situation, the true size of the test is smaller than the nominal size, .05, that we use.
2. *The variances are equal*. If the model is balanced, the procedures are not too sensitive to this assumption. This result is analogous to the result we derived for the two-sample model which showed that if the sample sizes are equal, then the two-sample t-test is asymptotically insensitive to the assumption of equal variances. (See Scheffé (1959), pp. 351–358, for further discussion.)
3. *The observations are independent*. The procedures are fairly sensitive to this assumption. Fortunately, in practice, we can often design the experiment in such a way that we can be comfortable with the assumption. However, in many experiments, we take several measurements on the same individual under different conditions. In such cases, it is not possible to assume that the obser-

vations are independent. To study such experiments, we use repeated measures models. (See Arnold (1981), pp. 209–242, 342–343, and 374–378 for further discussion.)

4. *The observations are normally distributed.* The procedures are asymptotically insensitive to this assumption for balanced experiments. (See Arnold (1981), pp. 141–158 for more details.) In addition, other procedures have been derived for some ANOVA models which do not depend on this assumption (although they do depend on the other three assumptions). (Some of these procedures are discussed in Hettmansperger (1984), Chapter 4.)

13.1 BASIC RESULTS

13.1.1 The One-Way Analysis-of-Variance Model

Consider the model in which we observe Y_{ij} independent, with

$$Y_{ij} \sim N(\mu_i, \sigma^2), \qquad j = 1, \ldots, n_i, \; i = 1, \ldots, k.$$

Note that, for each i,

$$Y_{i1}, \ldots, Y_{in_i}$$

are a sample from a normal distribution with mean μ_i and that the k samples are independent. Since the observations have been classified on one subscript (i), we call this model the *one-way analysis-of-variance* model. (We could also call the model the k-sample normal model.) When $k = 2$, the model is the same as the two-sample normal model with equal variances considered in Section 11.2.2, so that the results obtained here can be considered extensions of the results of that section.

We shall be concerned with testing that

$$\mu_1 = \mu_2 = \cdots = \mu_k$$

against $\mu_i \neq \mu_j$ for some i and j. In this section, we shall state procedures. In the next, we shall show that the results of this section can be derived from general results about orthogonal designs. These general results are themselves derived in Section 13.4.

We first consider testing that the μ_i are equal. Let N be the total number of observations, let $\overline{Y}_i$ be the average of the observations in the ith sample, and let $\overline{Y}$ be the average of all the observations. That is, let

$$N = \sum n_i, \qquad \overline{Y}_i = \frac{\sum_j Y_{ij}}{n_i}, \qquad \overline{Y} = \frac{\sum_i \sum_j Y_{ij}}{N} = \frac{\sum_i n_i \overline{Y}_i}{N}.$$

(Note that $\overline{Y}$ is not the average of the $\overline{Y}_i$, unless the n_i are equal.) Finally, let

$$T^2 = \sum_i n_i(\overline{Y}_i - \overline{Y})^2 = \sum_i n_i \overline{Y}_i^2 - N\overline{Y}^2, \qquad S^2 = \sum_i \sum_j (Y_{ij} - \overline{Y}_i)^2 = \sum_i \sum_j Y_{ij}^2 - \sum_i n_i \overline{Y}_i^2,$$

$$F = \frac{T^2/(k-1)}{S^2/(N-k)}.$$

If the μ_i are all equal, then

$$F \sim F_{k-1, N-k}.$$

(A derivation of this relation, using only properties of the one-sample normal model, is given in Exercise C1 for the balanced case.) The likelihood ratio test for testing that $\mu_i = \mu_j$ for all i and j against $\mu_i \neq \mu_j$ for some i and j rejects the null hypothesis if

$$F > F_{k-1, N-k}.$$

(See Exercise B4) Now, let $\overline{\mu} = \Sigma n_i \mu_i / N$. In Exercise B5, it is shown that

$$\frac{ET^2}{k-1} = \sigma^2 + \frac{\sum n_i(\mu_i - \overline{\mu})^2}{k-1} \quad \text{and} \quad \frac{ES^2}{n-p} = \sigma^2,$$

so it is not surprising that we should reject the hypothesis of equal means when F is too large.

Example 13–1

Suppose $k = 3$, $n_1 = 2$, $n_2 = 2$, and $n_3 = 3$. Suppose also that we observe 8 and 10 in the first sample, 3 and 5 in the second sample, and 2, 3, and 4 in the third sample. For these data, $N = 2 + 2 + 3 = 7$, $\overline{Y}_1 = 9$, $\overline{Y}_2 = 4$, and $\overline{Y}_3 = 3$. Therefore,

$$\overline{Y} = \frac{2(9) + 2(4) + 3(3)}{7} = 5, \qquad N\overline{Y}^2 = 175,$$

$$\sum\sum Y_{ij}^2 = 8^2 + 10^2 + 3^2 + 5^2 + 2^2 + 3^2 + 4^2 = 227,$$

$$\sum n_i \overline{Y}_i^2 = 2(9^2) + 2(4^2) + 3(3^2) = 221,$$

$$T^2 = \sum n_i \overline{Y}_i^2 - N\overline{Y}^2 = 221 - 175 = 46 \quad \text{and} \quad S^2 = \sum\sum Y_{ij}^2 - \sum n_i \overline{Y}_i^2 = 227 - 221 = 6.$$

Hence, for these data,

$$F = \frac{(4)(46)}{(2)(6)} = 15.33 > 6.94 = F_{2,4}^{.05},$$

so we can reject the hypothesis that the μ_i are equal with a size-.05 test.

We conclude the section with some notation which is useful in more complicated ANOVA models. Suppose we have some numbers a_{ij}. Then by $\overline{a}_{i.}$ we mean the average of all the a_{ij} whose first subscript is i, by $\overline{a}_{.j}$ we mean the average of all the a_{ij} whose second subscript is j, and by $\overline{a}_{..}$ we mean the average of all the a_{ij}. For the one-way ANOVA model given, $\overline{Y}_{i.}$ is what we have called $\overline{Y}_i$, and $\overline{Y}_{..}$ is what we have called $\overline{Y}$. From the definition of $\overline{Y}$, we see that $\overline{Y}_{..}$ is the average of all the Y_{ij}, not the average of the $\overline{Y}_{i.}$. Similarly, if we have c_{ijk}, then $\overline{c}_{i.k}$ is the average of all the c_{ijk} whose first component is i and third component is k, $\overline{c}_{.j.}$ is the average of all the c_{ijk} whose second component is j, etc.

Exercises—A

In this chapter, it is important not to round off the means when computing the sums of squares, because the computational formulas we are using are quite sensitive to round-off errors.

1. An experiment was conducted to determine the effect of alcohol on reaction time. Four students were given no alcohol, six were given two drinks, and eight were given four drinks. The students' reaction times to a visual stimulus were then measured, yielding the following data:

No drinks	4	5	3	7				
2 drinks	6	4	7	5	8	9		
4 drinks	8	12	6	11	7	13	14	9

Using these data with a .05 test, test that there is no significant difference in reaction times due to the number of drinks.

Exercises—B

1. Verify that $\Sigma n_i(\overline{Y}_i - \overline{Y})^2 = \Sigma n_i \overline{Y}_i^2 - N\overline{Y}^2$ and $\Sigma\Sigma(Y_{ij} - \overline{Y}_i)^2 = \Sigma\Sigma Y_{ij}^2 - \Sigma n_i \overline{Y}_i^2$.
2. (a) Show that when $k = 2$, $F = t^2$, where t is the two-sample t-statistic for testing the equality of means.
   (b) Show that when $k = 2$, the F-test of the equality of means rejects the null hypothesis if and only if the two-sided, two-sample t-test rejects it also.
3. For the one-way ANOVA model, show that
   (a) $(\overline{Y}_1, \dots, \overline{Y}_k, S^2)$ is a complete sufficient statistic.
   (b) $\overline{Y}_i$ and $S^2/(N-k)$ are the best unbiased estimators of μ_i and σ^2, respectively.
   (c) $(\overline{Y}_1, \dots, \overline{Y}_k, \hat{\sigma}^2)$ is the MLE of $(\mu_1, \dots, \mu_k, \sigma^2)$, where $\hat{\sigma}^2 = S^2/N$.
4. Consider testing that the μ_i are equal.
   (a) Show that $L(\overline{Y}_1, \dots, \overline{Y}_k, \hat{\sigma}^2) = (2\pi)^{-N/2}(\hat{\sigma}^2)^{-N/2}\exp(-N/2)$, where $\hat{\sigma}^2 = S^2/N$.
   (b) Show that the MLE under the null hypothesis is $\hat{\hat{\mu}}_i = \overline{Y}$, $\hat{\hat{\sigma}}^2 = \Sigma\Sigma(Y_{ij} - \overline{Y})^2/N$.
   (c) Show that $\hat{\hat{\sigma}}^2 = \hat{\sigma}^2 + T^2/N$.
   (d) Show that the likelihood ratio test statistic for this problem is $\Lambda = [1 + (k-1)F/(N-k)]^{N/2}$ and that the likelihood ratio test rejects the null hypothesis if $F > q$ for some q.
5. (a) Let $\overline{\mu} = \Sigma n_i \mu_i/N$. Show that $\overline{Y}_i \sim N(\mu_i, \sigma^2)$ and $\overline{Y} \sim N(\overline{\mu}, \sigma^2/N)$.
   (b) Show that $ET^2 = (k-1)\sigma^2 + \Sigma n_i(\mu_i - \overline{\mu})^2$ and $ES^2 = (n-p)\sigma^2$.

Exercises—C

1. Consider the one-way model when $n_i = n$ and $N = nk$. Let $V_i^2 = \Sigma_j(Y_{ij} - \overline{Y}_i)^2$.
   (a) Show that $V_i^2/\sigma^2 \sim \chi^2_{n-1}$ and are independent. (Note that V_i^2 is the sample variance of $(Y_{i1}, \dots, Y_{in_i})$.)
   (b) Show that $S^2/\sigma^2 = \Sigma V_i^2/\sigma^2 \sim \chi^2_{k(n-1)}$.
   (c) Show that $\overline{Y}_i \sim N(\mu_i, \sigma^2/n)$.
   (d) Show that if $\mu_i = \mu$, then $T^2/\sigma^2 \sim \chi^2_{k-1}$. (Note that $(k-1)T^2/n$ is the sample variance of the $\overline{Y}_{i\cdot}$.)

(e) Show that S^2 and T^2 are independent and, hence, that $F \sim F_{k-1,k(n-1)}$. (Note that $\overline{Y}_i$ and V_i^2 are independent. Why?)

13.1.2 Orthogonal Designs

In this section, we state the basic results on orthogonal designs and apply them to the one-way analysis-of-variance model. We use the letter q as a substitute for all the subscripts in the model. Thus, a sum over q is a sum over all the subscripts in the model. (In the one-way model, q stands for (i,j).)

Consider a model in which we observe N independent random variables Y_q, with

$$Y_q \sim N(\delta_q^{(1)} + \cdots + \delta_q^{(p)}, \sigma^2),$$

where the $\delta_q^{(i)}$ are unknown parameters. If

$$\sum_q \delta_q^{(i)} \delta_q^{(j)} = 0, \qquad i \neq j,$$

the model is an *orthogonal design* with p effects. As in previous chapters, we let $\boldsymbol{\delta}^{(i)}$ be the vector of $\delta_q^{(i)}$ (in any order). In orthogonal designs, we are primarily interested in testing that the $\boldsymbol{\delta}^{(1)} = 0$, in testing that the $\boldsymbol{\delta}^{(2)} = 0$, etc. We call $\boldsymbol{\delta}^{(i)}$ the *ith effect*.

The *ordinary least squares estimators* $\hat{\delta}_q^{(i)}$ of the $\delta_q^{(i)}$ are found by minimizing

$$Q(\boldsymbol{\delta}^{(1)}, \ldots, \boldsymbol{\delta}^{(p)}) = \sum_q (Y_q - \delta_q^{(1)} - \delta_q^{(2)} - \cdots - \delta_q^{(p)})^2.$$

Let

$$SS_i = \sum_q (\hat{\delta}_q^{(i)})^2, \qquad i = 1, \ldots, p, \quad \text{and} \quad SSE = \sum_q (Y_q - \hat{\delta}_q^{(1)} - \cdots - \hat{\delta}_q^{(p)})^2.$$

SS_i and SSE are respectively called the *sum of squares for the ith effect* and the *sum of squares for error*. Also,

$$SST = \sum_q Y_q^2$$

is called the *total sum of squares for the model*. In Section 13.5, it is shown that

$$SST = SS_1 + SS_2 + \cdots + SS_p + SSE$$

(i.e., that the sums of squares for the effects and the sum of squares for error add up to the total sum of squares). This result implies that

$$SSE = SST - \sum_i SS_i.$$

For $i = 1, \ldots, p$, let df_i be the number of independent parameters in $\boldsymbol{\delta}^{(i)}$, and let

$$dfe = N - \sum_i df_i.$$

The df_i and dfe are called the *degrees of freedom for the ith effect* and the *degrees of freedom for error*, respectively. Finally, let

$$MS_i = \frac{SS_i}{df_i}, \qquad MSE = \frac{SSE}{dfe}, \qquad F_i = \frac{MS_i}{MSE}.$$

MS_i and MSE are respectively called the *mean square for the ith effect* and the *mean square for error*, and F_i is called the *F-statistic for the ith effect*. (Note that there is no F-statistic for error.)

It can be shown that $(\hat{\boldsymbol{\delta}}^{(1)}, \ldots, \hat{\boldsymbol{\delta}}^{(p)}, MSE)$ is a complete sufficient statistic for the model, and $\hat{\boldsymbol{\delta}}^{(i)}$ and MSE are the best unbiased estimators of $\boldsymbol{\delta}^{(i)}$ and σ^2, and that $\hat{\boldsymbol{\delta}}^{(i)}$ and SSE/N are the MLEs of $\hat{\boldsymbol{\delta}}^{(i)}$ and σ^2, respectively. However, in analysis-of-variance models, there is not too much interest in point estimators of the parameters.

The following theorem is the main theorem for orthogonal designs and is proved in Section 13.5.1.

Theorem 13–1

a. SS_i and SSE are independent, and $SSE/\sigma^2 \sim \chi^2_{dfe}$. If $\boldsymbol{\delta}^i = \mathbf{0}$, then $SS_i/\sigma^2 \sim \chi^2_{df_i}$ and $F_i \sim F_{df_i, dfe}$.
b. $E(MSE) = \sigma^2$ and $E(MS_i) = \sigma^2 + \Sigma_q(\delta_q^{(i)})^2/df_i$.
c. The size-α likelihood ratio test rejects the null hypothesis if

$$F_i > F^{\alpha}_{df_i, dfe}.$$

Proof
See Theorems 13–6 and 13–7. □

Part b of Theorem 13–1 indicates why it is sensible to reject the null hypothesis if F_i is too large. If the null hypothesis is false and $\boldsymbol{\delta}^{(i)} \neq \mathbf{0}$, then we would expect the numerator (MS_i) of the F-statistic to be larger than the denominator (MSE).

The F-test given is unbiased and has other optimal properties. (See Arnold (1981), pp. 104–108.) It is UMP unbiased when $df_i = 1$. If $df_i > 1$, then there is no UMP unbiased test for this problem.

The results of the preceding calculations are often summarized in an ANOVA table in the following form:

Due to	df	SS	MS	F
$\delta^{(1)}$	df_1	SS_1	MS_1	F_1
$\delta^{(2)}$	df_2	SS_2	MS_2	F_2
.	.	.	.	.
.	.	.	.	.
.	.	.	.	.
$\delta^{(p)}$	df_p	SS_p	MS_p	F_p
Error	dfe	SSE	MSE	
Total	N	SST		

The MS column is found by taking the ratios of the associated entries in the SS and df columns, and the F column is found by taking the ratio of the entry in the MS

column to MSE. Therefore, when we give the formulas for ANOVA tables, we do not give the MS and F columns. When we do numerical examples, those columns are included. Note that the likelihood ratio test for testing that $\boldsymbol{\delta}^{(i)} = \mathbf{0}$ rejects the null hypothesis if F_i is too large, and the critical value is found in an F table with df_i and dfe degrees of freedom. F_i, df_i, and dfe can be read directly from the ANOVA table.

As an example of an orthogonal design, we consider the one-way analysis-of-variance model discussed in the previous section in which we observe Y_{ij} independent, with

$$Y_{ij} \sim N(\mu_i, \sigma^2), \qquad i = 1, \ldots, k, \quad j = 1, \ldots, n_i.$$

In order to apply Theorem 13–1 to this model, it is necessary to reparametrize it. Let

$$N = \sum_i n_i, \qquad \theta = \frac{\sum n_i \mu_i}{N}, \qquad \alpha_i = \mu_i - \theta, \tag{13–1}$$

so that

$$\mu_i = \theta + \alpha_i \quad \text{and} \quad \sum n_i \alpha_i = 0.$$

Then the μ_i are equal if and only if the α_i are zero. Now, let

$$\delta_{ij}^{(1)} = \theta \quad \text{and} \quad \delta_{ij}^{(2)} = \alpha_i.$$

Then

$$Y_{ij} \sim N(\delta_{ij}^{(1)} + \delta_{ij}^{(2)}, \sigma^2).$$

To verify that this model is an orthogonal design, we need to verify that

$$\sum_{i,j}\sum \delta_{ij}^{(1)} \delta_{ij}^{(2)} = 0.$$

In fact,

$$\sum_{i,j}\sum \delta_{ij}^{(1)} \delta_{ij}^{(2)} = \sum_{i=1}^{k} \sum_{j=1}^{n_i} \theta\alpha_i = \theta \sum_{i=1}^{k} n_i \alpha_i = 0.$$

(Note that the constraint $\Sigma n_i \alpha_i = 0$ is the only one that makes this model an orthogonal design.) Therefore, the model is an orthogonal design with $p = 2$ effects: $\delta_{ij}^{(1)} = \theta$ and $\delta_{ij}^{(2)} = \alpha_i$.

To find the ordinary least squares estimators, we must minimize

$$Q(\theta, \alpha_1, \ldots, \alpha_k) = \sum_{i=1}^{k} \sum_{j=1}^{n_i} (Y_{ij} - \delta_{ij}^{(1)} - \delta_{1j}^{(2)})^2 = \sum_{i=1}^{k} \sum_{j=1}^{n_i} (Y_{ij} - \theta - \alpha_i)^2.$$

If we differentiate Q with respect to θ and set the result equal to zero, we get

$$0 = \sum_{i=1}^{k} \sum_{j=1}^{n_i} (Y_{ij} - \theta - \alpha_i)(-2) = -2\left(\sum_{i=1}^{k} \sum_{j=1}^{n_i} Y_{ij} - \theta N - \sum_{i=1}^{k} n_i \alpha_i\right) = -2(N\overline{Y} - N\theta - 0)$$

(using the constraint on the α_i). Therefore, the ordinary least squares estimator for $\delta_{ij}^{(1)} = \theta$ is

$$\hat{\delta}_{ij}^{(1)} = \hat{\theta} = \overline{Y}_{..}.$$

If we now differentiate Q with respect to α_1 and set the result equal to zero, we obtain

$$0 = \sum_{j=1}^{n_1} (Y_{1j} - \theta - \alpha_1)(-2) = -2\left(\sum_{j=1}^{n_1} Y_{1j} - n_1\theta - n_1\alpha_1\right) = -2(n_1\overline{Y}_{1.} - n_1\theta - n_1\alpha_1)$$

(since the terms with $i > 1$ do not involve α_1). Therefore, the ordinary least squares estimator of $\delta_{1j}^{(2)} = \alpha_1$ is

$$\hat{\delta}_{1j}^{(2)} = \hat{\alpha}_1 = \overline{Y}_{1.} - \hat{\theta} = \overline{Y}_{1.} - \overline{Y}_{..}.$$

Similarly, $\hat{\delta}_{ij}^{(2)} = \hat{\alpha}_i = \overline{Y}_{i.} - \overline{Y}_{..}$. It is then easily verified that $\Sigma n_i\hat{\alpha}_i = 0$, so that the $\hat{\delta}_{ij}^{(2)} = \hat{\alpha}_i$ are ordinary least squares estimators of the α_i. (In later sections, we shall leave it to the reader to check that the estimators satisfy the constraints.) Consequently,

$$SS(\theta) = \sum_{i=1}^{k}\sum_{j=1}^{n_i} \hat{\theta}^2 = N\hat{\theta}^2 = N\overline{Y}_{..}^2$$

and

$$SS(\alpha) = \sum_{i=1}^{k}\sum_{j=1}^{n_i} \alpha_i^2 = \sum_{i=1}^{k} n_i\hat{\alpha}_i^2 = \sum_{i=1}^{k} n_i(\overline{Y}_{i.} - \overline{Y}_{..})^2 = \sum_{i=1}^{k} n_i\overline{Y}_{i.}^2 - N\overline{Y}_{..}^2$$

(where $SS(\theta)$ and $SS(\alpha)$ are what we have respectively called SS_1 and SS_2). In addition,

$$SS(\theta) + SS(\alpha) = \sum_i n_i\overline{Y}_{i.}^2 \quad \text{and} \quad SST = \sum_i\sum_j Y_{ij}^2.$$

Therefore,

$$SSE = \sum_i\sum_j Y_{ij}^2 - \sum_i n_i\overline{Y}_{i.}^2 = \sum_i\sum_j (Y_{ij} - \overline{Y}_{i.})^2.$$

The vector $\boldsymbol{\delta}^{(1)}$ has one parameter (θ), so that $df_1 = df(\theta) = 1$. $\boldsymbol{\delta}^{(2)}$ has k parameters (the α_i), but they are not linearly independent, since $\Sigma n_i\alpha_i = 0$. Since there are k parameters and one constraint, $\boldsymbol{\delta}^{(2)}$ has $k - 1$ linearly independent parameters, and hence, $df_2 = df(\alpha) = k - 1$. Finally,

$$dfe = N - (k - 1) - 1 = N - k.$$

Therefore, the table for the one-way ANOVA model is as follows:

Due to	df	SS
θ	1	$N\overline{Y}_{..}^2$
α	$k - 1$	$\Sigma n_i\overline{Y}_{i.}^2 - N\overline{Y}_{..}^2$
Error	$N - k$	$\Sigma\Sigma Y_{ij}^2 - \Sigma n_i\overline{Y}_{i.}^2$
Total	N	$\Sigma\Sigma Y_{ij}^2$

Note that T^2 and S^2 of the preview section are given here by

$$T^2 = SS(\alpha) \quad \text{and} \quad S^2 = SSE.$$

Hence, the F-test given here is the same as the F-test given in that section.

Example 13–2

Consider again Example 13–1, in which $k = 3$, $n_1 = 2$, $n_2 = 2$, $n_3 = 3$, $N = 7$, and

$$\overline{Y}_{1.} = 9, \quad \overline{Y}_{2.} = 4, \quad \overline{Y}_{3.} = 3, \quad \overline{Y}_{..} = 5, \quad SS(\theta) = 7(5)^2 = 175,$$
$$SS(\alpha) = T^2 = 46, \quad SSE = S^2 = 6.$$

We get the following ANOVA table for this model:

Due to	*df*	*SS*	*MS*	*F*
θ	1	175	175	116.67
α	2	46	23	15.33
Error	4	6	1.5	
Total	7	227		

(Recall that the *MS* column is computed by dividing *SS* for the effect by *df* for the effect, and the *F* column is computed by taking *MS* for the effect over *MSE*.)

We can test the hypothesis that $\theta = 0$ against $\theta \neq 0$ (although in practice we are rarely interested in inferences about the overall mean θ). The size-.05 test for this hypothesis rejects the hypothesis if

$$F(\boldsymbol{\theta}) = 116.67 > F_{1,4}^{.05} = 7.71.$$

Therefore, we reject the hypothesis. (Recall that the degrees of freedom for the numerator is the entry in the *df* column for the effect, while the degrees of freedom for the denominator is the entry in the *df* column for error.) Similarly, to test that the $\alpha_i = 0$ against the alternative that they are not all zero, we reject the null hypothesis if

$$F(\boldsymbol{\alpha}) = 15.33 > F_{2,4}^{.05} = 6.94.$$

So we also reject this hypothesis. (Note that this is the same test for this problem as in the previous section.) All the information we need for testing these hypotheses is contained in the ANOVA table.

In many ANOVA models, the first effect is an overall mean effect $\delta_{ij}^{(1)} = \theta$, as it is in the foregoing one-way model. If the model is an orthogonal design, then the least squares estimator of θ is $\hat{\theta} = \overline{Y}$, the average of all the observations, and $SS_1 = N\overline{Y}^2$ (see Exercise C2). Using the fact that the total sum of squares is the sum of the sums of squares for the effects, we see that

$$SST - N\overline{Y}^2 = SS_2 + \cdots + SS_p + SSE.$$

Therefore, in making an ANOVA table, we have "analyzed" the total variance in the observations ($SST - N\overline{Y}^2 = \Sigma_q(Y_q - \overline{Y})^2$) into pieces $SS_2, \ldots, SS_p$, and SSE, each representing a different effect. This may be the source of the name "analysis of variance" for these models. In models with an overall mean effect, the first row of the ANOVA table is often eliminated, and $SST - N\overline{Y}^2$ is called the total sum of squares. In this text, however, we shall always include a row for the overall mean if there is one in the model.

From the derivations given for the one-way ANOVA model, it is clear that it is often easy to calculate the ANOVA table for orthogonal designs. It is somewhat difficult to write down precisely the method we use, but we shall now give some rules which have been suggested by this model. We assume that we have N independent observations with equal variance σ^2 and that the means for the observations can be written as the sum of p different "effects" (e.g., the two effects θ and α_i). The rules are as follows:

1. Verify that the model is an orthogonal design. To do this, for each pair of effects, we must sum the products of the effects over all the subscripts in the model and verify that this sum is zero (e.g., verify that $\Sigma_i \Sigma_j \theta\alpha_i = 0$).
2. Find the ordinary least squares estimators of all the effects by minimizing the sum of the squares of the observations minus their means (e.g., minimize $\Sigma\Sigma(Y_{ij} - \theta - \alpha_i)^2$). These estimators can usually be determined fairly easily if we find the estimators for the parameters with fewer subscripts first. (If we do the minimization in another order, we often must use Lagrange multipliers or some such technique to assure that the estimators satisfy the constraints.)
3. Find the sum of squares for a particular effect by summing the squares of the ordinary least squares estimators for that effect over all the subscripts in the model (e.g., $SS(\boldsymbol{\alpha}) = \Sigma\Sigma\hat{\alpha}_i^2$).
4. Find the degrees of freedom for a particular effect by counting the number of linearly independent parameters contained in that effect (e.g., $k - 1$ linearly independent α_i).
5. Find SSE by summing the squares of the deviations of the observations from their estimated means (e.g., $SSE = \Sigma\Sigma(Y_{ij} - \hat{\theta} - \hat{\alpha}_i)^2$), or by subtracting the other sums of squares from the total sum of squares (e.g., $SSE = \Sigma\Sigma Y_{ij}^2 - SS(\boldsymbol{\theta}) - SS(\boldsymbol{\alpha})$).
6. Find the degrees of freedom for error by subtracting the degrees of freedom for the various effects from N, the total number of observations (e.g., $dfe = N - df(\boldsymbol{\theta}) - df(\boldsymbol{\alpha})$).

In later sections, we shall use these informal rules to derive ANOVA tables. In using the rules, it is very important that the sums always be over all the subscripts in the model, not just those occurring in the effect. (For example, $\Sigma\hat{\alpha}_i^2$ is *not* the sum of squares for α for the one-way model.)

It should be emphasized that the method suggested here for deriving ANOVA tables is valid only for orthogonal designs. (It is nontrivial to verify that it is even valid for such designs; see Section 13.5.1.) For nonorthogonal designs, the analysis is still more difficult. In particular, the sums of squares do not add up to the total sum of squares unless the design is orthogonal.

Exercises—A

1. **(a)** Write down the ANOVA table for the data in Exercise A1 of the last section.
   (b) Do a size-.05 test that $\theta = 0$ and a size-.01 test that $\alpha_i = 0$ for these data.
2. An experiment is performed to compare three drugs for treating blood pressure. Three people are given drug A, three are given drug B, and three are given drug C. After

several months, their blood pressures are recorded, as well as how many packs of cigarettes they smoke a day. The data are as follows (the numbers in parentheses are the number of packs of cigarettes smoked per day):

A	120 (0)	150 (1)	180 (2)
B	100 (0)	150 (1)	140 (2)
C	80 (0)	100 (1)	150 (2)

(For example, the second patient in each treatment smoked one pack of cigarettes a day.) Use the analysis-of-covariance model in exercise C1 (with x_{ij} as the number of packs per day and Y_{ij} as the blood pressure) to answer the following questions:

(a) Is there any significant difference between the medicines with $\alpha = .05$?

(b) Is there any significant effect due to smoking ($\lambda \neq 0$) for these data with a .05 test?

Exercises—B

1. Show that $\Sigma n_i \alpha_i = 0$, where the α_i are defined as in Equation (13–1).

Exercises—C

1. Consider the model in which we observe Y_{ij} independent, with $Y_{ij} \sim N(\theta + \alpha_i + \lambda(x_{ij} - \bar{x}_{i.}), \sigma^2), i = 1, \ldots, k, j = 1, \ldots, n$, where the x_{ij} are known constants and θ, α_i, λ, and σ^2 are all unknown parameters such that $\Sigma\alpha_i = 0$.
   (a) Show that this model is an orthogonal design with $\delta_{ij}^{(1)} = \theta$, $\delta_{ij}^{(2)} = \alpha_i$, and $\delta_{ij}^{(3)} = \lambda(x_{ij} - \bar{x}_{i.})$. (That is, show that $\Sigma\Sigma\theta\alpha_i = 0$, $\Sigma\Sigma\theta\lambda(x_{ij} - \bar{x}_{i.}) = 0$, and $\Sigma\Sigma\alpha_i\lambda(x_{ij} - \bar{x}_{i.}) = 0$.)
   (b) What is the ANOVA table for this model? ($SS(\theta)$, $df(\theta)$, $SS(\alpha)$, and $df(\alpha)$ are the same as for the one-way model, $\hat{\lambda} = \Sigma\Sigma(x_{ij} - \bar{x}_{i.})Y_{ij}/\Sigma\Sigma(x_{ij} - \bar{x}_{i.})^2$, $SS(\lambda) = \hat{\lambda}^2\Sigma\Sigma(x_{ij} - \bar{x}_{i.})^2$, and the number of degrees of freedom for λ is one. Why? Find SSE and dfe by subtraction.)

   (This model is a one-way *analysis-of-covariance* model, and the x_{ij} are called *covariates*.)
2. Suppose that we have an orthogonal design in which $\delta_q^{(1)} = \theta$.
   (a) Show that $\hat{\theta} = \bar{Y} = \Sigma_q Y_q/N$. (*Hint*: Since the $\boldsymbol{\delta}^{(i)}$ are orthogonal to $\boldsymbol{\delta}^{(1)}$, $\Sigma_q \delta_q^{(i)} = 0$ for $i > 1$.)
   (b) Show that $SS_1 = N\bar{Y}^2$.

13.2 CROSSED TWO-WAY ANOVA MODELS

13.2.1 Crossed Two-Way Analysis of Variance with No Replication

In this section, we consider the model in which we observe Y_{ij} independent, with

$$Y_{ij} \sim N(\mu_{ij}, \sigma^2), \qquad i = 1, \ldots, r, \quad j = 1, \ldots, c.$$

We think of i as representing a row of a matrix and j as representing a column. For this model, we are allowing the mean to depend on both the row and column. We call such a model a *crossed two-way analysis-of-variance* model, since we have a measurement for each possible pair (i, j).

If we make no assumption about the μ_{ij}, then SSE and dfe both turn out to be zero. Therefore, we assume that the μ_{ij} satisfy the additivity assumption

$$\mu_{ij} = \gamma_i + \xi_j.$$

That is, we assume that the mean of the observation in the ith row and the jth column is a sum of an effect from the ith row (γ_i) and an effect from the jth column (ξ_j). Such a model is called an *additive two-way analysis-of-variance* model.

In order to use the results on orthogonal designs, we must reparametrize the model. Let

$$\theta = \overline{\gamma} + \overline{\xi}, \qquad \alpha_i = \gamma_i - \overline{\gamma}, \qquad \beta_j = \xi_j - \overline{\xi}.$$

Then

$$\mu_{ij} = \theta + \alpha_i + \beta_j, \qquad \sum_i \alpha_i = 0, \qquad \sum_j \beta_j = 0.$$

We call θ the *overall mean*, the α_i the *row effects*, and the β_j the *column effects*. We shall derive tests for testing that the $\alpha_i = 0$ and that the $\beta_j = 0$. Testing that the $\alpha_i = 0$ is the same as testing that the μ_{ij} do not depend on i, i.e., that the rows have no effect on the distribution of the Y_{ij}. Similarly, $\beta_j = 0$ means that the μ_{ij} do not depend on j.

To show that the model is an orthogonal design, let $N = rc$, the total number of observations. We have three effects:

$$\delta_{ij}^{(1)} = \theta, \qquad \delta_{ij}^{(2)} = \alpha_i, \qquad \delta_{ij}^{(3)} = \beta_j.$$

The Y_{ij} are independent, with $Y_{ij} \sim N(\theta + \alpha_i + \beta_j, \sigma^2)$. To verify that the model is an orthogonal design, we must show that $\Sigma_i \Sigma_j \theta\alpha_i = 0$, $\Sigma_i \Sigma_j \theta\beta_j = 0$, and $\Sigma_i \Sigma_j \alpha_i \beta_j = 0$. But plainly,

$$\sum_i \sum_j \theta\beta_j = \theta r \sum_j \beta_j = 0 \quad \text{and} \quad \sum_i \sum_j \alpha_i \beta_j = \Big(\sum_i \alpha_i\Big)\Big(\sum_j \beta_j\Big) = 0.$$

The third equality is left as an exercise.

We now find the ordinary least squares estimators for the model. We must minimize

$$Q = \sum_i \sum_j (Y_{ij} - \theta - \alpha_i - \beta_j)^2.$$

Differentiating Q with respect to θ and setting the result equal to zero, we get

$$\begin{aligned} 0 &= -2\sum_i \sum_j (Y_{ij} - \theta - \alpha_i - \beta_j) \\ &= -2\Big(\sum_i \sum_j Y_{ij} - rc\theta - c\sum_i \alpha_i - r\sum_j \beta_j\Big) = -2rc(\overline{Y}_{..} - \theta - 0 - 0), \end{aligned}$$

and hence, $\hat{\theta} = \overline{Y}_{..}$. Differentiating Q with respect to α_i this time, and again setting the result equal to zero, we get

$$0 = -2\sum_j (Y_{ij} - \theta - \alpha_i - \beta_j) = -2c(\overline{Y}_{i.} - \theta - \alpha_i - 0)$$

(using the constraint on the β_j's). Therefore, $\hat{\alpha}_i = \overline{Y}_{i.} - \hat{\theta} = \overline{Y}_{i.} - \overline{Y}_{..}$, and similarly, $\hat{\beta}_j = \overline{Y}_{.j} - \overline{Y}_{..}$. Hence,

$$SS(\theta) = \sum_i \sum_j \hat{\theta}^2 = \sum_i \sum_j \overline{Y}^2_{..} = rc\overline{Y}^2_{..},$$

$$SS(\alpha) = \sum_i \sum_j \hat{\alpha}_i^2 = c\sum_i (\overline{Y}_{i.} - \overline{Y}_{..})^2 = c\sum_i \overline{Y}^2_{i.} - rc\overline{Y}^2_{..},$$

$$SS(\beta) = \sum_i \sum_j \hat{\beta}_j^2 = r\sum_j (\overline{Y}_{.j} - \overline{Y}_{..})^2 = r\sum_j \overline{Y}^2_{.j} - rc\overline{Y}^2_{..},$$

and

$$SSE = \sum_i \sum_j (Y_{ij} - \hat{\theta} - \hat{\alpha}_i - \hat{\beta}_j)^2 = \sum_i \sum_j (Y_{ij} - \overline{Y}_{i.} - \overline{Y}_{.j} + \overline{Y}_{..})^2$$

$$= \sum_i \sum_j Y^2_{ij} - c\sum_i \overline{Y}^2_{i.} - r\sum_j \overline{Y}^2_{.j} + rc\overline{Y}^2_{..}.$$

Now, there is only one parameter (θ) in θ, so that $df(\theta) = 1$. Also, there are $r - 1$ linearly independent α_i's, so that $df(\alpha) = r - 1$. And similarly, $df(\beta) = c - 1$. Finally,

$$dfe = N - df(\theta) - df(\alpha) - df(\beta) = rc - 1 - (r - 1) - (c - 1) = (r - 1)(c - 1).$$

Therefore, the table for the additive crossed two-way analysis-of-variance model is

Due to	*df*	*SS*
θ	1	$rc\overline{Y}^2_{..}$
α	$r - 1$	$c\Sigma\overline{Y}^2_{i.} - rc\overline{Y}^2_{..}$
β	$c - 1$	$r\Sigma\overline{Y}^2_{.j} - rc\overline{Y}^2_{..}$
Error	$(r - 1)(c - 1)$	$\Sigma\Sigma Y^2_{ij} - r\Sigma\overline{Y}^2_{.j} - c\Sigma\overline{Y}^2_{i.} + rc\overline{Y}^2_{..}$
Total	rc	$\Sigma\Sigma^2_{ij}$

Example 13–3

Suppose we have a two-way model with $r = 2$ and $c = 3$. Suppose also that we observe the following data:

		j	
	1	4	7
i	9	8	13

Then $\overline{Y}_{1.} = (1 + 4 + 7)/3 = 4$, $\overline{Y}_{2.} = 10$, $\overline{Y}_{.1} = (1 + 9)/2 = 5$, $\overline{Y}_{.2} = 6$, and $\overline{Y}_{.3} = 10$. Finally, $\overline{Y}_{..} = (1 + 4 + 7 + 9 + 8 + 13)/6 = 7$. Therefore,

$$T_1^2 = \sum\sum Y^2_{ij} = (1^2 + 4^2 + 7^2 + 9^2 + 8^2 + 13^2) = 380, \qquad T_2^2 = c\sum \overline{Y}^2_{i.} = 3(4^2 + 10^2) = 348,$$

$$T_3^2 = r\sum \overline{Y}^2_{.j} = 2(5^2 + 6^2 + 10^2) = 322, \qquad T_4^2 = rc\overline{Y}^2_{..} = 6(7^2) = 294.$$

Hence,

$$SS(\boldsymbol{\theta}) = T_4^2 = 294, \qquad SS(\boldsymbol{\alpha}) = T_2^2 - T_4^2 = 54, \qquad SS(\boldsymbol{\beta}) = T_3^2 - T_4^2 = 28,$$

$$SSE = T_1^2 - T_2^2 - T_3^2 + T_4^2 = 4.$$

Consequently, the ANOVA table for the given data is

Due to	*df*	*SS*	*MS*	*F*
θ	1	294	294	147
α	1	54	54	27
β	2	28	14	7
Error	2	4	2	
Total	6	380		

To test that the $\alpha_i = 0$ with a .05 test, we reject the null hypothesis if $F(\boldsymbol{\alpha}) > F_{1,2}^{;05} = 18.5$. Since $F(\boldsymbol{\alpha}) = 27$ for the data, we in fact reject the hypothesis. Similarly, to test that the $\beta_j = 0$, we reject the null hypothesis if $F(\boldsymbol{\beta}) > F_{2,2}^{;05} = 19.0$. Since $F(\boldsymbol{\beta}) = 7$, this time we accept the hypothesis.

Exercises—A

1. An experiment was conducted to compare the heat loss of four different makes of windows under each of three different temperature conditions. The following data were obtained:

	Type			
Temp	A	B	C	D
a	7	9	11	4
b	8	9	14	5
c	6	8	10	3

   (a) Find the ANOVA table for these data, and do a .05 test that there is no significant difference in heat loss for the four glass types.

   (b) Do a similar analysis comparing the heat losses for the three temperature conditions.

2. The data in Exercise A2 of the last section could also be analyzed as a two-way model (with treatments for rows and number of packs of cigarettes for columns). Using the two-way model in this section, answer the questions asked in Exercise A2 in the last section.

3. An experiment was done to assess the performance of two fertilizers, A and B, under two different temperature conditions, a and b, and two different moisture conditions, I and II (which are controlled in a laboratory). Each fertilizer was applied under each pair of temperature and moisture conditions, and the output from each plot was measured, yielding the following data:

	A			B	
	a	b		a	b
I	4	6	I	5	8
II	8	8	II	7	9

Use the three-way ANOVA model (Exercise B5) to answer the following questions with .05 tests:

(a) Is there any significant difference in the effects of the fertilizers?
(b) Is there any significant difference in the effects of the moisture?
(c) Is there any significant difference in the effects of the temperature?

Exercises—B

1. Show that $\Sigma_i \alpha_i = 0$, where $\alpha_i = \gamma_i - \overline{\gamma}$.
2. Show that $\Sigma_i \Sigma_j \theta\alpha_i = 0$.
3. Verify that $\hat{\beta}_j = \overline{Y}_{.j} - \overline{Y}_{..}$.
4. Verify that $\Sigma\Sigma(Y_{ij} - \overline{Y}_{i.} - \overline{Y}_{.j} + \overline{Y}_{..})^2 = \Sigma\Sigma Y_{ij}^2 - c\Sigma\overline{Y}_{i.}^2 - r\Sigma\overline{Y}_{.j}^2 + rc\overline{Y}_{..}^2$.
5. Consider a three-way additive model in which we observe Y_{ijk} independent, with $Y_{ijk} \sim N(\theta + \alpha_i + \beta_j + \delta_k, \sigma^2), i = 1, \ldots, r, j = 1, \ldots, c, k = 1, \ldots, d,$ where $\Sigma_i \alpha_i = 0$, $\Sigma_j \beta_j = 0$, and $\Sigma_k \delta_k = 0$.
   (a) Show that this model is an orthogonal design with $\delta_{ij}^{(1)} = \theta$, $\delta_{ij}^{(2)} = \alpha_i$, $\delta_{ij}^{(3)} = \beta_j$, and $\delta_{ij}^{(4)} = \delta_k$.
   (b) Find the ordinary least squares estimators of the parameters.
   (c) Find the ANOVA table for the model.
6. Consider the model presented in this section when $r = 2$. Let $U_j = Y_{1j} - Y_{2j}$, let $\overline{U}$ and V^2 be the sample mean and sample variance, respectively, of the U_j, and let $t = c^{1/2}(\overline{U})/V$. Show that $F(\alpha) = t^2$. (This implies that when $r = 2$, the F-test for row effects is the same as the two-sided paired t-test given in Section 11.2.4.)

13.2.2 Crossed Two-Way Analysis of Variance with Interaction

Consider the model in which we observe Y_{ijk} independent, with

$$Y_{ijk} \sim N(\mu_{ij}, \sigma^2), \qquad i = 1, \ldots, r, \qquad j = 1, \ldots, c, \qquad k = 1, \ldots, n.$$

We call i the row, j the column, and k the replication. (For fixed i and j, the Y_{ijk} are a sample from a normal distribution with mean μ_{ij}.) When $n = 1$, we have the model of the previous section. Here, we assume that $n > 1$, so that we have some replication. In this case, it is not necessary to make any assumption about the μ_{ij}.

To use the results on orthogonal designs, we reparametrize the model. Let

$$\theta = \overline{\mu}_{..}, \qquad \alpha_i = \overline{\mu}_{i.} - \overline{\mu}_{..}, \qquad \beta_j = \overline{\mu}_{.j} - \overline{\mu}_{..}, \qquad \gamma_{ij} = \mu_{ij} - \overline{\mu}_{i.} - \overline{\mu}_{.j} + \overline{\mu}_{..}.$$

Then

$$\mu_{ij} = \theta + \alpha_i + \beta_j + \gamma_{ij}, \qquad \sum_i \alpha_i = 0, \qquad \sum_j \beta_j = 0, \qquad \sum_i \gamma_{ij} = 0, \qquad \sum_j \gamma_{ij} = 0.$$

(That is, for each i, the sum on j of the γ_{ij} is zero, and for each j, the sum on i of the γ_{ij} is zero.) Note that in writing the model in this form, we are not making any assumption about the μ_{ij}; any set of μ_{ij} can be written this way. We call the α_i the *row effects*, the β_j the *column effects*, and the γ_{ij} the *interactions*. The model is known as the (crossed) *two-way analysis-of-variance-with-interaction* model and is *balanced*, since we have the same number n of measurements for each possible pair i and j.

We shall develop a test for testing that the $\alpha_i = 0$. Let $\overline{\mu}_{i.}$ be the average of the means in the ith row, averaged across the columns. Then testing that the $\alpha_i = 0$ is equivalent to testing that the $\overline{\mu}_{i.}$ are all equal. We think of this as saying that there is no "overall" effect due to the rows.

In a similar manner, we test that the $\beta_j = 0$, which is the same as testing that the $\overline{\mu}_{.j}$ are all equal. We shall also develop a test for testing that the $\gamma_{ij} = 0$, i.e., that the μ_{ij} are additive as defined in the previous section.

Example 13–4

To help make the meaning of the preceding parameters clear, suppose we have an experiment in which we are studying the effect of a new medicine on blood pressure. We divide the patients into smokers and nonsmokers, giving half of each group the new medicine and half the old medicine. Let μ_{NO} and μ_{NN} be the means for nonsmokers on the old medicine and for nonsmokers on the new medicine, and let μ_{SO} and μ_{SN} be the means for smokers on the old and new medicines, respectively. Finally, suppose that

$$\mu_{NO} = 120, \qquad \mu_{NN} = 100, \qquad \mu_{SO} = 130, \qquad \mu_{SN} = 170.$$

Then the overall mean for this experiment is

$$\theta = \overline{\mu}_{..} = 130.$$

The means for the nonsmokers and smokers are

$$\theta + \alpha_N = \overline{\mu}_{N.} = 110 \quad \text{and} \quad \theta + \alpha_S = \overline{\mu}_{S.} = 150,$$

respectively, so that the mean for smokers is 40 mm Hg higher than for nonsmokers (and $\alpha_N = -\alpha_S = -20$). Therefore, the hypothesis $\alpha_i = 0$ is not true, and there is an overall effect due to smoking. The means for the old medicine and the new medicine are

$$\theta + \beta_O = \overline{\mu}_{.O} = 125 \quad \text{and} \quad \theta + \beta_N = \overline{\mu}_{.N} = 135,$$

respectively (so that $\beta_O = -5$ and $\beta_N = 5$). Hence, the hypothesis $\beta_j = 0$ is also not true. In fact, the mean for the new medicine is 10 mm Hg higher than the mean for the old medicine, so that the overall effect of the new medicine is to increase the blood pressure, rather than decrease it. However, if we look at the means previously given, we see that the new medicine is better than the old medicine for the nonsmokers, but worse than the old medicine for the smokers. Therefore, we would recommend the new medicine for nonsmokers, but not for smokers. In this situation, we say that there is an interaction between the effects of smoking and medicine on blood pressure. In fact, the interactions for this setting are given by $\gamma_{ij} = \mu_{ij} - \theta - \alpha_i - \beta_j$ and are

$$\gamma_{NO} = 15, \qquad \gamma_{NN} = -15, \qquad \gamma_{SO} = -15, \qquad \gamma_{SN} = 15.$$

$\gamma_{NO} = 15$ implies that the effect of being a nonsmoker using the old medicine is 15 higher than the sum of the effect of being a nonsmoker and the effect of using the old medicine.

There are six equalities we must verify to show that the two-way analysis-of-variance-with-interaction model is an orthogonal design. We merely verify one here,

$$\sum_i \sum_j \sum_k \alpha_i \gamma_{ij} = n \sum_i \Big(\alpha_i \sum_j \gamma_{ij}\Big) = 0,$$

using the constraint on the γ_{ij}. In a similar manner, we can verify the other five equalities.

We now find the ordinary least squares estimators of the parameters. We have to minimize

$$Q = \sum_i \sum_j \sum_k (Y_{ijk} - \theta - \alpha_i - \beta_j - \gamma_{ij})^2.$$

By differentiating Q with respect to θ and using the constraints, we see that $\hat{\theta} = \overline{Y}_{...}$. Differentiating Q with respect to α_i and setting the result equal to zero, we get

$$0 = -2\left(\sum_j \sum_k (Y_{ijk} - \theta - \alpha_i - \beta_j - \gamma_{ij})\right) = -2(cn\overline{Y}_{i..} - cn\theta - cn\alpha_i - 0 - 0),$$

using the constraints on the β_j and γ_{ij}. (Note that α_1 does not occur in any of the terms with $i \neq 1$, α_2 does not occur in any of the terms with $i \neq 2$, etc.) Therefore,

$$\hat{\alpha}_i = \overline{Y}_{i..} - \hat{\theta} = \overline{Y}_{i..} - \overline{Y}_{...},$$

similarly, $\hat{\beta}_j = \overline{Y}_{.j.} - \overline{Y}_{...}$. Differentiating with respect to γ_{ij}, and setting the result equal to zero, we get

$$0 = -2\left(\sum_k (Y_{ijk} - \theta - \alpha_i - \beta_j - \gamma_{ij})\right) = -2n(\overline{Y}_{ij.} - \theta - \alpha_i - \beta_j - \gamma_{ij}).$$

(Note that, for example, γ_{13} only appears in terms with $i = 1$ and $j = 3$.) Therefore, $\hat{\gamma}_{ij} = \overline{Y}_{ij.} - \hat{\theta} - \hat{\alpha}_i - \hat{\beta}_j = \overline{Y}_{ij.} - \overline{Y}_{i..} - \overline{Y}_{.j.} + \overline{Y}_{...}$, and it follows that

$$SS(\theta) = \sum_i \sum_j \sum_k \hat{\theta}^2 = rcn\overline{Y}^2_{...},$$

$$SS(\alpha) = \sum_i \sum_j \sum_k \hat{\alpha}_i^2 = nc\sum_i (\overline{Y}_{i..} - \overline{Y}_{...})^2 = nc\sum_i \overline{Y}^2_{i..} - ncr\overline{Y}^2_{...},$$

$$SS(\beta) = \sum_i \sum_j \sum_k \hat{\beta}_j^2 = nr\sum_j (\overline{Y}_{.j.} - \overline{Y}_{...})^2 = nr\sum_j \overline{Y}^2_{.j.} - ncr\overline{Y}^2_{...},$$

$$SS(\gamma) = \sum_i \sum_j \sum_k \hat{\gamma}_{ij}^2 = n\sum_i \sum_j (\overline{Y}_{ij.} - \overline{Y}_{i..} - \overline{Y}_{.j.} + \overline{Y}_{...})^2$$

$$= n\sum_i \sum_j \overline{Y}^2_{ij.} - nc\sum_i \overline{Y}^2_{i..} - nr\sum_j \overline{Y}^2_{.j.} + ncr\overline{Y}^2_{...},$$

and

$$SSE = \sum_i \sum_j \sum_k (Y_{ijk} - \hat{\theta} - \hat{\alpha}_i - \hat{\beta}_j - \hat{\gamma}_{ij})^2$$

$$= \sum_i \sum_j \sum_k (Y_{ijk} - \overline{Y}_{ij.})^2 = \sum_i \sum_j \sum_k Y^2_{ijk} - n\sum_i \sum_j \overline{Y}^2_{ij.}.$$

To find the degrees of freedom of the effects, we note that θ has only one parameter, and so $df(\theta) = 1$. The α effect has $r - 1$ linearly independent parameters, so that $df(\alpha) = r - 1$, and similarly, $df(\beta) = c - 1$. Now, $\boldsymbol{\gamma}$ has rc parameters (the γ_{ij}). It also has a constraint for each row and a constraint for each column, so that there are $r + c$ constraints. However, if all the row constraints are satisfied and all but one of the column constraints are satisfied, then the last column constraint must be satisfied. Therefore, there are only $r + c - 1$ linearly independent constraints on the γ_{ij}, and it follows that

$$df(\gamma) = rc - (r + c - 1) = (r - 1)(c - 1).$$

(Another way to see this is that if we know that γ_{ij}, $i = 1, \ldots, r-1$ and $j = 1, \ldots, c-1$, then we can determine the γ_{ic} and the γ_{rj}.) Hence, there are $(r-1)(c-1)$ independent parameters.) Finally,

$$\begin{aligned} dfe &= N - df(\theta) - df(\alpha) - df(\beta) - df(\gamma) \\ &= nrc - 1 - (r-1) - (c-1) - (r-1)(c-1) = rc(n-1). \end{aligned}$$

The ANOVA table for the two-way model with interaction is thus as follows:

Due to	df	SS
θ	1	$nrc\overline{Y}^2_{...}$
α	$r-1$	$nc\Sigma\overline{Y}^2_{i..} - ncr\overline{Y}^2_{...}$
β	$c-1$	$nr\Sigma\overline{Y}^2_{.j.} - ncr\overline{Y}^2_{...}$
γ	$(r-1)(c-1)$	$n\Sigma\Sigma\overline{Y}^2_{ij.} - nc\Sigma\overline{Y}^2_{i..} - nr\Sigma\overline{Y}^2_{.j.} + ncr\overline{Y}^2_{...}$
Error	$rc(n-1)$	$\Sigma\Sigma\Sigma Y^2_{ijk} - \Sigma\Sigma\overline{Y}^2_{ij.}$
Total	rcn	$\Sigma\Sigma\Sigma Y^2_{ijk}$

Example 13–5

Suppose we have a two-way model with two rows, three columns, and two measurements in each cell, so that $r = 2$, $c = 3$, and $n = 2$. Suppose we observe the following data:

	1	2	3
1	1, 3	4, 8	18, 14
2	8, 16	14, 22	20, 16

Then, to take some examples, $\overline{Y}_{12.} = (4 + 8)/2 = 6$, $\overline{Y}_{.2.} = (4 + 8 + 14 + 22)/4 = 12$, $\overline{Y}_{1..} = (1 + 3 + 4 + 8 + 18 + 14)/6 = 8$, and $Y_{...} = (1 + 3 + 4 + \cdots + 20 + 16)/12 = 12$. Also,

$$T_1^2 = \sum\sum\sum Y^2_{ijk} = 2{,}266, \qquad T_2^2 = n\sum\sum\overline{Y}^2_{ij.} = 2{,}176, \qquad T_3^2 = nr\sum\overline{Y}^2_{.j.} = 1{,}928,$$

$$T_4^2 = nc\sum\overline{Y}^2_{i..} = 1{,}920, \qquad T_5^2 = ncr\overline{Y}^2_{...} = 1{,}728.$$

From these quantities, we can determine the entries in the ANOVA table:

Due to	df	SS	MS	F
θ	1	1,728	1,728	115.2
α	1	192	192	12.8
β	2	200	100	6.7
γ	2	56	28	1.9
Error	6	90	15	
Total	12	2,266		

Now, $F(\alpha) = 12.8 > F^{.05}_{1,6} = 5.99$, so we reject the hypothesis that there is no row effect with these data with a .05 test. Similarly, $F(\beta) = 6.7 > F^{.05}_{2,6} = 5.14$, so we reject the hypothesis that there is no column effect. And finally, $F(\gamma) = 1.9 < F^{.05}_{2,6} = 5.14$, so we accept the hypothesis that there is no interaction (i.e., we accept the hypothesis that the model is additive).

The two-way model is often used even when we are only interested in the row effect. For example, suppose we are interested in the effect of alcohol consumption on blood pressure. In doing an experiment to study this relationship, we should probably also consider smoking habits. We could design the experiment by classifying the people in the sample by smoking habits (e.g., no smoking, light smoking, and heavy smoking) and by drinking habits (e.g., no drinking, light drinking, moderate drinking, and heavy drinking). We would then measure the person's blood pressure. Such an experiment is a crossed two-way model. If we do not consider smoking in the design, it might happen that most of the light drinkers are heavy smokers and the heavy drinkers are light smokers. Then, if both smoking and drinking are harmful, but the harmful effects of smoking are stronger than the harmful effects of drinking, the light drinkers would have higher blood pressure than the heavy drinkers (because the light drinkers are the heavy smokers), and we would conclude, incorrectly, that heavy drinking reduces blood pressure. Therefore, even though we are not interested in smoking habits in the study, if we do not include them, we could be biasing our results. When we include a factor which we are not interested in studying in our experiment, we say it is a *blocking factor*, and we have *blocked* for that effect. So, in the foregoing example, we would say that we have blocked for smoking. (What we are discussing here is a variation of Simpson's paradox presented in Chapter 11; a numerical example of what can happen if we ignore smoking is given there also.)

It should be mentioned that there is some difficulty in interpreting the test for row effects (or column effects) in the presence of possible interactions. Arnold (1981, pp. 96–100) discusses this difficulty for the unbalanced case, which applies equally well to the balanced case.

The two-way model could be extended in a straightforward manner to higher way models. The formulas for the three-way model are derived in Exercise B6 below.

Exercises—A

1. An experiment was run to compare the gasoline mileage of two types of car and four types of gasoline. Two observations were made for each car-gasoline pair. The following data were obtained:

	Gas			
Car	A	B	C	D
1	12, 16	16, 15	22, 18	11, 13
2	18, 15	17, 23	27, 26	12, 15

(a) Compute the ANOVA table for these data.

(b) Is there a significant difference due to car? To gasoline? Is there any significant interaction? Use $\alpha = .05$.

2. An experiment was run to determine the effect of temperature and humidity on the size of potatoes. Since temperature and humidity could not be controlled for the experiment, it was necessary to use an unbalanced design. Suppose there were two temperature conditions A and B and two humidity conditions a and b. The following data were collected:

	A	B
a	2, 3, 2	4, 5
b	5, 6	7, 8, 9

(so that there are three measurements in the a, A cell, which are 2, 3, and 2). Use the model of Exercise C1 to answer the following questions:

a. Was there any significant difference due to temperature with a .05 test?

b. Was there any significant difference due to humidity with a .05 test?

Exercises—B

1. For Example 13–5, verify the entries for the ANOVA table.
2. Let $\gamma_{ij} = \mu_{ij} - \bar{\mu}_{i.} - \bar{\mu}_{.j} + \bar{\mu}_{..}$. Show that $\Sigma_i \gamma_{ij} = 0$.
3. (a) What are the six equalities which must be verified to show that the two-way model is an orthogonal design?
   (b) Verify that $\Sigma\Sigma\Sigma\alpha_i\beta_j = 0$ and $\Sigma\Sigma\Sigma\alpha_i\gamma_{ij} = 0$. (Assume that the constraints are satisfied.)
4. Verify the ordinary least squares estimators for β_j.
5. Consider the two-way model with replication but no interaction in which we observe Y_{ijk} independent, with $Y_{ijk} \sim N(\theta + \alpha_i + \beta_j, \sigma^2)$. Find the ANOVA table for this model.
6. Consider the three-way model in which we observe Y_{ijkm} independent, $i = 1, \ldots, r$, $j - 1, \ldots, c$, $k - 1, \ldots, d$, $m = 1, \ldots, n$, where

$$Y_{ijkm} \sim N(\theta + \alpha_i + \beta_j + \gamma_k + \delta_{ij} + \epsilon_{ik} + \varphi_{jk} + \eta_{ijk}, \sigma^2).$$

   (a) What constraints on the parameters make this model an orthogonal design?
   (b) Using these constraints, find the ANOVA table for the model.

Exercises—C

1. Suppose we observe $Y_{ijk} \sim N(\theta + \alpha_i + \beta_j, \sigma^2)$, independent, $i = 1, \ldots, r$, $j = 1, \ldots, c$, $k = 1, \ldots, n_{ij}$, so that the model is unbalanced. Suppose, however, that $\Sigma_j n_{ij} = m$ does not depend on i and $\Sigma_i n_{ij} = q$ does not depend on j. Let α_i and β_j satisfy the usual constraints, i.e., $\Sigma_i \alpha_i = 0$ and $\Sigma_j \beta_j = 0$.
   (a) Show that the effects due to θ, α, and β are orthogonal.
   (b) Find the ordinary least squares estimators of the parameters.
   (c) Find the ANOVA table for the model.

13.3 FURTHER ANOVA MODELS

* 13.3.1 The Latin Square Model

Consider the following square:

	1	2	3	4
1	4	2	1	3
2	3	1	2	4
3	1	3	4	2
4	2	4	3	1

Note that in each column the numbers 1, 2, 3, and 4 each appear exactly once, and in each row the numbers 1, 2, 3, and 4 also each appear exactly once. This is an example of a 4×4 Latin square. A general $r \times r$ *Latin square* is an $r \times r$ square filled with integers betwen 1 and r such that each integer occurs exactly once in each row and each column. The Latin square can sometimes be used to test for three effects in a two-way design.

Consider a particular $r \times r$ Latin square. Let k_{ij} be the number in the (i, j) cell of this square. Now suppose we observe Y_{ij} independent

$$Y_{ij} \sim N(\mu_{ij}, \sigma^2), \qquad i = 1, \ldots, r, \quad j = 1, \ldots, r,$$

where

$$\mu_{ij} = \theta + \alpha_i + \beta_j + \gamma_{k_{ij}}, \qquad \sum \alpha_i = 0, \qquad \sum \beta_j = 0, \qquad \sum \gamma_k = 0.$$

(As long as μ_{ij} can be written as a sum of an effect due to i, an effect due to j, and an effect due to k_{ij}, then μ_{ij} can be written in the preceding fashion.) We seek tests for testing that the $\alpha_i = 0$, that the $\beta_j = 0$, and that the $\gamma_k = 0$. Such a model is called a *Latin square model*.

Note that in order to use a Latin square model, we need the same number of levels for all three treatments. We also need to assume that the effects are additive. The benefit of the design is that we can analyze three effects, each with r levels, in r^2 measurements. If we used a three-way model of the type mentioned in Section 13.2.1, we would need r^3 measurements, even with no replication.

Note that the observations are independently normally distributed with constant variance σ^2. By the definition of a Latin square,

$$\sum_i \gamma_{k_{ij}} = \sum_j \gamma_{k_{ij}} = \sum \gamma_k = 0.$$

(Why?) We first verify that the model is an orthogonal design. We have

$$\sum_{ij}\sum \alpha_i \gamma_{k_{ij}} = \sum_i \left(\alpha_i \sum_j \gamma_{k_{ij}} \right) = 0,$$

so that the α_i and γ_k are orthogonal. In a similar manner, we can show that the α_i and

β_j are orthogonal, as are the β_j and the γ_k, θ and the α_i, θ and the β_j, and θ and the γ_k (see the exercises). Therefore, the model is an orthogonal design.

To find the least squares estimators, we must minimize

$$Q = \sum_i \sum_j (Y_{ij} - \theta - \alpha_i - \beta_j - \gamma_{k_{ij}})^2.$$

If we differentiate Q with respect to θ, set the result equal to zero, and use the constraints, we obtain $\hat{\theta} = \overline{Y}_{..}$. If we differentiate Q with respect to α_i and set the result equal to zero, we get

$$0 = -2\left(\sum_j (Y_{ij} - \theta - \alpha_i - \beta_j - \gamma_{k_{ij}})\right) = -2(r\overline{Y}_{i.} - r\theta - r\alpha_i - 0 - 0),$$

using the constraints on the β_j and the γ_k. Therefore, $\hat{\alpha}_i = \overline{Y}_{i.} - \hat{\theta} = \overline{Y}_{i.} - \overline{Y}_{..}$. Similarly, $\hat{\beta}_j = \overline{Y}_{.j} - \overline{Y}_{..}$. In order to find the least squares estimator for γ_k, let S_k be the set of (i,j) such that $k_{ij} = k$. (In the example given at the beginning of the section, $S_2 = \{(1,2), (2,3), (3,4), (4,1)\}$, since $k_{12} = k_{23} = k_{34} = k_{41} = 2$.) Let

$$\overline{Y}_{..k} = \frac{\sum\sum_{S_k} Y_{ij}}{r}.$$

That is, $\overline{Y}_{..k}$ is the average of the observations which get the kth γ treatment. (For the four-by-four Latin square shown at the beginning of the section, $Y_{..2} = (Y_{12} + Y_{23} + Y_{34} + Y_{41})/4$.) Now, as (i,j) runs through S_k, i runs through the integers $1, \ldots, r$, and so does j. Therefore,

$$\sum\sum_{S_k} \alpha_i = 0 \quad \text{and} \quad \sum\sum_{S_k} \beta_j = 0.$$

If we differentiate Q with respect to γ_2, use the previous constraints, and set the result equal to zero, we get

$$0 = -2\left(\sum\sum_{S_2} (Y_{ij} - \theta - \alpha_i - \beta_j - \gamma_k)\right) = -2(r\overline{Y}_{..2} - r\theta - 0 - 0 - r\gamma_2)$$

(in which we have contributions only from the terms in which $(i,j) \in S_2$). Therefore, $\hat{\gamma}_2 = \overline{Y}_{..2} - \hat{\theta} = \overline{Y}_{..2} - \overline{Y}_{..}$, and similarly, $\hat{\gamma}_k = \overline{Y}_{..k} - \overline{Y}_{..}$.

We can now find the sums of squares for the effects in the usual way. In particular,

$$SSE = \sum\sum (Y_{ij} - \hat{\theta} - \hat{\alpha} - \hat{\beta}_j - \hat{\gamma}_{k_{ij}})^2 = \sum\sum (Y_{ij} - \overline{Y}_{i.} - \overline{Y}_{.j} - \overline{Y}_{..k_{ij}} + 2\overline{Y}_{..})^2$$

$$= \sum\sum Y_{ij}^2 - r\sum \overline{Y}_{i.}^2 - r\sum \overline{Y}_{.j}^2 - r\sum \overline{Y}_{..k}^2 + 2r^2\, \overline{Y}_{..}^2.$$

We note that $df(\theta) = 1$, $df(\alpha) = r - 1$, $df(\beta) = r - 1$, $df(\gamma) = r - 1$, and that $N = r^2$. Therefore,

$$dfe = r^2 - 1 - 3(r - 1) = (r - 2)(r - 1).$$

We therefore get the following ANOVA table for the Latin square model:

Due to	*df*	*SS*
θ	1	$r^2\overline{Y}^2_{..}$
α	$r-1$	$r\Sigma\overline{Y}^2_{i.} - r^2\overline{Y}^2_{..}$
β	$r-1$	$r\Sigma\overline{Y}^2_{.j} - r^2\overline{Y}^2_{..}$
γ	$r-1$	$r\Sigma\overline{Y}^2_{..k} - r^2\overline{Y}^2_{..}$
Error	$(r-2)(r-1)$	$\Sigma\Sigma Y^2_{ij} - r\Sigma\overline{Y}^2_{i.} - r\Sigma\overline{Y}^2_{.j} - r\Sigma\overline{Y}^2_{..k} + 2r^2\overline{Y}^2_{..}$
Total	r^2	$\Sigma\Sigma Y^2_{ij}$

Example 13–6

Suppose we have the 3×3 Latin square

	1	2	3
1	3	2	1
2	1	3	2
3	2	1	3

with data

	1	2	3
1	4	6	5
2	11	4	9
3	9	11	4

(Note that the table for the data cannot be a Latin square, since it has integers greater than three.) Then $\overline{Y}_{1.} = (4 + 6 + 5)/3 = 5$, and $\overline{Y}_{2.}$ and $\overline{Y}_{3.}$ are computed similarly. Also, $\overline{Y}_{.2} = (6 + 4 + 11)/3 = 7$, and $\overline{Y}_{.1}$ and $\overline{Y}_{.3}$ are computed similarly. To find $\overline{Y}_{..1}$, we average the three numbers associated with the number one in the Latin square, to obtain $\overline{Y}_{..1} = (5 + 11 + 11)/3 = 9$. Similarly, $\overline{Y}_{..2} = 8$, $\overline{Y}_{..3} = 4$, and $\overline{Y}_{..} = 7$. Hence,

$$T_1^2 = \sum\sum Y_{ij}^2 = 513, \qquad T_2^2 = r\sum\overline{Y}^2_{i.} = 459, \qquad T_3^2 = r\sum\overline{Y}^2_{.j} = 447,$$

$$T_3^2 = r\sum\overline{Y}^2_{..k} = 3(9^2 + 8^2 + 4^2) = 483, \qquad T_4^2 = r^2\,\overline{Y}^2_{..} = 441.$$

We therefore get the following ANOVA table for the model given:

Due to	*df*	*SS*	*MS*	*F*
θ	1	441	441	147
α	2	18	9	3
β	2	6	3	1
γ	2	42	21	7
Error	2	6	3	
Total	9	513		

Since $F^{.05}_{2,2} = 19$, none of the effects are close to significance for the given data (except, of course, for the overall mean, which we do not care about.)

Exercises—A

1. Suppose we run an experiment to determine the effects of temperature, humidity, and fertilizer on potato yield. We use the 4×4 Latin square at the beginning of this section with rows (i) representing temperature, columns (j) representing humidity, and the third variable (k) representing fertilizer. We get the following data:

2	3	6	8
3	5	7	9
5	6	8	9
6	6	9	11

What is the ANOVA table for these data? Did temperature have a significant effect on potato yield? Did humidity? How about fertilizer? Use $\alpha = .05$.

Exercises—B

1. **(a)** Verify that the Latin square model is an orthogonal design.
   (b) Verify the formulas in the ANOVA table for the Latin square model.

* 13.3.2 *The Two-Fold Nested Analysis-of-Variance Model*

Consider the model in which we observe Y_{ijk} independent, with

$$Y_{ijk} \sim N(\mu_{ij}, \sigma^2), \qquad i = 1, \ldots, r, \quad j = 1, \ldots, c_i, \quad k = 1, \ldots, n.$$

Although this model looks much like the crossed two-way model discussed previously, in this section we interpret it differently. We assume that the j effect is nested in the i effect, so that there is no relationship between observations in the $(1, j)$ cell and observations in the $(2, j)$ cell. We also allow for different numbers of columns in the different rows; that is, we allow the c_i to be different.

We reparametrize this model in the following way. Let

$$\theta = \overline{\mu}_{..}, \qquad \alpha_i = \overline{\mu}_{i.} - \overline{\mu}_{..}, \qquad \delta_{ij} = \mu_{ij} - \overline{\mu}_{i.}.$$

Then

$$\mu_{ij} = \theta + \alpha_i + \delta_{ij}, \qquad \textstyle\sum_i c_i \alpha_i = 0, \qquad \sum_j \delta_{ij} = 0.$$

We shall develop tests that $\delta_{ij} = 0$ (μ_{ij} does not depend on j) and that the $\alpha_i = 0$ (that $\overline{\mu}_{i.}$ are equal). We call i the *class* and j the *subclass* for this model, so we call the α_i the *class effects* and δ_{ij} the *subclass effects*. The model is known as the (two-fold) *nested analysis-of-variance model.*

Example 13–7

Suppose we are interested in the effect of diet on blood pressure. We conduct a study with three types of diet: low salt, low fat, and normal. Suppose we have two low-salt diets, two low-fat diets, and three normal diets. Suppose also that the means for the low-salt diets are μ_{S1} and μ_{S2}, the means for the low-fat diets are μ_{F1} and μ_{F2}, and the means for the normal diets are μ_{N1}, μ_{N2}, and μ_{N3} and that the values of these means, in mm Hg, are as follows:

$$\mu_{S1} = 90, \quad \mu_{S2} = 100, \quad \mu_{F1} = 80, \quad \mu_{F2} = 110, \quad \mu_{N1} = 130, \quad \mu_{N2} = 120, \quad \mu_{N3} = 140.$$

Then the overall mean for the experiment is the average of these seven numbers, i.e.,

$$\theta = \bar{\mu}_{..} = 110,$$

and the average for the low-salt diets is 95, the average for the low-fat diets is also 95, and the average for the normal diets is 130. Therefore,

$$\alpha_S = \bar{\mu}_{S.} - \theta = 95 - 110 = -15, \qquad \alpha_F = 95 - 110 = -15, \qquad \alpha_N = 130 - 110 = 20.$$

(Note that $2\alpha_S + 2\alpha_F + 3\alpha_N = 0$.) We interpret $\alpha_S = -15$ to mean that the blood pressure for the low-salt diet is 15 mm Hg below the average blood pressure in the study.

We see that there is an effect due to the type of diet. In particular, both the low-salt and low-fat diet lead to lower blood pressure than the normal diet does. Now,

$$\delta_{S1} = 90 - 95 = -5, \quad \delta_{S2} = 5, \quad \delta_{F1} = -15, \quad \delta_{F2} = 15, \quad \delta_{N1} = 0, \quad \delta_{N2} = -10, \quad \delta_{N3} = -10$$

(Note that $\delta_{S1} + \delta_{S2} = 0$, $\delta_{F1} + \delta_{F2} = 0$, and $\delta_{N1} + \delta_{N2} + \delta_{N3} = 0$.) We interpret $\delta_{S1} = -5$ to mean that the blood pressure for the first low-salt diet is 5 mm Hg below the average blood pressure for low-salt diets in the experiment.

Note that in the nested analysis-of-variance model we are not assuming that the δ_{ij} sum to zero over i, and hence, the subclass effects δ_{ij} for the two-fold nested model are not the same as the interactions γ_{ij} for the two-way crossed model discussed in Section 13.2.2. In fact, if the c_i are equal, then $\delta_{ij} = \beta_j + \gamma_{ij}$. If the c_i are different, there is no relationship between the two-fold nested model of this section and the two-way crossed model of the previous one.

As usual, the observations in the two-fold nested model are independently normally distributed with common variance σ^2. In addition, the θ effect, the α_i effects, and the δ_{ij} effects are orthogonal (see the exercises). Therefore, the model is an orthogonal design. Furthermore, the ordinary least squares estimators of the parameters are

$$\hat{\theta} = \bar{Y}_{...}, \qquad \hat{\alpha}_i = \bar{Y}_{i..} - \bar{Y}_{...}, \qquad \hat{\delta}_{ij} = \bar{Y}_{ij.} - \bar{Y}_{i..}.$$

Let $C = \Sigma c_i$, i.e., the total number of subclasses. (Note that if the $c_i = c$, then $C = nc$). Then the ANOVA table for the model is

Due to	df	SS
θ	1	$nC\bar{Y}^2_{...}$
α	$r-1$	$n\Sigma c_i \bar{Y}^2_{i..} - nC\bar{Y}^2_{...}$
δ	$C-r$	$n\Sigma\Sigma \bar{Y}^2_{ij.} - n\Sigma c_i \bar{Y}^2_{i..}$
Error	$C(n-1)$	$\Sigma\Sigma\Sigma Y^2_{ijk} - n\Sigma\Sigma \bar{Y}^2_{ij.}$
Total	nC	$\Sigma\Sigma\Sigma Y^2_{ijk}$

Example 13–8

Suppose we have three classes, with one subclass in the first class, three subclasses in the second class, and two subclasses in the third class. Suppose also that we have three observations in each subclass. Then $r = 3$, $c_1 = 1$, $c_2 = 3$, $c_3 = 2$, and $n = 3$. Suppose next that we observe the following data:

Class	Subclass	Data
1	1	1, 3, 5
	1	1, 2, 3
2	2	4, 5, 6
	3	7, 8, 9
3	1	4, 7, 10
	2	8, 11, 14

For these data, we obtain, among other results, $\overline{Y}_{11.} = \overline{Y}_{1..} = 3$, $\overline{Y}_{21.} = (1+2+3)/3 = 2$, $\overline{Y}_{2..} = (1+2+\cdots+8+9)/9 = 5$, $\overline{Y}_{...} = (1+3+\cdots+11+14)/18 = 6$, and $C = 1 + 2 + 3 = 6$ (the total number of subclasses). We also get the following ANOVA table:

Due to	df	SS	MS	F
θ	1	648	648	155.52
α	2	90	45	10.80
δ	3	78	26	6.24
Error	12	50	4.17	
Total	18	866		

Now, $F(\boldsymbol{\alpha}) = 10.8 > 3.89 = F^{.05}_{2,12}$, so we reject the hypothesis that there is no class effect with a .05 test. Similarly, $F(\boldsymbol{\delta}) = 6.24 > 3.49 = F^{.05}_{3,12}$, so we also reject the hypothesis that there is no subclass effect with a .05 test.

In the preceding analysis, we have been assuming that we have the same number n of observations for each subclass, i.e., that the model is balanced. If the model is unbalanced, it is no longer an orthogonal design, but can still be analyzed by the regression approach. (See Section 13.5.2.)

The two-fold nested analysis-of-variance model can be extended in a straightforward way to three-fold and higher nested models. Note that we can also have models which have both crossed and nested effects. (See Exercise B5.)

Exercises—A

1. An experiment was done to compare rainfall in three different states. In the first and second states two sites were selected, while in the third state three sites were selected. At each site, two measurements were taken, yielding the following data:

State	1		2		3		
Site	1	2	1	2	1	2	3
	7	11	11	12	14	9	16
	8	13	10	13	15	8	17

Write down the ANOVA table for these data. Is there a significant difference between states? between sites within states? Use $\alpha = .05$.

2. An experiment was run to determine whether the amount of fat in a piece of beef is

affected by the steer's breed, by its herd, or by its sire. Two different breeds (A and B) of cattle were selected. Within each breed two herds (a and b) were selected, within each herd two sires (I and II) were selected, and for each sire two steers were selected, so that there were 16 steers in all. The steers were killed and their fat measured, yielding the following data:

A				B			
a		b		a		b	
I	II	I	II	I	II	I	II
8	10	12	10	16	12	20	16
6	10	8	10	16	14	18	18

(Note that there is no relationship between herd a or b in species A and herd a or b in species B, and no relationship between the various I and II studs, so that the design is that of a nested model.) Use the three-fold nested model given in Exercise B4 to find the ANOVA table for the model in this experiment. Which effects are significant with $\alpha = .05$?

Exercises—B

1. Verify the ANOVA table for Example 13–8.
2. Show that the balanced two-fold nested model is an orthogonal design.
3. **(a)** Verify the formulas in the text for the ordinary least squares estimators.
   (b) Verify the formulas given in the text for the ANOVA table.
4. Consider a model in which we observe $Y_{ijkm} \sim N(\theta + \alpha_i + \delta_{ij} + \epsilon_{ijk}, \sigma^2)$, independent, $i = 1, \ldots, r$, $j = 1, \ldots, c$, $k = 1, \ldots, s$, $m = 1, \ldots, n$, where θ, α_i, δ_{ij}, ϵ_{ijk}, and σ^2 are unknown constants such that $\Sigma_i \alpha_i = 0$, $\Sigma_j \delta_{ij} = 0$, and $\Sigma_k \epsilon_{ijk} = 0$. (Note, however, that $\Sigma_i \delta_{ij} \neq 0$, $\Sigma_j \epsilon_{ijk} \neq 0$, and $\Sigma_i \epsilon_{ijk} \neq 0$.)
   (a) Show that the θ, α_i, δ_{ij}, and ϵ_{ijk} effects are all orthogonal.
   (b) Find the ANOVA table for this model (called a *three-fold nested model*).
5. Suppose we observe Y_{ijk} independent, with $Y_{ijk} \sim N(\theta + \alpha_i + \beta_j + \delta_{ik}, \sigma^2)$, $i = 1, \ldots, r$, $j = 1, \ldots, c$, $k = 1, \ldots, n$, and with the constraints $\Sigma_i \alpha_i = 0$, $\Sigma_j \beta_j = 0$, and $\Sigma_k \delta_{ik} = 0$. (We are not assuming that the δ_{ik} sum to zero over i.) In this model, the α and β effects are crossed, but the δ effect is nested in the α effect.
   (a) Show that the θ, α, β, and δ effects are orthogonal.
   (b) Find the ordinary least squares estimators of θ, α_i, β_j, and δ_{ik}.
   (c) Find the ANOVA table for the model.

13.4 SIMULTANEOUS CONFIDENCE INTERVALS

* 13.4.1 Scheffé Simultaneous Confidence Intervals

Let us return to the general orthogonal design model described in Section 13.1.2 in which we observe Y_q independent, with

$$Y_q \sim N(\delta_q^{(1)} + \cdots + \delta_q^{(p)}, \sigma^2) \quad \text{and} \quad \sum_q \delta_q^{(i)} \delta_q^{(j)} = 0 \text{ for } i \neq j.$$

Often, when we have rejected the hypothesis that $\boldsymbol{\delta}^{(i)} = \mathbf{0}$, we want to establish what is causing the hypothesis to be rejected. We now present a procedure that is useful for that purpose.

For a particular i, let T_i be the set of possible $\boldsymbol{\delta}^{(i)}$, and let $\mathbf{d} \in T_i$. (Thus, the d_q have the same subscripts as the $\delta_q^{(i)}$ and satisfy the same constraints.) In the one-way model, $\delta_{ij}^{(2)} = \alpha_i$, $\Sigma n_i \alpha_i$, so that $d_{ij} = a_i$, $\Sigma n_i a_i = 0$.) A *contrast* in $\boldsymbol{\delta}^{(i)}$ is a function

$$C(\mathbf{d}) = \sum_q d_q \delta_q^{(i)}, \qquad \mathbf{d} \in T_i.$$

(Note that there are an infinite number of such contrasts.) We say that the set of intervals

$$L_C \leqq C(\mathbf{d}) \leqq R_C$$

is a set of *simultaneous* $(1 - \alpha)$ *confidence intervals* for the set of all contrasts $C(\mathbf{d})$ if

$$P(L_C \leqq C(\mathbf{d}) \leqq R_C \text{ for all contrasts } C(\mathbf{d})) = 1 - \alpha.$$

For the contrast $C(\mathbf{d}) = \Sigma_q d_q \delta_q^{(i)}$, let

$$\hat{C}(\mathbf{d}) = \sum_q d_q \hat{\delta}_q^{(i)}, \qquad L(\mathbf{d}) = \left(\sum_q d_q^2\right)^{1/2},$$

$$Q = [(MSE)(df_i)(F^\alpha)]^{1/2}, \qquad F^\alpha = F^\alpha_{df_i, dfe}.$$

(Note that df_i, dfe, and MSE can be read from the ANOVA table for the particular model and that F^α is the critical value for the size-α test that $\delta_i = 0$.) We call $\hat{C}(d)$ the *estimated contrast*. The *Scheffé* interval for the contrast $C(\mathbf{d})$ is given by

$$C(\mathbf{d}) \in \hat{C}(\mathbf{d}) \pm QL(\mathbf{d}). \tag{13–2}$$

Finally, a contrast $C(\mathbf{d})$ is *significant* if its Scheffé interval does not contain zero.

The following theorem is proved in Section 13.5.1.

Theorem 13–2

a. The set of Scheffé intervals is a set of simultaneous $(1 - \alpha)$ confidence intervals for the set of all contrasts in $\boldsymbol{\delta}^{(i)}$.
b. The hypothesis that the $\boldsymbol{\delta}^{(i)} = 0$ is rejected with the size-α F-test defined above if and only if at least one contrast is significant.
c. The hypothesis that the $\boldsymbol{\delta}^{(i)} = 0$ is rejected with the size-α F-test just mentioned if and only if the contrast $C(\hat{\boldsymbol{\delta}}^{(i)})$ is significant.

Proof

See Theorem 13–8. □

Part c of Theorem 13–2 says that the contrast with $\mathbf{d} = \hat{\boldsymbol{\delta}}^{(i)}$ is always significant when the hypothesis is rejected. This implies that the contrast is significant at a lower size than any other contrast. Therefore, we call the contrast with $\mathbf{d} = \hat{\boldsymbol{\delta}}^{(i)}$ the *most significant* contrast.

We now return to the one-way model discussed in Section 13.1 in which

$$\delta_{ij}^{(2)} = \alpha_i, \qquad \sum_i n_i \alpha_i = 0, \qquad \hat{\delta}_{ij}^{(2)} = \hat{\alpha}_i = \overline{Y}_{i.} - \overline{Y}_{..}, \qquad df_2 = df(\alpha) = k - 1.$$

A contrast in the α_i is a function $\Sigma d_{ij} \delta_{ij}^{(2)}$, where d_{ij} has the same form as $\delta_{ij}^{(2)}$. Therefore $d_{ij} = a_i$ for a_i such that $\Sigma n_i a_i = 0$. Hence, a contrast in the α_i is a function

$$C(\mathbf{d}) = \sum_i \sum_j d_{ij} \delta_{ij}^{(2)} = \sum_i \sum_j a_i \alpha_i = \sum_i n_i a_i \alpha_i,$$

where $\Sigma_i n_i a_i = 0$. Similarly,

$$\hat{C}(\mathbf{d}) = \sum_i n_i a_i \hat{\alpha}_i = \sum_i n_i a_i (\overline{Y}_i - \overline{Y}) = \sum_i n_i a_i \overline{Y}_i,$$

$$L(\mathbf{d}) = \left(\sum_i n_i a_i^2\right)^{1/2}, \qquad Q = [MSE(k-1)F_{k-1,N-k}^{\alpha}]^{1/2}.$$

Hence, the Scheffé simultaneous $(1 - \alpha)$ confidence intervals for contrasts in the α_i are

$$\sum n_i a_i \alpha_i \in \sum n_i a_i \overline{Y}_i \pm Q \left(\sum n_i a_i^2\right)^{1/2}, \qquad \sum n_i a_i = 0.$$

If we now let

$$c_i = n_i a_i,$$

so that

$$n_i a_i^2 = \frac{c_i^2}{n_i},$$

then the intervals become

$$\sum c_i \alpha_i \in \sum c_i \overline{Y}_i \pm Q \left(\frac{\sum c_i^2}{n_i}\right)^{1/2}, \qquad \Sigma c_i = 0. \tag{13–3}$$

The most significant contrast has $a_i = \hat{\alpha}_i$, or equivalently,

$$c_i = n_i a_i = n_i \hat{\alpha}_i = n_i(\overline{Y}_{i.} - \overline{Y}_{..}).$$

A *comparison* is a contrast of the form $\alpha_i - \alpha_j$, i.e., a contrast with $c_i = 1$, $c_j = -1$, and $c_k = 0$ for $k \neq i$ and $k \neq j$. Substituting into Equation (13–3), we obtain the Scheffé simultaneous confidence interval for the comparisons:

$$\alpha_i - \alpha_j \in \overline{Y}_i - \overline{Y}_j \pm (n_i^{-1} + n_j^{-1})^{1/2} Q.$$

Often, after rejecting with the F-test, at least one comparison is significant (i.e., its confidence interval does not contain zero). Therefore, after rejecting with the F-test, we look for significant comparisons. If we find any, we stop looking and consider those differences to be the ones which are causing the hypothesis to be rejected. If we do not find any significant comparisons, we look for more complicated significant contrasts. By Theorem 13–2, when we reject the null hypothesis, there must be at least one significant contrast. In particular, when we reject the null hypothesis, the most significant contrast must be significant.

Example 13–9

Consider again Example 13–2, in which $N = 7$, $k = 3$, $\overline{Y}_{1.} = 9$, $\overline{Y}_{2.} = 4$, $\overline{Y}_{3.} = 3$, $\overline{Y}_{.} = 5$, $MSE = 1.5$, and $F_{2,4}^{.05} = 6.94$. Since there is a significant difference in the α_i, we may want to investigate further to find the cause of it. Note first that

$$Q = \left[\frac{(6.94)(2)(1.5)}{(4)}\right]^{1/2} = 4.56.$$

Therefore, the Scheffé .95 intervals for comparisons are

$$\alpha_1 - \alpha_2 \in (9 - 4) \pm 4.56(.5 + .5)^{1/2} = 5 \pm 4.56,$$
$$\alpha_1 - \alpha_3 \in (9 - 3) \pm 4.56(.5 + .333)^{1/2} = 6 \pm 4.16,$$
$$\alpha_2 - \alpha_3 \in (4 - 3) \pm 4.56(.5 + .333)^{1/2} = 1 \pm 4.16.$$

The first two intervals do not contain zero for this problem, but the third interval does, so that we would conclude that the first mean is significantly different from the second and third means, but the second and third means are not significantly different from each other. The most significant contrast for the data is

$$2(9 - 5)\alpha_1 + 2(4 - 5)\alpha_2 + 3(3 - 5)\alpha_3 = 8\alpha_1 - 2\alpha_2 - 6\alpha_3.$$

The problem of deciding which if any means are different in a one-way ANOVA model is called the *multiple comparisons problem*. It is a difficult, but important, problem and one for which there is no definite answer at this time. Several of the procedures suggested (e.g., Fisher's least significant difference and Duncan's multiple range test) find too many significant differences, while the preceding Scheffé procedure finds too few. More powerful, but more complicated, procedures are the Ryan and modified Duncan multiple range tests. (For a more detailed discussion of the multiple comparisons problem, see Arnold (1981), pp. 180–200.)

As another example of Scheffé simultaneous confidence intervals, consider the crossed two-way ANOVA model with interaction discussed in Section 13.2.2. In that model, we observe

$$Y_{ijk} \sim N(\theta + \alpha_i + \beta_j + \gamma_{ij}, \sigma^2), \qquad i = 1, \ldots, r,\ j = 1, \ldots, c,\ k = 1, \ldots, n,$$

$$\sum_i \alpha_i = 0, \qquad \sum_j \beta_j = 0, \qquad \sum_i \gamma_{ij} = 0, \qquad \sum_j \gamma_{ij} = 0.$$

We first look at the contrasts in α_i. A contrast in the α_i is a function $\Sigma_i \Sigma_j \Sigma_k a_i \alpha_i$ such that the a_i satisfy the same constraint as the α_i, i.e., $\Sigma_i a_i = 0$. Now,

$$\sum_i \sum_j \sum_k a_i \alpha_i = nc \sum_i a_i \alpha_i, \qquad \sum_i \sum_j \sum_k a_i \hat{\alpha}_i = nc \sum_i a_i \hat{\alpha}_i = nc \sum_i a_i \overline{Y}_{i..},$$

$$\sum_i \sum_j \sum_k a_i^2 = nc \sum_i a_i^2.$$

Therefore, the Scheffé simultaneous $(1 - \alpha)$ confidence intervals for contrasts in the α_i are

$$nc \sum_i a_i \alpha_i \in nc \sum_i a_i \overline{Y}_{i..} \pm \left[MSE(r - 1)F^{\alpha} nc \sum_i a_i^2\right]^{1/2}, \qquad \sum_i a_i = 0$$

(where $F^\alpha = F^\alpha_{r-1, rc(n-1)}$). After dividing by nc, we get the intervals

$$\sum_i a_i \alpha_i \in \sum_i a_i \overline{Y}_{i..} \pm \left[\frac{MSE(r-1)F^\alpha \sum_i a_i^2}{nc}\right]^{1/2}, \qquad \sum_i a_i = 0.$$

Hence, the intervals for the comparisons are

$$\alpha_i - \alpha_m \in \overline{Y}_i \dot{-} \overline{Y}_m \pm \left[\frac{MSE(r-1)F^\alpha 2}{nc}\right]^{1/2}.$$

The confidence intervals for contrasts in the β_j can be derived analogously.

In a similar manner, a contrast in the γ_{ij} has the form

$$\sum\sum\sum_{ijk} g_{ij}\gamma_{ij} = n\sum_i\sum_j g_{ij}\gamma_{ij}, \qquad \sum_i g_{ij} = 0, \qquad \sum_j g_{ij} = 0.$$

Finally, by arguments like those given above, the set of simultaneous confidence intervals for the contrast in the γ_{ij} is given by

$$\sum\sum_{ij} g_{ij}\gamma_{ij} \in \sum\sum_{ij} g_{ij}\hat{\gamma}_{ij} \pm \left[\frac{MSE(r-1)(c-1)F^\alpha \sum\sum_{ij} g_{ij}^2}{n}\right]^{1/2},$$

where $F^\alpha = F^\alpha_{(r-1)(c-1), rc(n-1)}$ is the critical value for the size-α test for interactions.

Example 13–10

Consider again Example 13–5. There, we accepted the hypothesis that there was no interaction, so that we shall not compute simultaneous intervals for the interactions. Also, we rejected the hypothesis of equal rows; however, there are only two rows, so we do not need to check which rows are different. We have also rejected the hypothesis that the columns are different, so let us find which columns are significantly different, i.e., which column comparisons are significant. We note that $\overline{Y}_{.1.} = 7$, $\overline{Y}_{.2.} = 12$, $\overline{Y}_{.3.} = 17$, and $F^{.05} = F^{.05}_{2,6} = 5.14$. Hence

$$\left[\frac{MSE(c-1)F^{.05}2}{nr}\right]^{1/2} = \left[\frac{15(2)(5.14)2}{4}\right]^{1/2} = 8.8.$$

Therefore, the Scheffé .95 intervals for the comparisons are

$$\beta_1 - \beta_2 \in -5 \pm 8.8, \qquad \beta_1 - \beta_3 \in -10 \pm 8.8, \qquad \beta_2 - \beta_3 \in -5 \pm 8.8.$$

The only one of these intervals which does not contain zero is $\beta_1 - \beta_3$. (In fact, $4(\beta_3 - \beta_1)$ is the most significant contrast for the given data.) Therefore, based on the data, the mean for the third column is significantly higher than the mean for the first column, but there is no significant difference between the means of the first and second columns or between those of the second and third columns. Recall that accepting a null hypothesis is not proof of its truth, but merely lack of proof of its falsity. This explains the logical anomaly that we accept the hypothesis that the first and second columns are the same and that the second and third columns are the same, but reject the hypothesis that the first and third are the same. What we conclude from the data is that the first and third columns are different, but we do not know whether the second column is equal to the first or the third column (or perhaps different from both).

From the one-way and two-way models above, we see that it is often fairly easy to find the Scheffé simultaneous confidence intervals for contrasts in $\boldsymbol{\delta}_i$ for an orthogonal design. We first let $\mathbf{d}$ be a vector that has the same "form" as $\boldsymbol{\delta}^{(i)}$. That is, the components have the same subscripts as those in $\boldsymbol{\delta}^{(i)}$ and satisfy the same constraints as the components of $\boldsymbol{\delta}^{(i)}$ (e.g., $d_{ij} = a_i$ and $\Sigma_i n_i a_i = 0$). We then proceed according to the following rules:

1. Compute the contrast $C(\mathbf{d})$ and estimated contrast $\hat{C}(\mathbf{d})$ by summing (over all subscripts in the model) the products of the components of $\mathbf{d}$ and those of $\boldsymbol{\delta}^{(i)}$ and the products of the components of $\mathbf{d}$ and $\hat{\boldsymbol{\delta}}^{(i)}$ (e.g., $\Sigma\Sigma a_i \alpha_i = \Sigma n_i a_i \alpha_i$ and $\Sigma\Sigma a_i \hat{\alpha}_i = \Sigma n_i a_i \hat{\alpha}_i$).
2. Compute $L(\mathbf{d})$ by summing (over all subscripts in the model) the squares of the components of $\mathbf{d}$ (e.g., $\Sigma\Sigma a_i^2 = \Sigma n_i a_i^2$).
3. Substitute these quantities into the general formula (13–2).

Exercises—A

1. Find the Scheffé simultaneous 95% confidence intervals for comparisons for the data in Exercise A1, Section 13.1.1. Which comparisons are significant? What is the most significant contrast for these data?
2. Find the Scheffé simultaneous .95 confidence intervals for comparisons in the columns for the data in Exercise A1, Section 13.2.2. Which columns are significantly different? What is the most significant contrast for these data?
3. Use the data in Exercise A1, Section 13.3.2, and the results of Exercise B3 in this section to answer the following questions:
   (a) What are the .95 simultaneous confidence intervals for comparisons between states? Which states are significantly different?
   (b) What are the simultaneous .95 confidence intervals for comparisons of sites within states? Which sites within a state are significantly different? (There are five different contrasts, one in each of the first two states and three in the third state.)

Exercises—B

1. (a) Verify the simultaneous confidence intervals for contrasts in the γ_{ij} for the two-way crossed model.
   (b) Show that the comparisons $\gamma_{ij} - \gamma_{km}$ are not contrasts for this model.
2. Consider the Latin square model of Section 13.3.1.
   (a) Find the simultaneous confidence intervals for contrasts in the α_i.
   (b) Find the simultaneous confidence intervals for contrasts in the γ_k.
3. Consider the two-fold nested model of Section 13.3.2.
   (a) Find the Scheffé simultaneous confidence intervals for contrasts in the α_i.
   (b) Find the simultaneous confidence intervals for contrasts in the δ_{ij}.
   (c) Show that the comparison $\delta_{ij} - \delta_{km}$ is not a contrast unless $i = k$.

Exercises—C

1. Find the Scheffé simultaneous $(1 - \alpha)$ confidence intervals for the α_i in the one-way analysis-of-covariance model given in Exercise C1 of Section 13.1.2.

2. Let A_i be sets. Then $P(\cap_i A_i) \geqq 1 - \Sigma_i P(A_i)$, by Bonferroni's inequality. Consider the one-way ANOVA model of Section 13.1.1.
   (a) Let $\alpha^* = \alpha/\left[\binom{k}{2}\right]$. Show that
   $$P\left(\mu_i - \mu_j \in \overline{Y}_i - \overline{Y}_j \pm t^{\alpha^*}_{N-k}\left[\left(\frac{S^2}{(n-p)}\right)(n_i^{-1} + n_j^{-1})\right]^{1/2} \text{ for all } 1 \leqq i, j \leqq k\right) \geqq 1 - \alpha,$$
   so that these intervals are at least $(1-\alpha)$ simultaneous confidence intervals for the set of all comparisons.
   (b) Show that the foregoing intervals can be computed from the Scheffé intervals for comparisons by replacing $((k-1)F^{\alpha}_{k-1,N-k})^{1/2}$ by $t^{\alpha^*}_{N-k}$.
   (These simultaneous confidence intervals for comparisons are called the *Bonferroni* $(1-\alpha)$ *simultaneous confidence intervals*.)
3. Let $Z_1, \ldots, Z_k$ and U be independent, $Z_i \sim N(0,1)$, $U \sim \chi^2_n$. Let $Q = \max_{i,j} |Z_i - Z_j|/(U/n)^{1/2}$. We say that Q has a studentized range distribution with parameters k and n, and write $Q \sim q_{k,n}$. Let $q^{\alpha}_{k,n}$ be the upper α quantile of the distribution of Q.
   (a) In a balanced one-way ANOVA model with k classes and n observations in each class show that $\max_{i,j} |\overline{X}_{i.} - \overline{X}_{j.} - (\mu_i - \mu_j)|/(MSE/n)^{1/2} \sim q_{k,k(n-1)}$.
   (b) Show that $P(\mu_i - \mu_j \in \overline{X}_{i.} - \overline{X}_{j.} \pm q^{\alpha}_{k,k(n-1)}(MSE/n)^{1/2}$ for all $i, j) = 1 - \alpha$. (The intervals $\mu_i - M_j \in \overline{X}_{i.} - \overline{X}_{j.} \pm q^{\alpha}_{k,k(n-1)}\ (MSE/n)^{1/2}$ are called the Tukey simultaneous confidence intervals for the comparisons $\mu_i - \mu_j$.)
4. Let the $q_{k,n}$ distribution and $q^{\alpha}_{k,n}$ be as defined in Exercise C3.
   (a) In a balanced two-way model with r rows, c columns, and n observations per cell, show that $\max_{i,j} |\overline{X}_{i..} - \overline{X}_{j..} - (\alpha_i - \alpha_j)|/(MSE/cn)^{1/2} \sim q_{r,rc(n-1)}$.
   (b) Show that $\alpha_i - \alpha_j \in \overline{X}_{i..} - \overline{X}_{j..} \pm q^{\alpha}_{r,rc(n-1)}\ (MSE/cn)^{1/2}$ is a set of simultaneous $(1-\alpha)$ confidence intervals for the comparisons $\alpha_i - \alpha_j$.

13.5 THE THEORY OF ANOVA MODELS

13.5.1 Derivations for Orthogonal Designs

In this section, we prove the basic results used in previous sections to find optimal procedures for orthogonal designs. Although these results are fairly simple to use, to state them precisely and prove them, we need to talk about subspaces. A *subspace* of R^N is a set T such that if $\mathbf{u} \in T$ and $\mathbf{v} \in T$, then $a\mathbf{u} + b\mathbf{v} \in T$ for all real numbers a and b. Let $\mathbf{x}_1, \ldots, \mathbf{x}_r$ be vectors in T. Then the $\mathbf{x}_i$ are a *basis* for T if they are linearly independent and if every vector in T can be written as a linear combination of the $\mathbf{x}_i$. The *dimension* of a subspace is the number of vectors in a basis for T. (Note that although the basis is not unique, every basis has the same number of vectors.)

Two vectors $\mathbf{s}$ and $\mathbf{t}$ are *orthogonal* if $\mathbf{s}'\mathbf{t} = 0$, a vector $\mathbf{s}$ is *orthogonal* to a subspace T if $\mathbf{s}'\mathbf{t} = 0$ for all $\mathbf{t} \in T$, and two subspaces S and T are *orthogonal* if $\mathbf{s}'\mathbf{t} = 0$ for all $\mathbf{s} \in S, \mathbf{t} \in T$.

Let T be an r-dimensional subspace, and let $\mathbf{X}$ be an $n \times r$ matrix whose columns are a basis for T. Define

$$\mathbf{P}_T = \mathbf{X}(\mathbf{X}'\mathbf{X})^{-1}\mathbf{X}'.$$

$\mathbf{P}_T$ is the *projection matrix* for T. ($\mathbf{P}_T$ does not depend on what basis we have chosen for T; see the exercises.) We now derive the properties of $\mathbf{P}_T$ we shall be using in this section. (For some further properties, see Arnold (1981), pp. 32–40.)

Lemma 13–3

a. $\mathbf{P}_T$ is idempotent, and $\text{tr}(\mathbf{P}_T)$ is the dimension of T.
b. If $\mathbf{t} \in T$, then $\mathbf{P}_T\mathbf{t} = \mathbf{t}$. If $\mathbf{s}$ is orthogonal to T, then $\mathbf{P}_T\mathbf{s} = \mathbf{0}$.
c. For all $\mathbf{y} \in R^N$, $\mathbf{P}_T\mathbf{y} \in T$ and $\mathbf{y} - \mathbf{P}_T\mathbf{y}$ is orthogonal to T.
d. If S and T are orthogonal subspaces, then $\mathbf{P}_T\mathbf{P}_S = \mathbf{0}$.

Proof

a. See Exercise B1.
b. The vector $\mathbf{t}$ is a linear combination of the columns of $\mathbf{X}$ if and only if $\mathbf{t} = \mathbf{Xb}$ for some $\mathbf{b} \in R^p$. Therefore,

$$\mathbf{t} \in T \Leftrightarrow \mathbf{t} = \mathbf{Xb},$$

and it follows that if $t \in T$, then

$$\mathbf{P}_T\mathbf{t} = \mathbf{X}(\mathbf{X}'\mathbf{X})^{-1}\mathbf{X}'\mathbf{Xb} = \mathbf{Xb} = \mathbf{t}.$$

Also, if $\mathbf{s}$ is orthogonal to T, then $\mathbf{s}$ is orthogonal to the columns of $\mathbf{X}$, and hence, $\mathbf{X}'\mathbf{s} = \mathbf{0}$. But then, $\mathbf{P}_T\mathbf{s} = \mathbf{0}$.

c. $\mathbf{P}_T\mathbf{y} = \mathbf{X}[(\mathbf{X}'\mathbf{X})^{-1}\mathbf{X}'\mathbf{y}] = \mathbf{Xc}$ (where $\mathbf{c} = (\mathbf{X}'\mathbf{X})^{-1}\mathbf{X}'\mathbf{y}$), and hence, $\mathbf{Y} \in T$. Now, let $\mathbf{z} = \mathbf{Xb} \in T$. Then

$$\mathbf{z}'(\mathbf{y} - \mathbf{P}_T\mathbf{y}) = \mathbf{b}'\mathbf{X}'\mathbf{y} - \mathbf{b}'\mathbf{X}'\mathbf{X}(\mathbf{X}'\mathbf{X})^{-1}\mathbf{X}'\mathbf{y} = \mathbf{b}'\mathbf{X}'\mathbf{y} - \mathbf{b}'\mathbf{X}'\mathbf{y} = 0,$$

and therefore, $\mathbf{y} - \mathbf{P}_T\mathbf{y}$ is orthogonal to T.

d. Let $\mathbf{X}$ and $\mathbf{Z}$ be matrices whose columns form bases for T and S. Since S and T are orthogonal subspaces, the columns of $\mathbf{X}$ are orthogonal to those of $\mathbf{Z}$, and hence, $\mathbf{X}'\mathbf{Z} = \mathbf{0}$, which implies that $\mathbf{P}_T\mathbf{P}_S = \mathbf{0}$. □

Now, suppose we observe an N-dimensional random vector

$$\mathbf{Y} \sim N_N(\boldsymbol{\mu}, \sigma^2\mathbf{I}), \qquad \boldsymbol{\mu} = \sum_i \boldsymbol{\delta}_i, \qquad \boldsymbol{\delta}_i \in T_i,$$

where the T_i are orthogonal subspaces. Let df_i be the dimension of T_i. We say that this model is an *orthogonal design*. (This definition is just a more precise version of that given in Section 13.1.2; in the notation of that section, $\boldsymbol{\delta}_i = \boldsymbol{\delta}^{(i)}$. Note that $\mathbf{Y}$ has a spherical normal distribution.)

Let $\mathbf{P}_i$ be the projection matrix on T_i (i.e., $\mathbf{P}_i = \mathbf{P}_{T_i}$). Since T_i and T_j are orthogonal,

$$\mathbf{P}_i\mathbf{P}_j = 0 \quad \text{if } i \neq j.$$

Now, define

$$\mathbf{U}_i = \mathbf{P}_i\mathbf{Y} \quad \text{and} \quad \mathbf{Q} = \mathbf{I} - \sum_i \mathbf{P}_i,$$

where $\mathbf{I}$ is an $N \times N$ identity matrix. The following is the basic result that drives all other results on orthogonal designs.

Lemma 13–4

a. $\mathbf{Q}$ is idempotent, and $\mathbf{P}_i\mathbf{Q} = \mathbf{Q}\mathbf{P}_i = 0$.
b. $\mathbf{P}_i\boldsymbol{\mu} = \boldsymbol{\delta}_i$ and $\mathbf{Q}\boldsymbol{\mu} = \mathbf{0}$.
c. $\mathbf{Y}'\mathbf{Q}\mathbf{Y} = (\mathbf{Y} - \Sigma_i\mathbf{U}_i)'(\mathbf{Y} - \Sigma_i\mathbf{U}_i)$.
d. $(\mathbf{Y} - \boldsymbol{\mu})'(\mathbf{Y} - \boldsymbol{\mu}) = \mathbf{Y}'\mathbf{Q}\mathbf{Y} + \Sigma_i(\mathbf{U}_i - \boldsymbol{\delta}_i)'(\mathbf{U}_i - \boldsymbol{\delta}_i)$.

Proof

a. The proof of this part is left as an exercise.
b. Since $\boldsymbol{\delta}_i \in T_i$ and $\boldsymbol{\delta}_j$ is orthogonal to T_i, $\mathbf{P}_i\boldsymbol{\delta}_i = \boldsymbol{\delta}_i$ and $\mathbf{P}_i\boldsymbol{\delta}_j = \mathbf{0}$ if $i \neq j$, by part b of Lemma 13–3. Hence,

$$\mathbf{P}_i\boldsymbol{\mu} = \mathbf{P}_i\sum_j \boldsymbol{\delta}_j = \sum_j \mathbf{P}_i\boldsymbol{\delta}_j = \mathbf{P}_i\boldsymbol{\delta}_i = \boldsymbol{\delta}_i.$$

Similarly,

$$\mathbf{Q}\boldsymbol{\mu} = (\mathbf{I} - \sum_i \mathbf{P}_i)\boldsymbol{\mu} = \boldsymbol{\mu} - \sum_i \boldsymbol{\delta}_i = \boldsymbol{\mu} - \boldsymbol{\mu} = 0.$$

c. Using the idempotence of $\mathbf{Q}$, we have

$$\mathbf{Y}'\mathbf{Q}\mathbf{Y} = (\mathbf{Q}\mathbf{Y})'(\mathbf{Q}\mathbf{Y}) = ((\mathbf{I} - \sum_i \mathbf{P}_i)\mathbf{Y})'((\mathbf{I} - \sum_i \mathbf{P}_i)\mathbf{Y}) = (\mathbf{Y} - \sum_i \mathbf{U}_i)'(\mathbf{Y} - \sum_i \mathbf{U}_i).$$

d. From the definition of $\mathbf{Q}$ and the idempotence of $\mathbf{Q}$ and $\mathbf{P}_i$,

$$\mathbf{I} = \mathbf{Q} + \sum_i \mathbf{P}_i = \mathbf{Q}'\mathbf{Q} + \sum_i \mathbf{P}_i'\mathbf{P}_i.$$

Therefore,

$$(\mathbf{Y} - \boldsymbol{\mu})'\mathbf{I}(\mathbf{Y} - \boldsymbol{\mu}) = (\mathbf{Q}(\mathbf{Y} - \boldsymbol{\mu}))'(\mathbf{Q}(\mathbf{Y} - \boldsymbol{\mu})) + \sum_i(\mathbf{P}_i(\mathbf{Y} - \boldsymbol{\mu}))'(\mathbf{P}_i(\mathbf{Y} - \boldsymbol{\mu}))$$

$$= \mathbf{Y}'\mathbf{Q}\mathbf{Y} + \sum_i(\mathbf{U}_i - \boldsymbol{\delta}_i)'(\mathbf{U}_i - \boldsymbol{\delta}_i)$$

(using the fact that $\mathbf{Q}$ is idempotent and $\mathbf{Q}\boldsymbol{\mu} = 0$). □

Now, as in Section 13.1.2, let $\hat{\boldsymbol{\delta}}_i$ be the ordinary least squares estimator of $\boldsymbol{\delta}_i$, and let

$$SS_i = \hat{\boldsymbol{\delta}}_i'\hat{\boldsymbol{\delta}}_i, \qquad SST = \mathbf{Y}'\mathbf{Y}, \qquad SSE = \left(\mathbf{Y} - \sum_i \hat{\boldsymbol{\delta}}_i\right)'\left(\mathbf{Y} - \sum_i \hat{\boldsymbol{\delta}}_i\right).$$

Theorem 13–5. $\hat{\boldsymbol{\delta}}_i = \mathbf{U}_i = \mathbf{P}_i\mathbf{Y}$, $SS_i = \mathbf{Y}'\mathbf{P}_i\mathbf{Y}$, and $SSE = \mathbf{Y}'\mathbf{Q}\mathbf{Y}$.

Proof. By Lemma 13–4c and d

$$\left(\mathbf{Y} - \sum_i \boldsymbol{\delta}_i\right)'\left(\mathbf{Y} - \sum_i \boldsymbol{\delta}_i\right) = \mathbf{Y}'\mathbf{Q}\mathbf{Y} + \sum_i(\mathbf{U}_i - \boldsymbol{\delta}_i)'(\mathbf{U}_i - \boldsymbol{\delta}_i)$$

$$\geqq \mathbf{Y}'\mathbf{Q}\mathbf{Y} = \left(\mathbf{Y} - \sum_i \mathbf{U}_i\right)'\left(\mathbf{Y} - \sum_i \mathbf{U}_i\right),$$

with equality only if $\boldsymbol{\delta}_i = \mathbf{U}_i$. Since the $\mathbf{U}_i$ are in T_i, they must be the ordinary least squares estimators of the $\boldsymbol{\delta}_i$. Now,

$$SS_i = \mathbf{U}_i'\,\mathbf{U}_i = (\mathbf{P}_i\mathbf{Y})'(\mathbf{P}_i\mathbf{Y}) = \mathbf{Y}'\mathbf{P}_i\mathbf{Y},$$

using the idempotence of $\mathbf{P}_i$. Finally, by part c of Lemma 13.4, $SSE = \mathbf{Y}'\mathbf{Q}\mathbf{Y}$. □

The following identity follows from Theorem 13–5 and Lemma 13–4b:

$$\left(\mathbf{Y} - \sum_i \boldsymbol{\delta}_i\right)'\left(\mathbf{Y} - \sum_i \boldsymbol{\delta}_i\right) = SSE + \sum_i (\hat{\boldsymbol{\delta}}_i - \boldsymbol{\delta}_i)'(\hat{\boldsymbol{\delta}}_i - \boldsymbol{\delta}_i).$$

Since this identity is true for all $\boldsymbol{\delta}_i \in T_i$, it is true for $\boldsymbol{\delta}_i = \mathbf{0}$. Therefore, we also have

$$SST = SSE + \sum_i SS_i.$$

As in Section 13.1.2, let

$$MS_i = \frac{SS_i}{df_i}, \qquad dfe = N - \sum_i df_i, \qquad MSE = \frac{SSE}{dfe}, \qquad F_i = \frac{MS_i}{MSE}.$$

Theorem 13–6

a. SS_i and SSE are independent, and $SSE/\sigma^2 \sim \chi^2_{dfe}$. If $\boldsymbol{\delta}_i = \mathbf{0}$, then $SS_i/\sigma^2 \sim \chi^2_{df_i}$.
b. If $\boldsymbol{\delta}_i = 0$, then $F_i \sim F_{df_i, dfe}$.
c. $E(MSE) = \sigma^2$, $E(MS_i) = \sigma^2 + \boldsymbol{\delta}_i'\,\boldsymbol{\delta}_i/df_i$.

Proof

a. We use the basic results derived earlier for quadratic forms and spherical normal distributions. We have

$$\operatorname{tr}\mathbf{Q} = \operatorname{tr}\left(\mathbf{I} - \sum_i \mathbf{P}_i\right) = \operatorname{tr}\mathbf{I} - \sum_i \operatorname{tr}\mathbf{P}_i = N - \sum_i df_i = dfe \quad \text{and} \quad \boldsymbol{\mu}'\mathbf{Q}\boldsymbol{\mu} = 0.$$

Also, Q is idempotent; therefore,

$$\frac{SSE}{\sigma^2} = \frac{\mathbf{Y}'\mathbf{Q}\mathbf{Y}}{\sigma^2} \sim \chi^2_{dfe}.$$

Now, $\mathbf{P}_i$ is also an idempotent matrix, $\operatorname{tr}\mathbf{P}_i = df_i$, and

$$\boldsymbol{\mu}'\mathbf{P}_i\boldsymbol{\mu} = (\mathbf{P}_i\boldsymbol{\mu})'(\mathbf{P}_i\boldsymbol{\mu}) = \boldsymbol{\delta}_i'\,\boldsymbol{\delta}_i.$$

Therefore, if $\boldsymbol{\delta}_i = \mathbf{0}$, then

$$\frac{SS_i}{\sigma^2} = \frac{\mathbf{Y}'\mathbf{P}_i\mathbf{Y}}{\sigma^2} \sim \chi^2_{df_i}.$$

Finally, $\mathbf{P}_i\mathbf{Q} = \mathbf{0}$, so that SS_i and SSE are independent.

b. This result follows from part a by the definition of the F-distribution.
c. By part a,

$$E(MSE) = \frac{\sigma^2 E(SSE/\sigma^2)}{n-p} = \frac{\sigma^2(n-p)}{n-p} = \sigma^2.$$

But by Theorem 5–11,

$$ESS_i = E\mathbf{Y}'\mathbf{P}_i\mathbf{Y} = \boldsymbol{\mu}'\mathbf{P}_i\boldsymbol{\mu} + \text{tr}(\sigma^2\mathbf{P}_i) = \boldsymbol{\delta}_i'\boldsymbol{\delta}_i + \sigma^2 df_i;$$

hence,

$$E(MS_i) = \frac{E(SS_i)}{df_i} = \sigma^2 + \frac{\boldsymbol{\delta}_i'\boldsymbol{\delta}_i}{df_i}. \quad \square$$

Part b of Theorem 13–6 implies that the F-statistics that we have derived in earlier sections really do have the appropriate F-distributions under the null hypothesis. Part c was used in Section 13.1.2 to indicate why it is sensible to reject the null hypothesis if F_i is too large.

Theorem 13–7

a. The MLE for $(\boldsymbol{\delta}_1, \ldots, \boldsymbol{\delta}_p, \sigma^2)$ is $(\hat{\boldsymbol{\delta}}_1, \ldots, \hat{\boldsymbol{\delta}}_p, \hat{\sigma}^2)$, where $\hat{\sigma}^2 = SSE/N$.
b. The likelihood ratio test for testing that $\boldsymbol{\delta}_i = 0$ rejects the null hypothesis if F_i is too large.

Proof

a. From the model, $\mathbf{Y} \sim N_N(\boldsymbol{\mu}, \sigma^2\mathbf{I})$, so that the logarithm of the likelihood for the model is

$$\log[L_Y(\boldsymbol{\delta}, \sigma^2)] = -\left(\frac{N}{2}\right)\log(2\pi) - N\log\sigma - \frac{(\mathbf{Y}-\boldsymbol{\mu})'(\mathbf{Y}-\boldsymbol{\mu})}{2\sigma^2}$$

$$= -\left(\frac{N}{2}\right)\log(2\pi) - N\log\sigma - \frac{\left(SSE + \sum_i (\hat{\boldsymbol{\delta}}_i - \boldsymbol{\delta}_i)'(\hat{\boldsymbol{\delta}}_i - \boldsymbol{\delta}_i)\right)}{2\sigma^2}.$$

This expression is maximized when $\boldsymbol{\delta}_i = \hat{\boldsymbol{\delta}}_i$, so that $\hat{\boldsymbol{\delta}}_i$ is the MLE of $\boldsymbol{\delta}_i$. Routine differentiation of $\log[L_Y(\hat{\boldsymbol{\delta}}, \sigma^2)]$ with respect to σ then leads to the result that the MLE of σ^2 is SSE/N.

b. Let $\hat{\sigma}^2$ be the MLE of σ^2. Then

$$L_Y(\hat{\boldsymbol{\delta}}, \hat{\sigma}^2) = (2\pi)^{-N/2}\hat{\sigma}^{-N}\exp\left(\frac{-SSE}{2\hat{\sigma}^2}\right) = (2\pi)^{-N/2}(\hat{\sigma}^2)^{-N/2}e^{-N/2}.$$

Now, suppose that $\boldsymbol{\delta}_1 = 0$. Then by the same argument as in part a,

$$\hat{\hat{\boldsymbol{\delta}}}_1 = \mathbf{0}, \qquad \hat{\hat{\boldsymbol{\delta}}}_j = \hat{\boldsymbol{\delta}}_j, \qquad \hat{\hat{\sigma}}^2 = \frac{(SSE + \hat{\boldsymbol{\delta}}_1'\hat{\boldsymbol{\delta}}_1)}{N} = \frac{SSE + SS_1}{N},$$

and

$$L_Y(\hat{\hat{\boldsymbol{\delta}}}, \hat{\hat{\sigma}}^2) = (2\pi)^{-N/2}(\hat{\hat{\sigma}}^2)^{-N/2}e^{-N/2}.$$

Therefore, the likelihood ratio test statistic for the model is given by

$$\Lambda = \left(\frac{\hat{\hat{\sigma}}^2}{\hat{\sigma}^2}\right)^{-N/2} = \left(1 + \frac{SS_1}{SSE}\right)^{N/2} = \left[1 + \frac{(df_1)F_1}{dfe}\right]^{N/2}.$$

The likelihood ratio test rejects the null hypothesis when Λ is large, or equivalently, when F_1 is large. The proof for the other $\boldsymbol{\delta}_i$ is similar. $\square$

We next prove the theorem on Scheffé simultaneous confidence intervals.

Theorem 13–8. Let $F^\alpha = F^\alpha_{df_i, dfe}$. For any $\mathbf{d} \in T_i$, let I_d be the interval

$$\mathbf{d}'\boldsymbol{\delta}_i \in \mathbf{d}'\hat{\boldsymbol{\delta}}_i \pm (MSE\, df_i\, F^\alpha\, \mathbf{d}'\mathbf{d})^{1/2}.$$

Then:

a. $P(\mathbf{d}'\boldsymbol{\delta}_i \in I_\mathbf{d} \text{ for all } \mathbf{d} \in T_i) = 1 - \alpha$.

b. The hypothesis $\boldsymbol{\delta}_i = \mathbf{0}$ is rejected with the appropriate F-test if and only if $0 \notin I_d$ for at least one $\mathbf{d}$ in T_i.

c. If $0 \notin I_d$ for any $\mathbf{d} \in T_i$, then $0 \notin I_d$ for $\mathbf{d} = \hat{\boldsymbol{\delta}}^{(i)}$.

Proof. Let $\mathbf{v} \in T_i$, and let $\mathbf{d} \neq \mathbf{0}$ be another vector in T_i. Then by the Cauchy-Schwarz inequality for vectors,

$$\frac{(\mathbf{d}'\mathbf{v})^2}{\mathbf{d}'\mathbf{d}} \leqq \mathbf{v}'\mathbf{v}.$$

In addition, for $\mathbf{d} = \mathbf{v}$,

$$\frac{(\mathbf{d}'\mathbf{v})^2}{\mathbf{d}'\mathbf{d}} = \mathbf{v}'\mathbf{v}.$$

Therefore,

$$\mathbf{v}'\mathbf{v} = \sup_{\mathbf{d} \in T_i, \mathbf{d} \neq \mathbf{0}} \left[\frac{(\mathbf{d}'\mathbf{v})^2}{\mathbf{d}'\mathbf{d}}\right].$$

a. Let $\tilde{F} = (\hat{\boldsymbol{\delta}}_i - \boldsymbol{\delta}_i)'(\hat{\boldsymbol{\delta}}_i - \boldsymbol{\delta}_i)/(df_i\, MSE)$. Then $\tilde{F} \sim F_{df_i, dfe}$. (See Exercise B3.) Now, $\mathbf{0}'\boldsymbol{\delta}_i \in I_\mathbf{0}$. Hence,

$$P(\mathbf{d}'\boldsymbol{\delta}_i \in I_\mathbf{d} \text{ for all } \mathbf{d} \in T_i) = P\left(\frac{(\mathbf{d}'(\hat{\boldsymbol{\delta}}_i - \boldsymbol{\delta}_i))^2}{\mathbf{d}'\mathbf{d}} \leqq MSE\, df_i\, F^\alpha \text{ for all } \mathbf{d} \in T_i, \mathbf{d} \neq \mathbf{0}\right)$$

$$= P\left(\sup_{\mathbf{d} \in T_i, \mathbf{d} \neq \mathbf{0}} \left[\frac{(\mathbf{d}'(\boldsymbol{\delta}_i - \boldsymbol{\delta}))^2}{\mathbf{d}'\mathbf{d}}\right] \leqq MSE\, df_i\, F^\alpha\right)$$

$$= P((\hat{\boldsymbol{\delta}}_i - \boldsymbol{\delta}_i)'(\hat{\boldsymbol{\delta}}_i - \boldsymbol{\delta}_i) \leqq MSE\, df_i\, F^\alpha)$$

$$= P(\tilde{F} \leqq F^\alpha) = 1 - \alpha.$$

b. Zero is in all the intervals if and only if

$$\frac{(\mathbf{d}'\hat{\boldsymbol{\delta}}_i)^2}{\mathbf{d}'\mathbf{d}} \leqq MSE\, df_i\, F^\alpha \text{ for all } \mathbf{d} \in T_i, \mathbf{d} \neq \mathbf{0}$$

if and only if

$$\hat{\boldsymbol{\delta}}_i'\, \hat{\boldsymbol{\delta}}_i = \sup_{\mathbf{d} \in T_i, \mathbf{d} \neq \mathbf{0}} \frac{(\mathbf{d}'\hat{\boldsymbol{\delta}}_i)^2}{\mathbf{d}'\mathbf{d}} \leqq MSE\, df_i\, F^\alpha$$

if and only if

$$F_i = \frac{\hat{\boldsymbol{\delta}}_i'\, \hat{\boldsymbol{\delta}}_i}{df_i\, MSE} \leqq F^\alpha$$

if and only if we accept the null hypothesis with the F-test.

c. By the argument of part b, $0 \notin I_{\mathbf{d}}$ if and only if

$$\frac{(\mathbf{d}'\hat{\boldsymbol{\delta}}_i)^2}{(d'd)MSE(df_i)} > F^{\alpha}.$$

Now,

$$\frac{(\hat{\boldsymbol{\delta}}_i'\,\hat{\boldsymbol{\delta}}_i)^2}{\hat{\boldsymbol{\delta}}_i'\,\hat{\boldsymbol{\delta}}_i} = \hat{\boldsymbol{\delta}}_i'\,\hat{\boldsymbol{\delta}}_i = \sup_{\mathbf{d} \in T_1, \mathbf{d} \neq 0} \frac{(\mathbf{d}'\hat{\boldsymbol{\delta}}_i)^2}{\mathbf{d}'\mathbf{d}},$$

so that if any contrast C_d is significant, then the contrast C_d, $\mathbf{d} = \hat{\boldsymbol{\delta}}_i$, is significant. □

In concluding the section, we emphasize that the methods derived here are valid only for orthogonal designs and should be used only for such designs. For nonorthogonal designs, the analysis is much more difficult.

Exercises—B

1. Show that $\mathbf{P}_T$ is an idempotent matrix and $\text{tr}(\mathbf{P}_T)$ is equal to dimension of T. (*Hint*: $\text{tr}(\mathbf{P}_T) = \text{tr}[(\mathbf{X}'\mathbf{X})^{-1}\mathbf{X}'\mathbf{X}]$. Why?)
2. Show that $\mathbf{Q}$ is an idempotent matrix and that $\mathbf{QP}_i = \mathbf{0}$.
3. Let $\tilde{F}$ be defined as in the proof of Theorem 13–8. Show that $\tilde{F} \sim F_{df_i, dfe}$. (*Hint*: $\tilde{F} = (\mathbf{Y} - \boldsymbol{\mu})'\mathbf{P}_i(\mathbf{Y} - \boldsymbol{\mu})/df_i\,MSE$.)
4. Show that SS_i and SS_j are independent if $i \neq j$.

Exercises—C

1. Let T be a p-dimensional subspace, and let $\mathbf{X}$ and $\mathbf{Z}$ be two $n \times p$ matrices whose columns are a basis for T.
   - **(a)** Show that there exists a $p \times p$ matrix $\mathbf{A}$ such that $\mathbf{X} = \mathbf{ZA}$.
   - **(b)** Show that $\mathbf{A}$ is invertible. (*Hint*: Show that $\mathbf{A}$ has a rank of at least p.)
   - **(c)** Show that $\mathbf{X}(\mathbf{X}'\mathbf{X})^{-1}\mathbf{X}' = \mathbf{Z}(\mathbf{Z}'\mathbf{Z})^{-1}\mathbf{Z}'$.
2. **(a)** Show that $(\hat{\boldsymbol{\delta}}_1, \ldots, \hat{\boldsymbol{\delta}}_p, SSE)$ is a sufficient statistic for the model of this section. (*Hint*: The likelihood is given in the proof of part a of Theorem 13–5.)
   (b) Show that $\hat{\delta}_i$ and MSE are the best unbiased estimators of $\boldsymbol{\delta}_i$ and σ^2. (You may assume that $(\hat{\boldsymbol{\delta}}_1, \ldots, \hat{\boldsymbol{\delta}}_p, SSE)$ is a complete sufficient statistic.)

*# 13.5.2 Analysis of Variance as Regression

Analysis-of-variance models can be rewritten as regression models. Such rewriting is often quite useful for analysis-of-variance models which are not orthogonal designs, such as crossed or nested models which are unbalanced.

We illustrate how to write an analysis-of-variance model as a regression model by means of the two-way model with no replication presented earlier in which there are two rows and three columns. In that model, we observe Y_{ij} independent, with

$$Y_{ij} \sim N(\theta + \alpha_i + \beta_j, \sigma^2), \qquad i = 1, 2, j = 1, 2, 3.$$

To put this into a regression model, let

$$\mathbf{Y} = (Y_{11}, Y_{12}, Y_{13}, Y_{21}, Y_{22}, Y_{23})',$$

$$\boldsymbol{\mu} = E\mathbf{Y} = (\theta + \alpha_1 + \beta_1, \theta + \alpha_1 + \beta_2, \theta + \alpha_1 + \beta_3, \theta + \alpha_2 + \beta_1, \theta + \alpha_2 + \beta_2, \theta + \alpha_2 + \beta_3)',$$

$$\mathbf{x} = \begin{pmatrix} 1 & 1 & 0 & 1 & 0 & 0 \\ 1 & 1 & 0 & 0 & 1 & 0 \\ 1 & 1 & 0 & 0 & 0 & 1 \\ 1 & 0 & 1 & 1 & 0 & 0 \\ 1 & 0 & 1 & 0 & 1 & 0 \\ 1 & 0 & 1 & 0 & 0 & 1 \end{pmatrix}, \qquad \boldsymbol{\beta} = \begin{pmatrix} \theta \\ \alpha_1 \\ \alpha_2 \\ \beta_1 \\ \beta_2 \\ \beta_3 \end{pmatrix}.$$

Then

$$\boldsymbol{\mu} = \mathbf{x}\boldsymbol{\beta} \quad \text{and} \quad Y \sim N_6(\mathbf{x}\boldsymbol{\beta}, \sigma^2\mathbf{I}).$$

This model is not a multiple regression model, since the columns of $\mathbf{x}$ are not linearly independent. (The second and third columns sum to the first column, as do the last three columns.) To remove this multicollinearity, recall that

$$\alpha_1 + \alpha_2 = 0 \quad \text{and} \quad \beta_1 + \beta_2 + \beta_3 = 0,$$

so that

$$\alpha_2 = -\alpha_1 \quad \text{and} \quad \beta_3 = -\beta_1 - \beta_2.$$

Therefore, $\boldsymbol{\mu} = \mathbf{x}^*\boldsymbol{\beta}^*$, where

$$\mathbf{x}^* = \begin{pmatrix} 1 & 1 & 1 & 0 \\ 1 & 1 & 0 & 1 \\ 1 & 1 & -1 & -1 \\ 1 & -1 & 1 & 0 \\ 1 & -1 & 0 & 1 \\ 1 & -1 & -1 & -1 \end{pmatrix} \quad \text{and} \quad \boldsymbol{\beta}^* = \begin{pmatrix} \theta \\ \alpha_1 \\ \beta_1 \\ \beta_2 \end{pmatrix},$$

and hence,

$$\mathbf{Y} \sim N_6(\mathbf{x}^*\boldsymbol{\beta}^*, \sigma^2\mathbf{I}).$$

(Note that the columns of $\mathbf{x}^*$ are linearly independent.) Now, let

$$\mathbf{a} = (0, 1, 0, 0) \quad \text{and} \quad \mathbf{B} = \begin{pmatrix} 0 & 0 & 1 & 0 \\ 0 & 0 & 0 & 1 \end{pmatrix}.$$

Then the $\alpha_i = 0$ if and only if $\mathbf{a}\boldsymbol{\beta}^* = 0$, and the $\beta_j = 0$ if and only if $\mathbf{B}\boldsymbol{\beta}^* = \mathbf{0}$. Consequently, to test that the $\alpha_i = 0$, we use the test derived in Chapter 12 for testing that $\mathbf{a}\boldsymbol{\beta}^* = 0$, and to test that $\beta_j = 0$, we use the test derived in Section 12.3 for testing that $\mathbf{B}\boldsymbol{\beta} = \mathbf{0}$.

From the preceding example, it should be clear how to write a general analysis-of-variance model as a multiple regression model. First, we use the constraints to get a set of parameters that are linearly independent. Then we write the mean vector as a linear function of these parameters and put this linear function in matrix form. The resulting model is a multiple regression model. We can then test the hypotheses of interest using the tests developed for that model.

Exercises—A

1. An experiment was run to test the effect of traffic and humidity on a particular paint for marking streets. The wear was measured for samples of the paint under two traffic conditions (A and B) and two humidity conditions (a and b). The following data were obtained:

	A	B
a	7, 8, 6, 4	2, 3
b	5, 8, 3	5, 9, 6

(a) Using the two-way model with interaction (and the usual constraints), set up the **x** matrix and the **Y** vector for these data. What is the dimension of **x**′**x**? Write the hypothesis of no row effect in the form $\mathbf{A\beta} = \mathbf{0}$. Write the hypothesis of no column effect in the form $\mathbf{B\beta} = \mathbf{0}$. Write the hypothesis of no interaction effect in the form $\mathbf{C\beta} - \mathbf{0}$.

2. (a) Show how to write a two-fold nested model when $r = 3$, $c_1 = c_2 = c_3 = 2$, and $n = 2$ as a regression model. (The **x** matrix should be 12×6.)

(b) Write the hypothesis of no class effect in the form $\mathbf{A\beta} = \mathbf{0}$. Write the hypothesis of no subcolumn effect in the form $D\boldsymbol{\beta} = 0$.

3. (a) Show how to write a 3×3 Latin square model as a regression model. (The **x** matrix should be 9×7.) Use the Latin square in Example 13.6.

(b) Write the hypothesis $\gamma_k = 0$ in the form $\mathbf{C\beta} = \mathbf{0}$.

Chapter 14

INFERENCES FOR MULTINOMIAL DISTRIBUTIONS

In this chapter, we consider methods of drawing inferences for multinomial distributions. There are no pivotal quantities or test statistics for the models discussed; therefore, we concentrate on finding asymptotic tests.

14.1 THE ONE-SAMPLE MULTINOMIAL MODEL

14.1.1 The One-Sample Multinomial Model

Consider the model in which we observe $\mathbf{O} = (O_1, \ldots, O_p)$ having a multinomial distribution with parameters n and $\boldsymbol{\pi} = (\pi_1, \ldots, \pi_p)$, where n is known, but $\boldsymbol{\pi}$ is an unknown parameter vector. That is, we observe

$$\mathbf{O} \sim M_p(n, \boldsymbol{\pi}).$$

Note that when $p = 2$, this model is the same as the one-sample Bernoulli model discussed in Chapter 11. We call this model the *one-sample multinomial model*. (In writing the model in this form, we have reduced already to a sufficient statistic (see Exercise B1.)

As an example, suppose we want to find whether a die is fair. We roll the die n times. We let O_i be the number of times we get the ith face. Then

$$(O_1, \ldots, O_6) \sim M_6(n, (\pi_1, \ldots, \pi_6)),$$

where π_i is the probability that face i occurs on any particular roll. If the die is fair, then the π_i are all 1/6.

The following lemma is quite useful for multinomial models.

Lemma. Let $h(s_1, \ldots, s_p) = \Pi_i s_i^{a_i}$, where the a_i are constants. The maximum value of $h(s_1, \ldots, s_p)$ subject to $\Sigma b_i s_i = 1$ is

$$s_i = \frac{a_i}{b_i \sum_j a_j}.$$

Proof. We shall find the maximum of $\log(h(s_1, \ldots, s_p)) = \Sigma_i a_i \log(s_i)$ subject to the constraint given. We use the method of Lagrange multipliers. Let

$$g(s_1, \ldots, s_p) = \sum_i a_i \log(s_i) - \lambda \sum_i b_i s_i.$$

If we set the partial derivative of g with respect to s_i equal to 0, we get

$$0 = -\lambda b_i + \frac{a_i}{s_i}.$$

Therefore, the constrained maximum occurs when $s_i = a_i / \lambda b_i$. Now,

$$1 = \sum_j b_j s_j = \frac{\sum_j a_j}{\lambda},$$

so that $\lambda = \Sigma_j a_j$ and $s_i = a_i / b_i \Sigma_j a_j$. □

We are now ready to make inferences about the π_i.

Theorem 14–1

a. The likelihood for the one-sample multinomial model is

$$L_O(\boldsymbol{\pi}) = h(\mathbf{O}) \prod_i \pi_i^{O_i}.$$

b. $(O_1, \ldots, O_p)$ is a complete sufficient statistic.

c. O_i / n is the best unbiased estimator and MLE of π_i and is consistent.

Proof

a. From the multinomial density function, we see that

$$L_O(\boldsymbol{\pi}) = \begin{bmatrix} n \\ O_1, \ldots, O_p \end{bmatrix} \pi_1^{O_1} \pi_2^{O_2} \cdots \pi_p^{O_p} = h(\mathbf{O}) \Pi_i \pi_i^{O_i}.$$

b. To see that $\mathbf{O}$ is a complete sufficient statistic, we need to put the likelihood in exponential form with $p - 1$ dimensions, since $\boldsymbol{\pi}$ has $p - 1$ independent parameters. Because

$$O_p = n - \sum_i O_i$$

we have

$$L_O(\boldsymbol{\pi}) = h(\mathbf{O}) \exp\left[\sum_{i=1}^{p} O_i \log(\pi_i)\right] = h(\mathbf{O}) \pi_p^n \exp\left[\sum_{i=1}^{p-1} O_i c_i(\pi)\right],$$

where

$$c_i(\boldsymbol{\pi}) = \log\left(\frac{\pi_i}{\pi_p}\right).$$

Since $\boldsymbol{\pi}$ has $p - 1$ linearly independent components, $\mathbf{O}^* = (O_1, \ldots, O_{p-1})$ is a complete sufficient statistic. $\mathbf{O}$ is an invertible function of $\mathbf{O}^*$, and is therefore also a complete sufficient statistic.

c. In Section 5.1, we showed that $O_i \sim B(n, \pi_i)$. Therefore,

$$E\frac{O_i}{n} = \pi_i \quad \text{and} \quad \text{var}\left(\frac{O_i}{n}\right) = \frac{\pi_i(1-\pi_i)}{n}.$$

Hence, O_i/n is a consistent, unbiased estimator of π_i. By the Lehmann-Scheffe' theorem, O_i/n is the best unbiased estimator of π_i. To find the MLE of $\boldsymbol{\pi}$, we must maximize

$$L_O(\boldsymbol{\pi}) = h(\mathbf{O})\prod_i \pi_i^{O_i},$$

subject to the constraint that $\Sigma\pi_i = 1$. Therefore, by the lemma above, the MLE of π_i is

$$\frac{O_i}{\sum O_j} = \frac{O_i}{n}. \quad \square$$

Let us next consider the problem of testing that $\boldsymbol{\pi} = \mathbf{d}$, where $\mathbf{d}$ is a known p-dimensional vector whose components sum to one. If $\mathbf{d} = (d_1, \ldots, d_p)'$, then we are testing that

$$\pi_1 = d_1, \ldots, \pi_p = d_p.$$

Let

$$E_i = nd_i, \qquad i = 1, \ldots, p.$$

We think of the O_i as the number we actually observed in the ith class and the E_i as the number we would have expected to observe if the null hypothesis were true. Therefore, under the null hypothesis, we would expect the O_i and the E_i to be close.

Due to the discreteness of the observations, there is, in general, no exact test for this problem. Several asymptotic tests have been proposed. Let

$$U_P = \sum_i \frac{(O_i - E_i)^2}{E_i} \quad \text{and} \quad U_L = 2\sum_i O_i \log\left(\frac{O_i}{E_i}\right).$$

U_P is called *Pearson's* χ^2, and U_L is called the *likelihood ratio* χ^2.

Theorem 14–2

a. Let Λ be the likelihood ratio test (LRT) statistic for testing $\boldsymbol{\pi} = \mathbf{d}$ against $\boldsymbol{\pi} \neq \mathbf{d}$. Then

$$U_L = 2\log(\Lambda).$$

b. Under the null hypothesis $U_L - U_P \xrightarrow{P} 0$.

c. Under the null hypothesis

$$U_L \xrightarrow{d} V \sim \chi^2_{p-1} \quad \text{and} \quad U_P \xrightarrow{d} V \sim \chi^2_{p-1}.$$

Proof

a. Note that O_i/n is the MLE for π_i under the alternative hypothesis and

$E_i/n = d_i$ is the MLE under the null hypothesis. Therefore

$$\Lambda = \frac{h(\mathbf{O})\prod_i (O_i/n)^{O_i}}{h(\mathbf{O})\prod_i (E_i/n_i)^{O_i}} = \prod_i \left(\frac{O_i}{E_i}\right)^{O_i}.$$

Therefore,

$$2\,\log(\Lambda) = 2\sum O_i \log\left(\frac{O_i}{E_i}\right) = U_L.$$

b. Using Taylor's theorem, we see that

$$\log(1+t) = t - \frac{t^2}{2} + t^2 R(t) \quad \text{where } R(t) \to R(0) = 0 \text{ as } t \to \infty.$$

Let $T_i = (E_i - O_i)/O_i$. Then

$$U_L = -2\sum O_i \log(1+T_i) = -2\sum O_i T_i + \sum O_i T_i^2 - 2\sum O_i T_i^2 R(T_i).$$

Now,

$$\sum O_i T_i = \sum E_i - \sum O_i = n - n = 0.$$

Note that $(E_i - O_i)^2/E_i = T_i^2 O_i^2/E_i$. Therefore

$$U_L - U_P = \sum O_i T_i^2 H_i,$$

where $H_i = ((1 - O_i/E_i) - 2R(T_i))$. Now, under the null hypothesis,

$$H_i \xrightarrow{P} 0 \quad \text{and} \quad O_i T_i^2 \xrightarrow{d} (1-\pi_i)Z^2,$$

where $Z \sim N(0,1)$. (See Exercises C2 and C3.) Therefore, by Slutzky's theorem,

$$H_i O_i T_i^2 \xrightarrow{P} 0 \quad \text{and} \quad \sum H_i O_i T_i^2 \xrightarrow{P} 0.$$

(Note that there are only finitely many (p) terms in this sum.)

c. Note that $\pi_p = (1 - \pi_1 - \cdots - \pi_{p-1})$ is a redundant parameter. We are actually testing that $(\pi_1, \ldots, \pi_{p-1}) = (d_1, \ldots, d_{p-1})$. Also, $\mathbf{O}$ is a sufficient statistic from a sample of size n from a $M_p(1, \boldsymbol{\pi})$ distribution. Therefore, by Theorem 9–4, under the null hypothesis,

$$U_L \xrightarrow{d} V \sim \chi^2_{p-1}.$$

Now, by Slutzky's theorem, and part b,

$$U_P = U_L + (U_P - U_L) \xrightarrow{d} V + 0. \quad \square$$

We have not proved Theorem 9–4 in this text. Actually, the asymptotic distribution of U_P is often used in the proof of Theorem 9–4. For that reason, in Section 14.1.3, we give a derivation of the asymptotic distribution of U_P which does not depend on Theorem 9–4.

It seems clear that we want to reject the null hypothesis when the O_i and the E_i are far apart, so that a sensible test based on U_P would reject the hypothesis when U_P is too large. The likelihood ratio principle says that we should reject the null hypothesis when U_L is too large. Therefore, we have two possible asymptotic tests for this model, one that rejects the null hypothesis when U_P is greater than the upper α point of a χ^2_{p-1} distribution, and one which rejects it when U_L is greater than that point. By part b of Theorem 14–2, we would expect the two statistics to be about the same if the null hypothesis is true. Under the alternative hypothesis, the two statistics are different, even asymptotically. Pearson's χ^2 was the first test statistic suggested for this problem and is the statistic that is typically used in practice.

It is not clear from a theoretical perspective which procedure is better. In fact, it is difficult even to define what a better asymptotic procedure is. Is it one whose size converges faster or one that is asymptotically more powerful? Unfortunately, in most problems for which we can find only asymptotic tests, there are several possible asymptotic test statistics to choose from, and it is not obvious which one to choose.

One rule that is often given is that the two asymptotic tests involving U_P and U_L are applicable as long as all the $E_i \geqq 5$. Some authors feel that even this rule is too conservative, however, and that the tests may be applied as long as not too many cells have $E_i < 5$. In this book, we shall use the rule $E_i \geqq 5$ as the criterion for when these tests may be applied.

Before we look at examples, we recall that in this book all logarithms are natural (base e) logarithms.

Example 14–1

Let $p = 3$. Suppose we observe $O_1 = 4$, $O_2 = 13$, and $O_3 = 13$ and want to test that all three classes have the same probability (i.e., that $\pi_1 = \pi_2 = \pi_3 = \frac{1}{3}$). For this model, $n = 4 + 13 + 13 = 30$, and hence, $E_1 = E_2 = E_3 = \frac{30}{3} = 10$. Thus, the $E_i \geqq 5$. (Even though $O_1 < 5$, we can still use the tests as long as the $E_i \geqq 5$.) Now,

$$U_P = \frac{(4-10)^2}{10} + \frac{(13-10)^2}{10} + \frac{(13-10)^2}{10} = 5.4,$$

and

$$U_L = 8 \log\left(\frac{4}{10}\right) + 26 \log\left(\frac{13}{10}\right) + 26 \log\left(\frac{13}{10}\right) = 6.31.$$

Also, $p - 1 = 2$, and the upper .05 point of a χ^2_2 distribution is 5.99. Therefore, if we use U_P, we accept the null hypothesis with a .05 test, and if we use U_L, we reject the null hypothesis with a .05 test. In this situation, it is not clear which result to believe. In principle, we should compute only one statistic and make our decision based on that. In practice, the statistical computation programs would typically give both. Fortunately, situations like this one in which the statistics lead to different conclusions are fairly rare. Unfortunately, when they lead to different conclusions, it is not obvious what to conclude.

Example 14–2

Suppose that we observe $O_1 = 40$, $O_2 = 40$, $O_3 = 5$, $O_4 = 10$, and $O_5 = 5$ and that we want to test that $\pi_1 = .6$, $\pi_2 = .3$, $\pi_3 = .03$, $\pi_4 = .04$, and $\pi_5 = .03$. For these data, $n = 100$, $E_1 = 60$, $E_2 = 30$, $E_3 = 3$, $E_4 = 4$, and $E_5 = 3$. Since $E_3 < 5$, $E_4 < 5$, and $E_5 < 5$, we cannot use the tests based on U_L and U_P. (Note that all the $O_i \geqq 5$, but this is not good enough to use these tests.) When the E_i are too small, we often combine cells. So in this case, we combine the third, fourth, and fifth cells to get a new multinomial distribution with $O_1^* = 40$, $O_2^* = 40$, $O_3^* = O_3 + O_4 + O_5 = 20$, $\pi_1^* = .6$, $\pi_2^* = .3$, and $\pi_3^* = .03 + .04 + .03 = .1$. For the reduced data, $E_1^* = 60$, $E_2^* = 30$, and $E_3^* = 10$, and we get

$$U_P = \frac{(40-60)^2}{60} + \frac{(40-30)^2}{30} + \frac{(20-10)^2}{10} = 20.$$

For the reduced model, $p = 3$, so that the critical value for a .05 test is again 5.99, and we can reject the null hypothesis. We could test on the basis of U_L similarly.

We now return to the general multinomial model. Consider testing the more general null hypothesis that $\boldsymbol{\pi} \in A$, where A is a specified set. We again consider two different test statistics for this problem. Let $\hat{\pi}_i$ be the MLE of π_i under the null hypothesis, and let

$$E_i = n\hat{\pi}_i, \qquad \hat{U}_P = \sum\frac{(O_i - \hat{E}_i)^2}{\hat{E}_i}, \qquad U_L = 2\sum O_i \log\left(\frac{O_i}{\hat{E}_i}\right).$$

As before, $\hat{U}_P$ is called *Pearson's* χ^2 and $\hat{U}_L$ is called the *likelihood ratio* χ^2 for the problem. The following theorem summarizes results about $\hat{U}_P$ and $\hat{U}_L$.

Theorem 14–3

a. Let Λ be the LRT statistic for testing that $\pi \in A$ against $\pi \notin A$. Then

$$\hat{U}_L = 2\log(\Lambda).$$

b. Under the null hypothesis $\hat{U}_L - \hat{U}_P \xrightarrow{P} 0$.

c. Let a be the number of independent parameters under the null hypothesis, and let $b = p - 1 - a$. Then, under general regularity conditions,

$$\hat{U}_L \xrightarrow{d} W \sim \chi_b^2 \quad \text{and} \quad \hat{U}_P \xrightarrow{d} W \sim \chi_b^2.$$

Proof. Parts a and c are proved in the exercises. Part b is more difficult. See Serfling (1980), pp. 156–160 for a proof. □

The comments made earlier about U_P and U_L can be applied to $\hat{U}_P$ and $\hat{U}_L$. In particular, these asymptotic tests can be applied as long as all the $\hat{E}_i \geqq 5$.

Example 14–3

Suppose that $p = 3$ and we want to test that $\pi_1 = \theta^2$, $\pi_2 = 2\theta\delta$, and $\pi_3 = \delta^2$, for some unknown parameters θ and δ such that $\theta + \delta = 1$. (This model is useful in genetics when the first class consists of individuals that are AA, the second class of those that are Aa, and the third class of individuals that are aa. In such case, θ and δ represent the proportions of A and a genes, respectively, in the population.) To find the MLE of (π_1, π_2, π_3) under the null hypothesis, we must maximize

$$\frac{L_O(\boldsymbol{\pi})}{h(\mathbf{O})} = \pi_1^{O_1}\pi_2^{O_2}\pi_3^{O_3} = (\theta^2)^{O_1}(2\theta\delta)^{O_2}(\delta^2)^{O_3} = 2^{O_2}(\theta)^{2O_1+O_2}(\delta)^{O_2+2O_3}$$

subject to the constraint $\theta + \delta = 1$. By the lemma presented earlier, this quantity is maximized by

$$\hat{\theta} = \frac{2O_1 + O_2}{2O_1 + O_2 + O_2 + 2O_3} = \frac{2O_1 + O_2}{2n} \quad \text{and} \quad \hat{\delta} = \frac{O_2 + 2O_3}{2n}.$$

Therefore, $\hat{\pi}_1 = \hat{\theta}^2$, $\hat{\pi}_2 = 2\hat{\theta}\hat{\delta}$, and $\hat{\pi}_3 = \hat{\delta}^2$. We can then compute the $\hat{E}_i$ and either $\hat{U}_P$ or $\hat{U}_L$. The number of degrees of freedom for the model is $3 - 1 - 1 = 1$, since there are $p = 3$ classes and $d = 1$ independent parameter under the null hypothesis. For example, suppose we observe 10 AA's, 20 Aa's, and 20 aa's. Then $O_1 = 10$, $O_2 = O_3 = 20$, and $n = 10 + 20 + 20 = 50$, and it follows that

$$\hat{\theta} = .4, \qquad \hat{\delta} = .6, \qquad \hat{\pi}_1 = .16, \qquad \hat{\pi}_2 = .48, \qquad \hat{\pi}_3 = .36,$$
$$\hat{E}_1 = 50(.16) = 8, \qquad \hat{E}_2 = 24, \qquad \hat{E}_3 = 18.$$

Therefore,

$$\hat{U}_L = 20\log\left(\frac{10}{8}\right) + 40\log\left(\frac{20}{24}\right) + 40\log\left(\frac{20}{18}\right) = 1.38.$$

The critical value for a .05 test for a χ_1^2 is 3.84, so we would accept the null hypothesis with these data using the likelihood ratio test.

Exercises—A

1. Mendelian inheritance implies that the proportion of a particular type of pea which is round and yellow is $\frac{9}{16}$, the proportion which is wrinkled and yellow is $\frac{3}{16}$, the proportion round and green is $\frac{3}{16}$, and the proportion wrinkled and green is $\frac{1}{16}$. An experiment was conducted to test this theory. Of 200 peas studied, 120 were round and yellow, 40 were wrinkled and yellow, 35 were round and green, and 5 were wrinkled and green. Do an asymptotic size-.05 likelihood ratio test that the Mendelian model is satisfied.
2. According to Mendelian inheritance, the proportions of AA, Aa, and aa individuals in a population should be θ^2, $2\theta\delta$, and δ^2, respectively, for some θ and δ such that $\theta + \delta = 1$. Suppose that in a particular population we observe 30 AA individuals, 30 Aa individuals, and 40 aa individuals. Can we reject the Mendelian model with these data using an asymptotic size-.05 Pearson's χ^2 test?
3. We want to determine whether a coin is fair and whether the flips are independent. To test this hypothesis, we flip a coin five times and record the number of heads and then flip it five more times and record the number of heads. We replicate this experiment 100 times and get the following data: 5 times we get 0 heads, 17 times we get 1 head, 35 times we get 2 heads, 32 times 3 heads, 7 times 4 heads, and 4 times 5 heads. If the null hypothesis is true, the number of heads in five flips should have a $B(5, \frac{1}{2})$ distribution.
   (a) Use the tables of the binomial distribution to find the expected number of times we would expect zero heads, the number of times one head, etc.
   (b) Do an asymptotic size-.05 Pearson's χ^2 test of the null hypothesis that the data came from a $B(5, \frac{1}{2})$ distribution.

Exercises—B

1. Consider a model in which we classify each of n individuals into one of p classes. Let π_i be the probability that an individual is in the ith class, and let $\boldsymbol{\pi} = (\pi_1, \ldots, \pi_p)$. Let $X_{ij} = 1$ if the jth individual is in the ith class and 0 otherwise. Let $\mathbf{X}_j = (X_{1j}, X_{2j}, \ldots, X_{pj})$,

$O_i = \Sigma_j X_{ij}$, $\mathbf{O} = (O_1, \ldots, O_p) = \Sigma_j \mathbf{X}_j$. Assume that the $\mathbf{X}_j$ are independent random vectors.

(a) Argue that $\mathbf{X}_j \sim M_p(1, \boldsymbol{\pi})$.

(b) Show that the likelihood for this model is $L_X(\boldsymbol{\pi}) = \Pi_i \pi_i^{\Sigma_j X_{ij}}$, so that $\mathbf{O}$ is a sufficient statistic for this model.

(c) Show that $\mathbf{O} \sim M_p(n, \boldsymbol{\pi})$.

2. Verify that $\log(1+t) = t - t^2/2 + t^2 R_2(t)$ where $R_2(t) \to 0$ as $t \to 0$.

3. Prove parts a and c of Theorem 14–3. You may assume that part b is true, and that the hypothesis $\boldsymbol{\pi} \in A$ can be written in the form $\tau(\boldsymbol{\pi}) = \mathbf{a}$ for some b-dimensional function τ.

4. Suppose that $p = 5$ and we want to test that $\pi_1 = \theta^3$, $\pi_2 = 3\theta^2\delta$, $\pi_3 = 3\theta\delta^2$, and $\pi_4 = \delta^3$, where $\theta + \delta = 1$.

(a) Show that the MLEs under the null hypothesis are $\hat{\theta} = (3O_1 + 2O_2 + O_3)/3n$ and $\hat{\delta} = (O_2 + 2O_3 + 3O_4)/3n$.

(b) Suppose $O_1 = 25$, $O_2 = 30$, $O_3 = 30$, and $O_4 = 15$. Do a likelihood ratio χ^2 test for this model. Use an asymptotic .05 test.

5. Suppose that $p = 3$ and we want to test that $\pi_1 = \theta^3$, $\pi_2 = 3\theta\delta$, and $\pi_3 = \delta^3$, where $\theta + \delta = 1$.

(a) Find the MLEs under the null hypothesis.

(b) How many degrees of freedom are there for testing this hypothesis?

6. Suppose that $p = 5$, and consider testing that $\pi_1 = \pi_2$ and $\pi_3 = \pi_4 = \pi_5$.

(a) Show that the MLEs under the null hypothesis are $\hat{\pi}_1 = \hat{\pi}_2 = (O_1 + O_2)/2n$ and $\hat{\pi}_3 = \hat{\pi}_4 = \hat{\pi}_5 = (O_3 + O_4 + O_5)/3n$.

(b) If $O_1 = 10$, $O_2 = 20$, $O_3 = 20$, $O_4 = 20$, and $O_5 = 30$, do we accept or reject this hypothesis with an asymptotic size-.05 Pearson's χ^2 test? (There are three degrees of freedom for this test. Why?)

Exercises—C

1. Suppose that $p = 2$ and we are testing that $\pi_1 = d$ and $\pi_2 = 1 - d$. Let $Z = (O_1 - nd)/(d(1-d))^{1/2}$. Show that $Z^2 = U_P$, Pearson's χ^2 test statistic. (When $p = 2$, the multinomial distribution reduces to a binomial distribution. The result proved in this problem shows that when $p = 2$, the Pearson χ^2 test is the same as the two-sided Z-test in the one-sample Bernoulli model.)

2. Let O_i, E_i, T_i, and Q_i be defined as in the proof of part b of Theorem 14–2. Assume that the null hypothesis is true so that $E_i = n\pi_i$.

(a) Show that $O_i/E_i \xrightarrow{P} 1$. ($O_i/n$ is a consistent estimator of $\pi_i = E_i/n$.)

(b) Show that $T_i \xrightarrow{P} 0$ and hence $R(T_i) \xrightarrow{P} 0$.

(c) Show that $Q_i \xrightarrow{P} 0$.

3. Let O_i, E_i, and T_i be defined as in the proof of part b of Theorem 14–2. Assume that the null hypothesis is true so that $E_i = n\pi_i$.

(a) Show that $(O_i - E_i)/(E_i(1 - \pi_i))^{1/2} \xrightarrow{d} Z \sim N(0, 1)$. (This is just the normal approximation for the binomial distribution.)

(b) Show that $(O_i - E_i)/(O_i(1 - \pi_i))^{1/2} \xrightarrow{d} Z$. (Use part a of the previous exercise and Slutzky's theorem.)

(c) Show that $O_i T_i^2 \xrightarrow{d} (1 - \pi_i)Z^2$.

14.1.2 Contingency Tables

Consider now a model in which we observe a random $r \times c$ matrix $\mathbf{O}$ whose components have an rc-dimensional multinomial distribution with parameters n and $\boldsymbol{\pi}$, where n is a known integer and $\boldsymbol{\pi}$ is an unknown $r \times p$ matrix. Let O_{ij} and π_{ij} be the (i, j)th components of $\mathbf{O}$ and $\boldsymbol{\pi}$ respectively. Then O_{ij} is the number of observations in both the ith row and the jth column and π_{ij} is the probability an observation is in the ith row and jth column. We call the matrix $\mathbf{O}$ a two-dimensional *contingency table* and call this model the (two-dimensional) *contingency table model*.

Our main goal for this model is to test the hypothesis that the row classes and the column classes are independent, i.e., that

$$\pi_{ij} = \theta_i \delta_j$$

for some θ_i and δ_j such that

$$\sum \theta_i = 1 \quad \text{and} \quad \sum \delta_j = 1$$

(where θ_i is the probability of being in the ith row class and δ_j is the probability of being in the jth column class).

As an example of where a contingency table model might be appropriate, consider an election in which there are three candidates and 300 voters. Suppose we have divided the voters into four categories by ethnic background. We could analyze the data with a contingency table model with $r = 4$, $c = 3$, and $n = 300$. In this case, O_{ij} is the number of voters in the sample who are in the ith ethnic group and voted for the jth candidate. (Thus, if $O_{12} = 8$, then 8 of the 300 voters in the sample are in the first ethnic group and voted for the second candidate.)

Returning to the general contingency table model, note that $\mathbf{O}$ has an rc-dimensional multinomial distribution, so this model is just a rearrangement of the one-sample multinomial model in the previous section. Therefore, the following results follow from Theorem 14–1.

Theorem 14–4

a. $(O_{11}, \ldots, O_{rc})$ is a complete sufficient statistic for the contingency table model.
b. O_{ij}/n is the MLE and best unbiased estimator of π_{ij} and is consistent.

Now, consider testing that $\pi_{ij} = \theta_i \delta_j$, and let

$$O_{i.} = \sum_j O_{ij}, \qquad O_{.j} = \sum_i O_{ij}, \qquad \hat{E}_{ij} = O_{i.}\, O_{.j}/n.$$

That is, $O_{i.}$ is the total number of observations in the ith row class, and $O_{.j}$ is the total number of observations in the jth column class. The following theorem is the main result for contingency tables.

Theorem 14–5. Consider testing that the row and column classes are independent. Let

$$\hat{U}_P = \sum_i \sum_j \frac{(O_{ij} - \hat{E}_{ij})^2}{\hat{E}_{ij}} \quad \text{and} \quad \hat{U}_L = 2 \sum_i \sum_j O_{ij} \log\left(\frac{O_{ij}}{\hat{E}_{ij}}\right).$$

Then:

a. Under the null hypothesis,

$$\hat{U}_P \xrightarrow{d} \chi^2_{(r-1)(c-1)}, \qquad \hat{U}_L \xrightarrow{d} \chi^2_{(r-1)(c-1)}, \qquad \hat{U}_P - \hat{U}_L \xrightarrow{P} 0.$$

b. The likelihood ratio test rejects the null hypothesis if U_L is too large.

Proof. Under the null hypothesis, $\pi_{ij} = \theta_i \delta_j$, where $\Sigma\theta_i = 1$ and $\Sigma\delta_j = 1$. To find the MLEs of the θ_i and δ_j, we must maximize

$$\frac{L_O(\boldsymbol{\pi})}{h(\mathbf{O})} = \prod_i \prod_j \pi_{ij}^{O_{ij}} = \prod_i \prod_j (\theta_i \delta_j)^{O_{ij}} = \left(\prod_i \theta_i^{O_{i.}}\right)\left(\prod_j \delta_j^{O_{.j}}\right).$$

By the lemma proved in the previous section, this expression is maximized for

$$\hat{\theta}_i = \frac{O_{i.}}{\sum_i O_{i.}} = \frac{O_{i.}}{n}, \qquad \hat{\delta}_j = \frac{O_{.j}}{\sum_j O_{.j}} = \frac{O_{.j}}{n}.$$

Therefore, for the contingency table model,

$$\hat{E}_{ij} = n\hat{\theta}_i \hat{\delta}_j = \frac{O_{i.}\,O_{.j}}{n}.$$

Under the null hypothesis, there are $r - 1$ independent θ_i and $c - 1$ independent δ_j, and hence, there are $r - 1 + c - 1$ independent parameters. Therefore, the number of degrees of freedom for this model is

$$rc - 1 - (r - 1) - (c - 1) = (r - 1)(c - 1).$$

The desired result then follows from Theorem 14–3. □

As before, $\hat{U}_P$ is called Pearson's χ^2 statistic and $\hat{U}_L$ is called the likelihood ratio χ^2 statistic. To do a size-α test of the hypothesis that the rows and columns are independent for a set of O_{ij}, we first compute the $\hat{E}_{ij}$, then compute either Pearson's χ^2 or the likelihood ratio χ^2, and reject the null hypothesis if it is greater than the upper α point of a χ^2 distribution with $(r - 1)(c - 1)$ degrees of freedom.

Example 14–4

Suppose we obtain the 3×3 contingency table

$$O$$

$$\begin{Bmatrix} 14 & 16 & 10 \\ 4 & 10 & 16 \\ 2 & 4 & 24 \end{Bmatrix}$$

and want to test for independence. The expected cell frequencies are given by

$$\hat{E}$$

$$\begin{Bmatrix} 8 & 12 & 20 \\ 6 & 9 & 15 \\ 6 & 9 & 15 \end{Bmatrix}.$$

As an example,

$O_{1.} = 14 + 16 + 10 = 40, \quad O_{.3} = 10 + 16 + 24 = 50, \quad n = 14 + 16 + \cdots + 4 + 24 = 100,$

so that

$$\hat{E}_{13} = \frac{O_{1.}O_{.3}}{n} = \frac{(40)(50)}{100} = 20.$$

Note that the observations for this model are the O_{ij} and the $\hat{E}_{ij}$ are computed from the O_{ij}. Note also that for most data sets, the $\hat{E}_{ij}$ are not integers. Pearson's χ^2 for these observations is

$$\hat{U}_P = \frac{(14-8)^2}{8} + \frac{(16-12)^2}{12} + \cdots + \frac{(4-9)^2}{9} + \frac{(24-15)^2}{15} = 22.52.$$

The number of degrees of freedom for this test is given by $(3-1)(3-1) = 4$. The critical value for a .01 test is 13.3, so we can reject the hypothesis that the row and column classifications are independent with these data with a .01 test.

We now return to the general two-way contingency model. When $r = c$, another hypothesis which is often tested is symmetry, i.e., that

$$\pi_{ij} = \pi_{ji},$$

for all $i \neq j$. To test this hypothesis, we use either Pearson's χ^2 or the likelihood ratio χ^2, with

$$\hat{E}_{ij} = \frac{O_{ij} + O_{ji}}{2}$$

and degrees of freedom $c(c-1)/2$ (see the exercises).

When $r = c$, a third hypothesis which is sometimes tested is symmetry and independence, i.e., that

$$\pi_{ij} = \delta_i \delta_j,$$

for δ_i such that $\Sigma\delta_i = 1$. To test this hypothesis, we use either Pearson's χ^2 or the likelihood ratio χ^2 again, with

$$\hat{E}_{ij} = \frac{(O_{i.} + O_{.i})(O_{j.} + O_{.j})}{4n}$$

and degrees of freedom $c(c-1)$ (see the exercises).

Example 14–5

For the 3×3 table given in Example 14–4, the E_{ij} for the test of symmetry are given by

$$\begin{Bmatrix} 14 & 10 & 6 \\ 10 & 10 & 10 \\ 6 & 10 & 24 \end{Bmatrix}.$$

For example,

$$\hat{E}_{13} = \frac{(O_{31} + O_{13})}{2} = \frac{(10+2)}{2} = 6.$$

The likelihood ratio χ^2 for testing symmetry for the model is

$$\hat{U}_L = 2\left(14 \log\left(\frac{14}{14}\right) + 16 \log\left(\frac{16}{10}\right) + \cdots + 4 \log\left(\frac{4}{10}\right) + 24 \log\left(\frac{24}{24}\right)\right) = 21.24.$$

The number of degrees of freedom for this test is $3(3-1)/2=3$. Since $21.24>7.81=\chi_3^{2\ .05}$, we reject the hypothesis of symmetry with a .05 test for these data.

The $\hat{E}_{ij}$ for the test of symmetry and independence for these data are

$$\begin{Bmatrix} 9 & 9 & 12 \\ 9 & 9 & 12 \\ 12 & 2 & 16 \end{Bmatrix}.$$

For example, for the hypothesis of symmetry and independence,

$$\hat{E}_{13}=\frac{(O_{1.}+O_{.1})(O_{3.}+O_{.3})}{4n}=\frac{(40+20)(30+50)}{400}=12.$$

The Pearson χ^2 for testing symmetry and independence with this data is

$$\hat{U}_P=\frac{(14-9)^2}{9}+\frac{(16-9)^2}{9}+\cdots+\frac{(4-12)^2}{12}+\frac{(24-16)^2}{16}=30.44.$$

This test has $3(2)=6$ degrees of freedom. Since $30.44>12.6=\chi_6^{2\ .05}$, we reject the hypothesis of symmetry and independence with a .05 test for these data (which is not surprising, since we have already rejected the hypotheses of independence and symmetry separately).

Exercises—A

1. An experiment was conducted to determine whether automobile size is related to the fatality of accidents. The following data were collected:

	Small	Medium	Large
Fatal	90	70	20
Nonfatal	40	80	100

Do an asymptotic size-.05 Pearson's χ^2 test that fatality rate is independent of car size.

2. Two hundred patients are classified according to one of four blood types O, A, B, and AB. They are also classified by whether they are Rh positive or negative. The results are shown in the following table

	O	A	B	AB
+	56	60	34	12
−	8	18	6	6

Do an asymptotic size-.05 Pearson χ^2 that Rh factor is independent of blood type.

3. O blood type can be considered as an absence of both A and B, A blood type as a presence of A but an absence of B, B type as a presence of B but an absence of A, and AB as a presence of both A and B. If we collapse the table above over Rh factor, we get 64 patients with O blood, 78 with A blood, 40 with B blood, and 18 with AB blood.
   (a) Rearrange the data into a 2×2 table where the rows represent the absence or presence of A factor and the columns represent the presence or absence of B factor.
   (b) Do an asymptotic size-.05 likelihood ratio test that the A and B factors are independent.

4. The data in Exercise B2 can be considered a three-way contingency table in which the

patients have been classified by presence or absence of A factor, presence or absence of B factor, and Rh factor. Considering the data as a $2 \times 2 \times 2$ table, do an asymptotic size-.05 likelihood ratio χ^2 test that A factor, B factor, and Rh factor are independent. (See Exercise B4.)

5. A sample of 200 married couples was taken. The husbands and wives were asked separately whether they got their news from newspapers, radio, or television. The following data were recorded (so that, for example, in six couples the husband gets his news from the radio and the wife gets hers from the papers):

Husband / Wife	Papers	Radio	TV
Papers	15	6	10
Radio	11	10	20
TV	23	15	90

(a) Do an asymptotic size-.05 Pearson's χ^2 test of the null hypothesis that the news sources of the husbands and wives are symmetric.
(b) Do an asymptotic size-.05 likelihood ratio χ^2 test that the news sources of the husbands and wives are independent and symmetric.

Exercises—B

1. For the numerical example given in Examples 14–4 and 14–5, verify the entries for $\hat{E}_{23}$ for testing for independence, for testing for symmetry, and for testing for independence and symmetry.
2. For testing for symmetry, verify that $\hat{E}_{ij} = (O_{ij} + O_{ji})/2$ and that there are $c(c-1)/2$ degrees of freedom.
3. For testing for independence and symmetry, verify that $\hat{E}_{ij} = (O_{i.} + O_{.i})(O_{j.} + O_{.j})/4n$ and that the number of degrees of freedom is $c(c-1)$.
4. Suppose we have a three-way contingency table in which $\mathbf{O} = (O_{111}, \ldots, O_{rcd})$ and $\boldsymbol{\pi} = (\pi_{111}, \ldots, \pi_{rcd})$. The three classifications are independent if $\pi_{ijk} = \theta_i \delta_j \gamma_k$ for some parameters θ_i, δ_j, and γ_k such that $\Sigma\theta_i = 1$, etc.
   (a) What are the MLEs of the θ_i, δ_j, and γ_k under the hypothesis of independence?
   (b) Show that $\hat{E}_{ijk} = O_{i..} O_{.j.} O_{..k}/n^2$.
   (c) Show that the number of degrees of freedom for this problem is $rcd - r - c - d + 2$.

*# 14.1.3 Derivations

In this section, we derive the basic results used for testing simple hypotheses about multinomial parameters. Suppose we observe a sequence of p-dimensional multinomial random vectors

$$\mathbf{O}_n = (O_{n1}, \ldots, O_{np})' \sim M_p(n, \boldsymbol{\pi}), \qquad \boldsymbol{\pi} = (\pi_1, \ldots, \pi_p)'.$$

Let

$$E_{ni} = n\pi_i, \qquad U_{Pn} = \sum_{i=1}^{p} \frac{(O_{ni} - E_{ni})^2}{E_{ni}}.$$

Consider testing the simple hypothesis that $\boldsymbol{\pi} = \mathbf{d}$. Under the null hypothesis, U_{Pn} is Pearson's χ^2 statistic. (Under the alternative hypothesis U_{Pn} is not Pearson's χ^2, because we have used $E_i = n\pi_i$ instead of $E_i = nd_i$.) We show that $U_{Pn} \xrightarrow{d} V \sim \chi^2_{p-1}$. We have shown that $U_{Pn} - U_{Ln} \xrightarrow{P} 0$ under the null hypothesis (where U_{Ln} is the likelihood ratio χ^2 statistic). Therefore, the asymptotic distribution of the likelihood ratio χ^2 statistic follows directly from this result.

We use the following result of matrix algebra in the derivations.

Lemma. Let $\mathbf{A}$ be a $q \times q$ invertible matrix, let $\mathbf{b}$ and $\mathbf{c}$ be q-dimensional vectors, and let $d \neq 0$ be a number. Then

$$\mathbf{c}'(\mathbf{A} - d^{-1}\mathbf{b}\mathbf{b}')^{-1}\mathbf{c} = \mathbf{c}'\mathbf{A}^{-1}\mathbf{c} + \frac{(\mathbf{c}'\mathbf{A}^{-1}\mathbf{b})^2}{d - \mathbf{b}'\mathbf{A}^{-1}\mathbf{b}}.$$

Proof. See Exercise C1. □

We are now ready for the main theorem of this section.

Theorem 14–6. $U_{Pn} \xrightarrow{d} \chi^2_{p-1}$.

Proof. Let $\mathbf{E}_n = (E_{n1}, \ldots, E_{np})'$ and let $\mathbf{V}$ be the $p \times p$ matrix whose ith diagonal element is $v_{ii} = \pi_i(1 - \pi_i)$ and whose (i, j)th off-diagonal element is $v_{ij} = -\pi_i\pi_j$. In Chapter 6, we showed that

$$n^{-1/2}(\mathbf{O}_n - \mathbf{E}_n) \xrightarrow{d} N_p(\mathbf{0}, \mathbf{V}).$$

Unfortunately, $\mathbf{V}$ is not invertible (since the columns sum to zero). So let $\mathbf{O}_n^*$ and $\mathbf{E}_n^*$ be the $(p-1)$-dimensional vectors formed by eliminating the last elements of $\mathbf{O}_n$ and $\mathbf{E}_n$, and let $\mathbf{V}^*$ be the $(p-1) \times (p-1)$-dimensional matrix formed by eliminating the last row and column from $\mathbf{V}$. Then

$$n^{-1/2}(\mathbf{O}_n^* - \mathbf{E}_n^*) \xrightarrow{d} N_{p-1}(\mathbf{0}, \mathbf{V}^*).$$

Now, in Chapter 5, we showed that if $\mathbf{X} \sim N_q(\boldsymbol{\mu}, \boldsymbol{\Sigma})$ and $\boldsymbol{\Sigma} > 0$, then

$$(\mathbf{X} - \boldsymbol{\mu})'\boldsymbol{\Sigma}^{-1}(\mathbf{X} - \boldsymbol{\mu}) \sim \chi^2_q.$$

Therefore, by the theorem on continuous functions,

$$T_n = (\mathbf{O}_n^* - \mathbf{E}_n^*)'(n\mathbf{V}^*)^{-1}(\mathbf{O}_n^* - \mathbf{E}_n^*) \xrightarrow{d} \chi^2_{p-1}.$$

Now,

$$n\mathbf{V}^* = \mathbf{D}_n - n^{-1}\mathbf{E}_n^*\mathbf{E}_n^{*\prime},$$

where $\mathbf{D}_n$ is a $(p-1) \times (p-1)$ matrix whose ith diagonal element is $d_{ii} = E_i$ and whose off-diagonal elements are all zero. Hence, using the previous lemma with $\mathbf{A} = \mathbf{D}_n$, $\mathbf{b} = \mathbf{E}_n^*$, $\mathbf{c} = \mathbf{O}_n^* - \mathbf{E}_n^*$, and $d = n$, we obtain

$$T_n = (\mathbf{O}_n^* - \mathbf{E}_n^*)'\mathbf{D}_n^{-1}(\mathbf{O}_n^* - \mathbf{E}_n^*) + \frac{((\mathbf{O}_n^* - \mathbf{E}_n^*)'\mathbf{D}_n^{-1}\mathbf{E}_n^*)^2}{n - \mathbf{E}_n^{*\prime}\mathbf{D}_n^{-1}\mathbf{E}_n^*}.$$

By direct matrix multiplication, it follows that

$$(\mathbf{O}_n^* - \mathbf{E}_n^*)'\mathbf{D}_n^{-1}(\mathbf{O}_n^* - \mathbf{E}_n^*) = \sum_{i=1}^{p-1} \frac{(O_{ni} - E_{ni})^2}{E_{ni}}, \tag{14–1}$$

$$n - \mathbf{E}_n^{*\prime}\mathbf{D}_n^{-1}\mathbf{E}_n^* = n - \sum_{i=1}^{p-1} E_{ni} = E_{np}, \tag{14–2}$$

and

$$(\mathbf{O}_n^* - \mathbf{E}_n^*)'\mathbf{D}_n^{-1}\mathbf{E}_n^* = \sum_{i=1}^{p-1} (O_i - E_i) = E_{np} - O_{np}. \tag{14–3}$$

But then, plainly,

$$U_{Pn} = T_n \xrightarrow{d} \chi^2_{p-1}.$$

Exercises—B

1. Verify that the columns of $\mathbf{V}$ sum to zero.
2. Verify Equations (14–1), (14–2), and (14–3).

Exercises—C

1. **(a)** Under the conditions of the lemma given in this section, verify that

$$(\mathbf{A} - d^{-1}\mathbf{bb}')^{-1} = \mathbf{A}^{-1} + (d - \mathbf{b}'\mathbf{A}^{-1}\mathbf{b})^{-1}\mathbf{A}^{-1}\mathbf{bb}'\mathbf{A}^{-1}.$$

(Note that $\mathbf{B} = \mathbf{C}^{-1} \Leftrightarrow \mathbf{BC} = \mathbf{I}$ and that $\mathbf{b}'\mathbf{A}^{-1}\mathbf{b}$ is 1×1.)

(b) Verify the aforementioned lemma.

14.2 SOME k-SAMPLE MODELS

14.2.1 The k-Sample Multinomial Model

Consider the model in which we observe $\mathbf{O}_i$ independent p-dimensional random vectors,

$$\mathbf{O}_i \sim M_p(n_i, \boldsymbol{\pi}_i),$$

where the n_i are known integers and the $\boldsymbol{\pi}_i$ are unknown parameter vectors. We call this model the *k-sample multinomial model*. If $p = 2$, we call this model the *k-sample binomial model*. If $p = 2$ and $k = 2$, this model is the same as the two-sample Bernoulli model considered in Chapter 11. Our main goal for this model is to test the equality of the $\boldsymbol{\pi}_i$. We let O_{ij} be the jth component of $\mathbf{O}_i$ and let π_{ij} be the jth component of $\boldsymbol{\pi}_i$.

Consider again the example of Section 14.1.2 in which there are three candidates and four ethnic groups. Suppose we decide on a somewhat different sampling scheme from that described. We decide to find 50 persons in the first ethnic group and find out which candidate each chose, 100 in the second group and find out which candidate each chose, 50 in the third ethnic group, and 100 in the fourth. For this example, O_{ij} is the number of individuals in the ith ethnic group who chose candidate j, so that if $O_{23} = 6$, then six individuals in the second ethnic group supported the third candidate. The hypothesis that the $\boldsymbol{\pi}_i$ are equal for this experiment is the hypothesis that the proportion of individuals who support the jth candidate is the same for the different ethnic groups.

The primary difference between the model in Section 14.1.2 and the model in this section is that here we have sampled in a way that guarantees that there are 50 in the first group, 100 in the second, etc. In Section 14.1.2, we took a sample of size 300 from the whole population so that the number of individuals in each ethnic group would be random. Note that the O_{ij} are the same for both models, namely the number of people in the ith ethnic group who support candidate j, but that π_{ij} for this model is the probability that an individual who is in the ith ethnic group supports the jth candidate, while π_{ij} for the contingency table model is the probability that a randomly chosen individual is in the ith ethnic group and supports the jth candidate. However, the hypothesis of independence for contingency tables has essentially the same meaning as the hypothesis of equality of the $\boldsymbol{\pi}_i$ for the model in this section, namely, that the proportion of individuals who support the various candidates is the same in the different ethnic classes.

Theorem 14–7

a. $(\mathbf{O}_1, \ldots, \mathbf{O}_k)$ is a complete sufficient statistic for the k-sample multinomial model.
b. O_{ij}/n_i is the MLE and the best unbiased estimator of π_{ij} for this model and is consistent.

Proof. See the exercises. □

To test the equality of the $\boldsymbol{\pi}_i$, let

$$\hat{E}_{ij} = \frac{n_i O_{.j}}{N}, \qquad \hat{U}_P = \sum_i \sum_j \frac{(O_{ij} - \hat{E}_{ij})^2}{\hat{E}_{ij}}, \qquad \hat{U}_L = \sum_i \sum_j O_{ij} \log\left(\frac{O_{ij}}{\hat{E}_{ij}}\right).$$

where

$$O_{.j} = \sum_i O_{ij} \quad \text{and} \quad N = \sum_i n_i.$$

(In the example above $O_{.j}$ is the number of individuals who support candidate j). As before, $\hat{U}_P$ is called Pearson's χ^2, and $\hat{U}_L$ is called the likelihood ratio χ^2.

Theorem 14–8

a. The likelihood ratio test of the null hypothesis that the $\boldsymbol{\pi}_i$ are equal rejects the hypothesis if $\hat{U}_L$ is too large.
b. Under the null hypothesis that the $\boldsymbol{\pi}_i$ are all equal,

$$\hat{U}_P \xrightarrow{d} \chi^2_{(k-1)(p-1)}, \qquad \hat{U}_L \xrightarrow{d} \chi^2_{(k-1)(p-1)}, \qquad \hat{U}_P - \hat{U}_L \xrightarrow{P} 0.$$

Proof

a. See the exercises.

b. Proof of this part of the theorem is somewhat beyond the level of this book. Instead, we sketch a heuristic proof. Suppose that the $\boldsymbol{\pi}_i$ are equal, i.e.,

$$\boldsymbol{\pi}_i = \boldsymbol{\pi}_0 = (\pi_{01}, \ldots, \pi_{0p}).$$

Then the MLE of $\boldsymbol{\pi}_0$ is given by

$$\hat{\pi}_{0j} = \frac{O_{.j}}{N}.$$

(See the exercises.) Now, under the null hypothesis, $\mathbf{O}_i \sim M(n_i, \boldsymbol{\pi}_0)$. Therefore, a sensible estimator for the expected frequency for the jth cell in the ith sample is

$$\hat{E}_{ij} = n_i \hat{\pi}_{0j} = \frac{n_i O_{.j}}{N}.$$

The formulas for Pearson's χ^2 and the likelihood ratio χ^2 then follow from rules used earlier in the chapter. To compute the degrees of freedom in this problem, note that there are $p - 1$ degrees of freedom for each of the k populations, so that there are $k(p - 1)$ degrees of freedom for the whole model. Under the null hypothesis, we are estimating $p - 1$ independent parameters, the components of $\boldsymbol{\pi}_0$. (Note that $\Sigma\pi_{0j} = 1$.) Therefore, we would expect the degrees of freedom for this hypothesis to be

$$k(p - 1) - (p - 1) = (k - 1)(p - 1). \quad \square$$

As usual, we would test the hypothesis that the $\boldsymbol{\pi}_i$ are equal by first computing the $\hat{E}_{ij}$, then computing $\hat{U}_P$ or $\hat{U}_L$, and finally rejecting the null hypothesis if $\hat{U}_P$ or $\hat{U}_L$ is larger than the critical point, which we find in a χ^2 table for $(k - 1)(p - 1)$ degrees of freedom.

There is a close relationship between the problem of testing for independence in a two-way contingency table, as discussed in the previous section, and the problem of testing the equality of the $\boldsymbol{\pi}_i$, as discussed in this section. In the example of candidates and ethnic groups, we are testing that the support for the different candidates is the same across the ethnic groups with either test. In addition, the Pearson χ^2 test and the likelihood ratio χ^2 test for testing independence in a contingency table are the same as the Pearson χ^2 test and the likelihood ratio χ^2 test for testing the equality of multinomials derived in this section (see the exercises). Hence, it is often not important whether the model is a two-way contingency table or a k-sample multinomial model. (However, the test for symmetry and the test for symmetry and independence do not make sense for the k-sample multinomial model considered in this section.)

Exercises—A

1. Samples were drawn of 100 whites, 100 blacks, and 100 Hispanics. Of the 100 whites, 30 were nonsmokers, 50 were light smokers, and 20 were heavy smokers. Of the 100 blacks, 20 were nonsmokers, 50 were light smokers, and 30 were heavy smokers. Of the 100 Hispanics, 15 were nonsmokers, 40 were light smokers, and 45 were heavy smokers. Do

an asymptotic .05 Pearson χ^2 test to determine whether there is any significant difference between the smoking habits of the three populations.

2. Samples were drawn from each of three socioeconomic levels. Of the 40 low-income people, 30 passed their driver's tests on the first try. Out of 50 middle-income people, 45 passed their tests on the first try, and out of 40 high-income people, 35 passed on the first try. Do an asymptotic size-.05 likelihood ratio χ^2 test to determine whether there is any significant difference between the proportions who passed the driver's tests in the three levels.

Exercises—B

1. Prove Theorem 14–7.
2. **(a)** Show that the MLE of π_{0j} is $\hat{\pi}_{0j} = O_{.j}/N$.
   (b) Prove part a of Theorem 14–8.
3. **(a)** Show that the Pearson's χ^2 and the likelihood χ^2 test statistics given in this section are the same as the Pearson's χ^2 and likelihood ratio χ^2 test statistics given in the previous section for a contingency table with p rows and k columns.
   (b) Show that the number of degrees of freedom in this section is the same as the number of degrees of freedom in a $p \times k$ contingency table.
4. Suppose that $k = p = 2$, so that we are testing the equality of two binomial distributions. Let $Z = (\hat{\pi}_{11} - \hat{\pi}_{21})/(Q(1 - Q)(n_i^{-1} + n_2^{-1}))^{1/2}$ be the Z-statistic suggested in Section 11.3.2 for testing this hypothesis, where $\hat{\pi}_{ij} = O_{ij}/n_i$ and $Q = (O_{11} + O_{21})/(n_1 + n_2)$. Show that Z^2 is Pearson's χ^2 statistic U_P given in this section. (Therefore, in this case, Pearson's χ^2 test is the same as the two-sided Z-test given in Section 11.3.2.)

* 14.2.2 The k-Sample Poisson Model

Consider the model in which we observe X_{ij} independent, with

$$X_{ij} \sim P(\mu_i), \qquad i = 1, \ldots, k, j = 1, \ldots, n_i.$$

Note that for each i, the X_{ij} are a sample of size n_i from a Poisson distribution with mean μ_i. Since there are k such samples, we call this model the *k-sample Poisson model*. We are primarily interested in testing the equality of the μ_i for the model. Let

$$O_i = \sum_j X_{ij} = n_i \overline{X}_i \quad \text{and} \quad N = \sum_i n_i.$$

Then we have the following theorem.

Theorem 14–9

a. $(O_1, \ldots, O_k)$ is a complete sufficient statistic for the k-sample Poisson model.
b. The O_i are independent, with $O_i \sim P(n_i \mu_i)$.
c. $\overline{X}_i = O_i/n_i$ is the MLE and best unbiased estimator of μ_i and is consistent.

Proof. See the exercises. □

Since $(\overline{X}_{1.}, \ldots, \overline{X}_{k.})$ is an invertible function of $(O_1, \ldots, O_k)$, it is also a complete sufficient statistic for the k-sample Poisson model.

The following theorem is often used for testing the equality of the μ_i.

Theorem 14–10. Let $V = \Sigma_i O_i$. Then the conditional distribution of $(O_1, \ldots, O_k)$ given V is

$$(O_1, \ldots, O_k)|V \sim M_k(V, \boldsymbol{\pi}),$$

where $\boldsymbol{\pi} = (\pi_1, \ldots, \pi_k)$, $\pi_i = n_i \mu_i / \Sigma_j n_j \mu_j$.

Proof. The conditional density function is

$$f(o_1, \ldots, o_k|v) = P(O_1 = o_1, \ldots, O_k = o_k|V = v)$$
$$= \frac{P(O_1 = o_1, \ldots, O_k = o_k, V = v)}{P(V = v)}.$$

If $\Sigma o_i \neq v$, then the numerator is zero. Since the O_i are independent,

$$V = \sum_j O_j \sim P\left(\sum_j n_j \mu_j\right).$$

Therefore, if $\Sigma o_i = v$, then

$$\frac{P(O_1 = o_1, \ldots, O_k = o_k, V = v)}{P(V = v)} = \frac{\prod_i P(O_i = o_i)}{P(V = v)}$$
$$= \frac{\prod_i (\exp(-n_i \mu_i))(n_i \mu_i)^{o_i}/o_i!}{\left(\exp\left(-\sum n_j \mu_j\right)\right)\left(\sum n_j \mu_j\right)^{v}/v!}$$
$$= \left(\frac{v!}{\prod_i o_i!}\right) \prod_i (\pi_i^{o_i}),$$

which is the multinomial density function with parameters v and $\boldsymbol{\pi}$. □

We can use Theorem 14–10 to design asymptotic tests for testing that the μ_i are equal, which occurs if and only if

$$\pi_i = \frac{n_i}{N}.$$

Therefore, conditionally on V, the problem is equivalent to testing that $\boldsymbol{\pi}$ equals a known vector. (Note that V plays the role of n in the conditional multinomial distribution.) Now, let

$$E_i = \frac{V n_i}{N},$$

and as before, let

$$U_P = \sum \frac{(O_i - E_i)^2}{E_i} \quad \text{and} \quad U_L = 2 \sum O_i \log\left(\frac{O_i}{E_i}\right).$$

Using the results we have given for the multinomial distribution, we see that if the μ_i are equal (and hence, $\pi_i = n_i/N$), then as V goes to infinity,

$$U_P \xrightarrow{d} \chi^2_{k-1}, \qquad U_L \xrightarrow{d} \chi^2_{k-1}, \qquad U_P - U_L \xrightarrow{P} 0.$$

In Exercise C1, we show that as N goes to infinity, V also goes to infinity. Therefore, to do an asymptotic test of the equality of the μ_i, we compute either Pearson's χ^2 (U_P) or the likelihood ratio χ^2 (U_L) and reject the null hypothesis if it is larger than the critical value we find in a χ^2 table on $k - 1$ degrees of freedom. A common rule of thumb is that this asymptotic test is applicable as long as the $E_i \geqq 5$.

Exercises—A

1. A researcher wants to compare the number of calls coming into four different switchboards. For the first switchboard, he takes 8 periods and finds a sample mean of 13 calls. For another switchboard, he finds a sample mean of 10 calls in 6 periods. For the third and fourth switchboards, he finds sample means of 8 and 16 calls in samples of size 5 and 6. Do an asymptotic size-.05 Pearson's χ^2 test to test the null hypothesis that the means for the four switchboards are equal.

Exercises—B

1. Prove Theorem 14–9.
2. Show that the likelihood ratio test for testing equality of the means in the k-sample Poisson model rejects the null hypothesis if U_L is too large.
3. Show that when $k = 2$, the Pearson χ^2 reduces to the two-sample test described in Exercise B3 of Section 11.4.3 for testing the equality of means for two Poisson distributions.

Exercises—C

1. **a.** Let $V \sim P(N\mu)$, $\mu > 0$. Show that $V/n \xrightarrow{P} \mu$.
   b. Show that $P(V > k) \to 1$ for any fixed $k > 0$. (*Hint:* $P(V > k) = P((V - k)/n > 0)$ and $(V - k)/n \xrightarrow{P} \mu > 0$. Why?)

Chapter 15

NONPARAMETRIC STATISTICS

In Chapters 7–14, we have presented a theory of parametric statistics, together with some practical examples. In this chapter we present some results from nonparametric statistics.

We present results on rank tests, confidence intervals, and estimators. These procedures are frequently used as substitutes for the normal procedures discussed earlier. It should be emphasized that these models remove the normal assumption from these models, but not the remaining assumptions. In particular, it is necessary to assume that the observations in the samples are independent and have the same distribution. In the two-sample model, it is necessary to assume that the Y's have been shifted from the X's. In the normal model, this assumption reduces to assuming that the variances are the same in the two samples. The procedures developed in this chapter are for making inferences about the center of the distributions, so they reduce to making inferences about means in the normal case. Recall that in the normal model, such procedures are fairly robust against the normal assumption. (In fact, this assumption is the *only* one against which they are robust.)

Note that the procedures derived in earlier sections for binomial and multinomial models are often considered nonparametric procedures, since the assumed form for the distribution is the only one possible for data of that form. In particular, if a random variable can only take on the values zero and one, then it must be a Bernoulli random variable.

In parametric statistics, we assume that we have a statistical model, i.e., a random vector $\mathbf{X}$, a parameter vector $\boldsymbol{\theta}$, and a joint density function $f(\mathbf{x}; \boldsymbol{\theta})$ which is completely specified. In this chapter, we drop the assumption that the function $f(\mathbf{x}; \boldsymbol{\theta})$ is known.

Recall that the *median* θ of a continuous random variable X satisfies

$$P(X \leqq \theta) = P(X \geqq \theta) = \frac{1}{2}.$$

Any random variable has a median. In this chapter, we assume that the median is unique (i.e., that there is only one such θ). For the models in this chapter, the mean $\mu = EX$ may not exist. Therefore, we consider the median θ to be the center of the distribution.

We say that a random variable X has a distribution which is *symmetric* about θ (or X is symmetric about θ) if the distribution of $-(X-\theta)$ is the same as the distribution of $X-\theta$. If X is symmetric about θ, then θ is the median of X. If, in addition, X has a finite expectation, then $\theta = EX$.

In a *one-sample nonparametric model*, we observe a sample from an unspecified continuous distribution which is symmetric about its median θ. We make no other assumptions about the distribution. In Section 15.1, we derive size-α rank tests, $(1-\alpha)$ rank confidence intervals, and rank estimators for θ in the model. (Note that if we add the assumption that the X_i are normally distributed, this model is the one-sample normal model discussed in Section 10.2.1.)

In a *two-sample nonparametric* model, we observe independent samples $X_1, \ldots, X_m$ and $Y_1, \ldots, Y_n$ from continuous distributions such that $(Y_j - \delta)$ has the same distribution as X_i, where δ is an unknown parameter. (We make no additional assumption about the common distribution function of X_i and $Y_j - \delta$.) In other words, the Y_j have been shifted up or down from the X_i by a positive or negative δ. If $\delta = 0$, then X_i and Y_j have the same distribution. Also, if θ and ψ are the medians of X_i and Y_j, respectively, then $\delta = \psi - \theta$, and if the X_i and Y_j have finite expectations, then $\delta = EY_j - EX_i$. In Section 15.1.3, we derive rank tests, rank confidence intervals, and rank estimators for δ in this model. Note that we do not need to assume that the distributions are symmetric for the model. If the X_i and Y_j are normally distributed, the model is just the two-sample normal model with equal variances considered in Section 10.2.2. Nonparametric procedures have also been designed for analysis-of-variance and regression models, as well as for correlation analysis in bivariate models. We shall not discuss such models here, however. (See Hettmansperger (1984), Chapters 4–5, for a discussion of nonparametric procedures in these models.)

We define the *order statistics* $U_{[1]}, \ldots, U_{[k]}$ for a set of numbers $U_1, \ldots, U_k$ in the following way. $U_{[1]}$ is the smallest of the U_i, $U_{[2]}$ is the second smallest, ..., and $U_{[k]}$ is the largest of the U_i. The *rank* of U_i is the number of U_j less than or equal to U_i. For example, if the U_i are 1, 7, 4, and 3, then the $U_{[i]}$ are 1, 3, 4, and 7. It then follows that the rank of U_1 is 1 since it is the smallest observation, the rank of U_2 is 4 since it is the fourth smallest one, the rank of U_3 is 3, and the rank of U_4 is 2.

When nonparametric tests were first suggested, it was argued that the rank tests are simpler to understand than the t-tests. However, this argument is not very convincing. In particular, the rank tests are much slower to compute (even with computers) than the t-tests, primarily because computers order numbers more slowly than they add them. In addition, the distributions of the rank test statistics (even under the null hypothesis) are more complicated than the distributions of the t-tests. For example, the derivations of the asymptotic normal approximations for the rank tests are beyond the scope of the text.

A second argument that has been advanced for using rank tests is that rank tests should have more stable size than t-tests, since they are derived under more general assumptions. This argument, however, is also somewhat weak: The sizes of t-tests are actually quite robust against the normal assumption, as we have indicated in previous chapters.

Nonetheless, there is a very strong argument for using rank tests instead of

t-tests: Rank tests often have a superior power function. Recall that t-tests are optimal (UMP unbiased) tests for the normal model, so that no size-α test can have higher power than the t-test when the data are normally distributed. However, as we shall indicate in Section 15.2, the asymptotic power of some rank tests is nearly as good as the asymptotic power of the t-test when the data are normally distributed but may be much better when the data are not normally distributed. In fact, we shall discuss one rank test, the normal scores test, whose asymptotic power is just as high as the asymptotic power of the t-test when the data are normally distributed, and is much higher for distributions with infinite variance, such as the Cauchy distribution. Although these results are primarily asymptotic results, they seem to be fairly close, even for small samples. (See, e.g., Hettmansperger (1984), pp. 74–75.) We shall give similar asymptotic results for the rank confidence intervals and rank estimators.

The preceding argument for using rank procedures is quite compelling. However, it should be mentioned that those distributions for which t-tests do poorly are distributions which are unlikely to occur often in practice. There are sound reasons, based on the central limit theorem, for believing that many data sets should be nearly normally distributed, but the arguments are weak for the practicality of many of the distributions on which t-tests do poorly. Nonetheless, it seems likely that we often see data from models in which rank tests offer substantial improvement over t-tests.

15.1 ONE- AND TWO-SAMPLE RANK PROCEDURES

15.1.1 One-Sample Rank Tests

Consider the model in which we observe a sample $X_1, \ldots, X_n$ from a continuous distribution which is symmetric about its median θ. We assume that the distribution function of the X_i is unknown. In this section, we shall develop tests for testing the null hypothesis $\theta = d$ (where d is a specified constant) against one-sided and two-sided alternatives.

This model often arises when we observe a sample $(U_1, V_1), \ldots, (U_n, V_n)$ from a bivariate distribution. Let $X_i = U_i - V_i$. Then the procedures set out in this section can be used to test that the median of the X_i is zero against one-sided and two-sided alternatives, as long as the distribution of the X_i is symmetric about its median.

Suppose now that we want to test the null hypothesis that $\theta = d$ for some d. Let $Y_i = X_i - d$. We first order the $W_i = |Y_i|$. Let R_i be the rank of W_i, and let $S_i = 1$ if $Y_i > 0$ and $S_i = 0$ if $Y_i < 0$. (Since the distribution is continuous, $P(Y_i = 0) = 0$.) For example, if the Y_i are 4, -3, 1, and 6, then the S_i are 1, 0, 1, and 1, and the R_i are 3, 2, 1, and 4.

The one-sample *Wilcoxon test statistic* is defined to be

$$T_W = \sum S_i R_i$$

(i.e., T_W is the sum of the ranks associated with the positive Y_i's). The Wilcoxon test for testing that $\theta = d$ against $\theta > d$ rejects if $T_W > k$, where k is chosen so that the test has size α. (Note that the distribution of T_W is discrete, so that not all sizes may be attained.) The shape of the test is motivated by the following picture.

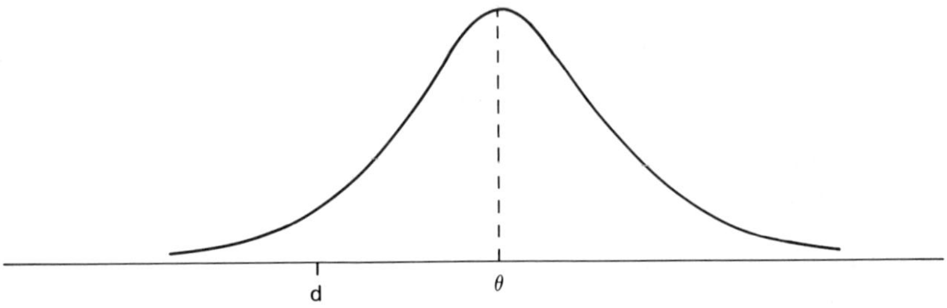

Figure 15–1 Graphs of $f(x)$.

If $\theta > d$, then most of the $Y_i = X_i - d$ should be positive, and the observations with large $|Y_i|$ should be associated observations with positive Y_i. Therefore, if $\theta > d$, then T_W should be large. Similarly, to test that $\theta = d$ against $\theta < d$, we reject if T_W is too small, and to test that $\theta = d$ against $\theta \neq d$, we reject if T_W is either too large or too small.

Before discussing some properties of T_W, we give a different, but equivalent, formula for it. Let $Q_i = 1$ if the Y_i with the ith smallest absolute value is positive and let $Q_i = 0$ if the ith smallest absolute value is negative. (In the previous example, the Y_i with the smallest absolute value (1) is positive, so that $Q_1 = 1$, and similarly, $Q_2 = 0$, $Q_3 = 1$, and $Q_4 = 1$.) Then

$$T_W = \sum R_i S_i = \sum i Q_i.$$

(In either definition given for T_W, T_W is the sum of the R_i over i such that $Y_i > 0$.) The advantage of the second form for T_W is that the Q_i are the only random elements. In the first form, both S_i and R_i are random.

The following lemma is the basic result of this section.

Lemma. If the X_i are a sample from a distribution which is symmetric about d, then the Q_i are independent, and $P(Q_i = 1) = P(Q_i = 0) = \frac{1}{2}$.

Proof. The $Y_i = X_i - d$ are independently identically distributed from a distribution which is symmetric about zero. Hence, S_i is independent of $|Y_i|$, and $P(S_i = 1) = P(S_i = 0) = .5$. (See Exercise B3.) Therefore, $(S_1, \ldots, S_n)$ and $(|Y_1|, \ldots, Y_n|)$ are all independent, and

$$P(S_1 = s_1, \ldots, S_n = s_n) = 2^{-n}$$

for any set $s_1, \ldots, s_n$ of zeros and ones. Let U_i be the observation Y_j such that $|Y_j|$ has rank i in the $|Y_i|$. (For example, if the Y's are 4, -5, and 3, then $U_1 = 3$, $U_2 = 1$, and $U_3 = 2$, since the third observation has the smallest absolute value, the first has the second smallest absolute value, and the second has the largest absolute value.) Since the U_i are computed from the $|Y_i|$, $\mathbf{S} = (S_1, \ldots, S_n)$ is independent of $\mathbf{U} =$

$(U_1, \ldots, U_n)$. Now,

$$Q_i = S_{U_i}.$$

Therefore,

$$\begin{aligned} P(Q_1 = q_1, \ldots, Q_n = q_n) &= P(S_{U_1} = q_1, \ldots, S_{U_n} = q_n) \\ &= E[P(S_{U_1} = q_1, \ldots, S_{U_n} = q_n|\mathbf{U})] \\ &= E(2^{-n}) = 2^{-n}, \end{aligned}$$

for any sequence $(q_1, \ldots, q_n)$ of zeros and ones (since the U_i are constants in the conditional expectation and the conditional distribution of **S** given **U** is the same as the marginal distribution of **S**). □

Theorem 15–1

a. T_W is a test statistic for testing $\theta = d$.
b. Under the null hypothesis that $\theta = d$,

$$ET_W = \mu_W = \frac{n(n+1)}{4} \quad \text{and} \quad \text{var}(T_W) = \sigma_W^2 = \frac{n(n+1)(2n+1)}{24}.$$

c. The null distribution of T_W is symmetric about μ_w.
d. Under the null hypothesis,

$$\frac{T_W - \mu_W}{\sigma_W} \xrightarrow{d} \mathbf{Z} \sim N(0,1).$$

Proof

a. By the previous lemma, the null distribution of $\mathbf{Q} = (Q_1, \ldots, Q_n)$ does not depend on any unknowns. Hence,

$$T_W = \sum iQ_i$$

is a test statistic.

b. $EQ_i = \frac{1}{2}$, $\text{var}(Q_i) = \frac{1}{4}$, and the Q_i are independent. Therefore,

$$ET_W = E\sum_{i=1}^{n} iQ_i = \sum_{i=1}^{n} iEQ_i = (\tfrac{1}{2}) \sum_{i=1}^{n} i = n(n+1)/4,$$

$$\text{var}(T_W) = \text{var}\left(\sum_{i=1}^{n} iQ_i\right) = \sum_{i=1}^{n} i^2 \text{var}(Q_i) = (\tfrac{1}{4}) \sum_{i=1}^{n} i^2 = n(n+1)(2n+1)/24.$$

c. Let $V_i = Q_i - \frac{1}{2}$. Under the null hypothesis $P(V_i = \frac{1}{2}) = P(V_i = -\frac{1}{2})$, and the V_i are independent. Therefore, the null distribution of $(-V_1, \ldots, -V_n)$ is the same as the null distribution of $(V_1, \ldots, V_n)$. Now,

$$T_W - \mu_W = \sum iQ_i - \frac{\sum i}{2} = \sum iV_i.$$

Consequently, the null distribution of $-(T_W - \mu_W) = \Sigma i(-V_i)$ is the same as the null distribution of $T_W - \mu_W = \Sigma iV_i$.

d. The proof of this part of the theorem is beyond the scope of the text. (See Hettmansperger (1984), pp. 47–47, for details.) □

Example 15–1

We indicate how to find the Wilcoxon density when $n = 5$. In that case, $T_W = 0$ if all the Y_i are negative, and $T_W = 15$ if all the Y_i are positive. Therefore, $0 \leqq T_W \leqq 15$. Let $\mathbf{Q} = (Q_1, \ldots, Q_5)$, where, as before, $Q_i = 1$ if the observation with the ith largest absolute value is positive, and $Q_i = 0$ otherwise. (If $\mathbf{Q} = (1, 0, 1, 1, 0)$, then $T_W = 1 + 0 + 3 + 4 + 0 = 8$.) Now,

$$T_W = 0 \Leftrightarrow \mathbf{Q} = (0, 0, 0, 0, 0), \qquad T_W = 1 \Leftrightarrow \mathbf{Q} = (1, 0, 0, 0, 0),$$

and

$$T_W = 2 \Leftrightarrow \mathbf{Q} = (0, 1, 0, 0, 0).$$

By the previous lemma, $P(Q_1 = q_1, Q_2 = q_2, \ldots, Q_5 = q_5) = (\frac{1}{2})^5 = \frac{1}{32}$. Therefore,

$$P(T_W = 0) = P(T_W = 1) = P(T_W = 2) = \tfrac{1}{32} = .03125.$$

By symmetry, we see that

$$P(T_W = 13) = P(T_W = 14) = P(T_W = 15) = .03125.$$

Now, T_W can equal three in two ways, viz., if $\mathbf{Q} = (0, 0, 1, 0, 0)$ or if $\mathbf{Q} = (1, 1, 0, 0, 0)$. Therefore,

$$P(T_W = 3) = P(T_W = 12) = \frac{2}{32} = .0625.$$

The remainder of the density function could be computed similarly. Therefore, a .09375 test that $\theta = d$ against $\theta > d$ would reject the null hypothesis when $T_W \geqq 13$, and a .15625 test would reject it if $T_W \geqq 12$.

Example 15–2

Let us apply the calculations in Example 15–1 to do a Wilcoxon test when we observe 3, 2, 6, 4, and 9 and want to test that $\theta = 2.3$ against $\theta > 2.3$. From Example 15–1,

$$P(T_W \geqq 14) = .0625.$$

Therefore, a .0625 Wilcoxon test for this problem rejects the null hypothesis if $T_W \geqq 14$. For the problem, the Y_i are

$$3 - 2.3 = .7, \qquad 2 - 2.3 = -.3, \qquad 6 - 2.3 = 3.7, \qquad 4 - 2.3 = 1.7, \qquad 9 - 2.3 = 6.7.$$

The ranks associated with the positive observations are 2, 3, 4, and 5, and therefore,

$$T_W = 2 + 3 + 4 + 5 = 14 \geq 14,$$

so that we reject the hypothesis that $\theta = 2.3$ in favor of $\theta > 2.3$ with a .0625 Wilcoxon test.

Several other tests similar to the Wilcoxon test have been suggested for this model. Let $0 \leqq c(1) \leqq c(2) \leqq \cdots \leqq c(n)$ be a set of constants, which we call scores. Then the *rank test statistic* associated with $\mathbf{c} = (c(1), \ldots, c(n))$ is defined to be

$$T_c = \sum S_i c(R_i) = \sum Q_i c(i)$$

(i.e., the sum of the scores associated with positive Y_i). If we are testing $\theta = d$ against $\theta > d$, we reject if T_c is too large, if we are testing $\theta = d$ against $\theta < d$, we reject if T_c

is too small, and if we are testing that $\theta = d$ against $\theta \neq d$, we reject if T_c is either too large or too small. The motivation for this shape is the same as for the Wilcoxon. (Note that the $c(i)$ are increasing in i.)

Theorem 15–2

a. T_c is a test statistic for testing that $\theta = d$.
b. Under the null hypothesis that $\theta = d$,

$$ET_c = \mu_c = \frac{\sum c(i)}{2} \quad \text{and} \quad \text{var}(T_c) = \sigma_c^2 = \frac{\sum (c(i))^2}{4}.$$

c. The null distribution of T_c is symmetric about μ_c.
d. Under fairly general conditions on $\mathbf{c}$,

$$\frac{(T_c - \mu_c)}{\sigma_c} \xrightarrow{d} Z \sim N(0, 1),$$

when the null hypothesis is true.

Proof. Parts a, b, and c are exercises. Part d is beyond the scope of this book and is proved in Hettmansperger (1984), pp. 84–86. □

The simplest rank test is the *sign test* in which all the $c(i) = 1$. In this case, $T_c = T_S$ is just the number of positive Y_i, i.e., the number of X_i which are greater than d. Under the null hypothesis, the probability that $X_i > d$ is .5, so that

$$T_S \sim B(n, .5),$$

when the null hypothesis is true. Hence we can use the binomial tables to find critical values for this test. For large n, we can use the normal approximation to the binomial distribution to find asymptotic critical values. (Note that the normal approximation to the binomial random variable T_S is the same as that given in Theorem 15–2.)

Example 15–3

As in Example 15–2, suppose we have $n = 5$, observe the X_i to be 3, 2, 6, 4 and 9, and we want to test that $\theta = 2.3$ against $\theta > 2.3$. From the binomial tables in the back of the book, we see that

$$P(T_S \geqq 5) = .03125.$$

Therefore, a .03125 test for this problem rejects if $T_S \geqq 5$. For the given data, there are four observations $(3, 6, 4, 9)$ above 2.3, so that $T_S = 4$ and we accept the null hypothesis that $\theta = 2.3$ with a .03125 sign test.

Another rank test that has been suggested is the one-sample *normal scores* test. Let $Z_1, \ldots, Z_n$ be a sample from an $N(0, 1)$ distribution, and let $U_i = |Z_i|$. Also, let

$$c(i) = EU_{[i]}$$

be the expected value of the ith-order statistic of the U_i. Then the $c(i)$ are called the *one-sample normal scores* and can be approximated by

$$c(i) \doteq 4.91(p_i^{0.14} - (1-p_i)^{0.14}), \qquad p_i = \frac{8n + 8i + 5}{16n + 10}. \tag{15–1}$$

In addition, it can be shown that the normal scores statistic is asymptotically normal. (See Hettmansperger, pp. 88–91.) In the past, this test has rarely been used because it is computationally messy. However, with computers, this issue should become less important, and the normal scores test may be used more often.

Example 15–4

Consider again the data (3, 2, 6, 4, 9) of Example 15–2, and suppose we wish to test that $\theta = 2.3$ against $\theta > 2.3$ with a normal scores test. Using Equation (15–1), we see that the normal scores for a sample of size 5 are

$$c_i = 4.91(p_i^{.14} + (1-p_i)^{.14}),$$

where $p_i = (45 + 8i)/90$, so that

$$c(1) = .22, \qquad c(2) = .46, \qquad c(3) = .71, \qquad c(4) = 1.01, \qquad c(5) = 1.59.$$

(Note that the true normal scores are .22, .45, .71, 1.04, and 1.57, so that the approximation is quite accurate when $n = 5$.) Recall that the only negative Y_i had $R_i = 1$. Therefore, the sum of the scores associated with positive R_i is

$$T_N = c(2) + c(3) + c(4) + c(5) = .46 + .71 + 1.04 + 1.59 = 3.82.$$

Note that

$$\mu_N = \frac{\sum c(i)}{2} = \frac{(.22 + .46 + .71 + 1.01 + 1.59)}{2} = 2.02,$$

$$\sigma_N^2 = \frac{\sum (c(i))^2}{4} = \frac{((.22)^2 + (.46)^2 + (.73)^2 + (1.04)^2 + (1.59)^2)}{4} = 1.10.$$

Now,

$$\frac{(T_N - \mu_N)}{\sigma_N} = 1.73 > 1.645 = z^{.05}$$

so that we (barely) reject the hypothesis that $\theta = 2.3$ in favor of $\theta > 2.3$ with an asymptotic size-.05 test. (The test which rejects if $T_N \geqq 3.82$ has exact size $\frac{2}{32} = .0625$ for this problem (see the exercises), so that we would reject with an exact .0625 test for these data.)

In this section, we are assuming that the observations come from continuous distributions. Therefore, $P(Y_i = 0) = 0$. However, in practice, it occasionally happens that a $Y_i = 0$. In that case, the observation is eliminated from the sample before doing any rank tests. In addition, because of the continuity, the probability of ties is zero. However, in practice, ties do occur. When there are ties, all the observations at that rank are assigned the average of the scores of the ranks with which they are tied. For example, if the Y's are 0, 3, 4, 2, −2, −3, 2, −1, 6, and −8, then the zero would be eliminated and the signed ranks of the remaining observations would be 5.5, 7, 3, −3, −5.5, 3, −1, 8, and −9. The Wilcoxon test statistic would then be

$$5.5 + 7 + 3 + 3 + 8 = 26.5.$$

Adjustments to the critical values of the Wilcoxon test have been suggested when there are ties. However, when there are not too many ties, such adjustments are not necessary. If there are many ties, then the experimenter should try to get more accurate readings. If he or she cannot, some other type of test should be considered.

The sign test has size α for testing hypotheses that the median $\theta = d$, even without assuming that the observations are symmetric. However, the sign test makes so little use of the observations (ignoring everything but the sign of the Y_i), that it seems inappropriate for most practical situations.

The Wilcoxon test seems preferable to the sign test when the distribution is symmetric. The idea of replacing the observations with the signed ranks is quite appealing, because it effectively moves unusually extreme observations back into the data. For example, suppose we observe the X's 3, -2, -4, 5, 6, -7, and 892 and want to test that $\theta = 0$ against $\theta \neq 0$. Then the signed ranks become 2, -1, -3, 4, 5, -6, and 7. The 892, which is probably some sort of mistake that would have damaged any parametric analysis, has been moved back into the data set.

One practical disadvantage of the Wilcoxon test, however, is that it separates close values. For example, suppose we observe the Y's -10.3, .7, 15.2, .8, and $-.6$. Then the signed ranks are -4, 2, 5, 3, and -1, and $T_W = 10$. Plainly, the Wilcoxon test is treating the difference between .6 and .7 in the same way as it treats the difference between 10.3 and 15.2 or .8 and 10.3. For example, if we change the $-.6$ to $-.9$, we change the signed ranks to -4, 1, 5, 2, and -3, and the signed rank statistic becomes 8. The Wilcoxon test is therefore unstable in the sense that small changes in close values can make large changes in the statistic. Fortunately, the problem occurs only when there is a cluster of close $|Y_i|$. In addition, if all the observations in that cluster have the same sign, then it does not matter how we assign the relevant ranks. Therefore, it is typically a problem only when the nearly tied Y_i are close to zero.

Often in an experiment, we may believe that the density function of the Y_i increases for $y < 0$, has a maximum at $y = 0$, and decreases for $y > 0$. In that case, the close observations should occur near the center of the distribution. A rank test will be less sensitive to the close observations in the center of the distribution if the scores c_i are close together when i is near zero and get further apart as i gets large. (Such scores should more nearly map what would happen if we get a "good" sample.) If there are ties, the ties are not so severe if the scores associated with the ranks that are tied are close together, and ties should occur near the center of the distribution. The normal scores have this property, but the Wilcoxon scores do not, which is one reason for preferring the normal scores to the Wilcoxon scores.

Exercises—A

1. A study is being made to see whether patients asking for cosmetic surgery are any heavier than the average weight of 115.3 for all patients. A sample of 20 such patients is collected, and their weights are recorded, giving the following data: 140, 125, 133, 117, 139, 145, 119, 121, 146, 153, 107, 102, 98, 104, 162, 109, 114, 137, 124, 111.
   - **(a)** Do an asymptotic .05 sign test on data.
   - **(b)** Do an asymptotic .05 Wilcoxon test on the data.
   - **(c)** Use the approximation to find the 5th and 8th normal scores for $n = 20$.

(d) The approximate normal scores for $n = 20$ are .06, .12, .18, .24, .30, .37, .43, .50, .57, .64, .72, .81, .89, .99, 1.09, 1.22, 1.35, 1.52, 1.76, 2.18. Use these normal scores to do an asymptotic .05 normal scores test on these data.

2. A test is run to see whether a blood pressure treatment is effective. Five patients are selected and given the treatment. Their before and after measurements are (140, 126), (154, 124), (125, 126), (115, 105), and (145, 120).
   (a) Do these data indicate a significant reduction in blood pressure with a size-$\frac{1}{32}$ sign test?
   (b) Is there a significant reduction with a $\frac{2}{32}$ Wilcoxon test?
   (Use the exact null distributions to determine the critical values here.)

Exercises—B

1. Suppose that the distribution of X is symmetric about θ.
   (a) Show that θ is a median of X.
   (b) Show that if EX exists, then $\theta = EX$. (Note that $E(X - \theta) = E(-(X - \theta))$.)
2. Find the density function of T_W under the null hypothesis when $n = 4$. What are the critical points for a .125 two-sided test in this case?
3. Suppose that the distribution of Y is symmetric about zero. Let $U = |Y|$, and let $S = 1$ if $Y > 0$ and $S = 0$ if $Y < 0$. Show that U and S are independent. (*Hint*: Show that $P(S = s, U \in A) = P(S = s)P(U \in A)$.)
4. Let $S_i^* = 1$ if $Y_i > 0$ and $S_i^* = -1$ if $Y_i < 0$. Let $T_c^* = \Sigma S_i^* c(R_i)$. Show that $T_c^* = 2(T_c - \mu_c)$. (*Hint*: What is the relationship between S_i^* and S_i?)
5. Suppose that $n = 2r$. Let $c(i) = 0$, $i = 0, \ldots, r$, and $c(i) = 1$, $i = r + 1, \ldots, 2r$. Find the mean and variance of T_c for these scores.
6. Let $X_1, \ldots, X_n$ be a sample from a (possibly nonsymmetric) distribution with median θ. Show that the sign test has size α for testing the null hypothesis that $\theta = d$ against one-sided and two-sided alternatives.
7. Show that when $n = 5$, the test which rejects when the normal scores test statistic $T_n \geqq 3.82$ has size $\frac{2}{32}$ for these data. (*Hint*: $T \geqq 3.82$ if and only if $T = 3.82$ or $T = 4.04$.)
8. Prove parts a, b, and c of Theorem 15–2.

Exercises—C

1. Find the null distribution of the normal scores statistic when $n = 4$. You may assume that the approximation given in Equation (15–1) is exact. (List all 16 possibilities for $\mathbf{Q}$, and evaluate the statistic for each possibility.)

15.1.2 Rank Confidence Intervals and Estimators

In the previous section, we gave a test statistic $T_c(d)$ for testing that $\theta = d$ for any d. Now, let $T_c(\theta)$ be that function evaluated at the true value θ. Since we have replaced the constant d with the unknown θ, $T_c(\theta)$ is no longer a statistic.

Theorem 15–3. $T_c(\theta)$ is a pivotal quantity for θ. The distribution of $T_c(\theta)$ is the same as the null distribution of $T_c(d)$ defined in the previous section.

Proof. For each fixed d, the distribution of $T_c(\theta)$ when $\theta = d$ is the same as the null distribution of $T_c(d)$, which does not depend on d. Therefore, for all θ, the distribution of $T_c(\theta)$ is the same as the null distribution of $T_c(d)$, and $T_c(\theta)$ is a pivotal quantity. □

Note that in proving Theorem 15–3, we have reversed the method used in earlier chapters of deriving test statistics from pivotal quantities. However, the reversed method works here only because the null distribution of $T_c(d)$ does not depend on d.

As in the previous section, let $\mu_c = ET_c(\theta)$. Since the distribution of $T_c(\theta)$ is symmetric about μ_c, choose a and $b = 2\mu_c - a$ such that $P(a \leqq T_c(\theta) \leqq b) = 1 - \alpha$. Now, suppose that

$$a \leqq T_c(\theta) \leqq b \Leftrightarrow L \leqq \theta < R.$$

Then the random interval $L \leqq \theta < R$ is the *rank* $(1 - \alpha)$ *confidence interval* associated with **c**. (The interval $L \leqq \theta < R$ is the set of all d such that we would accept the hypothesis $\theta = d$ with a two-sided size-α test based on T_c.) Because the distributions are continuous, $P(R = \theta) = 0$, so that $L \leqq \theta \leqq R$ would also be a $(1 - \alpha)$ confidence interval for θ.

Since the distribution of $T_c(\theta)$ is symmetric about μ_c, we define the *rank estimator* $\hat{\theta}_c$ of θ associated with **c** by

$$T_c(\hat{\theta}_c) = \mu_c.$$

For many scores, the associated confidence intervals and estimators have no simple formula and must be computed numerically. However, in this section and Section 15.1.3, we shall encounter several pivotal quantities of the form

$$Q(\gamma) = \#(V_k > \gamma),$$

where the right-hand expression denotes the number of V_k greater than or equal to γ for some random variables $V_1, \ldots, V_K$. Note that $0 \leqq Q(\gamma) \leqq K$. In such case, we can use the next theorem to find the confidence intervals and estimators.

Theorem 15–4. Suppose that $Q(\gamma) = \#(V_k > \gamma)$ is a pivotal quantity which is symmetric about $\mu = EQ(\gamma)$. Then

a. The estimator of γ associated with $Q(\gamma)$ is the sample median of the V_k.
b. If $b = K - a$, then

$$a \leqq Q(\gamma) \leqq b \Leftrightarrow V_{[a]} \leqq \theta < V_{[b+1]},$$

where the $V_{[i]}$ are the order statistics of the V_i.

Proof

a. Let $Q^* = \#(V_k < \theta) = K - Q(\theta)$. Then since $Q(\gamma)$ is symmetric, $\mu = EQ(\gamma) = K/2$. Therefore,

$$Q(\gamma) - Q^*(\gamma) = Q(\gamma) - (K - Q(\gamma)) = 2(Q(\gamma) - \mu).$$

Hence, $Q(\hat{\gamma}) - \mu = 0$ if and only if $Q(\hat{\gamma}) = Q^*(\hat{\gamma})$. If we now take $\hat{\gamma}$ to be the median of the V_k, the latter condition is satisfied, and the estimator of γ associated with $Q(\gamma)$ is the sample median of the V_k. (If K is odd, then this is the only possible choice; if K is even, then any number between the $K/2$ and $1 + K/2$ order statistics works.)

b. $Q(\gamma) \leqq b$ if and only if at most b of the V_k are greater than or equal to γ, which happens if and only if at least $K - b = a$ of the V_k are less than or equal to γ, which happens if and only if $\gamma \geqq V_{[a]}$. Similarly, $Q(\gamma) \geqq a$ if and only if at most $n - a = b$ of the observations are less than or equal to γ, which happens if and only if $\gamma < V_{[b+1]}$. □

For the sign scores, $c_i = 1$, so that

$$T_S(\theta) = \#(X_i > \theta).$$

By Theorem 15–4, the estimator associated with the sign scores is the sample median of the X_i.

Now, let a and $b = n - a$ be chosen from a binomial table, so that

$$P(a \leqq T_S(\theta) \leqq b) = 1 - \alpha.$$

By Theorem 15–4,

$$a \leqq T_S(\theta) \leqq b \Leftrightarrow X_{[a]} \leqq \theta < X_{[b+1]}.$$

Therefore, this interval is the $(1 - \alpha)$ rank confidence interval associated with the sign test.

For the Wilcoxon scores, $c_i = i$, so that

$$T_W(\theta) = \sum S_i R_i,$$

where R_i is the rank of $|Y_i(\theta)| = |X_i - \theta|$ in the set of $|Y_i(\theta)|$. In order to find the confidence interval and estimator associated with these scores, we put this statistic in a different form. Let

$$P_{ij} = \frac{X_i + X_j}{2}, \qquad 1 \leqq i \leqq j \leqq n$$

be the average of the observations X_i and X_j for all i and j. The P_{ij} are called the *Walsh averages* of the X_i. Note that we define these averages only for $1 \leqq i \leqq j \leqq n$, so that there are $n(n+1)/2$ of them. ($X_i = (X_i + X_i)/2$, so that the original observations are particular Walsh averages.)

Theorem 15–5. $T_W(\theta) = \#(P_{ij} > \theta)$.

Proof. Let $U_{ij} = (Y_i + Y_j)/2 = P_{ij} - \theta$ be the Walsh averages of the $Y_i = X_i - \theta$. Then $P_{ij} > \theta$ if and only if $U_{ij} > 0$. Observe that the set of $U_{ij}, 1 \leqq i \leqq j \leqq n$, is the same as the set of U_{ij}, $Y_i \leqq Y_j$. Now, if $Y_i > 0$ $(S_i = 1)$, then R_i is the number of Y_j such that $-Y_i \leqq Y_j \leqq Y_i$, or equivalently, the number of U_{ij} such that $Y_j < Y_i$ and $U_{ij} > 0$. Therefore, the sum T_W of the R_i with positive Y_i is the total number of U_{ij} such that $U_{ij} > 0$, i.e., $T_W(\theta) = \#(P_{ij} > \theta)$. □

By Theorems 15–4 and 15–5, the estimator associated with the Wilcoxon scores is the median of the Walsh averages. This estimator is often called the *Hodges-Lehmann* estimator of θ.

Now, let a and $b = 2\mu_W - a$ satisfy

$$P(a \leqq T_W(\theta) \leqq b) = 1 - \alpha.$$

Also, let $P_{[k]}$ be the kth smallest of the $n(n+1)/2$ P_{ij}'s. Then, by Theorem 15–4,

$$a \leqq \#(P_{ij} \geqq \theta) \leqq b \Leftrightarrow P_{[a]} \leqq \theta < P_{[b+1]},$$

so this interval is the Wilcoxon $(1 - \alpha)$ confidence interval.

Confidence intervals associated with the normal scores are more difficult and will not be discussed here.

Example 15–5

Suppose that $n = 5$, and that we observe 3, 2, 6, 4, and 10. Then the median is

$$\hat{\theta}_S = 4.$$

Using the binomial tables, we see that $P(1 \leqq T_S(\theta) \leqq 4) = .938$. Therefore, the .938 confidence interval associated with the sign test is

$$X_{[1]} \leqq \theta < X_{[5]} \Leftrightarrow 2 \leqq \theta < 10.$$

There are $5(6)/2 = 15$ Walsh averages, which are listed in the following table:

X_i	10	10	10	10	10	6	6	6	6	4	4	4	3	3	2
X_j	10	6	4	3	2	6	4	3	2	4	3	2	3	2	2
P_{ij}	10	8	7	6.5	6	6	5	4.5	4	4	3.5	3	3	2.5	2

Therefore, the Hodges-Lehmann estimator of θ is

$$\hat{\theta}_W = P_{[8]} = 4.5.$$

Now, from Example 15–1, $P(2 \leqq T_W \leqq 13) = .875$. Therefore, the .875 confidence interval associated with the Wilcoxon test is

$$P_{[2]} \leqq \theta < P_{[14]} \Leftrightarrow 2.5 \leqq \theta < 8.$$

Ties do not cause the same problems for the rank estimators and confidence intervals as for the rank tests. For example, suppose the confidence interval for θ is between the second and ninth order statistics of the data $6, 3, 4, 5, 7, 8, 3, 9, 8, 9$. The order statistics are then $3, 3, 4, 5, 6, 7, 8, 8, 9, 9$. Therefore, the second order statistic is 3 and the ninth one is 9; it does not matter that they are tied with other observations.

Exercises—A

1. Let θ be the median drop in blood pressure for the data of Exercise A2, Section 15.1.1.
   (a) Find the Wilcoxon (Hodges-Lehmann) estimator of θ.
   (b) Find the Wilcoxon .875 confidence interval for θ.
   (c) Find the sign estimator of θ.
   (d) Find the $\frac{15}{16}$ sign confidence interval for θ.

Exercises—B

1. Show that $\overline{X}$ is the point estimator of θ for the symmetric pivotal quantity $n^{1/2}(\overline{X}-\mu)/S$.

15.1.3 Two-Sample Rank Procedures

Consider the model in which we observe independent samples $X_1,\ldots,X_m$ and $Y_1,\ldots,Y_n$, where the $Y_j-\delta$ have the same continuous distribution as the X_i for some unknown δ. We make no assumptions about the common distribution of the X_i and $Y_j-\delta$, except that it is continuous. We are interested in inferences about δ, the shift parameter.

We first describe a collection of pivotal quantities for δ. Let $N=m+n$ be the total number of observations, and let $Z_j(\delta)=Y_j-\delta$. We rank the X_i and $Z_j(\delta)$ jointly and let $R_j=R_j(\delta)$ be the rank of $Z_j(\delta)$ in this joint ranking. (For example, if the X_i are 7 and 3 and the $Z_j(\delta)$ are 2, 5, and 4, then the joint order statistics are 2, 3, 4, 5, and 7, and hence, $R_1=1$ (since $Z_1=2$ has rank 1 in the joint sample), R_2 is 4 (the rank of $Z_2=5$), and $R_3=3$ (the rank of $Z_3=4$).)

Now let $c(1)\leqq c(2)\leqq\cdots\leqq c(N)$ be a set of constants which we call the *scores*. We say that the scores are *symmetric* if $c(k)+c(N+1-k)$ does not depend on k. Let

$$T_c(\delta)=\sum_{j=1}^{n}c(R_j(\delta)).$$

That is, $T_c(\delta)$ is the sum of the scores associated with the ranks of the Z_j's. The following theorem summarizes the basic properties of $T_c(\delta)$.

Theorem 15–6. Define $\bar{c}=\sum_{k=1}^{N}c(k)/N$, $Q=\sum_{k=1}^{N}(c(k)-\bar{c})^2$. Then

a. $T_c(\delta)$ is a pivotal quantity for δ.

b. $\mu_c=ET_c(\delta)=n\bar{c}$, and $\sigma_c^2=\text{var}(T_c(\delta))=nmQ/N(N-1)$.

c. If the scores are symmetric, then the distribution of $T_c(\delta)-\mu_c$ is symmetric about zero.

d. Under fairly general conditions, $(T_c(\delta)-\mu_c)/\sigma_c\xrightarrow{d}Z\sim N(0,1)$.

Proof

a. Note that the X_i and Z_j are all independent. In addition, $Z_j=Y_j-\delta$ has the same distribution as the X_i. Therefore, $X_1,\ldots,X_m,Z_1,\ldots,Z_n$ is a sample of size N from the distribution function F. Hence, any ordering of the observations is equally likely. But then, any ordered subset of $(1,\ldots,N)$ with n elements is equally likely for $(R_1,\ldots,R_n)$. Consequently, the joint distribution of the R_j does not depend on F or δ, so that $T_c(\delta)$ is a pivotal quantity for δ.

b. R_j is equally likely to be any number between one and N. Therefore,

$$P(R_j=k)=\frac{1}{N}\quad\text{and}\quad Ec(R_j)=\sum_{k=1}^{N}c(k)P(R_j=k)=\frac{\sum_{k=1}^{N}c(k)}{N}=\bar{c}.$$

Hence,

$$\mu_c = ET_c(\delta) = \sum_{j=1}^{n} Ec(R_j) = nEc(R_j) = n\bar{c}.$$

Now,

$$E[c(R_j)]^2 = \sum_{k=1}^{N} [c(k)]^2 P(R_j = k) = \frac{\sum_{k=1}^{N} [c(k)]^2}{N}$$

and

$$\text{var}[c(R_j)] = \sum_{k=1}^{N} \frac{[c(k)]^2}{N} - \bar{c}^2 = \frac{Q}{N}.$$

By a similar argument, we can show that

$$\text{cov}(c(R_j), c(R_k)) = \frac{-Q}{N(N-1)}.$$

But then,

$$\sigma_c^2 = \text{var}(T_c(\delta)) = \sum_{j=1}^{n} \text{var}(c(R_j)) + \sum_{j=1}^{n} \sum_{k \neq j} \text{cov}(c(R_j), c(R_k))$$

$$= \frac{nQ}{N} - \frac{n(n-1)Q}{N(N-1)} = \frac{n(N-n)Q}{N(N-1)} = \frac{nmQ}{N(N-1)}.$$

c. If $c(i) + c(N - i + 1) = d$, then $d = 2\bar{c}$, and hence,

$$c(i) - \bar{c} = -(c(N - i + 1) - \bar{c}).$$

Therefore,

$$-(T_c(\delta) - \mu_c) = -\sum_{i=1}^{n} (c(R_j) - \bar{c}) = \sum_{j=1}^{n} (c(N - R_j + 1) - \bar{c}).$$

Since any set of ranks is equally likely, $(N - R_1 + 1, \ldots, N - R_n + 1)$ has the same distribution as $(R_1, \ldots, R_n)$. But then $-(T_c(\delta) - \mu_c)$ has the same distribution as $T_c(\delta) - \mu$, and the distribution of $T_c(\delta) - \mu_c$ is symmetric about zero.

d. The proof of this part of the theorem is beyond the scope of the text. (See Hettmansperger (1984), pp. 147–157.) □

In the exercises, it is shown that $T_c(\delta)$ is a decreasing function of δ. Since $T_c(\delta)$ is a symmetric pivotal quantity for any symmetric scores, we can define confidence intervals, tests, and estimators associated with that quantity in the usual way.

We first look at the sign scores, $c_k = 1$. In this case, $T_c(\delta)$ is identically n, so that these scores do not lead to any interesting procedures.

We next look at the *Mood scores*. Suppose that $N = 2r$ is even, and let $c(k) = 1$ if $k > r$ and $c(k) = 0$ if $k \leqq r$. Note that $c(k) + c(N + 1 - k) = 1$, so that the scores are symmetric. The Mood pivotal quantity $T_M(\delta)$ is the number of $Z_j(\delta)$ above the joint median of the X_i and Z_j. The test statistic $T_M(0)$ for testing that $\delta = 0$ is the number of Y's which are above the joint median of X's and Y's, and the test using this statistic is often called the median test for this problem.

We can compute the distribution of $T_M(\delta)$ by considering the problem as one of sampling the n Z_j's at random from the set of r observations above the median and r below. Since all possible sets of rankings are equally likely for the Z_j, $T_M(\delta)$, the number of Z_j above the median has the hypergeometric distribution

$$P(T_M(\delta) = k) = \frac{\begin{bmatrix} r \\ k \end{bmatrix}\begin{bmatrix} r \\ n-k \end{bmatrix}}{\begin{bmatrix} 2r \\ n \end{bmatrix}}.$$

Note that $c(k) = c(k)^2$ is either zero or one, and that there are r zeros and r ones. Therefore, by part b of Theorem 15–6,

$$\bar{c} = \frac{1}{2}, \qquad \sum([c(k)]^2 - \bar{c}^2) = \frac{r}{2}, \qquad \mu_M = \frac{n}{2}, \qquad \sigma_M^2 = \frac{nm}{4(2r-1)}.$$

(These could also be derived from the formulas for the hypergeometric distribution.)

The rank estimator $\hat{\delta}_M$ of δ for these scores can be shown to be the sample median of the Y's minus the sample median of the X's. Let $X_{[i]}$ and $Y_{[j]}$ be the order statistics of the X's and Y's separately. Let a and $b = n - a$ satisfy $P(a \leqq T_M(\delta) \leqq b) = 1 - \alpha$. Then the $(1-\alpha)$ confidence interval associated with the Mood scores is

$$Y_{[a]} - X_{[r-a+1]} \leqq \delta \leqq Y_{[b+1]} - X_{[r-b]}.$$

(See Hettmansperger, p. 154.)

If $N = n + m$ is odd, then one of the observations is equal to the combined sample median, so it is neither greater nor less than it. What is often done in this situation is to eliminate the observation at the combined sample median and use the Mood procedures on the remaining data.

Example 15–6

Suppose we observe $X_i = 3.3$, 2.5, and 6.5, and $Y_j = 9.2$, 5.3, 7.7, 4.2, and 10.1. Then $T_M(\delta)$ must be 1, 2, 3, or 4, and

$$P(T_M = 1) = P(T_M = 4) = \frac{\begin{bmatrix} 4 \\ 1 \end{bmatrix}\begin{bmatrix} 4 \\ 4 \end{bmatrix}}{\begin{bmatrix} 8 \\ 5 \end{bmatrix}} = .07,$$

using the hypergeometric density function. Therefore, a .07 test that $\delta = 0$ against $\delta > 0$ rejects the null hypothesis if $T_M(0) \geqq 4$. With these data, $T_M(0)$ is the number of Y observations above the joint sample median $((6.5 + 5.3)/2 = 5.9)$, so that

$$T_M(0) = 3 < 4,$$

and we accept the null hypothesis with a .07 Mood test. The sample median for the Y's alone is 7.7 and for the X's alone is 3.3. Therefore, the Mood estimator for δ is

$$\hat{\delta}_M = 7.7 - 3.3 = 4.4.$$

Also, $P(2 \leqq T_M(\delta) \leqq 3) = .86$, so that the .86 Mood confidence interval for δ is

$$Y_{[2]} - X_{[3]} \leqq \delta \leqq Y_{[4]} - X_{[1]} \Leftrightarrow -1.2 \leqq \delta \leqq 6.7.$$

As in the one-sample model, the *Wilcoxon scores* are $c(k) = k$. Note that $c(k) + c(N + 1 - k) = N + 1$, so that these scores are symmetric. In this case,

$$T_W(\delta) = \sum_{j=1}^{n} R_j(\delta),$$

the sum of the ranks of the $Z_j = Y_j - \delta$ in the combined sample of X_i and Z_j.

Example 15–7

We illustrate how to find the distribution of T_W in the case where $m = 3$ and $n = 5$. Note that T_W must be at least $1+2+3+4+5=15$ and at most $4+5+6+7+8=30$. Recall that all possible sets of Z ranks are equally likely. There are

$$\begin{bmatrix} 8 \\ 5 \end{bmatrix} = 56$$

different choices for the ranks associated with the Z sample. There is only one choice $(1, 2, 3, 4, 5)$ which leads to $T_W = 15$, one choice $(1, 2, 3, 4, 6)$ which leads to $T_W = 16$, one choice $(3, 5, 6, 7, 8)$ for $T_W = 29$, and one choice for $T_W = 30$, so that

$$P(T_W = 15) = P(T_W = 16) = P(T_W = 29) = P(T_W = 30) = \frac{1}{56} = .018.$$

There are two choices $((1, 2, 3, 4, 7)$ and $(1, 2, 3, 5, 6))$ which lead to $T_W = 17$ and two choices $((2, 5, 6, 7, 8)$ and $(3, 4, 6, 7, 8))$ which lead to $T_W = 28$. Therefore,

$$P(T_W = 17) = P(T_W = 28) = \frac{2}{56} = .036.$$

The full density of T_W can be determined similarly.

The distribution function of T_W (or more often, $T_W - n(n + 1)/2$) for small m and n is given in many books on nonparametric statistics. For large m and n, the normal approximation may be used.

Observe that

$$\bar{c} = \frac{N+1}{2} \quad \text{and} \quad \sum_{k=1}^{N} (c_k - \bar{c})^2 = \frac{N(N+1)(N-1)}{12}.$$

Therefore, by part b of Theorem 15–6,

$$\mu_W = \frac{n(N+1)}{2} \quad \text{and} \quad \sigma_W^2 = \frac{mn(N+1)}{12}.$$

$T_W(0)$ is called the *two-sample Wilcoxon test statistic* and is used to test the null hypothesis $\delta = 0$ against one-sided and two-sided alternatives.

To find Wilcoxon confidence intervals and estimators, we need the following theorem.

Theorem 15–7. Let $D_{ij} = Y_j - X_i$, and let $U(\delta) = \#(D_{ij} \geqq \delta)$. Then $T_W(\delta) = U(\delta) + n(n + 1)/2$.

Proof. Clearly, $R_j(\delta) = \#(X_i \leqq Z_j) + \#(Z_k \leqq Z_j)$. Now, $\sum_{k=1}^{n} \#(Z_k \leqq Z_j) =$

$n(n+1)/2$, the number of possible pairs of Z_j. Therefore,

$$T_W(\delta) = \sum_{j=1}^{n} R_j(\delta) = \sum_{j=1}^{n} \#(X_i \leqq Z_j) + \frac{n(n+1)}{2} = \#(Y_j - X_i \geqq \delta) + \frac{n(n+1)}{2}$$
$$= U(\delta) + \frac{n(n+1)}{2}. \quad \square$$

The statistic $U(0)$ is called the two-sample *Mann-Whitney statistic*. Theorem 15–7 implies that $T_W(0) = U(0) + n(n+1)/2$. Also,

$$\mu_U = EU(\delta) = ET_W(\delta) - \frac{n(n+1)}{2} = mn,$$

$$\sigma_U^2 = \mathrm{var}(U(\delta)) = \mathrm{var}(T_W(\delta)) = \frac{mn(N+1)}{12},$$

and

$$U(\delta) - \mu_U = T_W(\delta) - \mu_W,$$

so that $U(\delta) - \mu_U$ is also symmetrically distributed about zero.

By Theorem 15–4, the Wilcoxon rank estimator is the median of the D_{ij}. This estimator is called the two-sample *Hodges-Lehmann estimator* of δ.

Now, let a and $b = nm - a$ satisfy

$$1 - \alpha = P(a \leqq U(\delta) \leqq b) = P\left(a + \frac{n(n+1)}{2} \leqq T_W(\delta) \leqq b + \frac{n(n+1)}{2}\right),$$

and let $D_{[i]}$ be the order statistics of the D_{ij}. Then, by Theorem 15–4 again,

$$a \leqq \#(D_{ij} \geqq \delta) \leqq b \Leftrightarrow D_{[a]} \leqq \delta < D_{[b+1]},$$

and hence, this interval is the $(1-\alpha)$ rank confidence interval associated with the Wilcoxon scores.

Example 15–8

Suppose, from Example 15–6, that $m = 3$, $n = 5$, the X_i are 3.3, 2.5, and 6.5, and the Y_j are 9.2, 5.3, 7.7, 4.2, and 10.1. From Example 15–7, we calculate

$$P(T_W \geqq 28) = .072.$$

Therefore, a .072 test that $\delta = 0$ against $\delta > 0$ rejects the null hypothesis if $T_W(0) \geqq 28$. Now, the ranks associated with the Y_j's in the joint sample are

$$(3, 4, 6, 7, 8),$$

so that

$$T_W(0) = 28 \geqq 28,$$

and we reject the null hypothesis that $\delta = 0$ against $\delta > 0$ with a .072 Wilcoxon test. There are 15 different D_{ij}, which are listed, together with their ranks, as follows:

Y	4.2	4.2	4.2	5.3	5.3	5.3	7.7	7.7	7.7	9.2	9.2	9.2	10.1	10.1	10.1
X	2.5	3.3	6.5	2.5	3.3	6.5	2.5	3.3	6.5	2.5	3.3	6.5	2.5	3.3	6.5
D	1.7	.9	−2.3	2.8	2.0	−1.2	5.2	4.4	1.2	6.7	5.9	2.7	7.6	6.8	3.6
R	5	3	1	8	6	2	11	10	4	13	12	7	15	14	9

Therefore, the Wilcoxon estimator for δ is the median of the D_{ij}, i.e.,

$$\hat{\delta}_W = d_{[8]} = 2.8.$$

(Note that the median of the differences is 2.8, but that the difference of the medians is 4.4.) Now, $P(17 \leqq T_W(\delta) \leqq 28) = P(2 \leqq U(\delta) \leqq 13) = .928$. Therefore, the .928 Wilcoxon interval for δ for these data is

$$D_{[2]} \leqq \delta \leqq D_{[14]} \Leftrightarrow -1.2 \leqq \delta \leqq 6.8.$$

Now, let $Z_1, \ldots, Z_n$ be a sample from an $N(0, 1)$ distribution, and let

$$c(k) = EZ_{[k]}$$

be the expectation of the kth order statistic of the Z_k. These $c(k)$ are called *two-sample normal scores*. A good approximation to the scores for the two-sample problem is

$$c(k) = 4.91(p_k^{0.14} - (1 - p_k)^{0.14}), \qquad p_k = \frac{8k - 3}{8N + 2}. \tag{15–2}$$

(see Hettmansperger, p. 145.) Since

$$c(N + 1 - k) = -c(k),$$

the $c(k)$ are symmetric. If we now let $T_N(\delta)$ be the pivotal quantity defined with these scores, then the statistic $T_N(0)$ can be used in the obvious way to test that $\delta = 0$ against one-sided and two-sided alternatives. Unfortunately, the rank estimators and tests based on the normal scores are quite complicated.

Example 15–9

Suppose again from Example 15–6 that $m = 3$, $n = 5$, the $X_i = 3.3$, 2.5, and 6.5, and the $Y_j = 9.2$, 5.3, 7.7, 4.2, and 10.1. Then, $N = 8$ and using Equation (15–3), we see that the normal scores are approximately

$$-1.43, \quad -.85, \quad -.47, \quad -.15, \quad .15, \quad .47, \quad .85, \quad 1.43.$$

(The true scores for $m + n = 8$ are −1.42, −.85, −.47, −.15, .15, .47, .85, 1.42, so the approximation formula is quite close again.) Since we do not have the critical values for the normal score statistic, we shall use the normal approximation. By Theorem 15–7,

$$\bar{c} = 0, \qquad \sum (c(k))^2 = 6.08, \qquad \mu_N = 0, \qquad \sigma_N^2 = \frac{(3)(5)(6.08)}{(8)(7)} = 1.63.$$

An approximate .05 normal scores test that $\delta = 0$ against $\delta > 0$ rejects the null hypothesis if

$$\frac{T_N - 0}{\sigma_N} \geqq 1.645.$$

Since the ranks associated with the Y_i are 3, 4, 6, 7, and 8, the normal scores statistic is

$$T_N = c(3) + c(4) + c(6) + c(7) + c(8) = -.47 - .15 + .47 + .85 + 1.43 = 2.28$$

and

$$\frac{T_N}{\sigma_N} = \frac{2.28}{(1.63)^{1/2}} = 1.78 > 1.645.$$

So we reject the null hypothesis with an asymptotic .05 normal scores test with these data.

Ties are treated in the same manner in the two-sample model as in the one-sample model, by assigning each of the tied observations the average of the scores associated with the ranks at which they are tied. Note that ties can be ignored when computing rank estimators or confidence intervals, as in the previous section.

Exercises—A

1. An experiment is done to compare temperatures in two cities in a particular season. Eight measurements are taken in the first city and seven in the second one, yielding the following data:

X	78	62	85	75	91	83	58	94
Y	55	47	61	43	64	44	50	

Let δ be the difference between the median temperatures for the two cities.
(a) Do an asymptotic size-.05 two-sample Wilcoxon test that $\delta = 0$ against $\delta < 0$.
(b) Find the asymptotic .95 Wilcoxon confidence interval for δ.
(c) Find the Wilcoxon estimator of δ.

2. In order to use the Mood procedures on the data in Exercise A1, it is necessary to drop the joint median 62, so that $m = n = r = 7$.
(a) Using the data in Exercise A1, do an exact .03 test that $\delta = 0$ against $\delta \neq 0$.
(b) Find the Mood estimator of δ.
(c) Find the exact .97 Mood confidence interval for δ.

3. Do an asymptotic .05 normal scores test that $\delta = 0$ against $\delta < 0$ for the data in Exercise A1.

4. An experiment is performed to see whether students from suburban high schools score higher than those from rural high schools on mathematics SAT tests. Scores are recorded from six suburban students and three rural students. The suburban students had mathematics SATs of 580, 640, 660, 800, 560, and 600, while the rural students had mathematics SATs of 610, 480, and 500. Let δ be the shift from suburban to rural scores.
(a) Do a $\frac{1}{21}$ Wilcoxon test that $\delta = 0$ against $\delta < 0$.
(b) Find a $\frac{1}{21}$ Wilcoxon confidence interval for δ.
(c) Find the Wilcoxon estimator of δ.
(d) Repeat parts a, b, and c if the score of 610 is changed to 570.

Exercises—B

1. Use the formulas derived earlier for the mean and variance of a hypergeometric distribution to verify the formula given in the text for the mean and variance of the Mood pivotal quantity.

2. Find the density of the Wilcoxon pivotal quantity when $m = 3$ and $n = 2$.
3. For the Wilcoxon scores, show that $\Sigma(c(k) - \bar{c})^2 = N(N + 1)(N - 1)/12$.
4. **(a)** Verify that $c(N + 1 - k) = -c(k)$ for the normal scores.
   (b) Verify this equality for the approximated normal scores given in Equation (15–2).
5. Verify that $\text{cov}(c(R_j), c(R_k)) = -Q/N(N - 1)$.

Exercises—C

1. **(a)** Show that the $R_j(\delta)$ are decreasing functions of δ.
   (b) Verify that $T_c(\delta) = \Sigma c(R_j(\delta))$ is a nondecreasing function as long as $c(k + 1) \geqq c(k)$.
2. Using the approximation given in Equation (15–2), find the density of the normal scores pivotal quantity when $n = 2$ and $m = 3$.

15.2 ASYMPTOTIC RELATIVE EFFICIENCY

* *15.2.1 Asymptotic Relative Efficiency*

In this section, we indicate that when the observations are normally distributed, the rank procedures discussed in previous sections are nearly as good asymptotically as the t-procedures based on the normal assumption. However, the rank procedures may be considerably better asymptotically than the t-procedures when the observations are not normally distributed.

We first look at one-sample models in which we observe a sample $\mathbf{X}_n = (X_1, \ldots, X_n)$ from an unknown density which is symmetric about its median θ. Let $Y_i = X_i - \theta$ have density $f(y)$.

Suppose that we have an asymptotic pivotal quantity $Q_n(\theta) = Q_n(\mathbf{X}_n, \theta)$ such that

$$\frac{Q_n(\theta) - \mu_n}{\sigma_n} \xrightarrow{d} Z \sim N(0, 1). \tag{15–3}$$

We assume that $Q_n(\theta)$ is symmetric about μ_n and decreasing in θ so that we can design confidence intervals, tests, and estimators based on $Q_n(\theta)$ as in previous sections. Let

$$\gamma_{n,f}(a) = E[Q_n(\theta - a)].$$

(Note that $\gamma_{n,f}(a)$ depends on f, although $\gamma_{n,f}(0) = \mu_n$ does not and that $\gamma_{n,f}(a)$ does not depend on θ, because θ is a location parameter.) Then the *efficacy* of Q_n for $f(y)$ is defined by

$$d_f = d_f(Q_n) = \lim_{n\to\infty} \frac{\gamma'_{n,f}(0)}{n^{1/2}\sigma_n},$$

where $\gamma'_{n,f}(0)$ is the derivative of $\gamma_{n,f}(a)$ at $a = 0$.

The following theorem is the main result of this section.

Theorem 15–8. Let $Q_n(\theta)$ be a symmetric asymptotic pivotal quantity which is decreasing in θ. Let d_f be the efficacy of $Q_n(\theta)$. Then, under general regularity conditions, the following are true:

a. Let $\hat{\theta}_n$ be the estimator of θ based on $Q_n(\theta)$. Then

$$n^{1/2} d_f(\hat{\theta}_n - \theta) \xrightarrow{d} U \sim N(0, 1).$$

b. Let (L_n, R_n) be the $(1 - \alpha)$ confidence interval for θ based on $Q_n(\theta)$. Then

$$n^{1/2}(R_n - L_n) \xrightarrow{P} \frac{2z^{\alpha}}{d_f}.$$

c. Let Φ_n be the size-α test for testing that $\theta = c$ against $\theta > c$. Let $K_{n,f}(\theta)$ be the power function of Φ_n, and let b_n satisfy $K_n(b_n) = \beta$. Then

$$n^{1/2}(b_n - c) \rightarrow \frac{z^{\beta} - z^{\alpha}}{d_f}.$$

Argument. A proof of this theorem is considerably beyond the level of this book. See Hettmansperger (1984), pp. 62–85 and 106 for details. We give a heuristic argument to indicate why these results should be true. The basic fact which we use is that for any sequence of constants a_n such that $n^{1/2}|a_n - \theta|$ is bounded,

$$\frac{(Q_n(a_n) - \mu_n)}{\sigma_n} - n^{1/2} d_f(\hat{\theta}_n - a_n) \xrightarrow{P} 0, \tag{15–4}$$

so that for large n and a_n near θ,

$$\frac{(Q_n(a_n) - \mu_n)}{\sigma_n} \doteq n^{1/2} d_f(\hat{\theta}_n - a_n) \tag{15–5}$$

and, in particular,

$$\frac{(Q_n(\theta) - \mu_n)}{\sigma_n} \doteq n^{1/2} d_f(\hat{\theta}_n - \theta). \tag{15–6}$$

a. By Equations (15–3) and (15–4) and Slutzky's theorem

$$n^{1/2} d_f(\hat{\theta}_n - \theta) = \frac{(Q_n(\theta) - \mu_n)}{\sigma_n} - \frac{(Q_n(\theta) - \mu_n)}{\sigma_n} - n^{1/2} d_f(\hat{\theta}_n - \theta) \xrightarrow{d} Z + 0.$$

b. Asymptotically, the $(1 - \alpha)$ rank confidence interval is given by

$$A_n = \left\{\theta\colon \frac{(Q_n(\theta) - \mu_n)}{\sigma_n} \in 0 \pm z^{\alpha/2}\right\}.$$

(Note that this set must be an interval because $Q_n(\theta)$ is a nonincreasing function of θ.) For large n, by Equation (15–6), this interval should approximately equal

$$B_n = \{\theta\colon n^{1/2} d_f(\hat{\theta}_n - \theta) \in 0 \pm z^{\alpha/2}\} = \left\{\theta\colon \theta \in \hat{\theta}_n \pm \frac{z^{\alpha/2}}{d_f n^{1/2}}\right\}.$$

Now, let $D_{1,n}$ and $D_{2,n}$ be the lengths of the intervals of the intervals A_n and B_n. Then, for large n,

$$n^{1/2} D_{1,n} \doteq n^{1/2} D_{2,n} = \frac{2z^{\alpha/2}}{d_f}.$$

c. Let b_n be a sequence of parameter values which converge to c, the value being tested. Note that the distribution of $Q_n(c)$ when $\theta = b_n$ is the same as the distribution of $Q_n(c - b_n)$ when $\theta = 0$. (Why?) Note also that when $\theta = 0$, $n^{1/2} d_f \hat{\theta}_n \stackrel{.}{\sim} N(0, 1)$ for large n. Therefore, using Equation (15–5), we see that

$$K_{n,f}(b_n) = P\left(\frac{(Q_n(c) - \mu_n)}{\sigma_n} > z^\alpha | \theta = b_n\right) = P\left(\frac{(Q_n(c - b_n) - \mu_n)}{\sigma_n} > z^\alpha | \theta = 0\right)$$

$$\doteq P(n^{1/2} d_f(\hat{\theta}_n - (c - b_n)) > z^\alpha | \theta = 0)$$

$$= P(n^{1/2} d_f \hat{\theta}_n > z^\alpha + n^{1/2} d_f(c - b_n) | \theta = 0)$$

$$\doteq P(Z > z^\alpha + n^{1/2} d_f(c - b_n)),$$

where $Z \sim N(0, 1)$. If we set the right-hand side of these approximations equal to β, we get

$$z^\alpha + n^{1/2} d_f(b_n - c) = z^\beta,$$

or equivalently

$$n^{1/2}(b_n - c) = \frac{(z^\beta - z^\alpha)}{d_f}. \quad \square$$

Part a of Theorem 15–8 implies that for large n,

$$\hat{\theta}_n \stackrel{.}{\sim} N(\theta, (nd_f^2)^{-1}),$$

so that the asymptotic performance of $\hat{\theta}_n$ is better when d_f is large. Part b implies that for large n, the approximate length of the confidence interval is

$$R_n - L_n \approx \frac{2z^\alpha}{nd_f},$$

and hence, the confidence intervals are asymptotically shorter when d_f is large. Part c of the theorem is often stated somewhat differently. For each θ, let $n(\theta)$ satisfy

$$K_{n(\theta)}(\theta) = \beta.$$

That is, $n(\theta)$ is the sample size necessary to get power β at θ. Then

$$(n(\theta))^{1/2}(\theta - c) \rightarrow \frac{z^\beta - z^\alpha}{d_f} \quad \text{as } \theta \rightarrow c. \tag{15–7}$$

This result implies that for large n and θ near c, the sample size necessary to get power β at θ is approximately

$$n(\theta) \doteq \frac{z^\beta - z^\alpha}{(\theta - c)\, d_f}.$$

Hence, the sample size necessary to get power β is smaller when d_f is large. (Although it seems fairly clear, intuitively, that Equation (15–7) is the same as part c of the theorem, it is somewhat difficult to prove the equivalence.) All these facts point to the desirability of finding procedures in which d_f is large.

Now, let $Q_n(\theta)$ and $Q_n^*(\theta)$ be two symmetric pivotal quantities for θ which are decreasing in θ with efficacies d_f and d_f^*, respectively. Because of Theorem 15–8, we define the (Pitman) *asymptotic relative efficiency* (ARE) of Q_n with respect to Q_n^* for the density $f(y)$ by

$$e_f(Q_n, Q_n^*) = \frac{(d_f)^2}{(d_f^*)^2}.$$

Now, suppose the ARE of Q_n with respect to Q_n^* is .6 for the density $f(y)$. Then part a of Theorem 15–8 says that for the density $f(y)$, the asymptotic variance of the estimator based on Q_n^* is 60% of the asymptotic variance of the estimator based on Q_n. Part b implies that the asymptotic length of the asymptotic $(1-\alpha)$ confidence interval based on Q_n^* is $(.60)^{1/2} = .77$ of the asymptotic length of the asymptotic $(1-\alpha)$ confidence interval based on Q_n, while part c says that, asymptotically, it takes $(.60)^{1/2} = .77$ as many measurements with Φ_n^* to get power β at a θ near the null hypothesis as it does for Φ_n (for the density f).

We next give formulas for the efficacies of some pivotal quantities studied in this chapter. Recall that the t pivotal quantity $t_n(\theta) = n^{1/2}(\overline{X} - \theta)/S$ is an exact pivotal quantity when the X_i are normally distributed, but is an asymptotic pivotal quantity as long as the X_i have a finite variance. Recall also that $\overline{X}$ is the estimator based on $t_n(\theta)$. The sign, Wilcoxon, and normal scores pivotal quantities are exact pivotal quantities for any symmetric, continuous distribution.

Theorem 15–9. Let $f(y)$ be a density function with distribution function $F(y)$ and variance τ^2. Then

a. The efficacy of the sign pivotal quantity is

$$d_{S,f} = 2f(0).$$

b. The efficacy of the Wilcoxon pivotal quantity is

$$d_{W,f} = (12)^{1/2} \int_{-\infty}^{\infty} f^2(y)dy.$$

c. The efficacy of the normal scores pivotal quantity is

$$d_{N,f} = \int_{-\infty}^{\infty} \frac{f^2(x)}{n(N^{-1}[F(x)])} dx,$$

where $n(z)$ and $N(z)$ are the density function and distribution function, respectively, of the standard normal distribution.

d. The efficacy of the t asymptotic pivotal quantity is

$$d_{t,F} = \tau^{-1}.$$

Proof

a. The sign statistic $T_S(\theta) = \#(X_i > \theta) \sim B(n, \tfrac{1}{2})$, so

$$\mu_n = ET_S(\theta) = \frac{n}{2} \quad \text{and} \quad \sigma_n^2 = \text{var}(T_S(\theta)) = \frac{n}{4}.$$

Now, $T_S(\theta - a) = \#(X_i \geqq \theta - a) = \#(Y_i \geqq -a) \sim B(n, 1 - F(-a))$. Therefore,

$$\gamma_{n,f} = ET_S(\theta - a) = n(1 - F(-a)),$$

$$\gamma_{n,f}'(a) = nf(-a), \qquad d_{S,f} = \lim_{n\to\infty} \frac{2nf(0)}{n} = 2f(0).$$

b. Recall that for the Wilcoxon pivotal quantity T_W,

$$\mu_n = ET_W(\theta) = \frac{n(n+1)}{4} \quad \text{and} \quad \sigma_n^2 = \frac{n(n+1)(2n+1)}{24}.$$

Recall also that $T_W(\theta - u) = \#(P_{ij} > \theta - a)$, where the P_{ij} are the $n(n+1)/2$ pairwise averages $(X_i + X_j)/2$. As before, let $Y_i = X_i - \theta$. Then

$$\begin{aligned} P(P_{ij} > \theta - a) &= P((Y_i + Y_j) > 2a) \\ &= \int_{-\infty}^{\infty}\int_{-2a-y_1}^{\infty} f(y_1)f(y_2)dy_2\,dy_1 \\ &= \int_{-\infty}^{\infty}(1 - F(-2a - y_1))f(y_1)dy_1. \end{aligned}$$

Hence,

$$\begin{aligned} \gamma_{n,f}(a) = ET_W(\theta - u) &= \sum\sum_{i<j} P(P_{ij} > \theta - a) \\ &= \begin{bmatrix} n \\ 2 \end{bmatrix}\int_{-\infty}^{\infty}(1 - F(-2a - v_1))f(y_1)dy_1, \end{aligned}$$

and

$$\gamma'_{n,f}(a) = \begin{bmatrix} n \\ 2 \end{bmatrix}\int_{-\infty}^{\infty} 2f(-2a - y_1)f(y_1)dy_1,$$

so that

$$\gamma'_{n,f}(0) = n(n+1)\int_{-\infty}^{\infty} f^2(y_1)dy_1$$

(since $f(-y) = f(y)$). Therefore,

$$d_{W,f} = \lim_{n\to\infty}\frac{\gamma'_{n,f}(0)}{n^{1/2}\sigma_n} = 12^{1/2}\int_{-\infty}^{\infty} f^2(y)dy.$$

c. The proof of this part of the theorem is beyond the scope of the text. (See Hettmansperger (1984), pp. 109–110.)

d. Let $t_n(\theta) = n^{1/2}(\overline{X} - \theta)/S$. We have shown that

$$n^{1/2}t_n(\theta) \xrightarrow{d} Z \sim N(0, 1),$$

so that $\mu_n = 0$ and $\sigma_n = 1$. Now,

$$\gamma_{n,f}(a) = E[t_n(\theta - a)] = n^{1/2}E[(\overline{X} - \theta + a)]E[S^{-1}] = n^{1/2}aES^{-1},$$

and it follows that

$$\gamma'_{n,f}(0) = n^{1/2}ES^{-1} \quad \text{and} \quad d_{t,f} = \lim_{n\to\infty} ES^{-1} = \tau^{-1}.$$

(The last equality, although fairly obvious intuitively, is not trivial to prove.) □

We now compute AREs when the observations are normally distributed.

Recall that the t-procedures are the optimal procedures for such a model, so that no procedure could be more efficient than those.

Example 15–10

Let $Y \sim N(0, \tau^2)$, so that

$$f(y) = \tau^{-1} n\left(\frac{y}{\tau}\right) = (2\pi)^{-1/2}\tau^{-1} \exp\left(\frac{-y^2}{2\tau^2}\right)$$

(where, as in Theorem 15–9, $n(z)$ is the standard normal density function).

a. The efficacy of the sign test is

$$d_{S,N} = 2f(0) = \left(\frac{2}{\pi}\right)^{1/2} \tau^{-1}.$$

b. Note that

$$f^2(y) = \tau^{-2} n^2\left(\frac{y}{\tau}\right) = (2\pi)^{-1/2}\tau^{-2} n\left(\frac{2^{1/2} y}{\tau}\right).$$

Therefore, the efficacy of the Wilcoxon test is

$$d_{W,N} = (12)^{1/2} \int_{-\infty}^{\infty} \tau^{-2} n^2\left(\frac{y}{\tau}\right) dy$$

$$= \left(\frac{3}{\pi}\right)^{1/2} \tau^{-1} \int_{-\infty}^{\infty} \left(\frac{2^{1/2}}{\tau}\right) n\left(\frac{2^{1/2} y}{\tau}\right) dy = \left(\frac{3}{\pi}\right)^{1/2} \tau^{-1}.$$

c. To find the efficacy of the normal scores test, note that $F(x) = N(x/\tau)$, and hence, $N^{-1}(F(x)) = x/\tau$, and $n(N^{-1}(F(x))) = n(x/\tau)$. Therefore, the efficacy of the normal scores test for normal data is

$$d_{N,N} = \int_{-\infty}^{\infty} \tau^{-2}\left[\frac{n^2(x/\tau)}{n(x/\tau)}\right] dx = \tau^{-1} \int_{-\infty}^{\infty} \tau^{-1} n\left(\frac{x}{\tau}\right) dx = \tau^{-1}.$$

d. The efficacy of the t-procedures is

$$d_{t,N} = \tau^{-1}.$$

Therefore, the ARE of the sign procedures to the t-procedures for normal data is

$$\frac{d_{S,N}^2}{d_{t,N}^2} = \frac{2}{\pi} = .64.$$

This reflects the unsurprising fact that if the observations really are normally distributed, then the procedures based on the sign pivotal quantity are inferior to those based on the t pivotal quantity. The ARE for the Wilcoxon procedures relative to the t-procedures for normal data is

$$\frac{d_{W,N}^2}{d_{t,N}^2} = \frac{3}{\pi} = .95.$$

This calculation indicates the rather surprising fact that even if the data are normally distributed, the Wilcoxon procedures are 95% as efficient asymptotically as are the t-procedures which are designed for normal distributions. Finally, the ARE of the normal scores test relative to the t-procedures is

$$\frac{d_{N,N}^2}{d_{t,N}^2} = 1.$$

Therefore, the normal scores procedures are asymptotically as efficient as the t-procedures, even when the observations are exactly normally distributed. We could also find the AREs for other pairs of tests. For example, the ARE of the sign test to the Wilcoxon test is $\frac{2}{3}$ for the normal model.

To indicate the interpretation of the results given in this example, we compare the asymptotic performances, under normality, of procedures based on the t pivotal quantity and the Wilcoxon pivotal quantity. Since the ARE of the Wilcoxon test to the t-test when the data are normally distributed is .95, the asymptotic variance of $\overline{X}$ is 95% of the variance of the Hodges-Lehmann estimator, and the squared length of the t confidence interval is 95% of the squared length of the Wilcoxon confidence interval. The sample size necessary to get a particular power for the t-test is $(.95)^{1/2} = .97$ as large as the sample size necessary for the Wilcoxon test.

The ARE of the Wilcoxon test to the t-test is .95, even when the observations are normally distributed. In fact, it can be shown that the ARE of the Wilcoxon test to the t-test must be at least .864 for any distribution (see Hettmansperger (1984), pp. 72–74), so that the Wilcoxon test can never be dramatically inefficient asymptotically as compared to the t-test. On the other hand, if the observations come from a Cauchy distribution, the efficacy of the t-test is zero (since the variance of a Cauchy distribution is infinite), but the efficacy of the Wilcoxon test is positive. Therefore, for the Cauchy distribution, the ARE of the t-test relative to the Wilcoxon test is zero, and we see that the t-test can be dramatically inefficient asymptotically, compared to the Wilcoxon test.

It can also be shown that the ARE of the normal scores test to the t-test is greater than one for any distribution but the normal distribution (for which the ARE is one). Similarly, if the observations are Cauchy distributed, then the efficiency of the t-test relative to the normal scores test is zero. Therefore, the normal scores test is asymptotically as efficient as the t-test for the normal distribution and more efficient asymptotically for any other distribution—dramatically more so for distributions with infinite variance.

We now look briefly at the two-sample model in which we observe $\mathbf{X}_m = (X_1, \ldots, X_m)$ and $\mathbf{Y}_n = (Y_1, \ldots, Y_n)$, independent samples from distributions such that $Y_j - \delta$ has the same density as the X_i. Let $f(z)$ be the common density of the X_i and the $Y_j - \delta$, and suppose we have a symmetric asymptotic pivotal quantity $Q_{m,n}(\delta)$ such that

$$\frac{Q_{m,n}(\delta) - \mu_{m,n}}{\sigma_{m,n}} \xrightarrow{d} Z \sim N(0,1).$$

Let

$$\gamma_{m,n,f}(a) = EQ_{m,n}(\delta - a).$$

Then the *efficacy* of $Q_{m,n}$, if it exists, is

$$d_f = \lim_{n\to\infty, m\to\infty} \left(\frac{\gamma'_{m,n,f}(0)}{h_{m,n}\sigma_{m,n}}\right),$$

where $h_{m,n} = (m^{-1} + n^{-1})^{-1/2}$. Now, let $\hat{\delta}_{m,n}$ and $(L_{m,n}, R_{m,n})$ respectively be the estimator and asymptotic confidence interval based on $Q_{m,n}$. Then

$$h_{m,n}\, d_f(\hat{\delta}_{m,n} - \delta) \xrightarrow{d} Z \sim N(0,1) \quad \text{and} \quad h_{m,n}(R_{m,n} - L_{m,n}) \xrightarrow{P} \frac{2z^{\alpha/2}}{d_f}. \qquad (15\text{–}8)$$

Similarly, if $b_{m,n}$ is the point at which the one-sided test based on $Q_{m,n}$ for testing $\delta = c$ against $\delta > c$ has power β for samples of size m and n, then

$$h_{m,n}(b_{m,n} - c) \to \frac{z^{\alpha} - z^{\beta}}{d_f}. \tag{15–9}$$

Now, let $Q_{m,n}(\delta)$ and $Q^*_{m,n}(\delta)$ be two asymptotic pivotal quantities with efficacies d_f and d_f^*, respectively. Then, as in the one-sample case, we define the (Pitman) *asymptotic relative efficiency* (ARE) of $Q_{m,n}(\delta)$ with respect to $Q^*_{m,n}(\delta)$ for the density f by

$$\text{eff}(Q_{m,n}, Q^*_{m,n}) = \left(\frac{d_f}{d_f^*}\right)^2.$$

It can be shown that the two-sample Wilcoxon, normal scores, and t-tests have the same efficacies for the two-sample model as their one-sample counterparts have for the one-sample model. In addition, it can be shown that the Mood test has the same efficacy for the two-sample model as the sign test has for the one-sample model. Therefore, all the comments made for the one-sample model can be extended immediately to the two-sample model.

Exercises—B

1. For the one-sample model, find the efficacies and AREs for the sign, Wilcoxon, and t-tests for the density $f(x) = 2^{-1}\exp(-|x|)$, $-\infty < x < \infty$.
2. Let $f(z) = [.85\exp(-z^2/2) + .05\exp(-z^2/18)]/(2\pi)^{1/2}$.
   a. Show that $f(z)$ is a density function.
   b. Find the efficacy of the one-sample Wilcoxon test for f.
   c. Find the efficacy of the t-test for this distribution.
   d. Find the AREs of the t-test to the Wilcoxon test for f.
3. Show that the efficacy of the two-sample t-test is the same as the efficacy of the one-sample t-test. (You may assume that $ES_p^{-1} \to \tau^{-1}$.)
4. a. For the two-sample model, show that $P(Y_i - \delta + a \geqq X_i) = \int_{-\infty}^{\infty} F(z + a)f(z)dz$.
   b. For the two-sample Wilcoxon test, show that $\gamma'_{m,n}(0) = mn\int_{-\infty}^{\infty}(f(z))^2\,dz$.
   c. Verify that the efficacy for the two-sample Wilcoxon test is $(12)^{1/2}\int_{-\infty}^{\infty}(f(z))^2\,dz$.

Exercises—C

1. Let $Q_n(\delta)$ be a symmetric pivotal quantity for δ in the two sample model such that $(Q_n(\delta) - \mu_n)/\sigma_n \xrightarrow{d} Z \sim N(0, 1)$, let d_f be its efficacy for a density $f(x)$ and let $\hat{\delta}_n$ be it associated estimator. For a_n such that $n^{1/2}|\delta - a_n|$ is bound, it can be shown that $(Q_n(a_n) - \mu_n)/\sigma_n - n^{1/2}d_f(\hat{\delta}_n - a_n) \xrightarrow{P} 0$. Use this fact to give a heuristic derivation of equations (15–8) and (15–9).

Chapter 16

BAYESIAN STATISTICS

The approach to statistics taken in the previous nine chapters is called the classical, or frequentist, approach. According to that approach, we observe a random vector $\mathbf{X}$ having joint density function $f(\mathbf{x};\boldsymbol{\theta})$, where $\boldsymbol{\theta}$ is an unobserved constant vector. We have discussed various methods for using the observed data $\mathbf{X}$ to draw conclusions about the unobserved vector $\boldsymbol{\theta}$.

In this chapter, we discuss a different approach to inference, called the *Bayesian* approach. In this approach, the vector $\boldsymbol{\theta}$ is still unobserved. However, we replace the assumption that $\boldsymbol{\theta}$ is constant with the assumption that $\boldsymbol{\Theta}$ is a random variable having a known distribution $\pi(\boldsymbol{\theta})$. (Note that we call the parameter $\boldsymbol{\Theta}$ instead of $\boldsymbol{\theta}$, since it is a random variable.) As we shall see, when we make this assumption, we get a very elegant approach to statistics. In addition, the interpretation of many concepts is clearer with the Bayesian approach.

A *Bayesian statistical model* consists of (1) an observed random variable or vector $\mathbf{X} \in \chi$, (2) an unobserved random variable or vector $\boldsymbol{\Theta} \in \Omega$, (3) the conditional density $f(\mathbf{x}|\boldsymbol{\theta})$ of $\mathbf{X}$ given $\boldsymbol{\Theta}$, and (4) the marginal density $\pi(\boldsymbol{\theta})$ of $\boldsymbol{\Theta}$. (Note that we have replaced $f(\mathbf{x};\boldsymbol{\theta})$ with $f(\mathbf{x}|\boldsymbol{\theta})$, since $\boldsymbol{\Theta}$ is now random.) We call $\pi(\boldsymbol{\theta})$ the *prior distribution* of $\boldsymbol{\Theta}$. As before, χ and Ω are the sample space and the parameter space, respectively. We can assume, without loss of generality, that $\pi(\boldsymbol{\theta}) > 0$ for all $\boldsymbol{\theta} \in \Omega$. (Otherwise, we can shrink the parameter space Ω to those $\boldsymbol{\theta}$ such that $\pi(\boldsymbol{\theta}) > 0$.) We make this assumption for the remainder of the chapter.

All the Bayesian procedures we shall discuss are defined in terms of the posterior distribution of $\boldsymbol{\Theta}$, which is the conditional distribution of $\boldsymbol{\theta}$ given the data $\mathbf{X}$. Section 16.1 begins with a discussion of the posterior distribution and how to compute it. This discussion then leads to a Bayesian definition of sufficiency, which we show is equivalent to the definition in classical statistics. We then consider Bayesian estimation, Bayesian interval estimators, Bayesian tests, and, finally, Bayesian inference for the one-sample normal model (with two parameters). In Section 16.2, we look at some additional Bayesian topics. In Section 16.3 we compare Bayesian and classical statistics.

Since the parameter $\boldsymbol{\Theta}$ is constant in classical statistics and a random vector in Bayesian statistics, when we consider classical properties of Bayes procedures, all of the expectations in classical statistics must be replaced by conditional expectations given $\boldsymbol{\Theta}$. In particular, if $T(\mathbf{X})$ is an estimator of $\tau(\boldsymbol{\Theta})$, then $T(\mathbf{X})$ is unbiased if

$$E(T(\mathbf{X})|\boldsymbol{\Theta}) = \tau(\boldsymbol{\Theta}),$$

and the mean squared error of $T(\mathbf{X})$ is

$$MSE_T(\boldsymbol{\Theta}) = E([T(\mathbf{X}) - \tau(\boldsymbol{\Theta})]^2|\boldsymbol{\Theta}).$$

If $\Phi(\mathbf{X})$ is a critical function, then the power function of Φ is

$$K_\Phi(\boldsymbol{\Theta}) = E(\Phi(\mathbf{X})|\boldsymbol{\Theta}).$$

16.1 BAYESIAN STATISTICS

16.1.1 The Posterior Distribution

A *Bayesian statistical model* consists of an observed random vector $\mathbf{X}$ called the data or the observations and an unobserved random vector $\boldsymbol{\Theta}$ called the parameter, together with the conditional density of $\mathbf{X}$ given $\boldsymbol{\Theta}$, $f(\mathbf{x}|\boldsymbol{\Theta})$, and the marginal density $\pi(\boldsymbol{\Theta})$ of $\boldsymbol{\Theta}$, called the *prior distribution*. From the latter two densities, we can find the joint density of $(\mathbf{X}, \boldsymbol{\Theta})$, and from that the conditional density $\pi(\boldsymbol{\theta}|\mathbf{x})$ of $\boldsymbol{\Theta}$ given $\mathbf{X}$. We call $\pi(\boldsymbol{\theta}|\mathbf{x})$ the *posterior distribution* of $\boldsymbol{\Theta}$ given the observations $\mathbf{X}$. (Note that the prior distribution of $\boldsymbol{\Theta}$ represents our knowledge of the distribution of $\boldsymbol{\Theta}$ before we look at the data, and the posterior distribution of $\boldsymbol{\Theta}$ represents our knowledge of the distribution of $\boldsymbol{\Theta}$ after we have observed the data.) A Bayesian statistician bases all of his or her statistical inference on the posterior distribution, since it represents the knowledge about $\boldsymbol{\Theta}$ at the conclusion of the experiment.

Now suppose that Θ is a discrete random variable taking values $1, \ldots, q$ and that $\mathbf{X}$ is a discrete random vector. Then the posterior distribution of Θ is given by

$$\pi(k|\mathbf{x}) = P(\Theta = k|\mathbf{X} = \mathbf{x}) = \frac{f(\mathbf{x}|k)\pi(k)}{\sum_k f(\mathbf{x}|k)\pi(k)} = \frac{P(\mathbf{X} = \mathbf{x}|\Theta = k)P(\Theta = k)}{\sum_k P(\mathbf{X} = \mathbf{x}|\Theta = k)P(\Theta = k)}.$$

This calculation is just Bayes rule, which explains how Bayesian statistics got its name.

Example 16–1

Let Θ be univariate. Suppose that conditionally on Θ, the X_i are a sample from an exponential distribution with mean $\mu = \Theta^{-1}$ and that Θ has a $\Gamma(a, b)$ distribution. (For Bayesian purposes, it is convenient to parametrize the exponential distribution by the inverse of the mean.) Then

$$f(\mathbf{x}|\theta) = \prod[\theta \exp(-x_i\theta)] = \theta^n \exp(-\theta n\bar{x}) \quad \text{and} \quad \pi(\theta) = k(a, b)\theta^{a-1} \exp\left(\frac{-\theta}{b}\right).$$

Therefore, the joint density of $(\mathbf{X}, \Theta)$ is

$$k(a, b)\theta^{n+a-1} \exp[-\theta(n\bar{x} + b^{-1})],$$

and the conditional density of Θ given $\mathbf{X}$ is

$$\pi(\theta|\mathbf{x}) = k(a,b)\theta^{n+a-1}\frac{\exp\left[\dfrac{-\theta}{b(bn\bar{x}+1)}\right]}{h(\mathbf{x})}$$

$$= k(a,b,\mathbf{x})\theta^{a^*-1}\exp\left(\frac{-\theta}{b^*}\right),$$

where

$$a^* = a+n, \qquad b^* = \frac{b}{bn\bar{x}+1},$$

and $h(\mathbf{x})$ is the marginal density of $\mathbf{X}$. Therefore,

$$\Theta|\mathbf{X} \sim \Gamma(A^*, B^*), \qquad A^* = a+n, \qquad B^* = \frac{b}{nb\bar{X}+1}$$

where we have used A^* and B^* in place of a^* and b^* to indicate that they are random variables. Note that $k(a,b,\mathbf{x})$ is a constant in this conditional distribution and is therefore the constant necessary to make the density integrate to one. Hence,

$$k(a,b,\mathbf{x}) = [(b^*)^{a^*}\Gamma(a^*)]^{-1}.$$

Note also that we really do not need to find this constant in order to find the posterior distribution.

Example 16–2

Let Θ be univariate, and suppose that, given Θ, the X_i are a sample from a Bernoulli distribution with mean Θ. Suppose also that the prior distribution of Θ is

$$\prod(\theta) = k(a,b)\theta^{a-1}(1-\theta)^{b-1}, \qquad 0<\theta<1.$$

Then we say that Θ has a *beta* distribution and write $\Theta \sim \beta(a,b)$. (In Exercise B6, it is shown that $(k(a,b))^{-1} = \Gamma(a)\Gamma(b)/\Gamma(a+b)$, which is often called the beta function of a and b.) It then follows that

$$f(\mathbf{x}|\theta) = \prod[\theta^{x_i}(1-\theta)^{1-x_i}] = \theta^{n\bar{x}}(1-\theta)^{n(1-\bar{x})} \quad \text{and} \quad \pi(\theta) = k(a,b)\theta^{a-1}(1-\theta)^{b-1},$$

and the joint density of $(\mathbf{X}, \Theta)$ is

$$f(\mathbf{x}|\theta)\pi(\theta) = k(a,b)\theta^{a+n\bar{x}-1}(1-\theta)^{b+n(1-\bar{x})-1}.$$

Hence, the conditional density of Θ given $\mathbf{X}$ is

$$\pi^*(\theta|\mathbf{x}) = \frac{k(a,b)\theta^{a+n\bar{x}-1}(1-\theta)^{b+n(1-\bar{x})-1}}{h(\mathbf{x},a,b)}$$

$$= k^*(\mathbf{x},a,b)\theta^{a^*-1}(1-\theta)^{b^*-1},$$

where

$$a^* = a+n\bar{x}, \qquad b^* = b+n(1-\bar{x}),$$

and $h(\mathbf{x},a,b)$ is the marginal density of $\mathbf{X}$. Note that π^* is the density function of a random variable which has a $\beta(a^*,b^*)$ distribution and, hence, that the posterior distribution of Θ given $\mathbf{X}$ is

$$\Theta|\mathbf{X} \sim \beta(A^*, B^*), \qquad A^* = a+n\bar{X}, \qquad B^* = b+n(1-\bar{X}),$$

where, again, we have used A^* and B^* instead of a^* and b^* to indicate that they are random variables. Since $k^*(\mathbf{x},a,b)$ is a constant in the conditional density of Θ given $\mathbf{X}$

and is just the constant that makes the density integrate to one, it follows that

$$k^*(\mathbf{x}, a, b) = \frac{\Gamma(a^*, b^*)}{\Gamma(a^*)\Gamma(b^*)}.$$

Again, we do not need to evaluate this constant to recognize the conditional density of Θ given $\mathbf{X}$.

Example 16–3

Suppose we observe (X, Y) having a trinomial distribution with known n and unknown parameter $\boldsymbol{\Theta} = (\Theta_1, \Theta_2)$. That is,

$$f((x, y)|(\theta_1, \theta_2)) = h(x, y)\theta_1^x \theta_2^y (1 - \theta_1 - \theta_2)^{n-x-y}, \qquad x, y = 1, 2, \ldots, x + y \leqq n.$$

Suppose also that (Θ_1, Θ_2) has the continuous joint density function

$$\pi(\theta_1, \theta_2) = k(a, b, c)\theta_1^{a-1}\theta_2^{b-1}(1 - \theta_1 - \theta_2)^{c-1},$$

$$\theta_1 > 0, \qquad \theta_2 > 0, \qquad \theta_1 + \theta_2 < 1.$$

We call this joint density function a *Dirichlet* density function with parameters (a, b, c). The constant $k(a, b)$ is given by

$$k(a, b, c) = \frac{\Gamma(a, b, c)}{\Gamma(a)\Gamma(b)\Gamma(c)}.$$

(See the exercises.) In this case, the posterior density of (Θ_1, Θ_2) given (X, Y) is

$$\begin{aligned}\pi((\theta_1, \theta_2)|(x, y)) &= g(x, y)k(a, b, c)\theta_1^{a+x-1}\theta_2^{b+y-1}(1 - \theta_1 - \theta_2)^{c+n-x-y-1} \\ &= k(a, b, c, \mathbf{x})\theta_1^{a^*-1}\theta_2^{b^*-1}(1 - \theta_1 - \theta_2)^{c^*-1},\end{aligned}$$

where

$$a^* = a + x, \qquad b^* = b + y, \qquad c^* = c + n - x - y.$$

Therefore, the posterior distribution of (Θ_1, Θ_2) given (X, Y) is a Dirichlet distribution with parameters

$$A^* = a + X, \qquad B^* = b + Y, \qquad C^* = c + n - X - Y.$$

From the preceding three examples, we see that when computing the joint density of $\mathbf{X}$ and $\boldsymbol{\Theta}$, we need only keep track of the parts that involve $\boldsymbol{\theta}$. Any parts that involve only constants or functions of $\mathbf{x}$ may be factored out in front and ignored. In addition, we do not need to find the marginal density of $\mathbf{X}$. These objects are merely constants in the conditional density of $\boldsymbol{\Theta}$ given $\mathbf{X}$ whose only purpose is to ensure that the density integrates to one.

A Bayesian statistician considers the posterior distribution of $\boldsymbol{\Theta}$ to summarize her present information about $\boldsymbol{\Theta}$. She would therefore use the posterior distribution from the previous experiment as the prior distribution for the present experiment. That is, for her first experiment, she would begin by somehow deciding on a prior distribution for the parameters. She would then conduct the first experiment, collecting some data $\mathbf{X}$, whose distribution she would model by $f(\mathbf{x}|\boldsymbol{\theta})$. She would then compute the posterior distribution of $\boldsymbol{\Theta}$. Then, when she did another experiment, she would use the posterior distribution from the first experiment as the prior distribution for the second. She would then collect some more data and update the posterior distribution on the basis of the data from both experiments. She would then use the new posterior distribution obtained as the prior distribution for the third experiment, etc. In this way, she can always use the present posterior distribution to describe her present state of knowledge about $\boldsymbol{\theta}$, conditionally on all the experiments that she has done.

Now, let $f(\mathbf{x}|\boldsymbol{\theta})$ be the conditional density of $\mathbf{X}$ given $\boldsymbol{\Theta}$. A family of prior distributions $\pi_c(\boldsymbol{\theta})$, $c \in C$ is called a *conjugate family of priors* for $f(\mathbf{x}|\boldsymbol{\theta})$ if whenever the prior distribution is in the family, the posterior distribution is also in the family. In Example 16–2, we have seen that if (conditionally on Θ) the X_i are a sample from a Bernoulli distribution with mean Θ, then the family of all beta priors is a conjugate family, since a beta prior leads to a beta posterior. Similarly, in Example 16–1, if (conditionally on Θ) the X_i are independently exponentially distributed with mean Θ^{-1}, then the family of gamma priors is a conjugate family. Finally, the family of Dirichlet distributions is a conjugate family for the trinomial distribution.

One motivation for using conjugate priors is the sequential approach to Bayesian inference suggested above. If we are going to be repeating an experiment many times, continually updating the prior, and if we start with a conjugate family of priors, then we shall merely move around that family as we update the prior after each experiment.

Another motivation for using conjugate families is that it is often easier to find the posterior for a prior from such a family than for other priors.

Theorem 16–1. Suppose that conditionally on $\boldsymbol{\Theta}$, $\mathbf{X}$ has the density function

$$f(\mathbf{x}|\boldsymbol{\theta}) = h(\mathbf{x}) \exp[T_1(\mathbf{x})d_1(\boldsymbol{\theta}) + \cdots + T_p(\mathbf{x})d_p(\boldsymbol{\theta})], \qquad \mathbf{x} \in \chi, \qquad \boldsymbol{\theta} \in \Omega.$$

Let $\mathbf{c} = (c_1, \ldots, c_p)$ be a p-dimensional vector, and suppose that $\boldsymbol{\Theta}$ has the prior distribution

$$\pi_c(\boldsymbol{\theta}) = k(\mathbf{c}) \exp[d_1(\boldsymbol{\theta})c_1 + \cdots + d_p(\boldsymbol{\theta})c_p].$$

Then the posterior distribution associated with π_c is just π_{C^*}, where

$$\mathbf{C}^* = (C_1^*, \ldots, C_p^*)', \qquad C_i^* = c_i + T_i(\mathbf{X}).$$

Proof. See Exercise B5. □

For Example 16–1,

$$f(\mathbf{x}|\theta) = \theta^{n\bar{x}}(1-\theta)^{n(1-\bar{x})} = \exp[n\bar{x}\log(\theta) + n(1-\bar{x})\log(1-\theta)],$$

so that the conjugate family for the model is

$$\pi_{c,d} = k(c,d) \exp[c \log(\theta) + d \log(1-\theta)] = k(c,d)\theta^c(1-\theta)^d.$$

Since the prior distribution that we chose in that example had this form with $c = a - 1$ and $d = b - 1$, by Theorem 16–1, the posterior distribution has

$$C^* = A^* - 1 = c + n\bar{X} = a + n\bar{X} - 1 \quad \text{and} \quad D^* = B^* - 1 = b + n(1-\bar{X}) - 1,$$

and hence,

$$A^* = a + n\bar{X} \quad \text{and} \quad B^* = b + n(1-\bar{X}),$$

which is the result we found in Example 16–1. The models in Examples 16–2 and 16–3 are also examples of this type.

We now discuss a Bayesian approach to sufficiency. A Bayesian statistician would say that a statistic $\mathbf{T} = T(\mathbf{X})$ is *sufficient* if the posterior distribution of $\boldsymbol{\Theta}$ given $\mathbf{X}$ is the same as the posterior distribution of $\boldsymbol{\Theta}$ given $\mathbf{T}$. The next theorem shows that this definition is the same as the classical definition given in Chapter 10.

Theorem 16–2. Let $\mathbf{T} = T(\mathbf{X})$ be a statistic with conditional density function $f^*(\mathbf{t}|\boldsymbol{\theta})$. Then the posterior distribution of $\boldsymbol{\Theta}$ given $\mathbf{X}$ is the same as the posterior distribution of $\boldsymbol{\Theta}$ given $\mathbf{T}$ if and only if

$$f(\mathbf{x}|\boldsymbol{\theta}) = k(\mathbf{x})f^*(\mathbf{T}(\mathbf{x})|\boldsymbol{\theta})$$

for some function $k(\mathbf{x}) \geqq 0$.

Proof. The posterior distributions of $\boldsymbol{\Theta}$ given $\mathbf{X}$ and Θ given $\mathbf{T}$ are

$$\pi(\boldsymbol{\theta}|\mathbf{x}) = \frac{f(\mathbf{x}|\boldsymbol{\theta})\pi(\boldsymbol{\theta})}{h(\mathbf{x})} \quad \text{and} \quad \pi^*(\boldsymbol{\theta}|\mathbf{t}) = \frac{f^*(\mathbf{t}|\boldsymbol{\theta})\pi(\boldsymbol{\theta})}{h^*(\mathbf{t})},$$

where $h(\mathbf{x})$ and $h^*(\mathbf{t})$ are the marginal density functions of $\mathbf{X}$ and $\mathbf{T}$, respectively. Therefore,

$$\pi(\boldsymbol{\theta}|\mathbf{x}) = \pi^*(\boldsymbol{\theta}|T(\mathbf{x})) \Leftrightarrow f(\mathbf{x}|\boldsymbol{\theta}) = \left[\frac{h(\mathbf{x})}{h^*(T(\mathbf{x}))}\right] f^*(T(\mathbf{x})|\boldsymbol{\theta}),$$

and the desired result follows immediately. □

Note that by the Fisher-Neyman criterion, $\mathbf{T}$ is sufficient for a Bayesian statistician if and only if $\mathbf{T}$ is sufficient for a classical statistician. (Note also that the derivation of the Fisher-Neyman criterion from the Bayesian definition is elementary, but the derivation of the Fisher-Neyman criterion from the classical definition involves measure theory, at least in the continuous case.)

Exercises—B

1. Let $X_1, \ldots, X_n$ be a sample from a Poisson distribution with mean Θ. Suppose that Θ has a prior distribution $\Gamma(a, b)$. Show that the posterior distribution of Θ given $\mathbf{X}$ is $\Gamma(A^*, B^*)$, where $A^* = a + n\overline{X}$ and $B^* = b/(1 + nb)$.
2. Let $X_1, \ldots, X_n$ be a sample from a normal distribution with mean Θ and variance one, and let $\boldsymbol{\Theta} \sim N(a, b^2)$. Show that the posterior distribution of Θ given $\mathbf{X}$ is $\Theta|\mathbf{X} \sim N(A^*, B^{*2})$, where $A^* = (a + nb^2\overline{X})/(1 + nb^2)$ and $B^{*2} = b^2/(1 + nb^2)$.
3. Let $X_1, \ldots, X_n$ be a sample from a geometric distribution with parameter Θ, and let Θ have a beta distribution with parameters a and b. What is the posterior distribution of Θ?
4. Let $X_1, \ldots, X_n$ be a sample from a normal distribution with mean zero and variance Θ^{-1}. Suppose that Θ has a $\Gamma(a, b)$ distribution. What is the posterior distribution of Θ?
5. Prove Theorem 16–1.
6. Let X and Y be independent, $X \sim \Gamma(a, 1)$, and $Y \sim \Gamma(b, 1)$. Let $U = X/(X + Y)$. Show that $U \sim \beta(a, b)$ and that $k(a, b) = \Gamma(a + b)/\Gamma(a)\Gamma(b)$.

Exercises—C

1. Suppose that the posterior distribution of $\boldsymbol{\Theta}$ given $\mathbf{X}$ depends on $\mathbf{X}$ only through $\mathbf{T} = T(\mathbf{X})$. Use the factorization criterion to show that $\mathbf{T}$ is sufficient.

2. Suppose that conditionally on $\mathbf{\Theta}$, $\mathbf{X}$ and $\mathbf{Y}$ are independent, with joint density functions $h(\mathbf{x}|\boldsymbol{\theta})$ and $g(\mathbf{x}|\boldsymbol{\theta})$, and that $\mathbf{\Theta}$ has prior density $\pi(\boldsymbol{\theta})$.
   (a) Find the posterior density of $\boldsymbol{\theta}$ given $(\mathbf{X}, \mathbf{Y})$.
   (b) Find the posterior density of $\boldsymbol{\theta}$ given $\mathbf{X}$.
   (c) Use the posterior density of $\boldsymbol{\theta}$ given $\mathbf{X}$ found in part b as the prior density when we observe $\mathbf{Y}$. Then find the posterior density after we observe $\mathbf{Y}$, and show that it is the same as the posterior density of $\boldsymbol{\theta}$ given $(\mathbf{X}, \mathbf{Y})$.

   (This problem shows that we get the same result by sequentially updating the prior density, as suggested in the text, as we would if we did a single Bayesian analysis using the original prior density and the joint density of $(\mathbf{X}, \mathbf{Y})$.
3. Let X, Y, and Z be independent, $X \sim \Gamma(a, 1)$, $Y \sim \Gamma(b, 1)$, and $Z \sim \Gamma(c, 1)$. Let $U = X/(X + Y + Z)$, $V = Y/(X + Y + Z)$, and $W = X + Y + Z$.
   (a) Use Jacobians to find the joint density of (U, V, W).
   (b) Show that (U, V) is independent of W.
   (c) Show that (U, V) has a Dirichlet distribution with parameters (a, b, c).
   (d) Verify the formula for $k(a, b, c)$ in Example 16–3. (Recall that $W \sim \Gamma(a + b + c, 1)$. Why?)
4. Let (U, V) have a Dirichlet distribution with parameters (a, b, c).
   (a) Show that U has a beta distribution with parameters a and $b + c$.
   (b) Show that $U + V$ has a beta distribution with parameters $a + b$ and c.

   (See Exercises B6 and C3.)

16.1.2 Bayes Estimators

Consider a Bayesian statistical model with parameter $\mathbf{\Theta}$ having posterior distribution $\pi(\boldsymbol{\theta}|\mathbf{X})$. Let $\tau(\mathbf{\Theta})$ be a univariate parameter. We define the *Bayes estimator* of $\tau = \tau(\mathbf{\Theta})$ by

$$T(\mathbf{X}) = E(\tau(\mathbf{\Theta})|\mathbf{X}).$$

That is, the Bayes estimator of τ is the expected value of τ computed from the posterior distribution of $\mathbf{\Theta}$ (which represents the statistician's knowledge about $\mathbf{\Theta}$ after the experiment has been completed).

Example 16–4

Conditionally on Θ, let $X_1, \ldots, X_n$ be a sample from a Bernoulli distribution with mean Θ. Suppose that $\Theta \sim \beta(a, b)$. In the previous section, we showed that the posterior distribution of Θ is

$$\Theta|\mathbf{X} \sim \beta(A^*, B^*), \qquad A^* = a + n\overline{X}, \qquad B^* = b + n(1 - \overline{X}).$$

Therefore, the Bayes estimator of $\tau(\Theta) = \Theta$ for this model is

$$T(\mathbf{X}) = E(\theta|\mathbf{X}) = \frac{A^*}{A^* + B^*} = \frac{a + n\overline{X}}{a + b + n}$$

(see Exercise B5), which we can write in the form

$$\frac{a}{a + b}\left(\frac{a + b}{a + b + n}\right) + \overline{X}\left(\frac{n}{a + b + n}\right).$$

Note that this estimator is a weighted average of the prior mean $a/(a + b)$ and the MLE $\overline{X}$. If n is large, the Bayes estimator puts more weight on the MLE, and if n is small, it puts more weight on the prior mean. Now,

$$ET(\mathbf{X}) = \frac{a + n\Theta}{a + b + n} \neq \Theta,$$

so that the Bayes estimator is not unbiased. Note, however, that

$$T(\mathbf{X}) = \frac{a}{a+b+n} + \frac{n\overline{X}}{a+b+n} \xrightarrow{P} 0 + \Theta,$$

so that the Bayes estimator is a consistent estimator of Θ. Now, suppose that $a = 5$, $b = 6$, $n = 10$, and $\overline{X} = .6$. Then $A^* = 11$, $B^* = 10$, and $T(\mathbf{X}) = \frac{11}{21} = .52$. Also, the MLE for these data is $\overline{X} = .6$, the prior mean is $\frac{5}{11} = .45$. We can also find the Bayes estimator of $\tau^*(\Theta) = \Theta(1-\Theta) = \text{var}(X_i)$ by

$$T^*(\mathbf{X}) = E(\Theta(1-\Theta)|\mathbf{X}) = \frac{A^*B^*}{(A^*+B^*)(A^*+B^*+1)}.$$

(See Exercise B5.)

Example 16–5

Let $X_1, \ldots, X_n$ be a sample from an exponential distribution with mean $\boldsymbol{\Theta}^{-1}$. Suppose that $\boldsymbol{\Theta} \sim \Gamma(a, b)$. In the previous section, we showed that the posterior distribution of $\boldsymbol{\Theta}$ is

$$\boldsymbol{\Theta}|\mathbf{X} \sim \Gamma(A^*, B^*), \qquad A^* = a + n, \qquad B^* = \frac{b}{nb\overline{X}+1}.$$

Therefore, the Bayes estimator for $\boldsymbol{\Theta}$ is

$$T(\mathbf{X}) = E(\boldsymbol{\Theta}|\mathbf{X}) = A^*B^* = \frac{(n+a)\,b}{nb\overline{X}+1}.$$

As in Example 16–4, $T(\mathbf{X})$ is consistent but not unbiased. Now, suppose that we want the Bayes estimator of $\mu = EX_i = \boldsymbol{\Theta}^{-1}$. This estimator is given by

$$T^*(\mathbf{X}) = E(\boldsymbol{\Theta}^{-1}|\mathbf{X}) = [(A^*-1)\,B^*]^{-1}$$
$$= \frac{nb\overline{X}+1}{b(n+a-1)} = \overline{X}\left(\frac{n}{n+a-1}\right) + ((a-1)\,b)^{-1}\left(\frac{a-1}{n+a-1}\right)$$

(see Theorem 4–26), which is again a weighted average of the MLE $\overline{X}$ and the prior mean $E\boldsymbol{\Theta}^{-1} = [b(a-1)]^{-1}$.

Although Bayes estimators are not always weighted averages of the MLE and the prior mean (and hence, do not always lie on the straight line between them), they are typically "somewhere between" these two quantities. Also, they are typically consistent and asymptotically efficient. (See Lehmann (1983), pp. 454–465, for more details on these properties.)

We next show that Bayes estimators can rarely be unbiased. Recall that if U and V are random variables, then $U \equiv V$ if and only if $P(U = V) = 1$.

Theorem 16–3. Let $T(\mathbf{X})$ be the Bayes estimator of $\tau(\boldsymbol{\Theta})$. Then $T(\mathbf{X})$ is unbiased only if $T(\mathbf{X}) \equiv \tau(\boldsymbol{\Theta})$.

Proof. Suppose that $T(\mathbf{X})$ is the Bayes estimator of $\tau(\boldsymbol{\Theta})$. Then

$$E(\tau(\boldsymbol{\Theta})|\mathbf{X}) = T(\mathbf{X}),$$

and therefore,

$$E(\tau(\boldsymbol{\Theta})\,T(\mathbf{X})) = E(E[T(\mathbf{X})\tau(\boldsymbol{\Theta})|\mathbf{X}]) = E(T(\mathbf{X})\,E(\tau(\boldsymbol{\Theta})|\mathbf{X})) = E(T(\mathbf{X}))^2.$$

Now, suppose that $T(\mathbf{X})$ is an unbiased estimator of $\tau(\boldsymbol{\Theta})$. Then

$$E(T(\mathbf{X})|\boldsymbol{\Theta}) = \tau(\boldsymbol{\Theta}) \Rightarrow E(T(\mathbf{X})\tau(\boldsymbol{\Theta})) = E(\tau(\boldsymbol{\Theta})\,E(T(\mathbf{X})|\boldsymbol{\Theta})) = E(\tau(\boldsymbol{\Theta}))^2.$$

Therefore,

$$E(T(\mathbf{X}) - \tau(\boldsymbol{\Theta}))^2 = E(T(\mathbf{X}))^2 - E(\tau(\boldsymbol{\Theta})\,T(\mathbf{X})) + E(\tau(\boldsymbol{\Theta}))^2 - E(T(\mathbf{X})\tau(\boldsymbol{\Theta})) = 0.$$

Since $(T(\mathbf{X}) - \tau(\boldsymbol{\Theta}))^2 \geqq 0$, Markov's inequality implies that

$$(T(\mathbf{X}) - \tau(\boldsymbol{\Theta}))^2 \equiv 0 \Rightarrow T(\mathbf{X}) \equiv \tau(\boldsymbol{\Theta}). \quad \square$$

Corollary 16–3. If $T(\mathbf{X})$ is an unbiased estimator of $\tau(\boldsymbol{\Theta})$ such that $\operatorname{var}(T(\mathbf{X})) > 0$, then $T(\mathbf{X})$ is not a Bayes estimator for any prior.

The mean squared error $MSE_T(\boldsymbol{\Theta})$ of an estimator $T(\mathbf{X})$ depends on $\boldsymbol{\Theta}$ and hence is a random variable for a Bayesian.

Theorem 16–4. Let $T(\mathbf{X})$ be the Bayes estimator of $\tau(\boldsymbol{\Theta})$, and let $T^*(\mathbf{X})$ be any other estimator. If $E(MSE_T(\boldsymbol{\Theta})) < \infty$, then

$$E(MSE_T(\boldsymbol{\Theta})) \le E(MSE_{T^*}(\boldsymbol{\Theta})).$$

Proof. We have

$$\begin{aligned} E(MSE_T(\boldsymbol{\Theta})) &= E(E[(T(\mathbf{X}) - \tau(\boldsymbol{\Theta}))^2|\boldsymbol{\Theta}]) = E(T(\mathbf{X}) - \tau(\boldsymbol{\Theta}))^2 \\ &= E(E[(T(\mathbf{X}) - \tau(\boldsymbol{\Theta}))^2|\mathbf{X}]). \end{aligned}$$

Similarly,

$$E(MSE_{T^*}(\boldsymbol{\Theta})) = E(E[(T^*(\mathbf{X}) - \tau(\boldsymbol{\Theta}))^2|\mathbf{X}]).$$

Now,

$$\begin{aligned} E[(T^*(\mathbf{X}) - \tau(\boldsymbol{\Theta}))^2|\mathbf{X}] &= E[(T^*(\mathbf{X}) - T(\mathbf{X}) + T(\mathbf{X}) - \tau(\boldsymbol{\Theta}))^2|\mathbf{X}] \\ &= E[(T^*(\mathbf{X}) - T(\mathbf{X}))^2|\mathbf{X}] + E[(T(\mathbf{X}) - \tau(\boldsymbol{\Theta}))^2|\mathbf{X}] \\ &\ge E[(T(\mathbf{X}) - \tau(\boldsymbol{\Theta}))^2|\mathbf{X}]. \end{aligned}$$

(See Exercise B8.) The result follows from taking the expected value of both sides. $\square$

In Exercise C2, Theorem 16–4 is extended slightly to show that the Bayes estimator $T(\mathbf{X})$ is an admissible estimator of $\tau(\boldsymbol{\Theta})$, i.e., that there is no other estimator $T^*(\mathbf{X})$ such that $MSE_{T^*}(\boldsymbol{\Theta}) \leqq MSE_T(\boldsymbol{\Theta})$ for all $\boldsymbol{\Theta}$, with strong inequality for some $\boldsymbol{\Theta}$. In other words, there is no other estimator that is better than $T(\mathbf{X})$ for all $\boldsymbol{\Theta}$.

Theorem 16–4 has an interesting interpretation. Suppose we want to estimate a parameter $\tau(\boldsymbol{\Theta})$. We cannot find any estimator that minimizes the mean squared error for all $\boldsymbol{\Theta}$. Suppose also that we have a weight function $w(\boldsymbol{\Theta}) \geqq 0$ and we want to minimize

$$r(T) = \int_\Omega w(\boldsymbol{\theta})\,MSE_T(\boldsymbol{\theta})\,d\boldsymbol{\theta}.$$

(Note that for each estimator $T(\mathbf{X})$, $r(T)$ is a number, so that we might reasonably hope to find a T that minimizes $r(T)$.) Finally, suppose that

$$w^* = \int_\Omega w(\boldsymbol{\theta})\, d\boldsymbol{\theta} < \infty.$$

Then $\pi(\boldsymbol{\theta}) = w(\boldsymbol{\theta})/w^*$ is a prior distribution for $\boldsymbol{\Theta}$. Furthermore, finding $T(\mathbf{X})$ that minimizes $r(T)$ is the same as finding $T(\mathbf{X})$ that minimizes $r(T)/w^*$, the expected mean squared error for the prior π. Therefore, to find $T(\mathbf{X})$ that minimizes $r(T)$, we find the Bayes estimator with respect to the prior distribution $\pi(\boldsymbol{\theta}) = w(\boldsymbol{\theta})/w^*$. Hence, the Bayes estimator can be considered as the estimator that minimizes a particular weighted average of the mean squared error.

It sometimes happens that we want to minimize the expected value of some other function $h(T - \tau)$ (averaged over both the sample space and the parameter space). To find the estimator $T(\mathbf{X})$ that minimizes $Eh(T(\mathbf{X}) - \tau(\boldsymbol{\Theta}))$, we find (if possible) for each $\mathbf{X}$ the value $T(\mathbf{X})$ that minimizes

$$r(T(\mathbf{X})) = E(h(T(\mathbf{X}) - \tau(\boldsymbol{\Theta}))|\mathbf{X}).$$

For example, if $h(T - \tau) = |T - \tau|$, we choose $T(\mathbf{X})$ to be the median of the posterior distribution of $\tau(\boldsymbol{\Theta})$ given $\mathbf{X}$. (See Exercise C1.) For this reason, the Bayes estimator, as we have defined it, is often called the *Bayes estimator with respect to squared error loss*, and the median of the posterior distribution is called the *Bayes estimator with respect to absolute error loss*.

Exercises—B

1. Find the Bayes estimators for Θ and Θ^2 for the model in Exercise B1, Section 16.1.1.

2. For the model of Exercise B2, Section 16.1.1.
  (a) Find the Bayes estimators of Θ and Θ^2.
  (b) Show that these estimators are consistent.

3. For the model of Exercise B3, Section 16.1.1
  (a) Find the Bayes estimator of Θ.
  (b) Find the Bayes estimator of $EX_i = (1 - \Theta)/\Theta$.
  (c) Show that the Bayes estimator of $(1 - \Theta)/\Theta$ is consistent.

4. For the model of Exercise B4, Section 16.1.1
  (a) Find the Bayes estimator of $\mathrm{var}(X_i) = \Theta^{-1}$, and show that it is consistent.
  (b) Find the Bayes estimator of $\Theta^{-1/2}$.

5. Let $\Theta \sim \beta(A^*, B^*)$. Show that $E\Theta = A^*/(A^* + B^*)$ and $E\Theta(1 - \Theta) = A^*B^*/(A^* + B^*)(A^* + B^* + 1)$.

6. For Example 16–5, show that the Bayes estimator of Θ is consistent.

7. **(a)** Let (U, V) have a Dirichlet distribution with parameters a, b, and c. Find a formula for $EU^h V^k (1 - U - V)^m$.
  (b) For Example 16–3, find the Bayes estimators of Θ_1, $\Theta_1\Theta_2$, and $\Theta_1\Theta_2(1 - \Theta_1 - \Theta_2)$.

8. Let $T(\mathbf{X})$ be the Bayes estimator of $\tau(\boldsymbol{\Theta})$.
  (a) Show that $E[(T^*(\mathbf{X}) - T(\mathbf{X}))(T(\mathbf{X}) - \tau(\boldsymbol{\Theta}))|\mathbf{X}] = 0$. (*Hint*: T^* and T are constants in the conditional distribution of $\boldsymbol{\Theta}|\mathbf{X}$.)
  (b) Verify that $E[(T^*(\mathbf{X}) - T(\mathbf{X}) + T(\mathbf{X}) - \tau(\boldsymbol{\Theta}))^2|\mathbf{X}] = E[(T^*(\mathbf{X}) - T(\mathbf{X}))^2|\mathbf{X}] + E[(T(\mathbf{X}) - \tau(\boldsymbol{\Theta}))^2|\mathbf{X}]$.

Exercises—C

1. (a) Let U be a random variable, and let m be the median of U. Show that $E|U-a| \geqq E|U-m|$ for any a.
   (b) Show that the $T(\mathbf{X})$ that minimizes $E|T(\mathbf{X})-\tau(\mathbf{\Theta})|$ is the posterior median.
2. (a) Let $T(\mathbf{X})$ be the Bayes estimator of $\tau(\mathbf{\Theta})$, and let $T^*(\mathbf{X})$ be any other estimator. Suppose that $E(MSE_{T^*}(\mathbf{\Theta})) = E(MSE_T(\mathbf{\Theta}))$. Show that $T(\mathbf{X}) \equiv T^*(\mathbf{X})$.
   (b) Suppose that $T(\mathbf{X})$ is the Bayes estimator of $\tau(\mathbf{\Theta})$. Show that there is no estimator $T^*(\mathbf{X})$ which satisfies $MSE_{T^*}(\mathbf{\Theta}) \leqq MSE_T(\mathbf{\Theta})$ for all $\mathbf{\Theta}$, with strong inequality for some $\mathbf{\Theta}$. (If so, then $E(MSE_{T^*}(\mathbf{\Theta})) \leqq E(MSE_T(\mathbf{\Theta}))$. Why?)
3. Consider the model in which we observe $X_1|\Theta \sim B(1,\Theta)$ and we have the discrete prior distribution given by $\pi(\Theta=0)=\frac{1}{3}$, $\pi(\Theta=1)=\frac{2}{3}$.
   (a) Show that the posterior distribution is given by $\pi(\Theta=0|X=0)=1$ and $\Pi(\Theta=1|X=1)=1$.
   (b) Show that $E(\Theta|X)=X$, an unbiased estimator.
   (c) Why does this example not violate Theorem 16–3?

16.1.3 Bayesian Intervals and Tests

Consider a Bayesian statistical model with parameter $\mathbf{\Theta}$. Let $\tau=\tau(\mathbf{\Theta})$ be a univariate parameter. A *Bayesian* $(1-\alpha)$ *interval* for τ is an interval $L(\mathbf{X}) \leqq \tau \leqq R(\mathbf{X})$ such that

$$P(L(\mathbf{X}) \leqq \tau(\mathbf{\Theta}) \leqq R(\mathbf{X})|\mathbf{X}) = 1-\alpha.$$

Theorem 16–5 gives a method which can often be used to find Bayesian $(1-\alpha)$ intervals.

Theorem 16–5. Let $Q=q(\mathbf{X},\tau)$ be a function such that the distribution of Q given $\mathbf{X}$ is known. Let $q^\alpha(\mathbf{X})$ be the upper α point of this conditional distribution. Let $\alpha=a+b$, and suppose that

$$q^{1-b}(\mathbf{X}) \leqq q(\mathbf{X},\tau) \leqq q^a(\mathbf{X}) \Leftrightarrow L(\mathbf{X}) \leqq \tau \leqq R(\mathbf{X}).$$

Then $L(\mathbf{X}) \leqq \tau \leqq R(\mathbf{X})$ is a Bayesian $(1-\alpha)$ interval for τ.

Proof

$$P(L(\mathbf{X}) \leqq \tau \leqq R(\mathbf{X})|\mathbf{X}) = P(q^{1-b}(\mathbf{X}) \leqq q(\mathbf{X},\tau) \leqq q^a(\mathbf{X})|\mathbf{X})$$
$$= a+b = \alpha. \qquad \square$$

If $a=b=\alpha/2$, we call the Bayesian interval an *equal-tailed* interval.

Example 16–6

Let $X_1,\ldots,X_n$ be a sample from an exponential distribution with mean Θ^{-1}, and let $\Theta \sim \Gamma(a,b)$. In Section 16.1.1, we showed that the posterior distribution of Θ given $\mathbf{X}$ is

$$\Theta|\mathbf{X} \sim \Gamma(A^*,B^*), \qquad A^* = a+n, \qquad B^* = \frac{b}{nb\bar{X}+1}.$$

Therefore,

$$\left(\frac{2\Theta}{B^*}\right)|\mathbf{X} \sim \chi^2_{2A^*}.$$

satisfies the conditions of Theorem 16–5. Moreover,

$$\chi_{2A^*}^{2\ 1-\alpha/2} \leqq \frac{2\Theta}{B^*} \leqq \chi_{2A^*}^{2\ \alpha/2}$$

if and only if

$$\frac{B^*}{2}\chi_{2A^*}^{2\ 1-\alpha/2} \leqq \Theta \leqq \frac{B^*}{2}\chi_{2A^*}^{2\ \alpha/2}$$

and hence, this interval is a Bayesian $(1-\alpha)$ interval for Θ. A Bayesian $(1-\alpha)$ interval for $\mu = \Theta^{-1}$ is

$$\frac{2}{B^*(\chi_{2A^*}^{2\ \alpha/2})} \leqq \mu \leqq \frac{2}{B^*(\chi_{2A^*}^{2\ 1-\alpha/2})}.$$

In particular, if $a = 5$, $b = 6$, $n = 10$, $\overline{X} = .2$, and $\alpha = .05$, then

$$A^* = 15, \qquad B^* = \frac{6}{13}, \qquad \chi_{30}^{2\ .975} = 16.791, \qquad \chi_{30}^{2\ .025} = 46.979,$$

the Bayesian interval for Θ is

$$\left(\frac{6}{26}\right)(16.791) \leqq \Theta \leqq \left(\frac{6}{26}\right)(46.979) \Leftrightarrow 3.87 \leqq \Theta \leqq 10.84,$$

and the Bayesian interval for μ is

$$.092 \leqq \mu \leqq .258.$$

For comparison, the classical $(1-\alpha)$ interval for μ is

$$\frac{2n\overline{X}}{\chi_{2n}^{2\ \alpha/2}} \leqq \mu \leqq \frac{2n\overline{X}}{\chi_{2n}^{2\ 1-\alpha/2}},$$

With the numbers given, we get

$$\chi_{20}^{2\ .025} = 34.170 \quad \text{and} \quad \chi_{20}^{2\ .975} = 9.591.$$

Hence, the classical confidence interval for μ in this situation is

$$.117 \leqq \mu \leqq .417.$$

Although the methods for finding classical confidence intervals and Bayesian intervals are quite similar, they typically lead to different answers and have different interpretations. In a $(1-\alpha)$ Bayesian interval,

$$P(L(\mathbf{X}) \leqq \tau \leqq R(\mathbf{X})|\mathbf{X}) = 1 - \alpha,$$

while in a classical interval,

$$P(L(\mathbf{X}) \leqq \tau \leqq R(\mathbf{X})|\boldsymbol{\Theta}) = 1 - \alpha.$$

(That is, for the Bayesian interval, the probability is computed conditionally on $\mathbf{X}$, but for the classical interval, the probability is computed conditionally on $\boldsymbol{\Theta}$.)

Since the Bayesian interval is computed from the posterior distribution, if we get a Bayesian .95 interval of $4 \leqq \tau \leqq 6$, then $P(4 \leqq \tau \leqq 6|\mathbf{X}) = .95$, so that conditionally on the data, the probability that τ is in the interval from 4 to 6 is .95. This interpretation is just the one many people incorrectly make for a classical interval.

Let N be a subset of the parameter space Ω. A *Bayes* α *test* of the null hypothesis that $\boldsymbol{\Theta} \in N$ against $\boldsymbol{\Theta} \notin N$ rejects the null hypothesis when

$$P(\Theta \in N) \leqq \alpha,$$

that is, when the posterior probability of the null hypothesis is at most α.

The next theorem is often useful in finding Bayes tests.

Theorem 16–6. Suppose that $q(\mathbf{X}, \tau)$ is an increasing function of τ satisfying the conditions of Theorem 16–5. Then

a. The Bayes α test of the null hypothesis that $\tau(\mathbf{\Theta}) \geqq c$ against $\tau(\mathbf{\Theta}) < c$ rejects the null hypothesis if and only if $q(\mathbf{X}, c) \geqq q^{\alpha}(\mathbf{X})$.
b. The Bayes α test of the null hypothesis that $\tau(\mathbf{\Theta}) \leqq c$ against $\tau(\mathbf{\Theta}) > c$ rejects the null hypothesis if and only if $q(\mathbf{X}, c) \leqq q^{1-\alpha}(\mathbf{X})$.

Proof

a. Since $q(\mathbf{X}, \tau)$ is increasing in τ,

$$\tau \geqq c \Leftrightarrow q(\mathbf{X}, \tau) \geqq q(\mathbf{X}, c).$$

In addition,

$$P(q(\mathbf{X}, \tau) \geqq q^{\alpha}(\mathbf{X}) | \mathbf{X}) = \alpha.$$

The Bayes test of the null hypothesis that $\tau \geqq c$ rejects the hypothesis if and only if

$$\alpha \geqq P(\tau(\mathbf{\Theta}) \geqq c | \mathbf{X}) = P(q(\mathbf{X}, \tau) \geqq q(\mathbf{X}, c) | \mathbf{X}),$$

which occurs if and only if $q(\mathbf{X}, c) \geqq q^{\alpha}(\mathbf{X})$.
b. See Exercise B6. □

Example 16–7

In Example 16–6, consider testing that $\Theta \geqq 8$ against $\Theta < 8$. From that example, we see that

$$q(\mathbf{X}, \Theta) = \frac{2\Theta}{b^*} \sim \chi^2_{2(a+n)}$$

satisfies Theorem 16–6. Therefore, the Bayes α test for this problem rejects the null hypothesis if

$$q(\mathbf{X}, 8) > \chi^{2}_{2A^*}{}^{\alpha}.$$

With the numbers given in this example, and with $\alpha = .05$,

$$q(\mathbf{X}, 8) = \frac{2(8)}{(6/13)} = 34.67 < 43.773.$$

Therefore, the Bayes .05 test for this problem accepts the null hypothesis that $\Theta \geqq 8$.

For most Bayesian models encountered, the posterior distribution is continuous. If so, then

$$P(\tau(\mathbf{\Theta}) = c) = 0$$

for any c, and hence,

$$P(\tau(\mathbf{\Theta}) = c | \mathbf{X}) = 0.$$

Therefore, a Bayesian would always reject the hypothesis that $\tau(\mathbf{\Theta}) = c$ in favor of $\tau(\mathbf{\Theta}) \neq c$ for any c, so he or she would rarely test two-sided problems of the type tested by classical statisticians.

The Bayesian rejects the null hypothesis that $\mathbf{\Theta} \in N$ if $P(\mathbf{\Theta} \in N|\mathbf{X}) \leqq \alpha$. That is, the Bayesian rejects the null hypothesis if, conditionally on the data observed, the probability of the null hypothesis is less than or equal to α. This assertion is the one that many people make about classical tests when it is not appropriate. We could define $P(\mathbf{\Theta} \in N|\mathbf{X})$ to be the Bayesian p-value for the experiment, since it is just the probability that the null hypothesis is true, given the data.

Example 16–8

Let $X_1, \ldots, X_n$ be a sample from a Bernoulli distribution with mean Θ, and let $\Theta \sim \beta(a, b)$. Then the posterior distribution of Θ given $\mathbf{X}$ is

$$\Theta|\mathbf{X} \sim \beta(A^*, B^*), \qquad A^* = a + n\overline{X}, \qquad B^* = b + n(1 - \overline{X}).$$

Since we do not have a table of the distribution function for the beta distribution, we use the fact that

$$U \sim \beta(a, b) \Rightarrow \frac{bU}{a(1-U)} \sim F_{2a, 2b},$$

(see the exercises), and hence,

$$F(\mathbf{X}, \Theta)|\mathbf{X} \sim F_{2A^*, 2B^*} \quad \text{where} \quad F(\mathbf{X}, \Theta) = \frac{(B^*)\Theta}{(A^*)(1-\Theta)},$$

which is an increasing function of Θ. Therefore, to test that $\Theta \leqq .2$ against $\Theta > .2$, we reject the null hypothesis if

$$F(\mathbf{X}, .2) = \frac{.2(B^*)}{.8(A^*)} = \frac{B^*}{4A^*} \leqq F^{1-\alpha}_{2A^*, 2B^*} = (F^{\alpha}_{2B^*, 2A^*})^{-1}.$$

In particular, if $\overline{X} = .6$, $n = 30$, $a = 2$, $b = 3$, and $\alpha = .05$, then

$$A^* = 20, \qquad B^* = 15, \qquad (F^{.05}_{30, 40})^{-1} = (1.74)^{-1} = .57.$$

Now,

$$F(\mathbf{X}, .2) = \frac{15}{4(20)} = .1875 \leqq .57,$$

so we reject the hypothesis that $\Theta \leqq .2$ with these data. To find a Bayesian $(1-\alpha)$ interval for Θ, we note that

$$F^{1-\alpha/2}_{2A^*, 2B^*} \leqq F(\mathbf{X}, \theta) \leqq F^{\alpha/2}_{2A^*, 2B^*} \Leftrightarrow \left[1 + \frac{B^*}{A^* F^{1-\alpha/2}_{2A^*, 2B^*}}\right]^{-1} \leqq \Theta \leqq \left[1 + \frac{B^*}{A^* F^{\alpha/2}_{2A^*, 2B^*}}\right]^{-1}$$

so that this interval is a Bayesian $(1-\alpha)$ interval for Θ. In particular, with the data given earlier in this example, and with $\alpha = .10$,

$$F^{.05}_{40, 30} = 1.79 \quad \text{and} \quad F^{.95}_{40, 30} = (F^{.05}_{30, 40})^{-1} = .56.$$

So the interval becomes

$$\left[1 + \frac{15}{20(.56)}\right]^{-1} \leqq \Theta \leqq \left[1 + \frac{15}{20(1.79)}\right]^{-1} \Leftrightarrow .42 \leqq \Theta \leqq .70.$$

Now, let $\Phi(\mathbf{X})$ be a test for testing the null hypothesis that $\mathbf{\Theta} \in N$ against $\mathbf{\Theta} \in A = N^c$. Let Φ have power function $K_\Phi(\mathbf{\Theta})$, and choose α. We define the *Bayes risk* of Φ by

$$r(\Phi) = (1-\alpha)\int_N K_\Phi(\boldsymbol{\theta})\pi(\boldsymbol{\theta})d\boldsymbol{\theta} + \alpha\int_A (1 - K_\Phi(\boldsymbol{\theta}))\pi(\boldsymbol{\theta})d\boldsymbol{\theta}.$$

If $\boldsymbol{\Theta} \in N$, then $K_\Phi(\boldsymbol{\Theta})$ is the probability of a type I error, so that the first integral is the probability of a type I error (averaged over the distribution of both $\mathbf{X}$ and $\boldsymbol{\Theta}$). Similarly, the second integral is the probability of a type II error. The Bayes risk $r(\Phi)$ therefore is a weighted average of the probabilities of type I and type II errors, with weights α and $1-\alpha$, respectively. Hence α and $1-\alpha$ represent the relative importance of type I and type II errors. Note that for each Φ, $r(\Phi)$ is a number, so that it is reasonable to ask whether there is a Φ that minimizes $r(\Phi)$.

Theorem 16–7. The Bayes α test minimizes $r(\Phi)$.

Proof. Note first that the Bayes test rejects the null hypothesis if and only if

$$(1-\alpha)P(\boldsymbol{\Theta} \in N|\mathbf{X}) \leqq \alpha P(\boldsymbol{\Theta} \in A|\mathbf{X}).$$

(See Exercise B7.) Let

$$h_\Phi(\mathbf{X}, \boldsymbol{\Theta}) = \begin{cases} (1-\alpha)\Phi(\mathbf{X}) & \text{if } \boldsymbol{\Theta} \in N \\ \alpha(1-\Phi(\mathbf{X})) & \text{if } \boldsymbol{\theta} \in A \end{cases}.$$

Then

$$r(\Phi) = E(E[h_\Phi(\mathbf{X}, \boldsymbol{\Theta})|\boldsymbol{\Theta}]) = E(E[h_\Phi(\mathbf{X}, \boldsymbol{\Theta})|\mathbf{X}]).$$

Therefore, to find a Φ that minimizes $r(\Phi)$, we find, for each $\mathbf{X}$, a $\Phi(\mathbf{X})$ that minimizes $E[h_\Phi(\mathbf{X}, \boldsymbol{\Theta})|\mathbf{X}]$. Now,

$$\begin{aligned} E[h_\Phi(\mathbf{X}, \boldsymbol{\Theta})|\mathbf{X}] &= (1-\alpha)\Phi(\mathbf{X})P(\boldsymbol{\Theta} \in N|\mathbf{X}) + \alpha(1-\Phi(\mathbf{X}))P(\boldsymbol{\Theta} \in A|\mathbf{X}) \\ &= \alpha P(\boldsymbol{\Theta} \in A|\mathbf{X}) + \Phi(\mathbf{X})[(1-\alpha)P(\boldsymbol{\Theta} \in N|\mathbf{X}) - \alpha P(\boldsymbol{\Theta} \in A|\mathbf{X})]. \end{aligned}$$

The first term does not involve Φ. To minimize this expression, we choose $\Phi(\mathbf{X}) = 1$ when the term inside the brackets is negative, and $\Phi(\mathbf{X}) = 0$ when that term is positive. That is, we let

$$\Phi(\mathbf{X}) = \begin{cases} 1 & \text{if } (1-\alpha)P(\boldsymbol{\Theta} \in N|\mathbf{X}) - \alpha P(\boldsymbol{\Theta} \in A|\mathbf{X}) < 0 \\ 0 & \text{if } (1-\alpha)P(\boldsymbol{\Theta} \in N|\mathbf{X}) - \alpha P(\boldsymbol{\Theta} \in A|\mathbf{X}) > 0 \end{cases}.$$

(We can randomize, if we choose, when $(1-\alpha)P(\boldsymbol{\Theta} \in N|\mathbf{X}) - \alpha P(\boldsymbol{\Theta} \in A|\mathbf{X}) = 0$.) Hence, the Bayes α test minimizes $r(\Phi)$. □

Exercises—A

1. Suppose that 20 smokers are observed for a particular period and that 9 had heart attacks during that period. Suppose also that, prior to the experiment, we believe that the true proportion Θ of smokers who have heart attacks in such a period has a beta distribution with parameters $a = 1$ and $b = 1$.
   (a) Find a 98% Bayesian interval for Θ.
   (b) Test the hypothesis that $\Theta \leqq .2$ against $\Theta > .2$ with a Bayes .05 test.
2. Suppose the accidents on a stretch of highway in a given period have a Poisson distribution with mean Θ. Suppose also that we believe that $\Theta \sim \Gamma(1, 10)$ before the

experiment. Finally, suppose we observe 14 accidents during the said period. Find a Bayesian .95 interval for Θ.

3. Suppose that the distance between potholes on a freeway has an exponential distribution with mean Θ^{-1} and that our prior distribution for Θ is $\Theta \sim \Gamma(2, .01)$. Suppose we observe 10 potholes and find that the average distance between them is 50 feet. Assume that the distances are independent. (Note that there are only 9 distances measured between the 10 potholes.)

(a) Find Bayesian .99 intervals for Θ and $\mu = \theta^{-1}$.

(b) Test the hypothesis that $\mu \leqq 40$ against $\mu > 40$ with a .05 test.

Exercises—B

1. Show that if $U \sim \beta(a, b)$, then $bU/a(1-U) \sim F_{2a,2b}$.

2. For Exercise B1, Section 16.1.1, find a Bayesian $(1-\alpha)$ interval for Θ.

3. For Exercise B2, Section 16.1.1, find the Bayes α test that $\Theta \geqq k$ against $\Theta < k$.

4. For Exercise B3, Section 16.1.1, find a $(1-\alpha)$ Bayesian interval for Θ.

5. For Exercise B4, Section 16.1.1, find the Bayes α test that $\Theta \leqq k$ against $\Theta > k$.

6. Prove part b of Theorem 16–6.

7. Show that the Bayes α test rejects the null hypothesis if and only if $(1-\alpha)P(\mathbf{\Theta} \in N|\mathbf{X}) \leqq \alpha P(\boldsymbol{\theta} \in A|\mathbf{X})$. (Note that $P(\mathbf{\Theta} \in A|\mathbf{X}) = 1 - P(\mathbf{\Theta} \in N|\mathbf{X})$.)

Exercises—C

1. Suppose the parameter space Θ consists of only 2 points, θ_0 and θ_1, with prior probabilities π_0 and π_1, and that $\mathbf{X}$ is a discrete random vector with likelihood function $L_X(\theta)$. Let $\lambda(\mathbf{X}) = L_X(\theta_1)/L_X(\theta_0)$ be the Neyman-Pearson test statistic for testing $\Theta = \theta_0$ against $\Theta = \theta_1$.

(a) Show that $P(\Theta = \theta_0|\mathbf{X}) = (1 + \pi_1\lambda(\mathbf{X})/\pi_0)^{-1}$.

(b) Show that the Bayes α test that $\Theta = \theta_0$ rejects if $\lambda(\mathbf{X}) > k$. What is k?

(This exercise shows that Bayes tests and the Neyman-Pearson tests are the same for simple vs. simple testing problems. Note that it is easier to determine the critical value k for the Bayes test.)

2. Let Θ be a one-dimensional parameter, and let $L(\mathbf{X}) \leqq \theta \leqq R(\mathbf{X})$ be a Bayes $(1-\alpha)$ confidence interval for θ. We say that this interval is the *highest posterior density* (HPD) interval for θ if there exists a k such that $L(\mathbf{X}) \leqq \theta \leqq R(\mathbf{X}) \Leftrightarrow \pi(\theta|\mathbf{X}) > k$. Show that for each $\mathbf{X}$, the HPD interval has shorter length than any other Bayes $(1-\alpha)$ interval. (*Hint*: Minimize $\int_{L(\mathbf{X})}^{R(\mathbf{X})} 1\,d\theta$ subject to $\int_{L(\mathbf{X})}^{R(\mathbf{X})} \pi(\theta|\mathbf{X})d\theta = 1-\alpha$. Follow the proof of the Neyman-Pearson theorem to show that this minimization occurs when $\pi(\theta|\mathbf{X})/1 > k$. Note that $\mathbf{X}$ is constant in this entire calculation.)

3. Suppose that Θ is a univariate parameter and that $L_X(\theta)$ has monotone likelihood ratio in $T(\mathbf{X})$.

(a) Show that $P(\Theta \leqq a|\mathbf{X})$ is a decreasing function of $T(\mathbf{X})$. (*Hint*:

$$P(\Theta \leqq a|\mathbf{X}) = (1 + \int_a^\infty [L_X(\theta)/L_X(a)]\pi(\theta)d\theta \Big/ \int_{-\infty}^a [L_X(a)/L_X(\theta)]^{-1}\pi(\theta)d\theta)^{-1}.)$$

(b) Show that the Bayes α test that $\Theta \leqq a$ rejects if $T(\mathbf{X})$ is too large. (This exercise shows that the Bayes tests are UMP tests for this setting.)

* 16.1.4 Bayesian Analysis of the Two-Parameter Normal Model

Consider the two-parameter normal model, in which we observe $X_1, \ldots, X_n$, a sample from a normal distribution with mean μ and variance δ^{-1}. We call $\delta = \sigma^{-2}$ the *precision* of the X_i. (Note that when the variance is small, the observations are precise, and hence, the precision is large. Similarly, when the variance is large, the precision is small.)

We say that (μ, δ) has a *normal-gamma* distribution with parameters a, $b > 0$, $c > 0$, and $d > 0$, and write

$$(\mu, \delta) \sim NG(a, b, c, d)$$

if

$$\mu|\delta \sim N(a, (b\delta)^{-1}), \qquad \delta \sim \Gamma(c, d^{-1}).$$

In this case, (μ, δ) has joint density

$$\begin{aligned}\pi(\mu, \delta) &= \pi_1(\mu|\delta)\pi_2(\delta) \\ &= h_1(a, b)\delta^{1/2} \exp\left[\frac{-(\mu - a)^2 b\delta}{2}\right] h_2(c, d)\delta^{c-1} \exp[-\delta d] \\ &= h(a, b, c, d)\delta^{c-1/2} \exp\left[\frac{-(\mu^2 \delta b - \mu\delta 2ab + \delta(a^2 b + 2d))}{2}\right].\end{aligned}$$

We first give some elementary results for this joint distribution.

Theorem 16–8. Let $(\mu, \delta) \sim NG(a, b, c, d)$. Then

a. $E\mu = a$, $E\delta = c/d$, and $E\delta^{-1} = d/(c-1)$.

b. If c is an integer or half-integer, then

$$t = \frac{(bc)^{1/2}(\mu - a)}{d^{1/2}} \sim t_{2c} \quad \text{and} \quad U = 2d\delta \sim \chi^2_{2c}.$$

Proof

a. $E\mu = E(E(\mu|\delta)) = E(a) = a$. The expressions for $E\delta$ and $E\delta^{-1}$ follow from Theorem 4–26.

b. By Theorems 4–26 and 4–33,

$$U \sim \Gamma\left(\frac{2c}{2}, 2\right) = \chi^2_{2c}.$$

By the definition of the normal distribution,

$$Z = (b\delta)^{1/2}(\mu - a)|\delta \sim N(0, 1).$$

Now, Z and δ are independent, since this distribution does not depend on δ. Hence, Z and U are also independent. But then, by the definition of the t-distribution,

$$\frac{Z}{(U/2c)^{1/2}} = \frac{(bc)^{1/2}(\mu - a)}{d^{1/2}} \sim t_{2c}. \quad \square$$

We now find the posterior distribution for the two-parameter normal model. Recall that $\overline{X}$ and S^2 are respectively the sample mean and sample variance of the X's.

Theorem 16–9. Let $X_1, \ldots, X_n$ be a sample from a normal distribution with mean μ and precision δ. Suppose that the prior distribution of μ and δ is given by

$$(\mu, \delta) \sim NG(a, b, c, d).$$

Then the posterior distribution of μ and δ is given by

$$(\mu, \delta) \sim NG(A^*, B^*, C^*, D^*),$$

where

$$A^* = \frac{ab + n\overline{X}}{b + n}, \qquad B^* = b + n, \qquad C^* = c + \frac{n}{2},$$

$$D^* = d + \frac{(n-1)S^2}{2} + \frac{nb(\overline{X} - a)^2}{2(b + n)}.$$

Proof. The conditional density of $\mathbf{X}$ given (μ, δ) is

$$f(\mathbf{x}; (\mu, \delta)) = k\delta^{n/2} \exp\left[\frac{-\sum(X_i - \mu)^2 \delta}{2}\right]$$

$$= k\delta^{n/2} \exp\left[\frac{-\left(\mu^2 \delta n - 2\mu\delta\left(\sum X_i\right) + \delta\left(\sum X_i^2\right)\right)}{2}\right].$$

Therefore, the posterior distribution of (μ, δ) given $\mathbf{X}$ is

$$\pi^*((\mu, \delta)|\mathbf{x}) = k\delta^{n/2} \exp\left[\frac{-(\mu^2 \delta n - 2\mu\delta\left(\sum x_i\right) + \delta\left(\sum x_i^2\right)}{2}\right]$$

$$\times h(a, b, c, d)\delta^{c - 1/2} \exp\left[\frac{-(\mu^2 \delta b - \mu\delta 2ab + \delta(a^2 b + 2d))/2}{2}\right] \Big/ q(\mathbf{x})$$

$$= h^*(a, b, c, d, \mathbf{x})\delta^{c + (n-1)/2}$$

$$\times \exp\left[\frac{-\left(\mu^2 \delta(n + b) - \mu\delta 2(ab + n\overline{x}) + \delta\left(a^2 b + 2d + \sum x_i^2\right)\right)}{2}\right]$$

$$= h^*(a^*, b^*, c^*, d^*)\delta^{c^* - 1/2}$$

$$\times \exp\left[\frac{-(\mu^2 \delta b^* - \mu\delta 2a^* b^* + \delta(a^{*2} b^* + 2d^*))}{2}\right],$$

where

$$c^* = \frac{c + n}{2}, \qquad b^* = n + b, \qquad a^* b^* = ab + n\overline{x},$$

$$a^{*2} b^* + 2d^* = a^2 b + 2d + \sum x_i^2. \tag{16–1}$$

Solving these equations, we obtain

$$a^* = \frac{ab + n\bar{x}}{b + n}, \qquad b^* = b + n, \qquad c^* = c + \frac{n}{2}, \tag{16–2}$$

$$d^* = d + \frac{(n-1)s^2}{2} + \frac{nb(\bar{x} - a)^2}{2(b+n)}, \tag{16–3}$$

and hence,

$$(\mu, \delta)|\mathbf{X} \sim NG(A^*, B^*, C^*, D^*). \quad \square$$

Theorem 16–9 implies that the normal-gamma family is a conjugate prior family for the two-parameter normal model.

From Theorem 16–8, we see that the Bayes estimator of μ is

$$T_1(\mathbf{X}) = E(\mu|\mathbf{X}) = A^* = \frac{ab + n\bar{X}}{b + n},$$

which is a weighted average of the prior mean and $\bar{X}$. The Bayes estimator of δ is

$$T_2(\mathbf{X}) = E(\delta|\mathbf{X}) = \frac{C^*}{D^*} = \frac{(2c+n)(b+n)}{2d(b+n) + (n-1)(b+n)S^2 + nb(X-a)^2},$$

and the Bayes estimator of the variance $\sigma^2 = \delta^{-1}$ is

$$T_3(\mathbf{X}) = E(\delta^{-1}|\mathbf{X}) = \frac{D^*}{C^* - 1} = \frac{2d}{2c + n - 2} + \frac{(n-1)S^2}{2c + n - 2} + \frac{nb(\bar{X} - a)^2}{(b+n)(2c+n-2)}.$$

The Bayes estimator of δ^{-1} can be written in the form

$$T_3(\mathbf{X}) = \left(\frac{d}{c-1}\right)\left(\frac{2(c-1)}{2c+n-2}\right) + S^2\left(\frac{n-1}{2c+n-2}\right) + \left(\frac{nb(\bar{X}-a)^2}{b+n}\right)\left(\frac{1}{2c+n-2}\right),$$

which is a weighted average of the prior mean $d/(c-1)$, the sample variance S^2, and $nb(\bar{X} - a)^2/(b+n)$. Note that

$$E\bar{X} = a, \qquad E\left(\left[\frac{nb(\bar{X}-a)^2}{b+n}\right]\Big|\delta\right) = \delta^{-1} = \sigma^2. \tag{16–4}$$

(See Exercise B2.) Therefore, the Bayes estimator for the variance for the two-parameter normal model uses not only the prior mean and the sample variance, but also how far $\bar{X}$ is from a.

By Theorem 16–8, if $2C^*$ is an integer, then

$$t(\mathbf{X}, \mu) = \frac{(B^*C^*)^{1/2}(\mu - A^*)}{D^{*\,1/2}} \sim t_{2C^*},$$

and therefore the interval

$$\mu \in A^* \pm t_{2C^*}^{\alpha/2}\left(\frac{D^*}{B^*C^*}\right)^{1/2}$$

is a Bayes $(1 - \alpha)$ interval for μ. Also, since $t(\mathbf{X}, \mu)$ is an increasing function of μ, a

Bayes α test that $\mu \leqq a$ against $\mu > a$ rejects the null hypothesis if

$$t(\mathbf{X}, a) < -t_{2C^*}^{\alpha}.$$

Similarly,

$$U(\mathbf{X}, \delta) = 2D^*\delta \sim \chi_{2C^*}^2,$$

so that a Bayes $(1-\alpha)$ interval for δ is

$$\frac{\chi_{2C^*}^{2\ \ 1-\alpha/2}}{2D^*} \leqq \delta \leqq \frac{\chi_{2C^*}^{2\ \ \alpha/2}}{2D^*}.$$

A Bayes $(1-\alpha)$ interval for $\sigma^2 = \delta^{-1}$ is

$$\frac{2D^*}{\chi_{2C^*}^{2\ \ \alpha/2}} \leqq \delta^{-1} = \sigma^2 \leqq \frac{2D^*}{\chi_{2C^*}^{2\ \ 1-\alpha/2}}.$$

$U(\mathbf{X}, \delta)$ is an increasing function of δ, so that a Bayes α test that $\delta \leqq c$ against $\delta > c$ (or $\sigma^2 \geqq 1/c$ against $\sigma^2 < 1/c$) rejects the null hypothesis if

$$U(\mathbf{X}, c) \leqq \chi_{2C^*}^{2\ \ 1-\alpha}.$$

Example 16–9

Suppose that we assume the prior in which

$$(\mu, \delta) \sim NG(2, 4, 5, 6)$$

and that we observe $n = 15$, $\overline{X} = 1$, and $S^2 = 3$. Then

$$A^* = \frac{23}{19} = 1.21, \qquad B^* = 19, \qquad C^* = 12.5,$$

$$D^* = \frac{6 + (14)(3)}{2} + \frac{60(-1)^2}{38} = 28.58.$$

Therefore, the Bayes estimators of μ, δ, and $\sigma^2 = \delta^{-1}$ are, respectively,

$$T_1(\mathbf{X}) = 1.21, \qquad T_2(\mathbf{X}) = \frac{12.5}{28.58} = .44, \qquad T_3(\mathbf{X}) = \frac{28.58}{11.5} = 2.49.$$

A Bayes 95% interval for μ is

$$\mu \in 1.21 \pm (2.06)\left[\frac{28.58}{(19)(12.5)}\right]^{1/2} \Leftrightarrow \mu \in 1.21 \pm .71.$$

If we now want to test that $\mu \geqq 1.9$ against $\mu < 1.9$ with a .05 Bayes test, then

$$t(\mathbf{X}, 1.9) = \left[\frac{(19)(12.5)}{28.58}\right]^{1/2}(1.9 - 1.21) = 1.99 > 1.708,$$

so we reject the hypothesis that $\mu \geqq 1.9$ in favor of the alternative hypothesis that $\mu < 1.9$. A Bayes 95% interval for δ is

$$\frac{13.12}{2(28.58)} \leqq \delta \leqq \frac{40.646}{2(28.58)} \Leftrightarrow .23 \leqq \delta \leqq .71.$$

A Bayes interval for $\sigma^2 = \delta^{-1}$ is

$$(.71)^{-1} \leqq \sigma^2 \leqq (.23)^{-1} \Leftrightarrow 1.41 \leqq \sigma^2 \leqq 4.36.$$

Exercises—A

1. An experiment was conducted to determine the average blood pressure of a group of people. Suppose we take a sample of 10 such people and assume their blood pressures are independently normally distributed with unknown mean μ and variance $\sigma^2 = \delta^{-1}$, and that $(\mu, \delta) \sim NG(110, 9, 5, 144)$. Suppose that we observe $\overline{X} = 120$, $S^2 = 36$.
   (a) What is the posterior distribution of (μ, δ)?
   (b) What are the Bayes estimators of μ, δ, and $\sigma^2 = \delta^{-1}$?
   (c) Find .95 Bayesian intervals for μ, δ, and σ.
   (d) Test the hypothesis that $\mu \leqq 100$ against $\mu > 100$ with a Bayes .10 test.

Exercises—B

1. Solve Equation (16–1) to verify Equations (16–2) and (16–3).
2. Verify Equation (16–4).
3. Find the formula for the Bayes estimator of $\sigma = \delta^{-1/2}$ for the two-parameter normal model.
4. **(a)** Let $(\mu, \delta) \sim NG(a, b, c, d)$. What is $E\mu\delta$? ($E\mu\delta = E(\delta E(\mu|\delta))$. Why?)
   (b) Find the Bayes estimator of $\mu\delta$ for the two-parameter normal model.

16.2 OTHER BAYESIAN TOPICS

* 16.2.1 Empirical Bayes Procedures

Suppose that we have a statistical model with a vector $\mathbf{X}$ of observations, a vector $\mathbf{\Theta}$ of parameters, and the joint density $f(\mathbf{x}|\boldsymbol{\theta})$ of $\mathbf{X}$ given $\mathbf{\Theta}$. Suppose also that we have a family $\pi(\boldsymbol{\theta}; \mathbf{a})$ of prior distributions. For each $\mathbf{a}$, we can compute a Bayes estimator $T(\mathbf{X}, \mathbf{a})$ for a parameter $\tau(\mathbf{\Theta})$. Now suppose we use the data to find a (classical) estimator $S(\mathbf{X})$ for $\mathbf{a}$. Then the estimator $T(\mathbf{X}, S(\mathbf{X}))$ is called an *empirical Bayes estimator* of $\tau(\mathbf{\Theta})$.

Example 16–10

Suppose that an insurance company insures n drivers. For each driver, it knows t_i, the time the driver has been driving, and X_i, the number of accidents he or she has had in that time. The company wants to estimate each driver's accident rate λ_i. The company considers the t_i to be known constants and the X_i to be observed random variables such that

$$X_i \sim P(\lambda_i t_i).$$

The MLE and best unbiased estimator of λ_i is

$$\hat{\lambda}_i = \frac{X_i}{t_i}.$$

However, this estimator has a major drawback: It estimates that if a person has an accident during the first week of driving, then that person will average one accident per week for the rest of his or her life. If the driver has no accident in the first week, $\hat{\lambda}_i$ estimates that he or she will never have any accidents.

Since the drivers can be considered a sample from the set of all possible drivers, the company believes that the λ_i are independently, identically distributed random variables, and it assumes that

$$\lambda_i \sim \Gamma(a, b).$$

Also, since the (X_i, λ_i) are independent random vectors, the conditional distribution of λ_i given $(X_1, \ldots, X_n)$ is the same as the conditional distribution of λ_i given X_i, which is

$$\lambda_i | X_i \sim \Gamma\left(a + X_i, \frac{b}{bt_i + 1}\right),$$

and the Bayes estimator of λ_i is

$$\hat{\lambda}_{iB} = \frac{ab + X_i b}{bt_i + 1}. \tag{16–5}$$

The insurance company has no idea how to choose a and b, so it decides to estimate them from the data. To find the MLE for (a, b), we would integrate out the λ_i to find the marginal density of the X_i. (Note that the X_i are still independent.) This marginal density is derived in the exercises, but is, unfortunately, not tractable. However, n, the number of drivers for the insurance company, is quite large, so that a consistent estimator of (a, b) should be adequate. Now,

$$EX_i = E(EX_i|\lambda_i) = E(\lambda_i t_i) = abt_i,$$

and

$$EX_i^2 = E(EX_i^2|\lambda_i) = E(\lambda_i^2 t_i^2 + \lambda_i t_i) = t_i^2(ab^2 + a^2b^2) + t_i ab.$$

Let

$$U = \frac{\sum X_i}{\sum t_i} \quad \text{and} \quad W = \frac{\sum X_i^2 - \sum X_i}{\sum t_i^2}.$$

Then

$$EU = E\left(\frac{\sum X_i}{\sum t_i}\right) = ab \quad \text{and} \quad EW = E\left(\frac{\sum X_i^2 - \sum X_i}{\sum t_i^2}\right) = a(a+1)b^2.$$

By similar, but more complicated arguments, it can be shown (under weak conditions on the t_i) that

$$\text{var}(U) \to 0 \quad \text{and} \quad \text{var}(W) \to 0$$

as $n \to \infty$. Therefore, U and W are consistent estimators of ab and $a(a+1)b^2$, respectively, and

$$Q = \frac{W - U^2}{U} \xrightarrow{P} \frac{a(a+1)b^2 - (ab)^2}{ab} = b,$$

by the continuous function theorem. Hence, Q is a consistent estimator of b, and therefore, we can make a sensible estimator for λ_i by substituting U for ab and Q for b into the Bayes estimator (Equation (16–5)), obtaining

$$\hat{\lambda}_{iE} = \frac{U + QX_i}{Qt_i + 1}.$$

This estimator is an empirical Bayes estimator because we have substituted classical estimators for a and b into the Bayes estimator for λ_i.

Note that

$$\hat{\lambda}_{iE} = \frac{U + Qt_i\hat{\lambda}_i}{Qt_i + 1}$$

is a weighted average of U, the estimated accident rate for the whole set of drivers, and $\hat{\lambda}_i$, the MLE of the accident rate for the ith driver. Note also that for small t_i, the empirical Bayes estimator puts most of its weight on the overall accident rate U, but as t_i gets large, the weight shifts to the accident rate for the individual driver.

In sum, the insurance company has used the Γ family of priors to get a Bayes estimator for λ_i and then used the data to select which member of the Γ family to use. Since the family of Γ priors is fairly large, this estimator should have desirable properties. In fact, simulations have indicated that the empirical Bayes estimator $\hat{\lambda}_{iE}$ defined here is better than the classical estimator $\hat{\lambda}_i$ in the sense that $E\Sigma(\hat{\lambda}_i - \lambda_i)^2 > E\Sigma(\hat{\lambda}_{iE} - \lambda_i)^2$, even for uniform and other nongamma priors.

It is fairly easy to motivate the use of Bayesian and empirical Bayesian procedures in the preceding example, because it is easy to imagine that the drivers are themselves a sample, and hence, the λ_i are random variables. However, it has been shown in many examples with nonrandom parameters that when the number of parameters in the original problem is much larger than the number of parameters in the prior, then empirical Bayes procedures have better classical properties than the "optimal" classical estimators such as MLEs and best unbiased estimators. In fact, some empirical Bayes estimators have been shown to be "better" than the MLEs and best unbiased estimators for estimating $\boldsymbol{\beta}$ in the multiple regression model as long as $\boldsymbol{\beta}$ has at least three parameters. (See Arnold (1981).) Similarly, some empirical Bayes estimators for estimating the vector of probabilities in the multinomial model have been shown to be better than the classical estimators, again, as long as there are at least three parameters being estimated.

Exercises—A

1. Suppose we have a collection of baseball players. For each player, we know X_i, the number of hits he has, and n_i, the number of times he had at bat. Suppose we believe that $X_i \sim B(n_i, \theta_i)$, independent, where θ_i represents the player's "true" batting average.
   (a) Show that the MLE for θ_i is X_i/n_i.
   (b) Assume that the θ_i are independent, with $\theta_i \sim \beta(a, b)$. Find the Bayes estimator of θ_i.
   (c) Let $Q = E\Sigma X_i/\Sigma n_i$ and $R = \Sigma(X_i^2 - X_i)/\Sigma n_i(n_i - 1)$. Find EQ and ER.
   (d) Use Q and R to find consistent estimators of a and b. (You may assume that var $Q \rightarrow 0$ and var $R \rightarrow 0$.)
   (e) Write down an empirical Bayes estimator of θ_i, and interpret it as a weighted average of Q, the overall proportion of hits, and $\hat{\theta}_i$, the player's proportion of hits. How do the weights vary with n_i?

Exercises—B

1. **(a)** Show that the conditional distribution of λ_i given $\mathbf{X} = (X_1, \ldots, X_n)$ is the same as the conditional distribution of λ_i given X_i.
   (b) Verify the formula for the posterior distribution of the λ_i.

2. Let X_i be independent, with $X_i \sim P(\lambda_i t_i)$.
   (a) Show that $\mathbf{X} = (X_1, \ldots, X_n)$ is a complete sufficient statistic for this model.
   (b) Show that $\hat{\lambda}_i$ is the best unbiased estimator of λ_i.
   (c) Show that $\hat{\lambda}_i$ is the MLE of λ_i.
3. (a) Under what conditions does $\operatorname{var}(U)$ go to zero as n goes to infinity?
   (b) Show that if $0 \leq a \leq t_i \leq b$ then $\operatorname{var} U \to 0$.
4. Suppose that $X|\lambda \sim P(t\lambda)$ and $\lambda \sim \Gamma(a, b)$. Find the marginal density of X.
5. Let $V = \Sigma X_i/nt_i$ and $R = \Sigma(X_i^2 - X_i)/nt_i^2$.
   (a) Show that $EV = ab$ and $ER = ab^2 + a^2b^2$.
   (b) Use V and R to give an alternative empirical Bayes estimator for the λ_i.

* 16.2.2 Bayesian Decision Theory

In a Bayesian statistical model we observe a random vector $\mathbf{X} \in \chi$, with known conditional density function $f(\mathbf{x}|\boldsymbol{\theta})$ given the unobserved parameter vector $\boldsymbol{\Theta} \in \Omega$, which is assumed to be a random vector with known density function $\pi(\boldsymbol{\theta})$. In a *Bayesian decision theory model*, we assume that we also have a set of possible actions $\mathbf{a} \in A$ and a function $L(\mathbf{a}, \boldsymbol{\theta})$ which measures the loss when we take action $\mathbf{a}$ and $\boldsymbol{\theta}$ is the value of $\boldsymbol{\Theta}$. (Throughout this section, the letter L will represent loss. We shall not use the likelihood function in this section.)

A *decision rule* is a function $d(\mathbf{X})$ from χ to A. (That is, for each possible outcome $\mathbf{X}$ of the experiment, d chooses an action $d(\mathbf{X})$.) The *risk function* $R(d, \boldsymbol{\theta})$ and *Bayes risk* $r(d)$ for a decision rule $d(\mathbf{X})$ are defined by

$$R(d, \boldsymbol{\theta}) = E(L(d(\mathbf{X}), \boldsymbol{\theta})|\boldsymbol{\Theta} = \boldsymbol{\theta}) \quad \text{and} \quad r(d) = ER(d, \boldsymbol{\Theta}).$$

(Note that the second expectation is with respect to prior distribution of $\boldsymbol{\Theta}$, and represents the average loss for $d(\mathbf{X})$ averaged over both $\mathbf{X}$ and $\boldsymbol{\Theta}$.) We say that $d_0(\mathbf{X})$ is a *Bayes rule* for this problem if $r(d_0) < \infty$ and

$$r(d_0) \leqq r(d)$$

for all possible rules d. That is, $d(\mathbf{X})$ is a Bayes rule if it minimizes the Bayes risk out of all possible rules.

Define the *posterior loss* of an action a given $\mathbf{X}$ by

$$L(\mathbf{a}|\mathbf{X}) = E(L(\mathbf{a}, \boldsymbol{\Theta})|\mathbf{X})$$

(i.e., the expectation of $L(\mathbf{a}, \boldsymbol{\Theta})$ taken over the posterior distribution of $\boldsymbol{\Theta}$). The following theorem is the basic result of Bayesian decision theory.

Theorem 16–10. Let $d_0(\mathbf{X})$ be a rule such that for each $\mathbf{X}$,

$$L(d_0(\mathbf{X})|\mathbf{X}) \leqq L(\mathbf{a}|\mathbf{X}) \quad \text{for all } \mathbf{a} \in A.$$

If $r(d_0) < \infty$, then $d_0(\mathbf{X})$ is a Bayes rule.

Proof. Let $d(\mathbf{X})$ be any rule. Using basic properties of conditional expectation, we see that

$$\begin{aligned} r(d) &= E(R(d, \boldsymbol{\Theta})) = E(E(L(d(\mathbf{X}), \boldsymbol{\Theta})|\boldsymbol{\Theta})) = E(L(d(\mathbf{X}), \boldsymbol{\Theta})) \\ &= E(E(L(d(\mathbf{X}), \boldsymbol{\Theta})|\mathbf{X})). \end{aligned}$$

By the definition of $d_0(\mathbf{X})$,

$$L(d_0(\mathbf{X})|\mathbf{X}) \leqq L(d(\mathbf{X})|\mathbf{X}),$$

and therefore,

$$r(d_0) = E(L(d_0(\mathbf{X})|\mathbf{X})) \leqq E(L(d(\mathbf{X})|\mathbf{X})) = r(d). \quad \square$$

This theorem implies that if for each $\mathbf{X}$ we choose $d_0(\mathbf{X})$ to be an action which minimizes $L(\mathbf{a}|\mathbf{X})$, then $d_0(\mathbf{X})$ is a Bayes rule, unless $r(d_0) = \infty$ (in which case there is no Bayes rule). If for some $\mathbf{X}$ there is a tie for the smallest posterior loss, we can choose any of the tied actions for $d_0(\mathbf{X})$.

Suppose we are estimating a parameter $\tau(\boldsymbol{\Theta})$ with loss function $L(a, \boldsymbol{\theta}) = (a - \tau(\boldsymbol{\theta}))^2$. If $d(\mathbf{X})$ is an estimator, then $R(d, \boldsymbol{\theta})$ is the mean squared error for d. In Section 16.2, we used the method of proof in this theorem to show that the estimator $d_0(\mathbf{X}) = E(\tau(\boldsymbol{\Theta})|\mathbf{X})$ minimizes the Bayes risk for this loss function, as long as $r(d_0) < \infty$.

Consider now testing that $\boldsymbol{\Theta} \in N$ against $\boldsymbol{\Theta} \notin N$. We have two possible actions: a_0—accept the null hypothesis; or a_1—reject the null hypothesis. Suppose we adopt the loss function

	N	N^c
a_0	0	L_0
a_1	L_1	0

(16–6)

(That is, we don't lose anything if we make a correct decision. We lose L_0 if we make a type I error and L_2 if we make a type II error.) Using Theorem 16–7 we can show that the Bayes rule for this problem rejects if

$$P(\Theta \in N|\mathbf{X}) \leqq \frac{L_0}{(L_0 + L_1)}.$$

That is, the Bayes rule is Bayes α test with $\alpha = L_0/(L_0 + L_1)$.
We now consider a nonstatistical example of Bayesian decision theory.

Example 16–11

Suppose that a piece of land may have oil under it. An oil magnate can buy the land for \$900,000. He believes that there is a .5 probability of no oil, a .3 chance of one well, and a .2 chance of two wells. He will make \$1,000,000 for each well on the land. He wants to decide whether or not to buy the land. Let Θ be the number of wells on the land, let b denote the event that he buys the land, and let c denote the event that he does not. Then

$$L(c, \theta) = 0 \quad \text{and} \quad L(b, \theta) = 900{,}000 - \theta(1{,}000{,}000).$$

Now

$$E\Theta = 0(.5) + 1(.3) + 2(.2) = .7.$$

Therefore, his expected loss if he does not buy the land is 0 and his expected loss if he buys it is

$$900{,}000 - 1{,}000{,}000\, E\Theta = 900{,}000 - 700{,}000 = 200{,}000.$$

Hence, he minimizes his Bayes risk by not buying the land.

Now, suppose that he can run an experiment for \$100,000. From this experiment he gets a Bernoulli random variable X such that $P(X = 1|\Theta = 0) = 0$, $P(X = 1|\Theta = 1) = .4$, and $P(X = 1|\Theta = 2) = .7$. In order to find his Bayes procedure for this situation, we need to find the posterior distribution of Θ given X.

$$\pi(\Theta = 2|X = 1) = \frac{P(\Theta = 2)P(X = 1|\Theta = 2)}{P(X = 1)} = \frac{(.2)(.7)}{(.2)(.7) + (.3)(.4)} = \frac{14}{26}.$$

Similarly, we find that

θ	0	1	2
$\pi(\theta\|X = 1)$	0	12/26	14/26
$\pi(\theta\|X = 0)$	50/74	18/74	6/74

Now, his loss function is

$$L(\theta, c) = 100{,}000 \quad \text{and} \quad L(\theta, b) = 1{,}000{,}000 - \theta(1{,}000{,}000)$$

(since he has paid \$100,000 to obtain X). Therefore, if he does not buy the land he loses \$100,000 (for either $X = 0$ or $X = 1$). Now, suppose that $X = 1$. Then

$$L(b|X = x) = 1{,}000{,}000(1 - E(\Theta|X = x)).$$

Now,

$$E(\Theta|X = 1) = 0(0) + \left(\frac{12}{26}\right) + 2\left(\frac{14}{26}\right) = \frac{40}{26}.$$

Therefore

$$L(b|X = 1) = 1{,}000{,}000\left(1 - \frac{40}{26}\right) = \left(\frac{-14}{26}\right)(1{,}000{,}000) < 100{,}000.$$

Hence, if he observes $X = 1$, he should buy the land. Similarly,

$$E(\Theta|X = 0) = \frac{50}{74} \quad \text{and} \quad L(b|X = 0) = 1{,}000{,}000\left(\frac{24}{74}\right) > 100{,}000.$$

Therefore he should not buy the land if he observes $X = 0$. Now,

$$P(X = 0) = \frac{74}{100} \quad \text{and} \quad P(X = 1) = \frac{26}{100}.$$

Therefore, the Bayes risk for this rule is

$$r(d) = E(L(d(\mathbf{X})|\mathbf{X})) = \left(\frac{74}{100}\right)(100{,}000) + \left(\frac{26}{100}\right)\left(\frac{-14}{26}\right)(1{,}000{,}000),$$

$$74{,}000 - 140{,}000 = -66{,}000.$$

Therefore, he can expect to make \$66,000 profit if he follows the rule of not buying when $X = 0$ and buying when $X = 1$. Note that before he observed, he would not have bought the land. Therefore, observing X at a cost of \$100,000 has made him a net profit of \$66,000 (so that he would still make money as long as X cost less than \$166,000.)

Typically the Bayes rule is fairly easy to find. By Theorem 16–10, we can determine $d(\mathbf{X})$ separately for each $\mathbf{X}$.

Exercises—A

1. In Example 16–11, suppose that the oil magnate can observe a random variable Y which takes on 3 values, 0, 1, and 2. Suppose that Y costs \$200,000 and has density

	0	1	2
$f(y\|\theta=0)$	.8	.2	0
$f(y\|\theta=1)$	.3	.4	.3
$f(y\|\theta=2)$	0	.1	.9

(a) Find the posterior distribution of θ given $y = 0$ and the Bayes decision when $y = 0$.
(b) Find the Bayes decision when $y = 1$ and when $y = 2$.
(c) Find the Bayes risk for this problem. How much could the magnate pay for Y and still break even?
(d) If the oil magnate had a choice between observing X of Example 16–11 for \$100,000 or Y of this problem for \$200,000, which should he choose?

2. Suppose a drug company is interested in studying the effectiveness of a new drug. Suppose that the company believes that the drug is ineffective, moderately effective, or very effective each with probability $\frac{1}{3}$. Finally, suppose that it would cost the company \$5,000,000 to study the drug, but that the company would get back \$8,000,000 if the drug is very effective, \$4,000,000 if it is moderately effective, and \$0 if the drug is ineffective. Should the company spend the \$5,000,000 to study the drug further?

3. In the setting of Exercise A2, suppose that for \$400,000 the drug company could do an experiment in which it finds a random variable $X \sim B(3, p)$, where $p = 0$ (i.e., $X \equiv 0$) if the drug is ineffective, $p = .4$ if the drug is moderately effective, and $p = .7$ if the drug is very effective.
(a) What is the Bayes action if $X = 3$?
(b) Find the Bayes rule for $X = 0, 1$, or 2.
(c) What is the Bayes risk for the Bayes rule?
(d) Should the drug company pay \$400,000 for the measurement X?
(e) How much money could the company pay for X and still break even?

Exercises—B

1. Consider the problem of testing that $\boldsymbol{\Theta} \in N$ with the loss function given in Equation (16–5). Use Theorem 16–7 to show that the rule which minimizes the Bayes risk rejects if $P(\boldsymbol{\Theta} \in N) \leq L_0/(L_0 + L_1)$. (Note that the Bayes risk of any test is finite. Why?)

2. Suppose that U is a random variable and $k(U)$ is a function such that $0 < E(k(U)) < \infty$, and $E(Uk(U)) < \infty$. Show that $g(a) = E(k(U)(U-a)^2)$ is minimized for $a = E(Uk(U))/E(k(U))$. (What is $g'(a)$?)

3. Consider estimating the parameter $\tau(\boldsymbol{\Theta})$ with the loss function $L(a,\theta) = k(\theta)(\tau(\theta) - a)^2$ where $k(\theta) > 0$. Let $d_0(\mathbf{X}) = E(k(\boldsymbol{\Theta})\tau(\boldsymbol{\Theta})|\mathbf{X})/E(k(\boldsymbol{\Theta})|\mathbf{X})$.
(a) Show that if $r(d_0) < \infty$, then $d_0(\mathbf{X})$ is the Bayes rule. (See Exercise B2.)
(b) Show that this rule is not unbiased unless $d(\mathbf{X}) \equiv \tau(\boldsymbol{\Theta})$. (*Hint*: Compute $E(k(\boldsymbol{\Theta})\tau(\boldsymbol{\Theta})d(\mathbf{X}))$ two ways.)

16.3 COMPARISON OF BAYESIAN AND CLASSICAL STATISTICS

16.3.1 The Likelihood Principle

Many principles have been suggested for statistical inference. We now present several such principles, show their relationships, and give some examples. These principles are sometimes quite deep, and it is necessary at the level of this book to give somewhat imprecise statements.

The *likelihood principle* has two parts:

a. Let $\mathbf{x}$ be a possible outcome of an experiment concerning a parameter $\mathbf{\Theta}$. Then any inference about $\mathbf{\Theta}$ should depend on the experiment only through $L_x(\mathbf{\Theta})$, the observed likelihood of $\mathbf{\Theta}$.
b. Let $\mathbf{x}$ and $\mathbf{y}$ be two outcomes of (possibly different) experiments about a common parameter $\mathbf{\Theta}$ such that the likelihood for $\mathbf{x}$ is proportional to the likelihood for $\mathbf{y}$. Then we should draw the same conclusions about $\mathbf{\Theta}$ for $\mathbf{x}$ and $\mathbf{y}$.

It is not immediately obvious why the likelihood principle should be followed. However, it can be derived from two other principles which we now mention. The first is the *sufficiency principle*, which says that any procedures for a model should depend only on a sufficient statistic for that model. This is a principle which most statisticians follow, as we have in this book.

Suppose in an experiment we will choose one of two sub-experiments to perform, each with probability .5. The *conditionality principle* says that in this case, any inference we make should depend only on the sub-experiment chosen. For example, suppose a homeowner is doing a test for radon gas in her house, and that her hardware store sells two different kits for such a test. She flips a fair coin to decide which kit to buy, takes it home, and uses it. The conditionality principle says that her conclusions should only depend on the analysis from the kit she has chosen, not what might have happened if she had chosen the other kit. In this setting, the conditionality principle is quite sensible. (The conditionality principle given in this paragraph is often called the weak conditionality principle.)

Theorem 16–11. The sufficiency principle and the conditionality principle imply the likelihood principle.

Proof. See Berger and Wolpert (1984), pp. 26–36. □

This theorem implies that if we believe the sufficiency principle and the conditionality principle, we are forced to believe the likelihood principle.

Any Bayesian inference based on the posterior distribution satisfies the likelihood principle, as we now show.

Theorem 16–12. Consider a Bayesian experiment with likelihood $L_X(\mathbf{\Theta})$ and prior distribution $\pi(\boldsymbol{\theta})$. Let $\pi^*(\boldsymbol{\theta}|\mathbf{x})$ be the posterior distribution of $\mathbf{\Theta}$ given $\mathbf{X}$. Then $\pi^*(\boldsymbol{\theta}|\mathbf{x})$ depends on $\mathbf{x}$ only through $L_x(\boldsymbol{\theta})$. If $L_x(\mathbf{\Theta})/L_y(\mathbf{\Theta})$ does not depend on $\mathbf{\Theta}$, then $\pi^*(\boldsymbol{\theta}|\mathbf{x}) = \pi^*(\boldsymbol{\theta}|\mathbf{y})$.

Proof. Suppose that $\mathbf{\Theta}$ is a continuous random vector. Then

$$\pi^*(\boldsymbol{\theta}|\mathbf{x}) = \frac{L_x(\boldsymbol{\theta})\pi(\boldsymbol{\theta})}{\int_\Omega L_x(\boldsymbol{\theta})\pi(\boldsymbol{\theta})d\boldsymbol{\theta}},$$

which depends on $\mathbf{x}$ only through $L_x(\mathbf{\Theta})$. Suppose that $L_y(\mathbf{\Theta}) = rL_x(\mathbf{\Theta})$. Then

$$\pi^*(\theta|\mathbf{y}) = \frac{rL_y(\boldsymbol{\theta})\pi(\boldsymbol{\theta})}{\int_\Omega rL_y(\boldsymbol{\theta})\pi(\boldsymbol{\theta})d\boldsymbol{\theta}} = \frac{L_x(\boldsymbol{\theta})\pi(\boldsymbol{\theta})}{\int_\Omega L_x(\boldsymbol{\theta})\pi(\boldsymbol{\theta})d\boldsymbol{\theta}} = \pi^*(\boldsymbol{\theta}|\mathbf{x}). \quad \square$$

We now state two very useful consequences of the likelihood principle. *The nonaveraging principle* says that any conclusions about $\boldsymbol{\Theta}$ should depend only on the data we actually observed, not on averages of data we might have observed but did not. (The nonaveraging principle is often called the weak likelihood principle.) The *stopping rule principle* says that any conclusions about $\boldsymbol{\Theta}$ should not depend on how it was decided to stop the experiment.

Theorem 16–13. The nonaveraging principle and the stopping rule principle follow from the likelihood principle.

Proof. The nonaveraging principle follows because the likelihood principle says that the inference should depend only on the likelihood for the observed data. A proof that the stopping rule principle follows from the likelihood principle is in Berger and Wolpert (1984), pp. 75–77, 86–88. □

We have seen how Bayesian analysis based on the posterior distribution satisfies the likelihood principle. However, Bayes risk, as defined in Section 16.2.2, averages the loss function over the sample space, and therefore does not satisfy the nonaveraging principle. Hence defining Bayes procedures as ones which minimize the Bayes risk is not compatible with the likelihood principle. That is why we have defined Bayes procedures in Section 16.1 in terms of the posterior distribution.

We now look at an example of how classical statistics often violates the nonaveraging principle.

Example 16–12

Suppose that 15 light bulbs are burned until they fail, and that the failure times, X_i, of the light bulbs are independently exponentially distributed with mean μ, the average lifetime for the particular brand. Then $\overline{X}$ is the best unbiased estimator for μ, and $30\overline{X}/\mu \sim \chi^2_{30}$, so that a .95 confidence interval for μ is $30\overline{X}/46.98 \leqq \mu \leqq 30\overline{X}/16.79$. The experimenter observes $\overline{X} = 100$, so that his point estimate is 100 and his .95 confidence interval is $63.85 \leqq \mu \leqq 178.68$. After analyzing the data, he learns the experiment would only have been conducted for a maximum of 250 hours and that any light bulb still burning after 250 hours would have gotten failure time 250 hours. He is not bothered by this, since all his light bulbs had burned out before 250 hours. However, if he really believed in unbiased estimators, he should be bothered. What he has observed is $U_i = \min(X_i, 250)$ and he is using $\overline{U}$ to estimate μ. Note that

$$EU_i = \int_0^{250} y\mu^{-1}\exp\left(\frac{-y}{\mu}\right)dy + \int_{250}^{\infty} 250\mu^{-1}\exp\left(\frac{-y}{\mu}\right)dy = \mu\left(1 - \exp\left[\frac{-250}{\mu}\right]\right).$$

Therefore, $E\overline{U} = \mu(1 - \exp[-250/\mu])$ and $\overline{U}$ is no longer an unbiased estimator of μ. Hence, if he insisted on using unbiased estimators, he could no longer use 100 as his estimate of μ. Similarly, $30\overline{U}/\mu$ does not have a χ^2 distribution, so that then the confidence interval defined above is no longer a 95% confidence interval.

The likelihood for μ depends only on the U_i observed, so that if all the $U_i \leqq 250$, then the likelihood is the same as if it had not been decided to terminate the experiment at 250. Hence any procedures compatible with the likelihood principle would draw the same conclusion as if the experiment had not been terminated. In particular, Bayes rules are unchanged by this termination at 250 hours, as long as all the bulbs burned out by 250 hours.

As another example of difficulties in classical statistics caused by averaging over the sample space, recall that in Section 11.5.2, we derived a size .05 test for μ which for some outcomes accepted a hypothesis which was known to be false.

What these examples illustrate is that to find a .95 confidence interval or unbiased estimator, it is necessary to determine what the estimator would be for values of the data which have not occurred. Similarly the size and power of a test or the variance and mean squared error of an estimator require averaging the procedure over **X**-values which have not actually been observed in the experiment, **X**-values which should be irrelevant. The nonaveraging principle seems even more appealing than the likelihood principle.

We now give an example of how classical statistics violates the stopping rule principle.

Example 16–13

Suppose we flip a coin with probability of heads Θ to see if $\Theta = .5$. Assume we flip it 4 times and get $X = 3$ heads. There are two different ways this experiment could have been performed. We could have flipped the coin 4 times and counted the heads, in which case $X \sim B(4, \Theta)$, or we could also have flipped until we got a tail, in which case X has a geometric distribution with parameter $1 - \Theta$. These are two different ways of stopping the experiment so that the stopping rule principle implies that we should come to the same conclusion about $\mathbf{\Theta}$ in either case. All that matters is that we observed 3 heads in 4 trials. For example, for any prior distribution on Θ, the posterior distribution is the same for the two models, so that Bayes procedures are the same for the two cases.

However, to a classical statistician, the two experiments should be analyzed differently. For example, in the first model, $P(X \geqq 3|\Theta = .5) = \frac{5}{16}$, while in the second model, $P(X \geqq 3|\theta = .5) = \frac{1}{8}$. Therefore, in the second model, we could reject the hypothesis that $\theta = .5$ with a size .2 test, but in the first model we could not reject the hypothesis with a .2 test.

To see that the stopping rule follows from the likelihood principle for this example, let $L_3(\Theta)$ and $\tilde{L}_3(\Theta)$ be the likelihoods for $X = 3$ for the binomial and geometric models. Then

$$L_3(\Theta) = \begin{bmatrix} 4 \\ 3 \end{bmatrix} \Theta^3(1 - \Theta), \qquad \tilde{L}_3(\Theta) = \Theta^3(1 - \Theta), \qquad \frac{L_3(\Theta)}{\tilde{L}_3(\Theta)} = \begin{bmatrix} 4 \\ 3 \end{bmatrix},$$

which does not depend on Θ.

In the exercises, inference for the rate λ in a Poisson process is considered. In this case, there are two obvious possibilities. One is to sample for a fixed time period t and observe the number X of accidents, so that $X \sim P(\lambda t)$. The other is to observe the time T until a fixed number of accidents x occur, so that $T \sim \Gamma(x, \lambda^{-1})$. According to the stopping rule principle, if $X = x$ and $T = t$, we should come to the same conclusions about λ. That is, the stopping rule principle implies that we should come to the same conclusion if we observed for 5 hours and had 10 arrivals or if we observed until the 10th arrival and waited 5 hours. According to the stopping rule principle, all that matters is that there were 10 arrivals in 5 hours.

One appealing aspect of the stopping rule principle is that it implies that we do not have to know the experimenter's true stopping rule, which is often difficult to determine. (What would we do if the experimenter disappeared, leaving only the information that there were 10 arrivals in 5 hours?)

One situation in which the stopping rule principle is especially appealing is in clinical trials. Suppose we have designed an experiment to compare a possible new drug with a control. Before the experiment is completed, however, it may be apparent that the new drug is much better. If we believe in the stopping rule principle, we can stop at that point and analyze the data as though we had planned to stop there all along. However, a classical statistician must complete the experiment as planned. All probabilities in classical statistics become irrelevant after we have looked at the data.

Some of classical statistics are compatible with the likelihood principle, as we now show.

Theorem 16–14. Consider a model with parameter $\boldsymbol{\theta}$.

a. The MLE $\hat{\boldsymbol{\theta}}$ satisfies the likelihood principle.
b. Let $\boldsymbol{\tau} = h(\boldsymbol{\theta})$ be a k-dimensional parameter and let $\Lambda(\mathbf{X})$ be the LRT statistic for testing $\boldsymbol{\tau} = \mathbf{0}$ against $\boldsymbol{\tau} \neq \mathbf{0}$. Then the test which rejects if

$$2 \log(\Lambda(\mathbf{X})) > \chi_k^{2.\,05}$$

satisfies the likelihood principle.

Proof

a. Let $\hat{\boldsymbol{\theta}}(\mathbf{x})$ be the MLE of $\hat{\boldsymbol{\theta}}$ when we observe the data $\mathbf{X} = \mathbf{x}$. Note first that $\hat{\boldsymbol{\theta}}(\mathbf{x})$ depends only on $L_x(\hat{\boldsymbol{\theta}})$, so that part a of the likelihood principle is satisfied. Suppose that $L_y(\boldsymbol{\theta}) = rL_x(\boldsymbol{\theta})$ for some $\mathbf{x}$, $\mathbf{y}$, and r. Note that $r > 0$. Let $\hat{\boldsymbol{\theta}}(\mathbf{x})$ be the MLE when $\mathbf{X} = \mathbf{x}$. Then

$$L_y(\hat{\theta}(\mathbf{x})) = rL_x(\hat{\theta}(\mathbf{x})) \geqq rL_x(\boldsymbol{\theta}) = L_y(\boldsymbol{\theta}) \quad \text{for all } \boldsymbol{\theta} \in \Omega$$

and hence $\hat{\theta}(\mathbf{x})$ is the MLE when $\mathbf{X} = \mathbf{y}$ also, by the definition of MLE. Therefore $\hat{\tau}$ is also the same when $\mathbf{X}$ is $\mathbf{x}$ or $\mathbf{y}$.
b. Note that $\chi_k^{2.\,05}$ depends only on the hypothesis tested, not the experiment or the data, and is hence compatible with the likelihood principle. If $\mathbf{X} = \mathbf{x}$, then $\Lambda(\mathbf{x})$ depends only on the observed likelihood $L_x(\boldsymbol{\theta})$. Suppose that for outcomes $\mathbf{y}$ and $\mathbf{x}$, $L_y(\boldsymbol{\theta}) = rL_x(\boldsymbol{\theta})$. Then

$$\hat{\theta}(\mathbf{x}) = \hat{\theta}(\mathbf{y}), \qquad \hat{\hat{\boldsymbol{\theta}}}(\mathbf{x}) = \hat{\hat{\boldsymbol{\theta}}}(\mathbf{y})$$

by part a. Therefore

$$\Lambda(\mathbf{y}) = \frac{L_y(\hat{\theta}(y))}{L_y(\hat{\hat{\theta}}(y))} = \frac{rL_x(\hat{\theta}(\mathbf{x}))}{rL_x(\hat{\hat{\theta}}(\mathbf{x}))} = \frac{L_x(\hat{\theta}(\mathbf{x}))}{L_x(\hat{\hat{\theta}}(x))} = \Lambda(\mathbf{x}),$$

and the test satisfies the likelihood principle. $\square$

Part b of Theorem 16–14 implies that the asymptotic size-.05 likelihood ratio test satisfies the likelihood criterion, at least in regular cases. (See Theorem 9–4.) Its motivation as a size .05 test does not satisfy the likelihood principle, since it involves averaging over the sample space. This phenomenon is similar to Bayesian analysis, in which a procedure defined in terms of the posterior distribution satisfies the likelihood principle, but its motivation in terms of Bayes risk does not. (The exact size α tests do not satisfy the likelihood principle because the critical values depend on the experiment chosen. See Example 16–13.)

Many asymptotic confidence intervals satisfy the likelihood principle, as we now indicate. For simplicity, suppose that θ is a univariate parameter, and let $\hat{\theta}$ and $I(\theta)$ be the MLE and Fisher information for θ. Let

$$i_X(\theta) = -\frac{\partial^2}{\partial\theta^2}\log(L_X(\theta)).$$

For many regular models

$$\frac{i_X(\hat{\theta})}{I(\theta)} \xrightarrow{P} 1.$$

(Some reasons for this are given in the exercises. Recall that for regular cases $\hat{\theta} \xrightarrow{P} \theta$.) By the discussion of Section 7.3.1, the interval

$$\theta \in \hat{\theta} \pm \frac{1.96}{(i_X(\hat{\theta}))} \tag{16–7}$$

is an asymptotic 95% confidence interval for θ in regular cases. Now, $\hat{\theta}$ and $i_X(\hat{\theta})$ satisfy the likelihood principle, so that this interval does also. (Its motivation as an asymptotic 95% confidence interval does not.) In the exercises, this interval is generalized to the case of vector $\boldsymbol{\theta}$. All the 95% or asymptotic 95% confidence intervals given in Chapters 11–14 (except those in Sections 11.1 and 11.5) equal this interval asymptotically. Therefore, these intervals satisfy the likelihood principle asymptotically. The exact confidence intervals do not satisfy the likelihood principle for the same reason that exact tests don't.

Exercises—A

1. Suppose that accidents on a stretch of highway occur as a Poisson process with unknown rate λ. Suppose that the highway is observed for t hours and X accidents occur so that $X \sim P(\lambda t)$.
   (a) Find the MLE, $\hat{\lambda}$, the best unbiased estimator, and the Bayes estimator for λ when $\lambda \sim \Gamma(3, 4)$.
   (b) Evaluate the estimators when the process is observed for 4.8 hours and 2 accidents occur.
   (c) Do an exact .048 test that $\lambda = 1$ against $\lambda < 1$ when $X = 2$, $t = 4.8$.
   (d) Show that an asymptotic 95% confidence interval for λ is $\lambda \in \hat{\lambda} \pm 1.96X^{1/2}/t$, as $t \to \infty$. Evaluate this interval when $X = 100$, $t = 200$.
2. In the setting of Exercise 1, suppose that the highway is observed until x accidents occur, so that the time $T \sim \Gamma(x, \lambda^{-1})$.
   (a) Find the MLE $\hat{\lambda}$ and best unbiased estimator for λ and the Bayes estimator when $\lambda \sim \Gamma(3, 4)$.

(b) Evaluate the estimators when the process is observed until 2 accidents occur and it takes 4.8 hours.

(c) Do an exact .05 test that $\lambda = 1$ against $\lambda < 1$ when $x = 2$, $T = 4.8$.

(d) Show that an asymptotic 95% confidence interval for λ is $\lambda \in \hat{\lambda} \pm 1.96x^{1/2}/T$, as $x \to \infty$. Evaluate this interval when $x = 100$, $T = 200$.

(Note that the MLE, Bayes estimator, and asymptotic confidence interval are the same, but the best unbiased estimator and the exact test are not.)

3. An experimenter wants to estimate the mean weight for male students. So he gets a sample of 16 students, weighs them, and finds a sample mean of 180 lbs. and a sample standard deviation of 30 lbs. Assume the X_i are a sample from a normal distribution with unknown mean and variance.

(a) Find a 95% confidence interval for the mean μ.

(b) Suppose his scale only goes to 250 lbs., but none of the students weighed that much. Is the confidence interval still valid? Why or why not?

Exercises—B

1. In Example 16–13, show that for any prior the posterior is the same when $X = 3$ for both the binomial and geometric models.

2. **(a)** Show that $i_X(\hat{\theta})$ satisfies the likelihood principle.

(b) Does $I(\theta)$ satisfy the likelihood principle? Why or why not?

3. Suppose θ is a univariate parameter and that $L_X(\theta) = h(\mathbf{X}) \exp[q(\theta) + d(\theta)T(\mathbf{X})]$, where $q(\theta)$ and $d(\theta)$ are twice differentiable and $d'(\theta) \neq 0$.

(a) By setting to 0 the derivative of $\log[L_X(\theta)]$, show that the MLE $\hat{\theta}$ satisfies $q'(\hat{\theta}) + d'(\hat{\theta})T(\mathbf{X}) = 0$.

(b) By recalling that the score function has mean 0, show that $E(q'(\theta) + d'(\theta)T(\mathbf{X})) = 0$.

(c) Show that $i(\theta) = -(q''(\theta) + d''(\theta)T(\mathbf{X}))$, and that $I(\theta) = -(q''(\theta) + d''(\theta)E(T(\mathbf{X}))$.

(d) Show that $i(\hat{\theta}) = I(\hat{\theta})$ in this case.

(Note that the $B(n, \theta)$, $NB(r, \theta)$, $P(t\theta)$, $\Gamma(a, \theta)$, $N(\theta, b^2)$, $N(a, \theta)$ densities are all in this form.)

Exercises—C

1. Let $X_1, \ldots, X_n$ be a sample from the density $f(x; \theta)$ (θ univariate). Let $I_n(\theta) = nI_1(\theta)$ be the Fisher information in $\mathbf{X}$, and let U_i be the second derivative of $\log(f(X_i; \theta))$ with respect to θ,

(a) Show that $\bar{U} = i_x(\theta)/n$, $EU_i = I_1(\theta)$.

(b) Use the weak law of large numbers to show that $i_x(\theta)/I_n(\theta) \xrightarrow{P} 1$.

2. Suppose that $\boldsymbol{\theta}$ is a vector parameter, with likelihood $L_X(\boldsymbol{\theta})$. Let $\hat{\boldsymbol{\theta}}$ be the MLE of $\boldsymbol{\theta}$, let $I(\boldsymbol{\theta})$ be the information matrix for this model and let $i_X(\theta)$ be minus the Hessian matrix of $\log(L_X(\boldsymbol{\theta}))$. Let $\tau = h(\boldsymbol{\theta})$ be a univariate parameter with MLE $\hat{\tau} = h(\hat{\theta})$. Let $\nabla h(\boldsymbol{\theta})$ be the gradient of $h(\boldsymbol{\theta})$ and let $v(\boldsymbol{\theta}) = (\nabla h(\boldsymbol{\theta}))'(i_X(\boldsymbol{\theta}))^{-1}(\nabla h(\boldsymbol{\theta}))$, $V(\boldsymbol{\theta}) = (\nabla h(\boldsymbol{\theta}))'(I(\boldsymbol{\theta}))^{-1}(\nabla h(\boldsymbol{\theta}))$. (Note that $'$ is transpose in this expression.) Suppose that $v(\hat{\boldsymbol{\theta}})/V(\boldsymbol{\theta}) \xrightarrow{P} 1$.

(a) Use the results of Section 7.3.2 to show that in regular case, an asymptotic 95% confidence interval for τ is $\tau \in \hat{\tau} \pm 1.96(v(\hat{\theta}))^{1/2}$.

(b) Show that this interval satisfies the likelihood principle.

3. Following Exercise C2, suppose that $L_X(\boldsymbol{\theta}) = h(\mathbf{X})\exp[q(\boldsymbol{\theta}) + (d(\boldsymbol{\theta}))'T(\mathbf{X})]$, where $q(\boldsymbol{\theta})$ and $d(\boldsymbol{\theta})$ are twice differentiable functions and the components of $\nabla d(\boldsymbol{\theta})$ are linearly independent. Show that $i(\hat{\theta}) = I(\hat{\theta})$ and $v(\hat{\theta})/V(\hat{\theta}) = 1$. (Note that all the nonuniform models in Chapters 11–14 are in this form.)

16.3.2 Further Comments

The classical and Bayesian approaches to statistics are completely different views of the subject. Even the words such as estimator, test, and parameter have different meanings in the two systems.

Classical statisticians often use Bayesian statistics, sometimes because they believe that the parameter is random (as in Example 16–10) and sometimes to suggest procedures whose classical properties are then determined. By a Bayesian statistician, we mean a statistician who claims that the Bayesian perspective on statistics is the only sensible one. We now discuss some relative advantages and disadvantages of the Bayesian and classical approaches to statistics.

One advantage of Bayesian analysis is that its concepts are easier to understand. Once we have the posterior distribution, Bayes estimators, intervals, and tests are defined in the obvious way, and it is dramatically easier to discuss testing from a Bayesian perspective than from a classical perspective. Also, it is very appealing to start with a prior distribution, observe some data, and use the data to update the prior to the posterior distribution, which is then used as the prior distribution for the next experiment.

When a Bayesian statistician computes a .95 interval for τ and gets $.3 \leqq \tau \leqq .4$, the probability that $.3 \leqq \tau \leqq .4$ after looking at the data is .95. This interpretation is the same as that most classical statisticians first make for a .95 confidence interval. However, the meaning of a classical .95 confidence interval is more problematic. All that can be said is that if we compute many .95 confidence intervals, then about 95% of them will be correct; we cannot say anything about any particular confidence interval. An illustration of the difficulty with classical analysis is given in Section 11.5.2, in which a .95 confidence interval is computed for τ which, for some $\mathbf{X}$, is sure to cover the true value of τ. An experimenter who brought data to a statistician for analysis might be somewhat surprised to be given an interval which the experimenter knows contains τ, but which the statistician can only say is 95% certain to contain τ. Although it seems unlikely that a statistician would often use the model in that section in practice, the fact that such anomalies occur for any model makes the concept of a classical confidence interval somewhat suspect.

Similarly, if a null hypothesis is rejected with a Bayes .05 test, then the probability that the null hypothesis is true (after observing the data) is .05. Again, this is the interpretation that many classical statisticians would like to make about classical size-.05 tests. However, the correct classical interpretation is that if we do many .05 tests, 95% will correctly accept the null hypothesis when it is true; no statement may be made about any particular .05 test. Again in Section 11.5.2, an example is given of a .05 test which accepts a hypothesis we know to be false from the data, which again might somewhat surprise the experimenter who brought the data to the statistician.

A second advantage of Bayesian statistics is that it is compatible with the likelihood principle, as discussed in the last section. One consequence of this is that a Bayes procedure does not depend on how the experiment was stopped. This advantage is especially useful in clinical trials, as discussed in Section 16.3.1.

The likelihood principle also implies that Bayes procedures are only dependent on what we have observed. Many classical concepts, such as unbiasedness, mean squared error, confidence coefficient, size, and power involve averages of the procedure over all possible observations. In practice, we typically want to discover what information there is in the observed data about the parameters. Therefore averaging a procedure over unobserved data points is often inappropriate. Example 16–12 in the last section illustrates how this averaging can sometimes be rather silly. In addition, it is this averaging that leads to the difficulty in explaining the classical concepts of size and confidence coefficient.

Although posterior distributions are often somewhat complicated to compute, especially for nonconjugate priors, Bayes procedures can often be computed numerically, because it is only necessary to compute the procedure for the particular realization $\mathbf{X}$ of the data which has actually occurred. For example, the Bayes estimator of $\tau(\mathbf{\Theta})$ is

$$T(\mathbf{X}) = (E\tau(\mathbf{\Theta})|\mathbf{X}) = \int_\Omega \tau(\boldsymbol{\theta})\pi^*(\boldsymbol{\theta}|\mathbf{X})d\boldsymbol{\theta} = \frac{\int_\Omega \tau(\boldsymbol{\theta})f(\mathbf{X}|\boldsymbol{\theta})\pi(\boldsymbol{\theta})d\boldsymbol{\theta}}{\int_\Omega f(\mathbf{X}|\boldsymbol{\theta})\pi(\boldsymbol{\theta})d\boldsymbol{\theta}},$$

and these two integrals can be computed numerically for a particular observation $\mathbf{X}$, since $f(\mathbf{X}|\boldsymbol{\theta})$, $\tau(\boldsymbol{\theta})$, and $\pi(\boldsymbol{\theta})$ are all known functions. Similarly,

$$P\left(\boldsymbol{\theta} \in N|\mathbf{X}\right) = \frac{\int_N f(\mathbf{X}|\boldsymbol{\theta})\pi(\boldsymbol{\theta})d\boldsymbol{\theta}}{\int_\Omega f(\mathbf{X}|\boldsymbol{\theta})\pi(\boldsymbol{\theta})d\boldsymbol{\theta}},$$

which can be again numerically computed.

As we have seen in the last section, MLEs, asymptotic likelihood ratio tests, and many asymptotic confidence intervals also satisfy the likelihood principle, and would share many of the advantages discussed in the last three paragraphs. In particular, since MLEs and likelihood ratio tests do not need to be determined for unobserved data, they may be computed numerically for the data observed. This fact allows computation of these procedures in many complicated models for which other procedures are unavailable.

A chief disadvantage of the Bayesian approach to statistical analysis is that it is often difficult to think of a parameter as a random variable. For example, if we perform an experiment to determine π, the ratio between the circumference and diameter of a circle, more accurately, it is hard to think of π as a random variable. For this reason, many Bayesians have embraced the subjective view of probability in which probability represents the observer's belief about the parameter. According to this view, where the prior is high is where the observer believes the parameter to "lie." For example, if it is known to five decimal places that π is 3.14159, then we might take as a prior for π a uniform distribution on the interval 3.141585 to 3.141595.

A second disadvantage of Bayesian statistics is that two people with different prior distributions have different posterior distributions. Therefore, one person's Bayesian analysis of a data set is of little help to another person who has different prior beliefs. Hence, if everyone did only Bayesian analysis, it would be difficult for scientists to assess the value of other scientists' work, and it might even be difficult for them to communicate.

Yet another disadvantage of Bayesian statistics is that it is "not objective." It would be hard to convince the Food and Drug Administration of the efficacy of a new medicine with a Bayesian analysis, which would be based on the company's prior belief that the medicine should be effective! Similarly, it is hard to imagine how journals would decide to publish papers of experiments based on Bayesian analysis, which could involve priors highly influenced by the experimenters' strong prior beliefs.

Several attempts have been made to determine "noninformative" priors that would be objective. The most obvious approach is to assume that the parameter is uniformly distributed over the parameter space. One difficulty with this approach is that it often leads to improper priors which are unacceptable to many people (e.g., priors like a uniform distribution of the whole real line, which cannot integrate to one). A second difficulty is illustrated by the one-sample normal model, in which it is not obvious whether to take a uniform prior on σ, on σ^2, or on $\delta = \sigma^{-2}$. Careful consideration of this example makes it clear that assuming a uniform distribution is actually a very strong assumption about the parameter. Another approach to finding objective priors is the Jeffreys prior, which lets $\pi(\theta) = (I(\theta))^{1/2}$, where $I(\boldsymbol{\theta})$ is the Fisher information. (If $\boldsymbol{\Theta}$ is a vector, then the Jeffreys prior takes $\pi(\boldsymbol{\theta}) = (\det(I(\boldsymbol{\theta})))^{1/2}$, where $I(\boldsymbol{\theta})$ is the Fisher information matrix.) However, the Jeffreys prior is typically improper and also leads to an unappealing choice even for the two-parameter normal model. (See Berger (1985), pp. 406–418, for an interesting discussion of where the Jeffreys prior goes wrong for this model.) For these reasons, it does not appear that there is any way to define objective priors for general statistical models.

An important advantage of classical statistics is that people have been using it for a long time, and many procedures have been developed which work reasonably well and which are understood by most scientists and evaluators of scientific work. Nonetheless, Berger (1985) has made the provocative argument that the only reason that classical procedures perform well in practice is because they are nearly Bayes procedures for fairly flat priors. However, these procedures may eventually need to be rejected in favor of Bayesian analyses as more complicated data sets are encountered, in the same way that Newtonian mechanics was eventually rejected for general relativity and quantum mechanics.

In practice, it seems that statistics is more of an art than a science. There are few experiments in which all the assumptions of an analysis (classical or Bayesian) are satisfied. Therefore, it is important to remain flexible, using whatever approach seems appropriate. If a person or organization is doing an experiment solely in order to learn something for him- or herself or for the organization, then a Bayesian analysis incorporating all of the person's or company's prior beliefs may be appropriate. If, on the other hand, the person or company wants to convince other people or organizations of the truth of his or her analysis, then a classical analysis would appear to be more appropriate.

APPENDIX

In this appendix, we review some of the mathematical background for the material in this text. We assume that readers are familiar with the basics of the algebra of real numbers and hence understand the notation

$$a + b, \qquad ab, \qquad a - b, \qquad \frac{a}{b}, \qquad a^b,$$

where a and b are elements of R, the set of real numbers, a/b is defined only if $b \neq 0$, and a^b is defined for all a when b is a positive integer and for all b when $a > 0$. Also,

$$0^b = 0 \quad \text{and} \quad a^0 = 1,$$

as long as $a \neq 0$ and $b \neq 0$ (0^0 is not defined), and

$$a^{1/2}$$

means the positive square root of a.

We also use the following notation. Let $a_1, \ldots, a_n$ be a finite set of real numbers. We write

$$\sum_{i=1}^{n} a_i \quad \text{and} \quad \prod_{i=1}^{n} a_i$$

for the numbers

$$\sum_{i=1}^{n} a_i = a_1 + a_2 + \cdots + a_n \quad \text{and} \quad \prod_{i=1}^{n} a_i = a_1 a_2 \ldots a_n.$$

When there is no ambiguity, we often use the simpler notation $\Sigma_i a_i$ and $\Pi_i a_i$. For example, if $n = 4$, $a_1 = 1$, $a_2 = 2$, $a_3 = 3$, and $a_4 = 4$, then

$$\sum_{i=1}^{4} a_i = \sum_i a_i = 1 + 2 + 3 + 4 = 10 \quad \text{and} \quad \prod_{i=1}^{4} a_i = \prod_i a_i = (1)(2)(3)(4) = 24.$$

We use the following facts about summations and products:

$$\sum_i ca_i = c \sum_i a_i, \qquad \sum_i (a_i + b_i) = \left(\sum_i a_i\right) + \left(\sum_i b_i\right),$$

$$\prod_i (a_i)^b = \left(\prod_i a_i\right)^b, \qquad \prod_i (a^{b_i}) = a^{(\Sigma_i b_i)}.$$

Now, let a and b be real numbers. We use the notation (a, b) for the interval of x's such that $a < x < b$, $(a, b]$ for the interval $a < x \leqq b$, $[a, b)$ for the interval $a \leqq x < b$, and $[a, b]$ for the interval $a \leqq x \leqq b$. We call these intervals *finite intervals*. We use $(-\infty, b)$ for the interval $x < b$, $(-\infty, b]$ for the interval $x \leqq b$, (a, ∞) for the interval $x > a$, and $[a, \infty)$ for the interval $x \geqq a$. We call these intervals *infinite intervals*.

A.1 COUNTING

A.1.1 Some Counting Formulas

If we have a k-stage experiment which has n_1 possible outcomes at the first stage, n_2 possible outcomes at the second stage for every possible outcome at the first stage, n_3 possible outcomes at the third stage for every pair of possible outcomes at the first two stages, etc., then the total number of possible outcomes for the experiment is

$$N = n_1 n_2 \ldots n_k = \prod_i n_i.$$

Let T be a set with n elements. Then the *number of ordered subsets* of T of size k is the number of distinct subsets of T with k points, where two subsets are distinct if they have different points or the same points in a different order. The *number of unordered subsets* of T of size k is the number of distinct subsets of T with k points, where two subsets are now distinct only if they have different points. For example, if T consists of the points 1, 4, and 7, and k is 2, then the distinct ordered subsets of T of size two are

$$\{1, 4\}, \quad \{1, 7\}, \quad \{4, 1\}, \quad \{4, 7\}, \quad \{7, 1\}, \quad \{7, 4\},$$

so that there are six distinct ordered subsets of T of size two. The distinct unordered subsets of T of size two are

$$\{1, 4\}, \quad \{1, 7\}, \quad \{4, 7\},$$

so that there are three distinct unordered subsets of T of size two. The distinct unordered subsets of T of any size are

$$\emptyset, \quad \{1\}, \quad \{4\}, \quad \{7\}, \quad \{1, 4\}, \quad \{1, 7\}, \quad \{4, 7\}, \quad \{1, 4, 7\}$$

(where $\emptyset$ is the set with no elements). There is therefore a total of eight unordered subsets of T.

Now, let $n > 0$ be a positive integer, and let *n factorial* ($n!$) and *zero factorial* ($0!$) be defined by

$$0! = 1, \qquad n! = n(n-1)! = n(n-1)(n-2)\cdots(2)(1) = \prod_{i \leqq n} i$$

(so that $3! = 3 \times 2 \times 1 = 6$).

Theorem A–1

a. The number of ordered subsets of size k of a set with n points is

$$_nP_k = n(n-1)\cdots(n-k+1) = \frac{n!}{(n-k)!}.$$

b. The number of unordered subsets of size k of a set with n points is

$$\begin{bmatrix} n \\ k \end{bmatrix} = \frac{n!}{k!(n-k)!}.$$

c. The total number of unordered subsets of any size of a set with n points is 2^n.

From Theorem A–1, the number of ordered subsets of size two of a set with three points is $_3P_2 = 6$, the number of unordered subsets is $\left[\begin{smallmatrix} 3 \\ 2 \end{smallmatrix}\right] = 3$, and the total number of unordered subsets is $2^3 = 8$. We call $_nP_k$ the number of *permutations* of n things taken k at a time. $\left[\begin{smallmatrix} n \\ k \end{smallmatrix}\right]$ is often written $_nC_k$ and is called the number of *combinations* of n things taken k at a time. $\left[\begin{smallmatrix} n \\ k \end{smallmatrix}\right]$ is also called a binomial coefficient, since it occurs in the binomial theorem, which says that

$$(a+b)^n = \sum_{k=0}^{n} \begin{bmatrix} n \\ k \end{bmatrix} a^k b^{n-k}.$$

Let S be a set, and let $B_1, B_2, \ldots, B_m$ be subsets of S. Then the B_i are a *partition* of S if each element in S is in exactly one of the B_i. In other words, a partition of S is a division of S into distinct subsets. Thus, if the set A consists of the four numbers 1, 2, 3, and 4, then the sets

$$\{1, 2\}, \qquad \{3\}, \qquad \{4\}$$

form a partition of A.

Theorem A–2. The number of ways of partitioning a set of n objects into m distinct subsets with k_1 objects in the first group, k_2 objects in the second subset, . . . , and k_m objects in the mth subset (where $n = k_1 + k_2 + \cdots + k_m$) is

$$\begin{bmatrix} & & n & \\ k_1 & k_2 & \cdots & k_m \end{bmatrix} = \frac{n!}{k_1!k_2!\ldots k_m!} = \frac{n!}{\prod_i k_i!}.$$

From Theorem A–2, if $m = 3$ and $n = i + j + k$, then

$$\begin{bmatrix} & n & \\ i & j & k \end{bmatrix} = \frac{n!}{i!j!k!},$$

and we obtain

$$\begin{bmatrix} & 9 & \\ 3 & 4 & 2 \end{bmatrix} = \frac{9!}{3!4!2!} = 1{,}260,$$

so that there are 1,260 ways to divide 9 objects into a set with 3 objects, a set with 4 objects, and a set with 2 objects.

The numbers

$$\begin{bmatrix} & & n & \\ k_1 & k_2 & \cdots & k_m \end{bmatrix}$$

are often called *multinomial coefficients*, since they occur in the multinomial theorem, which states that

$$(a_1 + a_2 + \cdots + a_m)^n = \sum \ldots \sum \begin{bmatrix} & & n & \\ k_1 & k_2 & \ldots & k_m \end{bmatrix} a_1^{k_1} a_2^{k_2} \ldots a_m^{k_m},$$

where this sum is taken over all non-negative integers $k_1, \ldots, k_m$ such that $k_1 + k_2 + \cdots + k_m = n$.

Finally, note that

$$\begin{bmatrix} & n \\ k & n-k \end{bmatrix} = \begin{bmatrix} n \\ k \end{bmatrix}.$$

A.2 CALCULUS

A.2.1 Functions

A *real-valued function* $f(x)$ on a set A is a rule which, to each point x in A, assigns the real number $f(x)$. (The set A is sometimes called the *domain* of $f(x)$.) When it causes no confusion, we often drop the term "real-valued" and call $f(x)$ a function.

For example, the function $f(x) = x^2$ is a real-valued function on the set R of all real numbers which assigns, to each number, its square. As another example, let Q be a finite set, and for each subset S of Q, let $h(S)$ be the number of points in S. Then $h(S)$ is a real-valued function on the set A of all subsets of Q. Further examples of functions and their properties are given in the next section.

One important property of functions is that the name for the variable is arbitrary. For example, the function $f(y) = y^2$ is the same as the function $f(x) = x^2$. In both cases, $f(7) = 49$.

Let $f(x)$ and $g(x)$ be two functions. We assume that the reader is also familiar with the functions $f(x) + g(x)$, $f(x) - g(x)$, $f(x)g(x)$, $f(x)/g(x)$, and $f(g(x))$. For example, if $f(x) = x^2$ and $g(x) = x^4 + 1$, then

$$f(x) + g(x) = x^2 + x^4 + 1, \qquad f(x)g(x) = x^2(x^4 + 1), \qquad \frac{f(x)}{g(x)} = \frac{x^2}{(x^4+1)},$$

$$g(f(x)) = (x^2)^4 + 1 = x^8 + 1, \qquad f(g(x)) = (x^4+1)^2 = x^8 + 2x^4 + 1.$$

Let A be a set and $f(x)$ be a real-valued function on A. We say that b is the *supremum* of $f(x)$ on A and write

$$b = \sup_{x \in A} f(x)$$

if b is the smallest number such that $f(x) \leqq b$ for all $x \in A$. If, in addition, there exists $a \in A$ such that $f(a) = b$, we say that b is the *maximum* of $f(x)$ on A. Similarly, c is the *infimum* of $f(x)$ on A ($c = \inf_{x \in A} f(x)$) if c is the largest number such that $f(x) \geqq c$ on A. And again, if, in addition, there exists $d \in A$ such that $f(d) = c$, we say that c is the *minimum* of $f(x)$ on A.

Example A–1

Let $f(x) = x^2$ and $A = (0, 2)$. Then $\sup_{x \in A} f(x) = 2^2 = 4$. Since there is no point $a \in (0, 2)$ such that $a^2 = 4$, 4 is not the maximum of $f(x)$ on A. (However, 4 is the maximum of $f(x)$ on $[0, 2]$.)

Let $a(n)$ be a real-valued function on the set of nonnegative integers. Then we say that $a(n)$ is a *sequence* and write a_n for $a(n)$. (Therefore, a sequence a_n is a rule which assigns, to each nonnegative integer n, the number a_n.) We often write

$$a_0, a_1, \ldots$$

for the sequence a_n. For example, the sequence $a_n = 2^n$ is $1, 2, 4, 8, 16, \ldots$. Sometimes a sequence has no a_0 term. For example, the sequence $a_n = n^{-1}, n = 1, 2, \ldots,$ is $1, \frac{1}{2}, \frac{1}{3}, \ldots$.

Let A be a set. If A has a finite number of elements, then A is a *finite set*. (Note that a finite interval is not a finite set.) If there is an invertible function from the positive integers to A, then A is a *countable set*. That is, A is countable if the elements of A can be written as a sequence with no duplications. We say that A is *uncountable* if A is not countable or finite. For example, the set consisting of the points 1, 2, and 3 is finite, the set consisting of the positive even integers is countable, and the set consisting of the points on the interval $(0, 1)$ is uncountable.

A.2.2 Limits and Continuity

Let $f(x)$ be a real-valued function from an interval of real numbers, and let a_n be a sequence of real numbers. We assume that the reader is familiar with the limits, if they exist, given by

$$\lim_{x \to a} f(x) = b \quad \text{and} \quad \lim_{n \to \infty} a_n = c.$$

We often say that $f(x)$ goes to b as x goes to a and that a_n goes to c as n goes to infinity and use the equivalent notations

$$f(x) \to b \text{ as } x \to a \quad \text{and} \quad a_n \to c \text{ as } n \to \infty.$$

We allow the possibility that a, b, or c may be infinity or minus infinity.

Example A–2

Suppose that $f(x) = (x+1)/x = 1 + x^{-1}$ for $x > 0$. Then

$$\lim_{x \to 1} f(x) = 2, \qquad \lim_{x \to 0} f(x) = 1 + \infty = \infty, \qquad \lim_{x \to \infty} f(x) = 1 + 0 = 1.$$

Now, let $a_n = (n+1)^{-1}$, $b_n = 2n$, and $c_n = a_n b_n = 2n(n+1)^{-1} = 2(1 + 1/n)^{-1}$. Then

$$\lim_{n \to \infty} a_n = 0, \qquad \lim_{n \to \infty} b_n = \infty, \qquad \lim_{n \to \infty} c_n = 2(1 + 0) = 2.$$

Let $f(x)$ be a function defined on an interval of the real line, and let a be a point in that interval. Then $f(x)$ is *continuous at a* if

$$f(x) \to f(a) \quad \text{as } x \to a.$$

We say that $f(x)$ is *continuous* if it is continuous at all points where it is defined. Most of the common functions that we encounter are continuous functions.

Let

$$P(x) = a_0 + a_1 x + \cdots + a_k x^k,$$

where $a_k \neq 0$. Then $P(x)$ is a continuous function on the entire real line. We say that

$P(x)$ is a *kth-degree polynomial*. If $k = 0$, then $P(x) = a_0$ is called a constant function. If $k = 1$, then $P(x) = a_0 + a_1x$ is called a *linear function*. If $k = 2$, then $P(x) = a_0 + a_1x + a_2x^2$ is called a *quadratic function*. (Third- and fourth-degree polynomials are called *cubic* and *quartic* functions, respectively.)

Let

$$e = \lim_{n\to\infty} (1 + n^{-1})^n = 2.71828\ldots,$$

and let

$$\exp(x) = e^x.$$

Then $\exp(x)$ is called the *exponential function* and is continuous on the entire real line. Some basic properties of the exponential function are as follows:

$$\exp(0) = 1, \qquad (\exp(a))^b = \exp(ba), \qquad \exp\left(\sum_i a_i\right) = \prod_i \exp(a_i),$$

$$\lim_{x\to-\infty} \exp(x) = 0, \qquad \lim_{x\to\infty} \exp(x) = \infty.$$

Now, let $\log(x)$ be the function defined by

$$y = \log(x) \Leftrightarrow x = \exp(y)$$

(where $\Leftrightarrow$ means "if and only if"). That is, $\log(x)$ is the number y such that $e^y = x$. Then $\log(x)$ is called the *natural logarithm* of x. Since the only logarithms we use in this book are natural logarithms, we call $\log(x)$ the *logarithm* of x. $\text{Log}(x)$ is defined only when x is positive and is continuous for all $x > 0$. Some basic properties of logarithms are the following:

$$\log(1) = 0, \qquad \log(a^b) = b\ \log(a), \qquad \log\left(\prod_i a_i\right) = \sum_i \log(a_i),$$

$$\lim_{x\to 0} \log(x) = -\infty, \qquad \lim_{x\to\infty} \log(x) = \infty.$$

Finally, we note that for any $a > 0$,

$$a^b = \exp[b\ \log(a)].$$

Therefore (by Theorem A–3, to follow),

$$f(x) = a^x = \exp[x\ \log(a)] \quad \text{and} \quad g(x) = x^a = \exp[a\ \log(x)]$$

are continuous functions wherever they are defined.

Occasionally, we use the trigonometric functions *sine* ($\sin(x)$), *cosine* ($\cos(x)$), and *tangent* ($\tan(x)$). We assume that the angle x is measured in radians. The functions $\sin(x)$ and $\cos(x)$ are continuous for all x in $(-\infty, \infty)$. $\text{Tan}(x)$ is continuous on the interval $(-\pi/2, \pi/2)$.

We also use the inverse trigonometric functions $\arcsin(x)$ and $\arctan(x)$. *Arcsin* (x) is the unique number y in the interval $[-\pi/2, \pi/2]$ such that $\sin(y) = x$ and is a continuous function on the interval $[-1, 1]$. *Arctan* (x) is the unique number y in the interval $(-\pi/2, \pi/2)$ such that $\tan(y) = x$ and is a continuous function on the interval $(-\infty, \infty)$. Note that $\arcsin(-x) = -\arcsin(x)$ and $\arctan(-x) = -\arctan(x)$.

We next give some basic results about continuous functions.

Theorem A–3. Let $f(x)$ and $g(x)$ be continuous functions, and let c be a real number. Then

$$cf(x), \qquad f(x)+g(x), \qquad f(x)-g(x), \qquad f(x)g(x), \qquad \frac{f(x)}{g(x)}, \qquad g(f(x))$$

are all continuous functions wherever they are defined.

Example A–3

Both $f(x) = x^2$ and $g(x) = \exp(x)$ are continuous functions if $x > 0$. Therefore,

$$f(x)g(x) = x^2 e^x, \qquad \frac{f(x)}{g(x)} = x^2 e^{-x}, \qquad g(f(x)) = \exp(x^2),$$

are also continuous functions. We can therefore conclude, among other things, that $x^2 e^x$ goes to $4e^2$ as x goes to 2.

Using Theorem A–3, we can see that many functions on the real line are continuous, so that many limits involving such functions can be evaluated using their continuity.

The following theorem is useful for evaluating limits of sequences.

Theorem A–4. Let a_n be a sequence such that $a_n \to a$ as $n \to \infty$, and let $f(x)$ be a function on an interval. Then

$$\lim_{n\to\infty} f(a_n) = \lim_{x\to a} f(x).$$

Example A–4

Let

$$a_n = \frac{4n^2+2n}{n^2+2} = \frac{4+2n^{-1}}{1+2n^{-2}} \to 4 \quad \text{as } n \to \infty,$$

and let $f(x) = \log(x)$. Then

$$\log(a_n) = \log\left(\frac{4n^2+2n}{n^2+2}\right) \to \log(4) \quad \text{as } n \to \infty,$$

since the logarithm function is continuous at $x = 4$.

A.2.3 Series

Let $a_0, a_1, \ldots,$ be a sequence of numbers. We assume that the student is familiar with the notion of the sum of an *infinite series*,

$$b = \sum_{k=0}^{\infty} a_k$$

if the limit b exists. (We allow the possibility that the limit b is infinite.)

Some basic properties of series are that

$$\sum_{k=0}^{\infty} (a_k + b_k) = \sum_{k=0}^{\infty} a_k + \sum_{k=0}^{\infty} b_k \quad \text{and} \quad \sum_{k=0}^{\infty} (ca_k) = c \sum_{k=0}^{\infty} a_k.$$

Also,

$$a_k \geqq 0 \Rightarrow \sum_{k=0}^{\infty} a_k \geqq 0 \quad \text{and} \quad b_k \geqq c_k \Rightarrow \sum_{k=0}^{\infty} b_k \geqq \sum_{k=0}^{\infty} c_k,$$

with equality only if $a_k = 0$ or $b_k = c_k$ for all k. When there is no confusion, we often use the simpler notation

$$\sum_k a_k = \sum_{k=0}^{\infty} a_k.$$

Note that

$$a_k = 0 \text{ for } k > n \Rightarrow \sum_{k=0}^{\infty} a_k = \sum_{k=0}^{n} a_k,$$

so that finite sums are special cases of series.

One important property of series is that the index of summation is irrelevant. That is,

$$\sum_k a_k = \sum_n a_n = \sum_i a_i.$$

Some sums and series used in this text are as follows:

1. $\sum_{k=0}^{\infty} a^k = 1/(1-a)$ if $|a| < 1$ (geometric series).
2. $\sum_{k=0}^{n} \binom{n}{k} a^k b^{n-k} = (a+b)^n$ for all a and b (binomial series).
3. $\sum_{k=0}^{\infty} \binom{r+k-1}{k} a^k = (1-a)^{-r}$ for $|a| < 1$ (negative binomial series).
4. $\sum_{k=0}^{\infty} a^k/k! = \exp(a) = e^a$ for all a (exponential series).
5. $\sum_{k=1}^{n} k = n(n+1)/2, \sum_{k=1}^{n} k^2 = n(n+1)/(2n+1)/6$.

To generalize somewhat, let $a_0, a_1, \ldots$ be a sequence of objects (possibly not in R) and let A be the set of the a_k. Let $f(x)$ be a real-valued function on A. Then we write

$$\textstyle\sum_A f(x) = \displaystyle\sum_{k=0}^{\infty} f(a_k) = f(a_0) + f(a_1) + \cdots.$$

If A is a finite set with elements $a_0, a_1, \ldots, a_n$, we take $a_k = 0$ for $k > n$ and get

$$\sum_A f(x) = \sum_{k=0}^{n} f(a_k) = f(a_0) + f(a_1) + \cdots + f(a_n).$$

A.2.4 Derivatives

Throughout this section and the next, we assume that $f(x)$ is a real-valued function on an interval of the real line. We also assume that the reader is familiar with the definition of $f'(b)$, the *derivative* of $f(x)$ at $x = b$ (if it exists). We say that $f(x)$ is *differentiable at* b if $f'(b)$ exists and that $f(x)$ is *differentiable* if $f'(x)$ exists for all x at which $f(x)$ is defined. We also use the notation

$$f'(x) = \frac{d}{dx} f(x) \quad \text{and} \quad f'(b) = \left. \frac{d}{dx} f(x) \right|_{x=b}.$$

That is,

$$f'(3) = \frac{d}{dx} f(x)\bigg|_{x=3}.$$

The derivative of $f(x)$ at b represents the "instantaneous slope" of $f(x)$ at b.

Theorem A–5. If $f(x)$ is differentiable at b, then $f(x)$ is continuous at b.

Theorem A–5 gives an easy way to show that many functions are continuous. The next theorem is quite helpful in finding derivatives.

Theorem A–6. Let $f(x)$ and $g(x)$ be differentiable, and let c be a real number.

a. Let $h(x) = cf(x)$. Then $h'(x) = cf'(x)$.
b. Let $s(x) = f(x) + g(x)$. Then $s'(x) = f'(x) + g'(x)$. (The sum rule).
c. Let $p(x) = f(x)g(x)$. Then $p'(x) = f(x)g'(x) + f'(x)g(x)$. (The product rule).
d. Let $q(x) = f(x)/g(x)$. If $g(x) \neq 0$, then $q'(x) = [f'(x)g(x) - f(x)g'(x)]/(g(x))^2$. (The quotient rule).
e. Let $c(x) = g(f(x))$. Then $c'(x) = g'(f(x))f'(x)$. (The chain rule).

The following table gives some commonly used derivatives:

$f(x)$	$f'(x)$
c	0
x^b	bx^{b-1}
$\exp(x)$	$\exp(x)$
$\log(bx)$	x^{-1}
$\sin(x)$	$\cos(x)$
$\cos(x)$	$-\sin(x)$
$\tan(x)$	$(\cos(x))^{-2}$
$\arctan(x)$	$(1+x^2)^{-1}$
$\arcsin(x)$	$(1-x^2)^{-1/2}$

Using the formulas in the table together with Theorem A–6, we can differentiate any differentiable function encountered in this book.

The *second derivative* $f''(b)$ of $f(x)$ at b (if it exists) is the derivative at b of the first derivative $f'(x)$. (Note that for $f''(b)$ to exist, $f'(x)$ must exist for x in an interval about b.) Similarly, the *kth derivative* of $f(x)$ at b, written $f^{(k)}(b)$, is the derivative at b of the $(k-1)$st derivative of $f(x)$. Therefore,

$$f''(x) = \frac{d}{dx} f'(x) \quad \text{and} \quad f^{(k)}(x) = \frac{d}{dx} f^{(k-1)}(x).$$

We also use the notation

$$f^{(k)}(b) = \frac{d^k}{dx^k} f(x)\bigg|_{x=b} \quad \text{and} \quad f^{(k)}(x) = \frac{d^k}{dx^k} f(x).$$

Example A–5

Let $k(x) = \exp(-x^2/2)$, $g(x) = \exp(x)$, and $f(x) = -x^2/2$. Then, by the chain rule,

$$k(x) = g(f(x)) \quad \text{and} \quad k'(x) = f'(x)g'(f(x)) = (-x)\exp(-x^2/2).$$

By the product rule,

$$k''(x) = \frac{d}{dx}(-x)\exp\left(\frac{-x^2}{2}\right) = (-x)\frac{d}{dx}\exp\left(\frac{-x^2}{2}\right) + \exp\left(\frac{-x^2}{2}\right)\frac{d}{dx}(-x)$$

$$(-x)(-x)\exp\left(\frac{-x^2}{2}\right) + (-1)\exp\left(\frac{-x^2}{2}\right) = (x^2 - 1)\exp\left(\frac{-x^2}{2}\right).$$

Let $f(x)$ be a function defined on an interval, and let M be a point in that interval. We say that $f(M)$ is a *maximum* of $f(x)$ on the interval if $f(M) \geqq f(x)$ for any x on the interval, and $f(M)$ is the *unique maximum* if $f(M) > f(x)$ for all $x \neq M$. We define a *minimum* and the *unique minimum* in a similar way. Note that there may be no maximum or minimum of $f(x)$. If $f(m)$ is a maximum or minimum of $f(x)$, we often say, somewhat imprecisely, that m is the maximum or minimum of $f(x)$.

Theorem A–7. If $f(x)$ is a differentiable function on the (possibly infinite) open interval (a, b), and if $f(m)$ is a minimum or maximum of $f(x)$ on that interval, then $f'(m) = 0$.

Points q such that $f'(q) = 0$ are often called *critical points* of f.

Example A–6

Let $f(x) = 3x^2 + 12x + 2$, $-\infty < x < \infty$. A brief sketch of this function indicates that $f(x)$ is unbounded as $x \to \pm\infty$ and, hence, that $f(x)$ has no maximum on the given interval. However, it is apparent that $f(x)$ has a minimum, which we now find. We set

$$0 = f'(x) = 6x + 12,$$

so that $m = -\frac{12}{6} = -2$ is the location of the minimum, and the minimum value of the function is $f(-2) = -10$.

The following result is often useful for evaluating limits of the form 0/0, and ∞/∞.

Theorem A–8 (L'Hospital's rule) Let $f(x)$ and $g(x)$ be functions such that $f(x) \to 0$ and $g(x) \to 0$ as $x \to a$, or $f(x) \to \pm\infty$ and $g(x) \to \pm\infty$ as $x \to a$. Then

$$\lim_{x \to a} \frac{f(x)}{g(x)} = \lim_{x \to a} \frac{f'(x)}{g'(x)}.$$

Example A–7

Both $\sin(x)$ and x go to zero as x goes to zero. Therefore,

$$\lim_{x \to 0} \frac{\sin(x)}{x} = \lim_{x \to 0} \frac{\cos(x)}{1} = 1.$$

We now use L'Hospital's to derive a result that is used several times in the text.

Theorem A–9. $\lim_{n\to\infty}(1 + a/n)^n = e^a = \exp(a)$.

Proof. As x goes to zero, $\log(1 + ax)$ goes to $\log(1) = 0$. Therefore, by L'Hospital's rule,

$$\lim_{x\to 0}\frac{\log(1+ax)}{x} = \lim_{x\to 0}\frac{(1+ax)^{-1}a}{1} = a.$$

Since the exponential function is continuous,

$$\lim_{x\to 0}(1+ax)^{1/x} = \lim_{x\to 0}\exp\left[\frac{\log(1+ax)}{x}\right]$$

$$= \exp\left[\lim_{x\to 0}\frac{\log(1+ax)}{x}\right] = \exp(a) = e^a.$$

Now, $1/n \to 0$ as $n\to\infty$. Hence,

$$\lim_{n\to\infty}\left(1+\frac{a}{n}\right)^n = \lim_{x\to 0}(1+ax)^{1/x} = e^a. \quad \square$$

We say that a function $f(x)$ is *increasing* if

$$a > b \Rightarrow f(a) > f(b)$$

and *nondecreasing* if

$$a > b \Rightarrow f(a) \geqq f(b).$$

(Thus, every increasing function is also nondecreasing.) We also say that $f(x)$ is *decreasing* if $-f(x)$ is increasing and is *nonincreasing* if $-f(x)$ is nondecreasing. We say that $f(x)$ is *monotone* if it is decreasing or increasing.

The function $f(x)$ is *invertible* if it has an inverse function $g(y)$, i.e., a function $g(y)$ which satisfies

$$f(x) = y \Leftrightarrow x = g(y).$$

We call $g(y)$ the *inverse* of $f(x)$. For example, $g(y) = \log(y)$ is the inverse function of $f(x) = \exp(x)$. Note that a continuous function is invertible if and only if it is monotone.

Example A–8

The function $f(x) = x^3$ is increasing and therefore invertible. If $h(x) = x^2$, then $h(x)$ is not monotone and is therefore not invertible on the whole line. (Of course, we already know this, because if $y = x^2$, then $x = \pm y^{1/2}$, which is not a function, since it assigns two x's to each y.) However, $h(x)$ is increasing on the set in which x is positive, and so it is invertible on that set. (This is again obvious, since if for $x > 0$, $y = x^2$, then $x = y^{1/2}$, the positive square root of y.)

A.2.5 Integrals

Let $f(x)$ be a real-valued function defined on the interval (a, b). Then the *integral* of $f(x)$ between the limits a and b (often called the *definite* integral of $f(x)$) is

$$I = \int_a^b f(x)dx.$$

(We allow the possibility that a and b are infinite and that the integral is infinite.)

Following are some basic properties of integrals:

$$\int_a^b cf(x)dx = c\int_a^b f(x)dx, \qquad \int_a^b (f(x)+g(x))dx = \int_a^b f(x)dx + \int_a^b g(x)dx,$$

$$f(x) \geqq 0 \Rightarrow \int_a^b f(x)dx \geqq 0, \qquad f(x) \geqq g(x) \Rightarrow \int_a^b f(x)dx \geqq \int_a^b g(x)dx.$$

Note that the variable of integration in the integral is irrelevant. That is,

$$\int_a^b f(x)dx = \int_a^b f(y)dy = \int_a^b f(t)dt.$$

Now, let $F(x)$ be a function on the set $[a, b]$. We define

$$F(x)|_a^b = F(b) - F(a).$$

If $F(x)$ is not defined at a or b or both, we use the same formula with

$$F(a) = \lim_{x\to a} F(x) \quad \text{and} \quad F(b) = \lim_{x\to b} F(x),$$

provided those limits exist. Again, we allow the possibility that a or b is infinite and that $F(b)$ or $F(a)$ or both are infinite. We also have

$$\infty - c = d - (-\infty) = \infty, \quad -\infty - c = d - \infty = -\infty, \quad \infty - (-\infty) = \infty, \quad -\infty - \infty = -\infty.$$

If $F(b)$ and $F(a)$ are both either infinity or minus infinity, then $F(b) - F(a)$ does not exist.

The following is the basic theorem on integration.

Theorem A–10 (The fundamental theorem)

a. Let $F(x)$ and $f(x)$ be defined on (a, b), and suppose that $f(x) = F'(x)$ on that interval. Then

$$\int_a^b f(x)dx = F(x)\Big|_a^b.$$

b. Suppose that $f(x)$ is continuous at a and

$$F(x) = \int_a^x f(t)dt.$$

Then $F'(a) = f(a)$.

If $F(b) - F(a)$ is finite, we say that the integral is *finite*, and if $F(b) - F(a) = \pm\infty$, we say that the integral is *infinite*. If $F(b) - F(a) = \infty - \infty$ or $(-\infty) - (-\infty)$, we say that the integral *does not exist*.

A function $F(x)$ such that $F'(x) = f(x)$ is often called an *indefinite integral* of $f(x)$ or a *primitive* of $f(x)$. If $F'(x) = f(x)$ and $G(x) = F(x) + c$, then $G'(x) = F'(x)$, so that the indefinite integral is not uniquely defined. However, the definite integral is still uniquely defined, since $G(b) - G(a) = F(b) - F(a)$. A short table of functions $F(x)$ and $f(x)$ such that $F'(x) = f(x)$ is given in the last section. We can use the fundamental theorem together with this table to evaluate many integrals.

Example A–9

Let $f(x) = 3x^2$ and $F(x) = x^3$. Then $F'(x) = f(x)$. Therefore,

$$\int_0^1 3x^2\,dx = x^3\Big|_0^1 = 1 - 0 = 1 \quad \text{and} \quad \int_{-\infty}^0 3x^2\,dx = x^3\Big|_{-\infty}^0 = 0 - (-\infty) = \infty,$$

so that integrals with infinite limits can be infinite. But also,

$$\int_0^\infty 2\exp(-2x)dx = -\exp(-2x)\Big|_0^\infty = 0 - (-1) = 1,$$

so that integrals with infinite limits can be finite. Moreover,

$$\int_0^1 x^{-1}\,dx = \log(x)\Big|_0^1 = 0 - (-\infty) = \infty,$$

so that integrals with finite limits can be infinite. Finally,

$$\int_{-\infty}^\infty 2x dx = x^2\Big|_{-\infty}^\infty = \infty - \infty,$$

which does not exist.

The following theorem is often helpful in evaluating integrals.

Theorem A–11

a. (Integration by parts) Let $F(x)$ and $G(x)$ be differentiable functions on (a, b). Then

$$\int_a^b F(x)G'(x)dx = F(x)G(x)\Big|_a^b - \int_a^b F'(x)G(x)dx.$$

b. (Substitution) Let $x = g(y)$ be a differentiable invertible function. Then

$$\int_a^b f(g(y))g'(y)dy = \int_{g(a)}^{g(b)} f(x)dx.$$

The next two examples illustrate how to apply the rules of Theorem A–11 to evaluate integrals.

Example A–10

Let $0 < a < b < \infty$. We first evaluate

$$\int_0^1 \log(x)dx,$$

using integration by parts. Let $F(x) = \log(x)$ and $G'(x) = 1$, so that $F'(x) = x^{-1}$ and $G(x) = x$. Then

$$I = x\ \log(x)\Big|_0^1 - \int_0^1 x(x^{-1})dx = (x\ \log(x) - x)\Big|_0^1 = -1 - \lim_{a\to 0} a(\log(a) - 1).$$

Now, $\log(a) - 1$ goes to minus infinity and a^{-1} goes to infinity as a goes to zero. Therefore, we can use L'Hospital's rule to evaluate the rightmost limit, getting

$$\lim_{a\to 0}\left(\frac{\log(a) - 1}{a^{-1}}\right) = \lim_{a\to 0}\left(\frac{(a^{-1})}{-a^{-2}}\right) = \lim_{a\to 0}(-a) = 0.$$

Hence, $I = -1 - 0 = -1$.

Example A–11

We now evaluate

$$I = \int_0^1 x \exp\left(\frac{-x^2}{2}\right)dx.$$

Let $y = g(x) = -x^2/2$, so that $dy = g'(x)dx = -xdx$. Then

$$I = -\int_0^1 \exp\left(\frac{-x^2}{2}\right)(-xdx) = -\int_0^{-1/2} \exp(y)dy = -\exp(y)\Big|_0^{-1/2} = -(e^{-1/2} - 1).$$

This problem shows that when we use substitution, it can happen that the lower value $g(a)$ is bigger than the upper value $g(b)$. However, we can still use the same method for evaluating the integral.

Integration is much more difficult than differentiation. In fact, it has been shown that one of the most important functions in this book,

$$N(x) = \int_{-\infty}^{x} (2\pi)^{-1/2} \exp\left(\frac{-x^2}{2}\right)dx,$$

cannot be written as any combination of powers, exponentials, logs, and trigonometric functions. That is, $N(x)$ has no formula. For this reason, $N(x)$ is in Table 1 in the back of the book.

The next theorem gives one more property of integrals which is often useful in computations.

Theorem A–12. Let $a_0 < a_1 < \cdots < a_n$. Then

$$\int_{a_0}^{a_n} f(x)dx = \sum_{k=1}^{n} \int_{a_{k-1}}^{a_k} f(x)dx.$$

As a particular example of Theorem A–12, we have

$$\int_1^4 f(x)dx = \int_1^2 f(x)dx + \int_2^3 f(x)dx + \int_3^4 f(x)dx.$$

Example A–12

$$\int_{-2}^{1} |x|dx = \int_{-2}^{0} |x|dx + \int_0^1 |x|dx = \int_{-2}^{0} -xdx + \int_0^1 xdx = \frac{-x^2}{2}\Big|_{-2}^{0} + \frac{x^2}{2}\Big|_0^1 = 2.5.$$

Now, let A be the set of all x such that $-1 < x < 2$ or $4 < x < 7$. Using the idea given in Theorem A–12, we define

$$\int_A f(x)dx = \int_{-1}^{2} f(x)dx + \int_4^7 f(x)dx.$$

In a similar way, we could define

$$\int_A f(x)dx$$

for any set A which is a disjoint union of finitely many intervals. In fact, this definition can be extended to a larger class, called *Borel sets*. In this text, we shall act as though we know what this definition means for sets in general, but shall only evaluate the integrals for intervals or disjoint unions of intervals.

A.2.6 Partial Derivatives and Multiple Integrals

In this section, we discuss some results in the calculus of several variables. A *bivariate vector* is an ordered pair (x, y) of real numbers. (Note that $(1, 3)$ is a different pair from $(3, 1)$.) We often think of (x, y) as representing a point in the plane, and we call the set of all bivariate vectors R^2.

Let $f(x, y)$ be a real-valued function on a set A in R^2. Then the *partial derivatives*

$$\frac{\partial^{j+k}}{\partial x^j \partial y^k} f(x, y) \quad \text{and} \quad \left.\frac{\partial^{j+k}}{\partial x^j \partial y^k} f(x, y)\right|_{(a, b)}$$

respectively represent the function we get when we differentiate $f(x, y)$ k times with respect to y and then j times with respect to x and the function we get when we substitute a and b for x and y in the first expression. (For all the functions in this book, the order of differentiation is irrelevant, so that we could differentiate j times with respect to x and then k times with respect to y.)

Example A–13

Let $f(x, y) = x^2 y \log(y)$. Then

$$\frac{\partial}{\partial y} x^2 y \log(y) = x^2(\log(y) + 1),$$

$$\frac{\partial^2}{\partial x \partial y} x^2 y \log(y) = \frac{\partial}{\partial x} x^2(\log(y) + 1) = 2x(\log(y) + 1),$$

and

$$\frac{\partial^3}{\partial x \partial y^2} x^2 y \log(y) = \frac{\partial}{\partial y} 2x(\log(y) + 1) = \frac{2x}{y}.$$

Therefore,

$$\left.\frac{\partial^3}{\partial x \partial y^2} x^2 y \log(y)\right|_{(7, 11)} = \frac{2(7)}{11} = \frac{14}{11}.$$

The value $f(a, b)$ is a *maximum* of $f(x, y)$ on a set A if

$$f(a, b) \geqq f(x, y)$$

for all (x, y) in A and is a *unique maximum* for the strict inequality $(x, y) \neq (a, b)$. A *minimum* and *unique minimum* are defined similarly.

Theorem A–13. Let A be an open set. If $f(a, b)$ is a maximum or minimum of $f(x, y)$ on A, then

$$\left.\frac{\partial}{\partial x} f(x, y)\right|_{(a, b)} = 0 \quad \text{and} \quad \left.\frac{\partial}{\partial y} f(a, b)\right|_{(a, b)} = 0.$$

Example A–14

Let $f(x, y) = x^2 + 2x + 3y + 4xy + y^2$. If we set

$$0 = \frac{\partial}{\partial x}(x^2 + 2x + 3y + 4xy + y^2) = 2x + 2 + 4y$$

and

$$0 = \frac{\partial}{\partial y}(x^2 + 2x + 3y + 4xy + y^2) = 3 + 4x + 2y$$

and solve, we see that the only solution is $(x, y) = (-\frac{2}{3}, -\frac{1}{6})$. Substitution of values around this point then indicates that $f(-\frac{2}{3}, -\frac{1}{6})$ is a minimum of $f(x, y)$. (Note that $f(x, y)$ goes to infinity as x or y goes to infinity.)

Let A be a subset of R^2. Then the *double integral* over A is given by

$$\iint_A f(x, y)dxdy = \iint_A f(x, y)dydx.$$

Example A–15

Let $f(x, y) = 8xy^2$ and $A = \{(x, y): 0 < x < y < 1\}$. Then x can be any value from zero to one, and for each x, y can be any value from x to one. Therefore,

$$\iint_A 8xy^2\,dxdy = \int_0^1 \left[\int_x^1 8xy^2\,dy\right]dx = \int_0^1 8x\left[\int_x^1 y^2\,dy\right]dx = \int_0^1 8x\left[\frac{y^3}{3}\bigg|_x^1\right]dx$$

$$= \left(\frac{8}{3}\right)\int_0^1 (x - x^4)dx = \left(\frac{8}{3}\right)\left(\frac{x^2}{2} - \frac{x^5}{5}\right)\bigg|_0^1 = \left(\frac{8}{3}\right)\left(\frac{3}{10}\right) = \frac{4}{5}.$$

In a similar way, y can take any value from zero to one, and for each fixed y, x can take any value from zero to y. Therefore,

$$\iint_A 8xy^2\,dxdy = \int_0^1 \left[\int_0^y 8xy^2\,dx\right]dy = \int_0^1 8y^2\left[\frac{x^2}{2}\bigg|_0^y\right]dy$$

$$= \int_0^1 4y^4\,dy = \frac{4y^5}{5}\bigg|_0^1 = \frac{4}{5}.$$

Notice that when we change the order of integration, we cannot just change the limits of integration; we have to go back to the original region and determine new limits.

We say that a real-valued function $f(x, y)$ is *continuous at* (a, b) if

$$\lim_{(x,y)\to(a,b)} f(x, y) = f(a, b).$$

Also, $f(x, y)$ is *continuous* if it is continuous at all points at which it is defined. Most common functions are continuous. In particular, the functions

$$ax + by, \qquad cx^a y^b, \qquad x^y$$

are all continuous functions.

A *k-dimensional vector* is a set of k ordered numbers $(x_1, \ldots, x_k)$, often written $\mathbf{x} = (x_1, \ldots, x_k)$. Throughout this book, we use boldface letters for vectors and italic letters for their components, which are real numbers or variables. R^k is the set of all k-dimensional vectors.

Let $f(\mathbf{x})$ be a real-valued function on a subset of R^k. We assume that the reader is familiar with partial derivatives of $f(\mathbf{x})$ and the notation for them analogous to that presented above for the case $k = 2$.

Theorem A–14 can be extended in the obvious way to functions of three or more variables.

Let A be a subset of R^k. Then multiple integration of a vector-valued function is given by

$$\int_A f(\mathbf{x})d\mathbf{x} = \int\int \ldots \int_A f(x_1, x_2, \ldots, x_k)dx_1\, dx_2 \ldots dx_k,$$

where the right-hand side is computed as an iterated integral.

Example A–16

Let $k = 3$, $f(x_1, x_2, x_3) = \exp(-x_1 - 2x_2 - 3x_3)$, and A be the set of $\mathbf{x}$ such that $0 < x_1 < x_2 < x_3$. Then $0 < x_1$, $x_1 < x_2$, and $x_2 < x_3$. Therefore,

$$\begin{aligned}\int_A f(\mathbf{x})d\mathbf{x} &= \int\int\int_A f(x_1, x_2, x_3)dx_1\, dx_2\, dx_3 \\ &= \int_0^\infty \exp(-x_1)\left[\int_{x_1}^\infty \exp(-2x_2)\left[\int_{x_2}^\infty \exp(-3x_3)dx_3\right]dx_2\right]dx_1 \\ &= \int_0^\infty \exp(-x_1)\left[\int_{x_1}^\infty \exp(-2x_2)\left[-\frac{1}{3}\exp(-3x_3)\Big|_{x_3=x_2}^{x_3=\infty}\right]dx_2\right]dx_1 \\ &= \int_0^\infty \exp(-x_1)\left[\int_{x_1}^\infty \frac{1}{3}\exp(-5x_2)dx_2\right]dx_1 \\ &= \int_0^\infty \exp(-x_1)\left(\frac{-1}{15}\right)\left[\exp(-5x_2)\Big|_{x_2=x_1}^{x_2=\infty}\right]dx_1 \\ &= \int_0^\infty \frac{1}{15}\exp(-6x_1)dx_1 = \frac{-1}{90}\exp(-6x_1)\Big|_{x_1=0}^{x_1=\infty} = \frac{1}{90}.\end{aligned}$$

A.3 MATRICES

A.3.1 Matrix Algebra

A *matrix* is a rectangular array of numbers whose *rows* go across the page and whose *columns* go down the page. We say that a matrix $\mathbf{A}$ is a $k \times p$ matrix if it has k rows and p columns. For example, the matrix

$$\mathbf{A} = \begin{Bmatrix} 2 & 3 & 4 \\ 1 & 5 & 6 \end{Bmatrix}$$

is a 2×3 matrix whose first row is $\{2, 3, 4\}$ and whose first column is $\{{2 \atop 1}\}$. We call the component of A in the ith row and jth column the (i, j)th component of $\mathbf{A}$ and write $\mathbf{A} = (a_{ij})$ to mean that $\mathbf{A}$ has (i, j)th component a_{ij}. For the preceding matrix $\mathbf{A}$, $a_{21} = 1$.

If $\mathbf{a}$ is a $k \times 1$ matrix, we say that $\mathbf{a}$ is a k-dimensional *column vector*, and if $\mathbf{b}$ is a $1 \times p$ matrix, we say that $\mathbf{b}$ is a p-dimensional *row vector*. We respectively define the *length* of the column vector **a** and the row vector **b** by

$$\|\mathbf{a}\| = (\mathbf{a}'\mathbf{a})^{1/2} \quad \text{and} \quad \|\mathbf{b}\| = (\mathbf{b}\mathbf{b}')^{1/2}.$$

(Note that $\|\mathbf{a}\|$ and $\|\mathbf{b}\|$ are just numbers.)

We call a $k \times p$ matrix **C**, all of whose elements are zero, a *zero matrix* and write

$$\mathbf{C} = \mathbf{0}.$$

We call a matrix **E** *square* if it has the same number of rows as columns. If $\mathbf{E} = (e_{ij})$ is square, we call the e_{ii} the *diagonal elements* and the $e_{ij}, i \neq j$, the *off-diagonal* elements. If the off-diagonal elements are all zero, we say that **E** is a *diagonal matrix*. If **D** is a $k \times k$ diagonal matrix whose diagonal elements are all one, we say that **D** is an *identity matrix* and write

$$\mathbf{D} = \mathbf{I}_k,$$

or $\mathbf{D} = \mathbf{I}$ if the dimensions are clear. For example,

$$\mathbf{I}_3 = \begin{Bmatrix} 1 & 0 & 0 \\ 0 & 1 & 0 \\ 0 & 0 & 1 \end{Bmatrix}.$$

If $\mathbf{A} = (a_{ij})$ is a $k \times p$ matrix, then the *transpose* of **A**, written

$$\mathbf{A}',$$

is the $k \times p$ matrix $\mathbf{B} = (b_{ij})$, where $b_{ij} = a_{ji}$. That is, to obtain the transpose of **A**, we interchange the rows and columns of **A**. For example, the transpose of the matrix

$$\mathbf{A} = \begin{Bmatrix} 2 & 3 & 4 \\ 1 & 5 & 6 \end{Bmatrix}$$

is

$$\mathbf{A}' = \begin{Bmatrix} 2 & 1 \\ 3 & 5 \\ 4 & 6 \end{Bmatrix}.$$

Note that

$$(\mathbf{A}')' = \mathbf{A}.$$

We say that a square matrix $\mathbf{S} = (s_{ij})$ is *symmetric* if $s_{ij} = s_{ji}$, or equivalently, if

$$\mathbf{S}' = \mathbf{S}.$$

Let $\mathbf{A} = (a_{ij})$ and $\mathbf{B} = (b_{ij})$ be $k \times p$ matrices. Then the *sum* of **A** and **B**, written

$$\mathbf{A} + \mathbf{B},$$

is the $k \times p$ matrix $\mathbf{C} = (c_{ij})$, where $c_{ij} = a_{ij} + b_{ij}$. For example, if

$$\mathbf{A} = \begin{Bmatrix} 2 & 3 & 4 \\ 1 & 5 & 6 \end{Bmatrix} \quad \text{and} \quad \mathbf{B} = \begin{Bmatrix} 1 & 2 & 4 \\ 3 & 5 & 7 \end{Bmatrix},$$

then

$$\mathbf{A} + \mathbf{B} = \begin{Bmatrix} 3 & 5 & 8 \\ 4 & 10 & 13 \end{Bmatrix}.$$

The basic properties of matrix addition are as follows:

$$\mathbf{A}+\mathbf{B}=\mathbf{B}+\mathbf{A}, \qquad (\mathbf{A}+\mathbf{B})+\mathbf{C}=\mathbf{A}+(\mathbf{B}+\mathbf{C}), \qquad (\mathbf{A}+\mathbf{B})'=\mathbf{A}'+\mathbf{B}'.$$

Note that the sum of **A** and **B** is defined only if **A** and **B** have the same dimensions.

If a matrix c is 1×1, then c is a *scalar*. Now, let $\mathbf{A} = (a_{ij})$ be a matrix, and let c be a scalar. (Typically, we use boldface letters for vectors and matrices and italic letters for scalars.) We define the *scalar product* of c and **A**, written

$$c\mathbf{A},$$

to be the matrix $\mathbf{B} = (b_{ij})$, where $b_{ij} = ca_{ij}$. For example, with **A** as defined earlier, and with $c = 2$, we get

$$2\mathbf{A} = 2\begin{Bmatrix} 2 & 3 & 4 \\ 1 & 5 & 6 \end{Bmatrix} = \begin{Bmatrix} 4 & 6 & 8 \\ 2 & 10 & 12 \end{Bmatrix}.$$

Some basic properties of scalar multiplication are as follows:

$$c' = c, \qquad (c\mathbf{A})' = c\mathbf{A}', \qquad c(\mathbf{A}+\mathbf{B}) = c\mathbf{A}+c\mathbf{B},$$

Let $\mathbf{A} = (a_{ij})$ be a $k \times p$ matrix and $\mathbf{B} = (b_{jk})$ be a $p \times r$ matrix. Then the *product* of **A** and **B**, written

$$\mathbf{AB},$$

is the matrix $\mathbf{C} = (c_{ik})$, where

$$c_{ik} = \sum_j a_{ij} b_{jk}.$$

For example, if

$$\mathbf{A} = \begin{Bmatrix} 2 & 3 & 4 \\ 1 & 5 & 6 \end{Bmatrix} \quad \text{and} \quad \mathbf{B} = \begin{Bmatrix} 1 & 2 & 3 \\ 4 & 5 & 6 \\ 7 & 8 & 9 \end{Bmatrix},$$

then

$$\mathbf{AB} = \begin{Bmatrix} 42 & 51 & 60 \\ 63 & 75 & 87 \end{Bmatrix}.$$

The basic properties of matrix multiplication are

$$(\mathbf{AB})\mathbf{C} = \mathbf{A}(\mathbf{BC}), \qquad (\mathbf{AB})' = \mathbf{B}'\mathbf{A}', \qquad \mathbf{A}(\mathbf{B}+\mathbf{C}) = (\mathbf{AB}) + (\mathbf{AC}),$$

$$(\mathbf{A}+\mathbf{B})\mathbf{C} = (\mathbf{AC}) + (\mathbf{BC}), \qquad c(\mathbf{AB}) = (c\mathbf{A})\mathbf{B} = \mathbf{A}(c\mathbf{B}), \qquad \mathbf{I}_k\mathbf{A} = \mathbf{AI}_p = \mathbf{A}$$

(where c is a scalar and **A** is $k \times p$). Note that **AB** is defined only if the number of columns of **A** equals the number of rows of **B**. Note also that typically, $\mathbf{AB} \neq \mathbf{BA}$.

We say that a $k \times k$ matrix **A** is *invertible* if there exists a $k \times k$ matrix **B** such that

$$\mathbf{AB} = \mathbf{BA} = \mathbf{I}_k.$$

We call **B** the *inverse* of **A** and write

$$\mathbf{B} = \mathbf{A}^{-1}.$$

(If **A** is square and $\mathbf{AB} = \mathbf{I}$, then $\mathbf{B} = \mathbf{A}^{-1}$.) If **A** and **B** are invertible matrices and c is a scalar, then

$$(c\mathbf{A})^{-1} = c^{-1}\mathbf{A}^{-1}, \qquad (\mathbf{A}')^{-1} = (\mathbf{A}^{-1})', \qquad (\mathbf{AB})^{-1} = \mathbf{B}^{-1}\mathbf{A}^{-1}.$$

(Note that a matrix **A** can be invertible only if it is square (although it need not be even then).)

Inverses are typically somewhat difficult to compute. However, if **A** is a diagonal matrix with diagonal elements $a_{ii} \neq 0$, then $\mathbf{A}^{-1}$ is a diagonal matrix with diagonal elements a_{ii}^{-1}. Another nice special case is when $k = 2$: If

$$\mathbf{A} = \begin{Bmatrix} a & b \\ c & d \end{Bmatrix},$$

then

$$\mathbf{A}^{-1} = (ad - bc)^{-1} \begin{Bmatrix} d & -b \\ -c & a \end{Bmatrix}.$$

Finally,

$$\begin{Bmatrix} \mathbf{C} & \mathbf{0} \\ \mathbf{0} & \mathbf{D} \end{Bmatrix}^{-1} = \begin{Bmatrix} \mathbf{C}^{-1} & \mathbf{0} \\ \mathbf{0} & \mathbf{D}^{-1} \end{Bmatrix}.$$

The columns of **A** are *linearly independent* if the only column vector **b** such that $\mathbf{Ab} = \mathbf{0}$ is the vector $\mathbf{b} = \mathbf{0}$. If **A** is a square matrix whose columns are linearly independent, then **A** is invertible.

If **A** is a square matrix, we write det(**A**) for the *determinant* of **A**. It is rather messy to define the determinant; however, determinants for the 2×2 and 3×3 cases are as follows:

$$\det\left(\begin{Bmatrix} a & b \\ c & d \end{Bmatrix}\right) = ad - bc, \qquad \det\left(\begin{Bmatrix} a & b & c \\ d & e & f \\ g & h & i \end{Bmatrix}\right) = aei + bfg + cdh - gec - hfa - idb.$$

If **A** has dimension greater than three, a nice special case occurs when $\mathbf{A} = (a_{ij}), a_{ij} = 0, i < j$ (i.e., **A** is lower triangular). Then $\det(\mathbf{A}) = \Pi_i a_{ii}$, the product of the diagonal elements of **A**. The same result holds if $a_{ij} = 0, i > j$ (i.e., **A** is upper triangular). In particular, if **A** is a diagonal matrix, then det(**A**) is the product of the diagonal elements. Some basic properties of determinants are

$$\det(\mathbf{A}') = \det(\mathbf{A}), \qquad \det(\mathbf{AB}) = \det(\mathbf{A})\det(\mathbf{B}),$$

$$\det(A^{-1}) = (\det(\mathbf{A}))^{-1}, \qquad \det(c\mathbf{A}) = c^k \det(\mathbf{A})$$

(where **A** is $k \times k$). Note that a square matrix **A** is invertible if and only if $\det(\mathbf{A}) \neq 0$.

If **A** is a square matrix, we define the *trace* of **A** (written tr(**A**)) to be the sum of the diagonal elements of **A**. It can be shown that

$$\mathrm{tr}(c\mathbf{A}) = c\,\mathrm{tr}(\mathbf{A}),\ \mathrm{tr}(\mathbf{A} + \mathbf{B}) = \mathrm{tr}(\mathbf{A}) + \mathrm{tr}(\mathbf{B}),\ \mathrm{tr}(\mathbf{AB}) = \mathrm{tr}(\mathbf{BA}).$$

(However, in general $\mathrm{tr}(\mathbf{AB}) \neq \mathrm{tr}(\mathbf{A})\mathrm{tr}(\mathbf{B})$ and $\mathrm{tr}(\mathbf{ABC}) \neq \mathrm{tr}(\mathbf{BAC})$.

A.3.2 Positive Definite and Idempotent Matrices

We say that a $p \times p$ symmetric matrix **S** is *nonnegative definite* and write $\mathbf{S} \geqq 0$ if

$$\mathbf{t}'\mathbf{S}\mathbf{t} \geqq 0$$

for all $\mathbf{t} \in R^p$. If

$$\mathbf{t}'\mathbf{S}\mathbf{t} > 0$$

for all $\mathbf{t} \in R^p, \mathbf{t} \neq 0$, then $\mathbf{S}$ is *positive definite*, written $\mathbf{S} > 0$. (Note that $\mathbf{t}'\mathbf{S}\mathbf{t}$ is a 1×1 matrix and hence is a number, so that $\mathbf{t}'\mathbf{S}\mathbf{t} \geqq 0$ is well defined.) We say that a symmetric matrix $\mathbf{P}$ is *idempotent* if $\mathbf{P}^2 = \mathbf{P}$. In this section, we derive the basic results used in the text for nonnegative definite, positive definite, and idempotent matrices. Note carefully that, for our purposes, any nonnegative definite, positive definite, or idempotent matrix is symmetric.

The following theorem gives some elementary properties of nonnegative definite matrices.

Theorem A–14. Let $\mathbf{X}$ be a $p \times q$ matrix and $\mathbf{Q}$ be $p \times p$. Then

a. $\mathbf{X}'\mathbf{X} \geqq 0$. If the columns of $\mathbf{X}$ are independent, then $\mathbf{X}'\mathbf{X} > 0$.
b. If $\mathbf{Q} \geqq 0$, then $\mathbf{X}'\mathbf{Q}\mathbf{X} \geqq 0$. If $\mathbf{Q} > 0$ and the columns of $\mathbf{X}$ are linearly independent, then $\mathbf{X}'\mathbf{Q}\mathbf{X} > 0$.

Proof

a. Note that

$$\mathbf{t}'(\mathbf{X}'\mathbf{X})\mathbf{t} = \|\mathbf{X}\mathbf{t}\|^2 \geqq 0,$$

with equality only if $\mathbf{X}\mathbf{t} = \mathbf{0}$. Hence, $\mathbf{X}'\mathbf{X} \geqq 0$, and if the columns of $\mathbf{X}$ are linearly independent, then $\mathbf{X}'\mathbf{X} > 0$. The proof of part b follows similarly from the fact that

$$\mathbf{t}'(\mathbf{X}'\mathbf{Q}\mathbf{X})\mathbf{t} = (\mathbf{X}\mathbf{t})'\mathbf{Q}(\mathbf{X}\mathbf{t}). \quad \square$$

The remaining results in this section are derived from Theorem A–15. This theorem has many names, including the principal axis theorem and the spectral decomposition theorem. It is appropriately called the fundamental theorem of matrix algebra.

We say that a $p \times p$ matrix $\mathbf{C}$ is *orthogonal* if

$$\mathbf{C}\mathbf{C}' = \mathbf{C}'\mathbf{C} = \mathbf{I},$$

or equivalently, if $\mathbf{C}^{-1} = \mathbf{C}'$.

Theorem A–15. Let $\mathbf{S}$ be a symmetric matrix. Then there exists an orthogonal matrix $\mathbf{C}$ and a diagonal matrix $\mathbf{D}$ such that

$$\mathbf{S} = \mathbf{C}\mathbf{D}\mathbf{C}'.$$

Proof. This theorem is proved at the end of most books on matrix algebra. It is also proved in Arnold (1981), pp. 462–463. $\square$

Interestingly, there does not appear to be any proof of Theorem A–15 using only matrix algebra. The two most common proofs use multivariable calculus or complex analysis.

Now, let

$$\mathbf{D} = \begin{pmatrix} d_1 & 0 & \cdot & \cdot & \cdot & 0 \\ 0 & d_2 & \cdot & \cdot & \cdot & 0 \\ \cdot & \cdot & \cdot & & & \cdot \\ \cdot & \cdot & & \cdot & & \cdot \\ \cdot & \cdot & & & \cdot & \cdot \\ 0 & 0 & \cdot & \cdot & \cdot & d_p \end{pmatrix} \quad \text{and} \quad \mathbf{C} = (\mathbf{C}_1, \ldots, \mathbf{C}_p).$$

Then the d_i are called the *eigenvalues* of $\mathbf{S}$, and $\mathbf{C}_i$ is called the *eigenvector* associated with d_i. (The d_i and $\mathbf{C}_i$ also have other names, e.g., latent roots and latent vectors, and characteristic roots and characteristic vectors.) If we assume that $d_i \geqq d_{i+1}$, then the eigenvalues are unique (but the eigenvectors are not). In this book, we do not use eigenvectors.

The next theorem characterizes some types of matrices by the properties of their eigenvalues.

Theorem A–16. Let $\mathbf{S}$ be a $p \times p$ symmetric matrix with eigenvalues $d_1, \ldots, d_p$. Then

a. $\mathbf{S}$ is invertible if and only if the $d_i \neq 0$.
b. $\mathbf{S} \geqq 0$ if and only if the $d_i \geqq 0$.
c. $\mathbf{S} > 0$ if and only if the $d_i > 0$.
d. $\mathbf{S}$ is idempotent if and only if the d_i are all zero or one.

Proof

a. Since $\mathbf{C}$ and $\mathbf{C}'$ are invertible, $\mathbf{S}$ is invertible if and only if $\mathbf{D}$ is invertible, which occurs if and only if the $d_i \neq 0$.
b. and c. Let $\mathbf{t} \in R^p$, and let $\mathbf{q} = \mathbf{C}'\mathbf{t}$. Then

$$\mathbf{t}'\mathbf{S}\mathbf{t} = \mathbf{q}'\mathbf{D}\mathbf{q}.$$

Therefore, $\mathbf{S} \geqq 0$ if and only if $\mathbf{D} \geqq 0$. Let q_i be the ith component of $\mathbf{q}$. Then

$$\mathbf{q}'\mathbf{D}\mathbf{q} = \sum q_i^2 d_i.$$

Therefore, $\mathbf{q}'\mathbf{D}\mathbf{q} \geqq 0$ for all $\mathbf{q}$ if and only if the $d_i \geqq 0$, and $\mathbf{q}'\mathbf{D}\mathbf{q} > 0$ for all $\mathbf{q} \neq \mathbf{0}$ if and only if the $d_i > 0$.

d. The orthogonality of $\mathbf{C}$ implies that

$$\mathbf{S}^2 = \mathbf{CDC}'\mathbf{CDC}' = \mathbf{CD}^2\mathbf{C}.$$

Therefore,

$$\mathbf{S}^2 = \mathbf{S} \Leftrightarrow \mathbf{D}^2 = \mathbf{D} \Leftrightarrow d_i^2 = d_i,$$

which occurs if and only if the d_i are all zero or one. □

The next theorem presents some additional results about nonnegative definite and positive definite matrices.

Theorem A–17. Let $\mathbf{Q}$ be $p \times p$.

a. If $\mathbf{Q} > 0$, then $\mathbf{Q}$ is invertible and $\mathbf{Q}^{-1} > 0$.
b. If $\mathbf{Q} \geqq 0$, then there exists $\mathbf{R} \geqq 0$ such that $\mathbf{Q} = \mathbf{R}^2$. If $\mathbf{Q} > 0$, then $\mathbf{R} > 0$.

Proof

a. Let $\mathbf{Q} = \mathbf{CDC}'$, and let the d_i be the eigenvalues of $\mathbf{Q}$. By parts a and c of Theorem A–15, if $\mathbf{Q} > 0$, then $\mathbf{Q}$ is invertible. Now,

$$\mathbf{Q}^{-1} = \mathbf{C}'^{-1}\mathbf{D}^{-1}\mathbf{C}^{-1} = \mathbf{CD}^{-1}\mathbf{C}'.$$

Therefore, the eigenvalues of $\mathbf{Q}^{-1}$ are d_i^{-1}, where the d_i are the eigenvalues of $\mathbf{Q}$. But then, if $\mathbf{Q} > 0$, then the $d_i > 0$, and hence, the $d_i^{-1} > 0$, so that $\mathbf{Q}^{-1} > 0$.

b. Since $\mathbf{Q} \geqq 0$, the $d_i \geqq 0$, and hence, $d_i^{1/2}$ exist. Let $\mathbf{D}^{1/2}$ be the diagonal matrix with diagonals $d_i^{1/2}$, and let

$$\mathbf{R} = \mathbf{CD}^{1/2}\mathbf{C}.$$

Then

$$\mathbf{R}^2 = \mathbf{CD}^{1/2}\mathbf{C}'\mathbf{CD}^{1/2}\mathbf{C} = \mathbf{CDC}' = \mathbf{Q}.$$

Now, the eigenvalues of $\mathbf{R}$ are the $d_i^{1/2} \geqq 0$, so that $\mathbf{R} \geqq 0$. But if $\mathbf{Q} > 0$, then $d_i^{1/2} > 0$, so that $\mathbf{R} > 0$. □

It can be shown that there is only one matrix $\mathbf{R} \geqq 0$ such that $\mathbf{R}^2 = \mathbf{Q}$. (See Arnold (1981), p. 453.) We call this matrix $\mathbf{Q}^{1/2}$. Now, suppose that $\mathbf{Q} > 0$. If $\mathbf{R}^2 = \mathbf{Q}$, then

$$(\mathbf{R}^{-1})^2 = \mathbf{R}^{-1}\mathbf{R}^{-1} = (\mathbf{RR})^{-1} = (\mathbf{R}^2)^{-1} = \mathbf{Q}^{-1}.$$

Therefore,

$$(\mathbf{Q}^{1/2})^{-1} = (\mathbf{Q}^{-1})^{1/2}.$$

We define $\mathbf{Q}^{-1/2}$ to be this matrix. (Note that $\mathbf{Q}^{-1/2}$ is defined only when $\mathbf{Q} > 0$.)

Theorem A–18. Let $\mathbf{Q} > 0$. Then

a. $\mathbf{Q}^{1/2}$ and $\mathbf{Q}^{-1/2}$ are symmetric, and

$$(\mathbf{Q}^{1/2})^2 = \mathbf{Q}, \qquad (\mathbf{Q}^{-1/2})^2 = \mathbf{Q}^{-1}, \qquad \mathbf{Q}^{-1/2}\mathbf{Q}\mathbf{Q}^{-1/2} = \mathbf{I}.$$

b. $\det(\mathbf{Q}) > 0$, and

$$\det(\mathbf{Q}^{1/2}) = (\det(\mathbf{Q}))^{1/2}, \qquad \det(\mathbf{Q}^{-1/2}) = (\det(\mathbf{Q}))^{-1/2}.$$

Proof

a. $\mathbf{Q}^{1/2}$ and $\mathbf{Q}^{-1/2}$ are symmetric because they are nonnegative definite. The first two equalities of the theorem follow directly from the definitions of $\mathbf{Q}^{1/2}$ and $\mathbf{Q}^{-1/2}$. Finally,

$$\mathbf{Q}^{-1/2}\mathbf{Q}\mathbf{Q}^{-1/2} = \mathbf{Q}^{-1/2}\mathbf{Q}^{1/2}\mathbf{Q}^{1/2}\mathbf{Q}^{-1/2} = \mathbf{I}.$$

b.

$$\det(\mathbf{Q}) = \det(\mathbf{Q}^{1/2}\mathbf{Q}^{1/2}) = \det(\mathbf{Q}^{1/2})\ \det(\mathbf{Q}^{1/2}) = (\det(\mathbf{Q}^{1/2}))^2 > 0.$$

Since $\mathbf{Q}^{1/2} > 0$, $\det \mathbf{Q}^{1/2} > 0$. Therefore

$$\det(\mathbf{Q}^{1/2}) = (\det(\mathbf{Q}))^{1/2}.$$

The last equality follows similarly. □

The next theorem gives the basic result about idempotent matrices.

Theorem A–19. Let $\mathbf{P}$ be a $p \times p$ idempotent matrix with trace k. Then there exists a $k \times p$ matrix $\mathbf{E}$ such that

$$\mathbf{EE}' = \mathbf{I}_k \quad \text{and} \quad \mathbf{E}'\mathbf{E} = \mathbf{P}.$$

Proof. Let $\mathbf{P} = \mathbf{CDC}'$, and let q be the number of eigenvalues of $\mathbf{P}$ which are one. Then, by Theorem A–16,

$$\mathbf{D} = \begin{Bmatrix} \mathbf{I}_q & \mathbf{0} \\ \mathbf{0} & \mathbf{0} \end{Bmatrix}.$$

Now, let $\mathbf{C} = (\mathbf{E}'\mathbf{F}')$, where $\mathbf{E}$ is $q \times p$ and $\mathbf{F}$ is $(p - q) \times p$. Then

$$\mathbf{P} = \mathbf{CDC}' = (\mathbf{E}'\ \mathbf{F}') \begin{Bmatrix} \mathbf{I}_q & \mathbf{0} \\ \mathbf{0} & \mathbf{0} \end{Bmatrix} \begin{Bmatrix} \mathbf{E} \\ \mathbf{F} \end{Bmatrix} = \mathbf{E}'\mathbf{E}.$$

Since $\mathbf{C}$ is orthogonal,

$$\mathbf{I} = \begin{Bmatrix} \mathbf{I}_q & \mathbf{0} \\ \mathbf{0} & \mathbf{I}_{p-q} \end{Bmatrix} = \mathbf{C}'\mathbf{C} = \begin{Bmatrix} \mathbf{EE}' & \mathbf{EF}' \\ \mathbf{FE}' & \mathbf{FF}' \end{Bmatrix},$$

and hence, $\mathbf{E}'\mathbf{E} = \mathbf{I}_q$. But

$$k = \mathrm{tr}(\mathbf{P}) = \mathrm{tr}(\mathbf{E}'\mathbf{E}) = \mathrm{tr}(\mathbf{EE}') = \mathrm{tr}(\mathbf{I}_q) = q,$$

so

$$\mathbf{E}'\mathbf{E} = \mathbf{I}_k. \quad \square$$

We close matters with a theorem that gives some other interesting properties of eigenvalues and eigenvectors which are not used in the text.

Theorem A–20. Let $\mathbf{S}$ be a symmetric matrix with eigenvalues d_i. Then

$$\det(S) = \prod_i d_i \quad \text{and} \quad \mathrm{tr}(S) = \sum_i d_i,$$

The d_i are the roots of the equation

$$h(t) = \det(\mathbf{S} - t\mathbf{I}) = 0.$$

Proof. First,

$$\det(\mathbf{S}) = \det(\mathbf{CDC}') = \det(\mathbf{C})\ \det(\mathbf{C}')\ \det(\mathbf{D}) = \det(\mathbf{CC}')\ \det(\mathbf{D}) = \det(\mathbf{D}) = \prod_i d_i.$$

Also,

$$\mathrm{tr}(\mathbf{S}) = \mathrm{tr}(\mathbf{CDC}') = \mathrm{tr}(\mathbf{DC}'\mathbf{C}) = \mathrm{tr}(\mathbf{D}) = \sum d_i.$$

Finally,

$$h(t) = \det(\mathbf{S} - t\mathbf{I}) = \det(\mathbf{CDC}' - t\mathbf{CC}') = \det(\mathbf{C})\det(\mathbf{C}')\det(\mathbf{D} - t\mathbf{I}) = \prod_i (d_i - t). \quad \square$$

As a final result, we mention that

$$\mathbf{SC} = \mathbf{CDC}'\mathbf{C} = \mathbf{CD},$$

which implies that

$$\mathbf{SC}_i = d_i \mathbf{C}_i,$$

where $\mathbf{C}_i$ is the eigenvector associated with the ith eigenvalue of $\mathbf{S}$.

REFERENCES AND FURTHER READING

In this section we briefly discuss some other textbooks related to the material included in this text. These books are ones which I have seen. There are certainly other excellent books of which I am unaware which have not been included here. I apologize to the authors of these books for their omission.

Other textbooks on probability and statistics at about the same level as this book include Brunk (1975), DeGroot (1986), Fraser (1976), Hogg and Craig (1978), Lindgren (1976), Mood, Graybill, and Boes (1974), and Rohatgi (1976). Larsen and Marks (1985) and Ross (1988) cover probability theory (Chapters 1–6 of this book) at about the same level as this book. Mosteller (1965) presents some interesting probability problems with their solutions, including problems of historical interest in the development of probability.

To make the definitions and derivations in probability theory precise, at least in the continuous case, it is necessary to use measure theory. Billingsley (1979) presents both measure and probability theory. Chung (1974) presents the probability theory, assuming that the reader already knows measure theory. Feller (1968) and (1971) present probability theory from a different perspective.

Chapter 6 presents some basic results in asymptotic theory. Serfling (1980) presents many more detailed asymptotic results.

Chapters 7–16 present an introduction to statistics. Bickel and Doksun (1977) present similar material at a somewhat more technical level. More advanced books in the theory of statistics include Ferguson (1967), Lehmann (1983), Lehmann (1986), and Rao (1973). Kendall and Stuart (1977 and 1979) and Kendall, Stuart, and Ord (1983) also present many detailed statistical results.

Chapters 12–13 present an introduction to two models, multiple regression and analysis of variance, which are special cases of linear models. Arnold (1981) presents a more detailed treatment of the theory of these linear models. Graybill (1976), Scheffe' (1959), Searle (1971), and Seber (1977) present treatments of both the theory and applications of these models. Books on the applications of these models include Draper and Smith (1980), Morrison (1983), and Neter, Wasserman, and Kutner (1985).

Chapter 14 contains an introduction to discrete models. Further reading on such models can be found in Bishop, Feinberg, and Holland (1975) and Haberman (1974).

Chapter 15 is an introduction to nonparametric models. Additional theory on nonparametric models is contained in Hettmansperger (1984) and Randles and Wolfe (1979). Connover (1971) gives an introduction to applied nonparametric models. Huber (1981) presents robust methods in statistics which are closely related to nonparametric procedures.

Chapter 16 gives a discussion of the Bayesian approach to statistics. Berger (1985), DeGroot (1970), and Savage (1972) contain further material on Bayesian inference. Berger and Wolpert (1984) present material on the likelihood principle. The fact that classical statistics violates this principle is one of the main arguments against classical statistics. De Finetti (1974) gives a discussion of subjective probability, which seems to be necessary for Bayesian inference.

REFERENCES

ARNOLD, S. F. (1981) *The Theory of Linear Models and Multivariate Analysis.* Wiley, New York.

BERGER, J. O. (1985) *Statistical Decision Theory and Bayesian Analysis.* 2nd ed. Springer, New York.

BERGER, J. O. and WOLPERT, R. L. (1984) *The Likelihood Principle.* Lecture notes No. 6, Inst. Math. Statist., Hayward, Calif.

BICKEL, P. J. and DOKSUM, K. A. (1977) *Mathematical Statistics: Basic Ideas and Selected Topics.* Holden-Day, San Francisco.

BILLINGSLEY, P. (1979) *Probability and Measure.* Wiley, New York.

BISHOP, Y. M., FEINBERG, S. E. and HOLLAND, P. W. (1975) *Discrete Multivariate Analysis.* M.I.T. Press, Cambridge, Mass.

BRUNK, H. D. (1975) *An Introduction to Mathematical Statistics.* 3rd ed. Xerox, Lexington, Mass.

CHUNG, K. L. (1974) *A Course in Probability Theory.* 2nd ed. Academic, New York.

CONNOVER, W. J. (1971) *Practical Non-parametric Statistics.* Wiley, New York.

DE FINETTI, B. (1974) *Theory of Probability.* Wiley, New York.

DEGROOT, M. H. (1970) *Optimal Statistical Decisions.* McGraw-Hill, New York.

DEGROOT, M. H. (1986) *Probability and Statistics.* 2nd ed. Addison-Wesley, Reading, Mass.

DRAPER, N. R. and SMITH, H. (1980) *Applied Regression Analysis,* 2nd ed. Wiley, New York.

FELLER, W. (1968) *An Introduction to Probability Theory and Its Applications,* 3rd ed. Wiley, New York.

FELLER, W. (1971) *An Introduction to Probability Theory and Its Applications.* 2nd ed. Wiley, New York.

FERGUSON, T. S. (1967) *Mathematical Statistics: A Decision Theoretic Approach.* Academic, New York.

FRASER, D. A. S. (1976) *Probability and Statistics, Theory and Application.* Duxbury, North Scituate, Mass.

GRAYBILL, F. A. (1976) *Theory and Application of the Linear Model.* Duxbury, North Scituate, Mass.

HABERMAN, S. J. (1974) *The Analysis of Frequency Data.* Univ. of Chicago Press, Chicago.

HETTMANSPERGER, T. P. (1984) *Statistical Inference Based on Ranks.* Wiley, New York.

HOGG, R.V. and CRAIG, A. T. (1978) *Introduction to Mathematical Statistics.* 4th ed. Macmillan, New York.

HUBER, P. J. (1981) *Robust Statistics.* Wiley, New York.

KENDALL, M. S. and STUART, A. (1977 and 1979) *The Advanced Theory of Statistics, v. 1 and 2.* 4th ed. Macmillan, New York.

KENDALL, M. S., STUART, A. and ORD, J. K. (1983) *The Advanced Theory of Statistics, v. 3.* 4th ed. Macmillan, New York.

LARSEN, R. J. and MARKS, M. L. (1985) *An Introduction to Probability and Its Applications.* Prentice-Hall, Englewood Cliffs, N.J.

LEHMANN, E. L. (1983) *Theory of Point Estimation.* Wiley, New York.

LEHMANN, E. L. (1986) *Testing Statistical Hypotheses.* 2nd ed. Wiley, New York.

LINDGREN, B. L. (1976) *Statistical Theory.* 3rd ed. Macmillan, New York.

MOOD, A. M., GRAYBILL, F. A. and BOES, D. C. (1974) *Introduction to the Theory of Statistics.* 3rd ed. McGraw-Hill, New York.

MORRISON, D. (1983) *Applied Linear Statistical Models.* Prentice Hall, Englewood Cliffs, N.J.

MOSTELLER, F. (1965) *Fifty Challenging Problems in Probability with Solutions.* Addison-Wesley, Reading, Mass.

NETER, J., WASSERMAN, W. and KUTNER, M. H. (1985) *Applied Linear Statistical Models.* 2nd ed. Irwin, Howard, Ill.

RAO, C. R. (1973) *Linear Statistical Inference and Its Applications.* 2nd ed. Wiley, New York.

RANDLES, R. H. and WOLFE, D. A. (1979) *Introduction to the Theory of Nonparametric Statistics.* Wiley, New York.

ROHATGI, V. K. (1976) *An Introduction to Probability Theory and Mathematical Statistics.* Wiley, New York.

ROSS, S. (1988) *A First Course in Probability.* 3rd ed. Macmillan, New York.

SAVAGE, L. J. (1972) *The Foundations of Statistics.* Dover, New York.

SCHEFFE', H. (1959) *The Analysis of Variance.* Wiley, New York.

SEARLE, S. R. (1971) *Linear Models.* Wiley, New York.

SEBER, G. A. F. (1977) *Linear Regression Analysis.* Wiley, New York.

SERFLING, R. J. (1980) *Approximation Theorems of Mathematical Statistics.* Wiley, New York.

ANSWERS TO ODD-NUMBERED EXERCISES

CHAPTER 1

1.1.1. A. 3 .00015 **5(a)** .052; **(b)** .0024; **(c)** .169 **7(b)** .125; **(c)** .375; **(d)** .5 **9** 32 **11(a)** .000039; **(b)** .00093; **(c)** 4.5×10^{-28}

2. A. 1(a) 103; **(b)** .437; **(c)** .460 **3** .067

1.2.1. B. 3(a) no; **(b)** $(-\infty,\infty)$, $\{0\}$, $\emptyset$, $(-\infty,0) \cup (0,\infty)$; **(c)** $(-\infty,n] \cup [n,\infty)$, $\emptyset$

2. A. 1(a) .57; **(b)** .29. **3** .9

B. 1(a) .9, .1; **(b)** .3, .7; **(c)** .4, .7

1.3.1. A. 1 .32 **3(a)** .12; **(b)** .44; **(c)** .26 **5(a)** .345; **(b)** .29 **7(a)** .5; **(b)** .75; **(c)** .6; **(d)** .33 **9(a)** .093, .011; **(b)** .118; **(c)** .0004. **11** 2/3

B. 1(a) .5, .25, .2; **(b)** .75, .375, .86

2. A. 3(a) .955; **(b)** .991; **5(a)** .87; **(c)** 7 **7(a)** .226, .226, .0547 **(b)** no **(c)** .242 **11** .53

CHAPTER 2

2.1.1. A. 1(a) $\begin{bmatrix}4\\x\end{bmatrix}\begin{bmatrix}48\\5-x\end{bmatrix}\Big/\begin{bmatrix}52\\5\end{bmatrix}$, $\begin{bmatrix}52\\5\end{bmatrix}$; **(b)** $\begin{bmatrix}5\\x\end{bmatrix}({}_4P_x)({}_{48}P_{5-x})/{}_{52}P_5$, ${}_{52}P_5$

3(a) 1/6(0), 1/2(1), 3/10(2), 1/30(3); **(b)** 3/10, 1/3; **(c)** 9/10, 1

5(a) $f(x) = 6^{-1}(5/6)^{x-1}$, $x = 1, 2, \ldots$; **(b)** $1-(5/6)^5$, $91(5^3)/6^6$; **(c)** 6/11

B. 1(a) .1; **(b)** .4, .6; **(c)** .5; **(d)** 1/3 **3(a)** 2/3; **(c)** .8

2. A. 1(a) 1/4; **(b)** 1/2; **(c)** 2/3; **(d)** 1/2

B. 3(a) 0, $(2e)^{-1}$, $1-(2e)^{-1}$, $1-e^{-1}$, $1-(2e)^{-1}(1-e^{-1})$, $1-e^{-2}$; **(b)** $1-e^{-1}$, $(2e^3)^{-1}$

3. B. 1 0, 3/16, 11/16 **3(a)** $F(x) = 0, .3, .6$ and 1 on the sets $x<0$, $0 \leqq x < 1$, $1 \leqq x < 2$, $2 \leqq x < 3$, $3 \leqq x < 4$ and $x \geqq 4$; **(b)** .5, .3 **5(b)** $f(x) = 1/4$, $x = 1, 2, 3, 4$

4. B. 1(a) .1(0), .3(1), .6(4); **(b)** .3(0), .7(1); **(c)** .1(0), .3(e), .6(e^2)

3(a) $(5/192)\,u^{2/3}$, $-8 \leqq u \leqq 8$; **(b)** $5(\log(v))^4/64v$, $e^{-2} \leqq v \leqq e^2$ **(c)** $5w^{5/4}/128$, $0 \leqq w \leqq 16$

2.2.1. A. 1(a) 25/72; **(b)** 50/72 **3(a)** .081
B. 1(a) 1/36; **(b)** 15/36, 5/36; **(c)** 1/6, 2/6, 3/6 **3(a)** 3/8; **(b)** 1/2, 1/2; **(c)** $3\pi/16$
2. A. 1 1/60; 1/60
B. 1 $f_1(x) = x/6, x = 1, 2, 3; f_2(y|x) = y/6, x = 1, 3, 2, y = 1, 2, 3$
3 $(3x^2+1)/2, 0 \leqq x \leqq 1; 3(x^2+y^2)/(3x^2+1), 0 \leqq y \leqq 1, 0 \leqq x \leqq 1$
7(a) $(2x)^{-1}, -x \leqq y \leqq x, 0 \leqq y \leqq 1$; **(b)** $-\log(|y|)/2, -1 \leqq y \leqq 1$; $(-x\log(y))^{-1}, |y| \leqq x \leqq 1, -1 \leqq y \leqq 1$
3. A. 1(a) 2/3; **(b)** .5
B. 1(a) independent **3** 1/5, 2/5
4. B. 1(a) 1/36 (2), 4/36 (2), 10/36 (4), 12/36 (5), 9/36 (8); **(b)** 1/36 (1), 4/36 (4), 6/36 (3), 4/36 (4), 12/36 (6), 9/36 (9) **3** $1/2\ (0 \leqq u \leqq 1), 1/2u^2\ (u \geqq 1)$.
5(a) $2(e^{-u} - e^{-2u}), u > 0$; **(b)** $2e^{-2v}, v > 0$ **7(a)** $2u\exp(-u^2), u > 0, 0 < v < u^2$; **(b)** $2^{-1}e^{2u}\exp[-e^{(u+v)/2} - e^{(u-v)/2}]$

2.3.1. B. 1(a) $f((x_2, x_3)|x_1) = \exp(2x_1 - x_2 - x_3), 0 < x_1 < x_2 < x_3$;
(b) $f(x_2|(x_1, x_3)) = \exp(-x_2)/(\exp(-x_1) - \exp(-x_3)), 0 < x_1 < x_2 < x_3$; $(\exp(-1) - \exp(-2))\ (\exp(-.4) - \exp(-2))$;
(c) $f(x_3|x_1) = 2(\exp(x_1 - x_3) - \exp(2x_1 - 2x_3), 0 < x_1 < x_3$ **3(a)** 1/4;
(b) $f(x_1, x_2|x_3 = 0) = 1/3$ ((0, 0), (0, 1), and (1, 0)); **(c)** $f(x_3|(0, 0) = 1/2$ (0 and 1); **(d)** $f(x_2|0) = 2/3\ (0), 1/3\ (1)$ **5(b)** $(4(x_2 + x_1))^{-1}, -x_1 \leqq x_3 \leqq x_2, 0 \leqq x_1 \leqq 2, 0 \leqq x_2 \leqq 2$
2. A. 1(a) .105; **(b)** $(3/200)\exp(-3u/200)$
B. 5(a) $f(x_i) = 1/2, x_i = 1, 2$; **(b)** $f(x_1, x_2) = 1/4, x_1 = 1, 2, x_2 = 1, 2$
3. B. 1 $u_1^2\exp(-u_1), u_1 > 0, 0 < u_2 < u_3 < 1$

CHAPTER 3

3.1.1. A. 1 25/72; **(b)** 1/2 **3(a)** 3/8000, 12/8000; **(b)** .82
B. 1 4/5, 4/6, 4/3, 8/9 **3(a)** 1/2, 17/12, 2/3, 3/4; **(b)** yes, yes; **(c)** 1 **5(a)** 5/3, 8/3, 4/3, 4/3; **(b)** no, yes
3. A. 1 9 **3** 1.3
B. 1(a) no; **(b)** −5, 156

3.2.1. A. 1 21/2, 105/12 **3(a)** 500, 1; **(b)** 2/3
B. 1 1.42, 4.03, 2.01 **3** 1, 1, 1 **7** $EU = -11$, $\text{var}(U) = 54$, $EW = 2$, $\text{var}(W) = 26$, $EV = -10$, $\text{var}(V) = 42$, $E(U + V + W) = -19$, $\text{var}(U + V + W) = 182$
2. A. 1 1, 1
B. 1 −11/13 **3** 0 **5(a)** .40; **(b)** −13, −.21

3.3.1. B. 1 $(e^t + 2e^{2t} + 3e^{3t} + 4e^{4t})/10$ **3** 10, 20 **5(a)** 0, 2; **(b)** $\exp(3t^2 + 1)$, $\exp(9t^2 + 5t)$; **7** $\exp(at)$
2. B. 1(a) $(1 + e^t + e^s + e^{2s} + e^{2t} + e^{s+t})/6$; **(b)** 2/3, 5/9, −5/18;
(c) $(3 + 2e^s + e^2s)/6$; **(d)** $(1 + 2e^t + 3e^{2t})/6$ **3(a)** 5, 5, −5; **(b)** $(1+r)^{-6}$, $(1-s)^{-5}, (1+t)^{-3}$, no; **(c)** $(1+r)^{-1}(1+2r+t)^{-2}(1+2r+2t)^{-3}$

3.4.1. B. 1(a) $(2+3y)/(3+6y)$, $(3+4y)/(6+12y)-((2+3y)/(3+6y))^2$; **(b)** $(2+3y)y^2/(3+6y)$ **3(a)** 20/9, 11/7, 1; **(b)** 50/81, 12/49, 0; **(c)** 50/9, 38/7, 3 **7(a)** 16/35, 201/4900; **(b)** $2y/3$, $y^2/18$; **(c)** $4z/5$, $2z^2/75$; **(d)** $8z^4/7$, $8z^3/7$

2. A. 1 16.8 **3** 40/3

3.5.1. B. 1 $2^{-1/2}$

CHAPTER 4

4.1.1. A. 1 .9974 **3** .3085

B. 1(a) .0228, .0228, 0, .9270, .9270, .9887; **(b)** 1.645, 1.96

4.2.1. A. 1(a) .678, .302, .624; **(b)** .0571, .9599, .0170 **3** .084

B. 1(a) .540, .097; **(b)** .5403, .0968 **3(a)** .189; **(b)** .322 **5(a)** 4; **(b)** 6 **7** 6

2. A. 1 .1711 **3** .4, .4 **5** .68

B. 1 $(1+p)^{-1}$

3. A. 1 .919, .998 **3** .0239, .063

B. 1 10 **3(a)** $(Np/(N-u))\begin{bmatrix}N(1-p)\\ u\end{bmatrix}\Big/\begin{bmatrix}N\\ u\end{bmatrix}$, $u=0,\ldots,N(1-p)$; **(b)** $((Np-r+1)/(N-v-r+1))\begin{bmatrix}N(1-p)\\ v\end{bmatrix}\begin{bmatrix}Np\\ 1-r\end{bmatrix}\Big/\begin{bmatrix}N\\ v+r-1\end{bmatrix}$, $v=0,\ldots,N(1-p)$

4.3.1. A. 1(a) .228, .371; **(b)** .830; **(c)** .2949 **3(a)** .007; **(b)** .258; **(c)** .1841, **(d)** .0165 **5** .2090

B. 1(a) $6^x(2/3)^y e^{-10}/y!(x-y)!$; $y=0,1,\ldots,x$; $x=0,1,\ldots$; **(b)** $4^x e^{-4}/x!$, $x=0,1,\ldots$.

3. A. 1 $e^{-5/4}$, $e^{-3/4}-e^{-5/4}$ **3** .6

B. 7(a) .715, .774; **(b)** .35, .22

C. 1(a) $y^x\exp(-3y/2)/2x!$, $y>0$, $x=0,1,\ldots$; **(b)** $(1/3)(2/3)^x$, $x=0,1,\ldots$; **(c)** 1/3, 19/27

4.4.1. A. 1(a) .9546; **(b)** .0573

B. 1(a) 3.29; **(b)** 5.56; **(c)** .7345; **(d)** .6611 **3** 2.65, 6.76 **5(a)** $(2\pi)^{-1/2}(\sigma y)^{-1}\exp(-(\log(y)-\mu)^2/2\sigma^2)$; **(b)** $\exp(\mu+\sigma^2/2)$, $\exp(2\mu+2\sigma^2)$.

2. B. 1(a) 4.87, 18.31; **(b)** 4.87, 15.99 **3(a)** .207, 4.82; **(b)** .23, 8.89 **9** $2k^2(k+m-2)/m(k-2)^2(k-4)$, $k>4$

C. 1(b) $w^{(k-1)/2}\exp(-w((t^2/2k)+1/2))/(k\pi)^{1/2}\Gamma(k/2)2^{(k+1)/2}$

3. B. 1 5, 26.5

C. 1(b) $(2\pi)^{-n/2}\sigma^{-n}|J|\exp(-(q(y_2,\ldots,y_n)+n(y_1-\mu)^2)2\sigma^2)$, where $|J|$ is the absolute value of the Jacobian of the linear function from the y's to the x's, which does not depend on the y's.

CHAPTER 5

5.1.1. A. **1(a)** .044; **(b)** .238, .672; **(c)** .354; **(d)** .260 **3(a)** .054; **(b)** .081; **(c)** .363; **(d)** .931 **5(a)** .969; **(b)** .988

B. **1(a)** $T(25, .3085, .1587)$; **(b)** .0008; **(c)** 29.38, 25.14
5(a) $e^{-q}m_1^{x_1}m_2^{x_2}m_3^{u-x_1-x_2}/x_1!x_2!(u-x_1-x_2)!$, $x_i = 0, 1, \ldots, x_1 + x_2 \leqq u$

C. **1(a)** $\begin{bmatrix}Np_1\\x_1\end{bmatrix}\begin{bmatrix}Np_2\\x_2\end{bmatrix}\begin{bmatrix}N(1-p_1-p_2)\\n-x_1-x_2\end{bmatrix}\Big/\begin{bmatrix}N\\n\end{bmatrix}$; **(b)** np_i, $np_i(1-p_i)(N-n)/(N-1)$; **(c)** $n(p_1+p_2)$, $n(p_1+p_2)(1-p_1-p_2)(N-n)/(N-1)$; **(d)** $-np_1p_2(N-n)/(N-1)$

2. A. **1(a)** .0004; **(b)** .015; **(c)** .015; **(d)** .15; **(e)** 12.38 **3(a)** .00006; **(b)** .023; **(c)** .097; **(d)** .006; **(e)** .8849 **5(a)** .07; **(b)** .49; **(c)** 1

B. **3(a)** $(n-1)!\,p_1^2(1-p_1-p_2)^{n-3}/(1-p_2)^{n-1}(2!)(n-3)!$;
(b) $B(n-X_2, p_1/(1-p_2))$;
(c) $(n-3)!(p_1)^2p_2(p_4)^{n-6}/(1-p_3)^{n-3}2!(n-6)!$
(d) $T(n-x_3, p_1/(1-p_3), p_2/(1-p_3))$ **5(a)** $n, q = p_1+p_2, r = p_3+p_4+p_5$;
(b) $B(n-V, q/(1-r))$; **(c)** $B(n-W, q/(1-p_6))$

C. **1(a)** $T(n, p_1, p_2+p_3)$

5.2.1. A. **1(a)** .4207, .5675; **(b)** .1056; **(c)** .4090; **(d)** .6554; **(e)** .3050

B. **1** $(7.2\pi)^{-1}\exp([-(x_1^2/4+(x_2-3)^2/9-1.6x_1(x_2-3)/6]/.72)$, $\exp(3t_2+(4t_1^2+9t_2^2+9.6t_1t_2)/2)$

5.3.1. B. **1(a)** $\begin{Bmatrix}2\\5\\7\\9\end{Bmatrix}, \begin{Bmatrix}4 & 2 & 0 & 0\\2 & 6 & 1 & 0\\0 & 1 & 12 & -1\\0 & 0 & -1 & 10\end{Bmatrix}$; **(b)** $\begin{Bmatrix}13\\-13\end{Bmatrix}, \begin{Bmatrix}98 & -47\\-47 & 138\end{Bmatrix}$

2. A. **1(a)** $\begin{Bmatrix}2.5\\2.8\\550\end{Bmatrix}, \begin{Bmatrix}.25 & .10 & 20\\.10 & .36 & 30\\20 & 30 & 10{,}000\end{Bmatrix}$; **(b)** $N_2\left(\begin{Bmatrix}2.5\\2.8\end{Bmatrix}, \begin{Bmatrix}.25 & .10\\.10 & .36\end{Bmatrix}\right)$; **(c)** .2946

B. **1(a)** $\begin{Bmatrix}6\\4\\2\end{Bmatrix}, \begin{Bmatrix}16 & 6 & 0\\6 & 25 & 0\\0 & 0 & 64\end{Bmatrix}$; **(b)** $(23{,}296)^{-1/2}\pi^{-3/2}\exp(-q(\mathbf{x})/2)$

$q(\mathbf{x}) = (364)^{-1}(25(x_1-6)^2+16(x_2-4)^2-12(x_1-6)(x_2-4)) + (64)^{-1}(x_3-2)^2$; **(c)** $\exp(6t_1+4t_2+2t_3+(16t_1^2+25t_2^2+64t_3^2+12t_1t_2)/2)$.
5(a) $\exp(\boldsymbol{\mu}'\mathbf{t})$

3. A. **1(a)** $N(1.23+.148X_2+.0016X_3, .204)$; **(b)** $N(226.25+52.5X_1 + 68.75X_2, 6887.5)$ **(c)** $N((222.8+123.4Y_1, 6913.5)$

B.1(a) $\begin{Bmatrix}2\\1\\3\\4\end{Bmatrix}, \begin{Bmatrix}1 & 1 & 1 & 1\\1 & 2 & 1 & 1\\1 & 1 & 3 & 1\\1 & 1 & 1 & 4\end{Bmatrix}$; **(b)** $N_2\left(\begin{Bmatrix}2\\1\end{Bmatrix}, \begin{Bmatrix}1 & 1\\1 & 2\end{Bmatrix}\right)$;

(c) $N_2\left((11)^{-1}\begin{Bmatrix}5+3X_3+2X_4\\-5+3X_3+2X_4\end{Bmatrix}, (11)^{-1}\begin{Bmatrix}6 & 6\\6 & 17\end{Bmatrix}\right)$;

(d) $N_2\left(5^{-1}\begin{Bmatrix} 5+2X_2+X_3 \\ 15+2X_2+X_3 \end{Bmatrix}, 5^{-1}\begin{Bmatrix} 2 & 2 \\ 2 & 17 \end{Bmatrix}\right)$;

(e) $N_2\left(3^{-1}\begin{Bmatrix} 3+X_3 \\ 9+X_3 \end{Bmatrix}, 3^{-1}\begin{Bmatrix} 2 & 2 \\ 2 & 11 \end{Bmatrix}\right.$;

(f) $N_2\left(\begin{Bmatrix} 6 \\ -6 \end{Bmatrix}, \begin{Bmatrix} 12 & -3 \\ -3 & 13 \end{Bmatrix}\right)$; **(g)** $N((96-3Y_2)/13, 147/13)$.

CHAPTER 6

6.1.1. B. 1(a) $(1-(1-v)^n\ 0 \leqq v \leqq 1$; **(c)** $1-(1-q/n)^n$, $q \geqq 0$. **3(a)** $(1-t/n)^n$ **5(a)** $1-(1-q/n^{1/2})^n$, $q \geqq 0$; **(c)** $1-(1-r/n^2)^n$, $n \geqq 0$

2. B. 1(a) $1+2x+2x^2$; **(b)** $1+x^2$

6.2.1. B. 1(a) $B(n,p)$

6.3.1. B. 5(a) 108, .5, .25; **(b)** 0, 15

2. B. 3(a) $2\mu\sigma Z \sim N(0, 4\mu^2\sigma^2)$; **(b)** $\sigma^2 Z^2 \sim \Gamma(.5, 2\sigma^2)$; **(c)** $0Z = 0$

C. 1 $g(X_n) - g(0) \xrightarrow{d} Z^2 \sim \chi_1^2$, $g'(0)\,Z = 0$

6.4.2. B. 1(a) $2X + 4Y \sim N(0,20)$, $2+4X+Y \sim N(2,17)$; **(b)** $2X + 0 \sim N(0,4)$, $2X+4Y \sim N(0,20)$ **5** $N(0, \sigma^4 + \mu\delta\sigma^3 + \mu^2(\gamma-1)\sigma^2/4)$, $N(0, 4\mu^2\sigma^{10} + 4\mu^3\delta\sigma^9 + (\gamma-1)\mu^4\sigma^8)$

CHAPTER 7

7.1.1. B. 3 $k\sigma^{-n}\exp(-\Sigma(X_i - i\theta)^2/2\sigma^2)$ **5** $k(\mathbf{X})\theta^{-a_n}\exp(-\Sigma X_i/\theta)$

2. B. 3 $\bar{X}$, $\exp(\bar{X})$ **5** $(1+\bar{X})^{-1}, (1+\bar{X})/\bar{X}$ **7** $(Q, ((n-1)\,T/n)^{1/2})$, $QT((n-1)/n)^{1/2}$, $QT^2(n-1)/n$

C. 1 $(-\bar{X} + (5\bar{X}^2 + 4W^2)^{1/2})/2$ if $\bar{X} > 0$ and $(-\bar{X} - (5\bar{X}^2 + 4W^2)^{1/2}/2$ if $\bar{X} < 0$, and either root if $\bar{X} = 0$, where $W^2 = (n-1)\,S^2/n$.

3. B. 1 $h\delta^{-2n}\exp(-\Sigma X_i/i\delta^2)$

7.2.1. B. 3 $Q^2 - b_n, nR^2/(n-1)$ **5** Q, T^2, QT^2 **9(b)** $n/\theta(n-1)$; **(c)** $(n-1)/nR$; **(d)** $(c^2n^2 - 2cn(n-2) + (n-1)(n-2)/(n-1)(n-2)\,\theta^2$; **(e)** $(n-2)/n$ **11(a)** $\Gamma((n-1)/2, 2\sigma^2/(n-1))$; **(c)** $\sigma^2(c^2 - 2kc + 1)$ $(k = ES/\sigma)$; **(d)** $c = k$ (note that for the unbiased estimator $c = k^{-1}$)

3. B. 1(b) $4\theta^2/b_n$ **3(b)** $\text{var}(\hat{\theta}) = \theta(1-\theta)/2n$, yes

7.3.2. B. 3 $\sigma^{-2}\begin{Bmatrix} b_n & 0 \\ 0 & 2n \end{Bmatrix}$; **(b)** σ^2/b_n, yes; **(c)** $2\sigma^4/n$ no; **(d)** $\sigma^4(\sigma^2/b_n + 2\theta^2/n)$, no

CHAPTER 8

8.1.1. B. 1(b) $\theta \in 9.47 \pm .26$ **3** $6.63 \leqq \theta \leqq 15.68$ **5(a)** $(n-1)\,T^2/\chi_{n-1}^{2\,\alpha/2} \leqq \sigma^2 \leqq (n-1)\,T^2/\chi_{n-1}^{2\,1-\alpha/2}$; **(c)** $\theta \in Q \pm t_{n-1}^{\alpha/2}\,T/(b_n)^{1/2}$; **(d)** $\theta \in 9.47 \pm 1.53$, $12 \leqq \sigma^2 \leqq 161$ **7(a)** $(1-\theta t/n)^{-n}$; **(c)** $1.76 \leqq \theta \leqq 6.26$

C. 1(a) $2n^{1/2}(\tilde{X} - \theta)/\pi \xrightarrow{d} Z \sim N(0,1)$; **(c)** $\theta \in \tilde{X} \pm z^{\alpha/2}\pi/2n^{1/2}$
3 $\theta \in P + z^{\alpha/2}/a_n \pm (4P(z^{\alpha/2})^2 a_n + (z^{\alpha/2})^4)^{1/2}/2a_n$

2. B. 1 reject if $Z = (Q - 11)(55)^{1/2} > 1.96$ or < -1.96, $Z = 11.3$ so reject **3** Reject if $U = 3P > 46.98$ or <16.79, $U = 29$ so accept **5(a)** Reject if $t = (55)^{1/2}(Q - 10)/T < -3.747$, $t = -.78$ so accept **(b)** Reject if $U = 4T^2/50 < .71$, $U = 2.27$ so accept **7** Reject if $U = 10\hat{\theta} > 31.41$, $U = 30$ so accept

C. 1 reject if $Z = 2n^{1/2}\tilde{X}/\pi > z^{\alpha}$

CHAPTER 9

9.1.1. B. 1 .08 **3(a)** $\Phi(X) = 1$ if $X \geqq 8$, .77 if $X = 7$, .0625 if $X = 0$ and 0 if $0 < X < 7$ **(b)** $\Phi(X) = 1$ if $X \geqq 6$, .074 if $X = 5$, 0 if $X < 5$
5 $(a + c)/(a + b + c + d)$.

9.2.1. B. 3 .43, .21, .08, .73, .99.

2. B. 1(a) $\Phi(X) = 1$ if $X = 1$; **(b)** $\Phi(X) = 1$ if $X = 1$, $=1/2$ if $X = 2$; **(c)** $\Phi(X) = 1$ if $X = 1$ or 2; **(d)** $\Phi(X) = 1$ if $X = 5$; **(e)** $\Phi(X) = 1$ if $X = 5$, $=1/5$ if $X = 4$ **3(a)** Reject if $U = 20R > 31.41$, $U = 44$, so reject; **(b)** same as **(a)** **7** Reject of $X = 0$, $Y = 2$

3. B. 1 Family has MLR in W, so reject if $W > k = 12.59$ **3** Family has MLR in $U = \Sigma X_i^2$, so reject if $U > k = 9.49$, $U = 18$ so reject **5(b)** $\Phi(U) = 1$ if $U > 7$, $=38/43$, $U = 7$; **(c)** $\Phi(U) = 1$ if $U < 2$, $=54/121$ if $U = 2$; **(d)** Reject if $(U - 20)/4 > 1.645$; **(d)** $\Phi(U) = 1$ if $U > 3$, $=31/61$ if $U = 3$

CHAPTER 10

10.1.1. B. 1(a) $\begin{bmatrix} 3 \\ 2 \end{bmatrix}(1/2)^3$; **(b)** $X|T \sim B(T, 1/2)$

C. 1(a) $T \sim P(a_n\theta)$; **(b)** $\begin{bmatrix} t \\ x_1\, x_2 \ldots x_n \end{bmatrix} \Pi(i/a_n)^{x_i}$ if $t = \Sigma x_i$, 0 if $t \neq \Sigma x_i$

10.3.1 B. 1 R, $nR^2/(n + 1)$, $(n - 1)R/n$ **3** $(2X + Y)/n$ **5** $\bar{X}$, $n\bar{X}(1 - \bar{X})/(n + 1)$
9(a) $1 - e^{-4\theta}$; **(b)** e^{-2P}; **(c)** $\exp(\theta(e^{-4} - 1)) - 2\exp(\theta(e^{-2} - 3)) + \exp(-4)$

C. 1 $I(\theta) = n/\theta^2$, $\text{var}((n - 1)/nR) = ((n - 2)\theta^2)^{-1} > (n\theta^2)^{-1}$, $\text{var}(nR^2/(n + 1)) = (4n + 6)\theta^4/n(n + 1) > 4\theta^4/n$ **3** $(1 - a_n^{-1})^{a_n P}$

CHAPTER 11

11.1.2. A. 1 $Z = -3.6 < -2.326$, so significantly lower, $\mu \in 3.9 \pm 1.29$
3 $Z = 5.57 > 1.645$, so significantly higher, $\delta \in 30 \pm 12.5$.
5 $Z = -2.5 < -1.645$, so significant decrease, $\delta \in 50 \pm 51.5$.

11.2.1. A. 1 $t = 1.11 < 1.711$, so accept
B. 1 $-4.303 \leqq t = 2.59 \leqq 4.303$, so accept
2. A. 1 $-2.756 \leqq t = 1.96 \leqq 2.756$, so accept, $\delta \in 5 \pm 5.2$, $32 \leqq \sigma^2 \leqq 91$
3. A. 1(a) $Z = 2.025 > 1.96$, so the difference is significant;
(b) $-1.96 \leqq t = 1.72 \leqq 1.96$, so the difference is not significant
4. A. 1 $t = 2.75 > 2.262$, so difference is significant, $\delta \in 1.3 \pm .87$

11.3.1. A. 1 Reject if $Y > 2$, $Y = 5$, so the proportion of defectives is significantly greater than .01 **3** Reject if $U \leqq 5$, $U = 6$ so the proportion of working bombs is not significantly less than .9
C. 1(c) $.818 \pm .078$, $.80 \pm .078$
2. A. 1 $-1.96 \leqq Z = 1.64 \leqq 1.96$, so no significant difference, $\delta \in .1 \pm .12$
3 $P(V \geqq 5 | W = 12) = .004$. $V = 5$, so a significant difference

11.4.1. A. 1 $-1.96 \leqq .45 \leqq 1.96$ so no significant difference for smokers, $-1.96 \leqq .67 \leqq 1.96$, so no significant difference for non-smokers
3(a) $Z = 7.017 \geqq 1.96$; **(b)** In both cases, $Z = -5.6 \leqq -1.96$
B. 1 .0000036
2. A. 1 $Z = 5.77 > 2.56$ so can declare winner, $\mu \in .55 \pm .02$ **3** $\mu \in .2 \pm .01$
3. A. 1 $\mu \in 4 \pm .55$ **3** Reject if $Z = 10(\bar{X} - 900)/900 < -2.326$, $Z = -1.67$, so accept, $\mu \in 750 \pm 147$

11.5.1 B. 1(b) $ET = n\theta/(n+1)$, $\text{var}(T) = n\theta^2/(n+1)^2(n+2)$
5(a) $f(p) = n\exp(-np)$, $p > 0$, $p^\alpha = -\log(\alpha)/n$;
(b) $S - p^{\alpha/2} \leqq \theta \leqq S - p^{1-\alpha/2}$ **6(e)** $h(T) = (T^3 - (T-1)^3)/(2T-1)$

CHAPTER 12

12.1.1. A. 1 $\bar{X} = 13$, $\bar{Y} = 51$, $S_{xx} = 54$, $S_{YY} = 824$, $S_{xY} = 220$; **(a)** $\hat{\gamma} = 3.7$, $\hat{\theta} = 2.9$;
(b) $S^2 = 20.81$
2. A. 1(a) $\theta \in 2.9 \pm 23.0$, $\gamma \in 3.7 \pm 1.7$, $7.5 \leqq \sigma^2 \leqq 173$; **(b)** $5.96 > 2.776$, reject;
(c) $\mu(12) \in 47.3 \pm 5.4$, $T \in 47.4 \pm 13.8$;
(d) $\mu(t) \in 2.9 + 3.7t \pm 2.3(9 + (t-13)^2)^{1/2}$

12.2.1. A. 1 $r = .92$, $t = 4.1 > 2.353$, so significant correlation. **3(a)** $.23 \leqq \rho \leqq .62$;
(b) $1.40 \leqq 1.645$ so accept

12.3.1. A. 1(b) $\beta_1 + 0\beta_2 + 150\beta_3 + 22500\beta_4 + 40\beta_5 + 6000\beta_6$

B. 1 $X'X = \begin{Bmatrix} 7 & 0 & 0 \\ 0 & 42 & -38 \\ 0 & -38 & 68 \end{Bmatrix}$, $(X'X)^{-1} = \begin{Bmatrix} 1/7 & 0 & 0 \\ 0 & 68/1412 & 38/1412 \\ 0 & 38/1412 & 42/1412 \end{Bmatrix}$,

$X'Y = \begin{Bmatrix} 2 \\ 100 \\ -77 \end{Bmatrix}$, $Y'Y = 254$; $\hat{\beta}' = (2/7, 3874/1412, 566/1412) = (.29, 2.74, .40)$,

$S^2 = 2.48$ **11** $\hat{\beta} = \Sigma x_i Y_i / \Sigma x_i^2$, $(n-1)S^2 = \Sigma(Y_i - x_i\hat{\beta})^2 = \Sigma Y_i^2 - \hat{\beta}^2 \Sigma x_i^2$

2. A. 1(a) $A = \begin{Bmatrix} 0 & 0 & 0 & 0 & 1 & 0 \\ 0 & 0 & 0 & 0 & 0 & 1 \end{Bmatrix}$; **(b)** $C = \begin{Bmatrix} 0 & 0 & 1 & 0 & 0 & 0 \\ 0 & 0 & 0 & 1 & 0 & 0 \\ 0 & 0 & 0 & 0 & 0 & 1 \end{Bmatrix}$

B. **1(a)** $t = 1.47 \leqq 2.776$ so accept; **(b)** $\beta_2 \in 2.74 \pm 1.29$ **(c)** $F = 49 > 6.94$, so reject **(d)** $\hat{\mu}(1,1,2) = 3.83$, $V^2(1,1,2) = .42$, $\mu(1,1,2) \in 3.83 \pm 2.84$, $T \in 3.83 \pm 5.21$, $\mu(1,1,2) \in 3.83 \pm 4.11$; **(e)** $R^2 = .96$, $F = 48 > 6.94$ so there is significant improvement in fit, $R^{*2} = .94$. (Note that the F for this part is slightly different from that in part c due to roundoff.)

C. **1(a)** $F = (A\hat{\beta}^*)'(A(x'v^{-1})x)^{-1}A')^{-1}A\hat{\beta}^*/kS^{*2}$;
(b) $\mu(t) \in \hat{\mu}^*(t) \pm t_{n-p}^{\alpha/2} S^*V^*$, where $\hat{\mu}^*(t) = t'\hat{\beta}^*$, $V^{*2}(t) = t'(x'v^{-1}x)^{-1}t$

CHAPTER 13

13.1.1. **A.** **1** $T^2 = 85.75$, $S^2 = 86.25$, $F = 7.46 > F_{2,15}^{.05} = 3.68$ so reject

2. **A.** **1**

Due to	df	SS	MS	F
θ	1	1058	1058	184
α	2	85.75	42.88	7.46
error	15	86.25	5.75	

13.2.1. **A.** **1(a)**

due to	df	SS	MS	F
θ	1	736.33	736.33	1389
α	2	10.17	5.08	9.6
β	3	92.34	30.78	58.1
error	6	3.16	.53	

$58.1 > F_{2,6}^{.05} = 5.99$, so reject hypothesis of no effect due to type
(b) $9.6 > F_{1,6}^{.05} = 5.14$, so reject hypothesis of no effect due to temperature
3(a) $1.28 \leqq 7.71$, so no significant difference between fertilizers;
(b) $11.51 > 7.71$ so a significant difference between moistures;
(c) $6.97 \leqq 7.71$, so no significant difference between temperatures

B. **5(b)** $\hat{\theta} = \overline{Y}_{...}$, $\hat{\alpha}_i = \overline{Y}_{i..} - \overline{Y}_{...}$, $\hat{\beta}_j = \overline{Y}_{.j.} - \overline{Y}_{...}$ $\hat{\gamma}_k = \overline{Y}_{..k} - \overline{Y}_{...}$;
(c) $SS(\theta) = rdc\overline{Y}^2_{...}$, $SS(\alpha) = dc\Sigma\overline{Y}^2_{i..} - rdc\overline{Y}^2_{...}$,
$SS(\beta) = dr\Sigma\overline{Y}^2_{.j.} - rdc\overline{Y}^2_{...}$, $SS(\gamma) = rc\Sigma\overline{Y}^2_{..k} - rdc\overline{Y}^2_{...}$,
$SSE = \Sigma\Sigma\Sigma Y^2_{ijk} - dc\Sigma\overline{Y}^2_{i..} - rd\Sigma\overline{Y}^2_{.j.} - rc\Sigma\overline{Y}^2_{..k} + 2rcd\overline{Y}^2_{...}$, $df(\theta) = 1$,
$df(\alpha) = r-1$, $df(\beta) = c-1$, $df(\gamma) = d-1$, $dfe = rcd - r - c - d + 2$

2. **A.** **1(a)**

due to	df	SS	MS	F
θ	1	4761	4761	828
α	1	56.25	56.25	9.78
β	3	242	80.67	14.03
γ	3	14.75	4.92	.86
error	8	46	5.75	

(b) $9.78 > 5.32$, so significant effect due to car, $14.03 > 4.07$, so significant effect due to gasoline, $.86 < 4.07$, so no significant interaction

B. **5** $SS(\theta) = nrc\overline{Y}^2_{...}$, $SS(\alpha) = nc\Sigma\overline{Y}^2_{i..} - ncr\overline{Y}^2_{...}$,
$SS(\beta) = nr\Sigma\overline{Y}^2_{.j.} - ncr\overline{Y}^2_{...}$, $SSE = \Sigma\Sigma\Sigma Y^2_{ijk} - nc\Sigma\overline{Y}^2_{i..} - nr\Sigma\overline{Y}^2_{.j.} + ncr\overline{Y}^2_{..}$,
$df(\theta) = 1$, $df(\alpha) = r-1$, $df(\beta) = c-1$, $dfe = rcn - r - c + 1$

13.3.1. **A.** **1**

Due to	df	SS	MS	F
θ	1	663.06	663.06	2138.90
α	3	23.19	7.73	24.94

β	3	68.19	22.73	73.32
γ	3	.69	.23	.74
error	6	1.87	.31	

$24.94 > 4.76$, so significant effect due to temperature, $73.32 > 4.76$, so significant effect due to humidity, $.74 \leqq 4.76$, so no significant effect due to fertilizer

2. A. 1

Due to	df	SS	MS	F
θ	1	1921.14	1921.14	2705.83
α	2	28.28	14.14	19.15
δ	4	93.58	23.40	33.00
error	7	5.00	.71	

$19.15 > 4.74$, so the effect of states is significant, $33.00 > 4.12$, so the effect of sites within states is also significant

B. 5(a) $\hat{\theta} = \overline{Y}\ldots, \hat{\alpha}_i = \overline{Y}_{i..} - \overline{Y}\ldots, \hat{\beta}_j = \overline{Y}_{.j.} - \overline{Y}\ldots, \hat{\delta}_{ik} = \overline{Y}_{i.k} = \overline{Y}_{i..}$;
(b) $SS(\theta) = rcn\overline{Y}^2_{...}, SS(\alpha) = cn\Sigma\overline{Y}^2_{i..} - rcn\overline{Y}^2_{...}$,
$SS(\beta) = rn\Sigma\overline{Y}^2_{.j.} - rcn\overline{Y}^2_{...}, SS(\gamma) = c\Sigma\Sigma\overline{Y}^2_{i.k} - cn\Sigma\overline{Y}^2_{i..}$,
$SSE = \Sigma\Sigma\Sigma Y^2_{ijk} - c\Sigma\Sigma\overline{Y}^2_{i.k} - rn\Sigma\overline{Y}^2_{.j.} + rcn\overline{Y}^2_{...}, df(\theta) = 1$,
$df(\alpha) = r - 1, df(\beta) = c - 1, df(\delta) = r(n - 1), dfe = (rn - 1)(c - 1)$

13.4.1 A. 1 $\alpha_1 - \alpha_2 \in -1.75 \pm 4.20, \alpha_1 - \alpha_3 \in -5.25 \pm 3.85, \alpha_2 - \alpha_3 \in -3.5 \pm 3.36$, $\alpha_1 - \alpha_2$ is only insignificant comparison, most significant contrast is $-11.67\alpha_1 - 7.00\alpha_2 + 18.67\alpha_3$ **3(a)** $\alpha_1 - \alpha_2 \in -1.75 \pm 1.83$, $\alpha_1 - \alpha_3 \in -3.42 \pm 1.67, \alpha_2 - \alpha_3 \in -1.667 \pm 1.674$, only $\alpha_1 - \alpha_3$ is significant (but $\alpha_2 - \alpha_3$ is sure close). **(b)** $\delta_{11} - \delta_{12} \in -4.5 \pm 3.42$, $\delta_{21} - \delta_{22} \in -2 \pm 3.42, \delta_{31} - \delta_{32} \in 6 \pm 3.42, \delta_{31} - \delta_{33} \in -2 \pm 3.42$, $\delta_{32} - \delta_{33} \in -8 \pm 3.42$, so there is a significant difference between the 2 sites in state 1, no significant difference between the two sites in state 2, and a significant difference between sites 1 and 2 and between sites 2 and 3 in state 3.

B. 1(a) $\Sigma b_i\alpha \in \Sigma b_i\overline{Y}_{i..} \pm ((r-1)\,MSE\,F^{\alpha}_{r-1,C(n-1)})^{1/2}(\Sigma b_i^2/nc_i)^{1/2}, \Sigma b_i = 0$;
(b) $\Sigma\Sigma d_{ij}\,\delta_{ij} \in \Sigma\Sigma d_{ij}\,\overline{Y}_{ij.} \pm ((\mathrm{C} - \mathrm{r})\,MSE\,F^{\alpha}_{C-r,C(n-1)})^{1/2}(\Sigma\Sigma d_{ij}^2/n)^{1/2}$, $\Sigma_j d_{ij} = 0$.

13.5.2. A. 1(a) $\mathbf{Y}' = (7,8,6,4,2,3,5,8,3,5,9,6), \boldsymbol{\beta}' = (\theta,\alpha_1,\beta_1,\gamma_{11})$,

$$\mathbf{x}' = \begin{pmatrix} 1 & 1 & 1 & 1 & 1 & 1 & 1 & 1 & 1 & 1 & 1 & 1 \\ 1 & 1 & 1 & 1 & 1 & 1 & -1 & -1 & -1 & -1 & -1 & -1 \\ 1 & 1 & 1 & 1 & -1 & -1 & 1 & 1 & 1 & -1 & -1 & -1 \\ 1 & 1 & 1 & 1 & -1 & -1 & -1 & -1 & -1 & 1 & 1 & 1 \end{pmatrix}$$

(b) $\mathbf{A} = (0,1,0,0), \mathbf{B} = (0,0,1,0), \mathbf{C} = (0,0,0,1)$.

CHAPTER 14

14.1.1. A. 1 $\chi^2_L = 6.7 \leqq 7.81$, so accept **3(a)** 3.1,15.6,31.3,31.3,15.6,3.1;
(b) Combine 1st and last cells, $\chi^2_P = 6.6 \leqq 9.49$ so accept

B. 5(a) $\hat{\theta} = (30_1 + O_2)/(3n - O_2), \hat{\delta} = (O_2 + 3O_3)/(3n - O_2)$; **(b)** 1

2. A. 1 $\chi_P^2 = 69.94 > 5.99$, so reject hypothesis of independence

3(a)

	A	NA
B	18	40
NB	78	64

; **(b)** $\chi_L^2 = 9.6 > 3.84$, so reject

5(a) $\chi_P^2 = 7.30 \leqq 7.81$, so accept symmetry; **(b)** $\chi_L^2 = 25 > 12.59$, so reject independence and symmetry

14.2.1. A. 1 $\chi_P^2 = 16.8 > 9.49$, so reject hypothesis of equal proportions

2. A. 1 $\chi_P^2 = 17.33 > 7.81$, so reject hypothesis of equal means

CHAPTER 15

15.1.1. A. 1(a) $(13 - 10)/5^{1/2} = 1.34 \leqq 1.645$, so accept null hypothesis; **(b)** $(159 - 105)/(717.5)^{1/2} = 2.02 > 1.645$, so reject null hypothesis; **(c)** .30, .50; **(d)** $(12.67 - 7.97)/(4.776)^{1/2} = 2.15 > 1.645$, so reject null

2. A. 1(a) 14.5; **(b)** $4.5 \leqq \theta \leqq 27.5$; **(c)** 14; **(d)** $-2 \leqq \theta \leqq 30$.

3. A. 1(a) $(31 - 56)/(74.67)^{1/2} = -2.89 < -1.645$, so reject; **(b)** $T_W \in 56 \pm 1.96(74.67)^{1/2} \Leftrightarrow 39 \leqq T_W \leqq 73 \Leftrightarrow 11 \leqq U \leqq 45$, $D_{[11]} \leqq \delta < D_{[46]} \Leftrightarrow -40 \leqq \delta \leqq -14$; **(c)** $(D_{28} + D_{29})/2 = 28$

3. The approximate normal scores for $N = 15$ are -1.74, -1.24, $-.94$, $-.71$, $-.51$, $-.33$, $-.16$, 0, .16, .33, .51, .71, .94, 1.24, 1.74, $(-5.14 - 0)/(3.37)^{1/2} = -2.8 < -1.645$ so reject

15.2.1. B. 1 $d_S = 1$, $d_W = (3/4)^{1/2}$, $d_t = 2^{-1/2}$, $e(T_W, T_S) = 3/4$, $e(t, T_S) = 1/2$, $e(t, T_W) = 2/3$

CHAPTER 16

16.1.1. B. 3 $\beta(a + n, b + n\overline{X})$

2. B. 1 $E\Theta|\mathbf{X} = A^*B^* = (a + n\overline{X})\, b/(1 + nb)$, $E\Theta^2|\mathbf{X} = A^*(A^* - 1)\, B^{*2} = (a + n\overline{X})(a + n\overline{X} - 1)/(1 + nb)^2$

3(a) $A^*/(A^* + B^*) = (a + n)/(a + b + n(1 + \overline{X}))$; **(b)** $B^*/(A^* - 1) = (b + n\overline{X})/(a + n - 1)$ **7(a)** $\Gamma(a + b + c + h + k + m)\,\Gamma(a)\,\Gamma(b)\,\Gamma(c)/\Gamma(a + b + c)\,\Gamma(a + h)\,\Gamma(b + k)\,\Gamma(c + m)$; **(b)** $E\Theta_1|(X, Y) = A^*/(A^* + B^* + C^*) = (a + X)/(a + b + c + n)$, $E\Theta_1\Theta_2|(X, Y) = (a + X)(b + Y)/(a + b + c + n)(A + b + c + n - 1)$, $E\Theta_1\Theta_2(1 - \Theta_1 - \Theta_2) = (a + X)(b + Y)(c + n - X - Y)/(a + b + c + n)$ $(a + b + c + n - 1)(a + b + c + n - 2)$

3. A. 1 $\Theta|\mathbf{X} \sim \beta(10,12)$, $12\Theta/10(1 - \Theta) \sim F_{20,24}$; **(a)** $.23 \leqq \Theta \leqq .70$; **(b)** $.3 < (2.08)^{-1}$ so reject **3** $\Theta|\mathbf{X} \sim \Gamma(11,(550)^{-1})$, $(1100)\Theta|X \sim \chi^2_{22}$; **(a)** $.088 \leqq \theta \leqq .039$, $25.7 \leqq \mu \leqq 127.3$; **(b)** $\mu \leqq 40 \Leftrightarrow \Theta \geqq .025$, $27.5 \leqq 33.92$, so accept

B. 3 $(\theta - A^*)/B^* \sim N(0,1)$, reject if $(k - A^*)/B^* > z^\alpha$ **5** $\theta|\mathbf{X} \sim \Gamma(A^*, B^*)$, $A^* = a + n/2$, $B^* = 2b/(b\Sigma X_i^2 + 2)$, reject if $2\theta/B^* \leqq \chi^{2\,1-\alpha}_{2A^*}$

4. A. 1(a) $NG(115.26,19,10,542.84)$; **(b)** 115.26, 0.18, 60.31;
(c) $\mu \in 115.26 \pm 3.53$, $.009 \leqq \delta \leqq .031$, $31.8 \leqq \sigma^2 \leqq 113.2$;
(d) $t(\mathbf{X},100) = 9.0 > 1.725$, so reject
B. 3 $(\Gamma(C^* - .5)/\Gamma(C^*))\, D^{*1/2}$

16.2.1 A. 1(b) $(a + X_i)/(a + b + n_i)$; **(c)** $EQ = a/(a + b)$,
$ER = (a^2 + a)/(a + b)(a + b + 1)$;
(d) Let $T = (Q - R)/(R - Q^2) \xrightarrow{P} a + b$,
$TQ \xrightarrow{P} a$, $T(1 - Q) \xrightarrow{P} b$; **(e)** $Q(T/(n_i + T)) + \hat{\theta}(n_i/(n_i + T))$
B. 3 $\text{var}(X_i) = t_i ab + t_i^2 ab^2$, $\text{var}(U) = \text{ab}/(\Sigma t_i) + ab^2(\Sigma t_i^2/(\Sigma t_i)^2$ goes to 0 if $\Sigma t_i^2/(\Sigma t_i)^2$ goes to 0.

2. A. 1(a) $P(\theta = 0|Y = 0) = 40/49$, $P(\theta = 1|Y = 0) = 9/49$,
$EL(b|Y = 0) = 100{,}000 - 1{,}000{,}000(1 - E(\theta|Y = 0)) = \$916{,}326.53$
$> \$200{,}000$ so don't buy; **(b)** don't buy when $Y = 1$, buy when
$Y = 2$; **(c)** $-\$7{,}000$, $\$207{,}000$; **(d)** X (which has Bayes risk $-\$66{,}000$)
3(a) study the drug; **(b)** study the drug unless $X = 0$;
(c) \$313,000; **(d)** yes; **(e)** \$713,000

16.3.1. A. 1(a) $X/t, X/t, 4(3 + X)/(4t + 1)$; **(b)** .42, .42, .99; **(c)** $X = 2 > 1$ so reject;
(d) $\mu \in .5 \pm .10$. **3(a)** $\mu \in 180 \pm 14.7$

TABLES

TABLE 1 THE STANDARD NORMAL DISTRIBUTION FUNCTION. $P(Z \leqq z), Z \sim N(0, 1)$.

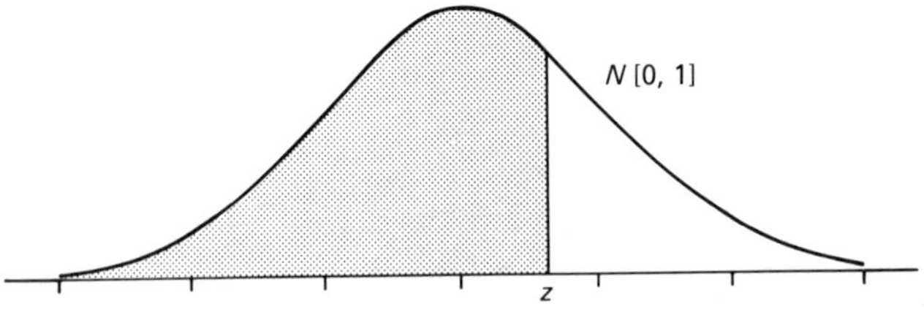

z	$P(Z \leqq z)$	z	$P(Z \leqq z)$	z	$P(Z \leqq z)$	z	$P(Z \leqq z)$
0.00	.5000	0.44	.6700	0.88	.8106	1.32	.9066
0.01	.5040	0.45	.6736	0.89	.8133	1.33	.9082
0.02	.5080	0.46	.6772	0.90	.8159	1.34	.9099
0.03	.5120	0.47	.6808	0.91	.8186	1.35	.9115
0.04	.5160	0.48	.6844	0.92	.8212	1.36	.9131
0.05	.5199	0.49	.6879	0.93	.8238	1.37	.9147
0.06	.5239	0.50	.6915	0.94	.8264	1.38	.9162
0.07	.5279	0.51	.6950	0.95	.8289	1.39	.9177
0.08	.5319	0.52	.6985	0.96	.8315	1.40	.9192
0.09	.5359	0.53	.7019	0.97	.8340	1.41	.9207
0.10	.5398	0.54	.7054	0.98	.8365	1.42	.9222
0.11	.5438	0.55	.7088	0.99	.8389	1.43	.9236
0.12	.5478	0.56	.7123	1.00	.8413	1.44	.9251
0.13	.5517	0.57	.7157	1.01	.8437	1.45	.9265
0.14	.5557	0.58	.7190	1.02	.8461	1.46	.9279
0.15	.5596	0.59	.7224	1.03	.8485	1.47	.9292
0.16	.5636	0.60	.7257	1.04	.8508	1.48	.9306
0.17	.5675	0.61	.7291	1.05	.8531	1.49	.9319
0.18	.5714	0.62	.7324	1.06	.8554	1.50	.9332
0.19	.5753	0.63	.7357	1.07	.8577	1.51	.9345
0.20	.5793	0.64	.7389	1.08	.8599	1.52	.9357
0.21	.5832	0.65	.7422	1.09	.8621	1.53	.9370
0.22	.5871	0.66	.7454	1.10	.8643	1.54	.9382
0.23	.5910	0.67	.7486	1.11	.8665	1.55	.9394
0.24	.5948	0.68	.7517	1.12	.8686	1.56	.9406
0.25	.5987	0.69	.7549	1.13	.8708	1.57	.9418
0.26	.6026	0.70	.7580	1.14	.8729	1.58	.9429
0.27	.6064	0.71	.7611	1.15	.8749	1.59	.9441
0.28	.6103	0.72	.7642	1.16	.8770	1.60	.9452
0.29	.6141	0.73	.7673	1.17	.8790	1.61	.9463
0.30	.6179	0.74	.7704	1.18	.8810	1.62	.9474
0.31	.6217	0.75	.7734	1.19	.8830	1.63	.9485
0.32	.6255	0.76	.7764	1.20	.8849	1.64	.9495
0.33	.6293	0.77	.7794	1.21	.8869	1.65	.9505
0.34	.6331	0.78	.7823	1.22	.8888	1.66	.9515
0.35	.6368	0.79	.7852	1.23	.8907	1.67	.9525
0.36	.6406	0.80	.7881	1.24	.8925	1.68	.9535
0.37	.6443	0.81	.7910	1.25	.8944	1.69	.9545
0.38	.6480	0.82	.7939	1.26	.8962	1.70	.9554
0.39	.6517	0.83	.7967	1.27	.8980	1.71	.9564
0.40	.6554	0.84	.7995	1.28	.8997	1.72	.9573
0.41	.6591	0.85	.8023	1.29	.9015	1.73	.9582
0.42	.6628	0.86	.8051	1.30	.9032	1.74	.9591
0.43	.6664	0.87	.8079	1.31	.9049	1.75	.9599

TABLE 1 (CONTINUED)

z	$P(Z \leqq z)$	z	$P(Z \leqq z)$	z	$P(Z \leqq z)$	z	$P(Z \leqq z)$
1.76	.9608	2.12	.9830	2.48	.9934	2.84	.9977
1.77	.9616	2.13	.9834	2.49	.9936	2.85	.9978
1.78	.9625	2.14	.9838	2.50	.9938	2.86	.9979
1.79	.9633	2.15	.9842	2.51	.9940	2.87	.9980
1.80	.9641	2.16	.9846	2.52	.9941	2.88	.9980
1.81	.9649	2.17	.9850	2.53	.9943	2.89	.9981
1.82	.9656	2.18	.9854	2.54	.9945	2.90	.9981
1.83	.9664	2.19	.9857	2.55	.9946	2.91	.9982
1.84	.9671	2.20	.9861	2.56	.9948	2.92	.9983
1.85	.9678	2.21	.9864	2.57	.9949	2.93	.9983
1.86	.9686	2.22	.9868	2.58	.9951	2.94	.9984
1.87	.9693	2.23	.9871	2.59	.9952	2.95	.9984
1.88	.9699	2.24	.9875	2.60	.9953	2.96	.9985
1.89	.9706	2.25	.9878	2.61	.9955	2.97	.9985
1.90	.9713	2.26	.9881	2.62	.9956	2.98	.9986
1.91	.9719	2.27	.9884	2.63	.9957	2.99	.9986
1.92	.9726	2.28	.9887	2.64	.9959	3.00	.9987
1.93	.9732	2.29	.9890	2.65	.9960	3.01	.9987
1.94	.9738	2.30	.9893	2.66	.9961	3.02	.9987
1.95	.9744	2.31	.9896	2.67	.9962	3.03	.9988
1.96	.9750	2.32	.9898	2.68	.9963	3.04	.9988
1.97	.9756	2.33	.9901	2.69	.9964	3.05	.9989
1.98	.9761	2.34	.9904	2.70	.9965	3.06	.9989
1.99	.9767	2.35	.9906	2.71	.9966	3.07	.9989
2.00	.9773	2.36	.9909	2.72	.9967	3.08	.9990
2.01	.9778	2.37	.9911	2.73	.9968	3.09	.9990
2.02	.9783	2.38	.9913	2.74	.9969	3.10	.9990
2.03	.9788	2.39	.9916	2.75	.9970	3.20	.9993
2.04	.9793	2.40	.9918	2.76	.9971	3.30	.9995
2.05	.9798	2.41	.9920	2.77	.9972	3.40	.9997
2.06	.9803	2.42	.9922	2.78	.9973	3.50	.9998
2.07	.9808	2.43	.9925	2.79	.9974	3.60	.9998
2.08	.9812	2.44	.9927	2.80	.9974	3.70	.9999
2.09	.9817	2.45	.9929	2.81	.9975	3.80	.9999
2.10	.9821	2.46	.9931	2.82	.9976	3.90	1.0000
2.11	.9826	2.47	.9932	2.83	.9977	4.00	1.0000

Adapted from R.N. Goldman and J.S. Weinberg, *Statistics—An Introduction.* (Englewood Cliffs, N.J.: Prentice Hall, 1985), 677–679.

TABLE 2 UPPER QUANTILES OF THE *t*-DISTRIBUTION

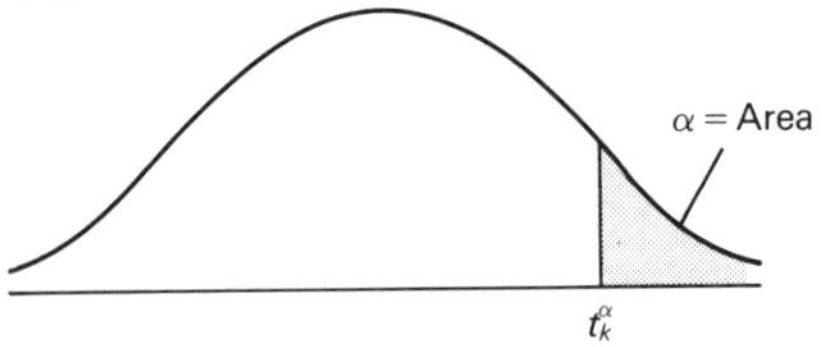

k	$t_k^{.25}$	$t_k^{.2}$	$t_k^{.15}$	$t_k^{.1}$	$t_k^{.05}$	$t_k^{.025}$	$t_k^{.01}$	$t_k^{.005}$
1	1.000	1.376	1.963	3.078	6.314	12.706	31.821	63.657
2	.817	1.061	1.386	1.886	2.920	4.303	6.965	9.925
3	.765	.978	1.250	1.638	2.353	3.183	4.541	5.841
4	.741	.941	1.190	1.533	2.132	2.776	3.747	4.604
5	.727	.920	1.156	1.476	2.015	2.571	3.365	4.032
6	.718	.906	1.134	1.440	1.943	2.447	3.143	3.707
7	.711	.896	1.119	1.415	1.895	2.365	2.998	3.500
8	.706	.889	1.108	1.397	1.860	2.306	2.896	3.355
9	.703	.883	1.100	1.383	1.833	2.262	2.821	3.250
10	.700	.879	1.093	1.372	1.813	2.228	2.764	3.169
11	.697	.876	1.088	1.363	1.796	2.201	2.718	3.106
12	.696	.873	1.083	1.356	1.782	2.179	2.681	3.055
13	.694	.870	1.079	1.350	1.771	2.160	2.650	3.012
14	.692	.868	1.076	1.345	1.761	2.145	2.624	2.977
15	.691	.866	1.074	1.341	1.753	2.132	2.602	2.947
16	.690	.865	1.071	1.337	1.746	2.120	2.583	2.921
17	.689	.863	1.069	1.333	1.740	2.110	2.567	2.898
18	.688	.862	1.067	1.330	1.734	2.101	2.552	2.878
19	.688	.861	1.066	1.328	1.729	2.093	2.539	2.861
20	.687	.860	1.064	1.325	1.725	2.086	2.528	2.845
21	.686	.859	1.063	1.323	1.721	2.080	2.518	2.831
22	.686	.858	1.061	1.321	1.717	2.074	2.508	2.819
23	.685	.858	1.060	1.319	1.714	2.069	2.500	2.807
24	.685	.857	1.059	1.318	1.711	2.064	2.492	2.797
25	.684	.856	1.058	1.316	1.708	2.060	2.485	2.787
26	.684	.856	1.058	1.315	1.706	2.056	2.479	2.779
27	.684	.855	1.057	1.314	1.703	2.052	2.473	2.771
28	.683	.855	1.056	1.313	1.701	2.048	2.467	2.763
29	.683	.854	1.055	1.311	1.699	2.045	2.462	2.756
30	.683	.854	1.055	1.310	1.697	2.042	2.457	2.750
31	.683	.854	1.054	1.310	1.696	2.040	2.453	2.744
32	.682	.853	1.054	1.309	1.694	2.037	2.449	2.739
33	.682	.853	1.053	1.308	1.692	2.035	2.445	2.733
34	.682	.852	1.053	1.307	1.691	2.032	2.441	2.728
35	.682	.852	1.052	1.306	1.690	2.030	2.438	2.724
36	.681	.852	1.052	1.306	1.688	2.028	2.434	2.720
37	.681	.852	1.051	1.305	1.687	2.026	2.431	2.716
38	.681	.851	1.051	1.304	1.686	2.024	2.428	2.712
39	.681	.851	1.050	1.304	1.685	2.023	2.426	2.708
40	.681	.851	1.050	1.303	1.684	2.021	2.423	2.705
50	.679	.849	1.047	1.299	1.676	2.009	2.403	2.678
60	.679	.848	1.046	1.296	1.671	2.000	2.390	2.660
70	.678	.847	1.044	1.294	1.667	1.995	2.381	2.648
80	.678	.846	1.043	1.292	1.664	1.990	2.374	2.639
90	.677	.846	1.043	1.291	1.662	1.987	2.368	2.632
100	.677	.845	1.042	1.290	1.660	1.984	2.364	2.626
∞	.674	.842	1.036	1.282	1.645	1.960	2.326	2.576

Adapted from D.B. Owen, *Handbook of Statistical Tables*, courtesy of Atomic Energy Commission. (Reading, Mass.: Addison-Wesley, 1962)

TABLE 3 UPPER QUANTILES OF THE χ^2 DISTRIBUTION

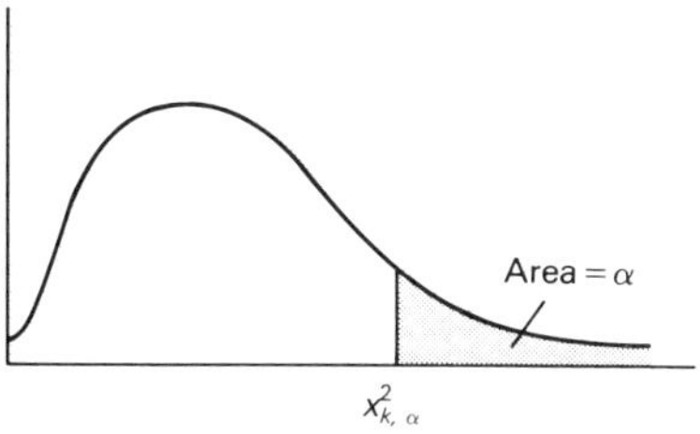

k	χ^2_{k} .995	χ^2_{k} .99	χ^2_{k} .975	χ^2_{k} .95	χ^2_{k} .90	χ^2_{k} .10	χ^2_{k} .05	χ^2_{k} .025	χ^2_{k} .01	χ^2_{k} .005
1	.00	.00	.00	.00	.02	2.71	3.84	5.02	6.63	7.88
2	.01	.02	.05	.10	.21	4.61	5.99	7.38	9.21	10.60
3	.07	.11	.22	.35	.58	6.25	7.81	9.35	11.34	12.84
4	.21	.30	.48	.71	1.06	7.78	9.49	11.14	13.28	14.86
5	.41	.55	.83	1.15	1.61	9.24	11.07	12.83	15.09	16.75
6	.68	.87	1.24	1.64	2.20	10.64	12.59	14.45	16.81	18.55
7	.99	1.24	1.69	2.17	2.83	12.02	14.07	16.01	18.48	20.28
8	1.34	1.65	2.18	2.73	3.49	13.36	15.51	17.54	20.09	21.96
9	1.73	2.09	2.70	3.33	4.17	14.68	16.92	19.02	21.67	23.59
10	2.16	2.56	3.25	3.94	4.87	15.99	18.31	20.48	23.21	25.19
11	2.60	3.05	3.82	4.57	5.58	17.28	19.68	21.92	24.72	26.76
12	3.07	3.57	4.40	5.23	6.30	18.55	21.03	23.34	26.22	28.30
13	3.57	4.11	5.01	5.89	7.04	19.81	22.36	24.74	27.69	29.82
14	4.07	4.66	5.63	6.57	7.79	21.06	23.68	26.12	29.14	31.32
15	4.60	5.23	6.26	7.26	8.55	22.31	25.00	27.49	30.58	32.80
16	5.14	5.81	6.91	7.96	9.31	23.54	26.30	28.85	32.00	34.27
17	5.70	6.41	7.56	8.67	10.09	24.77	27.59	30.19	33.41	35.72
18	6.26	7.01	8.23	9.39	10.86	25.99	28.87	31.53	34.81	37.16
19	6.84	7.63	8.91	10.12	11.65	27.20	30.14	32.85	36.19	38.58
20	7.43	8.26	9.59	10.85	12.44	28.41	31.41	34.17	37.57	40.00
21	8.03	8.90	10.28	11.59	13.24	29.62	32.67	35.48	38.93	41.40
22	8.64	9.54	10.98	12.34	14.04	30.81	33.92	36.78	40.29	42.80
23	9.26	10.20	11.69	13.09	14.85	32.01	35.17	38.08	41.64	44.18
24	9.89	10.86	12.40	13.85	15.66	33.20	36.42	39.36	42.98	45.56
25	10.52	11.52	13.12	14.61	16.47	34.38	37.65	40.65	44.31	46.93
26	11.16	12.20	13.84	15.38	17.29	35.56	38.89	41.92	45.64	48.29
27	11.81	12.88	14.57	16.15	18.11	36.74	40.11	43.19	46.96	49.65
28	12.46	13.56	15.31	16.93	18.94	37.92	41.34	44.46	48.28	50.99
29	13.12	14.26	16.05	17.71	19.77	39.09	42.56	45.72	49.59	52.34
30	13.79	14.95	16.79	18.49	20.60	40.26	43.77	46.98	50.89	53.67
50	27.99	29.71	32.36	34.76	37.69	63.17	67.50	71.42	76.15	79.49
100	67.33	70.06	74.22	77.93	82.36	118.5	124.3	129.6	135.8	140.2
500	422.3	429.4	439.9	449.1	459.9	540.9	553.1	563.9	576.5	585.2
1000	888.6	898.8	914.3	927.6	943.1	1058	1075	1090	1107	1119

Adapted from R.N. Goldman and J.S. Weinberg, *Statistics—An Introduction*. (Englewood Cliffs, N.J.: Prentice Hall, 1985), 681.

TABLE 4 UPPER QUANTILES OF THE *F*-DISTRIBUTION ($F^{\alpha}_{k,n}$)

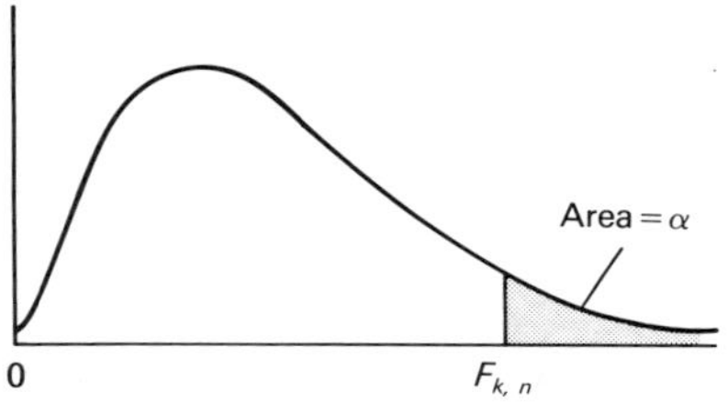

n	α	*k* = 1	2	3	4	5	6	7	8	9	10	11	12	15	20	24	30
(1)	.25	5.83	7.50	8.20	8.58	8.82	8.98	9.10	9.19	9.26	9.32	9.36	9.41	9.49	9.58	9.63	9.67
	.1	39.9	49.5	53.6	55.8	57.2	58.2	58.9	59.4	59.9	60.2	60.5	60.7	61.2	61.7	62.0	62.3
	.05	161	200	216	225	230	234	237	239	241	242	243	244	246	248	249	250
	.025	648	800	864	900	922	937	948	957	963	969	973	977	985	993	997	100^1
	.01	405^1	500^1	540^1	562^1	576^1	586^1	593^1	598^1	602^1	606^1	608^1	611^1	616^1	621^1	623^1	626^1
	.005	162^2	200^2	216^2	225^2	231^2	234^2	237^2	239^2	241^2	242^2	243^2	244^2	246^2	248^2	249^2	250^2
(2)	.25	2.57	3.00	3.15	3.23	3.28	3.31	3.34	3.35	3.37	3.38	3.39	3.39	3.41	3.43	3.43	3.44
	.1	8.53	9.00	9.16	9.24	9.29	9.33	9.35	9.37	9.38	9.39	9.40	9.41	9.42	9.44	9.45	9.46
	.05	18.5	19.0	19.2	19.2	19.3	19.3	19.4	19.4	19.4	19.4	19.4	19.4	19.4	19.4	19.5	19.5
	.025	38.5	39.0	39.2	39.2	39.3	39.3	39.4	39.4	39.4	39.4	39.4	39.4	39.4	39.4	39.5	39.5
	.01	98.5	99.0	99.2	99.2	99.3	99.3	99.4	99.4	99.4	99.4	99.4	99.4	99.4	99.4	99.5	99.5
	.005	198	199	199	199	199	199	199	199	199	199	199	199	199	199	199	199
(3)	.25	2.02	2.28	2.36	2.39	2.41	2.42	2.43	2.44	2.44	2.44	2.45	2.45	2.46	2.46	2.46	2.47
	.1	5.54	5.46	5.39	5.34	5.31	5.28	5.27	5.25	5.24	5.23	5.22	5.22	5.20	5.18	5.18	5.17
	.05	10.1	9.55	9.28	9.12	9.01	8.94	8.89	8.85	8.81	8.79	8.76	8.74	8.70	8.66	8.63	8.62
	.025	17.4	16.0	15.4	15.1	14.9	14.7	14.6	14.5	14.5	14.4	14.4	14.3	14.3	14.2	14.1	14.1
	.01	34.1	30.8	29.5	28.7	28.2	27.9	27.7	27.5	27.3	27.2	27.1	27.1	26.9	26.7	26.6	26.5
	.005	55.6	49.8	47.5	46.2	45.4	44.8	44.4	44.1	43.9	43.7	43.5	43.4	43.1	42.8	42.6	42.5
(4)	.25	1.81	2.00	2.05	2.06	2.07	2.08	2.08	2.08	2.08	2.08	2.08	2.08	2.08	2.08	2.08	2.08
	.1	4.54	4.32	4.19	4.11	4.05	4.01	3.98	3.95	3.94	3.92	3.91	3.90	3.87	3.84	3.83	3.82
	.05	7.71	6.94	6.59	6.39	6.26	6.16	6.09	6.04	6.00	5.96	5.94	5.91	5.86	5.80	5.77	5.75
	.025	12.2	10.6	9.98	9.60	9.36	9.20	9.07	8.98	8.90	8.84	8.79	8.75	8.66	8.56	8.51	8.46
	.01	21.2	18.0	16.7	16.0	15.5	15.2	15.0	14.8	14.7	14.5	14.4	14.4	14.2	14.0	13.9	13.8
	.005	31.3	26.3	24.3	23.2	22.5	22.0	21.6	21.4	21.1	21.0	20.8	20.7	20.4	20.2	20.0	19.9

(5)	.25	1.69	1.85	1.88	1.89	1.89	1.89	1.89	1.89	1.89	1.89	1.89	1.89	1.89	1.88	1.88	1.88
	.1	4.06	3.78	3.62	3.52	3.45	3.40	3.37	3.34	3.32	3.30	3.28	3.27	3.24	3.21	3.19	3.17
	.05	6.61	5.79	5.41	5.19	5.05	4.95	4.88	4.82	4.77	4.74	4.71	4.68	4.62	4.56	4.53	4.50
	.025	10.0	8.43	7.76	7.39	7.15	6.98	6.85	6.76	6.68	6.62	6.57	6.52	6.43	6.33	6.28	6.23
	.01	16.3	13.3	12.1	11.4	11.0	10.7	10.5	10.3	10.2	10.1	9.96	9.89	9.72	9.55	9.47	9.38
	.005	22.8	18.3	16.5	15.6	14.9	14.5	14.2	14.0	13.8	13.6	13.5	13.4	13.1	12.9	12.8	12.7
(6)	.25	1.62	1.76	1.78	1.79	1.79	1.78	1.78	1.78	1.77	1.77	1.77	1.77	1.76	1.76	1.75	1.75
	.1	3.78	3.46	3.29	3.18	3.11	3.05	3.01	2.98	2.96	2.94	2.92	2.90	2.87	2.84	2.82	2.80
	.05	5.99	5.14	4.76	4.53	4.39	4.28	4.21	4.15	4.10	4.06	4.03	4.00	3.94	3.87	3.84	3.81
	.025	8.81	7.26	6.60	6.23	5.99	5.82	5.70	5.60	5.52	5.46	5.41	5.37	5.27	5.17	5.12	5.07
	.01	13.7	10.9	9.78	9.15	8.75	8.47	8.26	8.10	7.98	7.87	7.79	7.72	7.56	7.40	7.31	7.23
	.005	18.6	14.5	12.9	12.0	11.5	11.1	10.8	10.6	10.4	10.2	10.1	10.0	9.81	9.59	9.47	9.36
(7)	.25	1.57	1.70	1.72	1.72	1.71	1.71	1.70	1.70	1.69	1.69	1.69	1.68	1.68	1.67	1.67	1.66
	.1	3.59	3.26	3.07	2.96	2.88	2.83	2.78	2.75	2.72	2.70	2.68	2.67	2.63	2.59	2.58	2.56
	.05	5.59	4.74	4.35	4.12	3.97	3.87	3.79	3.73	3.68	3.64	3.60	3.57	3.51	3.44	3.41	3.38
	.025	8.07	6.54	5.89	5.52	5.29	5.12	4.99	4.90	4.82	4.76	4.71	4.67	4.57	4.47	4.42	4.36
	.01	12.2	9.55	8.45	7.85	7.46	7.19	6.99	6.84	6.72	6.62	6.54	6.47	6.31	6.16	6.07	5.99
	.005	16.2	12.4	10.9	10.0	9.52	9.16	8.89	8.68	8.51	8.38	8.27	8.18	7.97	7.75	7.65	7.53
(8)	.25	1.54	1.66	1.67	1.66	1.66	1.65	1.64	1.64	1.64	1.63	1.63	1.62	1.62	1.61	1.60	1.60
	.1	3.46	3.11	2.92	2.81	2.73	2.67	2.62	2.59	2.56	2.54	2.52	2.50	2.46	2.42	2.40	2.38
	.05	5.32	4.46	4.07	3.84	3.69	3.58	3.50	3.44	3.39	3.35	3.31	3.28	3.22	3.15	3.12	3.08
	.025	7.57	6.06	5.42	5.05	4.82	4.65	4.53	4.43	4.36	4.30	4.24	4.20	4.10	4.00	3.95	3.89
	.01	11.3	8.65	7.59	7.01	6.63	6.37	6.18	6.03	5.91	5.81	5.73	5.67	5.52	5.36	5.28	5.20
	.005	14.7	11.0	9.60	8.81	8.30	7.95	7.69	7.50	7.34	7.21	7.10	7.01	6.81	6.61	6.50	6.40
(9)	.25	1.51	1.62	1.63	1.63	1.62	1.61	1.60	1.60	1.59	1.59	1.58	1.58	1.57	1.56	1.56	1.55
	.1	3.36	3.01	2.81	2.69	2.61	2.55	2.51	2.47	2.44	2.42	2.40	2.38	2.34	2.30	2.28	2.25
	.05	5.12	4.26	3.86	3.63	3.48	3.37	3.29	3.23	3.18	3.14	3.10	3.07	3.01	2.94	2.90	2.86
	.025	7.21	5.71	5.08	4.72	4.48	4.32	4.20	4.10	4.03	3.96	3.91	3.87	3.77	3.67	3.61	3.56
	.01	10.6	8.02	6.99	6.42	6.06	5.80	5.61	5.47	5.35	5.26	5.18	5.11	4.96	4.81	4.73	4.65
	.005	13.6	10.1	8.72	7.96	7.47	7.13	6.88	6.69	6.54	6.42	6.31	6.23	6.03	5.83	5.73	5.62
(10)	.25	1.49	1.60	1.60	1.59	1.59	1.58	1.57	1.56	1.56	1.55	1.55	1.54	1.53	1.52	1.52	1.51
	.1	3.28	2.92	2.73	2.61	2.52	2.46	2.41	2.38	2.35	2.32	2.30	2.28	2.24	2.20	2.18	2.16
	.05	4.96	4.10	3.71	3.48	3.33	3.22	3.14	3.07	3.02	2.98	2.94	2.91	2.85	2.77	2.74	2.70
	.025	6.94	5.46	4.83	4.47	4.24	4.07	3.95	3.85	3.78	3.72	3.66	2.62	3.52	3.42	3.37	3.31
	.01	10.0	7.56	6.55	5.99	5.64	5.39	5.20	5.06	4.94	4.85	4.77	4.71	4.56	4.41	4.33	4.25
	.005	12.8	9.43	8.08	7.34	6.87	6.54	6.30	6.12	5.97	5.85	5.75	5.66	5.47	5.27	5.17	5.07

TABLE 4 (CONTINUED)

		k															
n	α	1	2	3	4	5	6	7	8	9	10	11	12	15	20	24	30
(11)	.25	1.47	1.58	1.58	1.57	1.56	1.55	1.54	1.53	1.53	1.52	1.52	1.51	1.50	1.49	1.49	1.48
	.1	3.23	2.86	2.66	2.54	2.45	2.39	2.34	2.30	2.27	2.25	2.23	2.21	2.17	2.12	2.10	2.08
	.05	4.84	3.98	3.59	3.36	3.20	3.09	3.01	2.95	2.90	2.85	2.82	2.79	2.72	2.65	2.61	2.57
	.025	6.72	5.26	4.63	4.28	4.04	3.88	3.76	3.66	3.59	3.53	3.47	3.43	3.33	3.23	3.17	3.12
	.01	9.65	7.21	6.22	5.67	5.32	5.07	4.89	4.74	4.63	4.54	4.46	4.40	4.25	4.10	4.02	3.94
	.005	12.2	8.91	7.60	6.88	6.42	6.10	5.86	5.68	5.54	5.42	5.32	5.24	5.05	4.86	4.76	4.65
(12)	.25	1.46	1.56	1.56	1.55	1.54	1.53	1.52	1.51	1.51	1.50	1.50	1.49	1.48	1.47	1.46	1.45
	.1	3.18	2.81	2.61	2.48	2.39	2.33	2.28	2.24	2.21	2.19	2.17	2.15	2.11	2.06	2.04	2.01
	.05	4.75	3.89	3.49	3.26	3.11	3.00	2.91	2.85	2.80	2.75	2.72	2.69	2.62	2.54	2.51	2.47
	.025	6.55	5.10	4.47	4.12	3.89	3.73	3.61	3.51	3.44	3.37	3.32	3.28	3.18	3.07	3.02	2.96
	.01	9.33	6.93	5.95	5.41	5.06	4.82	4.64	4.50	4.39	4.30	4.22	4.16	4.01	3.86	3.78	3.70
	.005	11.8	8.51	7.23	6.52	6.07	5.76	5.52	5.35	5.20	5.09	4.99	4.91	4.72	4.53	4.43	4.33
(13)	.25	1.45	1.55	1.55	1.53	1.52	1.51	1.50	1.49	1.49	1.48	1.48	1.47	1.46	1.45	1.44	1.43
	.1	3.14	2.76	2.56	2.43	2.35	2.28	2.23	2.20	2.16	2.14	2.12	2.10	2.05	2.01	1.96	1.96
	.05	4.67	3.81	3.41	3.18	3.03	2.92	2.83	2.77	2.71	2.67	2.64	2.60	2.53	2.46	2.42	2.38
	.025	6.41	4.97	4.35	4.00	3.77	3.60	3.48	3.39	3.31	3.25	3.20	3.15	3.05	2.95	2.89	2.84
	.01	9.07	6.70	5.74	5.21	4.86	4.62	4.44	4.30	4.19	4.10	3.98	3.96	3.82	3.67	3.59	3.51
	.005	11.4	8.19	6.93	6.23	5.79	5.48	5.25	5.08	4.94	4.82	4.73	4.64	4.46	4.27	4.17	4.07
(14)	.25	1.44	1.53	1.53	1.52	1.51	1.50	1.49	1.48	1.47	1.46	1.46	1.45	1.44	1.43	1.42	1.41
	.1	3.10	2.73	2.52	2.39	2.31	2.24	2.19	2.15	2.12	2.10	2.07	2.05	2.01	1.96	1.94	1.91
	.05	4.60	3.74	3.34	3.11	2.96	2.85	2.76	2.70	2.65	2.60	2.58	2.53	2.46	2.39	2.35	2.31
	.025	6.30	4.86	4.24	3.89	3.66	3.50	3.38	3.29	3.21	3.15	3.10	3.05	2.95	2.84	2.79	2.73
	.01	8.88	6.51	5.56	5.04	4.69	4.46	4.28	4.14	4.03	3.94	3.87	3.80	3.66	3.50	3.43	3.35
	.005	11.1	7.92	6.68	6.00	5.56	5.26	5.03	4.86	4.72	4.60	4.52	4.43	4.25	4.06	3.96	3.86
(15)	.25	1.43	1.52	1.52	1.51	1.49	1.48	1.47	1.46	1.46	1.45	1.44	1.44	1.43	1.41	1.41	1.40
	.1	3.07	2.70	2.49	2.36	2.27	2.21	2.16	2.12	2.09	2.06	2.04	2.02	1.97	1.92	1.90	1.87
	.05	4.54	3.68	3.29	3.06	2.90	2.79	2.71	2.64	2.59	2.54	2.51	2.48	2.40	2.33	2.39	2.25
	.025	6.20	4.76	4.15	3.80	3.58	3.41	3.29	3.20	3.12	3.06	3.01	2.96	2.86	2.76	2.70	2.64
	.01	8.68	6.36	5.42	4.89	4.56	4.32	4.14	4.00	3.89	3.80	3.73	3.67	3.52	3.37	3.29	3.21
	.005	10.8	7.70	6.48	5.80	5.37	5.07	4.85	4.67	4.54	4.42	4.33	4.25	4.07	3.88	3.79	3.69
(16)	.25	1.42	1.51	1.51	1.50	1.48	1.47	1.46	1.45	1.44	1.44	1.43	1.43	1.41	1.40	1.39	1.38
	.1	3.05	2.67	2.46	2.33	2.24	2.18	2.13	2.09	2.06	2.03	2.01	1.98	1.94	1.89	1.87	1.84
	.05	4.49	3.63	3.24	3.01	2.85	2.74	2.66	2.59	2.54	2.49	2.46	2.42	2.35	2.28	2.24	2.19

	.025	6.12	4.69	4.08	3.73	3.50	3.34	3.22	3.12	3.05	2.99	2.94	2.89	2.79	2.68	2.62	2.57
	.01	8.53	6.23	5.29	4.77	4.44	4.20	4.03	3.89	3.78	3.69	3.62	3.55	3.41	3.26	3.18	3.10
	.005	10.6	7.51	6.30	5.64	5.21	4.91	4.69	4.52	4.38	4.27	4.18	4.10	3.92	3.73	3.64	3.54
(17)	.25	1.42	1.51	1.50	1.49	1.47	1.46	1.45	1.44	1.43	1.43	1.42	1.41	1.40	1.39	1.38	1.37
	.1	3.03	2.64	2.44	2.31	2.22	2.15	2.10	2.06	2.03	2.00	1.98	1.96	1.91	1.86	1.84	1.81
	.05	4.45	3.59	3.20	2.96	2.81	2.70	2.61	2.55	2.49	2.45	2.42	2.38	2.31	2.23	2.19	2.15
	.025	6.04	4.62	4.01	3.66	3.44	3.28	3.16	3.06	2.98	2.92	2.87	2.82	2.72	2.62	2.56	2.50
	.01	8.40	6.11	5.19	4.67	4.34	4.10	3.93	3.79	3.68	3.59	3.52	3.46	3.31	3.16	3.08	3.00
	.005	10.4	7.35	6.16	5.50	5.07	4.78	4.56	4.39	4.25	4.14	4.06	3.97	3.79	3.61	3.51	3.41
(18)	.25	1.41	1.50	1.49	1.48	1.46	1.45	1.44	1.43	1.42	1.42	1.41	1.40	1.39	1.38	1.37	1.36
	.1	3.01	2.62	2.42	2.29	2.20	2.13	2.08	2.04	2.00	1.98	1.96	1.93	1.89	1.84	1.81	1.78
	.05	4.41	3.55	3.16	2.93	2.77	2.66	2.58	2.51	2.46	2.41	2.38	2.34	2.27	2.19	2.15	2.11
	.025	5.98	4.56	3.95	3.61	3.38	3.22	3.10	3.01	2.93	2.87	2.81	2.77	2.67	2.56	2.50	2.44
	.01	8.29	6.01	5.09	4.58	4.25	4.01	3.84	3.71	3.60	3.51	3.44	3.37	3.23	3.08	3.00	2.92
	.005	10.2	7.21	6.03	5.37	4.96	4.66	4.44	4.28	4.14	4.03	3.94	3.86	3.68	3.50	3.40	3.30
(19)	.25	1.41	1.49	1.49	1.47	1.46	1.44	1.43	1.42	1.41	1.41	1.40	1.40	1.38	1.37	1.36	1.35
	.1	2.99	2.61	2.40	2.27	2.18	2.11	2.06	2.02	1.98	1.96	1.93	1.91	1.86	1.81	1.79	1.76
	.05	4.38	3.52	3.13	2.90	2.74	2.63	2.54	2.48	2.42	2.38	2.34	2.31	2.23	2.16	2.11	2.07
	.025	5.92	4.51	3.90	3.56	3.33	3.17	3.05	2.96	2.88	2.82	2.77	2.72	2.62	2.51	2.45	2.39
	.01	8.18	5.93	5.01	4.50	4.17	3.94	3.77	3.63	3.52	3.43	3.36	3.30	3.15	3.00	2.92	2.84
	.005	10.1	7.09	5.92	5.27	4.85	4.56	4.34	4.18	4.04	3.93	3.84	3.76	3.59	3.40	3.31	3.21
(20)	.25	1.40	1.49	1.48	1.47	1.45	1.44	1.43	1.42	1.41	1.40	1.39	1.39	1.37	1.36	1.35	1.34
	.1	2.97	2.59	2.38	2.25	2.16	2.09	2.04	2.00	1.96	1.94	1.91	1.89	1.84	1.79	1.77	1.74
	.05	4.35	3.49	3.10	2.87	2.71	2.60	2.51	2.45	2.39	2.35	2.31	2.28	2.20	2.12	2.08	2.04
	.025	5.87	4.46	3.86	3.51	3.29	3.13	3.01	2.91	2.84	2.77	2.72	2.68	2.57	2.46	2.41	2.35
	.01	8.10	5.85	4.94	4.43	4.10	3.87	3.70	3.56	3.46	3.37	3.29	3.23	3.09	2.94	2.86	2.78
	.005	9.94	6.99	5.82	5.17	4.76	4.47	4.26	4.09	3.96	3.85	3.76	3.68	3.50	3.32	3.22	3.12
(24)	.25	1.39	1.47	1.46	1.44	1.43	1.41	1.40	1.39	1.38	1.38	1.37	1.36	1.35	1.33	1.32	1.31
	.1	2.93	2.54	2.33	2.19	2.10	2.04	1.98	1.94	1.91	1.88	1.85	1.83	1.78	1.73	1.70	1.67
	.05	4.26	3.40	3.01	2.78	2.62	2.51	2.42	2.36	2.30	2.25	2.21	2.18	2.11	2.03	1.98	1.94
	.025	5.72	4.32	3.72	3.38	3.15	2.99	2.87	2.78	2.70	2.64	2.59	2.54	2.44	2.33	2.27	2.21
	.01	7.82	5.61	4.72	4.22	3.90	3.67	3.50	3.36	3.26	3.17	3.09	3.03	2.89	2.74	2.66	2.58
	.005	9.55	6.66	5.52	4.89	4.49	4.20	3.99	3.83	3.69	3.59	3.50	3.42	3.25	3.06	2.97	2.87
(30)	.25	1.38	1.45	1.44	1.42	1.41	1.39	1.38	1.37	1.36	1.35	1.35	1.34	1.32	1.30	1.29	1.28
	.1	2.88	2.49	2.28	2.14	2.05	1.98	1.93	1.88	1.85	1.82	1.79	1.77	1.72	1.67	1.64	1.61
	.05	4.17	3.32	2.92	2.69	2.53	2.42	2.33	2.27	2.21	2.16	2.13	2.09	2.01	1.93	1.89	1.84
	.025	5.57	4.18	3.59	3.25	3.03	2.87	2.75	2.65	2.57	2.51	2.46	2.41	2.31	2.20	2.14	2.07
	.01	7.56	5.39	4.51	4.02	3.70	3.47	3.30	3.17	3.07	2.98	2.91	2.84	2.70	2.55	2.47	2.39
	.005	9.18	6.35	5.24	4.62	4.23	3.95	3.74	3.58	3.45	3.34	3.25	3.18	3.01	2.82	2.73	2.63

TABLE 4 (CONTINUED)

		k															
n	α	1	2	3	4	5	6	7	8	9	10	11	12	15	20	24	30
(40)	.25	1.36	1.44	1.42	1.40	1.39	1.37	1.36	1.35	1.34	1.33	1.32	1.31	1.30	1.28	1.26	1.25
	.1	2.84	2.44	2.23	2.09	2.00	1.93	1.87	1.83	1.79	1.76	1.73	1.71	1.66	1.61	1.57	1.54
	.05	4.08	3.23	2.84	2.61	2.45	2.34	2.25	2.18	2.12	2.08	2.04	2.00	1.92	1.84	1.79	1.74
	.025	5.42	4.05	3.46	3.13	2.90	2.74	2.62	2.53	2.45	2.39	2.33	2.29	2.18	2.07	2.01	1.94
	.01	7.31	5.18	4.31	3.83	3.51	3.29	3.12	2.99	2.89	2.80	2.73	2.66	2.52	2.37	2.29	2.20
	.005	8.83	6.07	4.98	4.37	3.99	3.71	3.51	3.35	3.22	3.12	3.03	2.95	2.78	2.60	2.50	2.40
(60)	.25	1.35	1.42	1.41	1.38	1.37	1.35	1.33	1.32	1.31	1.30	1.29	1.29	1.27	1.25	1.24	1.22
	.1	2.79	2.39	2.18	2.04	1.95	1.87	1.82	1.77	1.74	1.71	1.68	1.66	1.60	1.54	1.51	1.48
	.05	4.00	3.15	2.76	2.53	2.37	2.25	2.17	2.10	2.04	1.99	1.95	1.92	1.84	1.75	1.70	1.65
	.025	5.29	3.93	3.34	3.01	2.79	2.63	2.51	2.41	2.33	2.27	2.22	2.17	2.06	1.94	1.88	1.82
	.01	7.08	4.98	4.13	3.65	3.34	3.12	2.95	2.82	2.72	2.63	2.56	2.50	2.35	2.20	2.12	2.03
	.005	8.49	5.80	4.73	4.14	3.76	3.49	3.29	3.13	3.01	2.90	2.82	2.74	2.57	2.39	2.29	2.19
(120)	.25	1.34	1.40	1.39	1.37	1.35	1.33	1.31	1.30	1.29	1.28	1.27	1.26	1.24	1.22	1.21	1.19
	.1	2.75	2.35	2.13	1.99	1.90	1.82	1.77	1.72	1.68	1.65	1.62	1.60	1.55	1.48	1.45	1.41
	.05	3.92	3.07	2.68	2.45	2.29	2.18	2.09	2.02	1.96	1.91	1.87	1.83	1.75	1.66	1.61	1.55
	.025	5.15	3.80	3.23	2.89	2.67	2.52	2.39	2.30	2.22	2.16	2.10	2.05	1.95	1.82	1.76	1.69
	.01	6.85	4.79	3.95	3.48	3.17	2.96	2.79	2.66	2.56	2.47	2.40	2.34	2.19	2.03	1.95	1.86
	.005	8.18	5.54	4.50	3.92	3.55	3.28	3.09	2.93	2.81	2.71	2.62	2.54	2.37	2.19	2.09	1.98

Table 4 is used with permission from Wilfred J. Dixon and Frank J. Massey, Jr.,

Introduction to Statistical Analysis, 4th ed., New York, McGraw-Hill, 1983, pp. 520–533.

[a] Read 562^1 as 5620, 231^2 as 23,100 and so on.

Adapted from Wilford J. Dixon and Frank J. Massey, Jr., *Introduction to Statistical Analysis*, 4th Ed. (New York: McGraw-Hill, 1983), 520–533.

TABLE 5 BINOMIAL DISTRIBUTION FUNCTIONS. $P(X \leqq x), X \sim B(n, p)$.

		p												
n	x	0.05	0.1	0.2	0.25	0.3	0.4	0.5	0.6	0.7	0.75	0.8	0.9	0.95
5														
	0	.774	.590	.328	.237	.168	.078	.031	.010	.002	.001	.000	.000	.000
	1	.977	.919	.737	.633	.528	.337	.188	.087	.031	.016	.007	.000	.000
	2	.999	.991	.942	.896	.837	.683	.500	.317	.163	.104	.058	.009	.001
	3	.999	.999	.993	.984	.969	.913	.813	.663	.472	.367	.263	.081	.023
	4	.999	.999	.999	.999	.998	.990	.969	.922	.832	.763	.672	.410	.226
	5	.999	.999	.999	.999	.999	.999	.999	.999	.999	.999	.999	.999	.999
10														
	0	.599	.349	.107	.056	.028	.006	.001	.000	.000	.000	.000	.000	.000
	1	.914	.736	.376	.244	.149	.046	.011	.002	.000	.000	.000	.000	.000
	2	.988	.930	.678	.526	.383	.167	.055	.012	.002	.000	.000	.000	.000
	3	.999	.987	.879	.776	.650	.382	.172	.055	.011	.004	.001	.000	.000
	4	.999	.998	.967	.922	.850	.633	.377	.166	.047	.020	.006	.000	.000
	5	.999	.999	.994	.980	.953	.834	.623	.367	.150	.078	.033	.002	.000
	6	.999	.999	.999	.996	.989	.945	.828	.618	.350	.224	.121	.013	.001
	7	.999	.999	.999	.999	.998	.988	.945	.833	.617	.474	.322	.070	.012
	8	.999	.999	.999	.999	.999	.998	.989	.954	.851	.756	.624	.264	.086
	9	.999	.999	.999	.999	.999	.999	.999	.994	.972	.944	.893	.651	.401
	10	.999	.999	.999	.999	.999	.999	.999	.999	.999	.999	.999	.999	.999
15														
	0	.463	.206	.035	.013	.005	.000	.000	.000	.000	.000	.000	.000	.000
	1	.829	.549	.167	.080	.035	.005	.000	.000	.000	.000	.000	.000	.000
	2	.964	.816	.398	.236	.127	.027	.004	.000	.000	.000	.000	.000	.000
	3	.995	.944	.648	.461	.297	.091	.018	.002	.000	.000	.000	.000	.000
	4	.999	.987	.836	.686	.515	.217	.059	.009	.001	.000	.000	.000	.000
	5	.999	.998	.939	.852	.722	.403	.151	.034	.004	.001	.000	.000	.000
	6	.999	.999	.982	.943	.869	.610	.304	.095	.015	.004	.001	.000	.000
	7	.999	.999	.996	.983	.950	.787	.500	.213	.050	.017	.004	.000	.000
	8	.999	.999	.999	.996	.985	.905	.696	.390	.131	.057	.018	.000	.000
	9	.999	.999	.999	.999	.996	.966	.849	.597	.278	.148	.061	.002	.000
	10	.999	.999	.999	.999	.999	.991	.941	.783	.485	.314	.164	.013	.001
	11	.999	.999	.999	.999	.999	.998	.982	.909	.703	.539	.352	.056	.005
	12	.999	.999	.999	.999	.999	.999	.996	.973	.873	.764	.602	.184	.036
	13	.999	.999	.999	.999	.999	.999	.999	.995	.965	.920	.833	.451	.171
	14	.999	.999	.999	.999	.999	.999	.999	.999	.995	.987	.965	.794	.537
	15	.999	.999	.999	.999	.999	.999	.999	.999	.999	.999	.999	.999	.999
20														
	0	.358	.122	.012	.003	.001	.000	.000	.000	.000	.000	.000	.000	.000
	1	.736	.392	.069	.024	.008	.001	.000	.000	.000	.000	.000	.000	.000
	2	.925	.677	.206	.091	.035	.004	.000	.000	.000	.000	.000	.000	.000
	3	.984	.867	.411	.225	.107	.016	.001	.000	.000	.000	.000	.000	.000
	4	.997	.957	.630	.415	.238	.051	.006	.000	.000	.000	.000	.000	.000
	5	.999	.989	.804	.617	.416	.126	.021	.002	.000	.000	.000	.000	.000
	6	.999	.998	.913	.786	.608	.250	.058	.006	.000	.000	.000	.000	.000
	7	.999	.999	.968	.898	.772	.416	.132	.021	.001	.000	.000	.000	.000
	8	.999	.999	.990	.959	.887	.596	.252	.057	.005	.001	.000	.000	.000
	9	.999	.999	.997	.986	.952	.755	.412	.128	.017	.004	.001	.000	.000
	10	.999	.999	.999	.996	.983	.872	.588	.245	.048	.014	.003	.000	.000
	11	.999	.999	.999	.999	.995	.943	.748	.404	.113	.041	.010	.000	.000
	12	.999	.999	.999	.999	.999	.979	.868	.584	.228	.102	.032	.000	.000

TABLE 5 (CONTINUED)

		p												
n	*x*	0.05	0.1	0.2	0.25	0.3	0.4	0.5	0.6	0.7	0.75	0.8	0.9	0.95
	13	.999	.999	.999	.999	.999	.994	.942	.750	.392	.214	.087	.002	.000
	14	.999	.999	.999	.999	.999	.998	.979	.874	.584	.383	.196	.011	.000
	15	.999	.999	.999	.999	.999	.999	.994	.949	.762	.585	.370	.043	.003
	16	.999	.999	.999	.999	.999	.999	.999	.984	.893	.775	.589	.133	.016
	17	.999	.999	.999	.999	.999	.999	.999	.996	.965	.909	.794	.323	.075
	18	.999	.999	.999	.999	.999	.999	.999	.999	.992	.976	.931	.608	.264
	19	.999	.999	.999	.999	.999	.999	.999	.999	.999	.997	.988	.878	.642
	20	.999	.999	.999	.999	.999	.999	.999	.999	.999	.999	.999	.999	.999
25														
	0	.277	.072	.004	.001	.000	.000	.000	.000	.000	.000	.000	.000	.000
	1	.642	.271	.027	.007	.002	.000	.000	.000	.000	.000	.000	.000	.000
	2	.873	.537	.098	.032	.009	.000	.000	.000	.000	.000	.000	.000	.000
	3	.966	.764	.234	.096	.033	.002	.000	.000	.000	.000	.000	.000	.000
	4	.993	.902	.421	.214	.090	.009	.000	.000	.000	.000	.000	.000	.000
	5	.999	.967	.617	.378	.193	.029	.002	.000	.000	.000	.000	.000	.000
	6	.999	.991	.780	.561	.341	.074	.007	.000	.000	.000	.000	.000	.000
	7	.999	.998	.891	.727	.512	.154	.022	.001	.000	.000	.000	.000	.000
	8	.999	.999	.953	.851	.677	.274	.054	.004	.000	.000	.000	.000	.000
	9	.999	.999	.983	.929	.811	.425	.115	.013	.000	.000	.000	.000	.000
	10	.999	.999	.994	.970	.902	.586	.212	.034	.002	.000	.000	.000	.000
	11	.999	.999	.998	.989	.956	.732	.345	.078	.006	.001	.000	.000	.000
	12	.999	.999	.999	.997	.983	.846	.500	.154	.017	.003	.000	.000	.000
	13	.999	.999	.999	.999	.994	.922	.655	.268	.044	.011	.002	.000	.000
	14	.999	.999	.999	.999	.998	.966	.788	.414	.098	.030	.006	.000	.000
	15	.999	.999	.999	.999	.999	.987	.885	.575	.189	.071	.017	.000	.000
	16	.999	.999	.999	.999	.999	.996	.946	.726	.323	.149	.047	.000	.000
	17	.999	.999	.999	.999	.999	.999	.978	.846	.488	.273	.109	.002	.000
	18	.999	.999	.999	.999	.999	.999	.993	.926	.659	.439	.220	.009	.000
	19	.999	.999	.999	.999	.999	.999	.998	.971	.807	.622	.383	.033	.001
	20	.999	.999	.999	.999	.999	.999	.999	.991	.910	.786	.579	.098	.007
	21	.999	.999	.999	.999	.999	.999	.999	.998	.967	.904	.766	.236	.034
	22	.999	.999	.999	.999	.999	.999	.999	.999	.991	.968	.902	.463	.127
	23	.999	.999	.999	.999	.999	.999	.999	.999	.998	.993	.973	.729	.358
	24	.999	.999	.999	.999	.999	.999	.999	.999	.999	.999	.996	.928	.723
	25	.999	.999	.999	.999	.999	.999	.999	.999	.999	.999	.999	.999	.999
30														
	0	.215	.042	.001	.000	.000	.000	.000	.000	.000	.000	.000	.000	.000
	1	.554	.184	.011	.002	.000	.000	.000	.000	.000	.000	.000	.000	.000
	2	.812	.411	.044	.011	.002	.000	.000	.000	.000	.000	.000	.000	.000
	3	.939	.647	.123	.037	.009	.000	.000	.000	.000	.000	.000	.000	.000
	4	.984	.825	.255	.098	.030	.002	.000	.000	.000	.000	.000	.000	.000
	5	.997	.927	.428	.203	.077	.006	.000	.000	.000	.000	.000	.000	.000
	6	.999	.974	.607	.348	.160	.017	.001	.000	.000	.000	.000	.000	.000
	7	.999	.992	.761	.514	.281	.044	.003	.000	.000	.000	.000	.000	.000
	8	.999	.998	.871	.674	.432	.094	.008	.000	.000	.000	.000	.000	.000
	9	.999	.999	.939	.803	.589	.176	.021	.001	.000	.000	.000	.000	.000
	10	.999	.999	.974	.894	.730	.291	.049	.003	.000	.000	.000	.000	.000
	11	.999	.999	.991	.949	.841	.431	.100	.008	.000	.000	.000	.000	.000
	12	.999	.999	.997	.978	.916	.578	.181	.021	.001	.000	.000	.000	.000
	13	.999	.999	.999	.992	.960	.715	.292	.048	.002	.000	.000	.000	.000

TABLE 5 (CONTINUED)

		p												
n	*x*	0.05	0.1	0.2	0.25	0.3	0.4	0.5	0.6	0.7	0.75	0.8	0.9	0.95
	14	.999	.999	.999	.997	.983	.825	.428	.097	.006	.001	.000	.000	.000
	15	.999	.999	.999	.999	.994	.903	.572	.175	.017	.003	.000	.000	.000
	16	.999	.999	.999	.999	.998	.952	.708	.285	.040	.008	.001	.000	.000
	17	.999	.999	.999	.999	.999	.979	.819	.422	.084	.022	.003	.000	.000
	18	.999	.999	.999	.999	.999	.992	.900	.569	.159	.051	.009	.000	.000
	19	.999	.999	.999	.999	.999	.997	.951	.709	.270	.106	.026	.000	.000
	20	.999	.999	.999	.999	.999	.999	.979	.824	.411	.197	.061	.000	.000
	21	.999	.999	.999	.999	.999	.999	.992	.906	.568	.326	.129	.002	.000
	22	.999	.999	.999	.999	.999	.999	.997	.956	.719	.486	.239	.008	.000
	23	.999	.999	.999	.999	.999	.999	.999	.983	.840	.652	.393	.026	.001
	24	.999	.999	.999	.999	.999	.999	.999	.994	.923	.797	.572	.073	.003
	25	.999	.999	.999	.999	.999	.999	.999	.998	.970	.902	.745	.175	.016
	26	.999	.999	.999	.999	.999	.999	.999	.999	.991	.963	.877	.353	.061
	27	.999	.999	.999	.999	.999	.999	.999	.999	.998	.989	.956	.589	.188
	28	.999	.999	.999	.999	.999	.999	.999	.999	.999	.998	.989	.816	.446
	29	.999	.999	.999	.999	.999	.999	.999	.999	.999	.999	.999	.958	.785
	30	.999	.999	.999	.999	.999	.999	.999	.999	.999	.999	.999	.999	.999

Adapted from R.N. Goldman and J.S. Weinberg, *Statistics—An Introduction*. (Englewood Cliffs, N.J.: Prentice Hall, 1985), 670–676.

TABLE 6 POISSON DISTRIBUTION FUNCTIONS. $P(X \leq x), X \sim P(m)$.

x \ m	.10	.20	.30	.40	.50	.60	.70	.80	.90
0	.905	.819	.741	.670	.607	.549	.497	.449	.407
1	.995	.982	.963	.938	.910	.878	.844	.809	.772
2	1.000	.999	.996	.992	.986	.977	.966	.953	.937
3		1.000	1.000	.999	.998	.997	.994	.991	.987
4				1.000	1.000	1.000	.999	.999	.998
5							1.000	1.000	1.000

x \ m	1.0	2.0	3.0	4.0	5.0	6.0	7.0	8.0	9.0	10.0	15.0
0	.368	.135	.050	.018	.007	.002	.001	.000	.000		
1	.736	.406	.199	.092	.040	.017	.007	.003	.001		
2	.920	.677	.423	.238	.125	.062	.030	.014	.006	.003	
3	.981	.857	.647	.433	.265	.151	.082	.042	.021	.010	
4	.996	.947	.815	.629	.440	.285	.173	.100	.055	.029	.001
5	.999	.983	.916	.785	.616	.446	.301	.191	.116	.067	.003
6	1.000	.995	.966	.889	.762	.606	.450	.313	.207	.130	.008
7		.999	.988	.949	.867	.744	.599	.453	.324	.220	.018
8		1.000	.996	.979	.932	.847	.729	.593	.456	.333	.037
9			.999	.992	.968	.916	.830	.717	.587	.458	.070
10			1.000	.997	.986	.957	.901	.816	.706	.583	.118
11				.999	.995	.980	.947	.888	.803	.697	.185
12				1.000	.998	.991	.973	.936	.876	.792	.268
13					.999	.996	.987	.966	.926	.864	.363
14					1.000	.999	.994	.983	.959	.917	.466
15						.999	.998	.992	.978	.951	.568
16						1.000	.999	.996	.989	.973	.664
17							1.000	.998	.995	.986	.749
18								.999	.998	.993	.819
19								1.000	.999	.997	.875
20									1.000	.998	.917
21										.999	.947
22										1.000	.967
23											.981
24											.989
25											.994
26											.997
27											.998
28											.999
29											1.000

TABLE 7 GAMMA DISTRIBUTION FUNCTIONS. $P(X \leqq x), X \sim \Gamma(a, 1)$.

x \ a	1	2	3	4	5	6	7	8	9	10
1	.632	.264	.080	.019	.004	.001	.000	.000	.000	.000
2	.865	.594	.323	.143	.053	.017	.005	.001	.000	.000
3	.950	.801	.577	.353	.185	.084	.034	.012	.004	.001
4	.982	.908	.762	.567	.371	.215	.111	.051	.021	.008
5	.993	.960	.875	.735	.560	.384	.238	.133	.068	.032
6	.998	.983	.938	.849	.715	.554	.398	.256	.153	.084
7	.999	.993	.970	.918	.827	.699	.550	.401	.271	.170
8	1.000	.997	.986	.958	.900	.809	.687	.547	.407	.283
9		.999	.994	.979	.945	.884	.793	.676	.544	.413
10		1.000	.997	.990	.971	.933	.870	.780	.667	.542
11			.999	.995	.985	.962	.921	.857	.768	.659
12			1.000	.998	.992	.980	.954	.911	.845	.758
13				.999	.996	.989	.974	.946	.900	.834
14				1.000	.998	.994	.986	.968	.938	.891
15					.999	.997	.992	.982	.963	.930

Adapted from Jay L. Devore, *Probability and Statistics for Engineering and the Sciences*. (Brooks-Cole, Monterey, Calif., 1982), 621.

INDEX

Entries marked with a * occur only in the exercises.